# Age-Structured Population Dynamics in Demography and Epidemiology

Hisashi Inaba

# Age-Structured Population Dynamics in Demography and Epidemiology

Springer

Hisashi Inaba
The University of Tokyo
Tokyo
Japan

ISBN 978-981-10-0187-1 ISBN 978-981-10-0188-8 (eBook)
DOI 10.1007/978-981-10-0188-8

Library of Congress Control Number: 2017930648



Printed on acid-free paper

This Springer imprint is published by Springer Nature
The registered company is Springer Nature Singapore Pte Ltd.
The registered company address is: 152 Beach Road, #21-01/04 Gateway East, Singapore 189721, Singapore

*To Yoko and Riki*

# Preface

The main purpose of this book was to explain deterministic, age-structured population dynamics models that appear in demography and epidemiology. The first part of this book addresses pure demographic models, while the second part considers epidemic models for infectious diseases.

It should be of essential importance to human beings to know the laws governing their own reproduction and associated results. In fact, the major problems with which we have been confronted relate to population dynamics, as was clearly recognized by Thomas Robert Malthus in the eighteenth century. From the beginning, the main concerns of demography have extended beyond the variations in population size and distribution to the age structure dynamics of the population. The starting point of modern demography was the life table technique developed to measure human life expectancy in the seventeenth century. In the eighteenth century, Euler developed a difference equation model for human populations to show that an age-structured population with constant fertility and mortality will grow geometrically. Moreover, Euler derived relations among demographic indices under this geometrical growth and suggested that these relations could be used to estimate incomplete data. Some 150 years later, Euler's idea was rediscovered by Sharpe and Lotka at 1911, and modern demography was born. As the human vital rates are under conscious control and strongly depend on historical, social, and economic environmental variations, mathematical studies of human populations have developed on the boundary between the social sciences and life sciences under problematic concerns that differ from those of population biology. It should be, however, noted that the basic mathematical tools used in demography (such as life tables, renewal equations, and the basic reproduction number) are universally applicable to the description of any self-renewing aggregates.

In Chap. 1, we introduce the most basic age-dependent population model, called the *stable population model*. We then explain many resulting facets of the Fundamental Theorem of Demography, because the stable population model is a central tenet in the development of modern demography. Indeed, its revision using functional analysis was the prelude to new, more general developments for structured population dynamics from the end of the 1970s. A crucial point of demographic applications is that the basic demographic indices cannot be interpreted without the stable population model. Although the operator semigroup theory is a powerful tool for studying structured population models, we here mainly adopt the classical integral equation approach, because the scalar integral equation is more elementary and is a most natural expression for the self-reproduction process of any population. Through the renewal equation, we can define the basic reproduction number, which enables the essential relation between individual vital rates and the Malthusian parameter to be established at the population level. The semigroup approach is, however, briefly introduced in Chap. 10.

In Chap. 2, we examine some linear extensions of the stable population model. First, we show that the Fundamental Theorem can be extended to the multistate stable population model, which was developed during the 1970s and the 1980s by various demographers. However, since the 1990s, the mathematical theory of the basic reproduction number has been developed in the context of mathematical epidemiology. Hence, we must reconsider classical results under the light of the modern $R_0$ theory. Next, we examine linear extensions that can recognize the marital status of individuals. The linear marriage models are mathematically simple, but practically important, because the marriage phenomenon is crucial to understanding the variation of human reproductivity. Subsequently, we consider the non-autonomous case, where the main tool is the *weak ergodicity* theorem. In particular, if the vital rates are time-periodic, the non-autonomous Lotka–McKendrick system has a periodic solution and, as time evolves, the age profile converges to a periodic age profile independent from the initial data, which reflects the strong ergodicity of the non-autonomous system.

In Chap. 3, one-sex nonlinear age-dependent models are examined. In 1974, Gurtin and MacCamy studied a nonlinear age-dependent population model with density dependence, which triggered a huge amount of research into age-structured population dynamics and motivated the functional analytic approach. However, we again use classical, elementary mathematics to examine the existence and uniqueness of the solution, the bifurcation of steady states and their stability and instability. In particular, we take up a cohort control model (the Easterlin model), because it is a unique nonlinear demographic model that has been empirically tested by demographers. Whether the Easterlin mechanism is a feasible explanation for real human population wave remains a controversial topic.

In Chap. 4, we consider demographic pair formation models. The two-sex problem is a fundamental, but difficult challenge in mathematical demography. Indeed, human population reproduction is essentially a result of mating and pair formation, which are nonlinear interactions between both sexes. Moreover, two-sex models are needed to study sexually transmitted diseases. First, we introduce Kendall's pair formation model without age structure and analyze it using homogeneous dynamical system theory. Next, we introduce the age-dependent pair formation models. Different from the age-independent model, we only have limited results on the age-dependent pair formation models. Although we can prove the existence of a Malthusian solution, its uniqueness and stability remain open.

Finally, we introduce a semigroup approach to the Fredrickson's age-structured pair formation model.

The focus in the second part of this book is on deterministic, continuous-time mathematical models for the spread of infectious diseases. It is well known that humans have repeatedly suffered from major outbreaks of infectious diseases, such as the Black Death pandemic in the 1300s, the Great Plague of London in 1665, the "Spanish" Flu of 1918, and the HIV/AIDS pandemic in the 1980s. Indeed, infectious diseases have played a role in changing the historical course of civilization, as we can see from the devastating decline in the native populations of the New World following contact with infectious diseases from the Old World. Corresponding to these considerable challenges, scientists have developed mathematical theories to understand and control epidemic phenomena since the eighteenth century. In 1760, Daniel Bernoulli submitted a paper to the Academy of Sciences in Paris in which the first age-dependent epidemic model was analyzed to calculate the gain in life expectancy after the elimination of a potentially lethal infectious disease. During the late nineteenth century, the earliest expression of a key idea, the basic reproduction number $R_0$, appeared in demography and epidemiology. Finally, between World War I and II, a seminal series of papers by Kermack and McKendrick formulated the basic ideas and modeling methods underlying the mathematical theory of infectious diseases.

Until the 1970s, however, very few researchers were interested in mathematical models for infectious diseases, because in developed countries, little attention was paid to infectious diseases as a result of the epidemiological transition in causes of death. There was increasing interest in cancer as well as heart, brain, and other non-communicable diseases. Indeed, in May 1980, smallpox was declared eradicated in the world. That was a very optimistic era: Many people thought that all infectious diseases would disappear in the near future as a result of progress in the medical sciences. The HIV/AIDS pandemic began in the 1980s, which I believe was the starting point for applying infectious disease modeling to policy decisions, and the study of mathematical epidemiology was strongly promoted from practical point of view. In 1990, we had entered a new epoch in mathematical epidemiology, because Diekmann, Heesterbeek, and Metz published a very influential paper about the definition of the basic reproduction number $R_0$. The paper gave the first mathematically rigorous definition of the basic reproduction number. Thereafter, the theoretical development of infectious disease epidemiology underwent marked acceleration, and that process was supported by the theory of structured population dynamics. As the 1990s progressed, it was no longer possible to talk about epidemic dynamics except in terms of $R_0$.

In Chap. 5, we thus begin by looking at the early Kermack and McKendrick model, from which the basic ideas and ingredients for epidemic modeling are introduced. The crucial insight is that we cannot precisely interpret the basic ideas and indices of infectious disease epidemiology without knowing the underlying nonlinear population dynamics. The early Kermack–McKendrick infection-age-dependent model was revived by Diekmann, Metz, and Thieme from a modern mathematical point of view in the late 1970s. The key idea of analysis in epidemic models is the basic reproduction number $R_0$ and its well-known *threshold principle*, that is, the final size of the epidemic is positive no matter how small the initial infected population is if $R_0 > 1$, whereas the final size becomes zero when the initial size of the infected population goes to zero if $R_0 < 1$. We try to show the threshold principle based on the original definition of the final size given by Kermack and McKendrick, although there are slightly different versions of the formulation. Next, we extend the original model to take into account the heterogeneity of individuals and to state a pandemic threshold theorem. Subsequently, we introduce the demography of the host population and prove the endemic threshold theorem, that is, there exists at least one endemic steady state if $R_0 > 1$, whereas there is no endemic steady state if $R_0 < 1$. This principle, however, does not always hold. We provide examples of the existence of subcritical endemic steady states, even when $R_0 < 1$, in the case that reinfection of the recovered individuals can occur, or if the disease is transmitted by vectors. This type of backward bifurcation is again discussed in Chaps. 7 and 8.

In Chap. 6, we introduce the age-structured SIR epidemic model, which is most useful when applied to real age-dependent data of common childhood infectious diseases. A crucial assumption here is that the host demography is not affected by the presence of diseases, so we can assume that the host population has already attained the stable age distribution. This is not the case if we cannot disregard the disease-induced mortality, as in the case of HIV/AIDS. We again establish the endemic threshold principle based on the basic reproduction number, which is defined by the spectral radius of the next-generation operator. Here, we sketch a functional analytic approach to the age-dependent model, because it is a natural way to deal with the infinite-dimensional problem.

In Chap. 7, we study structured population models for the HIV/AIDS epidemic. Since the 1980s, the HIV/AIDS epidemic has become a worldwide pandemic. This had a big impact on the developments of modern mathematical epidemiology for infectious diseases, because it was the first pandemic of a newly emerging virus since the 1970s—a time when people were very optimistic about the future control of infectious diseases—and it attacked the most scientifically developed countries. HIV/AIDS has a very long incubation period, so we have to take into account the demography of the host population. There also exists a true interaction between demography and epidemiology, because the disease-induced death rate is high and cannot be neglected. First, we summarize the epidemiological features of the HIV/AIDS epidemic and examine mathematical models for the initial invasion phase. We show that the invasion phase of HIV can be described by the stable population model, which can be used to estimate the size of the HIV-infected population based on the observable data. Next, we consider the age-duration-structured HIV epidemic model in a homosexual community and show that there can exist backwardly bifurcated endemic steady states even when $R_0 < 1$, which is a result of the homogeneous force of infection and the disease-induced death rate. Finally, we briefly sketch an age-structured model of in vivo HIV infection, by which we can show that the age-dependent model is a useful framework for studying the infection phenomena at the cell population level and considering effective treatments.

In Chap. 8, we take up three interesting epidemic models that suggest future extensions of classical epidemic models. It is now widely recognized that dynamic changes in host susceptibility due to evolutionary changes in infectious agents and changes in the host immunity distribution due to waning, boosting, or vaccination play important roles in the spread of infectious diseases. In fact, even for traditional common SIR-type childhood diseases such as measles, the natural decay of host immunity can be observed if the environmental virus disappears and the booster effect is lost. First, we consider the Pease model for type A influenza epidemic, which was the first attempt to account for the decay of host immunity due to antigenic changes in the virus population. A remarkable feature is that, for realistic parameter values, the antigenic drift of a dominant virus is a possible mechanism for recurrent outbreaks. Next, we formulate the Kermack–McKendrick reinfection model using the standard age-dependent population dynamics model and examine its basic properties. The potential importance of the Kermack–McKendrick reinfection model is that it can account for variable susceptibility and reinfection, so it is a useful starting point for considering the epidemiological life cycle of individuals. We show that the *reinfection threshold* is given as a threshold number of the fully vaccinated (by incomplete vaccine) system, so the disease cannot be controlled if $R_0$ exceeds the reinfection threshold. Moreover, subcritical endemic steady states may be created by a backward bifurcation. We show some realistic examples of the *reproductivity enhancement* that can create the backward bifurcation. Finally, we introduce the Aron model for acquired immunity boosted by exposure to infection. This was originally developed to understand the functional relation between the force of infection and the reversion rate and the age prevalence curve of malaria; however, our purpose here is not to give a realistic view of the malaria epidemic, but to show that the age-dependent model is a useful tool for describing the boosting mechanism.

In Chap. 9, we discuss a general theory for the basic reproduction number in heterogeneous environments. The basic reproduction number $R_0$ plays a central role in structured population dynamics. Although some roots of $R_0$ can be traced back to the nineteenth century, the specific concept was introduced to the demography literature in 1925. It took a further half century for this number to mature as a key concept in mathematical epidemiology, and it is only recently that the stable population theory has become a popular tool in the field. However, the progress of mathematical epidemiology over the past two decades has been remarkable, and the basic concept and applications of $R_0$ are now better developed in epidemiology than in demography. In particular, the successful introduction of a general definition of the basic reproduction number for structured populations in the context of epidemic models gave rise to a new epoch in our understanding. Since then, the theory of the basic reproduction number has been developed as a central tenet of both infectious disease epidemiology and general population dynamics. Recently, this basic idea has evolved considerably to allow its application to time-heterogeneous environments. In this chapter, first we formulate a general definition for the basic reproduction number $R_0$ of structured populations in time-heterogeneous environments. Based on the *generation evolution operator*, we show that the basic reproduction number can be calculated as the spectral radius of the *next-generation operator* in a constant environment or in a periodic environment. It is stressed that a biological reproduction process is described by a renewal equation, and its basic reproduction number should be uniquely determined. Subsequently, we define the *type-reproduction number* in a time-heterogeneous environment and examine some applications in demography and epidemiology. Finally, we discuss some methods to estimate $R_0$ from the available data.

In Chap. 10, we explain some mathematical tools commonly applied to demographic and epidemiological population models. As stated above, innovative functional analytic approaches have played a central role in modern developments of structured population dynamics. First, we explain a semigroup approach to consider the stable population model. Using the semigroup setting, the idea of strong ergodicity is extended as the *asynchronous exponential growth* of the semigroup, which can be applied to certain nonlinear problems. We then briefly discuss functional analytic methods for nonlinear population problems. Next, we introduce some results of the positive operator theory, which are useful for studying infinite-dimensional population dynamics. Finally, we summarize some basic results about the Laplace transformation and Volterra integral equation that are used in the previous chapters.

Tokyo, Japan
November 2016

Hisashi Inaba

# Acknowledgements

This project began as an English translation of my Japanese book "Suuri Jinkogaku" (Mathematical Demography, University of Tokyo Press, 2002) and eventually became a new version. It is a result of my studies since the 1980s in the Institute of Population Problems (Tokyo), the Institute of Theoretical Biology (Leiden), the Center for Mathematics and Computer Science (Amsterdam), and the University of Tokyo (Komaba). During this time, Prof. Odo Diekmann (Utrecht), who supervised my Ph.D. in Leiden, has been a guiding light to me. I was introduced to Odo by Prof. Masaya Yamaguti (Kyoto, 1925–1998), who was a generous mentor to me at Kyoto University and an extraordinary founder of the Japanese Society for Mathematical Biology. I would especially like to thank both professors, who had a strong and positive influence on my life and career.

For the idea of writing this book, I am deeply grateful to Prof. Mimmo Iannelli (Trento). As readers will surely understand, my book owes a considerable debt to his excellent monograph "Mathematical Theory of Age-Structured Population Dynamics" (1995). Furthermore, his gentle style of mathematics has always intrigued me. Thanks are also due to Prof. Horst R. Thieme (Tempe, ASU) and Prof. Glenn Webb (Vanderbilt), as they have continually stimulated and encouraged my studies. I would also like to thank Nicolas Bacaër (Bondy, IRD) for providing helpful comments on the early version of this manuscript. My young collaborators, Yoichi Enatsu (Tokyo, TUS), Shingo Iwami (Kyushu), Toshikazu Kuniya (Kobe), Shinji Nakaoka (Tokyo), Yukihiko Nakata (Shimane), Hiroshi Nishiura (Hokkaido), and Ryo Oizumi (MHLW and NIPSSR) also provided stimulating discussions on population mathematics, which are implicitly reflected in this book. Thanks also to my editors, Rika Tannai (University of Tokyo Press) and Yutaka Hirachi (Springer Japan).

During the years I spent completing this book, I was financially supported by the University of Tokyo, MEXT KAKENHI, the Japan Society for the Promotion of Science, and CREST, Japan Science and Technology Agency.

# Contents

# Chapter 1
# The Stable Population Model

**Abstract** In this chapter, we consider the most basic linear one-sex age-structured population model, known as the *stable population model*. In the eighteenth century, Euler developed a difference equation model to show that an age-structured population with constant fertility and mortality will grow geometrically. Moreover, Euler derived relations among various demographic indices under this geometrical growth and suggested that these relations could be used to estimate incomplete data. This brilliant discovery would have been the starting point of modern demography had it not been lost until the stable population theory was formulated in the twentieth century. Although the stable population model is very simple, it can be successfully applied to real populations that exhibit exponential growth. The stable population model has not only become a central tenet of modern demography, but has also stimulated a wide variety of mathematical studies in population biology, epidemiology, and social sciences. In this chapter, we formulate the stable population model as an initial-boundary value problem of the McKendrick partial differential equation. We investigate the basic properties of the model based on Lotka's integral equation, which also gives an alternative formulation of the basic model. Our main purpose is to prove the Fundamental Theorem of Demography (the Sharpe–Lotka–Feller theorem/the strong ergodicity theorem). By introducing the dual system, we then derive some stochastic interpretations of the Fundamental Theorem. Next, we present some applications in real human demography. Finally, we discuss the age-profile dynamics of quasi-stable populations.

## 1.1 Basic Model Ingredients

### *1.1.1 Introduction*

First, let us consider a homogeneous non-structured population in a constant environment without resource limitations. Let $P(t)$ be the population size at time $t$; $P(t)$ is assumed to be sufficiently large that it can be considered as a continuous function. Let $\beta$ and $\mu$ be the fertility rate and mortality rate per capita per unit time, respectively. The population growth can then be described by the following equation:



H. Inaba, *Age-Structured Population Dynamics in Demography and Epidemiology*, DOI 10.1007/978-981-10-0188-8_1

$$\frac{dP(t)}{dt} = (\beta - \mu)P(t) = \lambda P(t), \tag{1.1}$$

where $\lambda = \beta - \mu$ is the rate of population increase. Therefore, we have the exponential law $P(t) = P(0)e^{\lambda t}$. We call a population that obeys the exponential growth law a *Malthusian population* and call its growth rate $\lambda$ the *Malthusian parameter.* These names refer to Thomas R. Malthus, who published his famous book, *An Essay on the Principle of Population*, in 1798 [85]. This book led to the idea of geometrical population growth being widely accepted; however, such a concept had been mathematically established by Euler in 1760 [41, 118].

The Malthusian model (1.1) does not provide any information at the individual level. If we consider the population reproduction process as the replacement of successive generations and assume that there is no generational overlap and that the size of successive generations follows a constant ratio, the geometrical growth law of populations is self-evident. However, if we allow for overlapping generations, the relation between the Malthusian growth at the population level and the replacement process of successive generations is not clear.

Within biological populations, it can be observed that birth, death, and other demographic events have clear age-specific patterns. Thus, the age structure is an important characteristic in describing the population dynamics. A population living in a local habitat without migration is called a *closed population.* It is clear that the dynamics of a closed population are determined solely by births and deaths. The birth process and the aging process are essentially complex procedures, each of which would require a specific model for a detailed study. Here, however, we introduce the simplest phenomenological definitions of the birth rate and death rate at the population level to formulate age-dependent population models.

Consider a closed population in a local area and assume that the population is sufficiently large that its dynamics can be appropriately described by a continuous model. Let $p(t, a)$, $a \in [0, \omega]$, be the *age-density function* of the population at time $t$ where $a$ denotes chronological age and $0 < \omega \leq +\infty$ is the maximum attainable age. That is, we assume that the integral

$$\int_{a_1}^{a_2} p(t, a)da$$

gives the number of individuals in the age interval $[a_1, a_2]$ at time $t$ and

$$P(t) := \int_0^{\omega} p(t, a)da$$

is the total size of the population at time $t$. To guarantee $0 \leq P(t) < \infty$, we assume that $p(t, \cdot)$ is a nonnegative integrable function, that is, for any fixed $t$, $p(t, \cdot) \in L^1_+(0, \omega)$.

### 1.1.2 Mortality

The *force of mortality* or *age-specific death rate* $\mu(a)$ is defined as the instantaneous death rate at age $a$. To clarify the meaning of the instantaneous death rate, we give a probabilistic interpretation. Let $X$ be a random variable representing the age of death of an individual. Let $F(a)$ be the distribution function of $X$, and $f(a)$ be the probability density function. The force of mortality $\mu(a)$ can then be defined as follows:

$$\begin{aligned}\mu(a) &= \lim_{h\to 0} \frac{\Pr(a \le X < a+h | a \le X)}{h} \\ &= \lim_{h\to 0} \frac{\Pr(a \le X < a+h)}{\Pr(a \le X))h} = \frac{f(a)}{1-F(a)} = -\frac{d}{da}\log(1-F(a)),\end{aligned} \tag{1.2}$$

where $\Pr(A)$ denotes the probability of event $A$. The *survival probability* $\ell(a) := 1 - F(a)$[1] then describes the likelihood that an individual survives to age $a$. From (1.2), we have

$$\frac{d\ell(a)}{da} = -\mu(a)\ell(a).$$

Because $\ell(0) = 1$, we arrive at the expression:

$$\ell(a) = \exp\left(-\int_0^a \mu(\sigma)d\sigma\right),$$

and $f(a) = \mu(a)\ell(a)$. In particular, if the maximum attainable age $\omega$ is assumed to be finite, we also have that

$$\int_0^\omega \mu(a)da = \infty, \tag{1.3}$$

because $\ell(\omega) = 0$ (Figs. 1.1, 1.2).

It is assumed that $\mu(\cdot) \in L^1_{\text{loc},+}(0,\omega)$, that is, $\mu$ is a nonnegative, locally integrable function on $(0,\omega)$. Moreover, if $\omega = \infty$, $\mu$ is usually assumed to be essentially bounded. In particular, if we wish to treat constant parameters by which the basic model can be reduced to a system of ordinary differential equations (ODEs), we adopt the convention that $\omega = \infty$ and $\mu$ is bounded.

Under our deterministic interpretation, the number of deaths occurring in the age interval $[a_1, a_2]$ per unit time is given by $\int_{a_1}^{a_2} \mu(a)p(t,a)da$, and the total number of deaths at time $t$ is given by

$$D(t) := \int_0^\omega \mu(a)p(t,a)da. \tag{1.4}$$

[1] In traditional demography, $\ell(a)$ is called the *survival rate*, although it is not a "rate", but a dimensionless number.

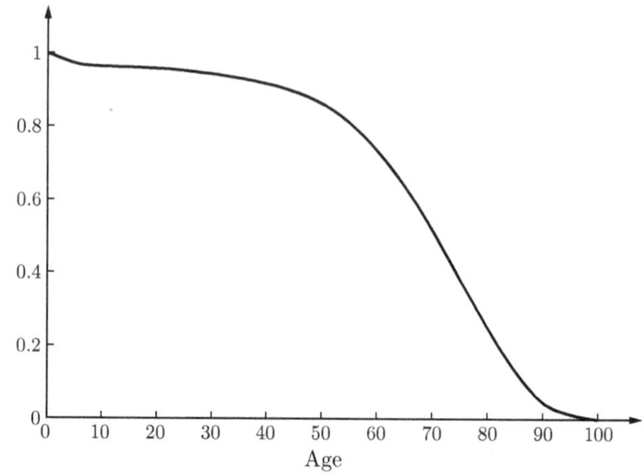


**Fig. 1.1** An example of a survival probability $\ell(a)$

**Fig. 1.2** An example of a force of mortality $\mu(a)$

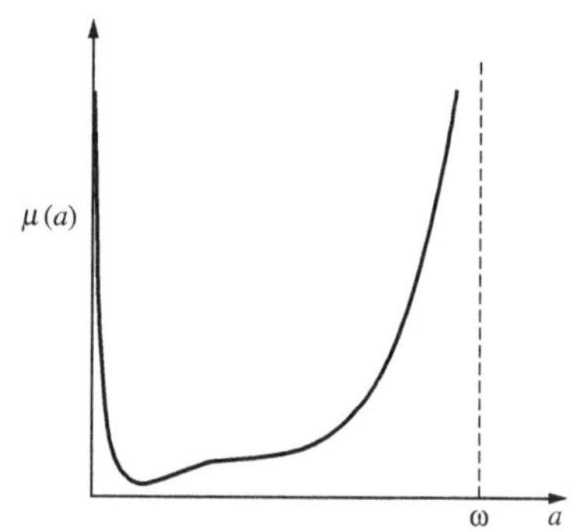


**Fig. 1.3** The age-specific birth rate $\beta(a)$ for Japanese women at 2004

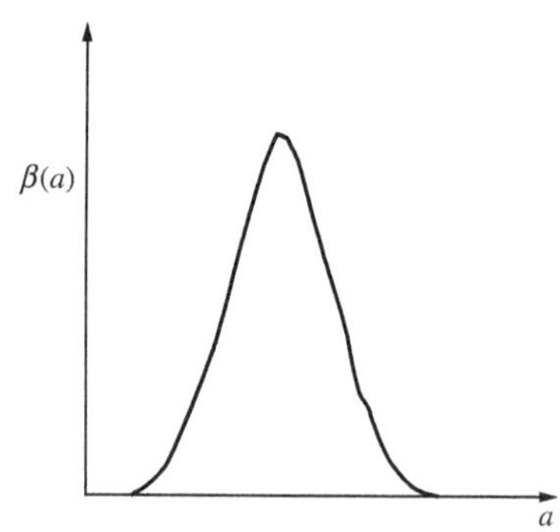

Note that if we adopt the singularity in (1.3), $p(t,\cdot) \in L^1_+(0,\omega)$ is not sufficient to ensure that $D(t) < \infty$; therefore, we must consider the problem on a subspace $\{\phi \in L^1(0,\omega) : \mu\phi \in L^1(0,\omega)\}$. Moreover, we often use the condition that $\lim_{a\to\infty} p(t,a) = 0$. This boundary condition naturally holds if $p(t,\cdot) \in W^{1,1}(\mathbb{R}_+) = \{\phi \in L^1(\mathbb{R}_+) : \phi \text{ is almost everywhere differentiable with } \phi' \in L^1(\mathbb{R}_+)\}$.

The *crude death rate* (CDR), defined as the death rate per unit time per capita, can be expressed as

$$\mathrm{CDR} := \frac{D(t)}{P(t)} = \frac{\int_0^\omega \mu(a)p(t,a)da}{\int_0^\omega p(t,a)da}$$

The average age of incidence of death, denoted by $e_0$, is called the *life expectancy* and is given by

$$e_0 := \int_0^\omega af(a)da = \int_0^\omega a\mu(a)\ell(a)da.$$

Integrating this expression partially, we have

$$\int_0^\omega a\mu(a)\ell(a)da = -\int_0^\omega a\frac{d\ell(a)}{da}da = \int_0^\omega \ell(a)da,$$

where we assume that $\lim_{a\to\omega} a\ell(a) = 0$. Moreover, we can define the *average remaining life expectancy* at age $a$, denoted by $e(a)$, as

$$e(a) := \frac{1}{\ell(a)}\int_a^\omega \ell(x)dx.$$

Note that $e(a)$ is not necessarily a monotone decreasing function. If infant mortality is particularly high, $e_0 = e(0)$ can be less than $e(a)$ for small $a$.

From the above, it is easy to see that the following equality holds:

$$\frac{1}{e_0}\int_0^\infty f(a)e(a)da = -\frac{1}{e_0}\int_0^\infty \ell(a)\log\ell(a)da, \tag{1.5}$$

where the right-hand side is called the *life table entropy*, which we denote by $H$. The above equality indicates that $H$ gives the ratio of the average remaining life expectancy lost by death to the life expectancy of newborns.

**Exercise 1.1** Define the *intrinsic age* by $\tau := a/e_0$ and the intrinsic survival probability by $\ell^*(\tau) := \ell(e_0\tau)$. Then, $\ell^*(\tau)$ can be interpreted as a probability density function. Calculate the probabilistic entropy $K = -\int_0^\infty \ell^*(\tau)\log\ell^*(\tau)d\tau$ to show that $K = H$ [36].

The life table entropy was introduced by Keyfitz [64] to examine the effect of changes in mortality on the life expectancy. Readers should note that the idea of *entropy* has been used in various aspects of demographic analysis [25–27, 47, 98, 128, 130]. Suppose that the mortality receives a small uniform perturbation $\delta\mu(a)$, $\delta \approx 0$. Then,

$$\exp\left(-(1+\delta)\int_0^a \mu(\sigma)d\sigma\right) \approx \ell(a)\left(1-\delta\int_0^a \mu(\sigma)d\sigma\right).$$

Hence, the relative change in life expectancy $e_0$ is calculated as

$$\frac{\Delta e_0}{e_0} = \frac{-\delta\int_0^\infty \ell(a)\int_0^a \mu(\sigma)d\sigma da}{\int_0^\infty \ell(a)da} = \delta\frac{\int_0^\infty \ell(a)\log\ell(a)da}{\int_0^\infty \ell(a)da}.$$

Therefore, we obtain

$$\frac{\Delta e_0}{e_0} = -\delta H,$$

which implies that $H$ describes the elasticity of the life expectancy with respect to a uniform change in mortality.

Let us consider two typical survival schedules. The *Type I* schedule is given as follows:

$$\ell(a) = \begin{cases} 1, & 0 \le a < \omega, \\ 0, & \omega \le a, \end{cases}$$

where $\omega$ is the maximum attainable age and $f(a) = \delta(a-\omega)$ ($\delta$ denotes the Dirac's delta function). The life table entropy of the Type I schedule is zero. If the force of mortality is a constant $\mu$, the schedule is called *Type II*. It is easy to see that the life table entropy of the Type II schedule is unity. Therefore, the life table entropy provides information about the convexity of the survival curve and the concentration of death events. From the above expression, we know that a decrease in entropy $H$ implies a decrease in the loss of remaining life due to death.

**Exercise 1.2** Suppose that a death is "repaired" with a probability $\pi \in [0, 1]$, or assume that $\pi$ is the recovery rate from any cause of death. Let $\ell_j(a)$ be the survival probability at age $a$ and state $j$ where the "state" denotes the number of repairs (resuscitations). We obtain the following *resuscitation model* [71, 90, 131, 132]:

$$\begin{aligned} \frac{d\ell_0(a)}{da} &= -\mu_0(a)\ell_0(a), \\ \frac{d\ell_n(a)}{da} &= \pi\mu_0(a)\ell_{n-1}(a) - \mu_0(a)\ell_n(a), \end{aligned} \tag{1.6}$$

where $\mu_0$ denotes the natural death rate and, for simplicity, we assume that $\mu_0$ can be applied to any state. The initial data are $\ell_0(0) = 1$ and $\ell_n(0) = 0$ ($n \ge 1$). Solving the above system of ODEs shows that the unconditional survival probability $\ell(a)$ is given by

$$\begin{aligned}\ell(a) &:= \sum_{n=0}^{\infty} \ell_n(a) = \ell_0(a) \sum_{n=0}^{\infty} \frac{(-\pi \log \ell_0(a))^n}{n!} \\ &= \ell_0(a) \exp(-\pi \log \ell_0(a)) = \ell_0(a)^{1-\pi}.\end{aligned} \tag{1.7}$$

The effective mortality is then given by

$$\mu(a) := -\frac{1}{\ell(a)} \frac{d\ell(a)}{da} = (1-\pi)\mu_0(a). \tag{1.8}$$

From the above exercise, it is apparent that a uniform decline in mortality can be realized by the resuscitation model. Let $e_0^{(1)}$ be the life expectancy under one-time resuscitation. Then, it follows that

$$e_0^{(1)} = \int_0^\infty \ell_0(a)(1 - \pi \log \ell_0(a))da = e_0(1 + \pi H), \tag{1.9}$$

where $e_0 = \int_0^\infty \ell_0(a)da$ is the life expectancy of the natural death rate and $H$ is the life table entropy of $\ell_0(a)$. The relative prolongation of life expectancy by one-time repair is then given by $\pi H$.

**Exercise 1.3** In a paper published in 1766, Daniel Bernoulli developed a mathematical model to estimate the increase in life expectancy after the elimination of smallpox (a potentially lethal infectious disease) [5, 8, 32]. Suppose that a host population is in an endemic steady state. Let $s(a)$ be the ratio of susceptible individuals at age $a$, $r(a)$ be the ratio of recovered individuals at age $a$, $\lambda(a)$ be the force of infection, and $\mu(a)$ be the natural mortality from other causes. The basic model can then be expressed as follows:

$$\begin{aligned}\frac{ds(a)}{da} &= -(\lambda(a) + \mu(a))s(a), \quad s(0) = 1, \\ \frac{dr(a)}{da} &= (1 - c(a))\lambda(a)s(a) - \mu(a)r(a), \quad r(0) = 0,\end{aligned} \tag{1.10}$$

where $1 - c(a)$ is the rate of recovery from infection with lifelong immunity and $c(a)$ is the age-dependent *case fatality*.

In this model, the infective period is neglected, because it is very short compared with the life expectancy.

1. Let $\ell(a) = s(a) + r(a)$ be the survival probability in the endemic steady state. Calculate the *catalytic curve* [93] $C(a) := r(a)/\ell(a)$.
2. Show that the following relation holds:

$$\ell_0(a) = \frac{\ell(a)}{1 - \int_0^a c(\tau)\lambda(\tau)e^{-\int_0^\tau \lambda(\sigma)d\sigma}d\tau}, \tag{1.11}$$

where $\ell_0(a)$ denotes the survival probability after the elimination of smallpox.

**Exercise 1.4** If the force of mortality depends on time $t \in \mathbb{R}$, the time-dependent survival probability $\ell(t,a)$ that an individual born at time $t-a$ survives to age $a$ is defined as a solution of the McKendrick equation

$$\left(\frac{\partial}{\partial t}+\frac{\partial}{\partial a}\right)\ell(t,a)=-\phi(t,a), \quad \ell(t,0)=1, \tag{1.12}$$

where $\phi(t,a) := \mu(t,a)\ell(t,a)$ is the probability density of death at time $t$ and age $a$.

1. Solve (1.12).
2. Show that the following relation holds:

$$\int_0^\infty \phi(t,a)da = 1-\frac{d}{dt}\int_0^\infty \ell(t,a)da, \tag{1.13}$$

   where $\int_0^\infty \phi(t,a)da$ is the period total mortality rate and $\int_0^\infty \ell(t,a)da$ is the *cross-sectional average length of life* [59].

**Exercise 1.5** Keyfitz [41] and Gyllenberg [49] pointed out that Euler [41] had given a formula for the present value of a whole-life annuity of one unit per year commencing at age $x$ (payable continuously) as

$$\bar{a}_x = \int_x^\infty e^{-r(s-a)}\frac{\ell(s)}{\ell(a)}ds,$$

where $r$ is the interest rate and $\bar{a}_x$ is the notation in actuarial mathematics. Prove the above formula by using the fact that

$$\bar{a}_x = \int_0^\infty \mu(x+\tau)\frac{\ell(x+\tau)}{\ell(x)}\int_0^\tau e^{-rs}dsd\tau,$$

and show its interpretation. Compare $\bar{a}_x$ with the reproductive value (1.73).

### 1.1.3 Fertility

The *age-specific birth rate* $\beta(a)$ is defined such that an average individual produces $\beta(a)da$ live births during the age interval $[a, a+da]$ per unit time. Therefore,

$$\int_{a_1}^{a_2} \beta(a)p(t,a)da$$

gives the number of newborns produced by the population $p(t,a)$ in age interval $[a_1, a_2]$ at time $t$. In this chapter, we consider a one-sex population model where $\beta(a)$

denotes the birth rate of newborns whose sex is the same as the adult population $p$, or we simply disregard the gender differences of individuals (Fig. 1.3).

The number of births produced from all individuals per unit time at time $t$, denoted by $B(t)$, is given by

$$B(t) = \int_0^\omega \beta(a)p(t,a)da. \tag{1.14}$$

The *crude birth rate* (CBR), defined as the number of births per capita per unit time, is given by

$$CBR := \frac{B(t)}{P(t)} = \frac{\int_0^\omega \beta(a)p(t,a)da}{\int_0^\omega p(t,a)da}.$$

We assume that the age-specific birth rate $\beta(a)$ is a nonnegative, bounded integrable function that can only be positive during the *reproductive age* interval $[\beta_1, \beta_2], 0 < \beta_1 < \beta_2 < \omega$. That is, we assume that $\beta(a) = 0, a \notin [\beta_1, \beta_2]$.

Attention should be paid to the difference between $\mu$ and $\beta$ from a statistical point of view. The force of mortality can be considered as the hazard rate of a non-repeatable event[2] for which the population at risk is the whole population. However, the age-specific birth rate is the incidence of births (the number of births per unit time per capita), and its population at risk depends on the *parity* (the number of children a woman has had) and mating chances. In this scenario, the stochastic interpretation of $\beta$ is not unique.

For example, let $X(a)$ be a probability variable that denotes the number of births at age $a$, and let $\beta(a)$ be the force of childbearing at age $a$, that is, the probability of the occurrence of a birth in the time interval $[a, a+h)$ is given by $\beta(a)h + o(h)$ and the probability that more than one births occur is given by $o(h)$. Let $P_n(a) := Pr(X(a) = n)$. The birth process can then be expressed by the time-inhomogeneous Poisson process

$$\begin{aligned} \frac{dP_0(a)}{da} &= -\beta(a)P_0(a), \\ \frac{dP_n(a)}{da} &= -\beta(a)P_n(a) + \beta(a)P_{n-1}(a), \quad (n \geq 1) \end{aligned}$$

where $P_0(0) = 1$, $P_n(0) = 0$ $(n \geq 1)$. By solving the above system, the family size distribution at age $a$ is determined to be

$$P_n(a) = \exp\left(-\int_0^a \beta(\sigma)d\sigma\right)\frac{1}{n!}\left(\int_0^a \beta(\sigma)d\sigma\right)^n.$$

Therefore, the average number of children at age $a$ is as follows:

[2]A demographic event is called *non-repeatable* if the population at risk is composed of individuals who have not yet experienced the event. A demographic event is called *repeatable* if the population at risk is not affected by the occurrence of the event [59].

$$\sum_{n=1}^{\infty} n P_n(a) = \exp\left(-\int_0^a \beta(\sigma)d\sigma\right)\sum_{n=1}^{\infty}\frac{1}{(n-1)!}\left(\int_0^a \beta(\sigma)d\sigma\right)^n$$
$$= \int_0^a \beta(\sigma)d\sigma \exp\left(-\int_0^a \beta(\sigma)d\sigma\right)\sum_{n=0}^{\infty}\frac{1}{n!}\left(\int_0^a \beta(\sigma)d\sigma\right)^n$$
$$= \int_0^a \beta(\sigma)d\sigma,$$

and $\int_0^\infty \beta(\sigma)d\sigma$ gives the lifelong expected total number of children (*total fertility rate*, TFR). Under such a stochastic interpretation, the final family size distribution is given by

$$P_n(\infty) = e^{-\alpha}\frac{\alpha^n}{n!},$$

where $\alpha$ is the TFR [67]. However, the above formula is not a good estimate for the real family size distribution of humans. This is because, in reality, the force of childbearing is controlled by the parity status of women; therefore, the simple Poisson process is not appropriate for modeling the human birth process [67]. To understand human fertility, we need far more elaborate mathematical models [115, 116].

### *1.1.4 Malthusian Populations*

Because an individual aged $a$ at time $t$ was born at time $t-a$ and has survived to age $a$, if we consider a closed population, we have the relations

$$p(t,a) = B(t-a)\ell(a), \quad t \in \mathbb{R}, \quad a \in [0,\omega] \tag{1.15}$$

and

$$P(t) = \int_0^\omega B(t-a)\ell(a)da. \tag{1.16}$$

Suppose that $B(t)$ is given by an exponential function as

$$B(t) = B_0 e^{\lambda t}, \quad t \in \mathbb{R}, \tag{1.17}$$

where $B_0$ is a positive number and $\lambda$ is the growth rate. Inserting (1.17) into (1.15), the age-density function is given by

$$p(t,a) = B_0 e^{rt} e^{-ra}\ell(a),$$

and so the *age profile* (the normalized age distribution), denoted by $w(a)$, becomes time-independent:

$$p(t, a) = P(t)w(a),$$

where

$$P(t) = \int_0^{\omega} p(t, a)da = B_0 e^{\lambda t} \int_0^{\omega} e^{-\lambda a} \ell(a)da,$$

$$w(a) := \frac{p(t, a)}{\int_0^{\omega} p(t, x)dx} = \frac{e^{-\lambda a} \ell(a)}{\int_0^{\omega} e^{-\lambda x} \ell(x)dx}.$$

If the survival probability is time-independent and the birth rate $B(t)$ obeys the exponential law, the population is said to be a *Malthusian population* and the growth rate $\lambda$ is called the *Malthusian parameter*. It follows from the above observations that, in a Malthusian population, the age profile is time-independent and its total size grows exponentially [73, 74].

The CBR of the Malthusian population is time-independent and given by

$$b := \frac{B(t)}{P(t)} = \frac{1}{\int_0^{\omega} e^{-\lambda a} \ell(a)da}.$$

Thus, we obtain

$$w(a) = be^{-\lambda a} \ell(a), \tag{1.18}$$

which we call the *Malthusian distribution*. As we shall discuss, if $\lambda$ is the intrinsic rate of natural increase, the Malthusian distribution is the stable age distribution.

If the survival rate and the CBR are time-independent, we can prove that the population converges to a Malthusian population over time. In fact, if the CBR $b$ is a constant, $B(t) = bP(t)$, and it follows from (1.16) that

$$P(t) = b \int_0^{\omega} P(t - a)\ell(a)da, \tag{1.19}$$

or equivalently that

$$B(t) = b \int_0^{\omega} B(t - a)\ell(a)da. \tag{1.20}$$

If we assume that $P(t)$ or $B(t)$ is known for $t < 0$, (1.19) and (1.20) become Volterra integral equations (renewal equations). As we shall see in the next section, (1.20) is a special case of Lotka's integral equation in which the age-specific birth rate is constant: $\beta(a) \equiv b$. Using a similar argument as in the Fundamental Theorem of Demography (Proposition 1.9), we know that $P(t)$ and $B(t)$ become asymptotically proportional to $e^{\lambda_0 t}$ as time evolves where $\lambda_0$ is the unique real root of the characteristic equation

$$1 = b \int_0^{\omega} e^{-\lambda_0 a} \ell(a)da. \tag{1.21}$$

Thus, we conclude that the population converges to a Malthusian population if the survival probability and the CBR are time-independent [108].

In the Malthusian population, the total number of deaths per unit time is given by

$$D(t) = \int_0^\omega \mu(a)p(t,a)da = B_0 e^{rt} \int_0^\omega e^{-ra}\ell(a)\mu(a)da.$$

Because $\mu(a)\ell(a) = -d\ell(a)/da$ and $\ell(\omega) = 0$, it follows from partial integration that

$$\int_0^\omega e^{-\lambda a}\ell(a)\mu(a)da = 1 - \lambda \int_0^\omega e^{-\lambda a}\ell(a)da.$$

Therefore, the CDR $d$ is calculated as

$$d := \frac{D(t)}{P(t)} = b - \lambda. \tag{1.22}$$

Because the survival probability $\ell(a)$ is a decreasing function, the age profile $w(a)$ also becomes a decreasing function with respect to age if $\lambda > 0$, which is the main reason why the age distribution of a growing population has a pyramid-like formula. Conversely, if $\lambda < 0$, the age profile becomes a unimodal distribution, which implies a highly aged population structure. In general, we expect an exponentially growing population to be a Malthusian population. The robust nature of the Malthusian distribution is guaranteed by the stable population theory, as long as changes in fertility are not overly large.

Let $A$ be the average age of a Malthusian population. Then, we have

$$A := \frac{\int_0^\omega ap(t,a)da}{\int_0^\omega p(t,a)da} = \int_0^\omega aw(a)da = b\int_0^\omega ae^{-\lambda a}\ell(a)da.$$

A simple calculation gives

$$\frac{dA}{d\lambda} = -\int_0^\omega (a-A)^2 w(a)da.$$

Thus, the average age of the Malthusian distribution is a decreasing function of the growth rate $\lambda$, so a decrease in the growth rate leads to an aging in the population structure. If $\lambda < 0$, the age profile $w(a)$ will reach an extrema in an old age class, which is the main reason for population aging.

**Exercise 1.6** Derive the following equality and describe the effect of changes in the Malthusian parameter on the age profile of the Malthusian population.

$$\frac{d\log w(a)}{d\lambda} = A - a.$$

If $\lambda = 0$, the size of the Malthusian population is constant, leading to the so-called *stationary population*. The age profile of the stationary population is given by

$$w(a) = \frac{\ell(a)}{\int_0^\omega \ell(x)dx} = \frac{\ell(a)}{e_0},$$

that is, the age distribution of the stationary population is proportional to the survival probability and its CBR is $1/e_0$. Moreover, the size of the stationary population is given by

$$P = \int_0^\omega p(a)da = B_0 \int_0^\omega \ell(a)da = B_0 e_0.$$

Hence, the total size of a stationary population is the product of the annual number of newborns and the average life expectancy.

**Exercise 1.7** Consider a closed population $p(t, a) = B(t-a)\ell(a), (t, a) \in \mathbb{R} \times \mathbb{R}_+$ where $\ell(a)$ denotes a given survival probability and $\ell'(a) = -\mu(a)\ell(a)$. Suppose that the age profile $w(t, a) = \frac{p(t,a)}{\int_0^\infty p(t,x)dx}$ is time-independent. Show that there exists a constant $\lambda$ such that $B(t) = B(0)e^{\lambda t}$ (Bortkiewicz's stable population, [92, 108, 113]).

## 1.2 Fundamental Theorem of Demography

### 1.2.1 The Stable Population Model

In the following, we assume that a female closed population reproduces under given time-independent birth and death rates. Although, in reality, pair formation between a male and a female is essential to produce newborns, we assume that the pair formation phenomena can be neglected as there exists a sufficient number of males that female fertility is guaranteed.

Let $F(t, a)$ be the cumulative population below the age of $a$ at time $t$:

$$F(t, a) := \int_0^a p(t, \sigma)d\sigma.$$

Let $B(t)$ be the number of newborns produced per unit time at time $t$, and let $\mu(a)$ be the force of mortality. If we consider the variation in the population during a time interval $h > 0$, we can obtain the following equation of balance (demographic accounting equation):

$$F(t+h, a+h) - F(t, a) = \int_t^{t+h} B(s)ds - \int_0^h \int_0^{a+s} \mu(\sigma)p(t+s, \sigma)d\sigma ds.$$

The first term on the right-hand side corresponds to the population increase due to newborns, and the second term is the decrease due to natural deaths. Differentiating the above equality with respect to $h$ and setting $h = 0$, we arrive at the continuity equation [69]:

$$p(t,a) + \int_0^a \frac{\partial p(t,\sigma)}{\partial t} d\sigma = B(t) - \int_0^a \mu(\sigma)p(t,\sigma)d\sigma.$$

In particular, if we set $a = 0$, we obtain $p(t,0) = B(t)$. Hence, if we introduce the age-specific (female) birth rate $\beta(a)$, it follows that

$$p(t,0) = B(t) = \int_0^\omega \beta(a)p(t,a)da, \tag{1.23}$$

where $\omega < +\infty$ is some fixed age below the upper bound of attainable age and above the maximum reproductive age so that $\mu(\cdot) \in L_+^\infty(0,\omega)$.

Differentiating the continuity equation with respect to $a$, we obtain

$$\frac{\partial p(t,a)}{\partial t} + \frac{\partial p(t,a)}{\partial a} = -\mu(a)p(t,a). \tag{1.24}$$

This is the *McKendrick equation*, which describes the aging process along the cohort lines (characteristic lines) in the $(t,a)$-plane (the *Lexis plane* in demographic terminology).

If we add the initial data $p(0,a) = p_0(a)$, the one-sex age-structured population dynamics can be formulated as an initial-boundary value problem:

$$\begin{aligned} &\frac{\partial p(t,a)}{\partial t} + \frac{\partial p(t,a)}{\partial a} = -\mu(a)p(t,a), \quad t > 0, \quad 0 < a < \omega, \\ &p(t,0) = \int_0^\omega \beta(a)p(t,a)da, \quad t > 0, \\ &p(0,a) = p_0(a). \end{aligned} \tag{1.25}$$

According to the terminology used in demography, we call this system (1.25) the *stable population model*. In the following, we consider the mathematical properties of the stable population model under reasonable conditions for human populations (as opposed to the most general conditions). The age-structured population described by the stable population model is often called a *stable population*, but note that, depending on the context, a stable population implies the exponential solution (Malthusian population) of the stable population model.

First, we solve the McKendrick equation along the characteristic line. Let us fix $(t,a)$ and define functions of $h$ as $p(t+h,a+h) \equiv \bar{p}(h)$ and $\mu(a+h) \equiv \bar{\mu}(h)$. Because

$$\frac{d\bar{p}(h)}{dh} = \frac{\partial p(t+h,a+h)}{\partial t} + \frac{\partial p(t+h,a+h)}{\partial a},$$

the McKendrick equation can be written as

$$\frac{d\bar{p}(h)}{dh} = -\bar{\mu}(h)\bar{p}(h). \tag{1.26}$$

Equation (1.26) implies that the size of a cohort (individuals born in the same period) can only decrease by the occurrence of deaths. Integrating (1.26), we obtain

$$\bar{p}(h) = \bar{p}(0)e^{-\int_0^h \bar{\mu}(\sigma)d\sigma}.$$

Replacing the variables by the original ones, we arrive at the following relation:

$$p(t+h, a+h) = p(t, a)e^{-\int_0^h \mu(a+\sigma)d\sigma}.$$

If $t > a$, we can replace $(t, a)$ by $(t-a, 0)$ and set $h = a$, which gives

$$p(t, a) = p(t-a, 0)e^{-\int_0^a \mu(\sigma)\sigma} = B(t-a)\ell(a).$$

However, if $t < a$, we can replace $(t, a)$ by $(0, a-t)$ and set $h = t$, giving

$$p(t, a) = p(0, a-t)e^{-\int_0^t \mu(a-t+\sigma)\sigma} = p_0(a-t)\frac{\ell(a)}{\ell(a-t)}.$$

Therefore, we arrive at the following expression:

$$p(t, a) = \begin{cases} B(t-a)\ell(a), & t-a > 0, \\ p_0(a-t)\frac{\ell(a)}{\ell(a-t)}, & a-t > 0. \end{cases} \tag{1.27}$$

Thus, if the initial data point $p_0$ and the unknown boundary value $B(t)$ are given, the age-density function is completely determined by (1.27).

Note that if $p(t, a)$ given by (1.27) is a solution of the McKendrick equation in a classical sense, it should be continuous and partially differentiable. To ensure continuity along the characteristic line $t - a = 0$, the initial data must satisfy the *consistency condition*

$$p_0(0) = B(+0) = \int_0^\omega \beta(a)p_0(a)da. \tag{1.28}$$

When $t \leq \omega$, inserting expression (1.27) into the boundary condition (1.23) and dividing the integral interval into $(0, t)$ and $(t, \omega)$ give

$$B(t) = G(t) + \int_0^t \Psi(a)B(t-a)da, \tag{1.29}$$

where $\Psi(a) := \beta(a)\ell(a)$ and $G(t)$ is a known function given by

$$G(t) := \int_t^{\omega} \beta(a)\frac{\ell(a)}{\ell(a-t)} p_0(a-t)da = \int_0^{\omega} \beta(a+t)\frac{\ell(a+t)}{\ell(a)} p_0(a)da, \quad (1.30)$$

where we use the convention that $\beta(a) = 0$ and $\ell(a) = 0$ for $a > \omega$. Therefore, we have $G(t) = 0$ for $t > \omega$, and (1.29) holds for all $t > 0$. In biological terms, $G(t)$ denotes the number of newborns produced by the initial female population per unit time. The special case $G(t) \equiv 0$ occurs if

$$\int_0^{\omega} \beta(a+t)p_0(a)\frac{\ell(a+t)}{\ell(a)}da = 0, \quad \forall t \geq 0.$$

Observe that $G(t) = 0$ for all $t \geq 0$ if and only if, for all $t \geq 0$, the following holds almost everywhere for $a \in [0, \omega]$:

$$\beta(a+t)p_0(a) = 0. \quad (1.31)$$

The condition in (1.31) occurs if and only if $p_0(a) = 0$ for almost all $a$ less than the maximum reproductive age—that is, the initial population is composed of individuals of post-reproductive age who can produce no children. Hence, the initial data satisfying condition (1.31) are called *trivial data*; otherwise, they are non-trivial data.

In the field of demography, the integral kernel $\Psi$ is called the *net reproduction function* or *net maternity function*. The integral equation (1.29) for an unknown function $B(t)$ is known to be a Volterra-type integral equation, or *renewal integral equation*. In the terminology of demography, this is sometimes called *Lotka's integral equation*. Readers are referred to [80–82, 84] for Lotka's integral equation approach. His personal history is summarized in [34, 66, 97].

*Remark 1.1* The McKendrick equation was first proposed by McKendrick in 1926 [87]. After being neglected for some 30 years, it was rediscovered independently by Scherbaum and Rasch [109] and Von Foerster [134] as a model for cellular systems. The age-structured cell population model is a special case of the age-size-dependent model [6, 117], which has been mathematically investigated by several authors [51, 140]. Bell and Anderson's cell population model is formulated as follows:

$$\frac{\partial p(t,a,x)}{\partial t} + \frac{\partial p(t,a,x)}{\partial a} + \frac{\partial v(a,x)p(t,a,x)}{\partial x} = -(b(a,x)+\mu(a,x))p(t,a,x),$$
$$p(t,0,x) = 2\int_0^{\infty}\int_0^{\infty} k(x,\sigma)b(a,\sigma)p(t,a,\sigma)d\sigma da,$$

where $p(t,a,x)$ is the density of cell population with age $a$ and size (or biomass, DNA, etc.) $x$, $v(a,x)$ is the growth rate of size $x$ at age $a$ and size $x$, $b(a,x)$ is the fission rate, $\mu(a,x)$ is the natural death rate, and $k(x,\sigma)$ is the fission probability that a daughter cell with size $x$ is produced by a mother cell with size $\sigma$. It is assumed that $k(x,\sigma) = 0$ for $x > \sigma$, $k(x,\sigma) = k(\sigma - x, \sigma)$, and $\int_0^{\infty} k(x,\sigma)dx = 1$. If the fission is symmetrical, we can assume that $k(x, \sigma) = \delta(x - \frac{\sigma}{2})$, and the boundary condition is reduced to

$$p(t, 0, x) = 4 \int_0^\infty b(a, 2x) p(t, a, 2x) da.$$

If we can assume that parameters are independent from size $x$, the age-density function $n(t, a) := \int_0^\infty p(t, a, x) dx$ satisfies the age-dependent model [105]:

$$\frac{\partial n(t,a)}{\partial t} + \frac{\partial n(t,a)}{\partial a} = -(b(a) + \mu(a)) n(t, a),$$
$$n(t, 0) = 2 \int_0^\infty b(a) n(t, a) da,$$

which is a kind of the stable population model. On the other hand, if we can assume that parameters are independent from age $a$ and the fission is symmetrical, the size distribution $m(t, x) := \int_0^\infty p(t, a, x) da$ satisfies the following model:

$$\frac{\partial m(t,x)}{\partial t} + \frac{\partial v(x) m(t,x)}{\partial x} = -(\mu(x) + b(x)) m(t, x) + 4b(2x) m(t, 2x)$$
$$m(t, \frac{x_0}{2}) = 0,$$

where $x_0$ is the minimum size at which the cell fission can occur. This size-dependent model is much more difficult to analyze [86]. Some historical aspects of the McKendrick equation were discussed by Dietz [31]. As we will see in Chap. 5, the stable population model provides a basic scheme for infection-age-structured epidemic models [137]. Although our primary concern in this text is human population dynamics, it is interesting and instructive to examine applications of the stable population theory to biological populations. Readers can find many applications of the stable population model in genetics, ecology, and population biology [14, 39, 53, 104].

### *1.2.2 Classical Solutions*

Roughly speaking, once the behavior of $B(t)$ has been determined by solving Lotka's integral equation (1.29), expression (1.27) gives a solution of the stable population model. However, we require a precise definition of the "solution" of the McKendrick system (1.25) before stating an existence and uniqueness theorem. First, we introduce the classical solution according to Iannelli [55]. Let $\Omega := \mathbb{R}_+ \times [0, \omega]$ and let $W^{1,p}(0, \omega) := \{\phi \in L^p(0, \omega) : \phi \text{ is differentiable a.e.}, \phi' \in L^p(0, \omega)\}$.

**Definition 1.1** We call $p$ the *classical solution* of the McKendrick system (1.25) if $p \in C_+(\Omega)$, $\partial p / \partial t$ exists for almost all $(t, a) \in \Omega$, $p(t, \cdot) \in W^{1,1}(0, \omega)$, $\mu(\cdot) p(t, \cdot) \in L^1(0, \omega)$ for $t \geq 0$ and $p$ satisfies the McKendrick system (1.25).

**Lemma 1.1** *Under our assumption, $\Psi(a) = 0$ for $a > \omega$ and $\Psi \in L^1_+(\mathbb{R}_+) \cap L^\infty_+(\mathbb{R}_+)$. Then, $G \in C_+(\mathbb{R}_+)$. Moreover, if $p_0 \in W^{1,1}(0, \omega)$ and $\mu p_0 \in L^1(0, \omega)$, then $G \in W^{1,\infty}(\mathbb{R}_+)$.*

**Exercise 1.8** Prove the above lemma.

**Proposition 1.1** *For the given initial data $p_0 \in L^1_+(0, \omega)$, (1.29) has a unique solution $B \in C_+(\mathbb{R}_+)$ such that*

$$|B(t)| \le \bar{\beta} e^{(\bar{\beta}-\underline{\mu})t} |p_0|_{L^1}, \tag{1.32}$$

*where $\bar{\beta} := \sup_{a\in[0,\omega]} \beta(a)$, $\underline{\mu} := \inf_{a\in[0,\omega]} \mu(a)$. If $p_0 \in W^{1,1}(0, \omega)$ and $\mu p_0 \in L^1(0, \omega)$, then $B \in W^{1,\infty}_{\mathrm{loc}}(\mathbb{R}_+)$ and the following holds:*

$$B'(t) = G'(t) + \Psi(t)B(0) + \int_0^t \Psi(a)B'(t-a)da. \tag{1.33}$$

*Proof* First, we can easily obtain the following estimates:

$$|G(t)| \le \bar{\beta} e^{-\underline{\mu}t} |p_0|_{L^1}, \quad |\Psi(a)| \le \bar{\beta} e^{-\underline{\mu}a}. \tag{1.34}$$

Then, if there exists a solution $B$ of (1.29), it follows that

$$|B(t)| \le \bar{\beta} e^{-\underline{\mu}t} |p_0|_{L^1} + \bar{\beta} \int_0^t |B(s)| e^{-\underline{\mu}(t-s)} ds.$$

Using Gronwall's inequality, we have (1.32). Next, for any fixed $t_0 > 0$, observe that there exists a sequence $\{B^k\}_{k=0,1,2,..} \subset C([0, t_0])$ such that

$$B^0(t) = G(t), \quad B^{k+1}(t) = G(t) + \int_0^t \Psi(t-s)B^k(s)ds.$$

It is then easy to show that

$$|B^k - B^{k-1}|_{C([0,t_0])} \le |\Psi|_{L^\infty(\mathbb{R}_+)} \frac{t_0^{k-1}}{(k-1)!} |B^1 - B^0|_{C([0,t_0])}. \quad (k \ge 2).$$

Thus, $\lim_{k\to\infty} B^k$ exists in $C_+([0, t_0])$ and gives a continuous solution of (1.29). If there are two solutions $B_1$ and $B_2$, the same iteration procedure gives that, for any integer $k$,

$$|B_1 - B_2|_{C([0,t_0])} \le |\Psi|_{L^\infty(\mathbb{R}_+)} \frac{t_0^{k-1}}{(k-1)!} |B_1 - B_2|_{C([0,t_0])},$$

which shows $B_1 \equiv B_2$; therefore, the solution is unique. Finally, suppose that $p_0 \in W^{1,1}(0, \omega)$ and $\mu p_0 \in L^1(0, \omega)$. It follows from Lemma 1.1 that $B^k \in W^{1,\infty}(\mathbb{R}_+)$. Let $V^k(t) := dB^k(t)/dt \in L^\infty(\mathbb{R}_+)$. We then have that

$$V^{k+1}(t) = G'(t) + \Psi(t)G(0) + \int_0^t \Psi(t-s)V^k(s)ds,$$

which yields

$$\left|V^{k+1} - V^k\right|_{L^\infty(\mathbb{R}_+)} \le |\Psi|_{L^\infty(\mathbb{R}_+)} \frac{t_0^k}{k!} \left|V^1 - V^0\right|_{L^\infty(\mathbb{R}_+)}.$$

Thus, the sequence $V^k$ again converges in $L^\infty(\mathbb{R}_+)$ to a function $V(t) \in L^\infty(\mathbb{R}_+)$ that satisfies

$$V(t) = G'(t) + \Psi(t)G(0) + \int_0^t \Psi(s)V(t-s)ds.$$

Integrating both sides of the above integral equation shows that $G(0) + \int_0^t V(s)ds$ satisfies the renewal equation (1.29). That is, $B(t) \in W^{1,\infty}_{\text{loc}}(\mathbb{R}_+)$, $B'(t) = V(t)$, and $B'$ satisfies (1.33). □

Let us define a subspace $\mathscr{D}(A) \subset L^1(0,\omega)$ by

$$\mathscr{D}(A) := \left\{\phi \in W^{1,1}(0,\omega) : \mu\phi \in L^1(0,\omega), \phi(0) = \int_0^\omega \beta(a)\phi(a)da\right\}. \quad (1.35)$$

**Lemma 1.2** *The set $\mathscr{D}(A)$ is dense in $L^1(0,\omega)$.*

*Proof* The set $\mathscr{D}(A)$ is the domain of the infinitesimal generator $A\phi := -\phi' - \mu\phi$ of the population semigroup. Proposition 10.4 implies that $\mathscr{D}(A)$ is dense in $L^1$. □

**Proposition 1.2** ([55], Proposition 4.2) *If $p_0 \in \mathscr{D}(A)$, the McKendrick system has a unique classical solution $p$ and $p(t,\cdot) \in \mathscr{D}(A)$ for all $t > 0$.*

*Proof* Let $p$ be a function defined by (1.27) with $p_0 \in \mathscr{D}(A)$. It follows from Proposition 1.1 that $p \in C_+(\Omega)$, $\partial p/\partial t$ exists for almost all $(t,a) \in \Omega$, $p(t,\cdot) \in W^{1,1}(0,\omega)$ and $p$ satisfies the McKendrick system (1.25). It is then sufficient to show that $\mu p(t,\cdot) \in L^1(0,\omega)$. For $t \ge \omega$, $p(t,a) = B(t-a)\ell(a)$, and it is clear that $\mu p(t,\cdot) \in L^1(0,\omega)$. For $t < \omega$, observe that

$$\int_0^\omega \mu(a)|p(t,a)|da \le \int_0^t \mu(a)\ell(a)|B(t-a)|da + \int_t^\omega \mu(a)|p_0(a-t)|\frac{\ell(a)}{\ell(a-t)}da,$$

where the first part is finite. Because $p_0$ is continuous on $[0,\omega]$, we have

$$\int_t^\omega \mu(a)|p_0(a-t)|\frac{\ell(a)}{\ell(a-t)}da \le \sup_{a\in[0,\omega]} |p_0(a)| e^{\int_0^{\omega-t}\mu(\sigma)d\sigma} \int_t^\omega \mu(a)\ell(a)da,$$

where the right-hand side is finite for $t < \omega$. □

### 1.2.3 Semigroup Solutions

Even when the right-hand side of (1.27) does not give a classical solution, it can be biologically meaningful in a weak sense. In fact, the partial differentiability is not needed to formulate the aging process along the characteristic line. To avoid the restriction caused by partial differentiability, the McKendrick equation can be formulated as follows:

$$Dp(t,a) = -\mu(a)p(t,a), \tag{1.36}$$

where $D$ denotes a differential operator along the characteristic line given by

$$Df(t,a) := \lim_{h\to 0} \frac{f(t+h, a+h) - f(t,a)}{h}. \tag{1.37}$$

**Definition 1.2** For a given $p_0 \in L^1(0,\omega)$, we call $p$ the *semigroup solution* for the McKendrick system (1.25) if $p$ satisfies (1.36) for almost all $(t,a) \in \Omega$, $p(t,0) = \int_0^\omega \beta(a)p(t,a)da$ for $t > 0$, $p(0,a) = p_0(a)$ and $p \in C_+(\mathbb{R}_+; L^1(0,\omega))$.

The expression in (1.27) can be seen as a linear operator in $L^1_+(0,\omega)$ that maps the initial data $p(0,a)$ to the age density $p(t,a)$ at some later time $t > 0$. Therefore, we define a family of linear operators $T(t), t \geq 0$, for $\phi \in L^1_+$ by

$$(T(t)\phi)(a) = \begin{cases} B(t-a;\phi)\ell(a), & t-a > 0, \\ \phi(a-t)\frac{\ell(a)}{\ell(a-t)}, & a-t > 0, \end{cases} \tag{1.38}$$

where $T(0) = I$ (the identity operator) and $B(t;\phi)$ is a solution of the renewal equation

$$B(t;\phi) = G(t;\phi) + \int_0^t \Psi(a)B(t-a;\phi)da, \tag{1.39}$$

and $G(t;\phi)$ is given by

$$G(t;\phi) := \int_0^\omega \beta(a+t)\frac{\ell(a+t)}{\ell(a)}\phi(a)da.$$

**Lemma 1.3** *For $\phi \in L^1(0,\omega)$, it follows that*

$$|T(t)\phi|_{L^1} \leq e^{(\bar{\beta}-\underline{\mu})t}|\phi|_{L^1}. \tag{1.40}$$

*Then, $T(t)$ is a bounded linear operator from $L^1(0,\omega)$ into itself for each $t > 0$.*

*Proof* From (1.32) and (1.38), we have

$$
\begin{aligned}
|T(t)\phi|_{L^1} &= \int_0^t |B(t-a;\phi)|\ell(a)da + \int_0^\omega \frac{\ell(a+t)}{\ell(a)}|\phi(a)|da \\
&\le \bar{\beta}\int_0^t e^{(t-a)(\bar{\beta}-\underline{\mu})}e^{-\underline{\mu}a}da + e^{-\underline{\mu}t}|\phi|_{L^1} = e^{t(\bar{\beta}-\underline{\mu})}|\phi|_{L^1}.
\end{aligned}
$$

Thus, we know that (1.40) holds. □

**Proposition 1.3** *For the given initial data $p_0 \in L^1_+(0,\omega)$, if a semigroup solution $p$ exists, then $p(t) = T(t)p_0$ and this solution is unique. Conversely, $T(t)p_0$ gives a semigroup solution for the McKendrick system* (1.25). *Moreover, if $p_0 \in \mathscr{D}(A)$, then $p(t) = T(t)p_0 \in \mathscr{D}(A)$ and*

$$\frac{dp(t)}{dt} = Ap(t) \tag{1.41}$$

*in $L^1(0,\omega)$ where $A\phi = -\phi' - \mu\phi$ for $\phi \in \mathscr{D}(A)$.*

*Proof* The first part is obvious from the construction of $T(t)$. Next, it is easy to see that $T(t)p_0$ satisfies (1.37) for almost all $(t,a) \in \Omega$, the boundary condition (1.23) for $t > 0$ and the initial data. Let us show that, for any $t_0 > 0$,

$$T(\cdot)\psi \in E := C_+([0,t_0];L^1(0,\omega)). \tag{1.42}$$

From Lemma 1.2, there exists a sequence $\{p_0^{(n)}\}_{n=1,2,\dots} \subset \mathscr{D}(A)$ such that

$$\lim_{n\to\infty} |p_0^{(n)} - p_0|_{L^1} = 0.$$

Let $p^{(n)} \in C_+(\Omega)$ be the classical solution corresponding to the initial data $p_0^{(n)}$, so we can write $p^{(n)}(t) = T(t)p_0^{(n)}$ and $p^{(n)}(\cdot) \in E$. From (1.40) and the linearity of the McKendrick system, we have

$$|T(\cdot)p_0 - p^{(n)}|_E \le e^{(\bar{\beta}-\underline{\mu})t_0}|p_0 - p_0^{(n)}|_{L^1}$$

and

$$|p^{(m)} - p^{(n)}|_E \le e^{(\bar{\beta}-\underline{\mu})t_0}|p_0^{(m)} - p_0^{(n)}|_{L^1},$$

which shows that $\{p^{(n)}\}$ is a Cauchy sequence in $E$ with the limit $T(\cdot)p_0 \in E$. Thus, we have (1.42) for any $t_0 > 0$, so we can conclude that $T(t)p_0$ is the semigroup solution. The final part is left as an exercise. □

**Exercise 1.9** Show that $p(t) = T(t)p_0 \in \mathscr{D}(A)$ and (1.41) holds if $p_0 \in \mathscr{D}(A)$.

**Proposition 1.4** *For any $t_1, t_2 \geq 0$, the following semigroup property holds:*

$$T(t_1)T(t_2) = T(t_1 + t_2). \tag{1.43}$$

*Proof* Let $p(t+t_1) = T(t+t_1)p_0, t \geq 0$. As a function of $t$, $p(t+t_1)$ is a semigroup solution with the initial data $T(t_1)p_0$, so we have $T(t+t_1)p_0 = T(t)T(t_1)p_0$. It then follows that $T(t_2+t_1)p_0 = T(t_2)T(t_1)p_0$ for any $p_0 \in L^1(0, \omega)$, which gives (1.43). □

**Exercise 1.10** Prove that $B(t; T(s)\phi) = B(t+s; \phi)$, $t, s > 0$. Using this fact and the definition of $T(t)$ (1.38), show that the semigroup property (1.43) holds.

From the above proposition, the one-parameter family of bounded linear operators $T(t), t > 0$, is a strongly continuous ($C_0$ class) semigroup of linear operators. We call $T(t), t \geq 0$, the *population semigroup*. The infinitesimal generator of the population semigroup is given by the population operator $A$ (see Chap. 10). For any $t > 0$, it follows that $T(t)(L^1_+) \subset L^1_+$. Moreover, we can show the following strict positivity:

**Proposition 1.5** *If $p_0$ is some non-trivial, nonnegative initial data, it follows that $|T(t)p_0|_{L^1} > 0$ for any $t > 0$.*

*Proof* If $|T(t_0)p_0|_{L^1} = 0$ for some $t_0 > 0$, it follows from the nonnegativity of $T(t)p_0$ that $T(t_0)p_0 = 0$ almost everywhere. For all $t \geq t_0$, we then have

$$|T(t)p_0|_{L^1} = |T(t-t_0)T(t_0)p_0|_{L^1} = 0.$$

If $p_0$ is some non-trivial, nonnegative initial data, it follows from Proposition 1.9 that $|T(t)p_0|_{L^1}$ becomes asymptotically positive, which is a contradiction. We can then conclude that, for any $t > 0$, $|T(t)p_0|_{L^1} > 0$. □

From expression (1.38), the population semigroup $T(t)$ can be decomposed as $T(t) = U(t) + W(t)$ where $U(t)$ and $W(t)$ are operators defined by

$$(U(t)\phi)(a) = \begin{cases} 0, & t-a > 0, \\ (T(t)\phi)(a), & a-t > 0, \end{cases} \tag{1.44}$$

$$(W(t)\phi)(a) = \begin{cases} (T(t)\phi)(a), & t-a > 0, \\ 0, & a-t > 0. \end{cases} \tag{1.45}$$

Thus, the following holds [50, 103, 139]:

**Proposition 1.6** *For any $t > 0$,* (i) *$W(t)$ is a compact operator and,* (ii) *$|U(t)\phi|_{L^1} \leq e^{-\nu t}|\phi|_{L^1}$ with $\nu = \inf \mu(\cdot)$.*

*Proof* Because (ii) is clear, let us prove (i). For a fixed $t > 0$, define $b_t(x; \psi)$, $x \in [0, t]$ as the solution of the renewal equation

$$b_t = G_t\phi + \Psi * b_t,$$

where $G_t$ is an operator from $L^1(0, \omega; \mathbb{R}^n)$ into $L^1(0, t; \mathbb{R}^n)$ given by

$$(G_t\phi)(x) = \int_x^{\omega} \frac{\Psi(a)}{\ell(a-x)}\phi(a-x)da, \quad x \in [0, t \wedge \omega],$$

and $G_t\phi = 0$ for $x \in (t \wedge \omega, t]$. Let $V_t$ be a convolution integral operator from $L^1(0, t; \mathbb{R}^n)$ into itself defined by

$$(V_t\phi)(x) = \int_0^x \Psi(a)\phi(x-a)da, \quad x \in [0, t].$$

$V_t$ is a bounded linear operator with a spectral radius of zero, and so we obtain the expression

$$b_t(x; \phi) = ((I - V_t)^{-1}G_t\phi)(x),$$

and

$$(W(t)\phi)(a) = \begin{cases} b_t(t-a; \phi)\ell(a), & t-a > 0, \\ 0, & a-t > 0. \end{cases}$$

It is easy to see that $G_t$ is a compact operator, so $W(t)$ is also a compact operator. □

**Corollary 1.1** *The population semigroup on a finite age space is compact for $t \geq \omega$ (eventually compact), so it is norm continuous on $t \in [\omega, \infty)$.*

Although we omit the spectral theory for population semigroups, the above property is a key for the asymptotic analysis of $T(t)$, $t \geq 0$, because the spectral mapping theorem holds for eventually norm continuous semigroups, so the growth bound of the population semigroup is given by the spectral bound of the population operator $A$ (see Chap. 10, [40, 94]). If $\omega = \infty$, however, we need the measure of non-compactness to observe the asymptotic behavior of $T(t)$ [139].

Finally, note that real populations have an infertile period before reaching reproductive age—that is, the minimum reproductive age is positive. In such a case, it would be remarkable if $T(t)$ could be expressed as the iteration of a simple, known operator without solving the integral equation. Suppose that $\beta$ has a compact support[3] $[\beta_1, \beta_2]$ with $\beta_1 > 0$. As newborns are produced only by the initial population for the time interval $0 < t < \beta_1$, we have

$$B(t; p_0) = \int_{\beta_1}^{\beta_2} \beta(a)p(t,a)da = \int_{\beta_1}^{\beta_2} \beta(a)\frac{\ell(a)}{\ell(a-t)}p_0(a-t)da.$$

Therefore, $T(t)$ for $0 < t \leq \beta_1$ is expressed as

[3]The support of a function $f$ is the closure of the set $\{x : f(x) \neq 0\}$ on $\mathbb{R}^n$.

$$(T(t)p_0)(a) = \begin{cases} \ell(a)\int_{\beta_1}^{\beta_2}\beta(s)\frac{\ell(s)}{\ell(s-t+a)}p_0(s-t+a)ds, & t-a>0, \\ p_0(a-t)\frac{\ell(a)}{\ell(a-t)}, & a-t>0. \end{cases} \tag{1.46}$$

For $t > \beta_1$, let $n = [t/\beta_1]$.[4] It then follows from the semigroup property that

$$T(t) = T(t-n\beta_1)T(n\beta_1) = T(t-n\beta_1)T^n(\beta_1).$$

$0 \le t - n\beta_1 < \beta_1$, $T(t-n\beta_1)$ and $T(\beta_1)$ are given by (1.46), and so $T(t)$ is expressed as the product of known simple, bounded linear operators.

For example, if we consider $T(t)p_0$ as a population projection, the future population up to $\beta_1$ years ahead can be calculated directly from the birth rate, death rate, and the initial data without solving the renewal equation. By repeating this projection, we can compute any future population, as long as it is described by the stable population model. This type of argument can also be applied to some nonlinear population models if there are no nonlinear interactions during the infertile age period $[0, \beta_1]$ [106, 107].

### 1.2.4 Generation Expansion and $R_0$

The well-known iteration procedure for constructing solutions of Lotka's integral equation has a biologically interesting interpretation [75–77]. Let us define the functions $B_n(t)$, $n = 1, 2, \ldots$, by the following iterative expression:

$$B_1(t) = G(t), \quad B_n(t) = \int_0^t \Psi(a)B_{n-1}(t-a)da, \quad (n \ge 2). \tag{1.47}$$

$B_n(t)$, $n \ge 1$, becomes a continuous function with a compact support. In fact, $G(t) = 0$ for $t \ge \beta_2$ and $B_n(t) = 0$ $(n \ge 1)$ for $t \notin [(n-1)\beta_1, n\beta_2]$. It is easy to see that $\sum_{n=1}^{\infty} B_n(t)$ is uniformly absolutely convergent in any finite time interval $[0, t_0] \subset \mathbb{R}_+$ and gives the continuous solution of the renewal equation (1.29).

The function $B_1(t)$ denotes the newborns (first generation) produced by the initial population per unit time, $B_2(t)$ denotes the newborns (the second generation, grandchildren of the initial population) produced by the first generation per unit time and so on. If $p_n(t,a)$ is the age-density function of the $n$th generation, we have

$$p_0(t,a) = \begin{cases} 0, & t-a>0, \\ p_0(a-t)\frac{\ell(a)}{\ell(a-t)}, & a-t>0, \end{cases}$$

$$p_n(t,a) = \begin{cases} B_n(t-a)\ell(a), & t-a>0, \\ 0, & a-t>0, \end{cases}, \quad n \ge 1.$$

[4] $[x]$ is the Gaussian notation for the largest integer that is not greater than $x$.

Therefore, the age-density function $p(t,a)$ is obtained as

$$p(t,a) = \sum_{n=0}^{\infty} p_n(t,a). \tag{1.48}$$

That is, the age-density function can be expressed as the sum of the age density of successive generations, which is called the *generation expansion* [106]. Because $p_n(t,a) = 0$ for $(t,a)$ such that $t - a \notin ((n-1)\beta_1, n\beta_2)$, we have $p_n(t,a) = 0$ for $n \geq 2 + [t/\beta_1]$. Thus, the generation expansion is a finite sum for a fixed $t$, so the number of generations that coexist at a given time is finite.

Let us calculate the total number of newborns produced in each generation. From (1.47), we have

$$\begin{aligned}
\int_0^\infty B_n(t)dt &= \int_0^\infty dt \int_0^t \Psi(a)B_{n-1}(t-a)da \\
&= \int_0^\infty \Psi(a)da \int_a^\infty B_{n-1}(t-a)dt \\
&= R_0 \int_0^\infty B_{n-1}(t)dt = R_0^{n-1} \int_0^\infty G(t)dt,
\end{aligned}$$

where

$$R_0 := \int_0^\infty \Psi(a)da$$

is called the *basic reproduction number*. From the above, it is apparent that the size of each generation in the stable population model expands or shrinks geometrically by the ratio $R_0$, so this is the per generation growth factor. The above argument implies that

$$\lim_{n\to\infty} \sqrt[n]{|B_n|_{L^1}} = R_0. \tag{1.49}$$

This generational interpretation of $R_0$ is clear for the scalar stable population model, but it is not self-evident (although still true) for the vector-valued stable population model (see Chap. 9).

The basic reproduction number in demography can also be interpreted as the average number of female children produced by a woman during her life. Traditionally, in the terminology of demography, $R_0$ is called the *net reproduction rate*, although $R_0$ is not a *rate* but a dimensionless number (ratio). Although it is certain that the true significance of $R_0$ was first clarified by Lotka's stable population theory, the origin of the idea of $R_0$ is controversial [108]. Contrary to the claims of Dublin and Lotka [35], Kuczynski [68] attributed its origin to Richard Böckh at the end of the nineteenth century, as also pointed out by Lotka himself [83]. Lewes [72] suggested that the concept of the net reproduction rate should actually be attributed to William Farr, whereas Nishiura and Inaba [96] found that the relevant conceptual frameworks for $R_0$ were developed 50–100 years earlier by French mathematicians. In the context of epidemiology, one of the original users of $R_0$ may have been Theophil Lotz in 1880 [95].

**Exercise 1.11** Let $f(a) = \mu(a)\ell(a)$ be the probability density function for the age at death. Show that

$$R_0 = \int_0^\infty f(a)M(a)da,$$

where $M(a) := \int_0^a \beta(x)dx$ denotes the lifetime reproduction of an individual who dies at age $a$ [110].

*Remark 1.2* Note that if $R_0 < 1$, the total number of future newborns is finite. In fact, we can observe that

$$\int_0^\infty B(t)dt = \int_0^\infty \sum_{n=1}^\infty B_n(t)dt$$
$$= \sum_{n=0}^\infty R_0^n \int_0^\infty G(t)dt = \frac{1}{1-R_0}\int_0^\infty G(t)dt.$$

For example, in 2005, the Japanese female population was calculated to have $R_0 = 0.61$. Thus, if this very low fertility rate were to persist, the total number of Japanese babies that will be born in the future is only 2.56 times the total number of babies born to women living in 2005.

**Exercise 1.12** Let us define a function $v(a)$ by

$$v(a) = \int_a^\infty \beta(x)\frac{\ell(x)}{\ell(a)}dx,$$

which gives the expected number of children produced by an individual at age $a$. Show that the average number of children produced by an individual of the initial population $p_0$ is

$$\frac{1}{P(0)}\int_0^\infty G(t)dt = \int_0^\infty v(a)w_0(a)da,$$

where $w_0(a) := p_0(a)/\int_0^\infty p_0(x)dx$ is the age profile of the initial population.

The *resolvent* with respect to the integral kernel $\Psi(t)$ is defined as the solution of the following integral equation (see Chap. 10):

$$\mathscr{R}(t) = \Psi(t) + \int_0^t \Psi(s)\mathscr{R}(t-s)ds.$$

Using the same argument as above, the resolvent can be calculated as follows:

$$\mathscr{R}(t) = \sum_{j=1}^{\infty} \Psi^{(j)}(t),$$

where $\Psi^{(j)}$, $j = 1, 2, \ldots$, are iteratively calculated as

$$\Psi^{(1)}(t) = \Psi(t), \quad \Psi^{(j+1)}(a) = (\Psi * \Psi^{(j)})(t) = \int_0^t \Psi(s)\Psi^{(j)}(t-s)ds,$$

where $*$ denotes the convolution operation and the above series is norm convergent. Using the resolvent, the solution of the integral equation (1.29) is given by

$$B(t) = G(t) + \int_0^t \mathscr{R}(a)G(t-a)ds,$$

because $B_{j+1} = \Psi^{(j)} * G$, $(j \geq 1)$.

If $p_0(a) = \delta(a)$, that is, the initial population is the unit population with age zero, we have $G(t) = \Psi(t)$. Therefore, the resolvent $\mathscr{R}(t)$ gives the birth rate at time $t$ of successive generations starting from one newborn at time zero.

Note that $\Phi(t) := \Psi(t)/R_0$ can be thought of as the probability density function of the childbearing age of the mothers. If we define

$$\Phi^{(1)}(t) = \Phi(t), \quad \Phi^{(j+1)}(t) = (\Phi * \Phi^{(j)})(t) = \int_0^t \Phi(s)\Phi^{(j)}(t-s)ds,$$

then $\Phi^{(j)}(t)$ gives the probability density function for the occurrence of the $j$th generation's birth. Observe that

$$\mathscr{R}(t) = \sum_{j=1}^{\infty} \Psi^{(j)}(t) = \sum_{j=1}^{\infty} R_0^j \Phi^{(j)}(t),$$

from which it is easy to see that the resolvent is integrable on $\mathbb{R}_+$ if and only if $R_0 < 1$, in which case

$$\int_0^{\infty} \mathscr{R}(t)dt = \sum_{j=1}^{\infty} R_0^j = \frac{R_0}{1 - R_0}$$

gives the average number of descendants of a newborn. It is clear that the average number of descendants of a newborn is $+\infty$ if $R_0 \geq 1$.

### 1.2.5 Fundamental Theorem of Demography

From the estimate in (1.32), the solution $B(t)$ of Lotka's integral equation has a Laplace transform for the complex number $\lambda$ such that $\Re\lambda > \bar{\beta} - \underline{\mu}$. The Laplace transforms of $G(\cdot)$ and $\Psi(\cdot)$ exist for all $\lambda$, as they have a compact support.

*Remark 1.3* For a locally integrable function $f \in L^1_{\text{loc}}(\mathbb{R}_+; \mathbb{R})$ and a complex parameter $\lambda \in \mathbb{C}$, if $\hat{f}(\lambda) = \int_0^\infty e^{-\lambda t} f(t)dt$ exists as an improper integral, that is, $\lim_{T\to+\infty} \int_0^T e^{-\lambda t} f(t)dt$ exists, $f$ is said to be *Laplace transformable* for $\lambda$. If this integral converges absolutely, $f$ is *absolutely Laplace transformable*. If $f$ is [absolutely] Laplace transformable for $\lambda_0$, it is also [absolutely] Laplace transformable for any $\lambda \in \mathbb{C}$ such that $\Re\lambda > \Re\lambda_0$. We can then define the coordinates (abscissa) of convergence as $\sigma := \inf\{\lambda \in \mathbb{R} : f \text{ is Laplace transformable at } \lambda\}$. The Laplace transformation defines a holomorphic function in the half plane $\Re\lambda > \sigma$, and the analytical function $\hat{f}(\lambda)$ is called the Laplace transformation of $f$. For further details on the Laplace transformation, readers may consult [33, 141].

The Laplace transform of the convolution of two functions is the product of the Laplace transforms of each individual function, and so the Laplace transform of the renewal equation (1.29) is

$$\hat{B}(\lambda) = \hat{G}(\lambda) + \hat{\Psi}(\lambda)\hat{B}(\lambda),$$

where

$$\hat{B}(\lambda) := \int_0^\infty e^{-\lambda t} B(t)dt$$

and so on. Let us define a set of complex numbers $\Lambda := \{\lambda \in \mathbb{C} : \hat{\Psi}(\lambda) = 1\}$ where

$$\hat{\Psi}(\lambda) = \int_0^\omega e^{-\lambda a}\Psi(a)da = 1 \tag{1.50}$$

is called *Lotka's characteristic equation*, and its roots $\lambda \in \Lambda$ are called the *characteristic roots*. If $\lambda \in \mathbb{C}\backslash\Lambda$, we have

$$\hat{B}(\lambda) = \frac{\hat{G}(\lambda)}{1 - \hat{\Psi}(\lambda)},$$

and $\hat{B}(\lambda)$ is analytic in $\mathbb{C}\backslash\Lambda$. Because $1 - \hat{\Psi}(\lambda)$ is an entire function for $\lambda$, the singular points of $\hat{B}(\lambda)$ are composed of poles of the meromorphic function $(1 - \hat{\Psi}(\lambda))^{-1}\hat{G}(\lambda)$. Taking the inverse Laplace transform, we obtain

$$B(t) = \frac{1}{2\pi i}\int_{x-i\infty}^{x+i\infty} e^{\lambda t}\hat{B}(\lambda)d\lambda, \quad x > \bar{\beta} - \underline{\mu}. \tag{1.51}$$

The asymptotic behavior of $B(t)$ is then determined by the distribution of singular points $\Lambda$. Thus, let us consider the structure of the set of characteristic roots $\Lambda$:

**Proposition 1.7** *Suppose that $R_0 > 0$. Then, the following statements hold:*

*(1) $\Lambda$ has a unique real root $\lambda_0$. $\lambda_0$ is a simple root satisfying $\lambda_0 \le \bar{\beta} - \underline{\mu}$. For any $\lambda \in \Lambda \setminus \{\lambda_0\}$, $\Re\lambda < \lambda_0$ holds. If $R_0 > 1$, $\lambda_0 > 0$; if $R_0 = 1$, $\lambda_0 = 0$; if $R_0 < 1$, $\lambda_0 < 0$.*
*(2) If $\lambda \in \Lambda$, then $\bar{\lambda} \in \Lambda$.*
*(3) For any real $\sigma$, the half plane $\Re\lambda > \sigma$ includes at most finitely many $\lambda \in \Lambda$.*

*Proof* Consider a real function such as

$$x \to \hat{\Psi}(x) = \int_0^{\omega} e^{-xa}\Psi(a)da, \quad x \in \mathbb{R}.$$

This is a decreasing continuous function, as $\Psi(a) \ge 0$ is nonnegative and takes a positive value on a finite interval. Therefore, we have $\lim_{x\to-\infty} \hat{\Psi}(x) = +\infty$, $\lim_{x\to+\infty} \hat{\Psi}(x) = 0$ and we can conclude that the characteristic equation $\hat{\Psi}(\lambda) = 1$ has a unique real root that we denote as $\lambda_0$. Moreover, observe that

$$\left.\frac{d}{dx}\hat{\Psi}(x)\right|_{x=\lambda_0} = -\int_0^{\omega} ae^{-\lambda_0 a}\Psi(a)da > 0,$$

from which it follows that $\lambda_0$ is a simple root. Because $\hat{\Psi}(0) = R_0$, we know that if $R_0 > 1$, then $\lambda_0 > 0$, if $R_0 = 1$, $\lambda_0 = 0$ and if $R_0 < 1$, $\lambda_0 < 0$. If $\lambda_0 + \underline{\mu} > 0$, it follows from (1.34) that

$$1 = \int_0^{\omega} e^{-\lambda_0 a}\Psi(a)da \le \frac{\bar{\beta}}{\lambda_0 + \underline{\mu}}.$$

It follows that $\lambda_0 \le \bar{\beta} - \underline{\mu}$. Conversely, if $\lambda_0 + \underline{\mu} \le 0$, the same inequality holds trivially. Moreover, if we let $\lambda_n \ne \lambda_0$ be a characteristic root, we have

$$\begin{aligned} 1 &= \int_0^{\omega} e^{-\lambda_0 a}\Psi(a)da = \Re\left(\int_0^{\omega} e^{-\lambda_n t}\Psi(a)da\right) \\ &= \int_0^{\omega} e^{-\Re\lambda_n a}\cos(\Im\lambda_n a)\Psi(a)da < \int_0^{\omega} e^{-\Re\lambda_n a}\Psi(a)da. \end{aligned}$$

Because $\hat{\Psi}(x)$ is monotone decreasing, we have $\Re\lambda_n < \lambda_0$. This completes the proof of (1). As (2) is clear, let us prove (3). Suppose that there exist infinitely many roots $\lambda_n = \alpha_n + i\beta_n$, $\sigma \le \alpha_n \le \bar{\beta} - \underline{\mu}$. As $\hat{\Psi}(\lambda) - 1$ is an analytic function that is not identically zero, its zeros have no accumulation point except for $\infty$. Thus, we can choose a subsequence $\lambda_{n(k)} = \alpha_{n(k)} + i\beta_{n(k)}$ such that when $k \to \infty$, $\alpha_{n(k)} \to \alpha^*$ and $\beta_{n(k)} \to \infty$. It follows that

$$\lim_{k\to\infty}\left|\hat{\Psi}(\lambda_{n(k)})-\int_0^\omega e^{-\alpha^* a-i\beta_{n(k)}a}\Psi(a)da\right|=0.$$

By contrast, the Riemann–Lebesgue lemma[5] implies that

$$\lim_{k\to\infty}\int_0^\omega e^{-\alpha^* a-i\beta_{n(k)}a}\Psi(a)da=0.$$

Hence, we have $\lim_{k\to\infty}\hat{\Psi}(\lambda_{n(k)})=0$, which contradicts our assumption that $\hat{\Psi}(\lambda_{n(k)})=1$. This completes the proof of (3). □

The dominant characteristic root $\lambda_0$ is called the *intrinsic rate of natural increase*, or simply the *intrinsic growth rate* or *stable growth rate*. It was pointed out by Feller [42] that the Lotka-type characteristic equation does not necessarily have infinitely many roots. However, this is not the case if the reproductive age interval is finite [125]:

**Proposition 1.8** *If the reproductive age interval is finite, Lotka's characteristic equation has infinitely many roots.*

*Proof* Let us define an entire function $f(\lambda)$ as

$$f(\lambda):=\int_0^\omega e^{-\lambda a}\Psi(a)da-1.$$

The characteristic roots are the zeros of $f(\lambda)$, and we can expand $f(\lambda)$ as follows:

$$f(\lambda)=\sum_{n=0}^{\infty}a_n\lambda^n,\quad a_n=\frac{f^{(n)}(0)}{n!}.$$

We know that the order of $f(\lambda)$ as an entire function is unity. In fact, the order of an entire function, denoted by $\rho$, is given by

$$\rho=\overline{\lim}_{n\to\infty}\frac{n\log n}{\log\left|\frac{1}{a_n}\right|}.$$

If we apply this formula to our case, we obtain

$$\frac{\log\left|\frac{1}{a_n}\right|}{n\log n}=1+\frac{\log\frac{(n!)^{\frac{1}{n}}}{n}}{\log n}-\frac{\log\left(\int_0^{\beta_2}a^n\Psi(a)da\right)^{\frac{1}{n}}}{\log n},$$

where $\beta_2$ is the upper bound of the reproductive age span. We can then observe that

[5] If $f\in L^1(0,T)$, it follows that $\lim_{|y|\to\infty}\int_0^T e^{-iyt}f(t)dt=0$.

$$\lim_{n\to\infty} \frac{\log\left|\frac{1}{a_n}\right|}{n\log n} = 1,$$

which means that the order of the entire function $f$ is one. It follows from the Hadamard theorem that $f(\lambda)$ can be expressed as

$$f(\lambda) = \lambda^m e^{Q(\lambda)} P(\lambda),$$

where $m$ denotes the order of the origin, $Q(\lambda)$ is at most first-order polynomial and $P(\lambda)$ is an entire function whose zeros are composed of zeros of $f$ except for the origin. If the set of characteristic roots is finite, $P(\lambda)$ is a polynomial whose roots are given by characteristic roots different from zero. Let $Q(\lambda) = u\lambda + v$. Then,

$$\int_0^{\beta_2} e^{-\lambda a}\Psi(a)da - 1 = \lambda^m e^{u\lambda+v} P(\lambda).$$

If $\Re u > 0$, the right-hand side goes to zero and the left-hand side goes to $+\infty$ if $\lambda \to -\infty$ on the real axis, which is a contradiction. Next, if we assume that $\Re u < 0$, the right-hand side goes to zero and the left-hand side goes to $-1$ if $\lambda \to +\infty$ on the real axis, giving another contradiction. Finally, if $\Re u = 0$, the absolute value of the right-hand side goes to $+\infty$ and the left-hand side goes to $-1$ if $\lambda \to +\infty$, which is again a contradiction. Therefore, $P(\lambda)$ cannot be a polynomial, and the number of characteristic roots must be infinite. Because $f(\lambda)$ is an entire function, the set of zeros is countable. □

Although the inverse Laplace transformation formula (1.51) is key when considering the asymptotic behavior of $B(t)$, it is more convenient to consider the Laplace transform of $B(t) - G(t)$ than that of $B(t)$ itself. We first state the following lemma:

**Lemma 1.4** *If we fix* $\sigma \in \mathbb{R}$, $\hat{G}(\sigma + iy)$ *and* $\hat{\Psi}(\sigma + iy)$ *are included in* $L^2(\mathbb{R})$ *as a function of* $y$ *and* $\hat{\Psi}(\sigma + iy)\hat{G}(\sigma + iy) \in L^1(\mathbb{R})$.

*Proof* If we define the functions

$$g_\sigma(t) = \begin{cases} e^{-\sigma t}G(t), & t > 0, \\ 0, & t < 0, \end{cases} \quad \psi_\sigma(t) = \begin{cases} e^{-\sigma t}\Psi(t), & t > 0, \\ 0, & t < 0, \end{cases}$$

then it is clear that $g_\sigma, \psi_\sigma \in L^1(\mathbb{R}) \cap L^2(\mathbb{R})$. It follows from Plancherel's theorem that their Fourier transforms $g_\sigma^*(y)$ and $\psi_\sigma^*(y)$ are included in $L^2(\mathbb{R})$ and

$$\sqrt{2\pi}\, g_\sigma^*(y) = \hat{G}(\sigma + iy) \in L^2(\mathbb{R}), \quad \sqrt{2\pi}\, \psi_\sigma^*(y) = \hat{\Psi}(\sigma + iy) \in L^2(\mathbb{R}).$$

Therefore, it follows from the Schwarz inequality that $\hat{\Psi}(\sigma + iy)\hat{G}(\sigma + iy) \in L^1(\mathbb{R})$. □

Under the above preparation, we prove the *Fundamental Theorem of Demography*, which was first proposed by Sharpe and Lotka [114] and subsequently proved by Feller [42]. Modern semigroup proofs have been given by Prüss [103], Song et al. [124] and Webb [138] (see Chap. 10). The following proof is based on the results of Heijmans [51] and Iannelli [55]:

**Proposition 1.9** (The Fundamental Theorem of Demography) *For Lotka's integral equation* (1.29), *there exists a positive number* $\eta > 0$ *such that*

$$B(t) = q_0 e^{\lambda_0 t}(1 + O(e^{-\eta t})), \tag{1.52}$$

*where*

$$q_0 = \frac{\int_0^\omega e^{-\lambda_0 t} G(t) dt}{\int_0^\omega a e^{-\lambda_0 a} \Psi(a) da}. \tag{1.53}$$

*Proof* Consider the following decomposition:

$$\hat{B}(\lambda) = \frac{\hat{G}(\lambda)}{1 - \hat{\Psi}(\lambda)} = \hat{G}(\lambda) + \frac{\hat{G}(\lambda)\hat{\Psi}(\lambda)}{1 - \hat{\Psi}(\lambda)}.$$

First, we prove that the following conditions hold for the second term of the above decomposition:

$$\lim_{|\lambda| \to +\infty, \Re\lambda > \delta} \frac{\hat{G}(\lambda)\hat{\Psi}(\lambda)}{1 - \hat{\Psi}(\lambda)} = 0, \tag{1.54}$$

$$\int_{-\infty}^{+\infty} \left| \frac{\hat{G}(\sigma + iy)\hat{\Psi}(\sigma + iy)}{1 - \hat{\Psi}(\sigma + iy)} \right| dy < +\infty, \tag{1.55}$$

where $\delta \in \mathbb{R}$ is any real number and $\sigma \in \mathbb{R}$ is a real number such that no element of $\Lambda$ lies on the line $\Re\lambda = \sigma$. For any half plane $\Re\lambda > \delta$, we have

$$\lim_{|\lambda| \to +\infty} \hat{\Psi}(\lambda) = \lim_{|\lambda| \to +\infty} \hat{G}(\lambda) = 0.$$

In fact, if we consider the region $S_1 = \{\Re\lambda > \delta; |\arg(\lambda - \delta)| \le \theta < \frac{\pi}{2}\}$, then $\lambda \in S_1$ and $|\lambda| \to \infty$ imply that

$$|\hat{\Psi}(\lambda)| \le \int_0^\omega e^{-xt} \Psi(t) dt \le \int_0^\omega e^{-t|\lambda| \cos\theta} \Psi(t) dt \to 0.$$

A similar estimate holds for $\hat{G}(\lambda)$. Therefore, we know that (1.54) follows. Next, note that

$$m_\sigma = \inf_{y \in \mathbb{R}} |1 - \hat{\Psi}(\sigma + iy)| > 0.$$

If we define the functions $g_\sigma(t)$ and $\psi_\sigma(t)$ as in Lemma 1.4, their Fourier transforms $g_\sigma^*(y)$, $\psi_\sigma^*(y)$ are included in $L^2(\mathbb{R})$ and

$$2\pi\psi_\sigma^*(y)g_\sigma^*(y) = \hat{\Psi}(\sigma+iy)\hat{G}(\sigma+iy) \in L^1(\mathbb{R}).$$

Therefore, we have

$$\frac{1}{2\pi}\left|\frac{\hat{G}(\sigma+iy)\hat{\Psi}(\sigma+iy)}{1-\hat{\Psi}(\sigma+iy)}\right| \le \frac{1}{m_\sigma}|g_\sigma^*(y)\psi_\sigma^*(y)| \in L^1(\mathbb{R}).$$

Thus, (1.55) holds. Let $\sigma > \lambda_0$ and define a function $J$ as

$$J(t) := \frac{1}{2\pi i}\int_{\sigma-i\infty}^{\sigma+i\infty}\frac{\hat{G}(\lambda)\hat{\Psi}(\lambda)}{1-\hat{\Psi}(\lambda)}e^{\lambda t}d\lambda.$$

From the well-known theorem for Laplace transforms [33], it follows that $J(t)$ has the Laplace transformation

$$\hat{J}(\lambda) = \frac{\hat{G}(\lambda)\hat{\Psi}(\lambda)}{1-\hat{\Psi}(\lambda)} = \hat{B}(\lambda) - \hat{G}(\lambda).$$

Because the original function in this Laplace transformation is unique, we can conclude that $B(t) = G(t) + J(t)$. Finally, we choose a number $\sigma_1 < \lambda_0$ such that any characteristic root except for $\lambda_0$ is located in the left-hand side of the line $\Re\lambda = \sigma_1$ and take the integral path shown in Fig. 1.4. It then follows from the Riemann–Lebesgue lemma that, along integral paths $i = 2$ and $i = 4$,

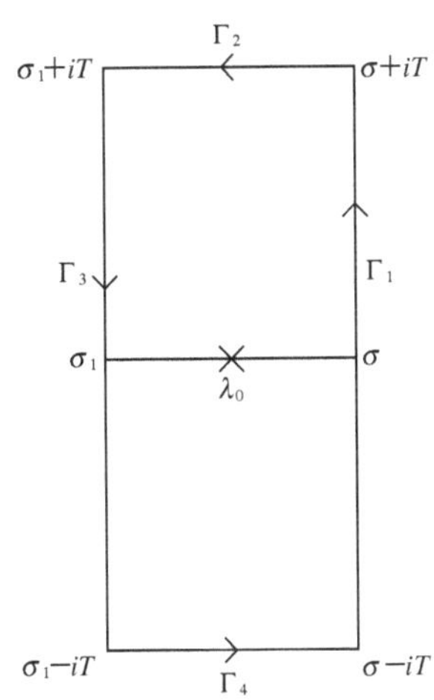


**Fig. 1.4** The integral path used in the proof of Proposition 1.9

$$\lim_{T\to\infty}\int_{\Gamma_i} e^{\lambda t}\hat{J}(\lambda)d\lambda = 0.$$

From Cauchy's integral theorem, it follows that

$$J(t) = \mathrm{Res}_{\lambda=\lambda_0}\{e^{\lambda t}\hat{J}(\lambda)\} + \frac{1}{2\pi i}\int_{\sigma_1-i\infty}^{\sigma_1+i\infty}\frac{\hat{G}(\lambda)\hat{\Psi}(\lambda)}{1-\hat{\Psi}(\lambda)}e^{\lambda t}d\lambda,$$

where $\mathrm{Res}_{\lambda=\lambda_0} f(\lambda)$ denotes the residue of a function $f(\lambda)$ at $\lambda = \lambda_0$. We can shift the integral path of $J(t)$ from $\Re\lambda = \sigma$ to $\Re\lambda = \sigma_1$ to obtain the expression

$$J(t) = e^{\lambda_0 t}(q_0 + \varepsilon_0(t)), \tag{1.56}$$

where

$$q_0 = \mathrm{Res}_{\lambda=\lambda_0}\left[\frac{\hat{G}(\lambda)\hat{\Psi}(\lambda)}{1-\hat{\Psi}(\lambda)}\right] = \frac{\int_0^\omega e^{-\lambda_0 t}G(t)dt}{\int_0^\omega ae^{-\lambda_0 a}\Psi(a)da},$$

$$|\varepsilon_0(t)| = \frac{e^{-\lambda_0 t}}{2\pi}\left|\int_{\sigma_1-i\infty}^{\sigma_1+i\infty}\frac{\hat{G}(\lambda)\hat{\Psi}(\lambda)}{1-\hat{\Psi}(\lambda)}e^{\lambda t}d\lambda\right| \le \frac{e^{-(\lambda_0-\sigma_1)t}}{m_{\sigma_1}}|g^*_{\sigma_1}|_{L^2}|\psi^*_{\sigma_1}|_{L^2}.$$

Note that $q_0 = 0$ if and only if $G(t) = 0$ for all $t \ge 0$. In this case, the renewal equation has a trivial solution $B(t) \equiv 0$. However, if $q_0 > 0$, it follows from (1.56) that

$$B(t) = q_0 e^{\lambda_0 t}\left(1 + \frac{e^{-\lambda_0 t}G(t)}{q_0} + \frac{1}{q_0}\varepsilon_0(t)\right).$$

If we let $\eta = \lambda_0 - \sigma_1$, then

$$\frac{e^{-\lambda_0 t}G(t)}{q_0} + \frac{1}{q_0}\varepsilon_0(t) = O(e^{-\eta t}).$$

This completes the proof. $\square$

From the Fundamental Theorem, we can obtain the asymptotic behavior of $B(t)$ by calculating the dominant characteristic root $\lambda_0$ and the coefficient $q_0$. If we calculate other characteristic roots and their coefficients, we can elicit the transient dynamics:

**Proposition 1.10** *Let $\lambda_i$, $i = 0, 1, 2, \ldots$, be poles of $\hat{J}(\lambda)$ with order $k_i$ that are ordered as $\lambda_0 > \Re\lambda_1 = \Re\lambda_2 > \Re\lambda_3 = \Re\lambda_4 > \ldots$. Then, the following asymptotic expansion holds:*

$$B(t) = \sum_{i=0}^{2n} e^{\lambda_i t}\left(\sum_{j=1}^{k_i}\frac{q^{(i)}_{-j}}{(j-1)!}t^{j-1}\right) + O(e^{\delta t}), \tag{1.57}$$

*where $\delta$ is some number such that $\Re\lambda_{2n} > \delta > \Re\lambda_{2n+1}$,*

$$q_{-1}^{(0)} = q_0, \quad q_{-j}^{(i)} = \frac{1}{2\pi i}\int_{\Gamma_i}(\lambda - \lambda_i)^{j-1}\hat{J}(\lambda)d\lambda, \quad (i \geq 1)$$

*and $\Gamma_i$ is a closed curve enclosing $\lambda_i$ but not enclosing other characteristic roots.*

*Proof* In the proof of the Fundamental Theorem, it is sufficient to shift the integral path from $\sigma + iy, -\infty < y < +\infty$ to $\delta + iy, -\infty < y < +\infty$ in the inverse Laplace expression of $J(t)$. □

From the Fundamental Theorem for the birth rate $B(t)$, we know that the asymptotic stable population behaves as a Malthusian population:

**Proposition 1.11** *Suppose that $p_0$ is some non-trivial initial data. The following holds uniformly for the age interval* $[0, \omega]$*:*

$$\lim_{t\to\infty} e^{-\lambda_0 t}p(t, a) = q_0 e^{-\lambda_0 a}\ell(a). \tag{1.58}$$

*Proof* For $t > \omega$, there exists a function $\varepsilon(t)$ such that $\varepsilon(t) = O(e^{-\eta t})$, $\eta > 0$ and

$$p(t, a) = e^{\lambda_0(t-a)}(q_0 + \varepsilon(t - a))\ell(a).$$

Hence, we can observe that

$$|e^{-\lambda_0 t}p(t, a) - q_0 e^{-\lambda_0 a}\ell(a)| \leq e^{-\lambda_0 a}\ell(a)\varepsilon(t - a).$$

However, it follows from Proposition 1.9 that there exist $M_{\sigma_1} > 0$ and $\sigma_1 < \lambda_0$ such that

$$|\varepsilon(t)| \leq M_{\sigma_1}e^{-(\lambda_0-\sigma_1)t}.$$

Therefore, we can conclude that

$$|e^{-\lambda_0 t}p(t, a) - q_0 e^{-\lambda_0 a}\ell(a)| \leq M_{\sigma_1}e^{-\sigma_1 a}\ell(a)e^{-(\lambda_0-\sigma_1)t}.$$

Because $e^{-\sigma_1 a}\ell(a) \leq \max\{1, e^{-\sigma_1\omega}\}$, the right-hand side converges to zero uniformly for $a \in [0, \omega]$ as $t \to \infty$. □

### *1.2.6 The Intrinsic Rate of Natural Increase*

Let us now introduce an approximate formula for the intrinsic rate of natural increase. Note that the normalized distribution $\psi(a) := \Psi(a)/R_0$ can be seen as the generation interval distribution, that is, the probability density function for the time from an individual's birth to their childbearing age. Epidemic applications have used the relation

$$\frac{1}{R_0} = \int_0^{\omega} e^{-\lambda_0 a}\psi(a)da,$$

as it is often necessary to estimate $R_0$ from observations of $\lambda_0$ and $\psi$. Readers can refer to [137] for epidemic applications of the demographic ideas explained here.

We replace the net reproduction function $\Psi(a)$ by a normal distribution with the same size, mean, and dispersion:

$$\Psi(a) \approx \frac{R_0}{\sigma\sqrt{2\pi}} e^{-\frac{(a-T_c)^2}{2\sigma^2}},$$

where $T_c$ denotes the *average age of childbearing in a cohort* (the *generation time* or *generation interval*) and $\sigma^2$ is its dispersion:

$$T_c := \int_0^{\omega} a\frac{\Psi(a)}{R_0}da, \quad \sigma^2 := \int_0^{\omega} (a-T_c)^2 \frac{\Psi(a)}{R_0}da.$$

Further, if we define

$$R_n := \int_0^{\omega} a^n \Psi(a)da,$$

then $T_c$ and $\sigma^2$ can be calculated as follows:

$$T_c = \frac{R_1}{R_0}, \quad \sigma^2 = \frac{R_2}{R_0} - \left(\frac{R_1}{R_0}\right)^2.$$

A normal distribution is defined on the real axis, and so we replace the integral interval of the characteristic equation by the real axis as follows:

$$\int_{-\infty}^{\infty} e^{-\lambda a} \frac{R_0}{\sigma\sqrt{2\pi}} e^{-\frac{(a-T_c)^2}{2\sigma^2}} da = R_0 e^{-\lambda T_c + \frac{\sigma^2\lambda^2}{2}} = 1.$$

The characteristic roots then satisfy the quadratic equation

$$\frac{\sigma^2}{2}\lambda^2 - T_c\lambda + \log R_0 = 0. \tag{1.59}$$

If (1.59) has two real roots, the smaller is positive if $R_0 > 1$ and negative if $R_0 < 1$, and the larger root is always positive. The smaller root is therefore a good approximation for the intrinsic rate of natural increase:

$$\lambda_0 \approx \frac{T_c - \sqrt{T_c^2 - 2\sigma^2 \log R_0}}{\sigma^2}. \tag{1.60}$$

Formula (1.60) was first proposed by Dublin and Lotka [35]. If we assume that the dispersion is zero in (1.59), we have

$$\lambda_0 \approx \frac{\log R_0}{T_c},$$

which is still an effective approximation formula; the error would be at most 5% for human populations [64]. Lotka defined the *average length of generation* $L$ as $L = (\log R_0)/\lambda_0$. It follows that $e^{\lambda_0 L} = R_0$, and so $L$ is the time needed for a population to grow to $R_0$ times the initial population. From the above argument, $T_c$ is approximated by $L$.

**Exercise 1.13** For $\lambda \in \mathbb{R}$, observe that

$$\frac{d\hat{\Psi}(\lambda)}{d\lambda} = -T(\lambda)\hat{\Psi}(\lambda),$$

where

$$T(\lambda) := \frac{1}{\hat{\Psi}(\lambda)} \int_0^{\omega} a e^{-\lambda a} \Psi(a) da$$

is the *average age of childbearing in a Malthusian population* with growth rate $\lambda$. Solving this differential equation, show that

$$1 = R_0 e^{-\int_0^{\lambda_0} T(\zeta) d\zeta}.$$

Prove that, if we adopt a linear approximation $T(\lambda) = T(0) + T'(0)\lambda$, we obtain the Dublin–Lotka approximation formula (1.60).

*Remark 1.4* Note that the basic reproduction number $R_0$ provides a threshold condition for the sign of the intrinsic growth rate, whereas a larger [smaller] $R_0$ does not necessarily imply a higher [lower] intrinsic growth rate. It is clear from (1.60) that the intrinsic growth rate is affected not only by $R_0$ but also by the timing of childbearing.[6] For example, if some *age shift* occurs in the reproduction schedule, that is, the net reproduction schedule is delayed [advanced] as $\varepsilon > 0$ [$\varepsilon < 0$], $\Psi(a)$ is replaced by $\Psi(a - \varepsilon)$. In this case, $R_0$ does not change, but $\lambda_0$ becomes smaller [larger].

If all $\lambda_j$ are simple roots of the characteristic equation, the meromorphic function $\hat{J}$ has first-order poles at $\lambda_j$. Hence, (1.57) can be reduced to the simple asymptotic expansion

$$B(t) = q_0 e^{\lambda_0 t} + \sum_{j=1}^{2n} q_j e^{\lambda_j t} + O(e^{\delta t}), \tag{1.61}$$

[6]It is also affected by the distribution of the age at death [110].

where

$$q_j := q_{-1}^{(j)} = \frac{\hat{G}(\lambda_j)}{-\hat{\Psi}'(\lambda_j)} \tag{1.62}$$

and $q_{2k} = \bar{q}_{2k-1}, k = 1, 2, \ldots$. Thus, the asymptotic expansion of $B(t)$ clearly has a term corresponding to a complex conjugate pair of characteristic roots:

$$q_{2k-1}e^{\lambda_{2k-1}t} + \bar{q}_{2k-1}e^{\bar{\lambda}_{2k-1}t} = 2|q_{2k-1}|e^{\Re\lambda_{2k-1}t}\cos(\Im\lambda_{2k-1}t + \arg q_{2k-1}),$$

which is the real oscillatory term. As the growth rate $\Re\lambda_{2k-1}$ of the oscillatory term is less than $\lambda_0$, the oscillations decay exponentially relative to the first exponential term $q_0e^{\lambda_0 t}$ as time evolves. However, if $\Re\lambda_1 > 0$, the absolute value of the oscillation (population wave) expands exponentially. If $\Re\lambda_1 < 0$, $B(t)$ converges to the exponential growth orbit, that is,

$$\lim_{t\to\infty} |B(t) - q_0e^{\lambda_0 t}| = 0. \tag{1.63}$$

For human populations, $\Re\lambda_1$ is usually negative. Hence, (1.63) holds, and it follows that

$$\frac{2\pi}{\Im\lambda_1} \approx T_c = \int_0^\omega a\frac{\Psi(a)}{R_0}da. \tag{1.64}$$

In fact, if $\lambda_1 = x + iy$ is a complex characteristic root, we have

$$\int_0^\omega e^{-iya}e^{-xa}\frac{\Psi(a)}{R_0}da = \frac{1}{R_0}.$$

Multiplying both sides by $e^{iyT_c}$,

$$\int_0^\omega e^{-i(a-T_c)y}e^{-xa}\frac{\Psi(a)}{R_0}da = \frac{1}{R_0}e^{iyT_c}. \tag{1.65}$$

Substituting the Taylor expansion

$$e^{-i(a-T_c)y} = 1 - i(a - T_c)y - \frac{(a - T_c)^2y^2}{2} - \ldots$$

and comparing imaginary parts, if the higher-order terms on the left-hand side in (1.65) can be neglected, we have

$$-\int_0^\omega (a - T_c)ye^{-xa}\frac{\Psi(a)}{R_0}da \approx \frac{1}{R_0}\sin(yT_c). \tag{1.66}$$

Because the left-hand side of (1.66) is zero if $x = 0$, the right-hand side of (1.66) must be close to zero if $x$ is small, which is expected for the second characteristic root $\lambda_1$. Hence, $T_c \Im\lambda_1 \approx 2\pi$, meaning that the shortest period $2\pi/\Im\lambda_1$ is close to $T_c$. In summary, the second oscillatory term is an attenuating oscillation whose period is close to the generation interval. As the effect of other higher-order oscillatory terms is relatively faint, their neglect has only a small influence [63].

Although $\lambda_0$ crosses the imaginary axis from left to right when $R_0$ crosses unity from below, whether $\Re\lambda_1$ crosses the imaginary axis to enter the right half plane as $R_0$ increases is a non-trivial question. The net reproduction function $\Psi$ is said to have the *rescaling property* if it can be multiplied by some positive number such that $\Re\lambda_1 = 0$. If the net reproduction function has the rescaling property, there exists a bifurcation point $R_0^*$ such that (1.63) holds if $R_0 < R_0^*$, whereas it does not hold if $R_0 > R_0^*$. It can be shown that the reproduction pattern usually has the rescaling property if the lower bound of the reproductive age is positive; that is, the net reproduction function for humans always has the rescaling property [135, 136].

## 1.3 The Dual System and the Reproductive Value

### *1.3.1 The Population Operator*

Let us apply the well-known separation of variables method to the stable population model (1.25). This enables us to consider the eigenvalue problem of the population operator. If we assume a separation-of-variable-type solution $p(t,a) = w(t)u(a)$ and insert this into the McKendrick equation in (1.25), we have

$$w'(t)u(a) + w(t)u'(a) = -\mu(a)w(t)u(a).$$

Dividing both sides by $w(t)u(a)$ gives

$$\frac{w'(t)}{w(t)} = \frac{1}{u(a)}\left(-\frac{d}{da} - \mu(a)\right)u(a),$$

which must be a constant because the left-hand side is a function of only $t$ and the right-hand side is a function of only $a$. If we assume that $\lambda$ is a constant of separation, we obtain a set of ODEs

$$w'(t) = \lambda w(t), \quad \left(-\frac{d}{da} - \mu(a)\right)u(a) = \lambda u(a),$$

from which it follows that

$$w(t) = e^{\lambda t}w(0), \quad u(a) = e^{-\lambda a}\ell(a)u(0).$$

Inserting the above expression into the boundary condition, we have

$$u(0) = \int_0^\omega \beta(a)u(a)da = u(0)\int_0^\omega e^{-\lambda a}\beta(a)\ell(a)da.$$

Therefore, there exists a non-trivial solution $u$ if and only if $\lambda$ is a root of Lotka's characteristic equation.

Let us now define a first-order differential operator $A$ in $X := L^1(0,\omega)$ as

$$(A\phi)(a) := \left(-\frac{d}{da} - \mu(a)\right)\phi(a), \quad \phi \in \mathscr{D}(A), \tag{1.67}$$

where the domain of $A$, denoted by $\mathscr{D}(A)$, is given by (1.35) and is dense in $L^1(0,\omega)$. Following Song et al. [122, 123, 125], we call $A$ the *population operator*. For the characteristic root $\lambda_i \in \Lambda$, $u_i(a) := e^{-\lambda_i a}\ell(a)$ becomes an eigenfunction[7] of the population operator, that is, Lotka's characteristic root $\lambda_i \in \Lambda$ is an eigenvalue of $A$ corresponding to the eigenfunction $u_i$. As shown in Chap. 10, the population operator $A$ is the infinitesimal generator of a strongly continuous semigroup $T(t) = e^{tA}$, $T(t)p_0$ is the solution of the Cauchy problem (1.41) and $e^{\lambda_0 t}u_0$ is an *exponential (persistent) solution* of (1.41).

Next, let us consider the adjoint eigenvalue problem of the population operator. The adjoint $A^*$ of the population operator $A$ is a linear operator defined on $L^\infty(0,\omega)$ that satisfies the relation[8]

$$\langle v, Au\rangle = \langle A^*v, u\rangle, \quad v \in \mathscr{D}(A^*),\ u \in \mathscr{D}(A),$$

where $\langle f, g\rangle := \int_0^\omega f(a)g(a)da$ denotes the value of the eigenfunctional $\langle f, \cdot\rangle$ (which is identified with $f \in L^\infty$) at $g \in L^1$. By a formal calculation, it is easy to see that

$$(A^*v)(a) = \frac{dv(a)}{da} - \mu(a)v(a) + \beta(a)v(0) \tag{1.68}$$

for $v \in \mathscr{D}(A^*) = \{v \in L^\infty(0,\omega) : v(\omega) = 0, \quad A^*v \in L^\infty(0,\omega)\}$. Thus, we can consider the adjoint eigenvalue problem as

$$A^*v = \lambda v, \quad v \in \mathscr{D}(A^*). \tag{1.69}$$

It is readily apparent that the adjoint eigenvector corresponding to the eigenvalue $\lambda = \lambda_j \in \Lambda$ can be calculated as follows:

$$v_j(a) = v_j(0)\int_a^\omega e^{-\lambda_j(s-a)}\frac{\ell(s)}{\ell(a)}\beta(s)ds. \tag{1.70}$$

Then, we can state the following:

---

[7] If $\omega = \infty$, it is unclear whether $u_i(a) = e^{-\lambda_i a}\ell(a)$ becomes an eigenvector in $L^1(\mathbb{R}_+)$. Let $\underline{\mu} = \inf_{a\in\mathbb{R}_+}\mu(a)$. If $\Re\lambda_i > -\underline{\mu}$, we have $u_i \in L^1(\mathbb{R}_+)$.

[8] For the precise definition of the adjoint (or dual) operator, readers are referred to [12, 52].

**Lemma 1.5** *The following relations between the eigenvectors $u_i$ and their adjoint eigenvectors $v_j$ hold:*

$$\langle v_j, u_i \rangle = \begin{cases} 0, & (i \neq j), \\ \int_0^\omega a e^{-\lambda_i a} \Psi(a) da, & (i = j), \end{cases} \tag{1.71}$$

$$q_j = \frac{\langle v_j, p_0 \rangle}{\langle v_j, u_j \rangle}, \tag{1.72}$$

*where $q_j$ is a coefficient in the asymptotic expansion* (1.61).

**Exercise 1.14** Prove (1.71).

### 1.3.2 The Reproductive Value

In particular, the adjoint eigenvector $v_0$ associated with the intrinsic rate of natural increase $\lambda_0$ is called the *reproductive value*. This was first introduced by Fisher to answer the question "To what extent will persons of this age, on the average, contribute to the ancestry of future generations?" [43]:

$$v_0(a) = v_0(0) \int_a^\omega e^{-\lambda_0 (s-a)} \frac{\ell(s)}{\ell(a)} \beta(s) ds, \tag{1.73}$$

where $v_0(0)$ is an arbitrary positive value. The reproductive value at age $a$ is the total prospective number of female children that would be born, with the prevailing net reproduction function $\beta(a)\ell(a)$, discounted at the intrinsic rate of natural increase $\lambda_0$. Thus, it is "the present value of future offspring." Moreover, the *total reproductive value*, denoted by $V(t)$, is defined as

$$V(t) := \langle v_0, p(t, \cdot) \rangle = \int_0^\omega v_0(a) p(t, a) da. \tag{1.74}$$

Then, the following holds:

**Proposition 1.12** *The total reproductive value of the stable population grows exponentially with the intrinsic rate of natural increase.*

*Proof* From the definition, we have

$$\begin{aligned} \frac{dV(t)}{dt} &= \langle v_0, \frac{\partial}{\partial t} p(t, \cdot) \rangle = \langle v_0, A p(t, \cdot) \rangle \\ &= \langle A^* v_0, p(t, \cdot) \rangle = \lambda_0 \langle v_0, p(t, \cdot) \rangle = \lambda_0 V(t). \end{aligned}$$

Thus, we obtain $V(t) = e^{\lambda_0 t} V(0)$. This completes the proof. □

**Exercise 1.15** Consider a heterogeneous population composed of $m$ separate subpopulations, each of which has its own set of demographic parameters and is described by the stable population model. Let $p_j(t,a)$ $(j = 1, 2, \dots, m)$ be the age-density function of the $m$th subpopulation with the Malthusian parameter $\lambda_{0j}$. Let $V_j(t)$ be the total reproductive value of the $j$th subpopulation. Then, we have

$$\frac{dV_j(t)}{dt} = \lambda_{0j} V_j(t).$$

Let $\zeta_j(t)$ be the frequency of the total reproductive value, defined by

$$\zeta_j(t) := \frac{V_j(t)}{\sum_{j=1}^m V_j(t)},$$

and $\lambda_0(t)$ be the average Malthusian parameter, defined by

$$\lambda_0(t) := \sum_{j=1}^m \zeta_j(t)\lambda_{0j}.$$

Show that

$$\frac{d\lambda_0(t)}{dt} = \sum_{j=1}^m \zeta_j(t)(\lambda_{0j} - \lambda_0(t))^2.$$

The rate of change of the average Malthusian parameter is equal to the variance among types in the Malthusian parameter, so the fitness increases as time evolves (Fisher's *Fundamental Theorem of Natural Selection* [14]).

Note that the total reproductive value still includes an arbitrary value $v_0(0)$. As was noted by Fisher himself [46], as long as we merely wish to compare the reproductive values of different age groups in a population, the convention that $v_0(0)$ be assigned a unit value is not open to objection; however, this is not suitable for the comparison of different populations. In his most famous book, Fisher defined the reproductive value at age $a$ only as a number relative to an arbitrary value of one at age zero [43]. However, Crow [21] pointed out that Fisher had earlier defined the reproductive value at age zero as

$$v_0(0) = \frac{1}{b_0 T_0}, \tag{1.75}$$

where

$$b_0 := \frac{1}{\int_0^\omega e^{-\lambda_0 a}\ell(a)da}, \quad T_0 := \int_0^\omega a e^{-\lambda_0 a}\Psi(a)da.$$

$b_0$ is the CBR of the stable population and $T_0$ is the *average age of childbearing in the stable population*.

We now introduce the normalized eigenvectors and adjoint eigenvectors associated with eigenvalues $\lambda_i$ ($i = 0, 1, 2, \dots$) of the population operator:

$$u_i(a) := b_i e^{-\lambda_i a}\ell(a), \quad v_i(a) := \frac{1}{b_i T_i}\int_a^\omega e^{-\lambda_i(s-a)}\frac{\ell(s)}{\ell(a)}\beta(s)ds,$$

where

$$b_i := \frac{1}{\int_0^\omega e^{-\lambda_i a}\ell(a)da}, \quad T_i := \int_0^\omega a e^{-\lambda_i a}\Psi(a)da.$$

Then, it follows from (1.71) that $\langle v_j, u_i\rangle = \delta_{ji}$ and $u_0$ is the stable age profile. The *normalized total reproductive value* is given by $V(t) := \langle v_0, p(t,\cdot)\rangle$.

Using the normalized reproductive value, (1.61) and (1.72), if all characteristic roots are simple, we have the following formal asymptotic expansion:

$$p(t,a) \sim \sum_{i=0}^\infty \langle v_i, p_0\rangle e^{\lambda_i t}u_i(a), \tag{1.76}$$

where the dominant exponential term can be written as

$$\langle v_0, p_0\rangle e^{\lambda_0 t}u_0(a) = V(0)e^{\lambda_0 t}u_0(a) = V(t)u_0(a).$$

Thus, we know that the normalized total reproductive value $V(t)$ converges to the total size of the stable population as time evolves:

$$\lim_{t\to\infty}\frac{V(t)}{P(t)} = 1.$$

If the population is described by the exponential solution, we have $P(t) = V(t)$. In particular, the ratio $V(0)/P(0)$ is called the *population momentum*, which can be interpreted as the ratio of the ultimate size of a stationary population to the size of the initial population if $\lambda_0 = 0$. This idea will be discussed again in Sect. 1.4. In other words, for any given initial population $p_0$, we expect that the future population will behave as a Malthusian population:

$$p(t,a) \sim V(0)e^{\lambda_0 t}u_0(a) = V(t)u_0(a), \quad t \to \infty. \tag{1.77}$$

In this sense, the total reproductive value $\langle v_0, p_0\rangle = V(0)$ (or $V(0)u_0(a)$) is also called the *stable equivalent* (with respect to $p_0$).

### 1.3.3 Fundamental Solutions

To examine the meaning of the reproductive value in more detail, let us consider a special solution $K(t, a|0, \zeta)$ of (1.25) such that

$$\left(\frac{\partial}{\partial t} + \frac{\partial}{\partial a} + \mu(a)\right) K(t, a|0, \zeta) = 0, \quad t > 0, \quad 0 < a < \omega,$$
$$K(t, 0|0, \zeta) = \int_0^\omega \beta(a) K(t, a|0, \zeta) da, \quad t > 0, \tag{1.78}$$
$$\lim_{t \downarrow 0} K(t, a|0, \zeta) = \delta(a - \zeta),$$

where $\delta(\cdot)$ denotes Dirac's delta function. The above initial condition then implies that

$$\lim_{t \downarrow 0} \int_0^\omega K(t, a|0, \zeta)\phi(a) da = \phi(\zeta), \quad \phi \in L^1(0, \omega).$$

Using (1.27), we can obtain the explicit expression

$$K(t, a|0, \zeta) = \begin{cases} B_\zeta(t - a)\ell(a), & t - a > 0, \\ \delta(a - t - \zeta)\frac{\ell(a)}{\ell(a-t)}, & a - t > 0, \end{cases} \tag{1.79}$$

where

$$B_\zeta(t) := \frac{\Psi(t + \zeta)}{\ell(\zeta)} + \int_0^t \mathscr{R}(t - s) \frac{\Psi(s + \zeta)}{\ell(\zeta)} ds$$

is the birth rate produced by the unit initial population aged $\zeta$.

Then, $K(t, a|0, \zeta)$ denotes the age-density function at time $t$ in the case that only one individual aged $\zeta$ is alive at the initial time. This solution $K$ is called the *fundamental solution*, and any solution with the initial data $p_0$ can be expressed as follows:

$$p(t, a) = \int_0^\omega K(t, a|0, \zeta) p_0(\zeta) d\zeta. \tag{1.80}$$

Because $K(t, a|0, \zeta)$ includes a delta function for $t \le \omega$, the expression in (1.80) is rather formal. For $t > \omega$, however, (1.80) has a rigorous meaning, because $K(t, a|0, \zeta)$ becomes a bounded integrable function. In fact, we can observe that, for $t > \omega$,

$$K(t, a|0, \zeta) = B_\zeta(t - a)\ell(a)$$
$$= \ell(a) \left[ \frac{\Psi(t - a + \zeta)}{\ell(\zeta)} + \int_0^{t-a} \mathscr{R}(t - a - s) \frac{\Psi(s + \zeta)}{\ell(\zeta)} ds \right].$$

Therefore, the time evolution of $p(t,\cdot)$ in the state space $L^1(0,\omega)$ is described by a positive linear integral operator for $t>\omega$. In fact, the solution orbit evolves according to a compact positive operator for $t>\omega$ (Corollary 1.1).

Because $\langle v_0, \delta_\zeta\rangle = v_0(\zeta)$, it follows from (1.76) that there exists some $\varepsilon>0$ such that

$$K(t,a|0,\zeta) = e^{\lambda_0 t}[v_0(\zeta)u_0(a)] + O(e^{(\lambda_0-\varepsilon)t}). \tag{1.81}$$

Therefore, a given unit population of age $\zeta$ at the initial time will add a population of $K(t,a|0,\zeta)\sim e^{\lambda_0 t}[v_0(\zeta)u_0(a)]$ asymptotically, the size of which is $e^{\lambda_0 t}v_0(\zeta)$.

Let $\Delta P(t)$ be the increment [decrement] in the total size of a stable population corresponding to an increase [decrease] of $\Delta p_0$ in the initial data. It then follows from (1.80) and (1.81) that

$$\Delta P(t) \approx e^{\lambda_0 t}\langle v_0, \Delta p_0\rangle, \tag{1.82}$$

which gives the long-term effect of one-time immigration (or emigration) on the total size of a closed population [62].

### *1.3.4 Backward System and Demographic Potential*

The *backward (dual) system* of the forward (primal) system (1.41) is formulated as follows:

$$\frac{dv(t)}{dt} = -A^*v(t), \quad t>0, \tag{1.83}$$

where $v(t,\cdot)\in\mathscr{D}(A^*)\subset X^* = L^\infty(0,\omega)$.

The *backward problem* of (1.83) can then be formulated in the dual space $X^*$ as a final-value problem:

$$\frac{dv(t)}{dt} = -A^*v(t), \quad v(t_0) = v^*\in X^*, \quad t\in[0,t_0), \tag{1.84}$$

where $v(t) = v(t,\cdot)\in\mathscr{D}(A^*)$ is an $X^*$-valued function and $v^*\in\mathscr{D}(A^*)$ is the *final* data. The weak* solution of the backward problem (1.84) is given by

$$v(t) = T^*(t_0-t)v^*, \quad t\in[0,t_0],$$

where $T^*(t) = T(t)^*$, $t>0$, is the *dual semigroup* and the dual $A^*$ is the weak* infinitesimal generator of the dual semigroup. The dual semigroup $T^*(t)$ is an important starting point for the *sun and star calculus* used to construct a solution semigroup for structured population models (see Chap. 10, [15, 16, 29]).

It is clear that the dual system (1.83) has a positive exponential solution $e^{-\lambda_0 t}v_0$ where $v_0\in X^*$ is the reproductive value. This positive exponential function is identified with the *importance functional* in the sense of Birkhoff (see Chap. 10). If we consider the dual system on the half line $[0, \infty)$, it follows from Proposition 10.26 that any persistent positive solution of (1.83) on $[0, \infty)$ is proportional to the exponential solution $e^{-\lambda_0 t} v_0$. Following Ediev [37, 38], we call the positive persistent solution (the importance functional) of the dual (stable population) system (1.83) for $t \in [0, \infty)$ the *demographic potential*. If $v(t)$ is the demographic potential, we call $\langle v(t), p(t) \rangle$ the *total demographic potential*. The dual system (1.83) has an essentially unique (up to a constant) positive persistent solution $e^{-\lambda_0 t} v_0$, so Fisher's reproductive value is the positive persistent age distribution of the dual system on $[0, \infty)$, and the demographic potential is a positive exponential solution that is proportional to the reproductive value at any time.

**Lemma 1.6** *The total demographic potential $\langle v(t), p(t) \rangle$ is time-independent.*

**Exercise 1.16** Prove the above lemma.

The above property is a special case of a more general result that holds for the time-inhomogeneous positive evolutionary system (see Chap. 10). From this point of view, Proposition 1.12 follows immediately, because if we set $v(t, a) = e^{-\lambda_0 t} v_0(a)$, we obtain

$$V(t) = \langle v_0, p(t) \rangle = e^{\lambda_0 t} \langle v(t), p(t) \rangle = e^{\lambda_0 t} \langle v(0), p(0) \rangle = e^{\lambda_0 t} V(0).$$

Integrating (1.83) along the characteristic line, we have

$$\begin{aligned} v(t, a) &= \int_a^\omega \frac{\ell(\sigma)}{\ell(a)} \beta(\sigma) v(t + \sigma - a, 0) d\sigma \\ &= \int_0^\omega \frac{\ell(a + x)}{\ell(a)} \beta(a + x) v(t + x, 0) dx. \end{aligned} \tag{1.85}$$

Thus, the boundary value $v(t, 0)$ must satisfy the *backward renewal equation*

$$v(t, 0) = \int_0^\omega \Psi(\sigma) v(t + \sigma, 0) d\sigma, \quad t > 0. \tag{1.86}$$

Equation (1.85) implies that the demographic potential at time $t$ and age $a$ is the summation of the demographic potential of newborns who will be produced by a mother at age $a$ and time $t$ in her remaining life. It is clear from Lotka's characteristic equation that the backward equation (1.86) has an exponential solution $e^{-\lambda_0 t}$, which is an essentially unique persistent positive solution for $t \in [0, \infty)$.

### 1.3.5 Stochastic Interpretations

Consider the forward system (1.41) and the dual system (1.83) on $\mathbb{R}_+$. The dual system has a (essentially unique) solution $v(t, a) = e^{-\lambda_0 t} v_0(a)$ for $t \in \mathbb{R}_+$. As the total demographic potential $\langle v(t,\cdot), p(t,\cdot)\rangle$ is constant, if the total reproductive value of the initial data is positive, the weighted distribution

$$\phi(t,a) := \frac{v(t,a)p(t,a)}{\langle v(t,\cdot), p(t,\cdot)\rangle} = \frac{v_0(a)p(t,a)}{V(t)} \tag{1.87}$$

is well defined and can be thought of as a probability density function, because its size is unity.

Moreover, it is easy to see that $\phi(t,a)$ satisfies the following McKendrick system:

$$\begin{aligned}
\frac{\partial \phi(t,a)}{\partial t} + \frac{\partial \phi(t,a)}{\partial a} &= -\frac{v_0(0)\beta(a)}{v_0(a)}\phi(t,a), \\
\phi(t,0) &= \int_0^\omega \frac{v_0(0)\beta(a)}{v_0(a)}\phi(t,a)da, \\
\phi(0,a) &= \frac{v_0(a)p(0,a)}{V(0)}
\end{aligned} \tag{1.88}$$

which conserves the total size:

$$\int_0^\omega \phi(t,a)da = 1.$$

Equation (1.88) generates a stochastic evolutionary system that evolves according to a Markov semigroup.[9] This system has a unique positive steady state $\phi^\dagger(a) = v_0(a)u_0(a)$, which is the stable age distribution weighted by the reproductive value. We can now study the asymptotic behavior of (1.88) based on the ideas developed in information theory and statistical mechanics. It is well known in the field of information theory that the *Kullback information* (Kullback divergence or relative entropy) or *information gain* of $\phi$ for the reference distribution $\phi^\dagger$ is given by

$$K(\phi,\phi^\dagger)(t) := \int_0^\omega \phi(t,a)\log\left(\frac{\phi(t,a)}{\phi^\dagger(t,a)}\right)da. \tag{1.89}$$

Conversely, Lasota and Mackey [70] define

$$H(\phi|\phi^\dagger) := -K(\phi,\phi^\dagger)$$

as the *conditional entropy* of $\phi$ with respect to $\phi^\dagger$. The *law of entropy increase* is therefore equivalent to the law of decreasing information gain.

More generally, if $G$ is an arbitrary convex function,

$$J(G;\phi,\phi^\dagger)(t) := \int_0^\omega \phi^\dagger(t,a)G\left(\frac{\phi(t,a)}{\phi^\dagger(t,a)}\right)da \tag{1.90}$$

---

[9] The matrix model version of (1.88) has been studied by Tuljapurkar [128].

is called the *generalized information gain* [111]. Note that $J = K$ if $G(x) = x \log x$. We can observe that

$$K(\phi, \phi^\dagger)(t) = \int_0^\omega \phi(t, a) \log \left( \frac{\phi(t, a)}{\phi^\dagger(t, a)} \right) da = \int_0^\omega \phi(t, a) G \left( \frac{\phi^\dagger(t, a)}{\phi(t, a)} \right) da,$$

where $G(x) = -\log x - 1 + x$. Thus, we have that the Kullback information is nonnegative and becomes zero if and only if $\phi = \phi^\dagger$.

Let us now prove that, for our system (1.88), $J(G; \phi, \phi^\dagger)(t)$ is monotone decreasing with respect to time $t$. First, observe that the time evolution of $\phi$ is described by a stochastic linear transformation. Using the fundamental solution of (1.88), the weighted distribution evolves in time according to the master equation

$$\phi(t, a) = \int_0^\omega K(t, a|s, \zeta) \phi(s, \zeta) d\zeta, \quad t > s, \tag{1.91}$$

where $K$ is a Green's function, that is, $K$ satisfies (1.88) and

$$\int_0^\omega K(t, a|s, \zeta) da = 1, \quad \lim_{t \downarrow s} K(t, a|s, \zeta) = \delta(a - \zeta).$$

The following proof is essentially that given by Schlögl [111]:

**Lemma 1.7** *If the distributions $\phi$ and $\phi^\dagger$ evolve according to the master equation* (1.91), *it follows that*

$$J(G; \phi, \phi^\dagger)(t) \le J(G; \phi, \phi^\dagger)(s), \quad t > s. \tag{1.92}$$

*Proof* For any convex function $G(x)$, we have

$$G(x) - G(y) \ge G'(y)(x - y).$$

Thus, it follows from (1.91) that

$$\begin{aligned}
&J(G; \phi, \phi^\dagger)(s) - J(G; \phi, \phi^\dagger)(t) \\
&= \int_0^\omega d\zeta \phi^\dagger(s, \zeta) G\left( \frac{\phi(s, \zeta)}{\phi^\dagger(s, \zeta)} \right) - \int_0^\omega da \phi^\dagger(t, a) G \left( \frac{\phi(t, a)}{\phi^\dagger(t, a)} \right) \\
&= \int_0^\omega da \int_0^\omega d\zeta K(t, a|s, \zeta) \phi^\dagger(s, \zeta) \left[ G\left( \frac{\phi(s, \zeta)}{\phi^\dagger(s, \zeta)} \right) - G \left( \frac{\phi(t, a)}{\phi^\dagger(t, a)} \right) \right] \\
&\ge \int_0^\omega da \int_0^\omega d\zeta K(t, a|s, \zeta) G' \left( \frac{\phi(t, a)}{\phi^\dagger(t, a)} \right) \phi^\dagger(s, \zeta) \left[ \frac{\phi(s, \zeta)}{\phi^\dagger(s, \zeta)} - \frac{\phi(t, a)}{\phi^\dagger(t, a)} \right].
\end{aligned}$$

In the last expression, we have

$$\int_0^\omega d\zeta K(t, a|s, \zeta)\phi^\dagger(s, \zeta) \left[ \frac{\phi(s, \zeta)}{\phi^\dagger(s, \zeta)} - \frac{\phi(t, a)}{\phi^\dagger(t, a)} \right] = \phi(t, a) - \phi(t, a) = 0.$$

Therefore, we have (1.92). □

Because the Kullback information $K(\phi, \phi^\dagger)(t)$ is a Lyapunov function of (1.88), it follows from LaSalle's invariance principle ([2], Sect. 18) and the relative compactness of the orbit $\phi(t, \cdot)$, $t \geq 0$ ([139], Sect. 3.4) that $\lim_{t\to\infty} \phi(t, a) = \phi^\dagger(a)$. This implies that

$$\lim_{t\to\infty} \frac{p(t, a)}{V(t)} = u_0(a),$$

that is, we can again obtain the strong ergodicity theorem for the stable population model. As shown by Lasota and MacKey [70], the convergence of a probability distribution evolved by a Markov semigroup to the stationary distribution can be proved as a general law of entropy increase under mild conditions. Schoen and Kim [112] called this phenomenon *a fundamental principle of population dynamics*. Vlad and Pop [133] stated this kind of result as an *H-theorem* for the multistate age-dependent master equation.

Finally, note that Michel, Mischler, and Perthame [88, 98] defined the *general relative entropy* as follows:

$$\mathcal{H}_v(p|n) := \int_0^\omega v(t, a)n(t, a)H\left(\frac{p(t, a)}{n(t, a)}\right) da,$$

where $v > 0$ is a solution of the dual system, $p$ and $n > 0$ are solutions of the primal system, $n$ is the reference distribution and $H : \mathbb{R}_+ \to \mathbb{R}_+$ is a convex function with $H(0) = 0$. If we choose a special set in which $v(t, a) = e^{-\lambda_0 t} v_0(a)$, $n(t, a) = e^{\lambda_0 t} u_0(a)$, and $G(x) = H(x) = x \log x$, it follows that

$$\begin{aligned}
\mathcal{H}_v(p|n) &= \int_0^\omega v(t, a)p(t, a) \log\left(\frac{v(t, a)p(t, a)}{v(t, a)n(t, a)}\right) da \\
&= V(0) \int_0^\omega \phi(t, a) \log\left(\frac{V(0)\phi(t, a)}{\phi^\dagger(a)}\right) da \\
&= V(0) \log V(0) + V(0)K(\phi, \phi^\dagger).
\end{aligned}$$

Thus, the general relative entropy also exhibits the monotone property. In fact, [88, 98] use a direct calculation to show that the generalized relative entropy plays the role of a Lyapunov function of the stable population model.

*Remark 1.5* The idea of entropy has been used in various contexts. Suppose that the number of female children produced by a woman aged $a$ per unit time at time $t$ is given by $\beta(a)p(t, a)$. The probability density that the mother's age is $a$ is then given by

$$\frac{\beta(a)p(t, a)}{\int_0^\omega \beta(x)p(t, x)dx} = \frac{\beta(a)\ell(a)B(t - a)}{B(t)}.$$

Let $\Psi(a) := \beta(a)\ell(a)$ be the female net reproduction function. Inserting $B(t) = q_0 e^{\lambda_0 t}$ into the above expression, we obtain $\psi(a) := e^{-\lambda_0 a}\Psi(a)$ as the probability density of the mother's age of female newborns under stable growth. The amount of uncertainty about the age of the mother is given by the *population entropy* $H$ introduced by Demetrius [23–28]:

$$H := -\frac{1}{T_0}\int_0^{\omega} \psi(a)\log\psi(a)da.$$

Let $\psi^*(a) := \Psi(a)/R_0$ be the probability density function for the age at which a woman produces a female child. Then, it follows that

$$\lambda_0 = \frac{\log R_0}{T_0} - \frac{K(\psi, \psi^*)}{T_0} = H - \Phi,$$

where $K$ denotes the Kullback distance and

$$\Phi := -\frac{1}{T_0}\int_0^{\omega} \psi(a)\log\Psi(a)da$$

is called the *reproductive potential*. Readers are referred to Emlen [39] and Smith [120, 121] for details of the use of the population entropy.

## 1.4 Some Demographic Applications

From the time of Lotka to the 1970s, the stable population model was mainly developed by demographers. Classical results in stable population theory, its applications and related population mathematics have been summarized by Coale [18], Impagriazzo [56], Pollard [99], Keyfitz [64, 65], and Smith and Keyfitz [119]. Recently, the stable population model has found a new role in mathematical epidemiology [30, 137]. Here, we consider some practical demographic applications of the stable population theory. Readers interested in discrete-time models or stochastic models, which are not considered here, can find a number of relevant reports in the literature [1, 13, 22, 91, 129].

### *1.4.1 Demographic Indices*

To apply the stable population model to real demographic data, we need to extend the basic one-sex model to distinguish both sexes. Let $p_m(t, a)$ be the age-density function of the male population and $p_f(t, a)$ be the age-density function of the female population. Let $\mu_m(a)$ [$\mu_f(a)$] be the force of mortality of males [females] and $\beta_m(a)$ [$\beta_f(a)$] be the age-specific birth rate of male [female] children. The *female-dominant model* is then formulated as follows:

$$\begin{aligned}
&\left(\frac{\partial}{\partial t}+\frac{\partial}{\partial a}\right)\begin{pmatrix}p_m(t,a)\\ p_f(t,a)\end{pmatrix}=\begin{pmatrix}-\mu_m(a) & 0\\ 0 & -\mu_f(a)\end{pmatrix}\begin{pmatrix}p_m(t,a)\\ p_f(t,a)\end{pmatrix},\\
&\begin{pmatrix}p_m(t,0)\\ p_f(t,0)\end{pmatrix}=\int_0^\infty\begin{pmatrix}0 & \beta_m(a)\\ 0 & \beta_f(a)\end{pmatrix}\begin{pmatrix}p_m(t,a)\\ p_f(t,a)\end{pmatrix}da,
\end{aligned}\tag{1.93}$$

where we use the convention that $\omega=\infty$.

In the female-dominant two-sex model, newborns are produced by the female population and the pair formation process is neglected. Because the female population dynamics are independent from those of the male population, their behavior is determined by the stable population theory.

Let $B_m(t)$ $[B_f(t)]$ be the number of births of male [female] children per unit time, let $\lambda_0$ be the intrinsic rate of natural increase for the female population and $\ell_f(a):=\exp(-\int_0^a\mu_f(\sigma)d\sigma)$. Then, we have

$$\int_0^\infty e^{-\lambda_0 a}\beta_f(a)\ell_f(a)da=1$$

and

$$\lim_{t\to\infty}e^{-\lambda_0 t}B_f(t)=V_f(0)b_{0f},$$

where $V_f(0)$ denotes the normalized total reproductive value of the initial female population and $b_{0f}$ is the CBR of the female stable age distribution. Therefore, it follows that

$$\begin{aligned}
\lim_{t\to\infty}e^{-\lambda_0 t}B_m(t)&=\lim_{t\to\infty}\int_0^\infty e^{-\lambda_0 a}\beta_m(a)\ell_f(a)e^{-\lambda_0(t-a)}B_f(t-a)da\\
&=V_f(0)b_{0f}\int_0^\infty e^{-\lambda_0 a}\beta_m(a)\ell_f(a)da.
\end{aligned}$$

The male population has the same intrinsic growth rate as the female population, and its evolution process is also strongly ergodic. Under the above setting, it is clear that the net reproduction rate (basic reproduction number) for the female population is given by

$$R_0=\int_0^\infty\beta_f(a)\ell_f(a)da,$$

which is a common threshold value for both sexes.

The reproduction index used more often than $R_0$ in statistical demographic arguments is the *total fertility rate* defined by

$$\mathrm{TFR}:=\int_0^\infty m(a)da,$$

where $m(a) := \beta_m(a) + \beta_f(a)$ is the age-specific fertility rate for both sexes. Thus, TFR can be interpreted as the average number of newborns produced by a woman during her reproductive age interval without termination by death. Though TFR is much easier to calculate, it does not provide a threshold condition because it does not take into account the effect of death and the sex ratio at birth. Hence, we define the *female critical fertility rate* (CFR) or the *population replacement level* (PRL) by

$$\text{CFR} := \frac{\text{TFR}}{R_0} = \frac{\int_0^\infty m(a)da}{\int_0^\infty \beta_f(a)\ell_f(a)da} = \frac{\int_0^\infty (1+k(a))\beta_f(a)da}{\int_0^\infty \beta_f(a)\ell_f(a)da},$$

where $k(a) := \beta_m(a)/\beta_f(a)$ denotes the age-specific sex ratio at birth. It then follows from the definition of CFR that

$$\text{sign}(R_0 - 1) = \text{sign}(\text{TFR} - \text{CFR}).$$

Therefore, we can use CFR as a threshold value for population growth. As we use some knowledge of $R_0$ to calculate CFR, it would appear that using CFR and TFR instead of $R_0$ is not particularly advantageous. However, if CFR exhibits a stable value, it is convenient to compare CFR with TFR, as TFR is calculated using only fertility data. In fact, although CFR varies with the sex ratio at birth and the mortality rate, it becomes almost constant if the death rate is sufficiently low until the end of the female reproductive period. For example, the Japanese female CFR has remained at approximately 2.08 since the 1980s, though it was 3.10 in 1925. Before World War II, Japanese infant mortality was very high and 1/3 of newborns died before reaching the reproductive period. Hence, women had to bear more than three children to maintain population replacement. Another reason why the population replacement level is greater than 2 is the effect of the sex ratio at birth. Under natural conditions, the number of male newborns exceeds that of female newborns by about 5%. The effect on CFR of the sex ratio at birth is stronger than that of the death rate if the death rate until the end of the reproductive period is sufficiently low.

For the female-dominant two-sex stable population, we define the asymptotic sex ratio at birth by

$$\gamma := \int_0^\infty \beta_m(a)\ell_f(a)e^{-\lambda_0 a}da = \lim_{t\to\infty} \frac{B_m(t)}{B_f(t)}.$$

If the age-specific sex ratio at birth is a constant $k$ that is independent of the mother's age, we have $\gamma = k$. The male and female stable age structures of the two-sex stable population, denoted by $w_m(a)$ and $w_f(a)$, are defined by

$$w_m(a) := \lim_{t\to\infty} \frac{p_m(t,a)}{P(t)} = \frac{\gamma e^{-\lambda_0 a}\ell_m(a)}{\int_0^\infty e^{-\lambda_0 x}(\gamma\ell_m(x) + \ell_f(x))dx},$$

$$w_f(a) := \lim_{t\to\infty} \frac{p_f(t,a)}{P(t)} = \frac{e^{-\lambda_0 a}\ell_f(a)}{\int_0^\infty e^{-\lambda_0 x}(\gamma\ell_m(x) + \ell_f(x))dx},$$

where

$$P(t) := \int_0^\infty (p_m(t,a) + p_f(t,a))da.$$

Moreover, the stable (crude) fertility rate is given by

$$b_0 := \lim_{t\to\infty} \frac{B_m(t) + B_f(t)}{P(t)} = \frac{\gamma + 1}{\int_0^\infty e^{-\lambda_0 a}(\gamma \ell_m(a) + \ell_f(a))da}.$$

Therefore, we can write

$$w_m(a) = \frac{\gamma}{\gamma+1} b_0 e^{-\lambda_0 a} \ell_m(a), \quad w_f(a) = \frac{1}{\gamma+1} b_0 e^{-\lambda_0 a} \ell_f(a),$$

and the stable age structure of the total population is

$$w(a) = w_m(a) + w_f(a) = \frac{e^{-\lambda_0 a}(\gamma \ell_m(a) + \ell_f(a))}{\int_0^\infty e^{-\lambda_0 x}(\gamma \ell_m(x) + \ell_f(x))dx}.$$

If the birth rate changes slowly and the population growth is almost Malthusian, as was observed in the Japanese population before World War II, the age structure is well approximated by the stable age structure calculated by the period data of the observation year (Fig. 1.5a). In contrast, as shown in Fig. 1.5b, the real age distribution differs markedly from the stable age distribution in 2000, because Japanese fertility rapidly decreased after World War II through the first demographic transition from the mid-1940s to mid-1950s and the second transition after the mid-1970s. The Japanese population fell below its replacement fertility in 1970, reaching an $R_0$ of 0.68 by 2012.

*Remark 1.6* In demographic statistics, TFR is usually calculated from data observed over a year, called the *period TFR* to distinguish it from the *cohort TFR* calculated

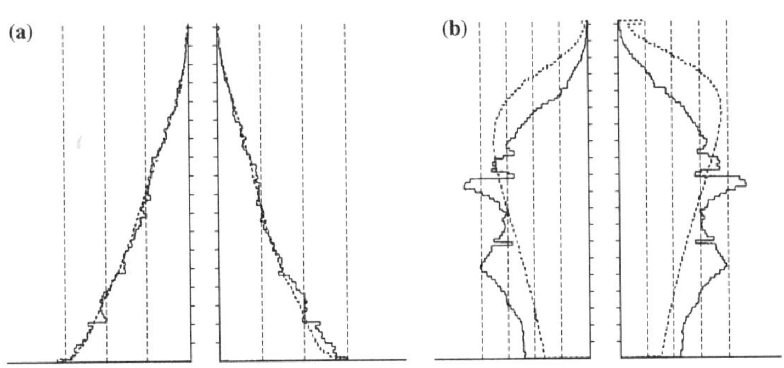


**Fig. 1.5** Age distributions in Japan (*solid line*) and stable age distributions (*dashed line*): **a** 1930, **b** 2000. *Source* National Institute of Population and Social Security

from real cohort data. Because the vital rates change with time, the period TFR is usually different from the cohort TFR. Understanding the relation between the period index and the cohort index is a basic concern in demographic analysis. As the cohort index reflects the real-life course of individuals, it has a biological regularity, whereas the period index is easily affected by environmental changes. However, very-long-time observations are needed to obtain cohort data. Let $m(t, a)$ be the age-specific fertility rate at time $t$ and age $a$. The period TFR at time $t$, denoted by $F(t)$, is given by $F(t) := \int_0^\infty m(t, a)da$, and the cohort TFR of a female cohort born at time $T$, denoted by $G(T)$, is given by $G(T) := \int_0^\infty m(a + T, a)da$. The period TFR denotes the average number of children for a hypothetical cohort. It does not necessarily indicate the real potential of population reproduction, because the period data are very sensitive to changes in the occurrence of demographic events. For example, suppose that the cohort fertility schedule is $m(a + T, a) = G(T)\psi(a; T)$ where $\psi(a; T)$ is a normalized fertility pattern of a cohort born at time $T$. Then, $\int_0^\infty \psi(a; T)da = 1$ and we have

$$F(t) = \int_0^\infty \psi(a; t - a)G(t - a)da.$$

If the cohort TFR is a constant $G$ and the fertility pattern is delayed with a speed $v$ along the cohort line such that $\psi(a; T) = \psi(a - vT)$ where $\psi(a)$ is a time-independent fertility schedule, then

$$F(t) = G \int_0^\infty \psi(a - v(t - a))da = \frac{G}{1 + v}.$$

Therefore, if there exists a persistent delay in the fertility schedule (postponement of childbearing), the observed period TFR will be smaller than the cohort TFR. Conversely, the advancement of childbearing makes the period TFR larger than the cohort TFR. This is a simple example of the well-known *tempo-distortion* of demographic indices [59].

### 1.4.2 The Population Momentum

To formulate the inertia of a population movement, Keyfitz [61] introduced the *momentum of population growth*, defined as the ratio of the ultimate stationary population size to the initial population size if the basic reproduction number was adjusted to unity for $t > 0$.

It follows from (1.58) that if $R_0 = 1$, a stable population converges to a stationary population

$$p_\infty(a) := \lim_{t \to \infty} p(t, a) = V(0)\frac{\ell(a)}{e_0},$$

where $V(0) = \langle v_0, p_0 \rangle$ is the normalized total reproductive value of the initial population, $v_0(a)$ is a reproductive value with $R_0 = 1$:

$$v_0(a) = \frac{e_0}{T_c} \int_a^{\infty} \frac{\ell(s)}{\ell(a)} \beta(s) ds,$$

where $T_c = \int_0^{\infty} a\beta(a)\ell(a)da$ is the average childbearing age in a cohort and $1/e_0$ gives the CBR of the stationary population.

Therefore, the momentum of population growth, denoted by $M_p$, is calculated as follows:

$$M_p = \frac{\int_0^{\infty} p_{\infty}(a)da}{\int_0^{\infty} p_0(a)da} = \langle v_0, w_0 \rangle = \int_0^{\infty} \frac{w_0(a)}{u_0(a)} z(a)da,$$

where

$$u_0(a) = \frac{\ell(a)}{e_0}, \quad w_0(a) := \frac{p_0(a)}{\int_0^{\infty} p_0(\sigma)d\sigma},$$

$$z(a) := v_0(a)u_0(a) = \frac{v_0(a)\ell(a)}{\int_0^{\infty} v_0(x)\ell(x)dx}.$$

According to Preston et al. [102], the distribution $z(a)$ (the stationary age profile weighted by the reproductive value) is a monotone decreasing function with a statistically stable pattern that is independent of regions and nations. Thus, if the age profile of the initial population is "younger" than the stationary population (that is, $w_0(a) > u_0(a)$ in young age classes), the momentum is greater than unity, whereas if the initial population is "older" than the stationary population ($w_0(a) < u_0(a)$ in young age classes), the momentum is less than unity.

If $\lambda_0 = 0$, the total reproductive value $V(0)$ gives the total size of the ultimate stationary population. Thus, we have $M_p = V(0)/P(0)$, which shows that the population momentum is the average reproductive value of the initial population. As we saw in Sect. 1.3.2, a unit population concentrated at age $\zeta$ at the initial time will asymptotically produce a population of $e^{\lambda_0 t} v_0(\zeta) u_0(a)$. In this case, it follows from $V(0) = v_0(\zeta)$ that

$$M_p = v_0(\zeta) = \frac{e_0}{T_c} \int_{\zeta}^{\infty} \frac{\ell(s)}{\ell(\zeta)} \beta(s) ds.$$

In other words, the reproductive value at age $\zeta$ with $\lambda_0 = 0$ is the population momentum of an individual aged $\zeta$. Because the momentum for prereproductive ages is approximately $e_0/T_c$, young people have a large momentum.

#### 1.4.2.1 Keyfitz's Formula

As a special case, let us consider the situation in which an initial population has a stable age distribution with a net reproduction function $\Psi(a)$ with the basic reproduction number $R_0 \neq 1$ and a fertility rate that changes from $\beta(a)$ to $\beta(a)/R_0$ at $t = 0$. The momentum of population growth is then calculated by *Keyfitz's momentum formula* as

$$M_p = \frac{b_0 e_0}{\lambda_0 T_c}\left(1 - \frac{1}{R_0}\right), \tag{1.94}$$

where $b_0$ is the CBR of the initial stable age distribution and $T_c$ is the mean age of childbearing in a cohort born at $t > 0$:

$$b_0 = \frac{1}{\int_0^\infty e^{-\lambda_0 a}\ell(a)da}, \quad T_c = \int_0^\infty a\frac{\Psi(a)}{R_0}da.$$

To derive this formula, suppose that $p_0(a) = p_0(0)e^{-\lambda_0 a}\ell(a)$. Then, the ultimate number of births per unit time is calculated as

$$\lim_{t\to\infty} B(t) = \frac{1}{T_c}\int_0^\infty G(t)dt,$$

where

$$G(t) = \int_t^\infty \frac{p_0(a-t)}{\ell(a-t)}\frac{\Psi(a)}{R_0}da = \frac{p_0(0)e^{\lambda_0 t}}{R_0}\int_t^\infty e^{-\lambda_0 a}\Psi(a)da.$$

Hence, it follows that

$$\begin{aligned}
\int_0^\infty G(t)dt &= \frac{p_0(0)}{R_0}\int_0^\infty dt\int_t^\infty e^{-\lambda_0(a-t)}\Psi(a)da \\
&= \frac{p_0(0)}{R_0}\int_0^\infty \Psi(a)da\int_0^a e^{-\lambda_0(a-t)}dt \\
&= \frac{p_0(0)}{R_0}\int_0^\infty \Psi(a)\left[\frac{1-e^{-\lambda_0 a}}{\lambda_0}\right]da.
\end{aligned}$$

Thus, we conclude that

$$\lim_{t\to\infty} B(t) = \frac{p_0(0)}{\lambda_0 T_c}\frac{R_0 - 1}{R_0}.$$

Because the size of the ultimate stationary population is $B(\infty)e_0$ and the size of the initial population is $\int_0^\infty p_0(a)da = p_0(0)/b_0$, we arrive at expression (1.94).

Although Keyfitz's momentum formula is useful for estimating the stationary population size as a result of a drastic population stabilization policy, it should be remarked that there are many other adjustment methods for changing $R_0$ to unity (see [89]). A weakness of this formula is that we do not know what kind of individual behavior changes could lead to such a period adjustment.

**Exercise 1.17** Consider again a situation in which the initial population has a stable age distribution with a net reproduction function $\Psi(a) = \beta(a)\ell(a)$ with the basic reproduction number $R_0 \neq 1$ and intrinsic growth rate $\lambda_0$. Suppose that the fertility rate changes from $\beta(a)$ to $e^{-\lambda_0 a}\beta(a)$ at $t = 0$, so that the net reproduction rate becomes unity for $t > 0$. Show that the population momentum is calculated as follows:

$$M_p = \frac{e_0 b_0}{T_0 \lambda_0}\left(1 - \int_0^\infty e^{-2\lambda_0 a}\Psi(a)da\right), \tag{1.95}$$

where $T_0 = \int_0^\infty a e^{-\lambda_0 a}\Psi(a)da$ is the average age of childbearing in the initial stable age distribution and is also the average age of childbearing in the final steady state [89].

#### 1.4.2.2 Frauenthal's Formula

In contrast to the above period adjustment, it would be more realistic to assume that the net reproduction function $\Psi(a)$ changes to $\Psi(a-\tau)/R_0$ for individuals born after the initial time, that is, individuals born after $t = 0$ adopt a two-child policy with time delay $\tau$ [44, 58, 127]. Again, we assume that the initial population is stable. Then, it follows that

$$\lim_{t\to\infty} B(t) = \frac{\int_0^\infty G(t)dt}{\int_0^\infty a\frac{\Psi(a-\tau)}{R_0}da}.$$

As the reproduction schedule of the initial population is given by $\Psi$, we can calculate $G(t)$ and its integral as

$$G(t) = p_0(0)e^{\lambda_0 t}\int_t^\infty e^{-\lambda_0 a}\Psi(a)da,$$

$$\int_0^\infty G(t)dt = p_0(0)\int_0^\infty \Psi(a)\left[\frac{1-e^{-\lambda_0 a}}{\lambda_0}\right]da.$$

From

$$\int_0^\infty a\frac{\Psi(a-\tau)}{R_0}da = T_c + \tau, \quad T_c = \int_0^\infty a\frac{\Psi(a)}{R_0}da,$$

we conclude that

$$\lim_{t\to\infty} B(t) = \frac{p_0(0)(R_0 - 1)}{\lambda_0(T_c + \tau)}.$$

Therefore, the momentum of population growth given by this cohort adjustment, denoted by $M_c$, can be calculated as

$$
\begin{aligned}
M_c &= \frac{p_0(0)e_0(R_0-1)}{\lambda_0(T_c+\tau)} \cdot \frac{1}{p_0(0)\int_0^\infty e^{-\lambda_0 a}\ell(a)da} \\
&= \frac{b_0 e_0(R_0-1)}{\lambda_0(T_c+\tau)} = \frac{R_0 T_c}{T_c+\tau} M_p.
\end{aligned} \tag{1.96}
$$

Although the definition of the momentum formula meant it was not originally applied to a stationary population, we can still use a tempo policy to regulate the stationary population. In the above cohort adjustment, if we let $R_0 = 1$ and $\lambda_0 \to 0$, then we have

$$
\lim_{t\to\infty} B(t) = \frac{\int_0^\infty G(t)dt}{\int_0^\infty a\Psi(a-\tau)da} = \frac{p_0(0)T_c}{T_c+\tau}.
$$

Thus, if childbearing is delayed ($\tau > 0$), the birth rate of the ultimate stationary population $B(\infty)$ is smaller than the birth rate of the initial population $p_0(0)$, whereas $B(\infty)$ is larger than $p_0(0)$ if childbearing is advanced ($\tau < 0$). Then, the momentum can be calculated as

$$
M_c = \frac{B(\infty)e_0}{p_0(0)e_0} = \frac{T_c}{T_c+\tau}.
$$

Therefore, if childbearing is delayed, the stationary population size becomes smaller.

In the case $R_0 > 1$, we have $M_c = M_p$ if we choose $\tau > 0$ such that $\tau = (R_0 - 1)T_c$. Therefore, if $|R_0 - 1|$ is not too large, the ultimate stationary population size given by Keyfitz's drastic assumption that $R_0$ is instantaneously adjusted to unity can be attained by a realistic family planning scenario in which individuals born after $t = 0$ have two children by delayed childbearing.

This example shows that both the final family size and the timing (tempo) of childbearing are important tools for regulating the population. Even though cultural norms or socioeconomic problems make it difficult to accept drastic reductions in the final family size, adjusting the timing of childbearing by delayed marriage and spacing between births would be a more acceptable, effective method of population control [9, 20]. That is, the population size is not only determined by the basic reproduction number, but also by the average interval between generations (generation time), so the tempo change in demographic events can regulate the population size without altering the family size of individuals.

**Exercise 1.18** Consider a situation in which an initial population is in a steady state with a survival probability $\ell(a)$ and fertility rate $\beta(a)$ such that $\int_0^\infty \beta(a)\ell(a)da = 1$. Suppose that the life cycle schedule of individuals born after $t = 0$ changes from $(\ell(a), \beta(a))$ to $(\ell^*(a), \beta^*(a))$ such that $\int_0^\infty \beta^*(a)\ell^*(a)da = 1$. Show that the population momentum $M_c$ is

$$
M_c = \frac{e_0^*}{T_c^*}\left[\frac{e_0}{T_c}\right]^{-1},
$$

**Table 1.1** Demographic indices of Japan: 1925–2010

| Year | TFR | $R_0$ | $\lambda_0$ (%) | CFR | Momentum |
|---|---|---|---|---|---|
| 1925 | 5.10 | 1.65 | 1.711 | 3.10 | – |
| 1930 | 4.70 | 1.52 | 1.423 | 3.09 | – |
| 1940 | 4.11 | 1.43 | 1.183 | 2.87 | – |
| 1950 | 3.65 | 1.50 | 1.388 | 2.43 | – |
| 1955 | 2.37 | 1.06 | 0.190 | 2.24 | 1.443 |
| 1960 | 2.00 | 0.92 | −0.301 | 2.18 | 1.385 |
| 1965 | 2.14 | 1.01 | 0.025 | 2.12 | 1.331 |
| 1970 | 2.13 | 1.03 | 0.014 | 2.13 | 1.284 |
| 1975 | 1.91 | 0.91 | −0.354 | 2.10 | 1.229 |
| 1980 | 1.75 | 0.83 | −0.650 | 2.09 | 1.166 |
| 1985 | 1.76 | 0.85 | −0.586 | 2.08 | 1.116 |
| 1990 | 1.54 | 0.74 | −1.026 | 2.08 | 1.066 |
| 1995 | 1.42 | 0.69 | −1.280 | 2.07 | 1.004 |
| 2000 | 1.36 | 0.65 | −1.423 | 2.08 | 0.945 |
| 2005 | 1.26 | 0.61 | −1.647 | 2.07 | 0.872 |
| 2010 | 1.39 | 0.67 | −1.311 | 2.07 | – |

where $e_0^*$ is the life span in the new schedule and $T_c^*$ is the average age of childbearing in a cohort of the new schedule [48].

Table 1.1 shows basic demographic indices of Japanese female from 1925 to 2010, taken from Latest Demographic Statistics 2014, NIPSSR, and *J. Pop. Problems* 62(4), 2006. The population momentum was calculated by Dr. Hutoshi Ishii (NIPSSR, Tokyo).

### 1.4.3 Preston–Coale System

We now introduce a statistical application of the McKendrick equation in demography that is known as the *variable r-method* or *Preston–Coale system*. Bennett and Horiuchi[10] [7] and Preston and Coale [100] found the following relation between the age-density function and the force of mortality:

$$\frac{1}{p(t,a)}\frac{\partial p(t,a)}{\partial a} = -\mu(t,a) - r(t,a), \tag{1.97}$$

where $r(t,a)$ is the *age-specific growth rate* of the population, defined by

[10] Prof. Horiuchi has informed me that they arrived at (1.97) without knowing the McKendrick equation.

$$r(t,a) := \frac{1}{p(t,a)} \frac{\partial p(t,a)}{\partial t}.$$

Though relation (1.97) is a time-dependent McKendrick equation, an important point is that (1.97) should hold for *any* closed population, as it exhibits a trivial balance. The age-specific growth rate can be measured using two successive censuses, so it is relatively easy to obtain data and a statistically robust parameter.

Integrating (1.97) with respect to the age variable, we have

$$p(t,a) = p(t,0)\ell_p(t,a) \exp\left(-\int_0^a r(t,\sigma)d\sigma\right), \tag{1.98}$$

where

$$\ell_p(t,a) := \exp\left(-\int_0^a \mu(t,\sigma)d\sigma\right)$$

is the period survival probability at time $t$. Equation (1.98) implies that the age distribution at time $t$ and the period life table function (survival probability) can be transformed into one another by means of the age-specific growth rate. If the parameters are time-independent, there exists an age distribution such that $r(t,a)$ is a constant, which gives the stable age distribution.

Let $w(t,a) = p(t,a)/\int_0^\infty p(t,a)da$ be the age profile and

$$b(t) = \frac{p(t,0)}{\int_0^\infty p(t,a)da}$$

be the CDR at time $t$. Then, (1.98) can be rewritten as

$$\frac{1}{\ell_p(t,a)} = \frac{b(t)e^{-\int_0^a r(\sigma,a)d\sigma}}{w(t,a)}. \tag{1.99}$$

Now consider the situation in which there are no vital statistics, but we can make use of two successive censuses. It is well known that a survival probability function can be transformed to another survival probability function by means of the logit transformation. Thus, if we choose a standard survival probability function $\ell_s(a)$, we obtain the logit transformation formula

$$\log\left(\frac{1-\ell_p(t,a)}{\ell_p(t,a)}\right) = \alpha + \beta \log\left(\frac{1-\ell_s(a)}{\ell_s(a)}\right), \tag{1.100}$$

where $\alpha$ and $\beta$ are unknown parameters. It follows from (1.99) and (1.100) that

$$\frac{e^{-\int_0^a r(\sigma,a)d\sigma}}{w(t,a)} = \frac{1}{b(t)} + \frac{e^\alpha}{b(t)}\left(\frac{1-\ell_s(a)}{\ell_s(a)}\right)^\beta. \tag{1.101}$$

Because the left-hand side of (1.101) is given by the census data, (1.101) can be thought of as a recurrence equation with unknown parameters $b(t)$, $\alpha$, and $\beta$. In particular, if we set $\beta = 1$ (for which the fitness of the logit transformation is still good), then (1.101) becomes a linear regression equation, and so $1/b(t)$ and $\alpha$ can be easily obtained. Therefore, from (1.99), we have $\ell_p(t, a)$, that is, the unknown mortality level is indirectly estimated [101].

The variable $r$-method is also useful for estimating unknown demographic parameters other than the death rate. For example, let $\beta(t, a)$ be the age-specific birth rate at time $t$ and $v(t, a)$ be the age distribution of childbearing for mothers:

$$v(t, a) = \frac{\beta(t, a)p(t, a)}{p(t, 0)} = \beta(t, a)\ell_p(t, a)e^{-\int_0^a r(\sigma,a)d\sigma}.$$

If we obtain data for $v(t, a)$ from a sample survey, the period basic reproduction number (net reproduction rate) can be estimated as follows:

$$R_0 = \int_0^\infty v(t, a)e^{\int_0^a r(\sigma,a)d\sigma} da.$$

Such indirect estimation methods have been applied to study anthropological groups without vital statistics [45]. Though the above assumes a closed population, the Preston–Coale system has been extended to allow migration. In the 1980s, the indirect estimation method, which is a classical theme in demography, was activated by the rediscovery of the McKendrick equation [4].

### 1.4.4 Perturbation Theory

A fundamental question in demography concerns how the age structure is influenced by changes in vital parameters. For example, the aging process of the population of France began in the nineteenth century and has become a hot issue among demographers seeking to understand the primary reason for population aging, fertility decline, or mortality decline. French demographers insisted there was an empirical law whereby population aging occurs mainly by a decline in fertility, rather than a decline in mortality. A point of issue here concerns the long-term effect of parameter changes on the population structure by eliminating the effect of initial data and applying a temporal perturbation. This argument was concluded in the 1950s when American and Japanese demographers independently proved the empirical aging law using comparative statistics to observe the effect of parameter changes on the stable population [17, 126].

Following Arthur [3], we now introduce a perturbation method to examine the effect of parameter changes on the stable age distribution. First, let us consider the following Lotka system:

$$
\begin{aligned}
&\int_0^\infty e^{-\lambda a}\ell(a)\beta(a)da - 1 = 0,\\
&b\int_0^\infty e^{-\lambda a}\ell(a)da - 1 = 0, \qquad (1.102)\\
&be^{-\lambda a}\ell(a) - w(a) = 0,
\end{aligned}
$$

where $\beta(a)$ is the age-specific fertility rate, $\ell(a)$ is the survival probability, $\lambda$ is the intrinsic rate of natural increase, $b$ is the CBR in the stable population and $w(a)$ is the stable age profile.

Let $\Delta\beta(a)$ be a small perturbation in the age-specific birth rate and let $\Delta\lambda$, $\Delta, b$ and $\Delta w(a)$ be corresponding perturbations in the intrinsic growth rate, the CBR and the stable age profile, respectively. Because $\Delta\lambda$ is sufficiently small that the approximation $e^{-\Delta\lambda a} = 1 - \Delta\lambda a$ holds, it follows from the above system that

$$
\begin{aligned}
&\int_0^\infty e^{-\lambda a}(1-\Delta\lambda a)\ell(a)(\beta(a)+\Delta\beta(a))da - 1 = 0,\\
&(b+\Delta b)\int_0^\infty e^{-\lambda a}(1-\Delta\lambda a)\ell(a)da - 1 = 0, \qquad (1.103)\\
&(b+\Delta b)e^{-\lambda a}(1-\Delta\lambda a)\ell(a) - (w(a)+\Delta w(a)) = 0.
\end{aligned}
$$

If we neglect the second-order small perturbation, (1.103) can be solved as follows:

$$
\begin{aligned}
\Delta\lambda &= \frac{1}{T_0}\int_0^\infty e^{-\lambda a}\ell(a)\Delta\beta(a)da,\\
\frac{\Delta b}{b} &= A\Delta\lambda = \frac{A}{T_0}\int_0^\infty e^{-\lambda a}\ell(a)\Delta\beta(a)da, \qquad (1.104)\\
\frac{\Delta w(a)}{w(a)} &= (A-a)\Delta\lambda = \frac{A-a}{T_0}\int_0^\infty e^{-\lambda a}\ell(a)\Delta\beta(a)da,
\end{aligned}
$$

where

$$
T_0 := \int_0^\infty ae^{-\lambda a}\ell(a)\beta(a)da, \quad A := \int_0^\infty aw(a)da,
$$

$T_0$ is the average age of childbearing in the stable population and $A$ is the average age of the stable population. Therefore, if the age-specific birth rate declines ($\Delta\beta < 0$), the intrinsic growth rate and the CBR decrease, while the population above the average age increases and the population below the average age decreases in the stable age distribution, that is, the population is aging.

Let us apply a similar analysis to the mortality rate. As mortality changes are expressed by changes in the survival probability, let $\Delta\ell(a)$ be a perturbation in the survival probability, and let $\Delta\lambda$, $\Delta b$, and $\Delta w(a)$ be the corresponding perturbations in the intrinsic growth rate, the CBR, and the stable age profile, respectively. It follows from (1.102) that

$$\begin{aligned}
&\int_0^\infty e^{-\lambda a}(1-\Delta\lambda a)\beta(a)(\ell(a)+\Delta\ell(a))da-1=0,\\
&(b+\Delta b)\int_0^\infty e^{-\lambda a}(1-\Delta\lambda a)(\ell(a)+\Delta\ell(a))da-1=0,\\
&(b+\Delta b)e^{-\lambda a}(1-\Delta\lambda a)(\ell(a)+\Delta\ell(a))-(w(a)+\Delta w(a))=0.
\end{aligned} \tag{1.105}$$

If we again neglect the second-order perturbations, we obtain

$$\begin{aligned}
\Delta\lambda &= \frac{1}{T_0}\int_0^\infty e^{-\lambda a}\beta(a)\Delta\ell(a)da,\\
\frac{\Delta b}{b} &= A\Delta\lambda - b\int_0^\infty e^{-\lambda a}\Delta\ell(a)da,\\
\frac{\Delta w(a)}{w(a)} &= (A-a)\Delta\lambda - b\int_0^\infty e^{-\lambda a}\Delta\ell(a)da+\frac{\Delta\ell(a)}{\ell(a)}.
\end{aligned} \tag{1.106}$$

From (1.106), we know that if the mortality rate decreases ($\Delta\ell > 0$) in the population below the reproductive age, the intrinsic growth rate increases; conversely, changes in mortality in the post-reproductive age population do not affect the intrinsic growth rate. In contrast, the directions of changes in the CBR and stable age profile are not uniquely determined. In the first term of the right-hand side of the third equation of (1.106), the population above the average age decreases and the population below the average age increases, that is, an anti-aging effect can be observed. The second and third terms reflect the aging effect of improved mortality on the age profile. If the mortality only improves in the post-reproductive age population, as observed among recently developed countries, the population structure becomes more aged, whereas the population becomes younger if the mortality improves among young people (in particular, infants), as was seen in the first stage of the population transition [102]. In summary, population aging in the modernization process is mainly caused by fertility decline, whereas the improvement of mortality among aged people is also a non-negligible cause of population aging in developed societies with sufficiently low fertility.

*Remark 1.7* We can consider the effect of fertility changes on the intrinsic growth rate using a concrete example [58]. Suppose that the perturbation in the age-specific birth rate is given by $\Delta\beta(a) = (1-k)\beta(a-\tau)-\beta(a)$. That is, the birth rate function retains its shape, but TFR is multiplied by a factor of $(1-k)$ and is delayed for $\tau$. Of course, this delay is less than the lower bound of the reproductive age. It then follows from (1.103) that

$$\Delta\lambda_0 = \frac{1}{T_0}\left((1-k)e^{-\lambda_0\tau}\int_0^\infty e^{-\lambda_0 a}\beta(a)\ell(a+\tau)da-1\right).$$

To calculate the integral term analytically, we assume that the force of mortality is a constant $\mu$ in the neighborhood of the reproductive age period. Then, we have $\ell(a+\tau)=e^{-\mu\tau}\ell(a)$ and so it holds that

$$\Delta\lambda_0 = \frac{1}{T_0}((1-k)e^{-(\lambda_0+\mu)\tau} - 1).$$

If there is no delay ($\tau = 0$), we have

$$\Delta\lambda_0 = -\frac{k}{T_0},$$

and so the intrinsic growth rate always decreases as TFR decreases, and its degree of decrease is inversely proportional to the mean age of childbearing [35]. If we consider the case in which $k > 0$ and $\tau > 0$, that is, TFR decreases and a delay in childbearing occurs, then $\Delta\lambda_0 > 0$ if and only if

$$\lambda_0 < \frac{\log(1-k)}{\tau} - \mu.$$

Therefore, the intrinsic growth rate can *increase* if the unperturbed fertility rate is below the replacement level and $|\lambda_0|$ is large. That is, for a population with below replacement fertility, a delay in childbearing does not necessarily lead to a decrease in the intrinsic growth rate.

Though the above arguments depend on an approximation when using the perturbation method, the following proposition can be proved without such a limitation [11]:

**Proposition 1.13** *Let $\beta_\tau(\cdot)$ be a shifted age-specific fertility function with delay $\tau$, that is, $\beta_\tau(a) := \beta(a-\tau)$. We adopt the convention that $\beta_\tau(a) = 0$ if $a - \tau < 0$. Let $\lambda_0(\tau)$ be the intrinsic growth rate corresponding to $\beta_\tau(\cdot)$. Then, $\lambda_0(\tau)$ is differentiable with respect to $\tau$ and $d\lambda_0(\tau)/d\tau < 0$ if and only if*

$$\lambda_0(\tau) + \int_0^\infty \mu(a)\ell(a)e^{-\lambda_0(\tau)a}\beta_\tau(a)da > 0, \tag{1.107}$$

*where $\mu(a)$ is the force of mortality. Moreover, the above condition is equivalent to*

$$M := \int_0^\infty \beta_\tau(a)\frac{dw^*(a)}{da}da < 0, \tag{1.108}$$

*where $w^*(a)$ denotes the stable age profile corresponding to $\lambda_0(\tau)$:*

$$w^*(a) := \frac{e^{-\lambda_0(\tau)a}\ell(a)}{\int_0^\infty e^{-\lambda_0(\tau)a}\ell(a)da}.$$

*Proof* The intrinsic growth rate $\lambda_0(\tau)$ satisfies the characteristic equation, as

$$F(\lambda_0, \tau) := \int_0^\infty e^{-\lambda_0 a}\beta_\tau(a)\ell(a)da - 1 = 0.$$

Because

$$F(\lambda_0, \tau) = e^{-\lambda_0 \tau} \int_0^\infty e^{-\lambda_0 a} \beta(a) \ell(a+\tau) da - 1,$$

$F$ is continuously differentiable in both $\lambda_0$ and $\tau$. We can observe that

$$\frac{\partial F}{\partial \tau} = -\lambda_0 (F(\lambda_0, \tau) + 1) - \int_0^\infty e^{-\lambda_0 a} \beta_\tau(a) \mu(a) \ell(a) da,$$

$$\frac{\partial F}{\partial \lambda_0} = -\int_0^\infty a e^{-\lambda_0 a} \beta_\tau(a) \ell(a) da.$$

It follows from $F(\lambda_0(\tau), \tau) = 0$ and the implicit function theorem that $\lambda_0(\tau)$ is differentiable and

$$\frac{d\lambda_0(\tau)}{d\tau} = -\frac{\partial F}{\partial \tau} \Big/ \frac{\partial F}{\partial \lambda_0} = -\frac{\lambda_0(\tau) + \int_0^\infty e^{-\lambda_0(\tau) a} \beta_\tau(a) \mu(a) \ell(a) da}{\int_0^\infty a e^{-\lambda_0 a} \beta_\tau(a) \ell(a) da}.$$

Thus, (1.107) is a necessary and sufficient condition for $\lambda_0(\tau)$ to become a decreasing function of $\tau$. Moreover, observe that

$$\begin{aligned}
&\int_0^\infty e^{-\lambda_0(\tau) a} \beta_\tau(a) \mu(a) \ell(a) da \\
&= \int_0^\infty \beta_\tau(a) \left[ \frac{d}{da} (e^{-\lambda_0(\tau) a} \ell(a)) + \lambda_0(a) e^{-\lambda_0(\tau) a} \ell(a) \right] da \\
&= -\lambda_0(\tau) - \int_0^\infty \beta_\tau(a) \frac{dp^*(a)}{da} da \left( \int_0^\infty e^{-\lambda_0(\tau) a} \ell(a) da \right).
\end{aligned}$$

Thus, we know that (1.107) and (1.108) are equivalent conditions. □

The index $M$ was called the *fertility momentum* by Busenberg and Iannelli [11]. That is, the intrinsic growth rate decreases corresponding to an age shift in the age-specific fertility rate if and only if the fertility momentum is negative. There is a possibility that the fertility momentum can be positive if the intrinsic growth rate is negative.

## 1.5 Age Profile Dynamics of Quasi-stable Populations

For the stable population model (1.25) with non-trivial initial data, for all $t \geq 0$ we define the *age profile* $w(t, a)$ as the normalized age-density function

$$w(t, a) := \frac{p(t, a)}{\int_0^\omega p(t, x) dx}. \tag{1.109}$$

One remarkable feature of the stable population model is that the time evolution of the age profile is *independent* of the growth in total population size.[11]

To observe the age-profile dynamics in a slightly more general setting, let us consider a *quasi-stable population* model[12]:

$$\begin{aligned} \frac{\partial q(t,a)}{\partial t} + \frac{\partial q(t,a)}{\partial a} &= -(\mu(a) + \eta(t))q(t,a), \\ q(t,0) &= \int_0^\omega \beta(a) q(t,a) da, \quad t > 0, \end{aligned} \tag{1.110}$$

where $\eta(t)$ is a given function denoting an age-independent, but time-dependent, perturbation to the force of mortality.

Let $w(t,a)$ be the age profile of the quasi-stable population in (1.110), and let

$$Q(t) := \int_0^\omega q(t,a) da$$

be its size. Inserting $q(t,a) = Q(t)w(t,a)$ into (1.110) and performing a simple calculation, we obtain the set of equations:

$$\begin{aligned} \frac{\partial w(t,a)}{\partial t} + \frac{\partial w(t,a)}{\partial a} &= -\mu(a) w(t,a) - \alpha(w(t,\cdot)) w(t,a), \\ w(t,0) &= \int_0^\omega \beta(a) w(t,a) da, \end{aligned} \tag{1.111}$$

$$\frac{dQ(t)}{dt} = [-\eta(t) + \alpha(w(t,\cdot))] Q(t), \tag{1.112}$$

where

$$\alpha(w(t,\cdot)) := \int_0^\omega [\beta(a) - \mu(a)] w(t,a) da.$$

System (1.111) shows that the age profile of a quasi-stable population is determined only by the birth rate $\beta$ and the standard death rate $\mu$ and is independent of the total population size $Q(t)$ and the death rate perturbation $\eta(t)$.

If $w$ and $Q$ satisfy (1.111) and (1.112), it can be easily checked that $Q(t)w(t,a)$ gives the solution to (1.110). Moreover, it is easy to see that

$$p(t,a) = Q(t) w(t,a) \exp\left( \int_0^t \eta(s) ds \right) = q(t,a) \exp\left( \int_0^t \eta(s) ds \right)$$

---

[11] The following results are given by [11, 55]. This type of separability of the age-profile dynamics is also seen in a class of nonlinear age-dependent models (*separable model*, see Chap. 3, [10, 54]).

[12] According to Coale [18], a population is called *quasi-stable* if its birth rate function is time-independent but the death rates rise or fall uniformly at all ages.

satisfies the stable population model (1.25) with initial data

$$p(0, a) = w(0, a)Q(0) = q(0, a).$$

Conversely, if we let $p(t, a)$ be the solution of the stable population model (1.25) with the initial data $p(0, a) = w(0, a)Q(0)$ and let $P(t)$ be its size $P(t) = \int_0^\omega p(t, a)da$, the solutions of (1.111) and (1.112) are given by

$$Q(t) = P(t)e^{-\int_0^t \eta(s)ds}, \quad w(t, a) = \frac{p(t, a)}{P(t)}.$$

That is, the age-profile dynamics (1.111) can be solved uniquely by reducing the system to the stable population model. Hence, the age-profile dynamics of quasi-stable populations are not influenced by the choice of the age-independent perturbation $\eta$.

**Proposition 1.14** *The age-profile dynamics* (1.111) *have a unique positive steady state.*

*Proof* Let us consider the stationary problem of (1.111):

$$\begin{aligned} \frac{dw(a)}{da} &= -\mu(a)w(a) - \alpha(w)w(a), \\ w(0) &= \int_0^\omega \beta(a)w(a)da, \\ \int_0^\omega w(t, a)da &= 1. \end{aligned} \tag{1.113}$$

Let $w^*(a)$ be a solution of (1.113), and define

$$\alpha^* := \alpha(w^*) = \int_0^\omega [\beta(\sigma) - \mu(\sigma)]w^*(\sigma)d\sigma.$$

Then, we have

$$w^*(a) = \frac{e^{-\alpha^* a}\ell(a)}{\int_0^\omega e^{-\alpha^* \sigma}\ell(\sigma)d\sigma}.$$

From the boundary condition, we obtain

$$\int_0^\omega e^{-\alpha^* \sigma}\beta(\sigma)\ell(\sigma)d\sigma = 1,$$

which is simply Lotka's characteristic equation, and its unique real root $\alpha^*$ is the intrinsic growth rate $\lambda_0$. The unique stationary solution of (1.113) is then given by

$$w^*(a) = \frac{e^{-\lambda_0 a}\ell(a)}{\int_0^\omega e^{-\lambda_0 \sigma}\ell(\sigma)d\sigma}.$$

This completes our proof. □

If the age profile is time-independent, $\alpha(w^*) = \lambda_0$, and it follows from (1.112) that

$$Q(t) = Q(0)\exp\left(\lambda_0 t - \int_0^t \eta(\sigma)d\sigma\right).$$

$Q(t)w^*(a)$ is called the *persistent solution*. In particular, if $\eta \equiv 0$, the stationary solution of the age-profile dynamics corresponds to an *exponential solution*: $p(t,a) = Q(0)e^{\lambda_0 t}w^*(a)$. It then follows from the Fundamental Theorem that the age profile converges to $w^*(a)$ in the $L^1$ norm as $t \to +\infty$:

**Proposition 1.15** *If $w_0$ is non-trivial, it follows that*

$$\lim_{t\to\infty}\int_0^\omega |w(t,a) - w^*(a)|da = 0. \tag{1.114}$$

*Proof* It follows from Proposition 1.9 that, for $t > \omega$,

$$p(t,a) = q_0 e^{\lambda_0(t-a)}\ell(a)(1+\varepsilon(t-a)),$$

where $\lim_{t\to\infty}\varepsilon(t) = 0$ and $q_0 > 0$. Then, we have

$$w(t,a) = \frac{e^{-\lambda_0 a}\ell(a)(1+\varepsilon(t-a))}{\int_0^\omega e^{-\lambda_0 a}\ell(a)(1+\varepsilon(t-a))da},$$

and (1.114) follows immediately. □

Proposition 1.15 shows that the age profile $w^*(a)$ is globally stable in the sense that any perturbed stable population system will ultimately recover the age profile $w^*$, as long as the age profile does not become a trivial profile. In this sense, $w^*(a)$ is called the *stable age distribution*. The stable distribution $w^*(a)$ is determined only by birth and death rates and is independent of the initial data. The population evolution process is said to be *strongly ergodic* if it has a limiting age distribution that is independent of the initial data. The term "ergodic" refers to the situation whereby the state of a system becomes asymptotically independent of its initial state. The Fundamental Theorem and related results are often called the *strong ergodicity theorem* [19]. If a population is described by the stable population model, the stable distribution is the unique persistent age distribution and, as pointed out by Lotka [79], the long-run exponential growth implies that the population age structure is given by the stable age distribution ([57, 78, 79], Chap. 10).

*Remark 1.8* In the stable population model (1.25), the *age-profile ratio* (APR) $r$ is defined by:

$$r(t,a) := \frac{p(t,a)}{P(t)w^*(a)}, \tag{1.115}$$

where $P(t) := \int_0^\omega P(t,a)da$, $p(t,a)/P(t)$ is the age profile at time $t$ and $w^*$ is the age profile of the stable age distribution. The age-profile ratio is well defined as long as the initial data $p_0$ are non-trivial. Then it is easy to see that APR $r$ satisfies the following equation:

$$\begin{aligned} \frac{\partial r(t,a)}{\partial t} + \frac{\partial r(t,a)}{\partial a} &= (\lambda_0 - h(r(t,\cdot))r(t,a), \\ r(t,0) &= \int_0^\omega \pi(\sigma)r(t,\sigma)d\sigma, \\ r(0,a) &= r_0(a) \in \Sigma, \end{aligned} \tag{1.116}$$

where $\pi(a) := e^{-\lambda_0 a}\beta(a)\ell(a)$, $\Sigma$ is the state space of $r$ given by

$$\Sigma := \left\{ \phi \in L^1_+(0,\omega) : \int_0^\omega w^*(a)\phi(a)da = 1 \right\},$$

and $h : \Sigma \to \mathbb{R}$ is a functional defined by

$$h(\phi) := \int_0^\omega (\beta(a) - \mu(a))w^*(a)\phi(a)da.$$

It is easy to see that $h(1) = \lambda_0$ and $h(r(t,\cdot))$ gives the growth rate of the total population at time $t$, that is, $N'(t)/N(t) = h(r(t,\cdot))$. Once the solution $r$ of (1.116) has been determined, the original population $p$ can be recovered as $p(t,a) = P(0)e^{\int_0^t h(r(\sigma,\cdot))d\sigma}w^*(a)r(t,a)$. In particular, $p(t,a)$ becomes the persistent solution $P(0)e^{\lambda_0 t}w^*(a)$ if and only if $r(t,a) = r^*(a) \equiv 1$, which is the unique non-trivial equilibrium solution of (1.116). From the stable population theory, the age-profile ratio $r$ converges to $r^* \equiv 1$ as $t \to \infty$. More precisely, we can prove that $|r - r^*|_{L^1}/|r_0 - r^*|_{L^1}$ decreases exponentially. To consider those normalization (age profile and age-profile ratio) for age distribution is often useful to discuss homogeneous nonlinear models (see Chap. 4, [60]).

## References


1. Alho, J.M., Spencer, B.D.: Statistical Demography and Forecasting. Springer, New York (2005)
2. Amann, H.: Ordinary Differential Equations: An Introduction to Nonlinear Analysis. Walter de Gruyter, Berlin (1990)
3. Arthur, W.B.: The analysis of linkages in demographic theory. Demography **21**(1), 109–129 (1984)

4. Arthur, W.B., Vaupel, J.W.: Some general relationships in population dynamics. Popul. Index **50**(2), 214–226 (1984)
5. Bacaër, N.: A Short History of Mathematical Population Dynamics. Springer, London (2011)
6. Bell, G.I., Anderson, E.C.: Cell growth division I. A mathematical model with applications to cell volume distributions in mammalian suspension cultures. Biophys. J. **7**, 329–351 (1967)
7. Bennett, N., Horiuchi, S.: Estimating the completeness of death registration in a closed population. Popul. Index **47**(2), 207–221 (1981)
8. Bernoulli, D.: Essai d'une nouvelle analyse de la mortalite causee par la petite verole et des avantages de l'incubation pout la prevenir. Mem. Math. Phys. Acad. R. Sci. Paris, 1–45 (1760/1766): English translation with review by Sally Blower. Rev. Med. Virol. **14**, 275–288 (2004)
9. Bongaarts, J., Greenhalgh, S.: An alternative to the one-child policy in China. Popul. Dev. Rev. **11**(4), 585–617 (1985)
10. Busenberg, S., Iannelli, M.: A class of nonlinear diffusion problems in age-dependent population dynamics. Nonl. Anal. Theory Meth. Appl. **7**(5), 501–529 (1983)
11. Busenberg, S., Iannelli, M.: Separable models in age-dependent population dynamics. J. Math. Biol. **22**, 145–173 (1985)
12. Butzer, P.L., Berens, H.: Semi-Groups of Operators and Approximation. Springer, Berlin (1967)
13. Caswell, H.: Matrix Population Models, 2nd edn. Sinauer, Sunderland (2001)
14. Charlesworth, B.: Evolution in Age-Structured Populations, 2nd edn. Cambridge University Press, Cambridge (1991)
15. Clément, Ph., Diekmann, O., Gyllenberg, M., Heijmans, H.J.A.M., Thieme, H.R.: Perturbation theory for dual semigroups I. The sun-reflexive case. Math. Ann. **277**, 709–725 (1987)
16. Clément, Ph., Heijmans, H.J.A.M., Angenent, S., van Duijn, C.J., de Pagter, B.: One-Parameter Semigroups, CWI Monograph 5. North-Holland, Amsterdam (1987)
17. Coale, A.J.: How the age distribution of a human population is determined. Cold Spring Harbor Symposia on Quantitative Biology **22**, 83–88 (1957)
18. Coale, A.J.: The Growth and Structure of Human Populations. Princeton University Press, Princeton (1972)
19. Cohen, J.E.: Ergodic theorems in demography. Bull. Amer. Math. Soc. **1**(2), 275–295 (1979)
20. Cohen, J.E.: Population system control. SIAM Rev. **32**(3), 494–500 (1990)
21. Crow, J.F.: Perspective: here's to Fisher, additive genetic variance, and the fundamental theorem of natural selection. Evolution **56**(7), 1313–1316 (2002)
22. Cushing, J.M.: An Introduction to Structured Population Dynamics. CBMS-NSF Regional Conference Series in Applied Mathematics, vol. 71. SIAM, Philadelphia (1998)
23. Demetrius, L.: Demographic parameters and natural selection. Proc. Natl. Acad. Sci. USA **71**(12), 4645–4647 (1974)
24. Demetrius, L.: Natural selection and age-structured populations. Genetics **79**, 535–544 (1975)
25. Demetrius, L.: Measures of fitness and demographic stability. Proc. Natl. Acad. Sci. USA **74**(1), 384–386 (1977)
26. Demetrius, L.: Adaptive value, entropy and survivorship curves. Nature **275**(21), 213–214 (1978)
27. Demetrius, L.: Relations between demographic parameters. Demography **16**(2), 329–338 (1979)
28. Demetrius, L.: The measurement of Darwinian fitness in human populations. Proc. R. Soc. Lond. B **222**, 33–50 (1984)
29. Diekmann, O., van Gils, S.A., Verduyn Lunel, S.M., Walther, H.-O.: Delay Equations: Functional-, Complex-, and Nonlinear Analysis. Appl. Math. Sci. vol. 110. Springer, Berlin (1995)
30. Diekmann, O., Heesterbeek, J.A.P., Britton, T.: Mathematical Tools for Understanding Infectious Disease Dynamics. Princeton University Press, Princeton (2013)
31. Dietz, K.: Introduction to McKendrick (1926) Applications of Mathematics to Medical Sciences. In: Kotz, S., Johnson, N.L. (eds.) Breakthroughs in Statistics, vol. 3, pp. 17–26. Springer, New York (1997)

32. Dietz, K., Heesterbeek, J.A.P.: Daniel Bernoulli's epidemiological model revisited. Math. Biosci. **180**, 1–21 (2002)
33. Doetsch, G.: Introduction to the Theory and Application of the Laplace Transformation. Springer, Berlin (1974)
34. Dublin, L.I.: Alfred James Lotka, 1880–1949. J. Amer. Stat. Assoc. **45**, 138–139 (1950)
35. Dublin, L.I., Lotka, A.J.: On the true rate of natural increase. J. Amer. Stat. Ass. New Series **20**(150), 305–339 (1925)
36. Eakin, T.: Intrinsic time scaling in survival analysis: application to biological populations. Bull. Math. Biol. **56**(6), 1121–1141 (1994)
37. Ediev, D.M.: On an extension of R.A. Fisher's result on the dynamics of the reproductive value. Theor. Popul. Biol. **72**, 480–484 (2007)
38. Ediev, D.M.: On the definition of the reproductive value: response to the discussion by Bacaër and Abdurahman. J. Math. Biol. **59**, 651–657 (2009)
39. Emlen, J.M.: Population Biology: The Coevolution of Population Dynamics and Behavior. Macmillan Publishing Company, New York (1984)
40. Engel, K.J., Nagel, R.: One-Parameter Semigroups for Linear Evolution Equations. Springer, Berlin (2000)
41. Euler, L.: Recherches générales sur la mortalité et la multiplication du genre humaine. Histoire de l'Academie Royale des Sciences et Belles Lettres **16**, 144–164 (1760) [A general investigation into the mortality and multiplication of the human species, translated by N. and B. Keyfitz, Theor. Popul. Biol. **1**, 307–314 (1970)]
42. Feller, W.: On the integral equation of renewal theory. Ann. Math. Stat. **12**, 243–267 (1941)
43. Fisher, R.A.: In: Bennett, J.H. (ed.) The Genetical Theory of Natural Selection: A Complete Variorum Edition. Oxford University Press, Oxford (1999)
44. Frauenthal, J.C.: Birth trajectory under changing fertility conditions. Demography **12**(3), 447–454 (1975)
45. Gage, T.B., Dyke, B., Riviere, P.G.: The population dynamics and fertility of the Trio of Surinam: an application of a two census method. Hum. Biol. **56**(4), 691–701 (1984)
46. Galindo, C.: On Fisher's reproductive value and Lotka's stable population, Paper presented at the 2007 Population Association of America (PAA) meeting, March 29–31, New York (2007)
47. Goldman, N., Lord, G.: A new look at entropy and the life table. Demography **23**(2), 275–282 (1986)
48. Goldstein, J.R., Schlag, W.: Longer life and population growth. Popul. Dev. Rev. **25**(4), 741–747 (1999)
49. Gyllenberg, M.: Mathematical aspects of physiologically structured populations: the contributions of J.A.J. Metz. J. Biol. Dyn. **1**(1), 3–44 (2007)
50. Hale, J.K.: Asymptotic Behavior of Dissipative Systems. Math. Survey Monogr. 25, Amer. Math. Soc., Providence (1988)
51. Heijmans, H.J.A.M.: The dynamical behaviour of the age-size-distribution of a cell population. In: Metz, J.A.J., Diekmann, O. (eds.) The Dynamics of Physiologically Structured Populations. Lecture Notes in Biomathematics, vol. 68, pp. 185–202. Springer, Berlin (1986)
52. Hille, E., Phillips, R.S.: Functional Analysis and Semi-Groups. American Mathematical Society, Providence, Rhode Island (1957)
53. Hoppensteadt, F.: Mathematical Theories of Populations: Demographics, Genetics and Epidemics. Society for Industrial and Applied Mathematics, Philadelphia (1975)
54. Iannelli, M.: Mathematical problems in the description of age structured populations. Mathematics in Biology and Medicine. Lecture Notes in Biomathematics, vol. 57, pp. 19–32. Springer, Berlin (1985)
55. Iannelli, M.: Mathematical Theory of Age-Structured Population Dynamics. Giardini Editori e Stampatori in Pisa (1995)
56. Impagriazzo, A.: Deterministic Aspects of Mathematical Demography. Springer, Berlin (1985)
57. Inaba, H.: Weak ergodicity of population evolution processes. Math. Biosci. **96**, 195–219 (1989)

58. Inaba, H.: Remarks on the effect of an age shift in fertility. Jinkogaku Kenkyu (J. Popul. Stud.) **26**, 21–27 (2000) [Japanese]
59. Inaba, H.: Effects of age shift on the tempo and quantum of non-repeatable events. Math. Popul. Stud. **14**(3), 131–168 (2007)
60. Inaba, H.: Age-structured homogeneous epidemic systems with application to the MSEIR epidemic model. J. Math. Biol. **54**, 101–146 (2007)
61. Keyfitz, N.: On the momentum of population growth. Demography **8**(1), 71–80 (1971)
62. Keyfitz, N.: Migration as a means of population control. Popul. Stud. **25**(1), 63–72 (1971)
63. Keyfitz, N.: Population waves. In: Greville, T.N.E. (ed.) Population Dynamics, pp. 1–38. Academic Press, New York (1972)
64. Keyfitz, N.: Introduction to the Mathematics of Population with Revisions. Addison-Wesley, Reading (1977)
65. Keyfitz, N., Caswell, H.: Applied Mathematical Demography, 3rd edn. Springer, New York (2005)
66. Kingsland, S.E.: Modeling Nature, 2nd edn. The University of Chicago Press, Chicago (1995)
67. Krishnamoorthy, S.: Family formation and the life cycle. Demography **16**(1), 121–129 (1979)
68. Kuczynski, R.R.: Fertility and Reproduction: Methods of Measuring the Balance of Births and Deaths. Falcon Press, New York (1932)
69. Langhaar, H.L.: General population theory in the age-time continuum. J. Franklin Inst. **293**, 199–214 (1972)
70. Lasota, A., Mackey, M.C.: Chaos, Fractals, and Noise: Stochastic Aspects of Dynamics. Applied Mathematical Sciences, vol. 97, 2nd edn. Springer, New York (1995)
71. Le Bras, H.: Mortality tempo versus removal of causes of mortality: opposite views leading to different estimations of life expectancy. Demogr. Res. **13**, 615–640 (2005)
72. Lewes, F.M.M.: A note on the origin of the net reproduction ratio. Popul. Stud. **38**, 321–324 (1984)
73. Lotka, A.J.: Relations between birth rates and death rates. Science N. S. **26**, 21–22 (1907)
74. Lotka, A.J.: Elements of Physical Biology. The Williams and Wilkins Co., Inc. (1925): Republished as Elements of Mathematical Biology. Dover, New York (1956)
75. Lotka, A.J.: The progeny of a population element. Am. J. Hyg. **8**, 875–901 (1928)
76. Lotka, A.J.: The spread of generations. Hum. Biol. **1**(3), 305–320 (1929)
77. Lotka, A.J.: The structure of a growing population. Hum. Biol. **3**, 459–493 (1931)
78. Lotka, A.J.: A historical error corrected. Hum. Biol. **9**, 104–107 (1937)
79. Lotka, A.J.: Population analysis: a theorem regarding the stable age distribution. J. Washington Acad. Sci. **27**(7), 299–303 (1937)
80. Lotka, A.J.: A contribution to the theory of self-renewing aggregates, with special reference to industrial replacement. Ann. Math. Stat. **10**(1), 1–25 (1939)
81. Lotka, A.J.: On an integral equation in population analysis. Ann. Math. Stat. **10**(2), 144–161 (1939)
82. Lotka, A.J.: Théorie Analytique des Associations Biologiques. Deuxième Partie: Analyse Démographique avec Application Particulière è l'Espèce Humaine, Actualités Scientifiques et Industrielles, No. 780. Hermann et Cie, Paris (1939)
83. Lotka, A.J.: Population analysis as a chapter in mathematical theory of evolution. In: LeGros Clark, W.E., Medawar, P.B. (eds.) Essays on Growth and Form, pp. 355–385. Oxford University Press, New York (1945)
84. Lotka, A.J.: Analytical Theory of Biological Populations. The Plenum Series on Demographic Methods and Population Analysis. Plenum Press, New York (1998)
85. Malthus, T.R.: An Essay on the Principle of Population, 1st edn. J. Johnson, London (1798)
86. Metz, J.A.J., Diekmann, O.: The Dynamics of Physiologically Structured Populations. Lecture Notes in Biomathematics, vol. 68. Springer, Berlin (1986)
87. McKendrick, A.G.: Application of mathematics to medical problems. Proc. Edinburgh. Math. Soc. **44**, 98–130 (1926)
88. Michel, P., Mischler, S., Perthame, B.: General relative entropy inequality: an illustration on growth models. J. Math. Pures Appl. **84**, 1235–1260 (2005)

89. Mitra, S.: Influence of instantaneous fertility decline to replacement level on population growth: an alternative model. Demography **13**(4), 513–519 (1976)
90. Mitra, S.: The effects of extra chances to live on life table functions. Theor. Popul. Biol. **16**, 315–322 (1979)
91. Mode, C.J.: Stochastic Processes in Demography and their Computer Implementation. Springer, Berlin (1985)
92. Morita, Y: Analysis of Population Growth, Nihon Hyoron Sha, Tokyo (1944) [Japanese]
93. Muench, H.: Catalytic Models in Epidemiology. Harvard University Press, Cambridge (1959)
94. Nagel, R. (ed.): One-Parameter Semigroups of Positive Operators. Lecture Notes in Mathematics, vol. 1184. Springer, Berlin (1986)
95. Nishiura, H., Dietz, K., Eichner, M.: The earliest notes on the reproduction number in relation to herd immunity: Theophil Lotz and smallpox vaccination. J. Theor. Biol. **241**, 964–967 (2006)
96. Nishiura, H., Inaba, H.: Discussion: Emergence of the concept of the basic reproduction number from mathematical demography. J. Theor. Biol. **244**, 357–364 (2007)
97. Notestein, F.W.: Alfred James Lotka 1880–1949. Popul. Index **16**, 22–29 (1950)
98. Perthame, B.: Transport Equations in Biology. Birkhäuser, Basel (2007)
99. Pollard, J.H.: Mathematical Models for the Growth of Human Populations. Cambridge University Press, Cambridge (1973)
100. Preston, S.H., Coale, A.J.: Age structure, growth, attrition, and accession: A new synthesis. Popul. Index **48**(2), 217–259 (1982)
101. Preston, S.H.: An integrated system for demographic estimation from two age distributions. Demography **20**(2), 213–226 (1983)
102. Preston, S.H., Heuveline, P., Guillot, M.: Demography: Measuring and Modeling Population Processes. Blackwell, Oxford (2001)
103. Prüss, J.: Equilibrium solutions of age-specific population dynamics of several species. J. Math. Biol. **11**, 65–84 (1981)
104. Roughgarden, J.: Theory of Population Genetics and Evolutionary Ecology: An Introduction. Prentice Hall, Upper Saddle River (1996)
105. Rubinow, S.I.: Age-structured populations in the theory of cell populations, In: Studies in Mathematical Biology, Part II: Population and Communities, Levin, S. A. (ed.), Studies in Mathematics Vol. 16, The Mathematical Association of America, 389–410 (1978)
106. Rubinow, S.I., Berger, R.O.: Time-dependent solution to age-structured equations for sexual populations. Theor. Popul. Biol. **16**, 35–47 (1979)
107. Sanchez, D.A.: Iteration and nonlinear equations of age-dependent population growth with a birth window. Math. Biosci. **73**, 61–69 (1985)
108. Samuelson, P.A.: Resolving a historical confusion in population analysis. Hum. Biol. **48**, 559–580 (1976)
109. Scherbaum, O., Rasch, G.: Cell size distribution and single cell growth in Tetrahymena pyriformis GL. Arch. Pathol. Microbiol. Scand. **41**, 161–182 (1957)
110. Schindler, S., Tuljapurkar, S., Gaillard, J.M., Coulson, T.: Linking the population growth rate and the age-at-death distribution. Theor. Pop. Biol. **82**, 244–252 (2012)
111. Schlögl, F.: Mixing distance and stability of steady states in statistical nonlinear thermodynamics. Zeitschrift für Physik B **25**, 411–421 (1976)
112. Schoen, R., Kim, Y.J.: Movement toward stability as a fundamental principle of population dynamics. Demography **28**(3), 455–466 (1991)
113. Schumpeter, J.A.: Ladislaus von Bortkiewicz. Econ. J. **XLII**, 338–340 (1932)
114. Sharpe, F.R., Lotka, A.J.: A problem in age-distribution. Philos. Mag. **6**(21), 435–438 (1911)
115. Sheps, M.C., Lapierre-Adamcyk, E. (eds.) On the Measurement of Human Fertlity: Selected Writings of Louis Henry. Elsevier, Amsterdam (1972)
116. Sheps, M.C., Menken, J.A.: Mathematical Models of Conception and Birth. The University of Chicago Press, Chicago (1973)
117. Sinko, J.W., Streifer, W.: A new model for age-size structure of a population. Ecology **48**(6), 910–918 (1967)

118. Smith, D.P.: An Euler contribution to stable theory from Süssmilch's Göttliche Ordnung. Theor. Pop. Biol. **12**, 246–251 (1977)
119. Smith, D., Keyfitz, N.: Mathematical Demography: Selected Papers. Springer, Berlin (1977)
120. Smith, J.D.H.: Demography and the canonical ensemble. Math. Biosci. **153**, 151–161 (1998)
121. Smith, J.D.H.: A macroscopic approach to demography. J. Math. Biol. **48**, 105–118 (2004)
122. Song, J., Tuan, C.H., Yu, J.Y.: Population Control in China: Theory and Applications. Praeger, New York (1985)
123. Song, J., Yu, J.: Population System Control. Springer, Berlin (1988)
124. Song, J., Yu, J.Y., Wang, X.Z., Hu, S.J., Zhao, Z.X., Liu, J.Q., Feng, D.X., Zhu, G.T.: Spectral properties of population operator and asymptotic behaviour of population semigroup. Acta Math. Sci. **2**(2), 139–148 (1982)
125. Song, J., Yu, J., Liu, C., Zhang, L., Zhu, G.: Spectral properties of population evolution and controllability of population system. Sci. Sinica (Series A), **XXIX**(8), 800–812 (1986)
126. Takagi, N.: A consideration on the stable-population theory. Jinko Mondai Kenkyu **63**, 42–49 (1956). [in Japanese]
127. Thieme, H.R.: Mathematics in Population Biology. Princeton University Press, Princeton (2003)
128. Tuljapurkar, S.D.: Why use population entropy? It determines the rate of convergence. J. Math. Biol. **13**, 325–337 (1982)
129. Tuljapurkar, S.: Population Dynamics in Variable Environments. Lecture Notes in Biomathematics, vol. 85. Springer, New York (1990)
130. Vaupel, J.W.: How change in age-specific mortality affects life expectancy. Popul. Stud. **40**, 147–157 (1986)
131. Vaupel, J.W.: Life expectancy at current rates vs. current conditions: a reflexion stimulated by Bongaarts and Feeney's "How long do we live?". Demogr. Res. **7**, 365–377 (2002)
132. Vaupel, J.W.: Lifesaving, lifetimes and lifetables. Demogr. Res. **13**, 597–614 (2005)
133. Vlad, M.O., Pop, A.: A new H-theorem for age-dependent dynamics. J. Phys. A: Math. Gen. **22**, 3945–3957 (1989)
134. Von Foerster, H.: Some remarks on changing populations. In: The Kinetics of Cellular Proliferation, pp. 382–407. Grune and Stratton, NY (1959)
135. Wachter, K.W.: Lotka's roots under rescalings. Proc. Natl. Acad. Sci. USA **81**, 3600–3604 (1984)
136. Wachter, K.E.: Pre-procreative ages in population stability and cyclicity. Math. Popul. Stud. **3**(2), 79–103 (1991)
137. Wallinga, J., Lipsitch, M.: How generation intervals shape the relationship between growth rates and reproductive numbers. Proc. R. Soc. B **274**, 599–604 (2007)
138. Webb, G.F.: A semigroup proof of the Sharpe-Lotka theorem. In: Kappel, F., Schappacher, W. (eds.) Infinite-Dimensional Systems. Lecture Notes in Mathematics, vol. 1076, pp. 254–268. Springer, Berlin (1984)
139. Webb, G.F.: Theory of Nonlinear Age-Dependent Population Dynamics. Marcel Dekker, New York (1985)
140. Webb, G.F.: Random transitions, size control, and inheritance in cell population dynamics. Math. Biosci. **85**, 71–91 (1987)
141. Widder, D.V.: The Laplace Transform. Princeton University Press, Princeton (1946)

# Chapter 2
# Extensions of the Linear Theory

**Abstract** For more than a century since it was first formulated, the stable population model has been used as a standard theory in demography. In fact, after World War II, the population explosions in developing countries became an important challenge to demographers working with prediction and control problems of Malthusian populations. Since the latter half of the 1970s, the onset of a *second demographic transition* (namely the persistence of below-replacement fertility) has been observed in developed countries, and the subsequent population problems require much more sophisticated tools to deal with the heterogeneities of individuals and environments. In this chapter, we introduce some linear extensions of the stable population model. These extensions are essentially multidimensional stable population models, so it is not surprising that the strong ergodicity theorem again holds. In contrast, we also examine non-autonomous systems to address the population dynamics in heterogeneous environments. In this case, we cannot expect a stable age distribution to exist, and thus introduce the idea of *weak ergodicity*. Using these extended models, we establish more elaborate demographic indices and characterizations for heterogeneous populations, and demonstrate their importance for applications in demography, epidemiology, and population biology.

## 2.1 Multistate Stable Population Model

In the previous chapter, we considered a population structured only by age and sex. Here, we deal with a model for a *multistate population*. This is an age-structured population divided according to some discrete state variables that indicate individual heterogeneity, such as the region of residence, parity status, labor status, and marital status. During the 1970s, Andrei Rogers and Herve Le Bras independently extended the stable population model to deal with multiregional populations and their migration between regions [70, 89]. In particular, Rogers and his collaborators developed numerical methods for parameter estimation from real data and computer simulations, and their results were successfully applied to both regional populations and other multistate populations [66, 98]. It is clear that extending the stable population model to the multistate model was an important first step in accounting for



H. Inaba, *Age-Structured Population Dynamics in Demography and Epidemiology*, DOI 10.1007/978-981-10-0188-8_2

the heterogeneity of populations. Because many quantitative problems in the social sciences, biology, and epidemiology can be formulated using multistate populations; the idea of multistate demography has a number of potential applications.

As we shall show in this chapter, the asymptotic behavior of the linear multistate population model is analogous to that of the stable population model. However, it is far more difficult to calculate the parameters of the multistate model from available data, so multistate demography has a greater focus on the parameter estimation problem. In this chapter, we present only a sketch of the mathematical properties of linear multistate population models.

Suppose that a one-sex population is divided into $n$ states, and denote each different state by the index $i = 1, 2, \ldots, n$. Each state of a population refers to any individual trait that could affect the life cycle parameters, for example, the place of residence, labor force status, marital status, or parity status (the number of children a woman has had). Let $p_i(t, a)$ be the age-density function of the $i$-th state, and let

$$p(t, a) = (p_1(t, a), \cdots, p_n(t, a))^{\mathrm{T}}$$

be the population vector, where T denotes the transpose of the vector. For simplicity, we assume that $\omega = \infty$ and the state space of a population density $p_i(t, \cdot)$ is $L^1_+(\mathbb{R}_+)$.

Let $q_{ij}(a)$, $i \neq j$, be the *force of transition* from state $j$ to state $i$, and let $\mu_j(a)$ be the force of mortality at state $j$. Moreover, we define

$$q_{jj}(a) = -\mu_j(a) - \sum_{i \neq j} q_{ij}(a), \quad 1 \leq j \leq n. \tag{2.1}$$

Let $Q(a)$ be an $n \times n$ matrix whose $(i, j)$-th entry is given by $q_{ij}(a)$, $1 \leq i, j \leq n$. Let $m_{ij}(a)$ be the age-specific fertility rate that gives rise to $i$-th state newborns from $j$-th state individuals. Let $M(a)$ be an $n \times n$ matrix whose $(i, j)$-th element is given by $m_{ij}(a)$, $1 \leq i, j \leq n$.

The *multistate stable population model* is then formulated by the following initial–boundary value problem of the vector-type McKendrick equation:

$$\begin{aligned}
&\frac{\partial p(t, a)}{\partial t} + \frac{\partial p(t, a)}{\partial a} = Q(a)p(t, a), \quad t > 0, \quad a > 0, \\
&p(t, 0) = \int_0^\infty M(a)p(t, a)da, \quad t > 0, \\
&p(0, a) = p_0(a),
\end{aligned} \tag{2.2}$$

where $p_0(a)$ is the initial population vector.

To derive the renewal equation for the multistate model, let us consider the evolutionary system (transition matrix) $\{L(a, \sigma) : a \geq \sigma \geq 0\}$ satisfying the forward equation

$$\frac{\partial}{\partial a} L(a, \sigma) = Q(a)L(a, \sigma), \quad L(\sigma, \sigma) = I, \tag{2.3}$$

where $I$ is the $n \times n$ identity matrix. Then, $L(a,\sigma)$ is the fundamental solution matrix for the master equation

$$\frac{d\ell(a)}{da} = Q(a)\ell(a),$$

where $\ell(a) := (\ell_1(a), \ldots, \ell_n(a))^{\tau}$; $\ell_j(a)$ denotes the probability that an individual is in state $j$ at age $a$, and $\ell(0)$ gives the state distribution at age zero, so $\sum_{j=1}^{N} \ell_j(0) = 1$. The evolutionary property

$$L(a,\tau)L(\tau,\sigma) = L(a,\sigma), \quad a \geq \tau \geq \sigma$$

holds, and the $(i, j)$-th entry $L(a,\sigma)$ denotes the probability that an individual of age $\sigma$ and state $j$ will survive to age $a$ and state $i$. For simplicity, $L(a,0)$ is often written as $L(a)$. This is a multidimensional extension of the survival probability $\ell(a)$ in the scalar model, because its $(i, j)$-th entry $\ell_{ij}(a)$ gives the probability that an individual born in state $j$ will survive to age $a$ and state $i$. From (2.1), $Q(a)$ is an *essentially non-negative matrix*[1]; that is, all off-diagonal elements are non-negative, and the fundamental solution matrix $L(a,\sigma)$ is a non-negative, non-singular matrix.[2]

Using the transition matrix, the McKendrick equation can be integrated along the characteristic line to obtain the following expression:

$$p(t,a) = \begin{cases} L(a,0)B(t-a), & t-a > 0, \\ L(a,a-t)p_0(a-t), & a-t > 0, \end{cases} \tag{2.4}$$

where $B(t) := p(t,0)$. Inserting expression (2.4) into the boundary condition in (2.2), we obtain the renewal integral equation

$$B(t) = G(t) + \int_0^t \Psi(a)B(t-a)da, \tag{2.5}$$

where $\Psi(a) := M(a)L(a,0)$ and

$$G(t) := \int_t^{\infty} M(a)L(a,a-t)p_0(a-t)da.$$

For our population problems, we can assume that $G(\cdot)$ and $\Psi(\cdot)$ are bounded and compactly supported integrable (vector-valued and matrix-valued) functions. In such a case, it is well known that the vector-type renewal integral equation (2.5) has

[1] An essentially non-negative matrix is also called *quasi-positive* if it is not zero [104].

[2] If a square matrix $A$ is irreducible and essentially non-negative, it is called *essentially positive*. $A$ is essentially positive if and only if $A + sI_d$ is a non-negative, irreducible, and primitive matrix for all sufficiently large $s > 0$. Moreover, a square matrix $A$ is essentially non-negative if and only if $e^{At} \geq 0$ for all $t \geq 0$, and $A$ is essentially positive if and only if $e^{At} > 0$ for all $t > 0$ (see Chap. 10, [17, 107]).

a unique locally integrable solution for $t \in \mathbb{R}_+$. If we know the behavior of $B(t)$, $p(t, a)$ is completely determined by (2.4).

Starting from the renewal equation (2.5), we can again determine the asymptotic behavior of the multistate model (2.2) as in Chap. 1. However, instead of repeating the same kind of argument, here we sketch another intuitive approach. A rigorous justification of the following is given by the semigroup approach [49, 50].

In the previous chapter, we took the Laplace transform of $B(t)$. Here, we take the Laplace transform of $p(t, a)$ directly:

$$\hat{p}(\lambda, a) := \int_0^\infty e^{-\lambda t} p(t, a) dt, \quad \lambda \in \mathbb{C}.$$

Because the norm of $p(t, a)$ grows at most exponentially, the Laplace transform of $p(t, a)$ exists for $\lambda$ with a sufficiently large real part. By taking the Laplace transform of both sides of (2.2), we have

$$\begin{aligned} \frac{d\hat{p}(\lambda, a)}{dt} &= p_0(a) + (-\lambda + Q(a))\hat{p}(\lambda, a), \\ \hat{p}(\lambda, 0) &= \int_0^\infty M(a)\hat{p}(\lambda, a) da. \end{aligned} \tag{2.6}$$

Solving the differential equation in (2.6), it follows that

$$\hat{p}(\lambda, a) = e^{-\lambda a} L(a)\hat{p}(\lambda, 0) + \int_0^a e^{-\lambda(a-s)} L(a, s) p_0(s) ds.$$

Inserting the above expression into the right-hand side of the boundary condition of (2.6) to solve for $\hat{p}(\lambda, 0)$, we obtain

$$\hat{p}(\lambda, 0) = (I - \hat{\Psi}(\lambda))^{-1} \int_0^\infty \int_s^\infty e^{-\lambda(a-s)} \Psi(a) da L^{-1}(s) p_0(s) ds.$$

Therefore, $\hat{p}(\lambda, a)$ is expressed as

$$\begin{aligned} \hat{p}(\lambda, a) = &\int_0^a e^{-\lambda(a-s)} L(a, s) p_0(s) ds \\ &+ e^{-\lambda a} L(a)(I - \hat{\Psi}(\lambda))^{-1} \int_0^\infty \int_s^\infty e^{-\lambda(a-s)} \Psi(a) da L^{-1}(s) p_0(s) ds. \end{aligned}$$

Taking the inverse transformation, we obtain the formal solution as

$$p(t, a) = \frac{1}{2\pi i} \int_{x-i\infty}^{x+i\infty} e^{\lambda t} \hat{p}(\lambda, a) d\lambda, \tag{2.7}$$

where the real number $x$ is sufficiently large that $\hat{p}(x + i\mathbb{R})$ converges absolutely. Let us define a set of complex numbers denoted by $\Omega$ such that

$$\Omega := \{\lambda \in \mathbb{C} : \det(I - \hat{\Psi}(\lambda)) = 0\} = \{\lambda \in \mathbb{C} : 1 \in \sigma(\hat{\Psi}(\lambda))\},$$

where $\sigma(A)$ denotes the set of spectra of matrix $A$. Then, as $\hat{p}(\lambda, a)$ is a meromorphic function with poles at $\Omega$, we can obtain its asymptotic expansion by shifting the integral path to the left, just as for the single-state stable population model in Chap. 1.

Let us again examine the structure of the set of characteristic roots $\Omega$. First, define $\underline{\mu} := \inf \mu_j(a)$, which leads to the estimate:

**Lemma 2.1**

$$|L(b, a)| \le e^{-\underline{\mu}(b-a)}, \tag{2.8}$$

*where the norm of a matrix* $A = (a_{ij})_{1 \le i,j \le n}$ *and of a vector* $x = (x_1, ..., x_n)$ *are given by:*

$$|x| = \sum_{i=1}^{n} |x_i|, \quad |A| = \max_{j} \sum_{i=1}^{n} |a_{ij}|.$$

*Proof* Let $\ell_{ij}(b, a)$ be the $(i, j)$-th element of $L(b, a)$. It follows that

$$\frac{\partial}{\partial b}\ell_{ij}(b, a) = \sum_{k=1}^{n} q_{ik}(b)\ell_{kj}(b, a), \quad \ell_{ij}(a, a) = \delta_{ij},$$

and

$$\begin{aligned} \frac{\partial}{\partial b}\sum_{i=1}^{n}\ell_{ij}(b, a) &= \sum_{i=1}^{n}\sum_{k=1}^{n} q_{ik}(b)\ell_{kj}(b, a) = \sum_{k=1}^{n}\sum_{i=1}^{n} q_{ik}(b)\ell_{kj}(b, a) \\ &= \sum_{k=1}^{n}(-\mu_k(b))\ell_{kj}(b, a) \le (-\underline{\mu})\sum_{i=1}^{n}\ell_{ij}(b, a). \end{aligned}$$

Therefore, we can conclude that

$$\sum_{i=1}^{n}\ell_{ij}(b, a) \le e^{-\underline{\mu}(b-a)},$$

which implies that (2.8) holds. □

From the above preparation, we can state the following proposition [49, 50]:

**Proposition 2.1** *Let* $\bar{m} := \sup_{0 \le a \le \omega} |M(a)|$. *Then, the following holds:*

*(1)* $\Omega$ *is included in the half-plane* $\Re\lambda \le \bar{m} - \underline{\mu}$.

*(2) If* $\lambda \in \Omega$, *then* $\bar{\lambda} \in \Omega$.

*(3) For any* $x \in \mathbb{R}$, *the half-plane* $\Re\lambda > x$ *contains at most finitely many* $\lambda \in \Omega$.

*Proof* We omit the proofs for (2) and (3), because they are the same as those for Proposition 1.7. Let us show (1). Because the matrix $\hat{\Psi}(\lambda)$ is non-negative for real $\lambda$, it has the Frobenius root $F(\lambda)$, $\lambda \in \mathbb{R}$. Let $\hat{\Psi}^*(\lambda)$, $\lambda \in \mathbb{C}$ be the matrix whose $(i, j)$-th element is given by the absolute value $|\hat{\psi}_{ij}(\lambda)|$ of the $(i, j)$-th element $\hat{\psi}_{ij}(\lambda)$ of $\hat{\Psi}(\lambda)$, and denote its Frobenius root as $F^*(\lambda)$. If we denote $r(A)$ as the *spectral radius*[3] of a matrix $A$, it can be proved that $r(\hat{\Psi}(\lambda)) \le F^*(\lambda)$. Moreover, it follows from $\hat{\Psi}^*(\lambda) \le \hat{\Psi}(\Re\lambda)$ that $F^*(\lambda) \le F(\Re\lambda)$. Hence, we have $r(\hat{\Psi}(\lambda)) \le F(\Re\lambda)$. Observe that for $\Re\lambda > -\underline{\mu}$,

$$|\hat{\Psi}(\Re\lambda)| \le \int_0^\infty e^{-\Re\lambda a}|\Psi(a)|da \le \int_0^\infty e^{-\Re\lambda a}|M(a)||L(a)|da \le \frac{\bar{m}}{\Re\lambda + \underline{\mu}}.$$

Thus, it follows that

$$F(\Re\lambda) \le \max_j \sum_{i=1}^n \hat{\psi}_{ij}(\Re\lambda) = |\hat{\Psi}(\Re\lambda)| \le \frac{\bar{m}}{\Re\lambda + \underline{\mu}}.$$

Therefore, if $\Re\lambda > \bar{m} - \underline{\mu}$, we have $r(\hat{\Psi}(\lambda)) \le F(\Re\lambda) < 1$. However, if $\lambda \in \Omega$, we have $r(\hat{\Psi}(\lambda)) \ge 1$. We can conclude that there is no element of $\Omega$ in the half-plane $\Re\lambda > \bar{m} - \underline{\mu}$. □

For the multistate stable population model, we define the *net reproduction matrix* (*next-generation matrix* in epidemic terminology, see Chap. 9) by

$$K := \hat{\Psi}(0) = \int_0^\infty \Psi(a)da, \tag{2.9}$$

and the *net reproduction rate* (basic reproduction number)[4] $R_0$ by its spectral radius

$$R_0 = r(\hat{\Psi}(0)). \tag{2.10}$$

Let $\hat{\psi}_{ij}(0)$ be the $(i, j)$-th entry of the net reproduction matrix:

$$\hat{\psi}_{ij}(0) = \sum_{k=1}^n \int_0^\infty m_{ik}(a)\ell_{kj}(a)da.$$

---

[3] $r(A) = \max_{\lambda \in \sigma(A)} |\lambda|$, where $\sigma(A)$ denotes the set of eigenvalues of $A$, and it follows that $r(A) = \lim_{n\to\infty} |A^n|^{1/n}$.

[4] Readers are referred to [58] for a historical review of the multistate net reproduction rate.

We then define the native-dependent reproduction number by

$$R_{0j} := \sum_{i=1}^{n} \hat{\psi}_{ij}(0),$$

which is the expected number of newborns produced by an individual born in state $j$ during her entire lifetime. As it is known that

$$\min \sum_{i=1}^{n} \hat{\psi}_{ij}(x) \le F(x) \le \max \sum_{i=1}^{n} \hat{\psi}_{ij}(x), \tag{2.11}$$

we can conclude that $R_0 < 1$ if $\max_{1 \le j \le n} R_{0j} < 1$ and $R_0 > 1$ if $\min_{1 \le j \le n} R_{0j} > 1$, whereas $R_0 = 1$ if $R_{0j} = 1$ for all of state $j$. An $n \times n$ matrix $A = (a_{ij})_{1 \le i,j \le n}$ is called *decomposable* (*reducible*) if the set of numbers $S = \{1, 2, \ldots, n\}$ is the sum of two disjoint subsets $S_1$ and $S_2$ such that $a_{ij} = 0$ if $i \in S_1$ and $j \in S_2$. If a matrix is not decomposable, it is called *indecomposable* (or *irreducible*). If $A$ is irreducible and $A^n = (a_{ij}^{(n)})_{1 \le i,j \le N}$, for any $i$ and $j$ there exists an integer $n_0$ such that $a_{ij}^{(n_0)} > 0$. Thus, any one state is accessible from any other state by multiple transitions [80].

If $\hat{\Psi}(0)$ is decomposable, it follows that for almost all $a \in \mathbb{R}_+$,

$$\sum_{k=1}^{n} m_{ik}(a)\ell_{kj}(a) = 0, \quad i \in S_1, \quad j \in S_2,$$

which implies that individuals born in state $S_2$ cannot produce children in state $S_1$. Thus, if we divide the population into two subpopulations, one composed of people born in state $S_1$ and the other composed of people born in state $S_2$, then the $S_1$ subpopulation can be reproduced only by the $S_1$ subpopulation and has become a closed subpopulation with respect to reproduction. In such a case, we cannot generally expect there to be a common Malthusian parameter for both subpopulations. Conversely, the indecomposability (irreducibility) of the net reproduction matrix implies the existence of the dominant Malthusian solution [49, 50]:

**Proposition 2.2** *If the net reproduction matrix $\hat{\Psi}(0)$ is indecomposable, the following holds:*

(1) *There exists $\lambda_0 \in \Omega \cap \mathbb{R}$ such that $\lambda_0 > \sup\{\Re\lambda : \lambda \in \Omega \setminus \{\lambda_0\}\}$ and $F(\lambda_0) = 1$. Moreover, it follows that*

$$\mathrm{sign}(\lambda_0) = \mathrm{sign}(R_0 - 1). \tag{2.12}$$

(2) *Let $R(\lambda) := (I - \hat{\Psi}(\lambda))^{-1}$. Then, $R(\lambda)$ has a pole of order one at $\lambda = \lambda_0$ and its residue $R_{-1}$ is a one-dimensional projection given by:*

$$R_{-1}\phi = \frac{v_0(0)^{\mathrm{T}}\phi}{v_0(0)^{\mathrm{T}}\Psi_1 u_0(0)} u_0(0), \tag{2.13}$$

*where $v_0(0)$ and $u_0(0)$ are the left and right positive eigenvectors of $\hat{\Psi}(\lambda_0)$ corresponding to the Frobenius root $F(\lambda_0) = 1$ and $\Psi_1$ is given by*

$$\Psi_1 = \int_0^\infty a\Psi(a)e^{-\lambda_0 a}da.$$

*Proof* Under the assumption, $\hat{\Psi}(x)$ $(x \in \mathbb{R})$ is a non-negative indecomposable matrix. Its Frobenius root $F(x) > 0$ therefore exists, is a strictly decreasing function of real $x$, and satisfies (2.11). Therefore, we have $\lim_{x\to-\infty} F(x) = +\infty$ and $\lim_{x\to\infty} F(x) = 0$. Thus, there exists a unique real root $\lambda_0$ of $F(x) = 1$ such that $\lambda_0 > 0$ if $F(0) > 1$, $\lambda_0 = 0$ if $F(0) = 1$ and $\lambda_0 < 0$ if $F(0) < 1$. From the definition, it is clear that $\lambda_0 \in \Omega$. For any $\lambda \in \Omega$, it follows from $1 \in \sigma(\hat{\Psi}(\lambda))$ and $\hat{\Psi}(\Re\lambda) \geq \hat{\Psi}^*(\lambda)$ that

$$r(\hat{\Psi}(\Re\lambda)) = F(\Re\lambda) \geq F^*(\lambda) \geq r(\hat{\Psi}(\lambda)) \geq 1.$$

As $F(x)$ is monotone decreasing, we have $\Re\lambda \leq \lambda_0$. Moreover, if $\Im\lambda \neq 0$, the equality of $\hat{\Psi}^*(\lambda) \leq \hat{\Psi}(\Re\lambda)$ does not hold, so we obtain

$$1 \leq r(\hat{\Psi}(\lambda)) \leq F^*(\lambda) < F(\Re\lambda).$$

Therefore, we have $\Re\lambda < \lambda_0$. We then know that $\lambda_0$ is a unique real root in $\Omega$, and $\lambda_0 > \sup\{\Re\lambda : \lambda \in \Omega \setminus \{\lambda_0\}\}$. Next we show (2). By differentiating the identity

$$(I - \hat{\Psi}(\lambda))\text{adj}(I - \hat{\Psi}(\lambda)) = \det(I - \hat{\Psi}(\lambda)) \cdot I,$$

we have

$$\begin{aligned} &\frac{d}{d\lambda}(I - \hat{\Psi}(\lambda))\bigg|_{\lambda=\lambda_0} \text{adj}(I - \hat{\Psi}(\lambda)) + (I - \hat{\Psi}(\lambda))\frac{d}{d\lambda}\text{adj}(I - \hat{\Psi}(\lambda))\bigg|_{\lambda=\lambda_0} \\ &= \frac{d}{d\lambda}\det(I - \hat{\Psi}(\lambda))\bigg|_{\lambda=\lambda_0} \cdot I. \end{aligned} \tag{2.14}$$

Observe that for any vector $\phi$, the following holds:

$$(I - \hat{\Psi}(\lambda_0))\text{adj}(I - \hat{\Psi}(\lambda_0))\phi = 0.$$

As the eigenspace of $\hat{\Psi}(\lambda_0)$ associated with its Frobenius root of unity is one-dimensional, there exists a scaler $c(\phi)$ such that

$$\text{adj}(I - \hat{\Psi}(\lambda_0))\phi = c(\phi)u_0(0).$$

Using $v_0(0)^{\mathrm{T}}(I - \hat{\Psi}(\lambda_0)) = 0$, we multiply (2.14) by $v_0(0)$ and $\phi$ from the left-hand side and right-hand side, respectively, and let $\lambda = \lambda_0$. Thus, we arrive at

$$c(\phi)v_0(0)^{\mathrm{T}}\Psi_1 u_0(0) = v_0(0)^{\mathrm{T}}\phi \, \frac{d}{d\lambda}\det(I - \hat{\Psi}(\lambda))\bigg|_{\lambda=\lambda_0}.$$

Therefore, we obtain

$$\lim_{\lambda\to\lambda_0}(\lambda-\lambda_0)(I-\hat{\Psi}(\lambda))^{-1}\phi = \left[\frac{d}{d\lambda}\det(I-\hat{\Psi}(\lambda))\bigg|_{\lambda=\lambda_0}\right]^{-1}\mathrm{adj}(I-\hat{\Psi}(\lambda_0))\phi$$
$$= \frac{v_0(0)^{\mathrm{T}}\phi}{v_0(0)^{\mathrm{T}}\Psi_1 u_0(0)} u_0(0),$$

which shows that $\lambda_0$ is a simple pole of $(I-\hat{\Psi}(\lambda))^{-1}$ and its residue at $\lambda=\lambda_0$ is given by (2.13). We conclude that (2) holds. □

From Proposition 2.2, for the indecomposable multistate stable population system, we can shift the integral path in (2.7) to the left so that $\lambda_0 > x > \sup\{\Re\lambda : \lambda \in \Omega \setminus \{\lambda_0\}\}$. It then follows from the residue theorem that

$$p(t,a) = \lim_{\lambda\to\lambda_0}(\lambda-\lambda_0)e^{\lambda t}\hat{p}(\lambda,a) + \frac{1}{2\pi i}\int_{y-i\infty}^{y+i\infty} e^{\lambda t}\hat{p}(\lambda,a)d\lambda, \tag{2.15}$$

where we can calculate the first term as

$$\lim_{\lambda\to\lambda_0}(\lambda-\lambda_0)e^{\lambda t}\hat{p}(\lambda,a) = e^{\lambda_0 t}\frac{\langle v_0, p_0\rangle}{\langle v_0, u_0\rangle}u_0(a).$$

Although we omit the proof, it can be shown that the growth bound of the second term of (2.15) is less than $\lambda_0$. Therefore, we have the following proposition [49, 50]:

**Proposition 2.3** *For the multistate stable population model with an irreducible net reproduction matrix, there exists some* $\varepsilon > 0$ *such that*

$$p(t,a) = e^{\lambda_0 t}\frac{\langle v_0, p_0\rangle}{\langle v_0, u_0\rangle}u_0(a) + O(e^{(\lambda_0-\varepsilon)t}), \tag{2.16}$$

*where* $\langle , \rangle$ *denotes the inner product of functions given by*

$$\langle f, g\rangle := \sum_{i=1}^{n}\int_0^\infty f_i(a)g_i(a)da$$

*for* $f = (f_1, \ldots, f_n)^{\mathrm{T}}$, $g = (g_1, \ldots, g_n)^{\mathrm{T}}$ *and* $u_0(a)$ *and* $v_0(a)$ *are positive vectors given by*

$$u_0(a) := e^{-\lambda_0 a}L(a)u_0(0), \quad v_0(a)^{\mathrm{T}} := v_0(0)^{\mathrm{T}}\int_a^\infty e^{-\lambda_0(s-a)}M(s)L(s,a)ds.$$

The normalized stable age distribution is given by $u_0(a)/|u_0|_{L^1}$. The vector $v_0(a)$ is called the *multistate reproductive value*, which is a multidimensional extension of the reproductive value in the scalar stable population theory.

**Corollary 2.1** *For the birth rate vector of* (2.2), *it follows that*

$$\lim_{t\to\infty} e^{-\lambda_0 t} B(t) = \frac{\langle v_0, p_0 \rangle}{\langle v_0, u_0 \rangle} u_0(0) = \frac{v_0(0)^{\mathrm{T}} \hat{G}(\lambda_0)}{v_0(0)^{\mathrm{T}} \Psi_1 u_0(0)} u_0(0). \tag{2.17}$$

From the above argument, and under the assumption of the irreducibility (indecomposability) of the net reproduction matrix, we know that there exists a dominant exponential solution for (2.2), so the sign of the intrinsic growth rate (Malthusian parameter) determines the stability of the zero solution of (2.2).

*Remark 2.1* In the multistate stable population model, we assume that the state transition process satisfies the *Markovian assumption*, that is, age-dependent forces of transition $q_{ij}(a)$, $1 \le i \le n$ are independent of the past history of individuals and depend only on their age and the present state. However, the Markovian assumption would not necessarily be satisfied for real human migration. In fact, the human migration pattern depends on individual migration histories, and especially the birth places of the individuals. Moreover, the duration in a region of residence could affect the force of migration. Non-Markovian models that can recognize these realistic aspects have been proposed by several authors [4, 72, 83], but it is very difficult to obtain real data for many related parameters. For Markovian models, parameter estimation methods are well developed, so there has been greater progress in their application and utilization [89–91].

*Remark 2.2* If the timescale of the migration dynamics is much faster than that of the birth and death process, the multistate system can be aggregated to a scalar stable population model, where the fast process has attained an equilibrium and the global variable is the total population size [5, 6, 19, 73].

## 2.2 Inhomogeneous Linear Problems

### 2.2.1 *Stable Population Model with Immigration*

We now extend the stable population model to allow immigration. For simplicity, the birth and death rates of immigrants are the same as those of the native population. This type of model has been investigated by several authors to consider the demographic effect of immigration [8, 20, 36, 81, 82].

If we let $f(t,\cdot) \in L^1_+(0,\omega;\mathbb{R}^n)$ be the age-density function of immigrants, we obtain an inhomogeneous problem of the McKendrick equation:

$$\begin{aligned}
&\frac{\partial p(t,a)}{\partial t} + \frac{\partial p(t,a)}{\partial a} = Q(a)p(t,a) + f(t,a), \quad t > 0, \quad 0 < a < \omega, \\
&p(t,0) = \int_0^{\omega} M(a)p(t,a)da, \quad t > 0, \\
&p(0,a) = p_0(a).
\end{aligned} \tag{2.18}$$

To address this inhomogeneous problem, a functional analytics approach is most effective, as shown in Chap. 10 of this text and by Inaba [51] under relaxed conditions. Here, however, we present a classical elementary calculation under restrictive conditions. For simplicity, we only deal with the autonomous case, that is, we assume that the migration term $f(t,a)$ is time-independent and given by a known function $f(a)$.

By integrating the McKendrick equation in (2.18) along characteristic lines, we have

$$p(t,a) = \begin{cases} L(a)B(t-a) + \int_0^a L(a,\rho)f(\rho)d\rho, & t-a > 0, \\ L(a,a-t)p_0(a-t) & \\ \quad + \int_0^t L(a,a-t+\rho)f(a-t+\rho)d\rho, & a-t > 0, \end{cases} \tag{2.19}$$

where $B(t) := p(t,0)$. Inserting this expression into the boundary condition in (2.18), we arrive at the renewal integral equation

$$B(t) = G(t) + H_1(t) + H_2(t) + \int_0^t \Psi(a)B(t-a)da, \tag{2.20}$$

where $\Psi(a) := M(a)L(a.0)$ and

$$\begin{aligned}
G(t) &:= \int_t^{\infty} M(a)L(a,a-t)p_0(a-t)da, \\
H_1(t) &:= \int_0^t M(a)\int_0^a L(a,\rho)f(\rho)d\rho da, \\
H_2(t) &:= \int_t^{\infty} M(a)\int_0^t L(a,a-t+\rho)f(a-t+\rho)d\rho da.
\end{aligned}$$

Note that $G(t) = H_2(t) = 0$ and $H_1(t) = H_1(\beta_2)$ for $t > \beta_2$, where $\beta_2$ is the upper bound of the reproductive age. We can divide (2.20) into two renewal equations:

$$\begin{aligned}
B_1(t) &= H_1(t) + \int_0^t \Psi(a)B_1(t-a)da, \\
B_2(t) &= G(t) + H_2(t) + \int_0^t \Psi(a)B_2(t-a)da.
\end{aligned} \tag{2.21}$$

It is then clear that the solution $B(t)$ of (2.20) is given as the sum of the two solutions to the renewal equation (2.21). Because the initial data $G(t)+H_2(t)$ of the second equation in (2.21) have a finite support, its asymptotic behavior is the same as the multistate Lotka integral equation (2.5), and is given by (2.17).

However, the initial data of the first equation in (2.21) are not integrable on $\mathbb{R}_+$, so we cannot directly apply result (2.17). If we assume $H_1 \in W^{1,\infty}(\mathbb{R}_+)$, then $B_1 \in W^{1,\infty}_{\mathrm{loc}}(\mathbb{R}_+)$, and it follows from $B_1(0)=0$ that

$$B_1'(t) = H_1'(t) + \int_0^t \Psi(a) B_1'(t-a)da. \tag{2.22}$$

It then follows that $H_1'(t) \in L^1_+(\mathbb{R}_+)$ and $\hat{H}_1'(0) = H_1(\infty)$, and we can apply Proposition 10.38 to (2.22). First, suppose that $\lambda_0 < 0$. Then, we have

$$B_1(\infty) = \int_0^\infty B_1'(t)dt = (I - \hat{\Psi}(0))^{-1} H_1(\infty).$$

It follows from $\lim_{t\to\infty} B_2(t) = 0$ that

$$\lim_{t\to\infty} B(t) = \lim_{t\to\infty} B_1(t) = (I - \hat{\Psi}(0))^{-1} H_1(\infty). \tag{2.23}$$

Next, suppose that $\lambda_0 = 0$. Then, we have

$$\lim_{t\to\infty} \frac{1}{t}\int_0^t B_1'(t)dt = \lim_{t\to\infty} \frac{B_1(t)}{t} = \frac{v_0(0)^{\mathrm{T}} H_1(\infty)}{v_0(0)^{\mathrm{T}} \Psi_1 u_0(0)} u_0(0),$$

where $v_0(0)$ and $u_0(0)$ are the left and right eigenvectors of $\hat{\Psi}(\lambda_0)$ associated with the eigenvalue unity. However, it follows from the renewal theorem that

$$\lim_{t\to\infty} B_2(t) = \frac{v_0(0)^{\mathrm{T}} (\hat{G}(0) + \hat{H}_2(0))}{v_0(0)^{\mathrm{T}} \Psi_1 u_0(0)} u_0(0).$$

Then, $\lim_{t\to\infty} B_2(t)/t = 0$ and we obtain

$$\lim_{t\to\infty} \frac{B(t)}{t} = \lim_{t\to\infty} \frac{B_1(t)}{t} = \frac{v_0(0)^{\mathrm{T}} H_1(\infty)}{v_0(0)^{\mathrm{T}} \Psi_1 u_0(0)} u_0(0). \tag{2.24}$$

Finally, suppose that $\lambda_0 > 0$. In this case, the initial data of (2.20) are bounded above for $t \ge 0$. Hence, for any positive number $\sigma > 0$, we have $e^{-\sigma t}(G(t) + H_1(t) + H_2(t)) \in L^1(\mathbb{R}_+) \cap L^2(\mathbb{R}_+)$. We can then repeat the same argument as for Proposition 1.9 for the vector-type renewal equation (2.20) to prove the following:

$$\lim_{t\to\infty} e^{-\lambda_0 t} B(t) = \frac{v_0(0)^{\mathrm{T}} (\hat{G}(\lambda_0) + \hat{H}_1(\lambda_0) + \hat{H}_2(\lambda_0))}{v_0(0)^{\mathrm{T}} \Psi_1 u_0(0)} u_0(0). \tag{2.25}$$

From (2.19), (2.23), (2.24) and (2.25), we can conclude the following [49]:

**Proposition 2.4** *Suppose that $\Psi$ is differentiable and $\hat{\Psi}(0)$ is indecomposable. For the stable population model of* (2.18) *with time-independent immigration term $f$, the following holds:*

*(1) If $\lambda_0 < 0$, then*

$$\lim_{t\to\infty} p(t,a) = L(a)(I - \hat{\Psi}(0))^{-1}H_1(\infty) + \int_0^a L(a,\rho)f(\rho)d\rho,$$

*where*

$$H_1(\infty) = \int_0^\infty d\rho \int_\rho^\infty M(a)L(a,\rho)daf(\rho)d\rho da.$$

*(2) If $\lambda_0 = 0$, then*

$$\lim_{t\to\infty} \frac{p(t,a)}{t} = \frac{\langle v_0, f\rangle}{\langle v_0, u_0\rangle} u_0(a),$$

*where $u_0(a)$ and $v_0(a)$ are given in Proposition 2.3 with $\lambda_0 = 0$.*

*(3) If $\lambda_0 > 0$, then*

$$\lim_{t\to\infty} e^{-\lambda_0 t} p(t,a) = \frac{\langle v_0, p_0\rangle + \lambda_0^{-1}\langle v_0, f\rangle}{\langle v_0, u_0\rangle} u_0(a),$$

*where $u_0(a)$ and $v_0(a)$ are given in Proposition 2.3.*

From Proposition 2.4, we know that if $\lambda_0 > 0$, the asymptotic Malthusian parameter is not affected by the existence of immigration and the asymptotic age structure is the stable age distribution associated with the net reproduction matrix $\Psi(a) = M(a)L(a)$. If $\lambda_0 = 0$, the size of the population is asymptotically growing in a linear manner, and the final age structure is the same as the stationary population structure $L(a)u_0(0)$ of a closed population. If $\lambda_0 < 0$, the population converges to a stationary population. Therefore, if the population has below-replacement fertility, a stationary population with an appropriate size can be achieved by a constant stream of immigration, but the final age structure will be more aged in comparison with the stationary state of a closed population (see Chap. 10).

### 2.2.2 *Population Dynamics of Marine Invertebrates*

We now introduce an ecological application of the stable population theory as an example of the inhomogeneous boundary value problem of the McKendrick equation. The McKendrick equation with inhomogeneous boundary conditions is investigated in [111].

The population dynamics of marine invertebrates such as *barnacles*, in which sessile adults and pelagic larvae are contained in a local area, are very much different from the population dynamics of vertebrates. Although the sessile adults can be viewed as living in a limited area, their larvae can freely move from one area to another, because each area (patch) is connected by the pelagic pool containing the larvae. That is, such a population system in a local area is essentially "open," because newly settled larvae can be carried from outside the region, while the whole multipatch system can be "closed" if the larvae are produced by the sessile adults in each patch [63]. Moreover, it has been observed in sessile marine populations that the space to be settled by the larvae is a principal limiting resource, and the number of settlements is approximately proportional to the free space available to larvae.

Under the observations mentioned above, Roughgarden et al. [92] proposed an age-structured population model for sessile invertebrates living in a local area. They derived a sufficient condition for the local stability of the uniquely existing steady state and used numerical examples to suggest that this steady state could be destabilized if the settlement rate were sufficiently high, leading to a limit cycle oscillation. The open marine population model has been further developed by Roughgarden and Iwasa [93, 94] and mathematically investigated by several authors [57, 63, 112, 113].

Originally, Roughgarden et al. recognized the absurd drawback of their model, whereby the population density may become negative for some initial conditions. The main reason for this shortcoming is that the demographic parameters of the size growth rate and mortality are assumed to be independent of the population density. In reality, these parameters will depend on the density of the population or the available free space. However, we overlook this limitation and simply sketch the mathematical formulation of the original linear model.

Let $p(t,a)$ denote the density of adults of age $a$ at time $t$, $A$ be the total area of available substrate, $F(t)$ be the size of the free space available to the larvae at time $t$, $k$ be the instantaneous settling rate per unit of free space, $\beta(a)$ be the size of individuals of age $a$, $\mu(a)$ be the age-specific death rate, and $\omega$ be the upper bound of the age of individuals. The Roughgarden–Iwasa–Baxter age-structured population model for sessile invertebrates living in a local area can then be formulated as

$$\begin{aligned}
&\frac{\partial p(t,a)}{\partial t}+\frac{\partial p(t,a)}{\partial a}=-\mu(a)p(t,a),\\
&p(t,0)=kF(t),\\
&F(t)=A-\int_0^\omega \beta(a)p(t,a)da,
\end{aligned} \tag{2.26}$$

where we assume that the maximum attainable age of individuals is finite, although $\omega=\infty$ in the original model.

We assume that $\beta\in L^\infty_+(0,\omega)$, $\mu(a)$ is positive for all $a\in[0,\omega]$, locally integrable on $[0,\omega)$ and $\int_0^\omega \mu(\sigma)d\sigma=\infty$. The survival probability (the proportion of newly settled larvae who can survive to age $a$) is given by $\ell(a):=\exp\left(-\int_0^a \mu(\sigma)d\sigma\right)$. The state space $\Omega$ of the age-density function is

$$\Omega = \left\{ p \in L^1_+(0,\omega) : \int_0^\omega \mu(a)p(a)da < \infty, \int_0^\omega \beta(a)p(a)da \le A \right\}.$$

Let us factor out the natural death rate in the basic model (2.26). If we define a new function $q(t,a)$ by $p(t,a) = \ell(a)q(t,a)$, then system (2.26) reduces to a simpler system for $q$ as follows:

$$\begin{aligned}
&\frac{\partial q(t,a)}{\partial t} + \frac{\partial q(t,a)}{\partial a} = 0,\\
&q(t,0) = k\left(A - \int_0^\omega \phi(a)q(t,a)da\right),\\
&q(0,a) = q_0(a) := \frac{p_0(a)}{\ell(a)},
\end{aligned} \tag{2.27}$$

where $\phi(a) := \beta(a)\ell(a)$ is the expected space size occupied by the population at age $a$ and we assume that $p_0/\ell \in L^1_+(0,\omega)$. If $\mu\ell \in L^\infty(0,\omega)$, we have $p = q\ell \in \Omega$ when $\int_0^\omega q(a)\phi(a)da \le A$.

For model (2.27), there always exists the unique positive steady state

$$q^*(a) = \frac{kA}{1 + k\int_0^\omega \phi(a)da}. \tag{2.28}$$

To rewrite the basic model, so as to have a homogeneous boundary condition, let us introduce a new variable $u(t,a)$ as $u(t,a) := q(t,a) - q^*(a)$. System (2.27) can then be rewritten as the following homogeneous system:

$$\begin{aligned}
&\frac{\partial u(t,a)}{\partial t} + \frac{\partial u(t,a)}{\partial a} = 0,\\
&u(t,0) = -k\int_0^\omega \phi(a)u(t,a)da,\\
&u(0,a) = q_0(a) - q^*(a).
\end{aligned} \tag{2.29}$$

Therefore, the open marine population model (2.26) can be reduced to a linear homogeneous age-dependent population system in $L^1$, and the method for the stable population model can be applied to this linear system.

Integrating (2.29) along the characteristic line, we obtain the following expression:

$$u(t,a) = \begin{cases} b(t-a), & t-a > 0,\\ u_0(a-t), & t-a < 0, \end{cases} \tag{2.30}$$

where $b(t) := u(t,0)$. Inserting (2.30) into the boundary condition of (2.29), we have the renewal integral equation

$$b(t) = -g(t) - k \int_0^t \phi(a)b(t-a)da, \tag{2.31}$$

where $g(t)$ is defined by

$$g(t) := k \int_t^\infty \phi(a)u_0(a-t)da.$$

Here, we extend the domain of $\phi(a)$ to $\phi(a) = 0$ for $a > \omega$. Let $\Lambda$ be the set of characteristic roots given by:

$$\Lambda := \{\lambda \in \mathbb{C} : 1 + k\hat{\phi}(\lambda) = 0\}, \tag{2.32}$$

where $\hat{\phi}$ denotes the Laplace transform of $\phi$. We can then state the following:

**Proposition 2.5** *$\Lambda \cap \mathbb{R} = \emptyset$ and $\Lambda$ is composed of a countably infinite number of complex conjugate pairs. For any real number $\alpha$, there are at most finitely many roots in the right half-plane $\Re\lambda > \alpha$, and there is a dominant pair whose real part is greater than the real part of any other characteristic root.*

Using the same kind of arguments as in Sect. 1.2, we obtain an asymptotic expansion of $b(t)$:

$$\begin{aligned} b(t) &= b_0 e^{\lambda_0 t} + \bar{b}_0 e^{\bar{\lambda}_0 t} + O(e^{(\Re\lambda_0 - \varepsilon)t}) \\ &= e^{\Re\lambda_0 t}[\Re b_0 \cos(\Im\lambda_0 t) - \Im b_0 \sin(\Im\lambda_0 t)] + O(e^{(\Re\lambda_0 - \varepsilon)t}), \end{aligned} \tag{2.33}$$

where $\lambda_0$ and $\bar{\lambda}_0$ are the dominant pair of characteristic roots, $\varepsilon > 0$ is a small number such that $\{\lambda : \lambda \in \Lambda \setminus \{\lambda_0, \bar{\lambda}_0\}\} \subset \{\lambda : \Re\lambda \le \Re\lambda_0 - \varepsilon\}$ and $b_0$ is given by:

$$b_0 := \frac{\int_0^\omega e^{-\lambda_0 t} g(t)dt}{k \int_0^\omega a e^{-\lambda_0 a}\phi(a)da}. \tag{2.34}$$

Hence, the asymptotically dominant part of the solution of the basic model (2.26) is given as

$$\ell(a)q^*(a) + |b_0| e^{\Re\lambda_0(t-a)}\ell(a)\cos(\Im\lambda_0(t-a) + \theta), \tag{2.35}$$

where $\theta := \arctan(\Im b_0/\Re b_0)$. Thus, there is no Malthusian solution, and the steady state is globally asymptotically stable if $\Re\lambda_0 < 0$, whereas it is unstable if $\Re\lambda_0 > 0$. In particular, if $\Re\lambda_0 > 0$, the positivity of the population density will be destroyed as time evolves.

An interesting point, however, is that the following *50% free space rule* for stability holds [92]:

**Proposition 2.6** *Let $\pi$ be the proportion of free space in the steady state:*

$$\pi := \frac{F^*}{A} = \frac{1}{1 + k\int_0^\omega \phi(a)da}. \tag{2.36}$$

*If $\pi > 1/2$, the steady state is globally asymptotically stable.*

*Proof* Suppose that $\pi > 1/2$ and there exists a characteristic root $\lambda = x + iy$ with $x \geq 0$. Then, it follows from (2.32) that

$$1 = k\left|\int_0^\omega e^{-\lambda a}\phi(a)da\right| \leq k\int_0^\omega e^{-xa}\phi(a)da \leq k\int_0^\omega \phi(a)da = \frac{1}{\pi} - 1.$$

This contradicts our assumption.

We will again encounter the *fifty percent stability rule* in the Pease influenza model (Chap. 8). Note that the condition $\pi > 1/2$ can be rewritten as

$$R := k\int_0^\omega \phi(a)da < 1, \tag{2.37}$$

where $R$ denotes the cumulative area occupied by settled larvae per unit area. As shown in the next chapter, there is a possibility that the characteristic Eq. (2.32) has a pair of characteristic roots with positive real parts if $k$ is sufficiently large, destabilizing the linear system. If there exists a characteristic root with a positive real part, the amplitude of the solution of (2.26) will grow without bound. In this case, the physical condition $0 \leq F(t) \leq A$ will be lost and the linear model breaks down.

As pointed out by Roughgarden et al., the above shortcoming can be overcome if we make the more realistic assumption that the mortality of the adult population increases as the free space decreases. Using numerical simulations, Roughgarden et al. found that the destabilization of the steady state in the density dependent model can lead to a limit cycle as the free space becomes exhausted.

*Remark 2.3* Note that in a formal sense, the open marine population dynamics are formulated by the Gurtin and MacCamy nonlinear model, which we investigate in Chap. 3. In fact, if we define the size-dependent fertility rate as

$$m(a, S(t)) := \frac{k(A - S(t))}{S(t)}\beta(a),$$

where $S(t) = \int_0^\omega \beta(a)p(t,a)da$, the open marine population model can be written as the Gurtin and MacCamy model [41], although it has a singularity at $S = 0$.

## 2.3 Linear Marriage Models

### 2.3.1 *First-Marriage Model*

In modern societies, fertility change is a most important factor in understanding population changes. From a demographic point of view, when considering the causal analysis of human reproduction, it is important to divide the population according to life stages that could affect the reproduction process. Thus, we need to model the birth process so as to clarify the relationship between the macrodynamics of populations and life cycle variables such as maturity, marriage, family building, and divorce. In this section, we consider a multistate stable population model that takes into account the marital status of individuals.

We consider a society in which there is a strong tradition of monogamous marriage, the divorce rate is still low and there is very little childbearing among the unmarried. That is, almost all newborns are produced by the legitimate first marriage, and the contribution of ex-nuptial phenomena to the total number of births is negligible. The reproductivity of such a population is determined by two factors: how many people will get married and how many children will be produced by married couples. Hence, changes in fertility should be factored into changes in nuptiality and marital fertility. Such a situation has been observed in Japan as well as in Mediterranean countries and other traditional societies. Understanding the fertility trends of those countries is one of the motivating factors behind the following theoretical framework [16, 56].

For simplicity, we assume that children are produced only by first-marriage couples. To formulate a dynamic population model based on nuptiality and marital fertility, we use the duration-specific marital fertility rate by age at marriage, instead of the age-specific fertility rate. This is because, for populations in which childbearing occurs predominantly within marriage, the duration of marriage is a more appropriate and influential variable than chronological age for describing controlled fertility [87, 96]. Of course, it should be noted that the duration-specific marital fertility by age at marriage implicitly accounts for the effect of chronological age, because the chronological age is given as the sum of the age at marriage and the duration of marriage. In addition, we assume that the force of marriage is a given function of age, that is, we disregard the nonlinear interaction between both sexes. The pair formation phenomena will be studied in Chap. 4.

Let us divide the population into three groups: $p_0$ (never married), $p_1$ (within first marriage), and $p_2$ (widowed or divorced population including remarried). Let $p_0(t, a)$ be the age density of the never-married population at time $t$ and age $a$, let $p_1(t, \tau; \zeta)$ be the density of married individuals at time $t$ and marital duration $\tau$ whose age at first marriage is $\zeta$ (that is, their chronological age is $\tau + \zeta$) and let $p_2(t, a)$ be the age density of widowed/divorced individuals at time $t$ and age $a$. Let $\lambda(a)$ be the *force of first marriage* at age $a$, $\mu(a)$ be the force of mortality (for simplicity, we do not assume differential mortality according to individuals' status), $m(\tau; \zeta)$ be the *marital fertility rate* of duration $\tau$ by age at first marriage $\zeta$, $\delta(\tau; \zeta)$

be the force of dissolution of couples with duration $\tau$ by age at marriage $\zeta$ and $\gamma$ be the proportion of female newborns. In the real reproduction process, the *parity* (the number of children that a woman has had) is also an important variable. However, we only use the marital fertility function aggregated with respect to the parity. A parity-structured model will be examined in the next section.

We can formulate the following one-sex model for the population reproduced by first marriage [56]:

$$\begin{aligned}
&\left(\frac{\partial}{\partial t}+\frac{\partial}{\partial a}\right)p_0(t,a)=-(\mu(a)+\lambda(a))p_0(t,a),\\
&\left(\frac{\partial}{\partial t}+\frac{\partial}{\partial \tau}\right)p_1(t,\tau;\zeta)=-\mu(\tau+\zeta)p_1(t,\tau;\zeta)-\delta(\tau;\zeta)p_1(t,\tau;\zeta),\\
&\left(\frac{\partial}{\partial t}+\frac{\partial}{\partial a}\right)p_2(t,a)=-\mu(a)p_2(t,a)+\int_0^a\delta(\tau;a-\tau)p_1(t,\tau;a-\tau)d\tau,\\
&p_0(t,0)=\gamma\int_0^\infty\int_0^\infty m(\tau;\zeta)p_1(t,\tau;\zeta)d\tau d\zeta,\\
&p_1(t,0;\zeta)=\lambda(\zeta)p_0(t,\zeta),\\
&p_2(t,0)=0,
\end{aligned} \tag{2.38}$$

where the initial data are given by:

$$p_0(0,a)=k_0(a),\ \ p_1(0,\tau;\zeta)=k_1(\tau;\zeta),\ \ p_2(0,a)=k_2(a).$$

If we define $n(t,a)$ as the age-density function of the total female population, it follows that

$$n(t,a)=p_0(t,a)+\int_0^a p_1(t,a-\zeta;\zeta)d\zeta+p_2(t,a),$$

and it is easy to see that $n(t,a)$ satisfies the McKendrick equation

$$\begin{aligned}
&\left(\frac{\partial}{\partial t}+\frac{\partial}{\partial a}\right)n(t,a)=-\mu(a)n(t,a),\\
&n(t,0)=p_0(t,0),\\
&n(0,a)=k_0(a)+\int_0^a k_1(a-\zeta;\zeta)d\zeta+k_2(a).
\end{aligned}$$

Let $B(t):=p_0(t,0)$ be the birth rate at time $t$. By integrating the McKendrick equation of (2.38) along the characteristic line, we obtain the following expression:

$$p_0(t,a) = \begin{cases} \Lambda(a)\ell(a)B(t-a), & t-a>0 \\ \frac{\Lambda(a)\ell(a)}{\Lambda(a-t)\ell(a-t)}k_0(a-t), & a-t>0 \end{cases}$$

$$p_1(t,\tau;\zeta) = \begin{cases} k_1(\tau-t;\zeta)\frac{\ell(\tau+\zeta)}{\ell(\tau+\zeta-t)}\exp\left(-\int_{\tau-t}^{\tau}\delta(s;\zeta)ds\right), & t<\tau \\ k_0(\tau+\zeta-t)\frac{\ell(\tau+\zeta)\phi(\zeta)}{\ell(\tau+\zeta-t)\Lambda(\tau+\zeta-t)}\exp\left(-\int_0^{\tau}\delta(s;\zeta)ds\right), & \tau<t<\tau+\zeta \\ B(t-\tau-\zeta)\phi(\zeta)\ell(\tau+\zeta)\exp\left(-\int_0^{\tau}\delta(s;\zeta)ds\right), & \tau+\zeta<t, \end{cases}$$

$$p_2(t,a) = n(t,a) - p_0(t,a) - \int_0^a p_1(t,a-\zeta;\zeta)d\zeta, \tag{2.39}$$

where $\ell(a)$ is the survival probability, $\Lambda(a)$ is the proportion of unmarried females at age $a$, and $\phi(a) := \lambda(a)\Lambda(a)$ is the (incomplete) probability density of first marriage:

$$\ell(a) := \exp\left(-\int_0^a \mu(\rho)d\rho\right), \quad \Lambda(a) := \exp\left(-\int_0^a \lambda(\sigma)d\sigma\right).$$

Inserting (2.39) into the boundary condition in (2.38) and changing the order of integrals, we obtain the renewal equation for $B(t)$:

$$B(t) = G(t) + \int_0^t \psi(a)B(t-a)da, \tag{2.40}$$

where $\beta(a)$ is the age-specific birth rate and $\psi(a)$ is the net reproduction function given by

$$\psi(a) = \gamma\beta(a)\ell(a), \quad \beta(a) = \int_0^a m(a-\zeta;\zeta)e^{-\int_0^{a-\zeta}\delta(s;\zeta)ds}\phi(\zeta)d\zeta$$

and

$$\begin{aligned} G(t) := &\int_t^\infty d\tau \int_0^\infty m(\tau;\zeta)k_1(\tau-t;\zeta)\frac{\ell(\tau+\zeta)}{\ell(\tau+\zeta-t)}e^{-\int_{\tau-t}^{\tau}\delta(s;\zeta)ds}d\zeta \\ &+ \int_0^t d\tau \int_{t-\tau}^\infty m(\tau;\zeta)\frac{k_0(\tau+\zeta-t)\ell(\tau+\zeta)\phi(\zeta)}{\ell(\tau+\zeta-t)\Lambda(\tau+\zeta-t)}e^{-\int_0^{\tau}\delta(s;\zeta)ds}d\zeta. \end{aligned}$$

If we let $\beta_2$ be the upper bound of reproductive age, we have $m(\tau;\zeta) = 0$ for $\tau+\zeta > \beta_2$ and $\psi(a) = 0$, $G(t) = 0$ for $t > \beta_2$. Thus, the strong ergodicity theorem holds for the renewal equation, and there is an intrinsic growth rate $\lambda_0$ such that

$$\lim_{t\to\infty} e^{-\lambda_0 t}B(t) = \frac{\int_0^\infty e^{-\lambda_0 s}G(s)ds}{\int_0^\infty a e^{-\lambda_0 a}\psi(a)da},$$

where $\lambda_0$ is the unique real root of the characteristic equation

$$\int_0^\infty e^{-\lambda_0 a}\psi(a)da = 1.$$

It is easy to calculate the stable age distributions by marital status as follows:

$$\lim_{t\to\infty}\frac{p_0(t,a)}{\int_0^\infty n(t,a)da} = w(a)\Lambda(a),$$
$$\lim_{t\to\infty}\frac{\int_0^a p_1(t,\tau;a-\tau)d\tau}{\int_0^\infty n(t,a)da} = w(a)\Gamma(a),$$
$$\lim_{t\to\infty}\frac{p_2(t,a)}{\int_0^\infty n(t,a)da} = w(a)(1-\Lambda(a)-\Gamma(a)),$$

where $w(a)$ is the age profile of the stable population given by

$$w(a) := \frac{e^{-r_0 a}\ell(a)}{\int_0^\infty e^{-r_0 a}\ell(a)da},$$

and $\Gamma(a)$ is the proportion of married population at age $a$ in the stable distribution defined by

$$\Gamma(a) := \int_0^a \phi(\zeta)e^{-\int_0^{a-\zeta}\delta(\sigma;\zeta)d\sigma}d\zeta.$$

From the renewal equation (2.40), we can calculate the basic reproduction number (net reproduction rate) $R_0$ and the total fertility rate as follows:

$$\begin{aligned} R_0 &= \int_0^\infty \psi(a)da = \gamma\int_0^\infty\int_0^\infty m(\tau;\zeta)\ell(\tau+\zeta)e^{-\int_0^\tau\delta(s;\zeta)ds}d\tau\phi(\zeta)d\zeta, \\ \mathrm{TFR} &= \int_0^\infty \beta(a)da = \int_0^\infty\int_0^\infty m(\tau;\zeta)e^{-\int_0^\tau\delta(s;\zeta)ds}d\tau\phi(\zeta)d\zeta. \end{aligned} \tag{2.41}$$

The period total fertility rate (TFR) observed at time $t$ is calculated as

$$\text{period TFR} = \int_0^\infty \frac{B(t,a)}{n(t,a)}da,$$

where $B(t,a)$ denotes the number of newborns produced by mothers aged $a$ at time $t$. For our first-marriage model, we have

$$B(t,a) = \int_0^a m(a-\zeta;\zeta)p_1(t,a-\zeta;\zeta)d\zeta.$$

Then, the period TFR can change as a function of time $t$ and coincides with the (time-independent) cohort TFR if and only if the population structure by marital status is consistent with the given marriage schedule, that is, $p_1$ is given as

$$p_1(t, a-\zeta; \zeta) = n(t, a)\phi(\zeta)e^{-\int_0^{a-\zeta}\delta(\sigma;\zeta)d\sigma}.$$

The above condition will be satisfied as the influence of the initial population disappears over time.

Let $T(\zeta)$ be the expected total number of children produced by a woman married at age $\zeta$ in the case that her reproduction process is not terminated by death. We then have

$$T(\zeta) = \int_0^\infty m(\tau;\zeta)e^{-\int_0^\tau \delta(s;\zeta)ds}d\tau.$$

Using $T(\zeta)$, TFR can be expressed as

$$\text{TFR} = \int_0^\infty T(\zeta)\phi(\zeta)d\zeta.$$

Let $\Phi(a)$ be the probability density of age at first marriage:

$$\Phi(a) = \frac{\phi(a)}{\int_0^\infty \phi(z)dz} = \frac{\phi(a)}{1-\Lambda(\infty)}.$$

We can then rewrite TFR as

$$\text{TFR} = (1-\Lambda(\infty))\int_0^\infty T(\zeta)\Phi(\zeta)d\zeta, \tag{2.42}$$

where $1-\Lambda(\infty)$ is the *proportion ever marrying* (PEM), so we know that TFR can be factored into the PEM and the average number of children produced per marriage in the case that the marital birth process is not terminated by death of female partner. Moreover, let us define $S(\zeta)$ by

$$S(\zeta) = \int_0^\infty m(\tau;\zeta)\frac{\ell(\tau+\zeta)}{\ell(\zeta)}e^{-\int_0^\tau \delta(s;\zeta)ds}d\tau.$$

Then, $S(\zeta)$ is the expected total number of children produced per marriage by age at marriage $\zeta$. Using $S(\zeta)$, the basic reproduction number is given by:

$$R_0 = \gamma\int_0^\infty S(\zeta)\ell(\zeta)\phi(\zeta)d\zeta. \tag{2.43}$$

As Henry [47] observed for *natural fertility*, if conception occurs randomly for each marital status, the *completed fertility* by age at first marriage given by

$$U(\zeta) := \int_0^\infty m(\tau;\zeta)d\tau$$

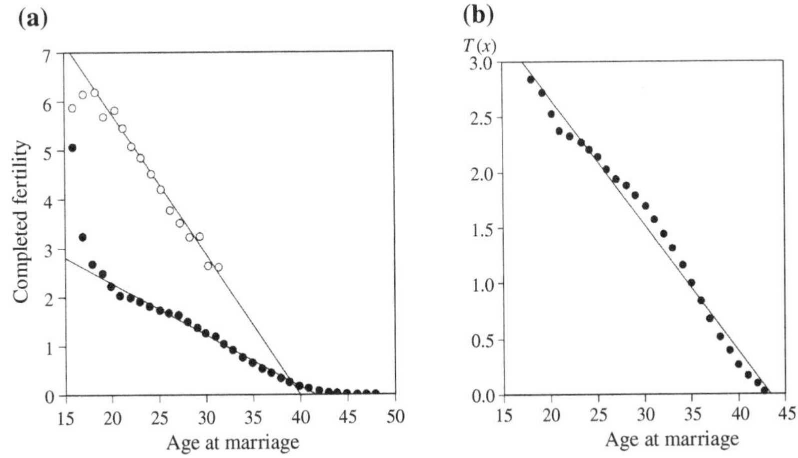


**Fig. 2.1** **a** Observed data for completed fertility rates in Japan (1940, ∘) and England (1939, •); **b** Observed data of $T(x)$ in Japan (1985) and the regression line [56, 62]

would become an almost linear function of the age at marriage. Even for the weakly controlled fertility historically observed in England and Japan before World War II, $U(\zeta)$ was approximately linear. More surprisingly, we can see that $T(\zeta)$ and $S(\zeta)$ for Japanese women in the 1980s, who had strong control over when they conceived, are also well approximated by linear functions (see Fig. 2.1 and [56]). Similar phenomena have been observed among modern Mediterranean countries such as Italy [16].

According to the above observations, we can write $T(\zeta)$ as

$$T(\zeta) = u + v\zeta + \xi(\zeta).$$

The coefficients of the linear part $u + v\zeta$ can be determined from a regression equation using real data. Inserting the above expression into (2.42), we then have

$$\text{TFR} = (1 - \Lambda(\infty))(u + va_0) + \int_0^\infty \xi(\zeta)\phi(\zeta)d\zeta, \tag{2.44}$$

where $\Lambda(\infty)$ is the proportion of lifelong unmarried, and $a_0$ is the mean age at first marriage. Though $T(\zeta)$ would not exhibit a linear pattern in the neighborhood of the upper and lower bounds of reproductive age, the residual term in (2.44) is very small because $\phi(\zeta)$ takes a very small value around the limits of reproductive age. As a result, the linear part of (2.44) is a sufficient approximation for TFR.

From the Japanese marital fertility table in 1985 (see [62] and Fig. 2.1), if we neglect the effect of remarriage, $T(\zeta)$ can be estimated as

$$T(\zeta) \approx 4.927 - 0.1136\zeta, \quad 18 \le \zeta \le 43,$$

where the coefficient of determination is 0.986. If we neglect the residual term in (2.44), we have

$$\text{TFR} \approx (1 - \Lambda(\infty))(4.927 - 0.1136a_0).$$

For Japanese women born between 1945 and 1950 (the Japanese baby boom cohort after World War II), $\Lambda(\infty)$ is less than 5% and the average age at marriage is 24 years. The TFR of this generation calculated by the above first approximation formula is 2.09, which is almost equal to the completed fertility rate observed from sample surveys and near to the population replacement level (critical fertility rate).

*Remark 2.4* Japanese demographers were concerned about the effect of nuptiality on fertility under pronatalist policy during World War II. After the war, however, the issue received less attention in Japan, because the first Japanese *demographic transition* occurred over the decade following the war, mainly because of a rapid decline in marital fertility, whereas the effect of nuptiality was relatively faint. Itoh [61] revived concern for the nuptiality–fertility relation to explain the *second demographic transition* that began in 1974 in Japan and derived "Itoh's formula" given in (2.41). In 1974, the period TFR of Japanese women was 2.05, which is below the replacement level ($R_0 = 0.97$). From 1974 to 2005, the period TFR in Japan monotonically decreased (except for a temporal increase from 1982 to 1984), falling to a postwar low of 1.26 in 2005. After 2005, TFR began to increase, reaching 1.41 in 2012. National fertility surveys, however, show that marital fertility is relatively stable. In fact, the completed marital fertility rate is 2.01 for cohorts born from 1960 to 1965, and the social norm of two children per couple is still accepted, even among younger generations. These observations indicate that a decline in period TFR since 1974 in Japan could be attributable to a change in nuptiality. In particular, delayed marriage and the increased proportion of lifelong unmarried are seen as major causes of fewer children being born and the declining population of Japan.

### 2.3.2 Reproduction by Non-persistent Unions

The assumption of reproduction only by first marriage gives a good approximation of real human reproduction if childbearing outside of marriage is exceptional and the union formation is persistent, a situation that has often been observed in traditional societies. However, this model is not so good when the pair formation process is unstable and temporary, as in post-modern societies.

We now consider the case in which each woman repeatedly (re)marries and divorces. This divides the female population into three states: never-married single ($p_0$), married ($p_1$), and divorced single ($p_2$). Alternatively, we can interpret $p_0$ as the unmatured state, $p_1$ as the susceptible state and $p_2$ as the non-susceptible state. Using the same notation as for the first-marriage model, the basic system is formulated as follows [55]:

$$\begin{aligned}
&\left(\frac{\partial}{\partial t}+\frac{\partial}{\partial a}\right)p_0(t,a)=-(\lambda(a)+\mu(a))p_0(t,a),\\
&\left(\frac{\partial}{\partial a}+\frac{\partial}{\partial a}+\frac{\partial}{\partial s}\right)p_1(t,a,s)=-(\mu(a)+\sigma(a,s))p_1(t,a,s),\\
&\left(\frac{\partial}{\partial t}+\frac{\partial}{\partial a}+\frac{\partial}{\partial s}\right)p_2(t,a,s)=-(\mu(a)+\delta(a,s))p_2(t,a,s),
\end{aligned} \tag{2.45}$$

$$\begin{aligned}
p_0(t,0)&=\int_0^\infty\int_0^a m(a,s)p_1(t,a,s)dsda,\\
p_1(t,a,0)&=\lambda(a)p_0(t,a)+\int_0^a\delta(a,s)p_2(t,a,s)ds,\\
p_2(t,a,0)&=\int_0^a\sigma(a,s)p_1(t,a,s)ds,\\
p_1(t,a,s)&=p_2(t,a,s)=0,\quad a\le s,
\end{aligned}$$

where $\lambda(a)$ is the force of first marriage, $\sigma(a,s)$ is the force of dissolution at age $a$ and marriage duration $s$ (by divorce or death of spouse), $\delta(a,s)$ is the force of remarriage at age $a$ and duration (since the last divorce) $s$, and $m(a,s)$ is the fertility rate at age $a$ for duration $s$. For simplicity, we assume that there is no differential mortality among the three states, state transitions and natural death are mutually independent phenomena, and newborns are only produced by married individuals.

To solve the basic system, let us consider the age–duration-dependent state transition process in a female cohort using survival probabilities. Let $q_0(a)$ be the survival probability at age $a$ in the never-married single state, $q_1(a,s)$ be the survival probability at age $a$ and duration (since the last marriage) $s$ in the marriage state, and $q_2(a,s)$ be the survival probability at age $a$ and duration (since the last dissolution) $s$ in the divorced single state. Neglecting the natural death rate, we obtain the following system:

$$\begin{aligned}
&\frac{dq_0(a)}{da}=-\lambda(a)q_0(a),\\
&\left(\frac{\partial}{\partial a}+\frac{\partial}{\partial s}\right)q_1(a,s)=-\sigma(a,s)q_1(a,s),\\
&\left(\frac{\partial}{\partial a}+\frac{\partial}{\partial s}\right)q_2(a,s)=-\delta(a,s)q_2(a,s),
\end{aligned} \tag{2.46}$$

where the boundary conditions are given by:

$$\begin{aligned}
q_0(0)&=1,\\
q_1(a,0)&=\lambda(a)q_0(a)+\int_0^a\delta(a,s)q_2(a,s)ds,\\
q_2(a,0)&=\int_0^a\sigma(a,s)q_1(a,s)ds,
\end{aligned}$$

and we adopt the convention that $q_1(a,s) = q_2(a,s) = 0$ for $a \le s$.

If we can determine the survival probabilities $q_j$ and assume that the number of newborns per unit time at time $t$, denoted by $B(t)$, is given for $t \in \mathbb{R}$, the age-density functions of each state can be expressed as follows:

$$
\begin{aligned}
p_0(t,a) &= \ell(a)q_0(a)B(t-a),\\
p_1(t,a,s) &= \ell(a)q_1(a,s)B(t-a),\\
p_2(t,a,s) &= \ell(a)q_2(a,s)B(t-a),
\end{aligned}
\tag{2.47}
$$

where $\ell(a) := \exp\left(-\int_0^a \mu(\sigma)d\sigma\right)$ is the survival probability with respect to natural death.

By integrating (2.46) along characteristic lines, we have the following expression:

$$
\begin{aligned}
q_0(a) &= e^{-\int_0^a \lambda(z)dz},\\
q_1(a,s) &= q_1(a-s,0)e^{-\int_0^s \sigma(a-s+z,z)dz}, \quad a > s,\\
q_2(a,s) &= q_2(a-s,0)e^{-\int_0^s \delta(a-s+z,z)dz}, \quad a > s.
\end{aligned}
\tag{2.48}
$$

The boundary values $q_1(a,0)$ and $q_2(a,0)$ are the incidences of marriage and divorce at age $a$, respectively.

Let $\phi(a) := q_1(a,0)$ and $\psi(a) := q_2(a,0)$. Inserting expression (2.48) into the boundary condition in (2.46), we obtain a Volterra integral equation system for the unknown vector of boundary values $X(a) := (\phi(a), \psi(a))^{\mathrm{T}}$ as follows:

$$
X(a) = G(a) + \int_0^a H(a,s)X(a-s)ds,
\tag{2.49}
$$

where

$$
G(a) := \begin{pmatrix} \lambda(a)q_0(a) \\ 0 \end{pmatrix},
$$

$$
H(a,s) := \begin{pmatrix} 0 & \delta(a,s)e^{-\int_0^s \delta(a-s+z,z)dz} \\ \sigma(a,s)e^{-\int_0^s \sigma(a-s+z,z)dz} & 0 \end{pmatrix}
$$

If we define $K(a,\rho) := H(a, a-\rho)$, then (2.49) can be rewritten as

$$
X(a) = G(a) + \int_0^a K(a,\rho)X(\rho)d\rho.
\tag{2.50}
$$

Define the resolvent $R(a,\rho)$ as the solution of

$$
R(a,\rho) = K(a,\rho) + \int_0^a K(a,\sigma)R(\sigma,\rho)d\sigma,
$$

in which case $X(a)$ can be expressed as

$$X(a) = G(a) + \int_0^a R(a,\rho)G(\rho)d\rho.$$

Using this solution, we have $\phi$ and $\psi$, and hence the age–duration-dependent survival probabilities by state can be calculated from (2.48).

From (2.47), omitting the initial data, we obtain a homogeneous renewal equation

$$B(t) = \int_0^\infty \left[\int_0^a m(a,s)q_1(a,s)ds\right]\ell(a)B(t-a)da. \tag{2.51}$$

The basic reproduction number and TFR are then given by:

$$\begin{aligned} R_0 &= \int_0^\infty \int_0^a m(a,s)q_1(a,s)ds\ell(a)da = \int_0^\infty S(a)\phi(a)\ell(a)da, \\ \text{TFR} &= \int_0^\infty \int_0^a m(a,s)q_1(a,s)dsda = \int_0^\infty T(a)\phi(a)da, \end{aligned} \tag{2.52}$$

where

$$\begin{aligned} S(a) &:= \int_0^\infty m(a+s,s)\frac{\ell(a+s)}{\ell(a)}e^{-\int_0^s \sigma(a+z,z)dz}ds, \\ T(a) &:= \int_0^\infty m(a+s,s)e^{-\int_0^s \sigma(a+z,z)dz}ds, \end{aligned}$$

$S(a)$ denotes the average number of newborns produced per marriage and $T(a)$ is the average number of newborns produced per marriage, provided that the marriage is not terminated by death. From (2.51), we can determine $B(t)$ for $t > 0$ and all age-density functions by (2.47), provided that the initial data $B(t)$, $t < 0$, are given.

Finally, let us consider the special case in which the forces of marriage and divorce do not depend on age, but only on duration. In such a case, $K(a,\sigma)$ depends only on $a-\sigma$, and (2.50) becomes a convolution integral equation with a kernel $K(a,\sigma) = K(a-\sigma)$. Integrating both sides, we have

$$\int_0^\infty X(a)da = \int_0^\infty G(a)da + \int_0^\infty K(a)da \int_0^\infty X(a)da,$$

from which it follows that

$$\int_0^\infty X(a)da = \left(I - \int_0^\infty K(a)da\right)^{-1} \int_0^\infty G(a)da,$$

where $I$ denotes the $2\times 2$ identity matrix. By calculating the matrices of the right-hand side of the above equation, we obtain

$$\int_0^\infty \phi(a)da = \frac{1 - \Lambda(\infty)}{1 - (1 - \Delta(\infty)(1 - \Sigma(\infty))},$$

$$\int_0^\infty \psi(a)da = \frac{(1 - \Delta(\infty))(1 - \Sigma(\infty))}{1 - (1 - \Delta(\infty)(1 - \Sigma(\infty))},$$

where $\Lambda(a) := \exp(-\int_0^a \lambda(z)dz)$ is the survival probability in the never-married state, $\Sigma(a) := \exp(-\int_0^a \sigma(z)dz)$ is the survival probability in the married state, and $\Delta(a) := \exp(-\int_0^a \delta(z)dz)$ is the survival probability in the divorced or widowed single state. The integrals of $\phi$ and $\psi$ give the average number of marriages and the average number of dissolutions per person, respectively, during their lifetime, provided that there is no termination by death.

In fact, the force of a third marriage is different from the force of a second marriage, the marital fertility of remarried couples differs from that of first-marriage couples, and so on. Thus, our model may need some extensions to capture the reality. If we introduce more fine-grained physiological aspects, model (2.46) could be extended as a model for the conception cycle. Moreover, if we take into account that the force of union formation $\lambda$ depends on the supply of the other sex, the basic model will become a nonlinear two-sex model. We will encounter such a nonlinear version as an epidemiological reinfection model in Sect. 8.4.

## 2.4 Parity Progression Model

In demography, *parity* refers to the number of children a woman has had. It is very important to divide the female population into subclasses according to their parity status, as women's childbearing decisions are strongly dependent on their parity status and the period since the last birth. In this section, we extend the stable population model to recognize the parity status and the duration of each parity status.

Let $p_i(t, a, s)$, $(i = 1, 2, \ldots, N)$ be the density of the female population whose parity is $i$ at age $a$, time $t$, and duration (since the last live birth) $s$. $N$ denotes the maximum parity, which is estimated as $N = 10 \sim 15$ for humans. For simplicity, we assume that $p_i$ are defined as zero for $i \geq N + 1$. Let $p_0(t, a)$ be the age density of women who have parity zero at age $a$ and time $t$, $\lambda_0(a)$ be the force of the first birth at parity zero and age $a$, and $\lambda_i(a, s)$ be the force of the $i + 1$-th birth at parity $i$, age $a$, and duration $s$. If $\beta_2$ is the upper bound of reproductive age, then $\lambda_0(a) = \lambda_i(a, s) = 0$ for $a > \beta_2$ or $i \geq N$.

Biologically, $\lambda_i(a, s)$ is defined only for $a \geq s$. However, we adopt the convention that $\lambda_i(a, s) = 0$ for $a \leq s$. $\lambda_i(a, s)$ and $\lambda_0(a)$ are the age–duration-dependent forces of parity progression. Let $\mu(a)$ be the natural force of mortality, which is, for simplicity, assumed to be independent of parity status. Under the above assumption, the *parity progression model* is formulated as follows:

$$\begin{aligned}
&\left(\frac{\partial}{\partial t}+\frac{\partial}{\partial a}\right)p_0(t,a)=-(\mu(a)+\lambda_0(a))p_0(t,a),\\
&p_0(t,0)=\gamma\sum_{i=1}^{\infty}\int_0^{\infty}p_i(t,a,0)da,\\
&\left(\frac{\partial}{\partial t}+\frac{\partial}{\partial a}+\frac{\partial}{\partial s}\right)p_i(t,a,s)=-(\mu(a)+\lambda_i(a,s))p_i(t,a,s),\quad (i\ge 1),\\
&p_1(t,a,0)=\lambda_0(a)p_0(t,a),\\
&p_i(t,a,0)=\int_0^a\lambda_{i-1}(a,s)p_{i-1}(t,a,s)ds,\quad (i\ge 2),\\
&p_0(0,a)=\phi_0(a),\\
&p_i(0,a,s)=\phi_i(a,s),\quad (i\ge 1),
\end{aligned}\tag{2.53}$$

where $\phi_0(a)$ and $\phi_i(a,s)$ denote the initial data and $\gamma$ is the proportion of female children among newborns. For simplicity, we assume that a woman produces at most one child per unit time, and we neglect stillbirths. The number of live births is then equal to the number of women whose parity increases.

The above model was first formulated by Inaba [53], although many authors have independently developed similar models [4, 21, 22, 30, 65]. This kind of stage structure is also observed in cell populations [42, 43, 84].

We define survival functions for the parity progression model as follows:

$$\Lambda_0(a):=\exp\left(-\int_0^a\lambda_0(\rho)d\rho\right),$$

$$\Lambda_i(h;a,s):=\exp\left(-\int_0^h\lambda_i(a+\rho,s+\rho)d\rho\right),\quad (i\ge 1).$$

Let $\ell(a):=\exp(-\int_0^a\mu(\sigma)d\sigma)$ be the survival probability with respect to natural death. We assume that $\ell(\infty)=\Lambda_0(\infty)=\Lambda(\infty;a,s)=0$. Let $\zeta_i(a)$ be the probability that the $i$-th birth occurs at age $a$. Then, $\zeta_i(a)$ satisfies the iterative relation

$$\zeta_1(a)=\lambda_0(a)\Lambda_0(a),$$

$$\zeta_{i+1}(a)=\int_0^a\lambda_i(a,s)\Lambda_i(s;a-s,0)\zeta_i(a-s)ds,\quad (i\ge 1).$$

Let $\Psi_i(a)$, $i\ge 0$ be the survival probability with respect to the parity progression rate at age $a$ and parity $i$, defined as

$$\Psi_0(a)=\Lambda_0(a),\quad \Psi_i(a)=\int_0^a\Lambda_i(s;a-s,0)\zeta_{i-1}(a-s)ds,\quad (i\ge 1).$$

It is then easy to see that the following holds for $0\le i$:

$$0 \le \Psi_i(a) \le 1, \quad \sum_{i=0}^{\infty} \Psi_i(a) = 1,$$

$$\Psi_0(a) = 1 - \int_0^a \zeta(\rho)d\rho, \quad \Psi_i(a) = \int_0^a [\zeta_{i-1}(\rho) - \zeta_i(\rho)]d\rho.$$

Using the above functions, we can induce useful demographic indices. For example, $\zeta_i(a)\ell(a)$ is the probability that the $i$-th birth occurs at age $a$ and

$$A_i := \int_0^\infty \zeta_i(a)\ell(a)da$$

is the probability that at least $i$ births occur. In the demographic literature,

$$\alpha_i := \frac{A_{i+1}}{A_i}$$

is called the *parity progression ratio*. If we let $A_0 = 1, \alpha_i$ $(i \ge 0)$ denotes the ratio of individuals who bear an $(i+1)$-th child among the individuals who have produced $i$ children. In this case, the basic reproduction number is given by:

$$R_0 = \gamma \sum_{i \ge 1} \int_0^\infty \zeta_i(a)\ell(a)da = \gamma \sum_{i \ge 1} i(A_i - A_{i+1}). \tag{2.54}$$

Let $T_i(a)$ be the expected sojourn time for individuals who have entered parity state $i$ $(i \ge 1)$ at age $a$. It follows that

$$\begin{aligned} T_i(a) &= \int_0^\infty s(\mu(a+s) + \lambda_i(a+s,s))\Lambda(s;a,0)\frac{\ell(a+s)}{\ell(a)}ds \\ &= \int_0^\infty \Lambda(s;a,0)\frac{\ell(a+s)}{\ell(a)}ds. \end{aligned}$$

Let us define the expected sojourn time at parity $i$ by

$$T_i := \int_0^\infty T_i(a)\zeta_i(a)\ell(a)da,$$

from which it is easy to see that

$$T_i = \int_0^\infty \Psi_i(a)\ell(a)da, \quad (i \ge 0).$$

Of course, $\sum_{i \ge 0} T_i = \int_0^\infty \ell(a)da$ gives the life expectancy at birth.

Using the survival functions, age–duration-density functions of each state can be calculated from the boundary values as

$$p_0(t,a) = \begin{cases} \Lambda_0(a)\ell(a)B(t-a), & t-a>0, \\ \dfrac{\Lambda_0(a)\ell(a)}{\Lambda_0(a-t)\ell(a-t)}\phi_0(a-t), & a-t>0, \end{cases}$$

$$p_i(t,a,s) = \begin{cases} \Lambda_i(s;a-s,0)\dfrac{\ell(a)}{\ell(a-s)}p_i(t-s,a-s,0), & t-s>0, \\ \Lambda_i(t;a-t,s-t)\dfrac{\ell(a)}{\ell(a-t)}\phi_i(a-t,s-t), & a \ge s \ge t, \end{cases} \tag{2.55}$$

where $B(t) := p_0(t,0)$. For the boundary value $p_i(t,a,0)$, it is easy to see that

$$p_i(t,a,0) = \zeta_i(a)\ell(a)B(t-a), \quad t-a>0.$$

To calculate $p_i(t,a,0)$ in the domain $a-t>0$, we need $g_i^j(t,a)$, which is the age density of women who have parity $i$ at the initial time and bear their $i+j$-th ($j \ge 1$) child at time $t$. $g_i^j(t,a)$ $(a-t>0)$ is calculated iteratively as follows:

$$g_0^j(t,a) = \frac{\zeta_j(a)}{\Lambda_0(a-t)}\phi_0(a-t), \quad j \ge 1,$$

$$g_i^1(t,a) = \int_t^a \lambda_i(a,s)\Lambda_i(t;a-t,s-t)\phi_i(a-t,s-t)ds,$$

$$g_i^{j+1}(t,a) = \int_0^t \lambda_{i+j}(a,s)\Lambda_{i+j}(s;a-s,0)g_i^j(t-s,a-s)ds.$$

By mathematical induction, we can conclude that:

**Lemma 2.2** *The boundary value $p_i(t,a,0)$ with parity $i \ge 1$ is given as*

$$p_i(t,a,0) = \begin{cases} \zeta_i(a)\ell(a)B(t-a), & t-a>0, \\ \left[\displaystyle\sum_{j=0}^{i-1} g_j^{i-j}(t,a)\right]\dfrac{\ell(a)}{\ell(a-t)}, & a-t>0. \end{cases} \tag{2.56}$$

Applying (2.56) to the relation

$$B(t) = \gamma\sum_{i=1}^{\infty}\int_0^{\infty} p_i(t,a,0)da,$$

we arrive at the renewal equation for the parity progression model:

$$B(t) = G(t) + \int_0^t \Phi(a)B(t-a)da, \tag{2.57}$$

where

$$\Phi(a) := \gamma \sum_{i\geq 1} \zeta_i(a)\ell(a), \quad G(t) := \gamma \sum_{j=0}^{i-1} \int_t^\infty g_j^{i-j}(t,a)\frac{\ell(a)}{\ell(a-t)}da.$$

Therefore, as in the stable population model, the strong ergodicity theorem also holds for the parity progression model. From the above consideration, we obtain the following result:

**Proposition 2.7** ([53]) *The solution of the parity progression model* (2.53) *is given as follows:*

$$p_0(t,a) = \begin{cases} \Lambda_0(a)\ell(a)B(t-a), & t-a>0 \\ \dfrac{\Lambda_0(a)\ell(a)}{\Lambda_0(a-t)\ell(a-t)}\phi_0(a-t), & a-t>0 \end{cases}$$

$$p_i(t,a,s) = \begin{cases} \Lambda_i(s;a-s,0)\zeta_i(a-s)\ell(a)B(t-a), & t>a>s, \\ \Lambda_i(s;a-s,0)\left[\sum_{j=0}^{i-1} g_j^{i-j}(t-s,a-s)\right]\dfrac{\ell(a)}{\ell(a-t)}, & a\geq t\geq s. \\ \Lambda_i(t;a-t,s-t)\dfrac{\ell(a)}{\ell(a-t)}\phi_i(a-t,s-t), & a\geq s\geq t, \end{cases}$$

*where* $B(t)$ *is the solution of the renewal equation* (2.57).

**Proposition 2.8** ([53]) *Let* $\lambda_0$ *be the intrinsic growth rate determined by the net reproduction function* $\Phi(a)$*. Then, the following holds uniformly on any finite age interval:*

$$\lim_{t\to\infty} e^{-\lambda_0 t} p_0(t,a) = q_0 e^{-\lambda_0 a}\ell(a)\Lambda_0(a),$$

$$\lim_{t\to\infty} e^{-\lambda_0 t} p_i(t,a,s) = q_0 e^{-\lambda_0 a}\ell(a)\Lambda_i(s;a-s,0)\zeta_i(a-s),$$

*where* $q_0$ *is given by*

$$q_0 = \frac{\int_0^\infty e^{-\lambda_0 t}G(t)dt}{\int_0^\infty a e^{-\lambda_0 a}\Phi(a)da}.$$

Finally, consider the special case in which the parity progression rate $\lambda_i(a,s)$ is age-independent—that is, it depends only on duration and parity status [37]. This type of parity progression model has been widely used in demography, because it depends on fewer parameters, and is therefore easier to apply to real data. We write $\lambda_i(a,s) = \lambda_i(s)$. Moreover, we neglect the age dependency of mortality or simply neglect the natural death rate. Even if we neglect the death rate, the model is still realistic for describing the populations of developed countries, which have very small death rates in the reproductive age classes.

In the following, we neglect the death rate. The survival probability at each parity status is then given by

$$\Lambda_i(s) = \exp\left(-\int_0^a \lambda_i(\rho)d\rho\right).$$

The probability of bearing an $i+1$-th child is given by $f_i(s) := \lambda_i(s)\Lambda_i(s)$, where $s$ denotes the duration since the $i$-th birth. Let us introduce the iterative functions $\zeta_i(a)$ as

$$\zeta_1(a) := f_0(a), \quad \zeta_{i+1}(a) := \int_0^a f_i(s)\zeta_i(a-s)ds, \, (i \geq 1).$$

$\zeta_i(a)$ gives the probability that the $i$-th birth occurs at age $a$. Then,

$$A_i = \int_0^\infty \zeta_i(a)da,$$

and the parity progression ratio is given by

$$\frac{A_{i+1}}{A_i} = \int_0^\infty f_i(s)ds = 1 - \Lambda_i(\infty).$$

In this case, the basic reproduction number is calculated as

$$R_0 = \gamma \sum_{i\geq 1} \int_0^\infty \zeta_i(a)da = \gamma \sum_{i \geq 1} A_i.$$

Let $B_i(t) := \gamma \int_0^\infty p_i(t,a,0)da$ be the number of $i$-th births per unit time. From the boundary condition in (2.53) and changing the order of integrals, we have

$$\begin{aligned}
B_i(t) &= \gamma \int_0^\infty p_i(t,a,0)da \\
&= \gamma \int_0^t \int_0^a \lambda_{i-1}(s)p_{i-1}(t,a,s)dsda + \gamma \int_t^\infty \int_0^a \lambda_{i-1}(s)p_{i-1}(t,a,s)dsda \\
&= \gamma \left[\int_0^t ds \int_s^t da + \int_0^t ds \int_t^\infty da + \int_t^\infty ds \int_s^\infty da\right] \lambda_{i-1}(s)p_{i-1}(t,a,s) \\
&= \gamma \int_0^t \lambda_{i-1}(s)\Lambda_{i-1}(s) \int_s^\infty p_{i-1}(t-s,a-s,0)dads + G_{i-1}(t) \\
&= \int_0^t f_{i-1}(s)B_{i-1}(t-s)ds + G_{i-1}(t),
\end{aligned}$$

where $B_0(t) := B(t)$ and

$$G_i(t) := \gamma \int_t^\infty \int_s^\infty f_i(s)\phi_i(a-t,s-t)dads.$$

$G_i(t)$ denotes the number of $i$-th births produced by the initial population at time $t$. Therefore, we arrive at the following renewal equation system:

$$B(t) = \sum_{i \geq 1} B_i(t),$$

$$B_i(t) = \int_0^t f_{i-1}(s) B_{i-1}(t-s) ds + G_{i-1}(t), \quad (i \geq 1).$$

Instead of giving the age–duration-density of the population at time $t = 0$ as initial data, if the past data for birth $B_i(t)$, $t < 0$ are given, we obtain the impressively simple system of renewal equations

$$B_i(t) = \int_0^\infty f_{i-1}(s) B_{i-1}(t-s) ds, \quad (i \geq 1).$$

To determine $B_i(t)$ for $t > 0$, we only need the (duration-dependent) parity progression rate $f_i(s)$. The parity progression model plays an important role in fertility analysis because it provides a theoretical framework from which to examine the effects of a delay or advance in the timing of reproduction and parity-dependent fertility decline in modern societies.

## 2.5 Growth and Diffusion in Continuous State Spaces

In the multistate population models considered so far, each state (trait) of the individuals is described by a finite set of natural numbers. We now consider the case in which individual states are indicated by a continuous (vector) parameter $\xi$, and the domain of $\xi$ (trait space) is a subset of $\mathbb{R}^n$, denoted by $\Omega$. The continuous state space may correspond to geographical position, social status, genetic traits, or any kind of individual heterogeneity. We present a formal exposition of how to induce the basic equation.

### 2.5.1 *McKendrick Equation with an Additional Structure*

First, we consider the survival process of individuals in the trait space $\Omega$. Let $\ell(a, x)$ be the survival probability of an individual at age $a$ and state $x \in \Omega$ under an initial distribution $\ell(0, \cdot)$ such that

$$\int_\Omega \ell(0, x) dx = 1.$$

Let $q(a; x, y)$ be the force of transition[5] from state $y \in \Omega$ to state $x \in \Omega$ at age $a$. Then, we obtain the *master equation*

$$\frac{\partial \ell(a,x)}{\partial a} = -\mu(a,x)\ell(a,x) + \int_\Omega [q(a;x,\eta)\ell(a,\eta) - q(a;\eta,x)\ell(a,x)]d\eta, \tag{2.58}$$

where $\mu(a, x)$ denotes the force of mortality at age $a$ and state $x$. Observe that

$$\frac{d}{da}\int_\Omega \ell(a,x)dx = -\int_\Omega \mu(a,x)\ell(a,x)dx.$$

Thus, we have

$$0 \le \int_\Omega \ell(a,x)dx \le 1$$

for all $a \ge 0$ if $\mu \in L^\infty(\mathbb{R}_+ \times \Omega)$.

The solution of the master equation can be expressed using a Green's function (transition operator) $W$ as

$$\ell(a,x) = \int_\Omega W(a,x|\sigma,\zeta)\ell(\sigma,\zeta)d\zeta, \tag{2.59}$$

where $W(a, x|\sigma, y)$ satisfies

$$\begin{aligned}\frac{\partial W(a,x|\sigma,\zeta)}{\partial a} &= -\mu(a,x)W(a,x|\sigma,\zeta)\\ &\quad + \int_\Omega [q(a;x,\eta)W(a,\eta|\sigma,\zeta) - q(a;\eta,x)W(a,x|\sigma,\zeta)]d\eta, \end{aligned} \tag{2.60}$$
$$\lim_{a\downarrow\sigma} W(a,x|\sigma,\zeta) = \delta(x-\zeta).$$

From (2.59), if $\ell(0, x) = \delta(x - \zeta)$, we have $\ell(a, x) = W(a, x|0, \zeta)$. Biologically speaking, $W(a, x|0, \zeta)$ is the survival probability for an individual born at $\zeta$ reaching age $a$ and state $x$.

Define a two-parameter family of operators $\{L(a, \sigma) : a \ge \sigma\}$ on $L^1(\Omega)$ as a solution operator of (2.58) such that $\ell(a, \cdot) = L(a, \sigma)\ell(\sigma, \cdot)$, that is,

$$(L(a,\sigma)f)(x) := \int_\Omega W(a,x|\sigma,\xi)f(\xi)d\xi. \tag{2.61}$$

Then, $L(a, \sigma)$, $a \ge \sigma$, forms an evolutionary system and $L(a, \sigma)$ maps the state-specific density function at age $\sigma$ to the density at age $a$, that is, $L(a, \sigma)$ describes the aging and state transition process in a birth cohort, so it is an infinite-dimensional

[5] That is, we only deal with the case in which the transition measure is absolutely continuous with density $q$. A more general formulation can be found in [33].

analog of the survival probability (transition) matrix introduced in Sect. 2.1. If an individual's traits do not change throughout their lifetime, there is no migration process in $\Omega$, and the survival probability at each state $x$ is simply given by

$$\ell(a,x) = e^{-\int_0^a \mu(\sigma,x)d\sigma}\ell(0,x).$$

In this case, the trait-specific survival probability is given by $e^{-\int_0^a \mu(\sigma,x)d\sigma}$, which is an analog of the survival probability $\ell(a)$ in the scalar model.

Under appropriate conditions, the master equation can be reduced to a reaction–diffusion equation. For example, let us consider a one-dimensional case such that $\Omega = \mathbb{R}$, and assume that the transition intensity can be decomposed as

$$q(a;x,y) = v(a,y)\phi(x-y),$$

where $\phi$ is the probability density function

$$\int_\Omega \phi(x)dx = 1.$$

Then, we have

$$\int_\Omega q(a;x,\eta)\ell(a,\eta)d\eta = \int_\Omega v(a,x-\xi)\ell(a,x-\xi)\phi(\xi)d\xi,$$
$$\int_\Omega q(a;\eta,x)\ell(a,x)d\eta = v(a,x)\ell(a,x).$$

A Taylor expansion gives

$$v(a,x-\xi)\ell(a,x-\xi) = v(a,x)\ell(a,x) - \xi\frac{\partial}{\partial x}v(a,x)\ell(a,x) + \frac{\xi^2}{2}\frac{\partial^2}{\partial x^2}v(a,x)\ell(a,x) + O(\xi^3).$$

Neglecting the higher-order terms, we obtain

$$\int_\Omega q(a;x,\eta)\ell(a,\eta)d\eta \sim v(a,x)\ell(a,x) - \langle\xi\rangle\frac{\partial}{\partial x}v(a,x)\ell(a,x) + \frac{\langle\xi^2\rangle}{2}\frac{\partial^2}{\partial x^2}v(a,x)\ell(a,x),$$

where

$$\langle\xi^n\rangle := \int_\Omega \xi^n\phi(\xi)d\xi.$$

Inserting the above expression into (2.58), we have

$$\frac{\partial \ell(a,x)}{\partial a} = -\mu(a,x)\ell(a,x) - \langle \xi \rangle \frac{\partial}{\partial x} \nu(a,x)\ell(a,x) + \frac{\langle \xi^2 \rangle}{2} \frac{\partial^2}{\partial x^2} \nu(a,x)\ell(a,x), \tag{2.62}$$

which is a *Fokker–Planck equation* or a *forward Kolmogorov equation* [86].

The Fokker–Planck equation can be written as a reaction–diffusion equation:

$$\frac{\partial \ell(a,x)}{\partial a} = -\mu(a,x)\ell(a,x) + \mathscr{L}(a)\ell(a,x), \tag{2.63}$$

where $\mathscr{L}(a)$ is a differential operator given by

$$\mathscr{L}(a) := -\frac{\partial}{\partial x} C(a,x) + \frac{\partial}{\partial x}\left( D(a,x) \frac{\partial}{\partial x} \right),$$

$$C(a,x) := \langle \xi \rangle \nu(a,x) - \frac{\langle \xi^2 \rangle}{2} \frac{\partial \nu(a,x)}{\partial x}, \quad D(a,x) := \frac{\langle \xi^2 \rangle}{2} \nu(a,x).$$

Conversely, if we start from a reaction–diffusion equation describing the survival process, $W(a,x|\sigma,\zeta)$ will be obtained as its fundamental solution such that

$$\begin{aligned} \frac{\partial W(a,x|\sigma,\zeta)}{\partial a} &= (-\mu(a,x) + \mathscr{L}(a)) W(a,x|\sigma,\zeta), \\ \lim_{a \downarrow \sigma} W(a,x|\sigma,\zeta) &= \delta(x-\zeta). \end{aligned} \tag{2.64}$$

Next, let $\beta(a,x,\xi)$ be the fertility function describing whether an individual at age $a$ and state $\xi \in \Omega$ produces a newborn of state $x \in \Omega$, and let

$$\psi(a,x,\xi) := \int_\Omega \beta(a,x,\eta) W(a,\eta|0,\xi) d\eta \tag{2.65}$$

be the *net reproduction function* depending on the birth state $\xi \in \Omega$.

We define a one-parameter operator $\Psi(a)$ on $L^1(\Omega)$, called the *net reproduction operator*, by

$$\begin{aligned} (\Psi(a)f)(x) :&= \int_\Omega \psi(a,x,\xi) f(\xi) d\xi \\ &= \int_\Omega \int_\Omega \beta(a,x,\eta) W(a,\eta|0,\xi) d\eta f(\xi) d\xi \\ &= \int_\Omega \beta(a,x,\eta)(L(a,0)f)(\eta) d\eta. \end{aligned} \tag{2.66}$$

$\Psi(a)$ maps the state vector $f$ of newborns to the state vector of their own newborns at age $a$.

Let $b(t,x)$ be the density of newborns of state $x$ at time $t$. It follows that

$$b(t,x)=\int_0^\infty\int_\Omega \beta(a,x,\xi)p(t,a,\xi)d\xi da, \tag{2.67}$$

where $p$ is the population density at age $a$ and state $\xi$. The population density is given by:

$$p(t,a,\xi)=\begin{cases}(L(a,0)b(t-a,\cdot))(\xi), & t-a>0,\\ (L(a,a-t)p(0,a-t,\cdot))(\xi), & a-t>0.\end{cases} \tag{2.68}$$

Inserting the above expression into (2.67), we obtain the renewal equation

$$b(t,x)=g(t,x)+\int_0^t\int_\Omega \psi(a,x,\xi)b(t-a,\xi)d\xi da, \tag{2.69}$$

where

$$g(t,x):=\int_t^\infty da\int_\Omega d\eta\int_\Omega d\xi\beta(a,x,\xi)W(a,\xi|a-t,\eta)p(0,a-t,\eta).$$

If we consider $b(t):=b(t,\cdot)$ as a vector-valued function in the state space $L^1(\Omega)$, (2.69) can be written as an abstract renewal integral equation in $L^1(\Omega)$:

$$b(t)=g(t)+\int_0^t \Psi(a)b(t-a)da. \tag{2.70}$$

Let us consider $W(a,x|\sigma,\zeta)$ to be given by the fundamental solution of the reaction–diffusion equation (2.63). For $t>a$, we have

$$p(t,a,x)=\int_\Omega W(a,x|0,\zeta)b(t-a,\zeta)d\zeta.$$

Observe that

$$\begin{aligned}\left(\frac{\partial}{\partial t}+\frac{\partial}{\partial a}\right)p(t,a,x)&=\int_\Omega \frac{\partial W(a,x|0,\zeta)}{\partial a}b(t-a,\zeta)d\zeta\\ &=\int_\Omega(-\mu(a,x)+\mathscr{L}(a))W(a,x|0,\zeta)b(t-a,\zeta)d\zeta\\ &=-\mu(a,x)p(t,a,x)+(\mathscr{L}(a)p(t,\cdot,\cdot))(a,x).\end{aligned}$$

Therefore, we obtain an age-dependent population problem as

$$\begin{aligned}\left(\frac{\partial}{\partial t}+\frac{\partial}{\partial a}\right)p(t,a,x)&=-\mu(a,x)p(t,a,x)+\mathscr{L}(a)p(t,a,x),\\ p(t,0,x)&=\int_0^\infty\int_\Omega \beta(a,x,\xi)p(t,a,\xi)d\xi da,\end{aligned} \tag{2.71}$$

which is the partial differential equation (PDE) formulation corresponding to the renewal equation (2.69). This type of PDE model (age-structured population models with an additional structure) has been studied by several authors, such as Tucker and Zimmerman [105] and Thieme [101, 102].

*Remark 2.5* Usually the basic Eq. (2.71) can be simply induced as follows: Let $\Omega_0$ be a subset with a smooth boundary $\partial\Omega_0$ in $\Omega$. Let $J(a,x)$ be the population flux vector and let $\mathbf{n}$ be the outward normal at $(a,x)$ on $\Omega_0$. Then, the outward flux at $(a,x)$ is given by $\mathbf{n}\cdot J(a,x)$, and the population balance on a cohort is expressed by

$$\left(\frac{\partial}{\partial t}+\frac{\partial}{\partial a}\right)\int_{\Omega_0} p(t,a,x)dx+\int_{\partial\Omega_0}\mathbf{n}\cdot Jd\sigma=-\int_{\Omega_0}\mu(a,x)p(t,a,x)dx,$$

where $\sigma$ denotes the Lebesgue measure on $\partial\Omega_0$. From the Gauss divergence theorem, we have

$$\int_{\partial\Omega_0}\mathbf{n}\cdot Jd\sigma=\int_{\Omega_0}\nabla Jdx.$$

Therefore, it follows that

$$\int_{\Omega_0}\left[\frac{\partial p(t,a,x)}{\partial t}+\frac{\partial p(t,a,x)}{\partial a}+\nabla J(a,x)+\mu(a,x)p(t,a,x)\right]dx=0.$$

Because $\Omega_0$ is arbitrary, we obtain

$$\frac{\partial p(t,a,x)}{\partial t}+\frac{\partial p(t,a,x)}{\partial a}+\nabla J(a,x)+\mu(a,x)p(t,a,x)=0.$$

The population flux is given by the product of density and velocity as $J=\mathbf{v}\cdot p$, so we arrive at (2.71).

As shown in Chap. 10, the asymptotic analysis in Chap. 1 for a scalar-type Volterra integral equation can be extended to the infinite-dimensional case of (2.70) in the same manner as long as the Laplace transformation $\hat{\Psi}(\lambda)$ is a non-supporting compact operator for $\lambda \in \mathbb{R}$.[6] That is, there exists a dominant characteristic root (Malthusian parameter) $\lambda_0$ and a number $\eta > 0$ such that

$$b(t)=g(t)+e^{\lambda_0 t}\left[\frac{\langle v_0,\hat{g}(\lambda_0)\rangle}{\langle v_0,\Psi_1 u_0\rangle}u_0+O(e^{-\eta t})\right], \tag{2.72}$$

where $v_0$ and $u_0$ are the positive eigenfunctional and the positive eigenfunction of the positive linear operator $\hat{\Psi}(\lambda_0)$ associated with the eigenvalue unity and

[6] Readers can find early attempts to analyze the abstract renewal equation (2.70) in [46, 77].

$$\Psi_1 = -\left.\frac{d}{d\lambda}\hat{\Psi}(\lambda)\right|_{\lambda=\lambda_0}.$$

Moreover, if we define

$$K := \int_0^\infty \Psi(a)da, \tag{2.73}$$

it follows that

$$\text{sign}(\lambda_0) = \text{sign}(r(K) - 1), \tag{2.74}$$

where $r(K)$ denotes the spectral radius of $K$. Therefore, it is reasonable to define $K$ as the *next-generation operator* (NGO), and its spectral radius $r(K)$ is the *basic reproduction number* (see Chap. 9).

Note that the NGO is expressed as an integral operator on $L^1(\Omega)$. In fact, it follows from (2.66) that

$$(Kf)(x) = \int_0^\infty da \int_\Omega \psi(a, x, \xi) f(\xi) d\xi = \int_\Omega \left[\int_0^\infty \psi(a, x, \xi) da\right] f(\xi) d\xi.$$

Therefore, we obtain

$$(Kf)(x) = \int_\Omega k(x, \xi) f(\xi) d\xi, \tag{2.75}$$

where

$$k(x, \xi) := \int_0^\infty \psi(a, x, \xi) da.$$

Biologically speaking, $k(x, \xi)$ is the total number of newborns at location $x$ produced by an individual born at $\xi$.

If the vital parameters are age-independent (non-age-structured population models), the survival rate $L(a, 0)$ is given by a strongly continuous semigroup $T(a) = e^{Qa}$ with a generator $Q$ whose growth bound is negative, that is, $\omega(Q) < 0$ and $\beta$ is a function of only $x$, and we have

$$K = \int_0^\infty MT(a)da = M(-Q)^{-1}, \tag{2.76}$$

where $M$ is a multiplication operator on $L^1(\Omega)$ defined by $(Mf)(x) := \beta(x)f(x)$ and $(Qf)(x) = -\mu(x)f(x) + (\mathscr{L}f)(x)$. In fact, if we use the well-known fact that $0 \in \rho(Q)$ and the resolvent operator is given by a Laplace transform of $T(a)$,

$$(\lambda - Q)^{-1} = \int_0^\infty e^{-\lambda a} T(a)da, \quad \lambda \in \rho(Q),$$

where $\rho(Q)$ denotes the resolvent set of $Q$, we have (2.76). The expression in (2.76) is an infinite-dimensional analog of the familiar framework for calculating the NGO (see Chaps. 9 and 10).

Note that the operator $(-Q)^{-1}$ is a positive integral operator on $L^1(\Omega)$ that gives the expected sojourn time at state $x$ for a typical individual. In fact, it follows from (2.61) that

$$\int_0^\infty (L(a,0)f)(x)da = \int_0^\infty da \int_\Omega W(a,x|0,\xi)f(\xi)d\xi$$
$$= \int_\Omega \left[\int_0^\infty W(a,x|0,\xi)da\right] f(\xi)d\xi,$$

so we obtain the expression

$$\int_0^\infty (L(a,0)f)(x)da = ((-Q)^{-1}f)(x) = \int_\Omega E(x,\xi)f(\xi)d\xi, \tag{2.77}$$

where

$$E(x,\xi) := \int_0^\infty W(a,x|0,\xi)da,$$

which gives the sojourn time at state $x$ for an individual born in state $\xi$.

Integrating the reaction–diffusion equation (2.64) with age-independent parameters from zero to $\infty$ with respect to $a$, we have

$$-\delta(x-\zeta) = (-\mu(x) + \mathscr{L})E(x,\zeta), \tag{2.78}$$

where

$$\mathscr{L} = -\frac{\partial}{\partial x}C(x) + \frac{\partial}{\partial x}\left(D(x)\frac{\partial}{\partial x}\right),$$

which is a differential equation satisfied by the integral kernel $E(x,\zeta)$. Using $E(a,\zeta)$, the NGO is given by:

$$(Kf)(x) = \beta(x)\int_\Omega E(x,\zeta)f(\zeta)d\zeta. \tag{2.79}$$

Because (2.78) is an ordinary differential equation, we can obtain the NGO without solving the master equation if the vital parameters are age-independent [79].

The age-dependent diffusion model was proposed by Gurtin [40] and subsequently investigated by many authors from the late 1970s to the 1990s [67, 77, 110]. Readers should note that the precise definition of $R_0$ for a structured population was first formulated by Diekmann et al. in 1990 [32]. Therefore, early studies on the age-dependent diffusion dynamics could not use the idea of $R_0$ and its threshold property.

Recently, some authors have incorporated $R_0$ into the study of reaction–diffusion models [1, 79, 88, 109].

### 2.5.2 Traveling Wave Solutions

An interesting phenomenon other than the asymptotic behavior in the renewal system (2.69) is the existence of *traveling wave solutions*. In fact, if the NGO $K$ is not a compact operator, our argument in the previous subsection does not necessarily hold, and we do not always expect the exponential invasion mode.

For simplicity, we assume that individuals do not change their traits during their lifetime, but newborns can exhibit different traits from their mothers, and the transmission rate from mother $\xi$ to daughter $x$ depends only on their distance $x - \xi$. Moreover, the survival process is independent of traits. That is, we assume that the net reproduction function is given as

$$\psi(a, x, \xi) = \beta_0(a)V(x - \xi)\ell_0(a) = \phi(a)V(x - \xi),$$

where $\ell_0$ is the survival rate, $\phi(a) := \beta_0(a)\ell_0(a)$ is a trait-independent net reproduction function, and $V(x)$ is a trait transmission probability from mother to daughter satisfying

$$\int_\Omega V(x)dx = 1, \quad V(x) = V(-x).$$

The basic reproduction number is then given by

$$R_0 = \int_0^\infty \phi(a)da.$$

The renewal equation (2.69) becomes as follows:

$$b(t, x) = \int_0^\infty \phi(a) \int_\Omega V(x - \xi)b(t - a, \xi)d\xi da, \tag{2.80}$$

where we assume that the initial population has died out.[7]

Inserting a trial solution $b(t, x) = w(x + ct)$ into (2.80), we have

$$w(x) = \int_{-\infty}^\infty V_c(x - \xi)w(\xi)d\xi, \tag{2.81}$$

where

$$V_c(z) := \int_0^\infty \phi(a)V(z - ca)da.$$

[7] A more general case is studied in [103].

From the symmetry of $V$, if $w(x)$ is a solution of (2.81), then $w(-x)$ becomes a solution of (2.81) by replacing $V_c$ with $V_{-c}$. Because our argument can proceed symmetrically, we assume, without loss of generality, that $c > 0$, that is, we only consider the traveling wave to the left-hand side.

Note that the characteristic equation of (2.81) is given by:

$$\hat{V}_c(\lambda) := \int_{-\infty}^{\infty} e^{-\lambda z} V_c(z) dz = 1, \tag{2.82}$$

where $\hat{V}_c(0) = R_0$. If the characteristic equation has a real root $\lambda_0$ with multiplicity $k \geq 1$, then $z^n e^{\lambda_0 z}$ $(n = 0, 1, \ldots, k-1)$ become real solutions of (2.81). Therefore, there exists a traveling wave solution if there is some $c > 0$ such that the characteristic equation has a real root. Suppose that $\hat{V}_c(\lambda)$ is defined in the right neighborhood of $\lambda = 0$, and $\hat{V}_c(0) = R_0 > 1$. Because

$$\hat{V}_c(\lambda) = \int_0^{\infty} e^{-\lambda c a} \phi(a) da \int_{-\infty}^{\infty} e^{-\lambda z} V(z) dz,$$

we obtain

$$\frac{d\hat{V}_c(0+)}{d\lambda} = -c \int_0^{\infty} a\phi(a) da < 0,$$

$$\frac{d^2 \hat{V}_c(\lambda)}{d\lambda^2} = \int_{-\infty}^{\infty} z^2 e^{-\lambda z} V_c(z) dz > 0.$$

As the graph of $\hat{V}_c(\lambda)$ is downwardly convex for $\lambda > 0$ and $\hat{V}_c(\lambda)$ is monotone decreasing for $c > 0$, for a sufficiently large $c > 0$, there must exist some $\lambda > 0$ such that $\hat{V}_c(\lambda) < 1$. Because $\hat{V}_c(0) > 1$, the characteristic equation (2.82) has a positive root. Consider the set defined by $\{c > 0 |$ there exists $\lambda > 0$ such that $\hat{V}_c(\lambda) < 1\}$. This set becomes a half line $(c_0, \infty)$. Then, if $R_0 > 1$, there exists a traveling wave solution for $c > c_0$ [31].

## 2.6 Ergodicity Theorems for Non-autonomous Systems

Although we have so far assumed time-independent vital rates for the stable population model and its linear extensions, this assumption is at best only approximately satisfied during a limited time interval in real-world applications. In fact, the essence of the autonomous models is to reveal the *potential power* of vital rates—they are not descriptive but normative, especially for modern populations. The vital rates of biological populations are always varying under socioeconomic and environmental changes, or as a result of interventions or controls.

We now examine the characteristics of the age profile dynamics for time-dependent vital rates. However, we do not consider any nonlinear interaction among the population parameters, but generally consider a linear one-sex population model with time-dependent vital rates. For simplicity, we consider only scalar models, although the main results can easily be extended to multistate or vector-valued models under appropriate additional assumptions.

A fundamental principle of non-autonomous positive linear evolutionary systems is *weak ergodicity*, which implies the asymptotic independence of the system state from its initial state. Although the idea of weak ergodicity can be traced back to the work of Norton [23, 85], our formulation essentially depends on the general theory of linear multiplicative processes that was mainly developed by Birkhoff in the context of lattice theory [18]. For further details, readers are referred to Chap. 10, [52, 69, 95].

For non-autonomous weakly ergodic population systems, we cannot generally expect a time-independent asymptotic age structure to exist. However, there are at least two exceptional cases, namely the asymptotically autonomous case and the time periodic case. We consider these two special cases later in this chapter.

### 2.6.1 Primary System and Ergodicity

A closed one-sex population with time-dependent vital rates is described by a boundary value problem of the non-autonomous McKendrick equation:

$$\begin{aligned}
&\frac{\partial p(t,a)}{\partial t} + \frac{\partial p(t,a)}{\partial a} = -\mu(t,a)p(t,a), \quad t \in \mathbb{R}, \quad 0 < a < \omega, \\
&p(t,0) = \int_0^\omega \beta(t,a)p(t,a)da, \quad t \in \mathbb{R}.
\end{aligned} \tag{2.83}$$

We consider the above system for $t \in \mathbb{R}$ and assume that $\omega < \infty$ is not the maximum attainable age, but the (time-independent) least upper bound of reproductive age. Thus, we can assume that for any given $t \in \mathbb{R}$, $\mu(t, \cdot), \beta(t, \cdot) \in L^\infty_+(0, \omega)$.

We define the cohort survival probability by

$$\ell(h; t, a) := \exp\left(-\int_0^h \mu(t+\sigma, a+\sigma)d\sigma\right), \quad h \geq 0.$$

Hence, $\ell(h; t, a)$ is the probability that individuals at time $t$ and age $a$ survive to age $a + h$ and time $t + h$. If we consider an age distribution $p(s, a) = \phi(a)$ at time $s$ as the initial data for system (2.83) and integrate the non-autonomous McKendrick equation (2.83) along characteristic lines for $t > s$, we obtain

$$p(t,a) = \begin{cases} \ell(a; t-a, 0)p(t-a, 0), & t-s > a, \\ \ell(t-s; s, a-t+s)\phi(a-t+s), & a > t-s. \end{cases} \tag{2.84}$$

Inserting expression (2.84) into the boundary condition in (2.83), we have

$$p(t,0) = \begin{cases} \int_0^{t-s} \beta(t,a)\ell(a; t-a, 0)p(t-a, 0)da \\ \quad + \int_{t-s}^{\omega} \beta(t,a)\ell(t-s; s, a-t+s)\phi(a-t+s)da, & t-s < \omega, \\ \int_0^{\omega} \beta(t,a)\ell(a; t-a, 0)p(t-a, 0)da, & t-s > \omega. \end{cases} \tag{2.85}$$

We now introduce the following functions for $h = t - s > 0$:

$$B(h; \phi, s) := p(s+h, 0),$$

$$\Psi(s+h, a) := \begin{cases} \beta(s+h, a)\ell(a; s+h-a, 0), & 0 < a < \omega, \\ 0, & a > \omega, \end{cases}$$

$$G(h; \phi, s) := \int_h^{\omega \vee h} \beta(s+h, a)\ell(h; s, a-h)\phi(a-h)da,$$

where $\Psi(t,a) = \beta(t,a)\ell(a; t-a, 0)$ is the net reproduction function for a cohort born at time $t - a$.

From (2.85), we know that $B(h; \phi, s)$ satisfies the following Volterra integral equation:

$$B(h; \phi, s) = G(h; \phi, s) + \int_0^h \Psi(s+h, a)B(h-a; \phi, s)da, \quad h > 0. \tag{2.86}$$

Let $R(h, x; s)$ be the resolvent kernel corresponding to the integral kernel $\Psi(h + s, h - x)$. That is, $R(h, x; s)$ is the solution of the resolvent equation

$$R(h, x; s) = \Psi(h+s, h-x) + \int_x^h \Psi(h+s, h-z)R(z, x; s)dz. \tag{2.87}$$

Then, (2.86) can be solved as follows:

$$B(h; \phi, s) = G(h; \phi, s) + \int_0^h R(h, \rho; s)G(\rho; \phi, s)d\rho. \tag{2.88}$$

Using the solution $B(h; \phi, s)$, the age distribution for $t > s$ is expressed as follows:

$$p(t,a) = \begin{cases} \ell(a; t-a, 0)B(t-s-a; \phi, s), & t-s > a, \\ \ell(t-s; s, a-t+s)\phi(a-t+s), & a > t-s, \end{cases} \tag{2.89}$$

where $p(s,a) = \phi(a)$.

To consider the time evolution problem of the non-autonomous system (2.83), let us introduce the *population evolution operator*, which is a time evolution operator acting on the state space of the age distributions $L^1(0,\omega)$ as follows:

$$(U(t,s)\phi)(a) := \begin{cases} \ell(a; t-a, 0)B(t-s-a; \phi, s), & t-s > a, \\ \ell(t-s; s, a-t+s)\phi(a-t+s), & a > t-s, \end{cases} \tag{2.90}$$

where $\phi \in L^1(0,\omega)$ and $B(h;\phi,s)$ are given by (2.88). Thus, $U(t,s)$, $t \ge s$, forms a positive exponentially bounded evolutionary family (multiplicative process) on $L^1_+(0,\omega)$:

**Definition 2.1** ([24]) On a Banach space $E$, a family of bounded linear operators $\{U(a,s)\}_{a \ge s}$ with $a, s \in \mathbb{R}$ or $a, s \in \mathbb{R}_+$ is called an exponentially bounded evolutionary family if

1. $U(a,s)U(s,\tau) = U(a,\tau)$ and $U(a,a) = I$ for all $a \ge s \ge \tau$,
2. for each $\phi \in E$, the function $(a,s) \to U(a,s)\phi$ is continuous for $a \ge s$,
3. there exist constants $M \ge 1$ and $\omega_0 > 0$ such that $\|U(a,s)\| \le Me^{\omega_0(a-s)}$ for all $a \ge s$.

Readers are referred to [25, 54, 101] for generation theorems regarding the evolutionary system based on the differential operator on $L^1_+(0,\omega)$. We can also apply the Z-space method [29] to obtain $U(t,s)$ as a continuous solution of the extended variation-of-constants formula. As shown in Chap. 1, for a given $\phi \in L^1(0,\omega)$, $U(t,s)\phi$ does not necessarily give a classical solution of the PDE system (2.83) with initial data $\phi$ at time $s$. However, if we replace $\frac{\partial}{\partial t} + \frac{\partial}{\partial a}$ by the directional derivative $D$ along the characteristic line defined by (1.37), $p(t) = U(t,s)\phi$ becomes a solution in the weak sense that $p$ satisfies $Dp(t,a) = -\mu(t,a)p(t,a)$ for almost all $(t,a) \in \mathbb{R} \times [0,\omega]$, $p(t,0) = \int_0^\omega \beta(t,a)p(t,a)da$ for $t > s$, $p(s,\cdot) = \phi(\cdot)$ and $p \in C_+([s,\infty); L^1(0,\omega))$.

Let $f : \mathbb{R} \to L^1$ be a function that takes values in the state space of age-density functions. The time series $\{f(t)\}_{t\in\mathbb{R}}$ is said to be *consistent* with the evolution system $\{U(t,s)\}_{s\le t}$ in $\Omega \subset \mathbb{R}$ if $f(t) = U(t,s)f(s)$ for all $s, t \in \Omega$ such that $s \le t$. The two-parameter family of non-negative linear operators $\{U(t,s)\}_{s\le t}$ has been extensively studied by Birkhoff in the theory of *multiplicative processes*. Here, we apply Birkhoff's theory to the population evolution operators. The key idea is a classical result given by Norton [85]. To formulate Norton's result, we adopt the following additional assumption for the vital rates:

**Assumption 2.9** (1) The vital rates $\mu$ and $\beta$ are uniformly bounded above, that is,

$$\bar{\mu} := \sup_{(t,a)\in\mathbb{R}\times[0,\omega]} \mu(t,a) < \infty, \quad \bar{\beta} := \sup_{(t,a)\in\mathbb{R}\times[0,\omega]} \beta(t,a) < \infty.$$

(2) There exist positive numbers $\varepsilon > 0$, $0 < \gamma_1 < \gamma_2 < \omega$ such that

$$\inf_{(t,a)\in\mathbb{R}\times[\gamma_1,\gamma_2]} \beta(t,a) > \varepsilon.$$

(3) For almost all $(s,x) \in \mathbb{R}\times[0,\omega]$, it follows that

$$\int_0^\omega \beta(s+z, x+z)\ell(x+z; s-x, 0)dz > 0.$$

Assumption (2) implies that there exists a "core" time-independent reproductive age period, which is not unnatural for real populations. Assumption (3) is satisfied if we assume that $\omega$ is the time-independent least upper bound of reproductive age, so the fertility function $\beta(t,a)$ is not zero in the left neighborhood of $a = \omega$ for any time $t$. In such a case, considering the age-density function on $[0,\omega]$ is sufficient to determine the whole population dynamics, because the post-reproductive age period is determined by the population on the age interval $[0,\omega]$. If condition (3) is satisfied, any non-zero population distribution on $[0,\omega]$ is *non-trivial*,[8] so the population does not become extinct within a finite time.

**Lemma 2.3** ([85]) *Under Assumption 2.9, for any $\delta$ such that $0 < \delta < (\gamma_2 - \gamma_1)/2$ and for any non-negative integers $n = 0, 1, 2, \ldots$, it follows that*

$$R(h,x;s) \ge \alpha^{n+1}\delta^n, \quad h - x \in [\gamma_1 + n(\gamma_1+\delta), \gamma_2 + n(\gamma_2 - \delta)], \tag{2.91}$$

*where $\alpha := \varepsilon \exp(-\bar{\mu}\gamma_2)$.*

*Proof* From the resolvent equation (2.87), if $h - x \in [\gamma_1, \gamma_2]$, we have

$$R(h,x;s) \ge \Psi(h+s, h-x) \ge \varepsilon e^{-\bar{\mu}\gamma_2} = \alpha.$$

Thus, (2.91) holds for $n = 0$. Suppose that (2.91) holds for some non-negative integer $n_0$, that is,

$$R(h,x;s) \ge \alpha^{n_0+1}\delta^{n_0}, \quad h - x \in [\gamma_1 + n_0(\gamma_1+\delta), \gamma_2 + n_0(\gamma_2-\delta)].$$

[8] A population age distribution is called *trivial* if it cannot produce any offspring.

First, we assume that $x \le h - \gamma_2$. Then,

$$\begin{aligned} R(h, x; s) &\ge \int_x^h \Psi(h+s, h-z) R(z, x; s) dz \\ &\ge \int_{h-\gamma_2}^{h-\gamma_1} \Psi(h+s, h-z) R(z, x; s) dz. \end{aligned} \tag{2.92}$$

Let $J$ be an interval given by

$$J := [h - \gamma_2, h - \gamma_1] \cap [x + \gamma_1 + n_0(\gamma_1 + \delta), x + \gamma_2 + n_0(\gamma_2 - \delta)].$$

From our assumption, the intervals $[h - \gamma_2, h - \gamma_1]$ and $[x + \gamma_1 + n_0(\gamma_1 + \delta), x + \gamma_2 + n_0(\gamma_2 - \delta)]$ are longer than $2\delta$. Therefore, if

$$h - x \in [\gamma_1 + (n_0 + 1)(\gamma_1 + \delta), \gamma_2 + (n_0 + 1)(\gamma_2 - \delta)], \tag{2.93}$$

it follows that

$$h - \gamma_1 - (x + \gamma_1 + n_0(\gamma_1 + \delta)) \ge \delta, \quad (x + \gamma_2 + n_0(\gamma_2 - \delta)) - (h - \gamma_2) \ge \delta.$$

Thus, the interval $J$ is not empty and its length is greater than $\delta$. Therefore, if (2.93) holds, it follows from (2.87) and the inductive assumption that

$$\begin{aligned} R(h, x; s) &\ge \int_{h-\gamma_2}^{h-\gamma_1} \Psi(h+s, h-z) R(z, x; s) dz \\ &\ge \int_J \Psi(h+s, h-z) R(z, x; s) dz \\ &\ge \alpha \alpha^{n_0+1} \delta^{n_0} \delta = \alpha^{n_0+2} \delta^{n_0+1}. \end{aligned}$$

Next, suppose that $h - \gamma_2 < x$. If we let

$$K := [x, h - \gamma_1] \cap [x + \gamma_1 + n_0(\gamma_1 + \delta), x + \gamma_2 + n_0(\gamma_2 - \delta)],$$

then the interval $K$ is longer than $\delta$ and

$$R(h, x; s) \ge \int_K \Psi(h+s, h-z) R(z, x; s) dz \ge \alpha^{n_0+2} \delta^{n_0+1}.$$

Therefore, (2.91) holds for $n = n_0 + 1$. Mathematical induction then implies that (2.91) holds for any natural number. □

**Lemma 2.4** *Under Assumption 2.9, there exists a positive number $\zeta > 0$ such that the population evolution operator $U(t, s)$ is uniformly positive for $t - s \ge \zeta$ in the sense of Birkhoff.*

*Proof* From the resolvent equation (2.87), we have

$$R(h, x; s) \leq \bar{\beta} + \bar{\beta} \int_x^h R(z, x; s) dz.$$

Therefore, it follows that

$$R(h, x; s) \leq \bar{\beta} e^{\bar{\beta}(h-x)}.$$

From (2.88) and (2.90), we can observe that for $t - s > 2\omega$,

$$\begin{aligned} (U(t,s)\phi)(a) &= \ell(a; t-a, 0) \int_0^\omega R(t-s-a, \rho; s) G(\rho; \phi, s) d\rho \\ &\leq \ell(a; t-a, 0) \bar{\beta} e^{\bar{\beta}(t-s)} \int_0^\omega G(\rho; \phi, s) d\rho. \end{aligned} \tag{2.94}$$

If we choose a natural number $n$ such that $n > (2\omega - \gamma_2 + \gamma_1)/(\gamma_2 - \gamma_1 - 2\delta)$, then there exists a positive number $\zeta$ such that $2\omega + \gamma_1 + n(\gamma_1 + \delta) < \zeta < \gamma_2 + n(\gamma_2 - \delta)$. If we let $t - s = \zeta$, for $a, \rho \in [0, \omega]$ we have $\zeta - a - \rho \in [\gamma_1 + n(\gamma_1 + \delta), \gamma_2 + n(\gamma_2 - \delta)]$. From (2.88), (2.90) and (2.91), we obtain

$$(U(s+\zeta, s)\phi)(a) \geq \ell(a; s+\zeta-a, 0) \alpha^{n+1} \delta^n \int_0^\omega G(\rho; \phi, s) d\rho. \tag{2.95}$$

Define a functional $\lambda_s(\phi)$, $\phi \in L^1(0, \omega)$ by

$$\lambda_s(\phi) := \int_0^\omega G(\rho; \phi, s) d\rho.$$

Changing the order of integrals, it is easy to see that

$$\lambda_s(\phi) = \int_0^\omega S(s, x) \phi(x) dx,$$

where

$$S(s, x) := \int_0^\omega \Psi(s+z, x+z) dz$$

is the total expected number of children who will be born in the remaining life of a mother at age $x$ and time $s$. From Assumption 2.9, $\lambda_s$ is a strictly positive functional, that is, $\lambda_s(\phi) > 0$ if $\phi \in L^1_+(0, \omega) \setminus \{0\}$. From (2.94) and (2.95), we have

$$\ell(a; s+\zeta-a, 0) \alpha^{n+1} \delta^n \lambda_s(\phi) \leq U(s+\zeta, s)\phi \leq \ell(a; s+\zeta-a, 0) \bar{\beta} e^{\bar{\beta}\zeta} \lambda_s(\phi).$$

Thus, the projective diameter of $U(s+\zeta, s)$ is estimated to be

$$\Delta(U(s+\zeta,s)) \le 2\log\left[\frac{\bar{\beta}e^{\bar{\beta}\zeta}}{\alpha^{n+1}\delta^n}\right].$$

That is, $U(s+\zeta,s)$ is uniformly positive, so $U(t,s)$ is uniformly positive for $t-s \ge \zeta$ in the sense of Birkhoff.[9] □

From Lemma 2.4, we obtain the following result (see Chap. 10):

**Proposition 2.10** ([52]) *Under Assumption 2.9, the population evolution system $\{U(t,s)\}_{s\le t}$ is a uniformly primitive multiplicative process for negative and positive time in the sense of Birkhoff, and is exponentially weakly ergodic.*

**Corollary 2.2** (The weak ergodicity theorem) *Suppose that $p_j(t,a)$, $t>0$, $(j = 1,2)$ is a solution of system (2.83) corresponding to the initial data $p_i(0,a) = \phi_i(a) \in L^1_+ \setminus \{0\}$. Under Assumption 2.9, it follows that*

$$\lim_{t\to\infty} |w_1(t,\cdot) - w_2(t,\cdot)|_{L^1} = 0, \tag{2.96}$$

*where $w_j(t,a) := p_j(t,a)/|p(t,\cdot)|_{L^1}$ is the age profile of population $p_j$.*

The result in (2.96) shows the *weak ergodicity* or the *asymptotic proportionality* of the non-autonomous age-structured population dynamics (2.83), whereby any two age distributions that evolve according to the same time-dependent vital rates become proportional to each other as time evolves, so the influences of the initial data disappear. In other words, for a uniformly primitive population evolution process, the long run-time variations in age profiles are uniquely determined by the vital rates.

In the weakly ergodic population evolution process, if the infinite time series of vital rates (birth and death rates) for $(-\infty, t_1]$ (where $t_1$ is a given time) are given, the time series of age profiles consistent with the given vital rates are uniquely determined. Even if the past vital rate data are given in a finite time interval $[t_0, t_1]$, the age profile at time $t_1$ is almost completely determined by the vital rate data if $t_1 - t_0$ is sufficiently large, because the convergence speed of the age profile is exponential. As Cohen pointed out [28], it is not necessary to know the prior age structures indefinitely far into the past to explain the current age structure of a population, because the weak ergodic theorem ensures that regardless of the age structure of a population some years previously, the vital rates that follow almost completely determine the current age structure.

If the vital rates are time-independent, the population evolution operator $U(t,s)$ depends only on the time interval $t-s$, and we can write $U(t,s) = T(t-s)$, where $T(t)$ is a population semigroup. In such a case, there exists an exponential solution $e^{\lambda_0(t-a)}\ell(a)$, and its age profile is time-independent. Therefore, for any solution $p(t,a) = (T(t)p_0)(a)$ corresponding to the positive initial data point $p_0 \in L^1 \setminus \{0\}$, it follows from (2.96) that

[9]The definition of uniform positivity is given in Chap. 10.

$$\lim_{t\to\infty}\left|\frac{p(t,\cdot)}{|p(t,\cdot)|_{L^1}}-w^*\right|_{L^1}=0,$$

where

$$w^*(a):=\frac{e^{-\lambda_0 a}\ell(a)}{\int_0^{\omega}e^{-\lambda_0\sigma}\ell(\sigma)d\sigma}$$

is the stable age profile. This result is simply the strong ergodicity theorem, which was derived in Chap. 1 using an integral equation approach.

Suppose that a population evolves according to a uniformly positive population semigroup for the time interval $t\in(-\infty,t_0]$, that is, $p(t,a)=(T(t-s)p(s,\cdot))(a)$, $t>s>-\infty$. Because the Malthusian solution $e^{\lambda_0(t-a)}\ell(a)$ is always consistent with $T(t)$ for this infinite time interval, it follows from the uniqueness of consistent time series of age profiles on an infinite time interval that for any $t\in(-\infty,t_0]$, $p(t,a)$ must have a stable age profile. Therefore, if we know that the birth and death rates have not changed for a sufficiently long time, we can expect the age distribution to have attained a stable age profile to within practical accuracy. This observation was first noticed by Lotka ([76, 97], and see Chap. 10).

*Remark 2.6* In demography, the weak ergodicity of the population evolution process was first pointed out by Coale [26], and a weak ergodicity theorem for the Leslie matrix model was proved by Lopez [74] (Coale–Lopez theorem). Readers can find early demographic discussions on the weak ergodicity theorem in [38, 71, 75]. Stochastic extensions were developed by Cohen [28].

### 2.6.2 Dual System and Ergodicity

The dual system of the non-autonomous problem in (2.83) is given as follows:

$$\begin{aligned}&\frac{\partial v(t,a)}{\partial t}+\frac{\partial v(t,a)}{\partial a}=\mu(t,a)v(t,a)-\beta(t,a)v(t,0),\\&v(t,\omega)=0.\end{aligned}\tag{2.97}$$

In an abstract setting, the backward problem in the dual space $X^*=L^\infty(0,\omega)$ is

$$\frac{dv(s)}{dt}=-A^*(s)v(s),\quad 0\le s<t,\quad v(t)=v^*\in X^*,\tag{2.98}$$

where

$$\begin{aligned}&(A^*(s)v)(a)=\frac{dv(a)}{da}-\mu(s,a)v(a)+\beta(s,a)v(0),\\&v\in\mathscr{D}(A^*(s))=\left\{v\in L^\infty(0,\omega):A^*(s)v\in L^\infty,\ v(\omega)=0\right\}.\end{aligned}$$

The weak* solution of the backward problem in (2.98) is then given by $U^*(s,t)v^*$ [25], where $U^*(s,t) = U(t,s)^*$ $(t > s > 0)$ is the dual evolutionary system of the forward system $U(t,s)$, $t > s$. As shown in Chap. 10 (Proposition 10.21), the dual system has an essentially unique solution (importance functional) $v^*(s) = U^*(s,t)v^*(t)$ for $0 < s < t < \infty$, which may be called the *demographic potential* [34].

Integrating (2.97) along the characteristic line, we have

$$v(t,a) = \int_a^\omega \beta(t-a+\sigma,\sigma)\ell(\sigma-a;t,a)v(t-a+\sigma,0)d\sigma, \tag{2.99}$$

which again implies that the demographic potential of a woman at age $a$ and time $t$ is equal to the total sum of the demographic potential of newborns produced in her remaining life. Therefore, the boundary value $v(t,0)$ must satisfy the *backward renewal equation*

$$v(t,0) = \int_0^\omega \Psi(t+\sigma,\sigma)v(t+\sigma,0)d\sigma, \tag{2.100}$$

where $\Psi(t,a) := \beta(t,a)\ell(a;t-a,0)$.

Let $p(t,a)$ be a positive solution of the primal system (2.83) and let $v(t,a)$ be a positive solution of the dual system. Again, it is easy to see that the total demographic potential $\langle v(t,\cdot), p(t,\cdot)\rangle$ is constant, because

$$\begin{aligned}\frac{d}{dt}\langle v(t,\cdot), p(t,\cdot)\rangle &= \langle \frac{\partial v(t,\cdot)}{dt}, p(t,\cdot)\rangle + \langle v(t,\cdot), \frac{\partial p(t,\cdot)}{dt}\rangle \\ &= -\langle A^*(t)v(t,\cdot), p(t,\cdot)\rangle + \langle v(t,\cdot), A(t)p(t,\cdot)\rangle = 0.\end{aligned}$$

Assume that $\langle v(0,\cdot), p(0,\cdot)\rangle > 0$. Then, the weighted distribution

$$\phi(t,a) := \frac{v(t,a)p(t,a)}{\langle v(t,\cdot), p(t,\cdot)\rangle} = \frac{v(t,a)p(t,a)}{\langle v(0,\cdot), p(0,\cdot)\rangle}$$

satisfies the following non-autonomous McKendrick equation:

$$\begin{aligned}\frac{\partial\phi(t,a)}{\partial t} + \frac{\partial\phi(t,a)}{\partial a} &= -\frac{v(t,0)\beta(t,a)}{v(t,a)}\phi(t,a), \\ \phi(t,0) &= \int_0^\omega \frac{v(t,0)\beta(t,a)}{v(t,a)}\phi(t,a)da, \\ \phi(0,a) &= \frac{v(0,a)p(0,a)}{\langle v(0,\cdot), p(0,\cdot)\rangle},\end{aligned} \tag{2.101}$$

which conserves the total size

$$\int_0^\omega \phi(t,a)da = 1,$$

because the total demographic potential is constant. The distribution $\phi(t, \cdot)$ then evolves as a Markov evolutionary system.

Let us fix the importance functional $v^*(t)$ for $0 < t < \infty$. Consider the two distributions

$$\phi_1(t,a) = \frac{v^*(t,a)p_1(t,a)}{\langle v^*(0,\cdot), p_1(0,\cdot)\rangle}, \quad \phi_2(t,a) = \frac{v^*(t,a)p_2(t,a)}{\langle v^*(0,\cdot), p_2(0,\cdot)\rangle},$$

satisfying (2.101), where $p_1$ and $p_2$ are two solutions of the forward system. As shown in Lemma 1.7, its Kullback information

$$K(\phi_1, \phi_2) = \int_0^\omega \phi_1(t,a) \log\left(\frac{\phi_1(t,a)}{\phi_2(t,a)}\right) da$$

is monotone decreasing as time evolves. Because this entropy-increasing (information gain decreasing) process does not stop as long as $\phi_1 \neq \phi_2$, the Kullback distance tends to zero as $t \to \infty$ ([68], Chap. 9). Therefore, in the long term, the weighted distribution (probability density function) $\phi$ becomes independent of the initial data, and the two age-density functions become proportional to each other:

$$p_1(t,a) \sim \frac{\langle v^*(0,\cdot), p_1(0,\cdot)\rangle}{\langle v^*(0,\cdot), p_2(0,\cdot)\rangle} p_2(t,a), \quad t \to \infty, \tag{2.102}$$

so the asymptotic proportional coefficient is the ratio of the two total demographic values at the initial time.

### 2.6.3 *Generalized Stable Populations*

We now consider the asymptotically autonomous case. If the population vital rates converge to time-independent values as time evolves, we could expect the age structure to converge to a stable age distribution calculated from the limiting vital rates. This type of result was first shown for a matrix population model by Artzrouni [7], who called such an asymptotically stable population the *generalized stable population*. To formulate this idea for the continuous time model, we introduce the following definition [54]:

**Definition 2.2** Let $p(t,\cdot) \in L^1_+(0,\omega)$ be the age distribution governed by (2.83). If there exist a real number $\lambda \in \mathbb{R}$, a positive functional $F : L^1_+(0,\omega) \to \mathbb{R}_+$ and an age distribution $q \in L^1_+(0,\omega)$ such that $\phi(a) = p(0,a)$ and

$$\lim_{t\to\infty} \int_0^\omega |e^{-\lambda t} p(t,a) - \langle F, \phi\rangle q(a)| da = 0, \tag{2.103}$$

then $p(t, a)$ is called the *generalized stable population* and the population evolution process is said to be strongly ergodic.

Even when there is a limiting system, it generally depends on the speed of convergence of the vital rates whether there exists an asymptotic time-independent age structure. We now introduce a sufficient condition:

**Assumption 2.11** In addition to Assumption 2.9, the following holds:

(1) There exists a time-independent survival probability $\ell(\cdot)$ such that

$$\lim_{t\to\infty}\int_0^{\omega}|\ell(a; t-a, 0)-\ell(a)|da = 0. \tag{2.104}$$

(2) There exists a real number $\lambda_0$ such that

$$\int_0^{\infty}\left|\int_0^{\omega}e^{-\lambda_0 a}\Psi(t, a)da - 1\right|dt < \infty, \tag{2.105}$$

where $\Psi(t, a) := \beta(t, a)\ell(a; t-a, 0)$.

(3) $\Psi(t, a)$ is uniformly Lipschitz continuous with respect to $t$ for all $a$.

Assumption (2) implies that $t \to \int_0^{\omega} e^{-\lambda_0 a}\Psi(t, a)da$ rapidly converges to unity as $t \to +\infty$. An important case in which (2) is satisfied is when the net reproduction kernel $\Psi(t, a)$ itself rapidly converges to a time-independent net reproduction kernel $\Psi(a)$ (that is, there exists a limiting fertility rate $\beta(a)$) as

$$\int_0^{\infty}\int_0^{\omega}|\Psi(t, a)-\Psi(a)|dadt < \infty.$$

In fact, in this case it is easy to see that there exists a unique real value $\lambda_0$ such that $\int_0^{\omega} e^{-\lambda_0 a}\Psi(a)da = 1$, and then (2.105) follows. Thus, the asymptotic growth rate $\lambda_0$ is the dominant characteristic root of the limiting net reproduction kernel. The following strong ergodicity theorem then holds [54]:

**Proposition 2.12** *Under Assumption 2.11, system* (2.83) *is strongly ergodic in the sense of Definition 2.2. That is, there exists a positive functional $F$ such that*

$$\lim_{t\to\infty}\int_0^{\omega}|e^{-\lambda_0 t}p(t, a)-\langle F, p_0\rangle e^{-\lambda_0 a}\ell(a)|da = 0.$$

The proof of the above statement is omitted, as it involves the theory of multiplicative processes and a generalized variation-of-constants formula for evolution operators, and these are beyond the scope of this chapter. However, we do provide a sketch of the proof. Under Assumption 2.11, we can decompose the net reproduction function $\Psi(t, a)$ to $\Psi(t, a) = \Psi_0(t, a) + (h(t)-1)\Psi_0(t, a)$, where

$$\Psi_0(t,a) = \frac{\Psi(t,a)}{h(t)}, \quad h(t) := \int_0^\omega e^{-\lambda_0 a}\Psi(t,a)da.$$

It then follows that

$$\int_0^\omega e^{-\lambda_0 a}\Psi_0(t,a)da = 1.$$

If we consider a non-autonomous Lotka–McKendrick system with fertility rate $\beta(t,a)/h(t)$ and survival probability $\ell(a; t-a, 0)$, we obtain the exponential solution $e^{\lambda_0(t-s)}\ell(a; t-a, 0)$, and all positive solutions are asymptotically proportional to the exponential solution because of the weak ergodicity. Hence, the evolutionary system produced by the fertility rate $\beta(t,a)$ and the survival probability $\ell(a; t-a, 0)$ is a perturbed system of a strongly ergodic evolutionary system, and we can prove that the original system itself also becomes strongly ergodic because $h(t)$ rapidly converges to unity by $\int_0^\infty |h(t)-1|dt < \infty$ [54, Proposition 5.6].

Finally, note that Iannelli [48] used a classical calculation to show that a renewal theorem holds for the non-autonomous Volterra integral equation with a rapidly convergent integral kernel.

### 2.6.4 Periodic Stable Populations

We now consider a weakly ergodic non-autonomous population system (2.83) whose vital rates $\mu$ and $\beta$ have a period $\theta > 0$ with respect to $t$:

$$\beta(t+\theta,a) = \beta(t,a), \quad \mu(t+\theta,a) = \mu(t,a), \quad \forall(t,a) \in \mathbb{R}\times[0,\omega].$$

In the periodic environment, the population evolution operator satisfies

$$U(t+\theta, s+\theta) = U(t,s). \tag{2.106}$$

In fact, $p(t) = U(t+\theta, s+\theta)p_0$ is the solution of (2.83) with the initial value $p(s) = p_0$, so it can be expressed as $p(t) = U(t,s)p_0$. Because $U(t+\theta, s+\theta)p_0 = U(t,s)p_0$ holds for any $p_0 \in L^1$, (2.106) holds.

In the following, we seek a positive *exponential solution* with exponent $\lambda$ for (2.83) such that $p(t,a) = e^{\lambda t}\phi(t,a)$, where $\lambda \in \mathbb{R}$ and $\phi(t,\cdot)$ is a $\theta$-periodic $L^1_+$-valued positive function. In fact, as shown in Chap. 10, if the population evolution process given by $U(t,s), t \geq s$, is weakly ergodic, then all positive solutions are asymptotically proportional to the positive exponential solution (if it exists), so its exponent can be seen as the intrinsic growth rate of the periodic system, and we can prove that the age profile converges to a periodic age profile. Moreover, as shown below, we can also define the basic reproduction number $R_0$ for the periodic stable population system. The exponential solution $e^{\lambda t}\phi(t,a)$ can then be called the *periodic stable population*.

If we insert $e^{\lambda t}\phi(t,a)$ into (2.83), we obtain the eigenvalue problem

$$\begin{aligned}(-D-\mu(t,a))\phi(t,a) &= \lambda\phi(t,a),\\ \phi(t,0) &= \int_0^{\omega}\beta(t,a)\phi(t,a)da,\end{aligned} \tag{2.107}$$

where $D$ denotes the directional derivative along the characteristic line defined in (1.37).

*Remark 2.7* From the perspective of functional analysis, the Malthusian parameter $\lambda$ is given as a real eigenvalue of the infinitesimal generator of the *evolution semigroup* $S(\sigma)$, $\sigma \geq 0$, associated with $U(t,s)$, which is defined as

$$(S(\sigma)f)(t) = U(t,t-\sigma)f(t-\sigma), \quad f \in \mathbb{F}_\theta,$$

where $\mathbb{F}_\theta = C_\theta(\mathbb{R}; L^1(0,\omega))$ or $\mathbb{F}_\theta = L^1_\theta(\mathbb{R}; L^1(0,\omega))$ [104]. In fact, the infinitesimal generator of the evolution semigroup $S(\sigma)$ is given by $-d/dt + A(t)$, where $A(t)$ is a generator of $U(t,s)$ such that $A(t) = -d/da - \mu(t,\cdot)$ with the domain $\mathscr{D}(A(t)) = \{\phi \in L^1(0,\omega) : A(t)\phi \in L^1, \phi(0) = \int_0^\omega \beta(t,a)\phi(a)da\}$.

Let $B(t)$ be a $\theta$-periodic function. It is then easy to see that $e^{-\lambda a}\ell(a;t-a,0)B(t-a)$ is the eigenfunction associated with the eigenvalue $\lambda$ for problem (2.107) if $B(t)$ satisfies the boundary condition

$$B(t) = \int_0^\infty e^{-\lambda a}\Psi(t,a)B(t-a)da, \tag{2.108}$$

where $\Psi(t,a) := \beta(t,a)\ell(a;t-a,0)$ is the net reproduction function. Here, we adopt the convention that $\Psi = 0$ for $a > \omega$ and apply a calculation method in [10, 59]. From the periodicity of the vital rates, we have $\Psi(t+\theta,a) = \Psi(t,a)$. Using the periodicity of $\Psi$, we can rewrite (2.108) as

$$B(u) = \int_0^\theta \Theta_\lambda(u,\sigma)B(\sigma)d\sigma, \quad u \in [0,\theta], \tag{2.109}$$

where

$$\Theta_\lambda(u,\sigma) := \begin{cases} \displaystyle\sum_{n=0}^{[\omega/\theta]+1} e^{-\lambda(u-\sigma+n\theta)}\Psi(u,u-\sigma+n\theta), & u-\sigma > 0,\\ \displaystyle\sum_{n=1}^{[\omega/\theta]+1} e^{-\lambda(u-\sigma+n\theta)}\Psi(u,u-\sigma+n\theta), & u-\sigma < 0.\end{cases}$$

In fact, for any $t$, there exists an integer $n$ such that $t = u + n\theta$, $u \in [0,\theta)$. It follows from the periodicity that (2.108) can be rewritten as

$$
\begin{aligned}
B(u) &= \int_0^\infty e^{-\lambda a}\Psi(u,a)B(u-a)da = \int_{-\infty}^{u} e^{-\lambda(u-\sigma)}\Psi(u,u-\sigma)B(\sigma)d\sigma, \\
&= \int_0^u e^{-\lambda(u-\sigma)}\Psi(u,u-\sigma)B(\sigma)d\sigma \\
&\quad + \sum_{n=1}^{\infty}\left\{\int_0^u d\sigma + \int_u^\theta d\sigma\right\} e^{-\lambda(u-\sigma+n\theta)}\Psi(u,u-\sigma+n\theta)B(\sigma).
\end{aligned}
$$

Thus, we have expression (2.109), because $\Psi(u, u-\sigma+n\theta) = 0$ for $n \geq [\omega/\theta]+2$.

For real $\lambda$, the right-hand side of (2.109) defines a positive linear operator on $L^1(0,\theta)$, denoted by $J(\lambda)$:

$$
(J(\lambda)f)(u) := \int_0^\theta \Theta_\lambda(u,\sigma)f(\sigma)d\sigma, \quad u \in [0,\theta].
$$

If we assume that $J(\lambda)$ is compact and non-supporting, then its spectral radius $r(J(\lambda))$ is a positive eigenvalue of $J(\lambda)$ and is strictly decreasing from $+\infty$ to zero with respect to $\lambda$. Therefore, there exists a unique $\lambda_0$ such that $r(J(\lambda_0)) = 1$ and a positive eigenfunction $B_0 \in L^1_+(0,\theta)$ such that $B_0 = J(\lambda)B_0$. The periodic extension $B(t) = B_0(t-[t/\theta]\theta)$ for $t \in \mathbb{R}$ then satisfies (2.109) with $\lambda = \lambda_0$, and $e^{\lambda_0(t-a)}\ell(a;t-a,0)B(t-a)$ becomes a positive exponential solution. Moreover, it follows that

$$
\operatorname{sign}(\lambda_0) = \operatorname{sign}(r(J(0)) - 1), \tag{2.110}
$$

and $\lambda_0$ can be seen as the asymptotic growth rate, because any positive solution of (2.83) is asymptotically proportional to the exponential solution $e^{\lambda_0(t-a)}\ell(a;t-a,0)B(t-a)$.

The above arguments suggest that the positive linear operator acting on the $\theta$-periodic locally integrable functions $f$ defined by

$$
(K_\theta f)(t) := \int_0^\infty \Psi(t,a)f(t-a)da \tag{2.111}
$$

can be seen as the NGO whose spectral radius gives the basic reproduction number $R_0$ for the periodic stable population system, because $r(K_\theta) = r(J(0))$. In fact, we can show that the spectral radius $r(K_\theta)$ gives the asymptotic per-generation growth factor of the periodic stable population model (see Chap. 9 and [60]). An early attempt to use $r(K_\theta)$ as a threshold value of population growth appeared in [2, 3], but the first precise recognition that $r(K_\theta)$ is the basic reproduction number was given in [9]. It is clear that the spectral radius of the monodromy operator $U(t+\theta,t)$ is a surrogate index for the basic reproduction number [44, 45].

Finally, to see the asymptotic behavior of the periodic stable population system, let us consider the *dual* eigenvalue problem

$$\begin{aligned} &\frac{\partial \phi(t,a)}{\partial t} + \frac{\partial \phi(t,a)}{\partial a} = (\lambda_0 + \mu(t,a))\phi(t,a) - \beta(t,a)\phi(t,0), \\ &\phi(t,\omega) = 0, \end{aligned} \tag{2.112}$$

where $\phi$ is a $\theta$-periodic $L^\infty$-valued positive function.

Let $p^\dagger(t,a) = e^{\lambda_0 t} u(t,a)$ be the exponential solution of (2.83), and assume that its population evolutionary system is uniformly primitive. From Proposition 10.30, the dual problem in (2.112) has a positive solution $\phi(t,a)$ corresponding to the eigenvalue $\lambda_0$, and $v^\dagger(t,a) := e^{-\lambda_0 t}\phi(t,a)$ is an exponential solution (importance functional) of the dual system (2.97) such that

$$\langle v^\dagger(t,\cdot), p^\dagger(t,\cdot)\rangle = \langle \phi(t,\cdot), u(t,\cdot)\rangle = 1.$$

It then follows from Propositions 10.19 and 10.30 that we can choose $p^\dagger(t,\cdot)$ as a reference distribution such that

$$\lim_{t\to\infty} \int_0^\omega |e^{-\lambda_0 t} p(t,a) - \langle v^\dagger(0,\cdot), p(0,\cdot)\rangle u(t,a)| da = 0, \tag{2.113}$$

where

$$\langle v^\dagger(0,\cdot), p(0,\cdot)\rangle = \langle \phi(0,\cdot), p(0,\cdot)\rangle$$

is the total demographic potential of the initial data $p(0,\cdot)$.

Bacaër and Abdurahman [12] defined $\phi(t,a)$ as the *reproductive value* at time $t$ and age $a$ in the periodic environment. With the total reproductive value defined by $V(t) := \langle \phi(t,\cdot), p(t,\cdot)\rangle$, it follows from Proposition 10.18 that $\langle v^\dagger(t,\cdot), p(t,\cdot)\rangle$ is constant, and so

$$V(t) = \langle \phi(t,\cdot), p(t,\cdot)\rangle = e^{\lambda_0 t}\langle v^\dagger(t,\cdot), p(t,\cdot)\rangle = e^{\lambda_0 t}\langle v^\dagger(0,\cdot), p(0,\cdot)\rangle = e^{\lambda_0 t} V(0),$$

which can be thought of as an extension of Fisher's theorem (Proposition 1.12) for the reproductive value in a constant environment. To date, although there is no consensus on how to define a reproductive value in a time-heterogeneous environment [35, 106], Bacaër and Abdurahman's definition is a natural extension of Fisher's reproductive value, because it mathematically describes a positive eigenfunction (corresponding to the Malthusian parameter as an eigenvalue) of the generator of the evolution semigroup associated with the population evolution system $U(t,s)$.

Using $V(t)$, it follows from (2.113) that

$$p(t,a) \sim V(t)u(t,a) = V(0)p^\dagger(t,a), \quad t \to \infty.$$

This shows that the periodic stable population model does not have a time-independent stable age distribution, but is strongly ergodic in the sense that its asymptotic age profile is independent of the initial data and has a Malthusian parameter $\lambda_0$ and basic reproduction number $R_0$, which satisfies the sign relation.

In a demographic sense, the periodic stable population was first observed by Coale [27], whereas the Volterra integral equation with periodic kernel and its renewal theorem was first mathematically established by Thieme [99]. Two decades later, based on Thieme's results, Bacaër and Guernaoui [9] defined $R_0$ for an age-structured periodic epidemic system using the spectral radius of an integral operator. For the essential progress of periodic population systems, readers should refer to a series of works by Bacaër and co-workers [9–15].

## References

1. Allen, L.J.S., Bolker, B.M., Lou, Y., Nevai, A.L.: Asymptotic profiles of the steady states for an SIS epidemic reaction-diffusion model. Disc. Cont. Dyn. Sys. **21**(1), 1–20 (2008)
2. Anita, S., Iannelli, M., Kim, M.-Y., Park, E.-J: Optimal harvesting for periodic age-dependent population dynamics. SIAM. J. Appl. Math. **58**(5), 1648–1666 (1998)
3. Anita, S.: Analysis and Control of Age-Dependent Population Dynamics. Kluwer, Dordrecht (2000)
4. Arino, O., Smith, W.V.: Migration in age structured population dynamics. Math. Mod. Meth. Appl. Sci. **8**(5), 905–925 (1998)
5. Arino, O., Sánchez, E., Bravo De La Parra, R.: A model for an age-structured population in a multipatch environment. Math. Comput. Model. **27**(4), 137–150 (1998)
6. Arino, O., Sánchez, E., Bravo De La Parra, R., Auger, P.: A singular perturbation in an age-structured population model. SIAM J. Appl. Math. **60**(2), 408–436 (1999)
7. Artzrouni, M.: Generalized stable population theory. J. Math. Biol. **21**, 363–381 (1985)
8. Bacaër, N.: The asymptotic behavior of the McKendrick equation with immigration. Math. Popul. Stud. **10**, 1–20 (2003)
9. Bacaër, N., Guernaoui, S.: The epidemic threshold of vector-borne diseases with seasonality. J. Math. Biol. **53**, 421–436 (2006)
10. Bacaër, N.: Approximation of the basic reproduction number $R_0$ for vector-borne diseases with a periodic vector population. Bull. Math. Biol. **69**, 1067–1091 (2007)
11. Bacaër, N., Ouifki, R.: Growth rate and basic reproduction number for population models with a simple periodic factor. Math. Biosci. **210**, 647–658 (2007)
12. Bacaër, N., Abdurahman, X.: Resonance of the epidemic threshold in a periodic environment. J. Math. Biol. **57**, 649–673 (2008)
13. Bacaër, N., Ait Dads, E.H.: Genealogy with seasonality, the basic reproduction number, and the influenza pandemic. J. Math. Biol. **62**, 741–762 (2011)
14. Bacaër, N.: The model of Kermack and McKendrick for the plague epidemic in Bombay and the type reproduction number with seasonality. J. Math. Biol. **64**(3), 403–422 (2012)
15. Bacaër, N., Ait Dads, E.H.: On the biological interpretation of a definition for the parameter $R_0$ in periodic population models. J. Math. Biol. **65**(4), 601–621 (2012)
16. Billari, F.C., Manfredi, P., Valentini, A.: Macro-demographic effects of the transition to adulthood: multistate stable population theory and an application to Italy. Math. Popul. Stud. **9**(1), 33–63 (2000)
17. Birkhoff, G., Varga, R.S.: Reactor criticality and nonnegative matrices. J. Soc. Indust. Appl. Math. **6**(4), 354–377 (1958)
18. Birkhoff, G.: Lattice Theory, 3rd edn. American Mathematical Society, Providence (1967)

19. Bravo De La Parra, R., Arino, O., Sánchez, E., Auger, P.: A model for an age-structured population with two time scales. Math. Comput. Model. **31**, 17–26 (2000)
20. Cerone, P.: On stable population theory with immigration. Demography **24**(3), 431–438 (1987)
21. Chan, W.L., Guo, B.Z.: Controlled age-dependent population dynamics based on parity progression. J. Math. Anal. Appl. **166**, 442–455 (1992)
22. Chan, W.L., Guo, B.Z.: Age-dependent population dynamics based on parity interval progression. Mathl. Comput. Model. **16**(4), 57–68 (1992)
23. Charlesworth, B.: Evolution in Age-Structured Populations, 2nd edn. Cambridge University Press, Cambridge (1994)
24. Chicone, C., Latushkin, Y.: Evolution Semigroups in Dynamical Systems and Differential Equations, Mathematical Surveys and Monographs, vol. 70. American Mathematical Society, Providence (1999)
25. Clément, P., Diekmann, O., Gyllenberg, M., Heijmans, H.J.A.M., Thieme, H.R.: Perturbation theory for dual semigroups II. Time-dependent perturbations in sun-reflexive case, Proc. Royal Soc. Edinb. **109A**, 145–172 (1988)
26. Coale, A.J.: How the age distribution of a human population is determined. Cold Spring Harb. Symp. Quant. Biol. **22**, 83–88 (1957)
27. Coale, A.J.: The Growth and Structure of Human Populations. Princeton University Press, Princeton (1972)
28. Cohen, J.E.: Ergodic theorems in demography. Bull. Amer. Math. Soc. **1**(2), 275–295 (1979)
29. Desch, W., Schappacher, W., Zhang K.P.: Semilinear evolution equations. Houston J. Math. **15**(4), 527–552 (1989)
30. Das Gupta, P.: Age-parity-nuptiality-specific stable population model that recognizes births to single woman. J. Amer. Stat. Ass. **71**(354), 308–314 (1976)
31. Diekmann, O.: Thresholds and travelling waves for the geographical spread of infection. J. Math. Biol. **6**, 109–130 (1978)
32. Diekmann, O., Heesterbeak, J.A.P., Metz, J.A.J.: On the definition and the computation of the basic reproduction ratio $R_0$ in models for infectious diseases in heterogeneous populations. J. Math. Biol. **28**, 365–382 (1990)
33. Diekmann, O., Gyllenberg, M., Metz, J.A.J., Thieme, H.R.: On the formulation and analysis of general deterministic structured population models I. Linear Theory. J. Math. Biol. **36**, 349–388 (1998)
34. Ediev, D.M.: On an extension of R.A. Fisher's result on the dynamics of the reproductive value. Theor. Popul. Biol. **72**, 480–484 (2007)
35. Ediev, D.M.: On the definition of the reproductive value: response to the discussion by Bacaer and Abdurahman. J. Math. Biol. **59**, 651–657 (2009)
36. Espenshade, T.J., Bouvier, L.F., Arthur, W.B.: Immigration and the stable population model. Demography **19**(1), 125–133 (1982)
37. Feeney, G.: Population dynamics based on birth intervals and parity progression. Popul. Stud. **37**, 75–89 (1983)
38. Feichtinger, G.: Demographische Analyse und populations-dynamische Modelle. Springer, Wien New York (1979)
39. Fujimoto, T., Krause, U.: Asymptotic properties for inhomogeneous iterations of nonlinear operators. SIAM J. Math. Anal. **19**(4), 841–853 (1988)
40. Gurtin, M.E.: A system of equations for age-dependent population diffusion. J. theor. Biol. **40**, 389–392 (1973)
41. Gurtin, M.E., MacCamy, R.C.: Non-linear age-dependent population dynamics. Arch. Rat. Mech. Anal. **54**, 281–300 (1974)
42. Gyllenberg, M.: The size and scar distributions of the yeast Saccharomyces cerevisiae. J. Math. Biol. **24**, 81–101 (1986)
43. Hamada, T., Kanno, S., Kano, E.: Stationary stage structure of yeast population with stage dependent generation time. J. Theor. Biol. **97**, 393–414 (1982)
44. Heesterbeek, J.A.P., Roberts, M.G.: Threshold quantities for helminth infections. J. Math. Biol. **33**, 415–434 (1995)

45. Heesterbeek, J.A.P., Roberts, M.G.: Threshold quantities for infectious diseases in periodic environments. J. Biol. Sys. **3**(3), 779–787 (1995)
46. Heijmans, H.J.A.M.: The dynamical behaviour of the age-size-distribution of a cell population. In: Metz, J.A.J., Diekmann, O. (eds.) The Dynamics of Physiologically Structured Populations, Lecture Notes in Biomathematics, vol. 68, pp. 185–202. Springer, Berlin (1986)
47. Henry, L.: Population: Analysis and Models. Edward Arnold, London (1976)
48. Iannelli, M.: Mathematical Theory of Age-Structured Population Dynamics, Giardini Editori e Stampatori in Pisa (1995)
49. Inaba, H.: Mathematical foundation of multidimensional stable population model I: Classical theory. Jinko Mondai Kenkyu (The Journal of Population Problems, in Japanese) **184**, 52–77 (1987)
50. Inaba, H.: A semigroup approach to the strong ergodic theorem of the multistate stable population process. Math. Popul. Stud. **1**(1), 49–77 (1988)
51. Inaba, H.: Asymptotic properties of the inhomogeneous Lotka-Von Foerster system. Math. Popul. Stud. **1**(3), 247–264 (1988)
52. Inaba, H.: Weak ergodicity of population evolution processes. Math. Biosci. **96**, 195–219 (1989)
53. Inaba, H.: Functional Analytic Approach to Age-Structured Population Dynamics, PhD Thesis, University of Leiden (1989)
54. Inaba, H.: Strong ergodicity for perturbed dual semigroups and application to age-dependent population dynamics. J. Math. Anal. Appl. **165**(1), 102–132 (1992)
55. Inaba, H.: A Mathematical Model for Human Population Reproduction by Iterative Marriage, Workin Paper Series 18. Institute of Population Problems, Tokyo (1993)
56. Inaba, H.: Human population reproduction via first marriage. Math. Popul. Studies **5**(2), 123–144 (1995)
57. Inaba, H.: Nonlinear dynamics of open marine population with space-limited recruitment: the case of mortality control. J. Math. Anal. Appl. **275**, 537–556 (2002)
58. Inaba, H.: The net reproduction rate and the type-reproduction number in multiregional demography. Vienna Yearb. Popul. Res. **2009**, 197–215 (2010)
59. Inaba, H.: The Malthusian parameter and $R_0$ for heterogeneous populations in periodic environments. Math. Biosci. Eng. **9**(2), 313–346 (2012)
60. Inaba, H.: On a new perspective of the basic reproduction number in heterogeneous environments. J. Math. Biol. **65**, 309–348 (2012)
61. Itoh, T.: Nuptiality, fertility and reproductivity in Japan: hypothetical studies on recent trend of fetility. J. Pop. Probl. **148**, 24–43 (1978) (Japanese)
62. Itoh, T., Bando, R.: Marital fertility taking account of marriage duration and age at marriage in the early 1980s of Japan. J. Pop. Probl. **189**, 51–69 (1989) (Japanese)
63. Kamioka, K.: Mathematical analysis of a metapopulation model with space-limited recruitment. Math. Biosci. **201**, 48–57 (2006)
64. Krause, U.: Positive Dynamical Systems in Discrete Time: Theory, Models, and Applications. de Gruyter, Berlin (2015)
65. Lamas, L.: Birth Intervals, Parity Specific Fertility, and Stable Populations, Ph.D. dissertation, University of Wisconsin (1985)
66. Land, K.C., Rogers, A. (eds.): Multidimensional Mathematical Demography. Academic Press, New York (1982)
67. Langlais, M.: Large time behavior in a nonlinear age-dependent population dynamics problem with spatial diffusion. J. Math. Biol. **26**(3), 319–346 (1986)
68. Lasota, A., Mackey, M.C.: Chaos, Fractals, and Noise: Stochastic Aspects of Dynamics, Applied Mathematical Sciences, vol. 97, 2nd edn. Springer, New York (1995)
69. Lasota, A., Yorke, J.A.: When the long-time behavior is independent of the initial density. SIAM J. Math. Anal. **27**(1), 221–240 (1996)
70. Le Bras, H.: Équilibre et croissance de populations soumises à des migrations. Theor. Popul. Biol. **2**, 100–121 (1971)

71. Le Bras, H.: Une formulation générale de la dynamique des populations, Population, numéro spécial, pp. 261–293 (1977)
72. Ledent, J., Rogers, A.: Srable growth in native-dependent multistate population dynamics. Math. Popul. Stud. **1**(2), 157–171 (1988)
73. Lisi, M., Totaro, S.: The Chapmann-Enskog procedure for an age-structured population model: initial, boundary and corner layer corrections. Math. Biosci. **196**, 153–186 (2005)
74. Lopez, A.: Problems in Stable Population Theory. Princeton University, Princeton, Office of Population Research (1961)
75. Lopez, A.: Asymptotic properties of a human age distribution under a continuous net maternity function. Demography **4**(2), 680–687 (1967)
76. Lotka, A.J.: Population analysis: a theorem regarding the stable age distribution. J. Wash. Acad. Sci. **27**(7), 299–303 (1937)
77. Marcati, P., Serafini, R.: Asymptotic behaviour in age dependent population dynamics with spatial spread. Boll. U. M. I. (5) **16-B**, 734–753 (1979)
78. Marcati, P.: Asymptotic behavior in age-dependent population dynamics with hereditary renewal law. SIAM J. Math. Anal. **12**(6), 904–916 (1981)
79. Mckenzie, H.W., Jin, Y., Jacobsen, J., Lewis, M.A.: $R_0$ analysis of a spatiotemporal model for a stream population. SIAM J. Appl. Dyn. Sys. **11**(2), 567–596 (2012)
80. Minc, H.: Nonnegative Matrices. Wiley, New York (1988)
81. Mitra, S.: Generalization of the immigration and the stable population model. Demography **20**(1), 111–115 (1983)
82. Mitra, S., Cerone, P.: Migration and stability. Genus **XLII-n**(1–2), 1–12 (1986)
83. Mode, C.J.: Stochastic Processes in Demography and Their Computer Implementation. Springer, Berlin (1985)
84. Nakaoka, S., Inaba, H.: Demographic modeling of transient amplifying cell population growth. Math. Bisci. Eng. **11**(2), 363–384 (2014)
85. Norton, H.T.J.: Natural selection and Mendelian variation. Proc. Lond. Math. Soc. **28**, 1–45 (1928)
86. Okubo, A., Levin, S.A.: Diffusion and Ecological Problems: Modern Perspectives, Second edn. Springer, New York (2001)
87. Page, H.J.: Patterns underlying fertility schedules: a decomposition by both age and marriage duration. Popul. Stud. **31**, 85–106 (1977)
88. Peng, R., Zhao, X.Q.: A reaction-diffusion SIS epidemic model in a time-periodic environment. Nonlinearity **25**, 1451–1471 (2012)
89. Rogers, A.: Introduction to Multiregional Mathematical Demography. Wiley, New York (1975)
90. Rogers, A., Willekens, F.J.: Migration amd Settlement: A Multiregional Comparative Study. D. Reidel, Dordrecht (1986)
91. Rogers, A.: Multiregional Demography: Principles, Methods and Extensions. Wiley, New York (1995)
92. Roughgarden, J., Iwasa, Y., Baxter, C.: Demographic theory for an open marine population with space-limited recruitment. Ecology **66**(1), 54–67 (1985)
93. Roughgarden, J., Iwasa, Y.: Dynamics of a metapopulation with space-limited subpopulations. Theor. Popul. Biol. **29**, 235–261 (1986)
94. Roughgarden, J., Iwasa, Y.: Interspecific competition among metapopulations with space-limited subpopulations. Theor. Popul. Biol. **30**, 194–214 (1986)
95. Rundnicki, R., Mackey, M.C.: Asymptotic similarity and Malthusian growth in autonomous and nonautonomous populations. J. Math. Anal. Appl. **187**, 548–566 (1994)
96. Ruzicka, L.T. (ed.): Nuptiality and Fertility, Liege, Ordina edn (1982)
97. Samuelson, P.A.: Resolving a historical confusion in population analysis. Hum. Biol. **48**, 559–580 (1976)
98. Schoen, R.: Modeling Multigroup Populations. Plenum Press, New York and London (1988)
99. Thieme, H.R.: Renewal theorems for linear periodic Volterra integral equations. J. Inte. Equ. **7**, 253–277 (1984)

100. Thieme, H.R.: Asymptotic proportionality (weak ergodicity) and conditional asymptotic equality of solutions to time-heterogeneous sublinear difference and differential equations. J. Diff. Equ. **73**, 237–268 (1988)
101. Thieme, H.R.: Semiflows generated by Lipschitz perturbations of non-densely defined operators. Differ. Integral Equ. **3**(6), 1035–1066 (1990)
102. Thieme, H.R.: Analysis of age-structured population models with additional structure. In: Arino, O., Axelrod, D.E., Kimmel, M. (eds.), Mathematical Population Dynamics, pp. 115–126. Marcel Dekker, New York (1991)
103. Thieme, H.R., Zhao, X.-Q.: Asymptotic speeds of spread and traveling waves for integral equations and delayed reaction-diffusion models. J. Diff. Equ. **195**, 430–470 (2003)
104. Thieme, H.R.: Spectral bound and reproduction number for infinite-dimensional population structure and time heterogeneity. SIAM J. Appl. Math. **70**(1), 188–211 (2009)
105. Tucker, S.L., Zimmerman, S.O.: A nonlinear model of population dynamics containing an arbitrary number of continuous structure variables. SIAM J. Appl. Math. **48**(3), 549–591 (1988)
106. Vandermeer, J.H.: Reproductive value in a population of arbitrary age distribution. Am. Nat. **102**(928), 586–589 (1968)
107. Varga, R.S.: Matrix Iterative Analysis, 2nd edn. Springer, Berlin (2000)
108. Wang, W., Zhao, X.-Q.: Threshold dynamics for compartmental epidemic models in period environments. J. Dyn. Diff. Equat. **20**, 699–717 (2008)
109. Wang, W., Zhao, X.Q.: Basic reproduction numbers for reaction-diffusion epidemic models. SIAM J. Appl. Dyn. Sys. **11**(4), 1652–1673 (2012)
110. Webb, G.F.: Diffusive age-dependent population models and an application to genetics. Math. Biosci. **61**, 1–16 (1982)
111. Webb, G.F.: Theory of Nonlinear Age-Dependent Population Dynamics. Marcel Dekker, New York and Basel (1985)
112. Zhang, S., Freedman, H.I.: On an unbounded linear operator arising in the demographic theory for an open population with space limited recruitment. Dyn. Contin. Discret. Impuls. Syst. **5**, 317–325 (1999)
113. Zhang, S., Freedman, H.I., Liu, X.Z.: Analysis of a population model with space-dependent recruitment in continuous time. J. Math. Anal. Appls. **232**, 99–118 (1999)

# Chapter 3
# Nonlinear One-Sex Models

**Abstract** The stable population model is an age-structured version of the Malthus model. Therefore, any criticism of the Malthus model can also be applied to the stable population model. In reality, if the intrinsic growth rate is positive, asymptotic exponential growth cannot continue forever. In the long term, we should take into account the interactions that exist between the population and its environment. Population parameters such as fertility and mortality could be affected by environmental parameters (resources, seasons, life conditions and so on), while population growth itself will affect the environmental conditions. Therefore, through these feedback loops, demographic parameters generally depend on the population size and structure, resulting in a necessarily nonlinear model. Although there are many nonlinear effects that should be taken into account to reflect the self-regulation of the population, in this chapter, we only consider the mechanism of density-dependence in a one-sex age-structured population. In the first half of this chapter, we discuss the *period control model*, and in the latter half, we consider the *cohort control model*.

## 3.1 Period Control Model

### *3.1.1 Basic Model and Its Well-Posedness*

If we take into account that the population has some self-regulation mechanism, the Malthus model (1.1) may be replaced by a nonlinear equation

$$\frac{dP(t)}{dt} = \alpha(P(t))P(t), \tag{3.1}$$

where $\alpha(x) : [0, +\infty) \to \mathbb{R}$ describes the effect of population size on birth and death rates. From observations in population biology, we assume that there exists a critical size $x_0 \geq 0$ such that $\alpha(x)$ satisfies the following properties [23]:



H. Inaba, *Age-Structured Population Dynamics in Demography and Epidemiology*, DOI 10.1007/978-981-10-0188-8_3

$$\begin{cases} (1) \quad \alpha'(x) > 0, & 0 < x < x_0, \\ (2) \quad \alpha'(x) < 0, & x > x_0, \\ (3) \quad \lim_{x\to\infty}\alpha(x) < 0. \end{cases}$$

If $x_0 = 0$, only (2) and (3) are meaningful. $x_0$ denotes the critical population size (density) such that the density effect switches from positive to negative. Property (1) is called the *Allee effect* and describes the phenomenon whereby, when the population density is low, the population growth rate increases as the population size increases; this is because the mating chances will increase and the protection effect of collective life will develop. Properties (2) and (3) describe the *logistic effect* whereby, when the population density is relatively high, the population size has a negative effect on population growth because of resource constraints and environmental destruction. As a result of the logistic effect, the population cannot grow indefinitely.

In the following, we consider a one-sex closed, age-structured population whose age-specific birth and death rates are controlled by one variable (the weighted total size of the population) given by

$$S(t) := \int_0^{\omega} \gamma(a)p(t,a)da,$$

where $\gamma(a)$ denotes a weight for the age class. If the vital rates simply depend on the total population size, we can choose $\gamma(a) \equiv 1$.

More generally, we can consider population dynamics controlled by many control variables, although for simplicity, we focus on the case in which the vital rates are controlled by only one variable $S(t)$. Readers are referred to a study by Iannelli [23] for the case of vector control. If we replace $\beta(a)$ and $\mu(a)$ by $\beta(a, S(t))$ and $\mu(a, S(t))$, the stable population model becomes the following nonlinear one-sex model:

$$\begin{aligned} &\frac{\partial p(t,a)}{\partial t} + \frac{\partial p(t,a)}{\partial a} = -\mu(a, S(t))p(t,a), \\ &p(t,0) = \int_0^{\omega} \beta(\sigma, S(t))p(t,\sigma)d\sigma, \\ &S(t) = \int_0^{\omega} \gamma(\sigma)p(t,\sigma)d\sigma, \\ &p(a,0) = p_0(a). \end{aligned} \tag{3.2}$$

System (3.2) is an age-structured version of the nonlinear ordinary differential equation (ODE) in (3.1) (the Allee–Logistic model). A special case with $\gamma(a) \equiv 1$ was first rigorously examined by Gurtin and MacCamy [19]. Their study made a considerable impact and promoted various systematic studies of nonlinear age-structured population dynamics. System (3.2) is also called the *period control model* in demography, because the vital rates at time $t$ are controlled by the size of the population $S(t)$ at the same time (period) $t$. In Chap. 8, we will show that a classical Kermack–McKendrick endemic model can be considered as the Gurtin and MacCamy model.

First, observe that the nonlinear system (3.2) can be reduced to an integral equation system as well as the stable population model. If we consider the control variable $S(t)$ as a given function, (3.2) can be regarded as a nonautonomous linear problem and can be integrated along the characteristic lines. Then, we have

$$p(t,a) = \begin{cases} p_0(a-t)\ell(t,a,t;S), & a \geq t, \\ B(t-a;S)\ell(t,a,a;S), & a < t, \end{cases} \tag{3.3}$$

where $S(t) \in C_+([0,T];\mathbb{R})$,[1] $T > 0$ is a fixed positive number and $\ell$ is the survival probability depending on $S$ given by

$$\ell(t,a,x;S) := \exp\left[-\int_0^x \mu(a-\sigma, S(t-\sigma))d\sigma\right], \quad 0 \leq x \leq \min\{a,t\},$$

which is the probability that individuals at age $a-x$ and time $t-x$ will survive to age $a$ and time $t$ under the given condition $S$.

If we insert expression (3.3) into the boundary condition in (3.2), we obtain an integral equation for the birth rate $B(t;S) := p(t,0)$:

$$B(t;S) = G(t;S) + \int_0^t \Psi(t,a;S)B(t-a;S)da, \tag{3.4}$$

where

$$\Psi(t,a;S) := \beta(a,S(t))\ell(t,a,a;S),$$
$$G(t;S) := \int_t^\infty \beta(a,S(t))\ell(t,a,t;S)p_0(a-t)da,$$

and we adopt the convention that all functions are extended to be zero for $a > \omega$. If we insert expression (3.3) into the definition of $S(t)$, it follows that

$$S(t) = H(t;S) + \int_0^t \gamma(a)\ell(t,a,a;S)B(t-a;S)da, \tag{3.5}$$

where

$$H(t;S) := \int_t^\infty \gamma(a)\ell(t,a,t;S)p_0(a-t)da.$$

Therefore, we arrive at a Volterra nonlinear integral equation system (3.4) and (3.5) for the unknown functions $(B(t;S), S(t))$. It is clear that if we can determine $B$ and $S$ from (3.4) and (3.5), the solution $p(t,a)$ of the original system can be recovered from (3.3). Although the existence and uniqueness of continuous solutions of the simultaneous integral equation system for $(B,S)$ can be studied using the classical fixed point argument [19], we instead sketch an $L^1$-scheme for the existence and

[1] $C_+([0,T];\mathbb{R})$ denotes the set of nonnegative continuous functions on $[0,T]$.

uniqueness of solutions of the nonlinear problem (3.2) developed by Iannelli [23]. We adopt the following technical assumption for $\beta$, $\mu$, and $\gamma$:

**Assumption 3.1** For any $x \in \mathbb{R}_+$, $\beta(\cdot, x) \in L^1_+(0, \omega)$, $\mu(\cdot, x) \in L^1_{\mathrm{loc},+}(0, \omega)$ and there exists a number $\beta_+ > 0$ such that $0 \le \beta(a, x) \le \beta_+$, $\forall (a, x) \in [0, \omega] \times \mathbb{R}_+$. Moreover, $\mu(a, x) \ge 0$, $\forall (a, x) \in [0, \omega] \times \mathbb{R}_+$ and

$$\int_0^{\omega} \mu(\sigma, x) d\sigma = +\infty.$$

For all $M > 0$, there exists a constant $L(M) > 0$ such that $|\beta(a, x) - \beta(a, \bar{x})| \le L(M)|x - \bar{x}|$, $|\mu(a, x) - \mu(a, \bar{x})| \le L(M)|x - \bar{x}|$ if $|x| \le M$ and $|\bar{x}| \le M$. In addition, $\gamma(\cdot) \in L^\infty_+(0, \omega)$.

**Exercise 3.1** Prove that if $|S| \le M$ and $|\bar{S}| \le M$, we have

$$|\ell(t, a, x; S) - \ell(t, a, x; \bar{S})| \le L(M) \int_{t-x}^{t} |S(\zeta) - \bar{S}(\zeta)| d\zeta. \tag{3.6}$$

The above condition implies that the vital rates $\beta$ and $\mu$ satisfy the same condition as the stable population model when we fix the control variable $x$. Moreover, they are uniformly Lipschitz continuous with respect to $x$ for $a \in [0, \omega]$. Under the above condition, for a given $S \in C_+([0, T]; \mathbb{R})$, $G(\cdot, S)$ and $B(\cdot, S)$ become continuous functions. Thus, we know that $t \to p(t, \cdot)$ defined by (3.3) belongs to $C_+([0, T]; L^1(0, \omega))$.

**Lemma 3.1** *For a given $S \in C([0, T]; \mathbb{R})$, it holds that*

$$|B(t; S)| \le \beta_+ e^{\beta_+ t} |p_0|_{L^1}, \quad t \in [0, T]. \tag{3.7}$$

*Moreover, suppose that $S, \bar{S} \in C([0, T]; \mathbb{R})$ with $|S(t)| \le M$ and $|\bar{S}(t)| \le M, \forall t \in [0, T]$. Then, there exists $\tilde{L}(M) > 0$ such that*

$$|B(t; S) - B(t; \bar{S})| \le \tilde{L}(M) |p_0|_{L^1} \left( |S(t) - \bar{S}(t)| + \int_0^t |S(\sigma) - \bar{S}(\sigma)| d\sigma \right). \tag{3.8}$$

**Exercise 3.2** Prove Lemma 3.1.

Define a function space $E := C([0, T]; L^1(0, \omega))$ with norm $|q|_E = \sup_{t \in [0,T]} |q(t, \cdot)|_{L^1}$ and its closed subset $K := \{q \in E | \ q(t, a) \ge 0, \ |q(t, \cdot)|_{L^1} \le M\}$. For $q \in K$, let

$$Q(t) = \int_0^{\omega} \gamma(a) q(t, a) da,$$

and define a map $\mathscr{F} : K \subset E \to E$ by

$$(\mathscr{F}q)(t,a) := \begin{cases} p_0(a-t)\ell(t,0,a-t;Q), & a \geq t, \\ B(t-a;Q)\ell(a,t-a,0;Q), & a < t, \end{cases} \tag{3.9}$$

where $p_0 \in L^1(0,\omega)$ is a given initial data. If this map has a fixed point $p \in K$, $p = p(t,a)$ becomes a solution of system (3.2) if the differential operator $\partial_t + \partial_a$ is replaced by the directional derivative $D$.

**Lemma 3.2** *For a fixed $T > 0$ and a given $p_0 \in L^1_+(0,\omega)$, choose $M > 0$ such that $e^{\beta_+ T}|p_0|_{L^1} < M$. Then, $\mathscr{F}(K) \subset K$ and there exists a number $C(M,T) > 0$ such that, for all $n = 1, 2, \ldots,$*

$$|\mathscr{F}^n q - \mathscr{F}^n \bar{q}|_E \leq \frac{C(M,T)^n T^n}{n!}|q - \bar{q}|_E. \tag{3.10}$$

*Proof* Let $q \in K$, and observe that

$$|(\mathscr{F}q)(t,\cdot)|_{L^1} = \int_0^t B(t-a;Q)\ell(a,t-a,0;Q)da + \int_t^\omega p_0(a-t)\ell(t,0,a-t;Q)da$$
$$\leq \int_0^t B(\sigma;Q)d\sigma + \int_0^\omega p_0(a)da \leq e^{\beta_+ t}|p_0|_{L^1} < M,$$

where we have used (3.7). Then, $\mathscr{F}(K) \subset K$. Next, we can show that there exists $C(M,T) > 0$ such that

$$|(\mathscr{F}q)(t,\cdot) - (\mathscr{F}\bar{q})(t,\cdot)|_{L^1} \leq C(M,T)\int_0^t |q(\sigma,\cdot) - \bar{q}(\sigma,\cdot)|_{L^1} d\sigma. \tag{3.11}$$

Thus, (3.10) holds for $n = 1$. By mathematical induction, we can show that, for all $n = 1, 2, \ldots,$

$$|(\mathscr{F}^n q)(t,\cdot) - (\mathscr{F}^n \bar{q})(t,\cdot)|_{L^1} \leq \frac{C(M,T)^n t^n}{n!}|q - \bar{q}|_E. \tag{3.12}$$

Therefore, we have (3.10). □

**Exercise 3.3** Prove that (3.11) and (3.12) hold.

From the above, we can state the following:

**Proposition 3.2** ([23]) *Let $p_0 \in L^1(0,\omega)$. Then, there exists a unique $p \in K$ such that*

$$p(t,a) = \begin{cases} p_0(a-t)\ell(t,a,t;S), & a \geq t, \\ B(t-a;S)\ell(t,a,a;S), & a < t, \end{cases}$$
$$S(t) = \int_0^\omega \gamma(a)p(t,a)da. \tag{3.13}$$

*Moreover, $p(t,a)$ satisfies*

$$Dp(t,a) = -\mu(a, S(t))p(t,a), \text{ a.e. } (t,a) \in \mathbb{R}_+ \times (0,\omega), \tag{3.14}$$

$$|p(t,\cdot)|_{L^1} \le e^{\bar{\beta}t}|p_0|_{L^1}, \tag{3.15}$$

$$|p(t,\cdot) - \bar{p}(t,\cdot)|_{L^1} \le e^{C(M,T)t}|p_0 - \bar{p}_0|_{L^1}, \tag{3.16}$$

*where $C(M,T)$ is a constant depending on $M$ and $T$ and $\bar{p}(t,\cdot)$ denotes the solution corresponding to the initial data $\bar{p}_0$.*

*Proof* For a given $p_0 \in L^1(0,\omega)$, we choose $M$ such that $e^{\bar{\beta}T}|p_0|_{L^1} < M$. From (3.10), $\mathscr{F}^n$ becomes a contraction mapping for a sufficiently large integer $n$, so it has a unique fixed point $p$ in $K$ that satisfies (3.13). Because (3.14) is clear, we show (3.15). From (3.7), $B(t;S) \le \bar{\beta}e^{\bar{\beta}t}|p_0|_{L^1}$ and (3.15) follows immediately. Finally, let $\mathscr{F}_{p_0}$ be the operator defined by (3.9) corresponding to the initial value $p_0$. Then, we have

$$p(t,\cdot) = (\mathscr{F}_{p_0}p)(t,\cdot), \quad \bar{p}(t,\cdot) = (\mathscr{F}_{\bar{p}_0}\bar{p})(t,\cdot).$$

From (3.9) and (3.11), we observe that

$$\begin{aligned}
&|p(t,\cdot) - \bar{p}(t,\cdot)|_{L^1} \\
&\le |(\mathscr{F}_{p_0}p)(t,\cdot) - (\mathscr{F}_{\bar{p}_0}p)(t,\cdot)|_{L^1} + |(\mathscr{F}_{\bar{p}_0}p)(t,\cdot) - (\mathscr{F}_{\bar{p}_0}\bar{p})(t,\cdot)|_{L^1} \\
&\le C(M,T)\int_0^t |p(\sigma,\cdot) - \bar{p}(\sigma,\cdot)|_{L^1}d\sigma + \int_t^\infty |p_0(a-t) - \bar{p}_0(a-t)|da.
\end{aligned}$$

Equation (3.16) follows from Gronwall's inequality. □

The above result shows that the age-density function $p(t,a)$ given by (3.13) is a solution of the nonlinear problem (3.2) with the directional derivative $D$ instead of the partial differential operator $\partial_t + \partial_a$, which can be regarded as a generalized solution of (3.2) without losing the original biological meaning. In other words, the map $p_0 \to p(t,\cdot)$ defined by (3.13) forms a dynamical system (nonlinear semigroup) in $L^1(0,\omega)$ [46, 47]. It is not difficult to show that the solution of (3.13) becomes a classical solution of (3.2) when we assume appropriate regularities of the initial data and vital rates.

**Exercise 3.4** Suppose that there exists an age interval $[a_1, a_2]$ such that $\beta(a,x) = 0$ and $\gamma(a) = 0$ for $a \notin [a_1, a_2]$ and all $x$. Moreover, assume that $\mu$ is independent of the size $S$. Let $T(t)$, $0 \le t \le a_1$, be a nonlinear operator on $L^1(0,\omega)$ defined by

$$(T(t)\phi)(a) = \begin{cases} \phi(a-t)\frac{\ell(a)}{\ell(a-t)}, & a-t > 0, \\ B(t-a)\ell(a), & t-a > 0, \end{cases}$$

where $\ell(a) = \exp(-\int_0^a \mu(x)dx)$ is the size-independent survival probability and

$$B(t) = \int_{a_1}^{a_2} \beta(a, S(t))\phi(a-t)\frac{\ell(a)}{\ell(a-t)}da, \quad S(t) = \int_{a_1}^{a_2} \gamma(a)\phi(a-t)\frac{\ell(a)}{\ell(a-t)}da.$$

Show that the solution $p(t, a), t > 0$, of (3.2) is given by $T(t - na_1)T^n(a_1)p_0$, where $n = [\frac{t}{a_1}]$, that is, we can obtain the solution of (3.2) without solving the nonlinear integral equations [36].

### 3.1.2 Stationary Solutions and Their Stability

Let $p^*(a)$ be the stationary solution of (3.2). Then, we have

$$\begin{aligned} \frac{dp^*(a)}{da} &= -\mu(a, S^*)p^*(a), \\ p^*(0) &= \int_0^\omega \beta(a, S^*)p^*(a)da, \\ S^* &= \int_0^\omega \gamma(a)p^*(a)da. \end{aligned} \tag{3.17}$$

It is clear that there exists a trivial solution $p^*(a) \equiv 0$, and so our problem is to seek non-trivial solutions of (3.17). From the differential equation part of (3.17), it follows that

$$p^*(a) = p^*(0)e^{-\int_0^a \mu(\sigma, S^*)d\sigma} = p^*(0)\ell(a; S^*), \tag{3.18}$$

where

$$\ell(a; S^*) := \exp\left(-\int_0^a \mu(\sigma, S^*)d\sigma\right),$$

is the survival probability depending on the stationary (weighted) population size $S^*$. Inserting (3.18) into the boundary condition in (3.17), we obtain

$$\begin{aligned} p^*(0) &= p^*(0)\int_0^\omega \beta(\sigma, S^*)\ell(\sigma; S^*)d\sigma, \\ S^* &= p^*(0)\int_0^\omega \gamma(\sigma)\ell(\sigma; S^*)d\sigma. \end{aligned}$$

Therefore, if there exists some $S^* > 0$ such that

$$\int_0^\omega \beta(\sigma, S^*)\ell(\sigma; S^*)d\sigma = 1, \tag{3.19}$$

then the positive stationary state solution is given by

$$p^*(a) = p^*(0)\ell(a; S^*) = \frac{S^*\ell(a; S^*)}{\int_0^\omega \gamma(\sigma)\ell(\sigma; S^*)d\sigma}. \tag{3.20}$$

If we define the net reproduction rate depending on the population size $S$ by

$$R(S) := \int_0^\omega \beta(\sigma, S)\ell(\sigma; S)d\sigma,$$

then the stationary condition (3.19) implies that the equation $R(S) = 1$ has a positive root $S^*$, that is, the net reproduction rate at the steady state must be unity. This is an intuitively clear and expected result.

The behavior of the reproduction number $R(S)$ as a function of $S$ depends on the population growth mechanism. Here, we assume that $R(S)$ satisfies the Allee–Logistic assumption, as follows:

$$\begin{cases} (1)\ \ R'(S) > 0, \quad 0 < S < S_0, \\ (2)\ \ R'(S) < 0, \quad S > S_0, \\ (3)\ \ \lim_{S\to\infty} R(S) = 0, \end{cases} \tag{3.21}$$

where $S_0 \geq 0$ is the critical population size at which the feedback effect on the reproductivity changes from positive to negative. If $S_0 = 0$, then only (2) and (3) are meaningful, and the model becomes a pure logistic model (Fig. 3.1). The Allee–Logistic assumption is realized if $\beta(a, S)$ is increasing [decreasing] and $\mu(a, S)$ is decreasing [increasing] for $0 < S < S_0$ [$S > S_0$]. Considering the positive solution of the equation $R(S) = 1$, it is easy to see that the following holds:

**Proposition 3.3** *Let $R_0 = R(0)$ be the basic reproduction number. Under the assumptions in* (3.21), *there exist three cases provided that $S_0 > 0$:*

*(1) If $R_0 > 1$, there exists a unique non-trivial stationary solution.*
*(2) If $R_0 < 1 < R(S_0)$, there exist exactly two non-trivial stationary solutions.*
*(3) If $R(S_0) < 1$, there is no non-trivial stationary solution.*

*If $S_0 = 0$, only two cases are possible:*

*(1) If $R_0 > 1$, there exists a unique non-trivial stationary solution.*
*(2) If $R_0 \leq 1$, there is no non-trivial stationary solution.*

Subsequently, let us consider the stability of the stationary solutions. We commence by confirming the definition of stability.

**Definition 3.1** The stationary solution $p^*(\cdot)$ is said to be *stable* if, for any $\varepsilon > 0$, there exists $\delta > 0$ such that the solution $p(t, \cdot)$ corresponding to the initial data $p_0(\cdot)$ satisfying $|p_0 - p^*|_{L^1} \leq \delta$ exists for all $t \geq 0$ and $|p(t, \cdot) - p^*(\cdot)|_{L^1} \leq \varepsilon$ for all $t \geq 0$. If $p^*$ is stable and we can take $\delta$ such that $\lim_{t\to\infty} |p(t, \cdot) - p^*(\cdot)|_{L^1} = 0$, then the solution is said to be *asymptotically stable*. If $p$ is not stable, it is called *unstable*.

To analyze the local stability, let us consider the linearized equation at the steady state $p^*(\cdot)$. We assume that the parameters have sufficient smoothness with respect to each variable. The stationary solution is given by (3.20), where its size $S^*$ is a

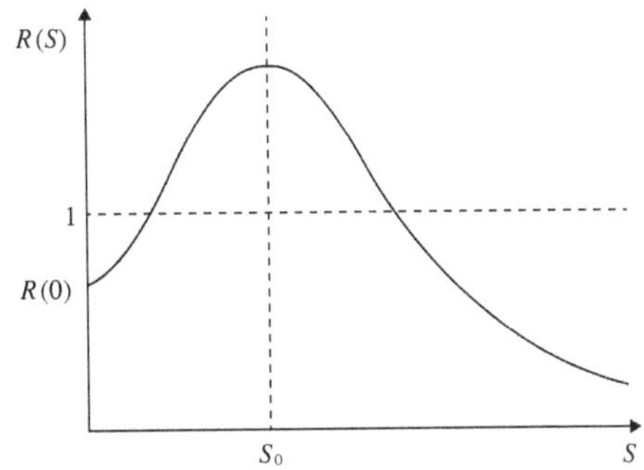


**Fig. 3.1** A case that there exist two non-trivial stationary solutions

nonnegative solution of the equation $R(S) = 1$. We define the perturbation from the stationary state as follows:

$$\xi(t,a) := p(t,a) - p^*(a), \quad \eta(t) := S(t) - S^*.$$

Inserting the above expressions into (3.2) and neglecting the second-order terms of $\xi(a,t)$ and $\eta(t)$, we obtain the following *linearized equation*:

$$\begin{aligned}
&\frac{\partial \xi(t,a)}{\partial t} + \frac{\partial \xi(t,a)}{\partial a} = -\mu(a,S^*)\xi(t,a) - \mu'_x(a,S^*)p^*(a)\eta(t),\\
&\xi(t,0) = \int_0^\omega \beta(\sigma,S^*)\xi(t,\sigma)d\sigma + \kappa\eta(t),\\
&\eta(t) = \int_0^\omega \gamma(\sigma)\xi(t,\sigma)d\sigma,\\
&\xi(0,a) = p_0(a) - p^*(a),\\
&\kappa := \int_0^\omega \beta'_x(\sigma,S^*)p^*(\sigma)d\sigma.
\end{aligned} \tag{3.22}$$

To consider the asymptotic behavior of the linearized equation, let us consider the exponential solutions $\xi(t,a) = e^{\lambda t}u_\lambda(a)$. Then, $u_\lambda(a)$ must satisfy the following system:

$$\begin{aligned}
&u'_\lambda(a) = -(\lambda + \mu(a,S^*))u_\lambda(a) - \mu'_x(a,S^*)p^*(a)\int_0^\omega \gamma(\sigma)u_\lambda(\sigma)d\sigma,\\
&u_\lambda(0) = \int_0^\omega [\beta(\sigma,S^*) + \kappa\gamma(\sigma)]u_\lambda(\sigma)d\sigma.
\end{aligned} \tag{3.23}$$

By solving the ODE in (3.23), we have

$$u_\lambda(a) = u_\lambda(0)e^{-\lambda a}\ell(a,S^*) - p^*(a)\int_0^a e^{-\lambda(a-z)}\mu'_x(z,S^*)dz\int_0^\omega \gamma(\sigma)u_\lambda(\sigma)d\sigma.$$

Therefore, it follows that

$$\int_0^\omega \gamma(\sigma)u_\lambda(\sigma)d\sigma = \frac{u_\lambda(0)\int_0^\omega e^{-\lambda a}\gamma(a)\ell(a,S^*)da}{1+\int_0^\omega \gamma(a)p^*(a)\int_0^a e^{-\lambda(a-z)}\mu_x'(z,S^*)dzda}.$$

Inserting the above relation into the boundary condition in (3.22), we arrive at the following equation:

$$\begin{aligned} 1 = &\int_0^\omega e^{-\lambda a}\beta(a,S^*)\ell(a,S^*)da \\ &+ H(\lambda,S^*)\frac{\int_0^\omega e^{-\lambda a}\gamma(a)p^*(a)da}{1+\int_0^\omega \gamma(a)p^*(a)\int_0^a e^{-\lambda(a-z)}\mu_x'(z,S^*)dzda}, \end{aligned} \tag{3.24}$$

where

$$\begin{aligned} H(\lambda,S^*) :=& \int_0^\omega \beta_x'(a,S^*)\ell(a,S^*)da \\ &- \int_0^\omega \beta(a,S^*)\ell(a,S^*)\int_0^a \mu_x'(z,S^*)e^{-\lambda(a-z)}dzda \\ =& R'(S^*) + \int_0^\omega \beta(a,S^*)\ell(a,S^*)\int_0^a \mu_x'(z,S^*)(1-e^{-\lambda(a-z)})dzda. \end{aligned}$$

Moreover, if we use (3.20), we can rewrite the above equation as $F(\lambda)=1$, where

$$\begin{aligned} F(\lambda) :=& \int_0^\omega e^{-\lambda a}\beta(a,S^*)\ell(a,S^*)da \\ &+ \frac{H(\lambda,S^*)S^*\int_0^\omega e^{-\lambda a}\gamma(a)\ell(a,S^*)da}{\int_0^\omega \gamma(a)\ell(a,S^*)da + S^*\int_0^\omega \gamma(a)\ell(a,S^*)\int_0^a e^{-\lambda(a-z)}\mu_x'(z,S^*)dzda}. \end{aligned}$$

We call $F(\lambda)=1$ the *characteristic equation*, and its roots are called *characteristic roots*. Let $\Lambda := \{\lambda \in \mathbb{C} : F(\lambda)=1\}$.

Let us consider the linearized equation (3.22) as an abstract ODE on the state space $L^1$:

$$\frac{d\zeta(t)}{dt} = A\zeta(t), \quad \zeta(0)=\zeta_0 = p_0 - p^*, \tag{3.25}$$

where $\zeta(t)$ is a vector-valued function $\zeta(t)=\zeta(t,\cdot)\in L^1(0,\omega)$, and $A$ is a differential operator given by

$$\begin{aligned} (Af)(a) &:= -\frac{df(a)}{da} - \mu(a,S^*)f(a) - \mu_x'(a,S^*)p^*(a)\int_0^\omega \gamma(\sigma)f(\sigma)d\sigma, \\ \mathscr{D}(A) &:= \left\{ f\in L^1(0,\omega) : f(0) = \int_0^\omega [\beta(\sigma,S^*)+\kappa\gamma(a)]f(\sigma)d\sigma, \quad Af \in L^1(0,\omega)\right\}. \end{aligned}$$

As for the stable population model, we can then expect the perturbation $\xi$ to admit the asymptotic expansion

$$\xi(t,a) \approx \sum_{\lambda \in \Lambda} \alpha_\lambda e^{\lambda t} u_\lambda(a), \tag{3.26}$$

where $\alpha_\lambda$ is given by

$$\alpha_\lambda = \frac{\langle v_\lambda, \xi(0,\cdot)\rangle}{\langle v_\lambda, u_\lambda\rangle},$$

$\langle v_\lambda, u_\lambda\rangle = \int_0^\omega v_\lambda(a) u_\lambda(a) da$ and $v_\lambda$ denotes the adjoint eigenfunction associated with the eigenvalue $\lambda$ of the linearized generator $A$.

If (3.26) holds and all real parts of the characteristic roots are negative, it follows that $\lim_{t\to\infty} \zeta(t,a) = 0$, and we can conclude that the stationary solution $p^*$ is locally asymptotically stable. Although the classical proof for this argument can be given using the integral equation approach, we here sketch another intuitive justification for the expansion (3.26).

Let us denote the Laplace transform as

$$\hat{\zeta}(\lambda) := \int_0^\infty e^{-\lambda t} \zeta(t) dt.$$

Taking the Laplace transform of (3.25) with respect to the time variable, we have, for sufficiently large $\lambda > 0$,

$$\int_0^\infty e^{-\lambda t} \frac{d\zeta(t)}{dt} dt = -\zeta_0 + \lambda \hat{\zeta}(\lambda) = A\hat{\zeta}(\lambda).$$

Thus, if there exists an inverse transformation $(\lambda - A)^{-1}$, we obtain $\hat{\zeta}(\lambda) = (\lambda - A)^{-1}\zeta_0$. The set of complex numbers $\lambda$ such that the bounded linear operator $(\lambda - A)^{-1}$ exists is called the *resolvent set*, denoted by $\rho(A)$, and $(\lambda - A)^{-1}$ is called the *resolvent* of the linear operator $A$. The complement of the resolvent set is called the *spectrum*, denoted by $\sigma(A) = \mathbb{C} \setminus \rho(A)$. For the population operator $A$, we can generally prove that $\{\lambda \in \mathbb{C} : \Re\lambda > \alpha\} \subset \rho(A)$ for a large $\alpha$.

Using the inverse Laplace transformation, for a large $\alpha > 0$, we have

$$\zeta(t) = \frac{1}{2\pi i} \int_{\alpha - i\infty}^{\alpha + i\infty} e^{\lambda t} (\lambda - A)^{-1} \zeta_0 d\lambda,$$

where $\{\alpha + ix : x \in \mathbb{R}\} \subset \rho(A)$. Therefore, as for the Fundamental Theorem of Demography, the asymptotic behavior of $\zeta(t)$ is determined by the location of the eigenvalues of $A$ if the resolvent is compact, and we obtain the asymptotic expansion (3.26) if the eigenvalues are all simple roots.

By solving the equation $(\lambda - A)\phi = \psi$ formally, we obtain the following expression:

$$\begin{aligned}\phi(a) &= (\lambda - A)^{-1}\psi(a)\\ &= \phi(0)e^{-\lambda a}\ell(a,S^*) - p^*(a)\int_0^a e^{-\lambda(a-z)}\mu_x'(z,S^*)dz\int_0^\omega \gamma(\sigma)\phi(\sigma)d\sigma\\ &\quad + \int_0^a \psi(\sigma)e^{-\lambda(a-\sigma)}\frac{\ell(a,S^*)}{\ell(\sigma,S^*)}d\sigma.\end{aligned}$$

The unknown terms $\int_0^\omega \gamma(\sigma)\phi(\sigma)d\sigma$ and $\phi(0)$ are calculated as follows:

$$\begin{aligned}&\int_0^\omega \gamma(\sigma)\phi(\sigma)d\sigma\\ &\quad= \frac{\phi(0)\int_0^\omega e^{-\lambda a}\gamma(a)\ell(a,S^*)da + \int_0^\omega \gamma(a)da\int_0^a \psi(\sigma)e^{-\lambda(a-\sigma)}\frac{\ell(a,S^*)}{\ell(\sigma,S^*)}d\sigma}{1+\int_0^\omega \gamma(a)p^*(a)da\int_0^a e^{-\lambda(a-z)}\mu_x'(z,S^*)dz},\end{aligned}$$

$$\begin{aligned}\phi(0) &= (1-F(\lambda))^{-1}\left[\int_0^\omega [\beta(a,S^*)+\kappa\gamma(a)]da\int_0^a \psi(\sigma)e^{-\lambda(a-\sigma)}\frac{\ell(a,S^*)}{\ell(\sigma,S^*)}d\sigma\right.\\ &\quad + \frac{\int_0^\omega \gamma(a)da\int_0^a \psi(\sigma)e^{-\lambda(a-\sigma)}\frac{\ell(a,S^*)}{\ell(\sigma,S^*)}d\sigma}{1+\int_0^\omega \gamma(a)p^*(a)da\int_0^a e^{-\lambda(a-z)}\mu_x'(z,S^*)dz}\\ &\quad\left.\times\int_0^\omega [\beta(a,S^*)+\kappa\gamma(a)]p^*(a)da\int_0^a e^{-\lambda(a-z)}\mu_x'(z,S^*)dz\right].\end{aligned}$$

Therefore, we know that if $\omega < \infty$ and all parameter functions are bounded, the spectrum of $(\lambda - A)^{-1}$ is composed of characteristic roots of $F(\lambda) = 1$ (eigenvalues of $A$). Hence, a necessary and sufficient condition for the asymptotic stability of the zero solution of the linearized equation is that all eigenvalues are included in the left half plane $\Re z < 0$. The *principle of linearized stability* for the nonlinear system (3.2) is that this stability condition of the zero solution of the linearized system implies the local asymptotic stability of the stationary solution $p^*$. Then, we can state the following:

**Proposition 3.4** *Let $p^*(a) = p^*(0)\ell(a;S^*)$ be the stationary solution of* (3.2). *If all real parts of the characteristic roots of the equation $F(\lambda) = 1$ are negative, $p^*(\cdot)$ is locally asymptotically stable, whereas it is unstable if there exists a characteristic root with a positive real part.*

The rigorous proof of the principle of linearized stability was given by Desch and Schappacher [10] and Webb [47] under a more general setting, where it was also shown that the stationary solution is unstable if there exists a characteristic root with a positive real part. For the nonlinear system of (3.2), using the integral equation approach and the Laplace transformation technique, Gurtin and MacCamy [19] provided a classical proof for the sufficient condition of the local stability of the stationary solution. Because their calculation is long and cumbersome, we do not repeat it here (see also [23]). However, we note that the essential point of their proof is the same as the results for the cohort control model, which is discussed in Sect. 3.3.

**Exercise 3.5** Consider renewal equations (3.4) and (3.5). Define the perturbation variables $(U_0, U_1)$ by $B(t; S) = p^*(0) + U_0(t)$ and $S(t) = S^* + U_1(t)$. Show that (3.4) and (3.5) can be rewritten as

$$U(t) = CU(t) + \int_0^t A(\sigma)U(t-\sigma)d\sigma + F(t),$$

where

$$U(t) = \begin{pmatrix} U_0(t) \\ U_1(t) \end{pmatrix}, \quad C = \begin{pmatrix} 0 & c_{01} \\ 0 & 0 \end{pmatrix}, \quad A(\sigma) = \begin{pmatrix} A_{00}(\sigma) & A_{01}(\sigma) \\ A_{10}(\sigma) & A_{11}(\sigma) \end{pmatrix},$$

and $F(t)$ denotes the second-order term. Determine matrices $C$ and $A(\sigma)$ and the characteristic equation

$$\det(I - (I - C)^{-1}\hat{A}(\lambda)) = 0,$$

where $\hat{A}(\lambda)$ denotes the Laplace transformation of $A(\cdot)$. From Proposition 10.39, if there is no characteristic root with a nonnegative real part, the zero solution $U = 0$ is stable, which implies the local stability of the steady state $(p^*(0), S^*)$ [23].

### 3.1.3 *Exchange of Stability*

We now consider the stability of the stationary state under some specialized assumptions for the Allee–Logistic model. As $F(\lambda)$ is not necessarily positive for real $\lambda$, we cannot use the same argument as for Lotka's characteristic equation. Therefore, we first consider a simple case in which the death rate is independent of the population size. In this case, because $\mu'_x = 0$, the characteristic equation is drastically simplified to

$$F(\lambda, S^*) = \int_0^\omega e^{-\lambda a}\beta(a, S^*)\ell(a)da + R'(S^*)S^* \frac{\int_0^\omega e^{-\lambda a}\gamma(a)\ell(a)da}{\int_0^\omega \gamma(a)\ell(a)da} = 1, \quad (3.27)$$

where $\ell(a)$ denotes the survival probability independent of the population size. Define a function $\phi(a, x)$ as follows:

$$\phi(a, x) := \beta(a, x)\ell(a) + xR'(x)\frac{\gamma(a)\ell(a)}{\int_0^\omega \gamma(a)\ell(a)da}.$$

The characteristic equation is then written as

$$F(\lambda, S^*) = \int_0^\omega e^{-\lambda a}\phi(a, S^*)da = 1.$$

If $R'(S^*) \geq 0$, then $\phi$ is positive, so we can apply the same argument as for Lotka's characteristic equation. Observe that

$$\int_0^{\omega} \phi(a, S^*)da = \begin{cases} 1 + S^* R'(S^*), & S^* > 0, \\ R_0, & S^* = 0. \end{cases}$$

Thus, we can immediately conclude that:

**Proposition 3.5** *If the death rate is independent of the population size and the Allee–Logistic assumption* (3.21) *is satisfied, it follows that*

*(1) If there is no non-trivial stationary solution, the trivial stationary solution is locally asymptotically stable.*
*(2) If there exists a unique non-trivial stationary solution, the trivial stationary solution is unstable.*
*(3) If there exist two non-trivial stationary solutions, one of them is unstable and the trivial stationary solution is locally asymptotically stable.*

*Proof* If there is no non-trivial stationary solution, it holds that $R_0 < R(S_0) < 1$. Then, we have $F(0, 0) = R_0 < 1$, so any real part of the root of the characteristic equation for the trivial stationary solution is negative. If there exists a unique non-trivial stationary solution, it follows that $R_0 > 1$ and the characteristic equation at the trivial stationary solution has a positive root. Finally, if there exist two non-trivial stationary solutions, we have $F(0, 0) = R_0 < 1 < R(S_0)$, so again all real parts of the roots of the characteristic equation at the trivial stationary solution are negative. For the non-trivial stationary solution with the smaller population size, it follows that $R'(S^*) > 0$. Thus, the characteristic equation at the solution $F(\lambda, S^*) = 1$ has a positive root because $\phi(a, S^*) > 0$ and $F(0, S^*) = 1 + S^* R_0'(S^*) > 1$. □

In the above proposition, we have no result for the stability of a non-trivial stationary solution for which $R'(S^*) < 0$. However, this case is most interesting for real applications, as this type of stationary solution appears as a result of the logistic effect. We will take up this problem again in Sect. 3.3; here, we give only intuitive explanations.

Suppose that the age-specific birth rate is written as

$$\beta(a, x) = R_0 \psi(x) \beta_0(a),$$

where $\beta_0(a)$ is the normalized age-specific birth rate such that

$$\int_0^{\omega} \beta_0(a) \ell(a) da = 1.$$

To reflect the pure logistic effect, we assume that the function $\psi : [0, +\infty) \to (0, +\infty)$ satisfies

$$\psi(0) = 1, \quad \psi'(x) < 0, \quad \lim_{x \to +\infty} \psi(x) = 0.$$

Then, $R_0\psi(x)$ gives the reproduction number at population size $x$. It is clear from the logistic assumption that if $R_0 \leq 1$, there is no non-trivial stationary solution, whereas a unique non-trivial stationary solution exists if $R_0 > 1$. From Proposition 3.5, we know that the trivial stationary solution is stable if $R_0 < 1$ and is unstable if $R_0 > 1$. In such a case, the non-trivial stationary solution forwardly bifurcated from the trivial stationary solution at $R_0 = 1$ is called *supercritical*, and the non-trivial stationary solution backwardly bifurcated from the trivial stationary solution at $R_0 = 1$ is called *subcritical*. The subcritical bifurcation does not occur for the pure logistic model.

From the stationarity condition (3.19), it follows that $R_0\psi(S^*) = 1$. Therefore, we have

$$S^* = \psi^{-1}(R_0^{-1}), \quad \phi(a, S^*) = \beta_0(a)\ell(a) + R'(S^*)S^* \frac{\gamma(a)\ell(a)}{\int_0^\omega \gamma(a)\ell(a)da}.$$

Let us introduce a bifurcation parameter $\varepsilon$ by

$$\varepsilon = S^*R'(S^*) = \psi^{-1}(R_0^{-1})R'(\psi^{-1}(R_0^{-1})).$$

Then, $\varepsilon \leq 0$ ($\varepsilon = 0$ corresponds to $R_0 = 1$), and the characteristic equation is written as

$$F(\lambda, \varepsilon) := \int_0^\omega e^{-\lambda a} \left[ \beta_0(a)\ell(a) + \varepsilon \frac{\gamma(a)\ell(a)da}{\int_0^\omega \gamma(a)\ell(a)da} \right] da = 1.$$

Let $\lambda(\varepsilon)$ be the dominant root—that is, the characteristic root whose real part is the largest among all characteristic roots. We know that such a root is uniquely determined for small $\varepsilon$, because $\lambda(0) = 0$ is an isolated characteristic root, and all other roots are isolated and located in the left half plane. If $\lambda(\varepsilon)$ is a smooth function, it follows from the Implicit Function Theorem that

$$\left.\frac{d\lambda}{d\varepsilon}\right|_{\varepsilon=0} = \frac{\frac{\partial F}{\partial \varepsilon}(0, 0)}{\int_0^\omega a\beta_0(a)\ell(a)da} > 0.$$

Thus, the orbit of $\lambda(\varepsilon)$ starting from $\lambda(0) = 0$ moves to the right if $\varepsilon$ increases from 0 and moves to the left if $\varepsilon$ decreases. If $\varepsilon$ is sufficiently small, we can show that other characteristic roots remain in the left half plane using Rouché's theorem.

*Remark 3.1* According to Iannelli [23, Chap. IV], we here sketch some details of the above argument. Consider a characteristic equation $\hat{K}_0(\lambda) + \varepsilon\hat{K}_1(\lambda) = 1$, where $\hat{K}_j$ denotes the Laplace transform of nonnegative functions $K_j$. Suppose that $K_j$ has a compact support and $\hat{K}_j(0) = 1$. The unperturbed equation $\hat{K}_0(\lambda) = 1$ has the real root $\lambda = 0$ which is the unique one in the half plane $\Re\lambda \geq \alpha$ for some $\alpha < 0$. Let us choose a positive number $m$ as $m := \inf_{y \in (-\infty, \infty)} |1 - \hat{K}_0(\alpha + iy)| > 0$. From the

Riemann–Lebesgue lemma that we can take a large number $L > 0$ such that for $|\lambda| > L$, $\Re\lambda \geq \alpha$, $\frac{1}{2} < |1 - \hat{K}_0(\lambda)|$. Moreover, there exists a $M > 0$ such that $|\hat{K}_1(\lambda)| < M$ for $\Re\lambda \geq \beta$ for some negative number $\beta < \alpha$. Then if $\varepsilon > 0$ is sufficiently small such that $|\varepsilon| < (1/M) \min\left(m, \frac{1}{2}\right)$, then we obtain $|\varepsilon||\hat{K}_1(\lambda)| < |1 - \hat{K}_0(\lambda)|$ on the domain $\{\lambda \in \mathbb{C} : |\lambda| > L, \Re\lambda > \alpha\}$ and on the vertical line $\{\lambda \in \mathbb{C} : \Re\lambda = \alpha\}$. Therefore it follows from the Rouché theorem, the perturbed equation $(1 - \hat{K}_0(\lambda)) - \varepsilon\hat{K}_1(\lambda) = 0$ has one and only one root in the half plane $\Re\lambda \geq \alpha$, since equation $1 - \hat{K}_0(\lambda) = 0$ has only one zero in this half plane. Then, for small $\varepsilon$, there exists a unique (dominant) root $\lambda(\varepsilon)$ of the perturbed equation in the half plane $\Re\lambda \geq \alpha$ such that $\lambda(0) = 0$.

Then, for the pure logistic model, when $R_0 > 1$ and $|R_0 - 1|$ is sufficiently small, the supercritically bifurcated non-trivial stationary solution $S^* > 0$, which is also small, becomes locally stable because all of its characteristic roots are located in the left half plane. This is a typical example of the well-known *principle of exchange of stability*, whereby as a parameter changes, a stable stationary solution (which is the trivial steady state in our case) becomes unstable, and a supercritically bifurcated stationary solution becomes stable.

To examine what happens if $|\varepsilon|$ becomes larger, let us consider the special case $\beta_0(a) = \gamma(a)$. In such a case, the characteristic equation is simplified to

$$F(\lambda, \varepsilon) = (1 + \varepsilon) \int_0^{\omega} e^{-\lambda a} \beta_0(a) \ell(a) da = 1.$$

It is then clear that, for $-1 < \varepsilon < 0$, all characteristic roots have a negative real part, and the non-trivial stationary solution is stable. However, as we see in Sect. 3.3, if $\varepsilon < -2$, a conjugate pair of characteristic roots cross the imaginary axis into the right half plane. That is, the non-trivial stationary solution becomes unstable and may bifurcate into a periodic solution (Hopf bifurcation, see [2, 9, 30]). This mechanism is again studied in Sect. 3.3 in the context of the Easterlin model.

## 3.2 Global Behavior: Illustrative Examples

In the previous section, we examined the local stability of stationary solutions. In general, however, it is far more difficult to study the global behavior of solutions of the nonlinear system (3.2). Readers may refer to [23, 24, 37, 47] for some global results. Here, according to [23], we introduce two well-known examples in which the global behavior can be determined analytically. Note that those examples are, more generally, investigated in Metz and Diekmann [29, Chap. IV, Sect. 5] as the problem of finite dimensional representability for age-dependent population models.

### 3.2.1 Existence of Periodic Solutions

We assume that the age space is the infinite interval $[0, \infty)$ and the age-specific birth and death rates are given by

$$\beta(a, x) = \beta(x)e^{-\alpha a}, \quad \mu(a, x) = \mu(x), \tag{3.28}$$

where $\alpha > 0$ is a constant. That is, the birth rate is decreasing as the population ages and the death rate is independent of age. Moreover, we assume that $\gamma \equiv 1$, that is, the vital rates depend on the total size of the population. Under this special assumption, the nonlinear population model (3.2) can be written as follows [19, 23]:

$$\begin{aligned} &\frac{\partial p(t,a)}{\partial t} + \frac{\partial p(t,a)}{\partial a} = -\mu(P(t))p(t,a), \\ &p(t,0) = \beta(P(t)) \int_0^\infty e^{-\alpha\sigma} p(t,\sigma)d\sigma, \\ &p(0,a) = p_0(a), \\ &P(t) = \int_0^\infty p(t,\sigma)d\sigma. \end{aligned} \tag{3.29}$$

We introduce a new variable $Q(t)$ as

$$Q(t) := \int_0^\infty e^{-\alpha\sigma} p(t,\sigma)d\sigma.$$

As long as $\alpha \neq 0$, $P(t)$ and $Q(t)$ are different variables. It follows that

$$\begin{aligned} \frac{dP(t)}{dt} &= \int_0^\infty p_t(t,a)da = -\int_0^\infty p_a(t,a)da - \bar{\mu}(P(t))P(t) \\ &= \beta(P(t))Q(t) - \mu(P(t))P(t), \\ \frac{dQ(t)}{dt} &= \int_0^\infty e^{-\alpha a} p_t(t,a)da \\ &= -\int_0^\infty e^{-\alpha a} p_a(t,a)da - \mu(P(t)) \int_0^\infty e^{-\alpha a} p(t,a)da \\ &= \beta(P(t))Q(t) - \alpha Q(t) - \mu(P(t))Q(t). \end{aligned}$$

Therefore, (3.29) can be reduced to the following system of ODEs:

$$\begin{aligned} \frac{dP(t)}{dt} &= \beta(P(t))Q(t) - \mu(P(t))P(t), \\ \frac{dQ(t)}{dt} &= [\beta(P(t)) - \alpha - \mu(P(t))]Q(t), \end{aligned} \tag{3.30}$$

where the initial data are given by

$$P(0) = P_0 := \int_0^\infty p_0(a)da, \quad Q(0) := Q_0 = \int_0^\infty e^{-\alpha a} p_0(a)da.$$

If $(P(t), Q(t))$ is the solution of (3.30), the solution of the original system (3.29) is

$$p(t,a) = \begin{cases} p_0(a-t)e^{-\int_0^t \mu(P(\sigma))d\sigma}, & a \ge t, \\ B(t-a)e^{-\int_{t-a}^t \mu(P(\sigma))d\sigma}, & a < t, \end{cases}$$

where

$$B(t) := p(t,0) = \beta(P(t))Q(t).$$

Therefore, we mainly focus on the behavior of the ODE system (3.30). To retain its biological meaning, we only consider the initial data $(P_0, Q_0)$ such that $P_0 \ge Q_0 \ge 0$. The natural state space of (3.30) is then given by $\Omega := \{(P, Q) | \, 0 \le Q \le P\} \subset \mathbb{R}^2$, and we can prove that the following statement holds:

**Proposition 3.6** *$\Omega$ is positively invariant with respect to the flow defined by* (3.30).

*Proof* Let $(P(t), Q(t))$ be the solution of (3.30) corresponding to the initial data point $(P_0, Q_0) \in \Omega$. It is sufficient to show that $P(t) \ge Q(t) \ge 0$, $\forall t > 0$. Because $Q_0 \ge 0$, it follows from the second equation of (3.30) that, for all $t > 0$, $Q(t) \ge 0$. Define $W(t) := P(t) - Q(t)$. Then, we have

$$\frac{d}{dt} W(t) = -\mu(P(t))W(t) + \alpha Q(t), \quad W(0) > 0.$$

Using the variation-of-constants formula, we obtain

$$W(t) = W(0)e^{-\int_0^t \mu(P(\sigma))d\sigma} + \int_0^t e^{-\int_s^t \mu(P(\sigma))d\sigma} \alpha Q(s)ds,$$

which shows that $W(t) \ge 0$ for all $t > 0$. □

Let $(P^*, Q^*)$ be the steady state of (3.30). Then, we can observe that

$$\begin{cases} \beta(P^*)Q^* - \mu(P^*)P^* = 0, \\ [\beta(P^*) - \alpha - \mu(P^*)]Q^* = 0. \end{cases} \tag{3.31}$$

If a non-trivial stationary solution exists, $P^* > 0$ must satisfy

$$\beta(P^*) = \alpha + \mu(P^*).$$

This condition implies that the net reproduction rate of (3.29) equals unity at $P(t) = P^*$. In fact, the reproduction number at the population size $x$ is given by

$$R(x) = \int_0^\infty \beta(x) e^{-\alpha\sigma - \mu(x)\sigma} d\sigma = \frac{\beta(x)}{\alpha + \mu(x)}.$$

From (3.31), once $P^*$ has been determined, $Q^*$ is calculated as

$$Q^* = \frac{\mu(P^*)P^*}{\beta(P^*)}.$$

Because $\beta(P^*) > \mu(P^*)$, we know that any non-trivial stationary state $(P^*, Q^*)$ belongs to the state space $\Omega$.

To reflect the logistic effect, we introduce the following assumption:

$$\begin{cases} \beta'(x)(\alpha + \mu(x)) - \beta(x)\mu'(x) < 0, \quad \forall x \geq 0, \\ \lim_{x \to +\infty} \beta(x) = 0, \\ \lim_{x \to +\infty} \mu(x) > 0. \end{cases} \tag{3.32}$$

The above assumption guarantees that the reproduction number $R(x)$ is a decreasing function of the population size, and the following holds:

$$R'(x) < 0, \quad \lim_{x \to +\infty} R(x) = 0, \tag{3.33}$$

which shows that the population growth rate $[\beta(\cdot) - \mu(\cdot)]$ ultimately becomes negative. Therefore, if $P(t)$ becomes sufficiently large, $P'(t) < 0$ and the solution orbit remains in a bounded set. That is, if we take a large number $P^+ > 0$, the compact set $\Omega^+ := \{(P, Q) : 0 \leq Q \leq P \leq P^+\}$ becomes positively invariant with respect to the flow defined by (3.30). Moreover, if

$$R_0 = R(0) = \frac{\beta(0)}{\alpha + \mu(0)} > 1,$$

then $R(x) = 1$ has a unique positive root and system (3.30) has a unique stationary solution in $\Omega^+$, whereas if $R_0 \leq 1$, only the trivial stationary solution exists.

Let us now consider the stability of stationary solutions. First, the following holds for the trivial steady state $(0, 0)$:

**Proposition 3.7** *If $R_0 < 1$, only the trivial steady state exists and it is globally asymptotically stable. If $R_0 > 1$, the trivial steady state becomes unstable and there exists a unique non-trivial steady state.*

*Proof* The equation for $Q(t)$ can be rewritten as

$$\frac{dQ(t)}{dt} = (\mu(P(t)) + \alpha)(R(P(t)) - 1) \leq (\bar{\mu} + \alpha)(R_0 - 1)Q(t),$$

where $\bar{\mu} := \max_{0 \leq P \leq P^+} \mu(P)$. Then, we have

$$Q(t) \le e^{(\bar{\mu}+\alpha)(R_0-1)t}. \tag{3.34}$$

Therefore, if $R_0 < 1$, then $\lim_{t\to\infty} Q(t) = 0$. Conversely, for $P(t)$, observe that

$$\frac{dP(t)}{dt} = \beta(P(t))Q(t) - \mu(P(t))P(t) \le \bar{\beta}Q(t) - \underline{\mu}P(t),$$

where $\bar{\beta} := \max_{0\le P\le P^+} \beta(P)$ and $\underline{\mu} := \min_{0\le P\le P^+} \mu(P) > 0$. Thus, we obtain

$$P(t) \le P(0)e^{-\underline{\mu}t} + \bar{\beta}\int_0^t e^{-\underline{\mu}(t-s)}Q(s)ds.$$

Using (3.34), it is easy to show that $\lim_{t\to\infty} P(t) = 0$ if $R_0 < 1$. The Jacobian matrix at the origin is given by

$$J(0,0) = \begin{pmatrix} -\mu(0) & \beta(0) \\ 0 & \beta(0) - \alpha - \mu(0) \end{pmatrix}.$$

If there exists a non-trivial steady state, then $R_0 > 1$, so $\beta(0) - \alpha - \mu(0) > 0$ and the origin becomes the saddle point. $\square$

For the non-trivial stationary solution $(P^*, Q^*)$, its Jacobian matrix has two eigenvalues given by

$$\lambda_\pm = \frac{1}{2}(A \pm \sqrt{A^2 + 4B}),$$

where

$$A := \beta'(P^*)Q^* - \mu'(P^*)P^* - \mu(P^*),$$
$$B := \beta(P^*)[\beta'(P^*) - \mu'(P^*)]Q^*.$$

It follows from (3.31) and (3.32) that $B < 0$. Thus, the stability of $(P^*, Q^*)$ depends on the sign of $A$. If $A < 0$, $(P^*, Q^*)$ becomes a sink point, whereas if $A > 0$, it becomes a source. If $A > 0$ and $Q(0) > 0$, the omega limit set of any orbit whose initial point is not the stationary state does not include the stationary state, so it is a periodic orbit (Poincaré–Bendixon Theorem). Thus, we can conclude the following:

**Proposition 3.8** *Suppose that $R_0 > 1$. If $A < 0$, the non-trivial stationary solution $(P^*, Q^*)$ is locally asymptotically stable. If $A > 0$, it is unstable and there exists at least one periodic solution.*

The existence of periodic solutions for system (3.30) can also be proved by the fact that the non-trivial stationary solution becomes unstable and a Hopf bifurcation occurs as the parameter changes when $\beta'(x) < 0$ and $\mu'(x) < 0$ [38, 41]. However, if we use the pure logistic assumption as $\beta'(x) < 0$ and $\mu'(x) > 0$, then although the reproduction number is again a decreasing function, $A < 0$ always holds, so the non-trivial stationary solution is locally stable. Therefore, we require the Allee effect on the birth rate or the death rate to destabilize the non-trivial steady state. Gurtin and MacCamy [17, 18] showed that there is no periodic solution if $\beta(x)$ is constant, but their result can be considered as a special case of the following separable model derived by Busenberg and Iannelli [5]. The simple example of (3.29) shows that the age structure can lead to periodic behavior in a population. If the birth rate is concentrated in one fertile age class, as for semelparous species, periodic solutions appear [35].

**Exercise 3.6** Instead of (3.28), suppose that

$$\beta(a,x) = \sum_{j=0}^{n} \beta_j(x) a^j e^{-\alpha a}, \quad \mu(a,x) = \mu(x),$$

and $\gamma \equiv 1$. Show that (3.29) reduces to a system of ordinary differential equations of $n+2$ variables, $P(t)$ and

$$E_j(t) = \int_0^\infty a^j e^{-\alpha a} p(t,a) da, \quad j = 0, 1, 2, \ldots, n.$$

This kind of reduction is called *linear chain trickery* [29].

### 3.2.2 *Separable Models*

In contrast to the above model, we now consider a nonlinear age-structured population model that has a globally stable non-trivial steady state, that is, periodic solutions are excluded. This example is a simple case of the *separable model* studied by Busenberg and Iannelli [5] and Webb [47]. Readers are referred to [11, 23, 26, 32, 33] for global stability results in systems that are not necessarily separable.

Suppose that the birth rate is independent of the population size, $\beta(a,x) = \beta_0(a)$ and the size dependency of the death rate is given by

$$\mu(a,x) = \mu_0(a) + \eta(x),$$

where $\beta_0$ and $\mu_0$ are the basic birth and death rates, which satisfy the same assumption as in the stable population model. We then have the following nonlinear model:

$$\begin{aligned}
\frac{\partial p(t,a)}{\partial t} + \frac{\partial p(t,a)}{\partial a} &= -\mu_0(a)p(t,a) - \eta(S(t))p(t,a), \\
p(t,0) &= \int_0^\omega \beta_0(a)p(t,a)da, \\
p(0,a) &= p_0(a), \\
S(t) &= \int_0^\omega \gamma(a)p(t,a)da.
\end{aligned} \tag{3.35}$$

The most important aspect of (3.35) is that, if we introduce the total size of the population $P(t) = \int_0^\omega p(t,a)da$, the age profile $w(t,a) := p(t,a)/P(t)$ satisfies the age profile equation of the stable population model with birth rate $\beta_0$ and death rate $\mu_0$ (see Sect. 1.5). That is, $w(t,a)$ satisfies the following system:

$$\begin{aligned}
\frac{\partial w(t,a)}{\partial t} + \frac{\partial w(t,a)}{\partial a} &= -(\alpha(t) + \mu_0(a))w(t,a), \\
w(t,0) &= \int_0^\omega \beta_0(a)w(t,a)da, \\
w(0,a) &= w_0(a), \\
\alpha(t) &= \int_0^\omega [\beta_0(a) - \mu_0(a)]w(t,a)da,
\end{aligned} \tag{3.36}$$

where $\int_0^\omega w(t,a)da = 1$. The above fact implies that the time evolution of the age profile is not affected by the existence of the additional age-independent death rate $\eta(S(t))$. Because we have already examined the behavior of the solution $w(t,a)$ in Sect. 1.5, it is sufficient to consider the behavior of the total size $P(t)$.

It is easy to see that the total population size $P(t)$ satisfies the following nonautonomous ODE:

$$\frac{dP(t)}{dt} = F(t, P(t)), \quad P(0) = \int_0^\omega p_0(a)da, \tag{3.37}$$

where

$$F(t,x) := [\alpha(t) - \eta(\Gamma(t)x)]x, \quad \Gamma(t) := \int_0^\omega \gamma(a)w(t,a)da.$$

Note that $S(t) = \Gamma(t)P(t)$. Let $w^*(a)$ be the stationary solution of (3.36). We then have

$$\begin{aligned}
\lim_{t\to+\infty} \alpha(t) &= \lambda_0 = \int_0^\omega [\beta_0(\sigma) - \mu_0(\sigma)]w^*(\sigma)d\sigma, \\
\lim_{t\to+\infty} \Gamma(t) &= \Gamma^* = \int_0^\omega \gamma(\sigma)w^*(\sigma)d\sigma,
\end{aligned}$$

where

$$w^*(a) = \frac{e^{-\lambda_0 a}\ell_0(a)}{\int_0^\omega e^{-\lambda_0 \sigma}\ell_0(\sigma)d\sigma}, \quad \ell_0(a) := \exp\left(-\int_0^a \mu_0(\sigma)d\sigma\right), \tag{3.38}$$

and $\lambda_0$ is the intrinsic growth rate satisfying

$$1 = \int_0^\omega e^{-\lambda_0\sigma}\beta_0(\sigma)\ell_0(\sigma)d\sigma. \tag{3.39}$$

Therefore, (3.37) has the limiting equation

$$\frac{dQ(t)}{dt} = F_\infty(Q(t)), \tag{3.40}$$

where

$$F_\infty(x) := [\lambda_0 - \eta(\Gamma^* x)]x = \lim_{t\to\infty} F(t, x).$$

Thus, we can state the following:

**Proposition 3.9** *$Q^*$ is the non-trivial stationary solution of the limiting equation* (3.40) *if and only if $p^*(a) = Q^*w^*(a)$ is the non-trivial stationary solution of* (3.35). *Let $S^* = \Gamma^*Q^*$ be the size of the stationary solution. If $\eta'(S^*) > 0$, the stationary solution of size $S^*$ is locally stable, whereas it is unstable if $\eta'(S^*) < 0$.*

*Proof* If $Q^*$ is the non-trivial stationary solution of the limiting equation (3.40), then $\lambda_0 = \eta(\Gamma^*Q^*)$. If we let $S^* = \Gamma^*Q^*$, (3.39) implies that

$$R(S^*) = \int_0^\omega \beta_0(a)\ell_0(a)e^{-a\eta(S^*)}da = 1. \tag{3.41}$$

As seen in the previous section, (3.35) has a stationary solution

$$p^*(a) = \frac{S^*\ell_0(a)e^{-\lambda_0 a}}{\int_0^\omega \gamma(a)\ell_0(a)e^{-\lambda_0 a}da}$$

whose size is $S^*$. As $w^*$ is given by (3.38), it follows that

$$p^*(a) = \frac{S^*w^*(a)}{\int_0^\omega \gamma(a)w^*(a)da} = Q^*w^*(a).$$

Conversely, if $p^*(a) = Q^*w^*(a)$ is the stationary solution of (3.35), its size

$$S^* = \int_0^\omega \gamma(a)p^*(a)da = Q^*\Gamma^*$$

must satisfy the stationarity condition of (3.41). Thus, we have $\lambda_0 = \eta(Q^*\Gamma^*)$ and conclude that $Q^*$ is the stationary solution of (3.40). Subsequently, note that the linearized form of the limiting equation at $Q^*$ is given by

$$\frac{dz(t)}{dt} = -\eta'(S^*)Q^*z(t).$$

Therefore, the size $Q^*$ is stable if $\eta'(S^*) > 0$, whereas it is unstable if $\eta'(S^*) < 0$. Finally, let us examine the stability of $p^*(a)$. If $\lambda \neq 0$, the characteristic equation $F(\lambda) = 1$ can be rewritten as

$$\left(1 - \int_0^\omega \beta_0(a)\ell_0(a)e^{-(\lambda+\lambda_0)a}da\right)(\lambda + S^*\eta'(S^*))G(\lambda) = 0, \tag{3.42}$$

where

$$G(\lambda) := \frac{\int_0^\omega \gamma(a)\ell_0(a)e^{-\lambda_0 a}da}{(\lambda + S^*\eta'(S^*))\int_0^\omega \gamma(a)\ell_0(a)e^{-\lambda_0 a}da - S^*\eta'(S^*)\int_0^\omega \gamma(a)\ell_0(a)e^{-(\lambda+\lambda_0)a}da}.$$

If $\eta'(S^*) > 0$, we immediately have that $\lambda = 0$ is not a characteristic root, and it follows from (3.42) that all characteristic roots have negative real parts. Conversely, if $\eta'(S^*) < 0$, we know that (3.42) has a positive root $\lambda = -S^*\eta'(S^*)$. This completes the proof. □

Although we omit the proof here, it is known that the following holds for the global behavior of the separable model:

**Proposition 3.10** ([23]) *Suppose that the limiting equation* (3.40) *has exactly* $k+1$ *isolated stationary states* $0 = Q_0^* < Q_1^* < \cdots < Q_k^* < +\infty$ *and, for sufficiently large* $x$, $F_\infty(x) < 0$. *Then, there exists* $j \in \{0, 1, \ldots, k\}$ *such that*

$$\lim_{t\to\infty} P(t) = Q_j^*.$$

*Therefore, if we set* $p_j^*(a) = Q_j^* w^*(a)$,

$$\lim_{t\to\infty}\int_0^\omega |p(t,a) - p_j^*(a)|da = 0.$$

The above proposition shows that the separable model has no periodic solution. If we only know the initial size $P(0)$, we cannot determine which $Q_j^*$ is reached by $P(t)$, because $\alpha(t)$ and $\Gamma(t)$ depend on the initial distribution $p_0(a)$.

*Remark 3.2* There is another method to deal with the separable model (3.35) [29, p. 177]. Let us introduce a new distribution $u$ as

$$u(t,a) := \exp\left(\int_0^t \eta(S(\sigma))d\sigma\right)p(t,a).$$

Then, we know that $u(t, a)$ satisfies the stable population model:

$$\frac{\partial u(t,a)}{\partial t} + \frac{\partial u(t,a)}{\partial a} = -\mu_0(a)u(t,a),$$
$$u(t,0) = \int_0^\omega \beta_0(a)u(t,a)da,$$
$$u(0,a) = p_0(a).$$

Therefore, from the stable population theory, there exists a positive number $q > 0$ (corresponding to a non-trivial initial data) such that

$$u(t,a) = qe^{\lambda_0 t}(u_0(a) + v(t,a)),$$

where $u_0(a) = e^{-\lambda_0 a}\ell_0(a)$ is the stable age distribution and there exists a number $\varepsilon > 0$ such that $|v(t,\cdot)|_{L^1} = O(e^{-\varepsilon t})$. Then, the weighted size $S(t)$ is calculated as

$$S(t) = qe^{\lambda_0 t - \int_0^t \eta(S(\sigma))d\sigma} \int_0^\omega \gamma(\sigma)[u_0(a) + v(t,a)]d\sigma,$$

and the limiting equation is given by

$$\frac{dS(t)}{dt} = (\lambda_0 - \eta(S(t)))S(t), \tag{3.43}$$

which is satisfied by $\Gamma^* Q(t)$ as (3.40), i.e., we can expect the asymptotic behavior of $S(t)$ to be determined by (3.43).

**Exercise 3.7** Show that $v(t,a)$ satisfies

$$\frac{\partial v(t,a)}{\partial t} + \frac{\partial v(t,a)}{\partial a} = -(\lambda_0 + \mu_0(a))v(t,a),$$
$$v(t,0) = \int_0^\omega \beta_0(a)v(t,a)da.$$

*Example 3.1* Let us consider an example in which we can determine the asymptotic behavior of the separable model. Let $\Pi(t)$ be the weighted size of

$$u(t,a) = e^{\int_0^t \eta(S(\sigma))d\sigma} p(t,a).$$

Then

$$\Pi(t) := \int_0^\omega \gamma(a)u(t,a)da = e^{\int_0^t \eta(S(\sigma))d\sigma} S(t).$$

Suppose that the additional death rate is given by $\eta(S(t)) = KS(t)$ for some positive constant $K$. It follows that

$$\int_0^t K\Pi(z)dz = \int_0^t e^{\int_0^z KS(\sigma)d\sigma} KS(z)dz = e^{K\int_0^t S(\sigma)d\sigma} - 1.$$

Therefore, we obtain

$$p(t,a) = \frac{u(t,a)}{1 + K\int_0^t \Pi(z)dz}. \tag{3.44}$$

It follows from the Fundamental Theorem of Demography (Chap. 1) that if the initial data are non-trivial, there exists some $q_0 > 0$ such that

$$\lim_{t\to\infty} e^{-\lambda_0 t} u(t,a) = q_0 e^{-\lambda_0 a} \ell_0(a).$$

It is then clear from expression (3.44) that if $\lambda_0 < 0$, we have $\lim_{t\to\infty} p(t,a) = 0$. However, if $\lambda_0 > 0$,

$$\lim_{t\to\infty} p(t,a) = \lim_{t\to\infty} \frac{e^{-\lambda_0 t} u(t,a)}{e^{-\lambda_0 t} + K e^{-\lambda_0 t}\int_0^t \Pi(z)dz},$$

where

$$e^{-\lambda_0 t}\int_0^t \Pi(z)dz = \int_0^t e^{-\lambda_0(t-s)}\phi(s)ds, \quad \phi(s) := e^{-\lambda_0 s}\int_0^\omega \gamma(a)u(s,a)da.$$

Observe that

$$\lim_{s\to\infty} \phi(s) = q_0 \int_0^\omega \gamma(a) e^{-\lambda_0 a}\ell_0(a)da =: \phi(\infty),$$

$$\lim_{t\to\infty} \int_0^t e^{-\lambda_0(t-s)}\phi(s)ds = \frac{\phi(\infty)}{\lambda_0}.$$

Thus, if $\lambda_0 > 0$, we can conclude that

$$\lim_{t\to\infty} p(t,a) = \frac{\lambda_0}{K} \frac{e^{-\lambda_0 a}\ell_0(a)}{\int_0^\omega e^{-\lambda_0 a}\ell_0(a)\gamma(a)da} =: v(a),$$

where the stationary solution $v(a)$ is globally stable. In this case, the size of the stationary population is proportional to the intrinsic growth rate $\lambda_0$ and is inversely proportional to $K$, which is the elasticity of the additional death rate with respect to the weighted population size.

**Exercise 3.8** In the above example, derive an equation satisfied by $P(t)$ and solve it for $\gamma \equiv 1$, i.e., $P(t) = S(t)$ [47, Sect. 5.4].

Note that even for the nonlinear model in which only the death rate is size-dependent, if the perturbation term $\eta$ has some age-dependency, periodic solutions could appear from stationary solutions via a Hopf bifurcation [31].

**Exercise 3.9** As another example of the separable model, consider an age-dependent Lotka–Volterra system. Let $v(t,a)$ be the age-density function of a prey population,

$P(t)$ be the total size of the prey population and $Q(t)$ be the density of the predator population at time $t$. The basic system is as follows [28]:

$$\begin{aligned}
\frac{\partial v(t,a)}{\partial t} + \frac{\partial v(t,a)}{\partial a} &= -(\mu(a) + f_1(P(t)) + f_2(Q(t)))v(t,a),\\
v(t,0) &= \int_0^\infty \beta(a)v(t,a)da,\\
P(t) &= \int_0^\infty v(t,a)da,\\
\frac{dQ(t)}{dt} &= -bQ(t) + g(P(t))Q(t),
\end{aligned}$$

where $b$ is the predator death rate, $\beta(a)$, $\mu(a)$ are the birth rate and death rate of the prey, respectively, $f_1(P)$ is the density-dependent death rate of the prey, $f_2(Q)$ is the (predator-)density-dependent death rate of the prey and $g(P(t))$ is the growth rate of the predator population. We assume that $f_1(x)$, $f_2(x)$ and $g(x)$ are nonnegative, strictly increasing continuous functions for $x \in \mathbb{R}_+$ and $f_1(0) = f_2(0) = g(0) = 0$. Suppose that the net reproduction rate of the prey is greater than unity:

$$\int_0^\infty \beta(a) \exp\left(-\int_0^a \mu(\sigma)d\sigma\right) da > 1.$$

(1) Derive a partial differential equation and its boundary condition satisfied by the age profile of the prey population $w(t,a) = v(t,a)/P(t)$, and prove that it has a unique positive steady state.
(2) Let $w^*(a)$ be the steady state solution obtained in (1). Derive a system of ODEs that is satisfied by $P(t)$ and $Q(t)$ when $w(t,a) = w^*(a)$.
(3) Derive a necessary and sufficient condition such that the ODEs obtained in (2) have a positive steady state, and examine its stability.

*Remark 3.3* Bocharnov and Hadeler [3, 21] discuss that a certain class of ordinary differential equations and delay differential equations served as population models can be interpreted as special cases of the Gurtin–MacCamy model, while Barbarossa, Hadeler and Kuttler [1] show that state-dependent neutral delay equations can be derived from the Gurtin–MacCamy model not as a special case but as an extension. For example, suppose that the size-dependent weight function is given as $\gamma(a, S(t)) = H(a - \tau(S(t)))$, where $H(\cdot)$ denotes the Heaviside function and $\tau(S)$ is the state-dependent delay. Then, the model (3.2) expresses a competition of juveniles and adults such that a larger adult population leads to a longer juveniles period if we assume that $\tau$ is non-decreasing. It is not a special case of the basic system (3.2).

## 3.3 Cohort Control Model

We now consider a nonlinear model in which the vital rates at each age class depend on the size of the cohort to which the age class belongs. This type of model is called the *cohort control model*. Because a population at a given instant can be seen as a superposition of many cohorts, the period control model is controlled by the weighted average of many cohort sizes. Conversely, in the cohort control model, vital rates are controlled by the cohort to whom the vital rates belong. Hence, the cohort control model is a special case of the period control model. However, it is worth examining this special case separately.

The cohort control model is also called the *Easterlin model*, because it reflects the *Easterlin hypothesis* in economic demography which states that human fertility is mainly determined by the relative difference in economic status between successive generations. If the economic status of a generation is improved in comparison with their parents' generation, they are likely to have more children; otherwise, they will have fewer children. We assume that a large cohort size reflects good economic status in the parent generation, although it will lead to greater competition among individuals belonging to the large-size cohort. As a result, the economic status of the large cohort becomes relatively worse, causing their fertility to decrease. Therefore, under the Easterlin hypothesis, fertility is assumed to be controlled by negative feedback with respect to the cohort size.

*Remark 3.4* Just like the cohort model, the period model also has an economic interpretation. For example, if we consider the weight function $\gamma(\cdot)$ as the age-specific labor force rate in (3.2), we obtain a period control model with the period labor force population as a control variable. If we adopt a neoclassical economic assumption that the labor force population size is negatively correlated with fertility via wages and welfare level, we obtain a logistic-type period control model. For economic demographic interpretations of nonlinear population models, readers are referred to [6, 13, 27].

### 3.3.1 Basic Model

In the following section, we discuss a cohort control model based on the integral equation approach, because this is easier to analyze using elementary calculus. Let us consider a closed population in which $B(t)$ is the number of births per unit time at time $t$. For simplicity, we assume that the death rate is time-independent. If the size of a cohort is $x > 0$, its age-specific birth rate is given by $\beta(a)\psi(x)$, where $\beta(a)$ is a normalized age-specific birth rate such that

$$\int_0^\omega \beta(a)\ell(a)da = 1,$$

where $\ell(a)$ is the survival probability. Therefore, the nonnegative function $\psi(x)$ gives the net reproduction rate of a cohort of size $x$ at birth.

The Easterlin model is then formulated as the following nonlinear Volterra integral equation:

$$B(t) = G(t) + \int_0^t \phi(a)\psi(B(t-a))B(t-a)da, \quad t > 0, \tag{3.45}$$

where $\phi(a) := \beta(a)\ell(a)$ is the normalized net reproduction function whose domain is extended, if necessary, such that $\phi(a) = 0$ for $a > \omega$. The starting function $G(t)$ is the number of births per unit time given by the initial population. If we assume that $x\psi(x)$ is Lipschitz continuous, the iteration method can be used to show that (3.45) has a unique nonnegative solution corresponding to nonnegative initial data.

For the net reproduction function $\psi(x)$, let us adopt the Allee–Logistic-type differentiable positive function

$$\begin{cases} (1) \quad \psi'(x) > 0, & 0 < x < x_0, \\ (2) \quad \psi'(x) < 0, & x > x_0, \\ (3) \quad \lim_{x\to\infty}\psi(x) = 0, \end{cases} \tag{3.46}$$

where $x_0 \geq 0$. If $x_0 = 0$, condition (1) is neglected. The value of $x_0$ denotes the cohort size that maximizes the basic reproduction number. The case in which $x_0 = 0$ is a pure Easterlin model that takes into account only negative feedback.

First, we consider some basic global properties of model (3.45) stated by Brauer [4]:

**Proposition 3.11** *Under Assumption* (3.46), *the nonnegative solution of* (3.45) *is upper bounded for* $0 \leq t < \infty$.

*Proof* From Assumption (3.46), if we take a sufficiently large $K > 0$, it follows that for $0 < \rho < 1$, $\psi(x) < \rho \ \forall x \in [K, \infty)$. Let us decompose the integral interval into two parts as $I_1 := \{t \geq 0 : B(t) > K\}$ and $I_2 := \{t \geq 0 : B(t) \leq K\}$. Then, we have

$$\begin{aligned} B(t) &= G(t) + \int_{I_1\cap[0,t]} \phi(t-a)\psi(B(a))B(a)da + \int_{I_2\cap[0,t]} \phi(t-a)\psi(B(a))B(a)da \\ &\leq |G|_{L^\infty} + \rho \int_{I_1\cap[0,t]} \phi(t-a)B(a)da + K\psi(x_0) \\ &\leq |G|_{L^\infty} + \rho \sup_{0\leq x\leq t} B(x) + K\psi(x_0). \end{aligned}$$

Because the above inequality holds for all $t \geq 0$, for any $T > 0$ it holds that

$$\sup_{0\leq x\leq T} B(x) \leq |G|_{L^\infty} + \rho \sup_{0\leq x\leq t} B(x) + K\psi(x_0).$$

Therefore, we have

$$\sup_{0\leq x\leq T} B(x) \leq \frac{|G|_{L^\infty} + K\psi(x_0)}{1-\rho}.$$

Because the right-hand side of the above inequality is independent of $T > 0$, it follows that

$$\sup_{0 \le x < \infty} B(x) \le \frac{|G|_{L^\infty} + K\psi(x_0)}{1-\rho}.$$

Therefore, we know that $B(t)$ is bounded above for $t \ge 0$. □

This result reflects the fact that population growth is suppressed by the Logistic effect. However, even if the population density is very small, the population governed by (3.45) is not eradicated if the basic reproduction number is greater than unity:

**Proposition 3.12** *Suppose that $\psi(x) \ge 1$ for sufficiently small $x \ge 0$. Then, there is no solution that is not a trivial solution and $\lim_{t\to\infty} B(t) = 0$.*

*Proof* Let $f(x) := \psi(x)x - x$. It follows from our assumption that, for sufficiently small $x \ge 0$, $f(x) \ge 0$. Suppose that $B(t)$ is not a solution associated with trivial initial data and $\lim_{t\to\infty} B(t) = 0$. Then, $B(t) > 0$ for all $t \ge 0$. From the assumptions, we can take some $\tau > 0$ such that $B(t) > 0$ for all $t \le \tau$ and $f(B(t)) \ge 0,\ t \ge \tau$, because $B(t),\ t \ge \tau$ is sufficiently small. For $t \ge \tau$, it follows that

$$\begin{aligned} B(t) &= G(t) + \int_{t-\tau}^{t} \phi(a)\psi(B(t-a))B(t-a)da \\ &\quad + \int_0^{t-\tau} \phi(a)\psi(B(t-a))B(t-a)da \\ &= G(t) + \int_{t-\tau}^{t} \phi(a)\psi(B(t-a))B(t-a)da \\ &\quad + \int_0^{t-\tau} \phi(a)f(B(t-a))da + \int_0^{t-\tau} \phi(a)B(t-a)da. \end{aligned}$$

Therefore, we can rewrite (3.45) as

$$B(t) = F(t) + \int_0^{t-\tau} \phi(a)B(t-a)da,$$

where

$$F(t) := G(t) + \int_{t-\tau}^{t} \phi(a)\psi(B(t-a))B(t-a)da + \int_0^{t-\tau} \phi(a)f(B(t-a))da.$$

Hence, $B(t+\tau)$ satisfies the renewal equation as

$$B(t+\tau) = F(t+\tau) + \int_0^t \phi(a)B(t+\tau-a)da, \quad t > 0.$$

Let $C(t)$ be the solution of a linear integral equation

$$C(t) = H(t) + \int_0^t \phi(a)C(t-a)da,$$

where

$$H(t) := G(t+\tau) + \int_t^{t+\tau} \phi(a)\psi(B(t+\tau-a))B(t+\tau-a)da.$$

Then, $F(t+\tau) \geq H(t)$, and $H(t)$ is not identically zero but $H(t) = 0$ for $t > \omega$. Thus, it follows from the Fundamental Theorem of Demography in Chap. 1 that $C(t)$ converges to a positive constant as $t \to \infty$. Using a comparison argument, it is easily seen that $B(t+\tau) \geq C(t)$, which contradicts the assumption that $\lim_{t\to\infty} B(t) = 0$. □

### 3.3.2 Easterlin Cycle

The model in (3.45) becomes a homogeneous equation for $t > \omega$ as

$$B(t) = \int_0^\omega \phi(a)\psi(B(t-a))B(t-a)da. \tag{3.47}$$

Let $B^*$ be a positive stationary solution of (3.47). It must then hold that

$$\psi(B^*) = 1. \tag{3.48}$$

Conversely, if there exists $B^* > 0$ satisfying (3.48), we have a positive stationary solution. From Assumption (3.46), it is easy to see that the following holds [40]:

**Proposition 3.13** *Suppose that $x_0 > 0$. If $\psi(0) > 1$, there exists a unique positive stationary state. If $\psi(0) < 1 < \psi(x_0)$, there exist exactly two positive stationary solutions. If $\psi(x_0) < 1$, there is no positive stationary solution. In the case that $x_0 = 0$, there exists a unique positive stationary solution if $\psi(0) > 1$, whereas there is no positive stationary solution if $\psi(0) \leq 1$.*

In the following, we assume that there exists a positive stationary solution $B^* > 0$ for which we can consider the stability. Let us define a small perturbation from the equilibrium solution by

$$\zeta(t) = B(t) - B^*,$$

and consider the linearized equation around $B = B^*$. Observe that the cohort birth rate $x\psi(x)$ is expressed by a Taylor expansion

$$x\psi(x) = B^* + (1 + B^*\psi'(B^*))(x - B^*) + g(x - B^*),$$

where $g(x)$ is the second-order term with $g(0) = g'(0) = 0$. Inserting the above expression into (3.47), we obtain a nonlinear equation for the perturbation $\zeta(t)$:

$$\zeta(t) = (1-\gamma)\int_0^{\omega} \phi(a)\zeta(t-a)da + \int_0^{\omega} \phi(a)g(\zeta(t-a))da, \tag{3.49}$$

where $\gamma := -B^*\psi'(B^*)$ is a parameter whose absolute value denotes the strength (elasticity) of feedback. If the feedback is positive (a larger cohort has higher fertility), then $\gamma < 0$, whereas if the feedback is negative (a larger cohort has lower fertility), $\gamma > 0$. The characteristic equation of the linear part of the perturbation equation (3.49) is then given by

$$(1-\gamma)\int_0^{\omega} e^{-\lambda a}\phi(a)da = 1. \tag{3.50}$$

If $\gamma \neq 1$, all of the characteristic roots are included in the half plane $\Re\lambda \le |1-\gamma|\bar{\beta}$. In fact, if $\lambda$ is a characteristic root with $\Re\lambda > 0$, we have

$$1 = \left|(1-\gamma)\int_0^{\omega} e^{-\lambda a}\phi(a)da\right| \le |1-\gamma|\bar{\beta}\int_0^{\omega} e^{-(\Re\lambda)a}da \le |1-\gamma|\frac{\bar{\beta}}{\Re\lambda}.$$

Therefore, it follows that $\Re\lambda \le |1-\gamma|\bar{\beta}$ for any characteristic root $\lambda$. In particular, there is no real characteristic root if $\gamma > 1$.

**Proposition 3.14** *If all real parts of the roots of the characteristic equation* (3.50) *are negative, the stationary solution $B^*$ is locally asymptotically stable.*

*Proof* Let $Y = C[-\omega, 0]$ (the set of continuous functions on the interval $[-\omega, 0]$) be the state space of the initial data with norm $|\zeta|_Y := \sup_{t\in[-\omega,0]}|\zeta(t)|$. We prove that, for any $\varepsilon > 0$, there exists a number $\delta > 0$ such that for any initial data satisfying $|\zeta_0|_Y = \sup_{t\in[-\omega,0]}|B(t) - B^*| < \delta$, the perturbation equation (3.49) has a unique solution for all $t > 0$ such that

$$|\zeta(t)| = |B(t) - B^*| \le \varepsilon, \quad \forall t \ge 0,$$

and $\lim_{t\to\infty}\zeta(t) = 0$. The perturbation equation (3.49) can be written as a nonlinear Volterra integral equation:

$$\zeta(t) = f(\zeta, \zeta_0)(t) + (1-\gamma)\int_0^{t} \phi(a)\zeta(t-a)da, \tag{3.51}$$

where

$$\begin{aligned} f(\zeta, \zeta_0)(t) :=& \int_{t\wedge\omega}^{\omega} [(1-\gamma)\zeta_0(t-a) \\ &+ g(\zeta_0(t-a))]\phi(a)da + \int_0^{t\wedge\omega} \phi(a)g(\zeta(t-a))da, \end{aligned}$$

and $\zeta(t) = \zeta_0(t), \quad t \in [-\omega, 0]$ is the given initial data. Let $R(t)$ be the resolvent kernel associated with $(1-\gamma)\phi(a)$. It follows from our assumption that $R(t)$ is integrable (Paley–Wiener's Theorem). Equation (3.51) is then solved as follows:

$$\zeta(t) = f(\zeta, \zeta_0)(t) + \int_0^t R(t-s) f(\zeta, \zeta_0)(s) ds. \tag{3.52}$$

Let $X := C_0(\mathbb{R}_+)$ be the set of continuous functions vanishing at $+\infty$. We use $X$ as the state space of solutions $\zeta(t)$ with norm defined by $|\zeta|_X := \sup_{0 \le t < \infty} |\zeta(t)|$. For the given initial data, the nonlinear operator $f$ is locally Lipschitz continuous as an operator from $X$ into itself. That is, there exists a nonnegative function $L(r)$, $r > 0$ such that $\lim_{r \to 0} L(r) = 0$ and, if $|\zeta|_X \le r$, $|\xi|_X \le r$, then

$$|f(\zeta, \zeta_0) - f(\xi, \zeta_0)|_X \le L(r) |\zeta - \xi|_X,$$

and the following inequality holds for $\zeta_0 \in Y$, $|\zeta_0|_Y \le r$:

$$|f(0, \zeta_0)|_X \le (|1-\gamma| + L(r)) |\zeta_0|_Y.$$

In fact, it is sufficient to choose $L(r)$ such that $|g(x)| \le L(r)|x|$ for $|x| \le r$, which is possible because $g(x) = o(x^2)$. For any $\varepsilon > 0$, we can choose $\eta < \varepsilon$ such that

$$L(\eta) < \min \left\{ 1, \frac{1}{(2 + |1-\gamma|)(1 + |R|_{L^1})} \right\}.$$

Let $\delta = L(\eta)\eta$ and let $X_0$ be a subset of $X$ defined by $X_0 := \{\zeta \in X : |\zeta|_X \le \eta\}$. For any initial data $\zeta_0 \in Y$ such that $|\zeta_0|_Y \le \delta = L(\eta)\eta$, define a map $F$ by

$$(F\zeta)(t) := f(\zeta, \zeta_0)(t) + \int_0^t R(t-s) f(\zeta, \zeta_0)(s) ds.$$

For any $\zeta \in X_0$, we can observe that

$$\begin{aligned} |F\zeta|_Y &\le (1 + |R|_{L^1})(L(\eta)|\zeta|_X + (|1-\gamma| + L(\eta))|\zeta_0|_Y) \\ &\le (1 + |R|_{L^1})(1 + |1-\gamma| + L(\eta)) L(\eta)\eta \le \eta, \end{aligned}$$

where we have used the relation $|f(\zeta, \zeta_0)|_X \le |f(0, \zeta_0)|_X + L(r)|\zeta|_X$. It follows from the integrability of the resolvent $R$ that $\lim_{t \to \infty} (F\zeta)(t) = 0$ if $\zeta \in X$. Thus, $F$ becomes a mapping from $X_0$ into itself. Moreover, for $\zeta, \xi \in X_0$, we have

$$|F\zeta - F\xi|_X \le (1 + |R|_{L^1}) L(\eta) |\zeta - \xi|_X < \frac{1}{2} |\zeta - \xi|_X.$$

$F$ then becomes a contraction mapping from $X_0$ to itself, so it has a unique fixed point in $X_0$. This shows that, for any initial data $\zeta_0 \in Y$ such that $|\zeta_0|_Y \le \delta$, the perturbation equation (3.41) has a unique solution such that $|\zeta|_X \le \eta < \varepsilon$ and $\lim_{t\to\infty} \zeta(t) = 0$. □

From the above proposition, we obtain a sufficient condition for the local stability of steady state solutions. As the principle of linearized stability holds for the Volterra integral equation [12], the stationary solution is unstable if there exists a characteristic root with a positive real part. The stability of stationary solutions changes as the elasticity parameter $\gamma$ varies. We examine this point in the following discussion.

**Proposition 3.15** *If $\gamma < 0$, the stationary solution is unstable, whereas if $0 < \gamma < 1$ or $1 < \gamma < 2$, it is locally asymptotically stable.*

*Proof* If $\gamma < 1$, (3.50) becomes the Euler–Lotka characteristic equation and

$$R_0 = (1-\gamma)\int_0^\omega \phi(a)da = 1-\gamma.$$

If $\gamma < 0$, the characteristic equation has a positive root, and hence, the stationary solution is unstable, whereas if $0 < \gamma < 1$, all real parts of the characteristic roots are negative, and hence, the stationary solution is locally asymptotically stable. Finally, consider the case in which $\gamma > 1$. If there exists a characteristic root $\lambda$ whose real part is nonnegative, it follows that

$$1 = \left|(1-\gamma)\int_0^\omega e^{-\lambda a}\phi(a)da\right| \le (\gamma-1)\int_0^\omega e^{-\Re\lambda a}\phi(a)da \le \gamma - 1,$$

which shows that $\gamma \ge 2$. Therefore, if $1 < \gamma < 2$, all real parts of the characteristic roots are negative, and hence, the stationary solution $B^*$ is locally asymptotically stable. □

Whether the stationary solution will become unstable as the parameter $\gamma$ grows to above two is an important problem. In fact, in such a situation, it is strongly suggested that a periodic solution for the basic nonlinear system would appear via a Hopf bifurcation. However, we first consider some examples in which destabilization does not occur.

**Proposition 3.16** *For any real number $y$, if*

$$\int_0^\omega \phi(a)\cos(ya)da \ge 0,$$

*then the stationary solution is always locally asymptotically stable if $\gamma > 1$.*

*Proof* Let us define a number $\gamma^*$ as

$$\gamma^* := \sup\{\delta : \text{if } \gamma \in (1,\delta], \text{ all real parts of the characteristic roots are negative}\}.$$

We then have $\gamma^* \geq 2$. If we assume that $\gamma^* < \infty$, the characteristic equation must have purely imaginary characteristic roots at $\gamma = \gamma^*$, which we denote by $iy$. It follows from the real part of the characteristic equation that

$$(1-\gamma^*)\int_0^\omega \phi(a)\cos(ya)da = 1,$$

which contradicts our assumption because $1-\gamma^* \leq -1$. Thus, we know that $\gamma^* = \infty$, and the real parts of all characteristic roots are negative if $\gamma > 1$. $\square$

**Proposition 3.17** ([34]) *Suppose that $\gamma > 1$ and the reproduction period is given by an infinite interval $[0, \infty)$. If $\phi(a)$, $a \geq 0$, is monotone decreasing, the stationary solution is always locally asymptotically stable.*

*Proof* From the assumption, there is no real characteristic root. Let $\lambda = x + iy, \quad y > 0$ be a characteristic root. Then, it must hold that

$$J := \int_0^\infty \phi(a)e^{-xa}\sin(ya)da = 0.$$

Let $v(a) := \phi(a)e^{-xa}$. $J$ can then be calculated as

$$\begin{aligned} J &= \sum_{n=0}^{\infty}\left\{\int_{\frac{2n\pi}{y}}^{\frac{(2n+1)\pi}{y}} v(a)\sin(ya)da + \int_{\frac{(2n+1)\pi}{y}}^{\frac{(2n+2)\pi}{y}} v(a)\sin(ya)da\right\} \\ &= \sum_{n=0}^{\infty}\left\{\int_0^\pi v(\frac{2n\pi+a}{y})\frac{\sin a}{y}da + \int_0^\pi v(\frac{(2n+1)\pi+a}{y})\frac{\sin(a+\pi)}{y}da\right\} \\ &= \sum_{n=0}^{\infty}\frac{1}{y}\int_0^\pi\left(v(\frac{2n\pi+a}{y}) - v(\frac{(2n+1)\pi+a}{y})\right)\sin a da. \end{aligned}$$

If $x \geq 0$ and $\phi(a)$ is monotone decreasing, then $e^{-xa}\phi(a)$ is also monotone decreasing. It follows from the last equation above that $J > 0$, which contradicts the fact that $x + iy$ is a characteristic root. Then, under the assumption, all characteristic roots have negative real parts. $\square$

As suggested by the above proposition, the situation whereby bifurcated solutions are always stable is likely to be observed if the pre-reproductive period (that is, the period needed for maturity) is negligibly short and the reproductive period is infinite or sufficiently large. Conversely, as observed in real human populations, if the length of the pre-reproductive period is similar to that of the reproductive period, as the elasticity parameter $\gamma$ becomes greater than unity, the stability of the steady state solution for the Easterlin model (3.47) will be lost. Although we have not yet derived a sufficiently general, precise result, the existence of a non-negligible pre-reproductive period would be essential for the destabilization of stationary solutions, and it is predicted that destabilization would be likely to occur as the reproductive period becomes shorter [45].

Consider the extreme case in which the reproductive age is concentrated at one age $a_0 > 0$, that is, $\phi(a) = \delta(a - a_0)$ (where $\delta$ denotes the delta function). From (3.47), we have $B(t) = F(B(t - a_0))$, where $F(x) := x\psi(x)$ is a unimodal function. It has been proved that, under such a function $F$, this difference equation could lead to very complex, chaotic solutions [16]. Therefore, if the net reproduction function is concentrated on a short age interval, we can expect the Easterlin model to also exhibit very complex behavior [20, 22, 29]. We now derive a sufficient condition for destabilization:

**Proposition 3.18** ([25]) *Let $[\beta_1, \beta_2]$ be the reproductive period. If $\beta_2 < \frac{3}{2}\beta_1$, there exists a number $\gamma^* > 2$ such that the characteristic equation (3.50) has purely imaginary roots when $\gamma = \gamma^*$. Moreover, when $\gamma$ becomes larger than $\gamma^*$, characteristic roots with positive real parts appear.*

*Proof* Let $\lambda = x + iy$. The characteristic equation (3.50) can then be reduced to a pair of equations

$$\begin{aligned} F(x, y, \gamma) &:= \int_{\beta_1}^{\beta_2} e^{-xa}\cos(ya)\phi(a)da + \frac{1}{\gamma - 1} = 0, \\ G(x, y) &:= \int_{\beta_1}^{\beta_2} e^{-xa}\sin(ya)\phi(a)da = 0. \end{aligned} \tag{3.53}$$

Define an interval $I$ by $I = [\pi/\beta_2, \alpha\pi/\beta_2]$, where $\alpha$ is such that $\beta_2/\beta_1 < \alpha < 3/2$. It follows that, for $a \in [\beta_1, \beta_2]$,

$$\frac{\pi}{\beta_2}a \in \left[\frac{\beta_1}{\beta_2}\pi, \pi\right], \quad \frac{\alpha\pi}{\beta_2}a \in \left[\frac{\alpha\beta_1}{\beta_2}\pi, \alpha\pi\right] \subset \left(\pi, \frac{3}{2}\pi\right).$$

Therefore, for any fixed $x$, we have $G(x, \pi/\beta_2) > 0$ and $G(x, \alpha\pi/\beta_2) < 0$. If $y \in I$ and $a \in [\beta_1, \beta_2]$, then $ya \in (\frac{\pi}{2}, \frac{3}{2}\pi)$. Hence, it holds that

$$\frac{\partial G}{\partial y} = \int_{\beta_1}^{\beta_2} ae^{-xa}\cos(ya)\phi(a)da < 0.$$

Because $G(x, y)$ is monotone decreasing with respect to $y \in I$ and changes its sign on both sides of the interval, $G(x, y) = 0$ has a unique solution $y = y(x) \in I$. Observe that

$$y'(x) = -\frac{G_x}{G_y} = \frac{\int_{\beta_1}^{\beta_2} ae^{-xa}\sin(y(x)a)\phi(a)da}{\int_{\beta_1}^{\beta_2} ae^{-xa}\cos(y(x)a)\phi(a)da}. \tag{3.54}$$

Inserting the solution $y(0)$ corresponding to $x = 0$ into (3.53) and solving for $\gamma$, we obtain the solution

$$\gamma^* = 1 - \frac{1}{\int_{\beta_1}^{\beta_2} \cos(y(0)a)\phi(a)da}.$$

That is, for $\gamma = \gamma^*$, there exists a purely imaginary root. Moreover, if we use (3.54) in

$$F(x, y(x), \gamma) = 0,$$

we obtain

$$\frac{\partial F}{\partial x} = -\int_{\beta_1}^{\beta_2} ae^{-xa}\cos(y(x)a)\phi(a)da - \int_{\beta_1}^{\beta_2} ae^{-xa}\sin(y(x)a)y'(x)\phi(a)da$$

$$= -\frac{\left(\int_{\beta_1}^{\beta_2} ae^{-xa}\cos(y(x)a)\phi(a)da\right)^2 + \left(\int_{\beta_1}^{\beta_2} ae^{-xa}\sin(y(x)a)\phi(a)da\right)^2}{\int_{\beta_1}^{\beta_2} ae^{-xa}\cos(y(x)a)\phi(a)da} > 0.$$

It follows from the Implicit Function Theorem that there exists a root $x = x(\gamma)$ such that

$$x'(\gamma) = -\frac{F_\gamma}{F_x} = \frac{1}{(\gamma - 1)^2 F_x} > 0.$$

Then, if $\gamma$ becomes larger, the real part of the characteristic root increases. Therefore, the purely imaginary root at $\gamma = \gamma^*$ becomes a characteristic root with a positive real part as $\gamma$ increases beyond $\gamma^*$. □

For $\gamma > 2$, all characteristic roots exist as conjugate pairs of complex roots. The above observation tells us that, just like a Hopf bifurcation in ODEs, pairs of characteristic roots for the steady state solution move from the left half plane to the right half plane across the imaginary axis as $\gamma$ increases. In such a case, the nonlinear Volterra integral equation theory tells us that, under appropriate conditions, an asymptotically stable periodic solution (limit cycle) bifurcates from the stationary solution [7–9, 12, 42]. Although we omit the proof, we should mention the following result:

**Proposition 3.19** ([39]) *Suppose that the characteristic equation* (3.50) *has a pair of purely imaginary roots* $\pm y^* i$ *at* $\gamma = \gamma^*$, *and that* $ny^* i$, $n \neq \pm 1$ *(n is an integer) are not characteristic roots. If*

$$\int_0^\omega a\phi(a)\cos(y^* a)da \neq 0, \tag{3.55}$$

*then the stationary solution* $B^*$ *of* (3.47) *bifurcates to a periodic solution at* $\gamma = \gamma^*$ *with period close to* $2\pi/y^*$. *The period is almost twice the average age of childbearing.*

Condition (3.55) implies that the speed component along the real axis of a conjugate pair of complex roots crossing the imaginary axis is not zero, which is shown by the same kind of argument as used in the proof of Proposition 3.18.

Following Frauenthal [14], we present a simple proof for the latter part of the above proposition. Suppose that $y^*i$ is a purely imaginary root of the characteristic equation. Then, we have

$$\int_0^{\omega} e^{-iy^*a}\phi(a)da = \frac{1}{1-\gamma^*}. \tag{3.56}$$

Let

$$\mu := \int_0^{\omega} a\phi(a)da,$$

where $\mu$ denotes the average age of childbearing. Suppose that $\phi(a)$ is a distribution concentrated around $a = \mu$. Multiplying both sides of (3.56) by $e^{iy^*\mu}$ and expanding $\exp(-iy^*(a-\mu))$ to its second-order Taylor series, we obtain that

$$\int_0^{\omega} \left[1 - iy^*(a-\mu) - \frac{(y^*)^2}{2}(a-\mu)^2\right]\phi(a)da \approx \frac{e^{iy^*\mu}}{1-\gamma^*}.$$

Dividing the above equation into its real and imaginary parts, it follows that

$$-y^*\int_0^{\omega}(a-\mu)\phi(a)da = 0 \approx \frac{\sin y^*\mu}{1-\gamma^*}, \tag{3.57}$$

$$\int_0^{\omega}\left[1 - \frac{(y^*)^2}{2}(a-\mu)^2\right]\phi(a)da \approx \frac{\cos y^*\mu}{1-\gamma^*}. \tag{3.58}$$

From (3.57), we have $y^*\mu = n\pi$ $(n = 0, \pm 1, \pm 2, \ldots)$. Inserting $y^*\mu = n\pi$ into (3.58) to calculate $\gamma^*$,

$$\gamma^* \approx 1 - (-1)^n\left(1 - \frac{y^*\sigma^2}{2}\right)^{-1}, \tag{3.59}$$

where $\sigma^2$ denotes the variance of the age of childbearing:

$$\sigma^2 := \int_0^{\omega}(a-\mu)^2\phi(a)da.$$

From (3.59), if $\sigma$ is sufficiently small, $n$ should be an odd number, because $\gamma^* > 2$. Thus, the smallest period corresponds to $n = 1$, that is, $y^* = \pi/\mu$ and the period is given by

$$T = \frac{2\pi}{y^*} \approx 2\mu,$$

and $\gamma^*$ is estimated as

$$\gamma^* \approx 2 + \frac{\pi\sigma^2}{2\mu}.$$

Frauenthal and Swick [15] applied the Easterlin model to data from the United States, and concluded that the elasticity parameter $\gamma$ was in the domain for which periodic solutions occur, although Wachter [43, 44] criticized this conclusion. In fact, it should be remarked that the bifurcation of periodic solutions for the Easterlin type model (3.47) is only valid in the neighborhood of the demographic steady state, whereas real populations are not in the stationary state. Readers are referred to the work of Chu [6] for a statistical test of the validity of the Easterlin hypothesis.

## References


1. Barbarossa, M.V., Hadeler, K.P., Kuttler, C.: State-dependent neutral delay equations from population dynamics. J. Math. Biol. **69**, 1027–1056 (2014)
2. Bertoni, S.: Periodic solutions for non-linear equations of structured populations. J. Math. Anal. Appl. **220**, 250–267 (1998)
3. Bocharnov, G., Hadeler, K.: Structured population models, conservation laws, and delay equations. J. Diff. Equ. **168**, 212–237 (2000)
4. Brauer, F.: On a nonlinear integral equation for population growth problems. SIAM J. Math. Anal. **6**(2), 312–317 (1975)
5. Busenberg, S., Iannelli, M.: Separable models in age-dependent population dynamics. J. Math. Biol. **22**, 145–173 (1985)
6. Cyrus Chu, C.Y.: Population Dynamics: A New Economic Approach. Oxford University Press, New York (1998)
7. Cushing, J.M.: Nontrivial periodic solutions of integrodifferential equations. J. Integral Equ. **1**, 165–181 (1979)
8. Cushing, J.M.: Nontrivial periodic solutions of some Volterra integral equations. Volterra Equations. Lecture Notes in Mathematics, vol. 737, pp. 50–66. Springer, Berlin (1979)
9. Cushing, J.M.: Bifurcation of periodic solutions of nonlinear equations in age-structured population dynamics. Nonlinear Phenomena in Mathematical Sciences, pp. 279–288. Academic Press, New York (1982)
10. Desch, W., Schappacher, W.: Spectral properties of finite-dimensional perturbed linear semigroups. J. Diff. Equ. **59**, 80–102 (1985)
11. Di Blasio, G., Iannelli, M., Sinestrari, E.: Approach to equilibrium in age structured populations with increasing recruitment process. J. Math. Biol. **13**, 371–382 (1982)
12. Diekmann, O., van Gils, S.A.: Invariant manifolds for Volterra integral equations of convolution type. J. Diff. Equ. **54**, 139–180 (1984)
13. Fanti, L., Iannelli, M., Manfredi, P.: Neoclassical growth with endogenous age distribution. Poverty vs low-fertility traps as steady states of demographic transition. J. Popul. Econ. **26**, 1457–1484 (2013)
14. Frauenthal, J.C.: A dynamical model for human population growth. Theor. Popul. Biol. **8**, 64–73 (1975)
15. Frauenthal, J.C., Swick, K.E.: Limit cycle oscillations of the human population. Demography **20**(3), 285–298 (1983)
16. Guckenheimer, J., Oster, G., Ipaktchi, A.: The dynamics of density dependent population models. J. Math. Biol. **4**, 101–147 (1977)
17. Gurtin, M.E., MacCamy, R.C.: Some simple models for nonlinear age-dependent population dynamics. Math. Biosci. **43**, 199–211 (1979)

18. Gurtin, M.E., MacCamy, R.C.: Population dynamics with age dependence. In: Knops, R.J. (ed.) Nonlinear Analysis and Mechanics: Heriot-Watt Symposium, vol. III, pp. 1–35. Pitman, London (1979)
19. Gurtin, M.E., MacCamy, R.C.: Non-linear age-dependent population dynamics. Arch. Rat. Mech. Anal. **54**, 281–300 (1974)
20. Gurtin, M.E.: Some questions and open problems in continuum mechanics and population dynamics. J. Diff. Equ. **48**, 293–312 (1983)
21. Hadeler, K.: Neutral delay equations from and for population dynamics. Electronic Journal of Qualitative Theory of Differential Equations, Proceedings of 8th Coll. QTDE, No. 11, 1–18 (2008)
22. Hoppensteadt, F.C.: A nonlinear renewal equation with periodic and chaotic solutions. SIAM-AMS Proc. **10**, 51–60 (1976)
23. Iannelli, M.: Mathematical Theory of Age-Structured Population Dynamics, Giardini Editori e Stampatori in Pisa (1995)
24. Iannelli, M., Kim, M.-Y., Park, E.-J., Pugliese, A.: Global boundedness of the solutions to a Gurtin-MacCamy system. Nonlinear Differ. Equ. Appl. **9**, 197–216 (2002)
25. Inaba, H.: Nonlinear dynamics of open marine population with space-limited recruitment: the case of mortality control. J. Math. Anal. Appl. **275**, 537–556 (2002)
26. Lamberti, L., Vernole, P.: Existence and asymptotic behaviour of solutions of an age structured population model. Bollettino U.M.I. Analisi Funzionale e Applicazioni, Series V, Vol. XVIII -C,N.1, 119–139 (1981)
27. Lee, R.: The formal dynamics of controlled populations and the echo, the boom and the bust. Demography **11**(4), 563–585 (1974)
28. Li, J.: Dynamics of age-structured predator-prey population models. J. Math. Anal. Appl. **152**, 399–415 (1990)
29. Metz, J.A.J., Diekmann, O.: The Dynamics of Physiologically Structured Populations. Lecture Notes in Biomathematics, vol. 68. Springer, Berlin (1986)
30. Prüss, J.: Stability analysis for equilibria in age-specific population dynamics. Nonl. Anal., Theo. Meth. Appl. **7**(12), 1291–1313 (1983)
31. Prüss, J.: On the qualitative behaviour of populations with age-specific interactions. Comp. Maths. Appls. **9**(3), 327–339 (1983)
32. Rorres, C.: Stability of an age specific population with density dependent fertility. Theor. Popul. Biol. **10**, 26–46 (1976)
33. Rorres, C.: A nonlinear model of population growth in which fertility is dependent on birth rate. SIAM J. Appl. Math. **37**(2), 423–432 (1979)
34. Roughgarden, J., Iwasa, Y., Baxter, C.: Demographic theory for an open marine population with space-limited recruitment. Ecology **66**(1), 54–67 (1985)
35. Rudnicki, R., Wieczorek, R.: On a nonlinear age-structured model of semelparous species. Disc. Conti. Dyna. Sys. Ser. B **19**(8), 2641–2656 (2014)
36. Sánchez, D.A.: Iteration and nonlinear equations of age-dependent population growth with a birth window. Math. Biosci. **73**, 61–69 (1985)
37. Smith, H.L., Thieme, H.R.: Dynamical Systems and Population Persistence, Graduate Studies in Mathematics 118. Amer. Math. Soc, Providence, Rhode Island (2011)
38. Swick, K.E.: Periodic solutions of a nonlinear age-dependent model of single species population dynamics. SIAM J. Math. Anal. **11**(5), 901–910 (1980)
39. Swick, K.E.: A nonlinear model for human population dynamics. SIAM J. Appl. Math. **40**(2), 266–278 (1981)
40. Swick, K.E.: Stability and bifurcation in age-dependent population dynamics. Theor. Popul. Biol. **20**, 80–100 (1981)
41. Swick, K.E.: Some reducible models of age dependent dynamics. SIAM J. Appl. Math. **45**(2), 256–267 (1985)
42. Tuljapurkar, S.: Cycles in nonlinear age-structured models I. Renewal equations. Theor. Popul. Biol. **32**, 26–41 (1987)
43. Wachter, K.E.: U.S. births and limit cycle models. Demography **26**(1), 99–115 (1989)

44. Wachter, K.E.: Elusive cycle: are there dynamically possible Lee-Easterlin models for U.S. births? Popul. Stud. **45**, 109–135 (1991)
45. Wachter, K.E.: Pre-procreative ages in population stability and cyclicity. Math. Popul. Stud. **3**(2), 79–103 (1991)
46. Webb, G.F.: Compactness of bounded trajectories of dynamical systems in infinite dimensional spaces. Proc. Roy. Soc. Edinburgh **84A**, 19–33 (1979)
47. Webb, G.F.: Theory of Nonlinear Age-Dependent Population Dynamics. Marcel Dekker, New York (1985)

# Chapter 4
# Pair Formation Models

**Abstract** The stable population model can be extended to take into account marriage phenomena. However, this proves unsatisfactory as these linear models are essentially one-sex models that neglect the process of *pair formation* between the sexes. In reality, the mating chances depend on the existence of the other sex. It is not self-evident why persistent population growth, as observed in the real world, is possible under this model of pair formation between the sexes. Various questions cannot be answered by linear population models, such as those regarding the laws of pair formation that make persistent growth possible, the two-sex Malthusian growth rate, and the result of an imbalanced sex ratio. Although two-sex population dynamics models without age structures have been developed by several authors, it is very difficult to study age-structured two-sex models. Thus, there is not yet a satisfactory two-sex nonlinear demographic theory. Temporary or persistent pair formation also plays an important role in understanding the spread of sexually transmitted diseases, so there are a number of studies about pair formation phenomena in the context of epidemic models. In this chapter, however, we focus on pure continuous-time demographic theories.

## 4.1 The Two-Sex Problem in Demography

In modern demography, many reproduction indices are calculated based on the stable population model for the female population, because the female-dominant model naturally corresponds to the biological process of childbearing. However, formally speaking, we can develop a male-dominant stable population model in which males reproduce male children and calculate reproduction indices based on the male-dominant theory. In general, these one-sex models are inconsistent, and from a formal, theoretical point of view, it is difficult to determine which model should be used in the scope of linear theory. Before World War II, Kuczynski pointed out that the net reproduction rate of French males was 1.194 from 1920 to 1923, whereas that of French females over the same period was 0.977. Thus, it follows from the



H. Inaba, *Age-Structured Population Dynamics in Demography and Epidemiology*, DOI 10.1007/978-981-10-0188-8_4

stable population theory that the French male population would increase, whereas the female population would decrease. It is clear that these results are not consistent in the long term. Such inconsistencies derived from one-sex theory are called the *two-sex problem* in demography [25]. The two-sex problem does not imply that stable population theory includes a contradiction, but does suggest that the assumption of a time-independent net maternity function for males and females at the same time is inconsistent, because the real reproduction function depends on the interaction of both sexes.

Pollard [24] first proposed a two-sex linear extension of the stable population model to give a common intrinsic growth rate for males and females. Let $p_m(t,a)$ be the age-density function of males at time $t$, and let $p_f(t,a)$ be the female age-density function. The Pollard model is then formulated as follows:

$$\begin{aligned}\left(\frac{\partial}{\partial t}+\frac{\partial}{\partial a}\right)\begin{pmatrix}p_m(t,a)\\ p_f(t,a)\end{pmatrix}&=\begin{pmatrix}-\mu_m(a) & 0\\ 0 & -\mu_f(a)\end{pmatrix}\begin{pmatrix}p_m(t,a)\\ p_f(t,a)\end{pmatrix},\\ \begin{pmatrix}p_m(t,0)\\ p_f(t,0)\end{pmatrix}&=\int_0^\infty\begin{pmatrix}0 & \beta_{mf}(a)\\ \beta_{fm}(a) & 0\end{pmatrix}\begin{pmatrix}p_m(t,a)\\ p_f(t,a)\end{pmatrix}da,\end{aligned}\tag{4.1}$$

where $\beta_{mf}(a)$ denotes the female birth rate of male children, $\beta_{fm}(a)$ is the male birth rate of female children, $\mu_m$ is the force of mortality of male, and $\mu_f$ is the force of mortality of female. The two McKendrick equations are not independent, but are coupled by the boundary condition.

Let $B_m(t) := p_m(t,0)$ and $B_f(t) := p_f(t,0)$ be the male and female birth rates at time $t$. The Pollard model can then be reduced to the following integral equation system:

$$B_m(t)=G_m(t)+\int_0^t\beta_{mf}(a)\ell_f(a)B_f(t-a)da,$$

$$B_f(t)=G_f(t)+\int_0^t\beta_{fm}(a)\ell_m(a)B_m(t-a)da,$$

where $\ell_m$ denotes the male survival probability, $\ell_f$ is the female survival probability and

$$G_m(t)=\int_t^\infty p_f(0,a-t)\frac{\ell_f(a)}{\ell_f(a-t)}\beta_{mf}(a)da,$$

$$G_f(t)=\int_t^\infty p_m(0,a-t)\frac{\ell_m(a)}{\ell_m(a-t)}\beta_{fm}(a)da.$$

Inserting the renewal equation for males into the equation for females, we can derive a single integral equation for $B_f(t)$ as

$$B_f(t)=F(t)+\int_0^t\phi(a)B_f(t-a)da,$$

where

$$F(t) := G_f(t) + \int_0^t \beta_{fm}(a)\ell_m(a)G_m(t-a)da,$$

$$\phi(a) := \int_0^a \beta_{fm}(s)\ell_m(s)\beta_{mf}(a-s)\ell_m(a-s)ds.$$

The same integral equation holds for $B_m(t)$, so we obtain a common net maternity function $\phi(a)$ as the convolution of $\beta_{mf}(a)\ell_f(a)$ and $\beta_{fm}(a)\ell_m(a)$. Therefore, the two-sex basic reproduction number, which is also called the *joint reproduction rate*, is given by

$$R_{\text{joint}} = \int_0^\infty \phi(a)da = \int_0^\infty \beta_{mf}(a)\ell_f(a)da \int_0^\infty \beta_{fm}(a)\ell_m(a)da, \tag{4.2}$$

and the common intrinsic growth rate (the *joint growth rate*) $\lambda_0$ is given as the unique real root of the characteristic equation

$$\int_0^\infty e^{-\lambda a}\phi(a)da = \int_0^\infty e^{-\lambda a}\beta_{fm}\ell_m(a)da \int_0^\infty e^{-\lambda a}\beta_{mf}(a)\ell_f(a)da = 1.$$

Although its cross-reproduction assumption makes the Pollard model rather artificial as a model for human reproduction, it is appropriate as an epidemic model of sexually transmitted diseases. We can formulate a more general linear extension by replacing the boundary condition in (4.1) with

$$\begin{pmatrix} p_m(t,0) \\ p_f(t,0) \end{pmatrix} = \frac{1}{2}\int_0^\infty \begin{pmatrix} \beta_{mm}(a) & \beta_{mf}(a) \\ \beta_{fm}(a) & \beta_{ff}(a) \end{pmatrix} \begin{pmatrix} p_m(t,a) \\ p_f(t,a) \end{pmatrix} da, \tag{4.3}$$

where $\beta_{mm}(a)$ $[\beta_{fm}(a)]$ denotes the fertility rate of male [female] newborns from the male population at age $a$ and $\beta_{mf}(a)$ $[\beta_{ff}(a)]$ denotes the fertility rate of male [female] newborns from the female population at age $a$. An early study on the linear two-sex model in (4.3) was conducted by Bartlett [1].

We can then define the net reproduction matrix as

$$\Phi(a) = \frac{1}{2}\begin{pmatrix} \beta_{mm}(a)\ell_m(a) & \beta_{mf}(a)\ell_f(a) \\ \beta_{fm}(a)\ell_m(a) & \beta_{ff}(a)\ell_f(a) \end{pmatrix}.$$

As shown in Chap. 2, if the next-generation matrix (NGM)

$$K = \int_0^\infty \Phi(a)da$$

is indecomposable, the linear two-sex model (4.3) has a Malthusian (intrinsic) growth rate. It is clear that the Pollard model satisfies the indecomposability condition. The net reproduction rate (basic reproduction number) of (4.3) is given by the spectral radius $r(K)$ of the NGM $K$, and its intrinsic growth rate is given as the real number $\lambda_0$ such that the Laplace transform of the net reproduction matrix

$$\hat{\Phi}(\lambda) = \int_0^\infty e^{-\lambda a}\Phi(a)da$$

has the unit spectral radius at $\lambda = \lambda_0$, and hence, the sign relation

$$\text{sign}(\lambda_0) = \text{sign}(r(K) - 1),$$

holds.

In particular, if the sex ratio at birth, denoted by $s$, is constant, the fertility rate of males is $\beta_m(a)$ and the fertility rate of females is given by $\beta_f(a)$, we have

$$\Phi(a) = \frac{1}{2}\begin{pmatrix} \gamma\beta_m(a)\ell_m(a) & \gamma\beta_f(a)\ell_f(a) \\ (1-\gamma)\beta_m(a)\ell_m(a) & (1-\gamma)\beta_f(a)\ell_f(a) \end{pmatrix},$$

where $\gamma := \frac{s}{1+s}$ is the proportion of male newborns among the total newborns. The NGM is then given by

$$K = \frac{1}{2}\begin{pmatrix} \gamma g_m & \gamma g_f \\ (1-\gamma)g_m & (1-\gamma)g_f \end{pmatrix},$$

where

$$g_m := \int_0^\infty \beta_m(a)\ell_m(a)da, \quad g_f := \int_0^\infty \beta_f(a)\ell_f(a)da,$$

are the gross reproduction rates of males and females, respectively. In this case, $K$ is rank one, so we can easily calculate its spectral radius as

$$R_0 = r(K) = \frac{1}{2}(\gamma g_m + (1-\gamma)g_f),$$

and the characteristic equation is given by

$$1 = \frac{1}{2}\int_0^\infty e^{-\lambda a}(\gamma\beta_m(a)\ell_m(a) + (1-\gamma)\beta_f(a)\ell_f(a))da.$$

If we consider the Pollard model as a special case of the Bartlett model, its basic reproduction number is given by the spectral radius of the NGM as

$$K = \begin{pmatrix} 0 & \int_0^\infty \beta_{mf}(a)\ell_f(a)da \\ \int_0^\infty \beta_{fm}(a)\ell_m(a)da & 0 \end{pmatrix}.$$

Therefore, it follows that $R_0 = \sqrt{R_{\rm joint}}$, because the male [female] population needs two generations of male–female–male [female–male–female] to reproduce themselves, so the unit of generation is twice that of the Bartlett model.

It is evident that the fault in the Bartlett model is that the population can be reproduced by only one sex, so it cannot reflect the mating of both sexes and reproduction by pair formation, which is an essentially nonlinear phenomenon. In the following, we examine some nonlinear pair formation models to overcome the two-sex problem. Here, we only consider continuous-time models; however, discrete two-sex models have been considered by several authors [22, 23].

## 4.2 Kendall's Marriage Model

### 4.2.1 *Basic Model and Its Preliminary Analysis*

Kendall [16] proposed a nonlinear two-sex model that takes into account pair formation. Though not well analyzed for some time, Hadeler and his collaborators successfully clarified the general mathematical characteristics of Kendall's model some 40 years later [5, 29].

Neglecting the age structure, we define $x(t)$, $y(t)$, and $p(t)$ as the density of female singles, the density of male singles, and the density of couples at time $t$, respectively. Let $\mu_x$ and $\mu_y$ be the death rates of females and males, and let $\sigma$ be the divorce rate of couples. In original Kendall's model, the divorce rate was not taken into account. Let $\kappa_x$ and $\kappa_y$ be the birth rate of females and the birth rate of males per unit time and per couple. Let $\psi(x, y)$ be the *marriage function*, which gives the number of couples produced by $x$ single females and $y$ single males per unit time. Kendall's marriage model can then be formulated as a system of ordinary differential equations (ODEs):

$$\begin{aligned} \dot{x} &= (\kappa_x + \mu_y + \sigma)p - \mu_x x - \psi(x, y), \\ \dot{y} &= (\kappa_y + \mu_x + \sigma)p - \mu_y y - \psi(x, y), \\ \dot{p} &= -(\mu_x + \mu_y + \sigma)p + \psi(x, y). \end{aligned} \tag{4.4}$$

In this model, it is assumed that all newborns are produced by couples and that no children are produced by temporary intercourse.

We assume that the marriage function (pair formation function) $\psi$ satisfies the following axioms:

$$\begin{aligned} &(1)\quad \psi(x, y) \geq 0, \quad \forall x, y \geq 0, \\ &(2)\quad \psi(x, 0) = \psi(0, y) = 0, \quad \forall x, y \geq 0, \\ &(3)\quad u \geq 0, v \geq 0 \Rightarrow \psi(x + u, y + v) \geq \psi(x, y), \\ &(4)\quad \alpha \geq 0 \Rightarrow \psi(\alpha x, \alpha y) = \alpha\psi(x, y). \end{aligned} \tag{4.5}$$

Conditions (1)–(3) are self-evident from the meaning of the marriage function. Condition (4) (homogeneity of degree one) may not be essential, but it reflects the fact that the force of marriage would become scale independent in a very large population. In fact, under the homogeneity condition, we can observe that

$$\psi(x, y) = \psi\left(1, \frac{y}{x}\right) x = \psi\left(\frac{x}{y}, 1\right) y,$$

which shows that the force of marriage on females or males depends only on the ratio of singles. That is, the chance of marriage depends not on the number of people of the other sex, but on their relative frequency.

**Exercise 4.1** A typical homogeneous marriage function is the generalized average (CES production function)

$$\psi(x, y) = \rho(\beta x^\alpha + (1 - \beta) y^\alpha)^{\frac{1}{\alpha}},$$

where $\rho > 0$, $1 > \beta > 0$, and $\alpha < 1$ are given parameters. Show that if $\alpha = -1$, this becomes a harmonic mean-type marriage function such as

$$\psi(x, y) = \frac{\rho x y}{\beta x + (1 - \beta) y},$$

that if $\alpha \to 0$, we obtain the geometric mean (Cobb–Douglas production function)

$$\psi(x, y) = \rho x^\beta y^{1-\beta},$$

and that if $\alpha \to -\infty$, we have the minimum value function

$$\psi(x, y) = \rho \min(x, y).$$

Before considering the general system (4.4), let us consider a special case. If there is no difference between the female and male birth and death rates, we can write $\kappa_x = \kappa_y = \kappa$ and $\mu_x = \mu_y = \mu$. From the equations for singles, it follows that

$$\dot{x} - \dot{y} = -\mu(x - y).$$

Then, we have

$$x(t) - y(t) = e^{-\mu t}(x(0) - y(0)),$$

so the difference between the number of females and males converges to zero as time goes to infinity. Hence, if we assume that the number of females and males is the same in advance, $x(t) = y(t)$ for all $t \geq 0$ and the three-dimensional system (4.4) reduces to the two-dimensional linear system

$$\begin{pmatrix} \dot{x} \\ \dot{p} \end{pmatrix} = \begin{pmatrix} -\mu - \rho & \kappa + \mu + \sigma \\ \rho & -2\mu - \sigma \end{pmatrix} \begin{pmatrix} x \\ p \end{pmatrix}, \tag{4.6}$$

where $\rho = \psi(1, 1)$. That is, if there is no difference between the vital rates of females and males, system (4.4) becomes an asymptotically linear system whose asymptotic behavior is determined by the eigenvalues of the coefficient matrix $A$:

$$A := \begin{pmatrix} -\mu - \rho & \kappa + \mu + \sigma \\ \rho & -2\mu - \sigma \end{pmatrix}.$$

Because

$$\det(A - \lambda I) = \lambda^2 + (3\mu + \sigma + \rho) + (\mu + \rho)(2\mu + \sigma) - \rho(\kappa + \mu + \sigma),$$

the eigenvalues are given by

$$\lambda = \frac{1}{2}\left[-(3\mu + \sigma + \rho) \pm \sqrt{D}\right],$$

where

$$D := (3\mu + \sigma + \rho)^2 - 4H,$$
$$H := (\mu + \rho)(2\mu + \sigma) - \rho(\kappa + \mu + \sigma).$$

If $H > 0$, two eigenvalues have negative real parts, so the population becomes extinct. In particular, if $D < 0$, the population exhibits a damping oscillation. In contrast, if $H < 0$, one eigenvalue is positive, and the population grows exponentially. If $H = 0$, the population reaches a stationary state, because the eigenvalues are zero and negative. The threshold condition for population growth $H = 0$ can be rewritten as

$$\frac{\rho\kappa}{\mu(2\mu + \sigma + \rho)} = 1. \tag{4.7}$$

To see the demographic implication of condition (4.7), consider individuals who move between single status and married status with given vital rates. Let $\ell_1(a)$ [$\ell_2(a)$] be the probability that an individual remains in the single [married] state at age $a$. Then, the following holds:

$$\begin{pmatrix} \dot{\ell}_1 \\ \dot{\ell}_2 \end{pmatrix} = \begin{pmatrix} -\mu - \rho & \mu + \sigma \\ \rho & -2\mu - \sigma \end{pmatrix} \begin{pmatrix} \ell_1 \\ \ell_2 \end{pmatrix}, \quad \begin{pmatrix} \ell_1(0) \\ \ell_2(0) \end{pmatrix} = \begin{pmatrix} 1 \\ 0 \end{pmatrix} \tag{4.8}$$

where the coefficient matrix is given by eliminating the birth rate from $A$. Integrating (4.8), we have

$$\ell_1(a) = \frac{(\mu + \sigma)e^{-\mu a}}{\mu + \sigma + \rho} + \frac{\rho e^{-(2\mu+\rho+\sigma)a}}{\mu + \sigma + \rho},$$
$$\ell_2(a) = \frac{\rho}{\mu + \sigma + \rho}(e^{-\mu a} - e^{-(2\mu+\rho+\sigma)a}).$$

The duration in the married state is then calculated as

$$\int_0^\infty \ell_2(a)da = \frac{\rho}{\mu(2\mu + \sigma + \rho)}.$$

Because $\kappa$ is the number of children produced by a married individual per unit time, the left-hand side of (4.7) gives the number of children whose gender is the same as the individual considered here produced by that individual during their married status. This is none other than the basic reproduction number

$$R_0 = \frac{\kappa\rho}{\mu(2\mu + \sigma + \rho)}, \tag{4.9}$$

and hence, the threshold condition for population growth is formulated by setting the basic reproduction number to unity.

**Exercise 4.2** Split the coefficient matrix of the linear system (4.6) as

$$M + Q = \begin{pmatrix} 0 & \kappa \\ 0 & 0 \end{pmatrix} + \begin{pmatrix} -\mu - \rho & \mu + \sigma \\ \rho & -2\mu - \sigma \end{pmatrix}$$

Show that the basic reproduction number $R_0$ given by (4.9) is calculated as the spectral radius of a matrix $K = M(-Q)^{-1}$ (for this recipe, see Chap. 9).

### 4.2.2 *Exponential Solutions*

Let us now consider the general case in which females and males have different vital rates within Kendall's marriage model (4.4). For the homogeneous system, we are mainly interested in persistent (exponential) solutions, which play a role as stationary solutions in non-homogeneous nonlinear dynamical systems. Thus, if we assume that

$$x(t) = e^{\lambda t} x_0, \quad y(t) = e^{\lambda t} y_0, \quad p(t) = e^{\lambda t} p_0,$$

and insert the above expressions into (4.4), we have

$$\begin{aligned} \lambda x_0 &= (\kappa_x + \mu_y + \sigma) p_0 - \mu_x x_0 - \psi(x_0, y_0), \\ \lambda y_0 &= (\kappa_y + \mu_x + \sigma) p_0 - \mu_y y_0 - \psi(x_0, y_0), \\ \lambda p_0 &= -(\mu_x + \mu_y + \sigma) p_0 + \psi(x_0, y_0), \end{aligned} \tag{4.10}$$

which formulate a nonlinear eigenvalue problem for the unknown number (growth rate) $\lambda$ and unknown vector $(x_0, y_0, p_0)$. It is easy to see that the nonlinear eigenvalue problem (4.10) has the trivial solution

$$(x_0, y_0, p_0, \lambda_0) = (1, 0, 0, -\mu_x), \quad (0, 1, 0, -\mu_y), \tag{4.11}$$

where $\lambda_0$ denotes the eigenvalue associated with the eigenvector $(x_0, y_0, p_0)$. This trivial solution corresponds to the extinction process of one sex without the other sex.

If we sum the equations in (4.10), we obtain

$$\begin{aligned}(\lambda + \mu_x)(x_0 + p_0) &= \kappa_x p_0, \\ (\lambda + \mu_y)(y_0 + p_0) &= \kappa_y p_0.\end{aligned} \tag{4.12}$$

Therefore, it follows that

$$x_0 = p_0 \left( \frac{\kappa_x}{\lambda + \mu_x} - 1 \right), \quad y_0 = p_0 \left( \frac{\kappa_y}{\lambda + \mu_y} - 1 \right). \tag{4.13}$$

To ensure there is a positive eigenvector, we require

$$\lambda > \underline{\lambda} := \max(-\mu_x, -\mu_y).$$

Similarly, it follows from (4.12) that

$$\lambda = \frac{\kappa_x p_0}{x_0 + p_0} - \mu_x, \quad \lambda = \frac{\kappa_y p_0}{y_0 + p_0} - \mu_y.$$

The eigenvalue associated with a positive eigenvector should satisfy

$$\lambda < \bar{\lambda} := \min(\kappa_x - \mu_x, \kappa_y - \mu_y).$$

If $\bar{\lambda} \leq \underline{\lambda}$, there is no such positive eigenvector.

Because only the trivial solution exists if $p_0 = 0$, we assume that $p_0 \neq 0$. Inserting (4.13) into the third equation of (4.10) and dividing by $p_0$, we arrive at the characteristic equation

$$F(\lambda) := \psi \left( \frac{\kappa_x}{\lambda + \mu_x} - 1, \frac{\kappa_y}{\lambda + \mu_y} - 1 \right) = \mu_x + \mu_y + \sigma + \lambda. \tag{4.14}$$

If $\bar{\lambda} > \underline{\lambda}$, $F(\lambda)$ goes to zero monotonically as $\lambda$ moves from $\underline{\lambda}$ to $\bar{\lambda}$. The right-hand side of (4.14) is a straight line, and the necessary and sufficient condition that a unique positive eigenvalue exists is given by

$$F(\underline{\lambda}) = \lim_{\lambda \downarrow \underline{\lambda}} F(\lambda) > \mu_x + \mu_y + \sigma + \underline{\lambda}. \tag{4.15}$$

Let $x = 1/h$. Then, if $x \to \infty$,

$$\psi(x, y) = x\psi(1, \frac{y}{x}) = \frac{1}{h}\psi(1, yh) \to \psi_y(1, 0)y, \quad (h \to 0),$$

whereas if $y \to \infty$, we have $\psi(x, y) \to \psi_x(0, 1)x$, $(1/y \to 0)$, where we include the case $\psi_y(1, 0) = \psi_x(0, 1) = \infty$. From this fact, we know that

$$F(\underline{\lambda}) = \begin{cases} \psi_y(1, 0)\left(\frac{\kappa_y}{\mu_y - \mu_x} - 1\right), & (\mu_y > \mu_x) \\ \psi_x(0, 1)\left(\frac{\kappa_x}{\mu_x - \mu_y} - 1\right), & (\mu_x > \mu_y) \end{cases}.$$

If $\mu_y > \mu_x$, (4.15) holds if and only if

$$\psi_y(1, 0)\left(\frac{\kappa_y}{\mu_y - \mu_x} - 1\right) > \mu_y + \sigma.$$

On the other hand, if $\mu_y < \mu_x$, (4.15) holds if and only if

$$\psi_x(0, 1)\left(\frac{\kappa_x}{\mu_x - \mu_y} - 1\right) > \mu_x + \sigma.$$

Finally, if $\mu_y = \mu_x$, (4.15) always holds. It is easy to see that these conditions can be expressed as (Fig. 4.1)

$$\mu_y + \frac{\kappa_x \psi_x(0, 1)}{\mu_x + \sigma + \psi_x(0, 1)} > \mu_x > \mu_y - \frac{\kappa_y \psi_y(1, 0)}{\mu_y + \sigma + \psi_y(1, 0)}. \tag{4.16}$$

From the above argument, we conclude that

**Proposition 4.1** ([5]) *The nonlinear eigenvalue problem* (4.10) *always has a trivial solution* (4.11). *If $\bar{\lambda} > \underline{\lambda}$, there exists a unique positive eigenvector if and only if* (4.16) *holds. In such a case, for the eigenvalue $\lambda_0$ associated with the positive eigenvector, the following holds if $\bar{\lambda} > 0$:*

$$\begin{cases} F(0) > \mu_x + \mu_y + \sigma \Rightarrow \lambda_0 > 0, \\ F(0) = \mu_x + \mu_y + \sigma \Rightarrow \lambda_0 = 0, \\ F(0) < \mu_x + \mu_y + \sigma \Rightarrow \lambda_0 < 0, \end{cases}$$

*where*

$$F(0) = \psi\left(\frac{\kappa_x}{\mu_x} - 1, \frac{\kappa_y}{\mu_y} - 1\right).$$

*If $\bar{\lambda} \le 0$, the eigenvalue (growth rate) is always negative (as long as it exists).*

Finally, note the relation between condition (4.16) and the basic reproduction number. Consider the case whereby infinitely many individuals of the opposite sex exist. Suppose that the size of the male population is so large that the marriage rate for females is saturated. It follows from $\psi(x, y) \to \psi_x(0, 1)x$, $(y \to \infty)$ that the female population and the couple population satisfy the linear system

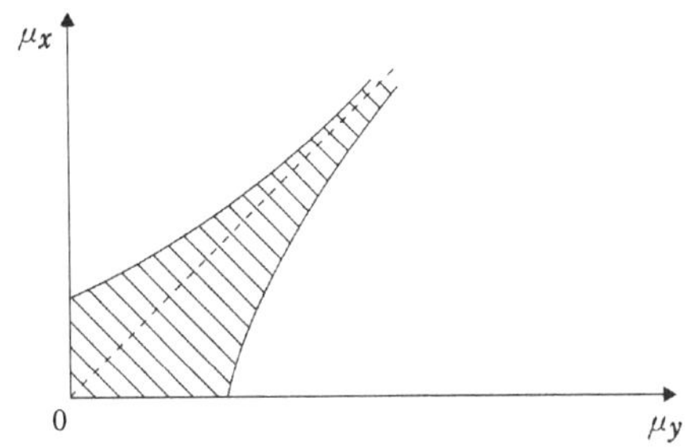


**Fig. 4.1** Schematic representation of domain of existence for a bisexual stationary state [5]

$$\dot{x} = (\kappa_x + \mu_y + \sigma)p - \mu_x x - \rho_x x,$$
$$\dot{p} = -(\mu_x + \mu_y + \sigma)p + \rho_x x,$$

where $\rho_x := \psi_x(0, 1)$ is the marriage rate for females. By a similar argument in the previous subsection, the basic reproduction number for females is calculated as

$$R_0^f = \frac{\kappa_x \rho_x}{\mu_x(\mu_x + \mu_y + \sigma + \rho_x)}.$$

Under the saturation assumption for the marriage rate for males, the basic reproduction number for males is calculated as

$$R_0^m = \frac{\kappa_y \rho_y}{\mu_y(\mu_x + \mu_y + \sigma + \rho_y)},$$

where $\rho_x := \psi_y(1, 0)$. It is easy to see that condition (4.16) is satisfied if $R_0^m > 1$ and $R_0^f > 1$. That is, if both males and females can invade a large-scale population of the opposite sex, then there exists a balanced growth orbit with a positive Malthusian parameter in the pair formation model.

### 4.2.3 Stability of the Homogeneous System

We now consider the orbital stability of non-trivial persistent solutions. For this purpose, we introduce the principle of linearized stability for the homogeneous dynamical system.

Let $f : \mathbb{R}^n \to \mathbb{R}^n$ be a locally Lipschitz continuous function that is homogeneous of degree one and continuously differentiable except at the origin. That is, it follows that

$$\forall \alpha \in \mathbb{R}, \quad f(\alpha x) = \alpha f(x).$$

Let us consider a system of ODEs

$$\dot{x} = f(x). \tag{4.17}$$

The origin is always in a stationary state, and solutions starting from any points other than the origin will not become zero. In particular, it is most important for many applications that the system makes the positive cone $\mathbb{R}^n_+$ invariant. Thus, we assume that

$$x \geq 0, \; x_i = 0 \Longrightarrow f_i(x) \geq 0.$$

The initial data are then in the cone $\mathbb{R}^n_+$, and the solution $x = x(t)$ stays in $\mathbb{R}^n_+$ for all $t > 0$.

Let $e = (1, 1, \ldots, 1)^{\mathrm{T}}$, and let us define a vector $z$ as

$$z = \frac{x}{\langle e, x \rangle}, \quad x \in \mathbb{R}^n_+ \setminus \{0\},$$

where $\langle , \rangle$ denotes the inner product of vectors. If $x$ satisfies system (4.17), we can obtain a dynamical system on the set $S := \{z \geq 0 : \langle e, z \rangle = 1\}$:

$$\dot{z} = f(z) - \langle e, f(z) \rangle z. \tag{4.18}$$

Conversely, if $z$ satisfies (4.18), we can construct a solution of (4.17) by

$$x(t) = z(t) \exp\left( \int_0^t \langle e, f(z(s)) \rangle ds \right) \langle e, x(0) \rangle. \tag{4.19}$$

Note that this kind of transformation can also be applied for the non-autonomous case [7] and the infinite-dimensional case [11].

If $z^* \in S$ is a stationary solution of (4.18), $z^*$ is the solution of the following nonlinear eigenvalue problem:

$$f(z^*) = \lambda^* z^*, \quad \lambda^* = \langle e, f(z^*) \rangle. \tag{4.20}$$

The corresponding solution in (4.17) is

$$x(t) = z^* e^{\lambda^* t} \langle e, x(0) \rangle.$$

Therefore, the stability of the exponential solution of the homogeneous system (4.17) is equivalent to the stability of the corresponding stationary solution of (4.18) on the set $S$.

To clarify the stability of the stationary solution $z$, let us calculate the Jacobian matrix

$$J(z) = f'(z) - z e^{\mathrm{T}} f'(z) - \langle e, f(z) \rangle I_d,$$

where $I_d$ is the identity matrix. Let $f(x) = (f_1(x), \ldots, f_n(x))^{\mathrm{T}}$. It follows from Euler's theorem for homogeneous functions that

$$f'(z)z = f(z). \tag{4.21}$$

From (4.20) and (4.21), we have

$$f'(z^*)z^* = \lambda^* z^*.$$

Using the fact that

$$J(z^*) = f'(z^*) - z^* e^{\mathrm{T}} f'(z^*) - \lambda^* I_d,$$

we obtain

$$J(z^*)z^* = \lambda^* z^* - \lambda^* z^* - \lambda^* z^* = -\lambda^* z^*,$$
$$e^{\mathrm{T}} J(z^*) = e^{\mathrm{T}} f'(z^*) - e^{\mathrm{T}} f'(z^*) - \lambda^* e^{\mathrm{T}} = -\lambda^* e^{\mathrm{T}}.$$

That is, $e^{\mathrm{T}}$ and $z^*$ are the left and the right eigenvectors of the Jacobian matrix $J(z^*)$ with respect to the eigenvalue $-\lambda^*$. This eigenvector corresponds to the movement from the origin to the stationary point $z^*$ on $S$ in the vector field of the cone $\mathbb{R}^n_+$.

Let $\lambda \in \sigma(f'(z^*))$ have the left and right eigenvectors $v$ and $u$.[1] Therefore, $v^{\mathrm{T}} f'(z^*) = \lambda v^{\mathrm{T}}$, $f'(z^*)u = \lambda u$. Note that if $\lambda \neq \lambda^*$, it follows that $\langle v, z^* \rangle = 0$ and $\langle e, u \rangle = 0$. We then obtain

$$v^{\mathrm{T}} J(z^*) = v^{\mathrm{T}} f'(z^*) - \langle v, z^* \rangle e^{\mathrm{T}} f'(z^*) - \lambda^* v^{\mathrm{T}} = (\lambda - \lambda^*) v^{\mathrm{T}},$$

$$J(z^*)u = (\lambda - \lambda^*)u.$$

Hence, if $\sigma(f'(z^*)) = \{\lambda^*, \lambda_1, \ldots, \lambda_{n-1}\}$, it follows that

$$\sigma(J(z^*)) = \{-\lambda^*, \lambda_1 - \lambda^*, \ldots, \lambda_{n-1} - \lambda^*\},$$

and eigenvector $u_j$ of $J(z^*)$ associated with $\lambda_j - \lambda^*$ is seen as a vector on the set $S$, because $\langle e, u_j \rangle = 0$.

The flow around $z^*$ on $S$ is then determined by eigenvalues $\lambda - \lambda^*$ and their eigenvectors, so we conclude that

**Proposition 4.2** ([5]) *Suppose that a function $f : \mathbb{R}^n \to \mathbb{R}^n$ is locally Lipschitz continuous, continuously differentiable except at the origin and homogeneous of degree one. If $z^* e^{\lambda^* t}$, $z^* \in S$, is an exponential solution of the homogeneous system* (4.17), *then $\lambda^*$ is an eigenvalue of $f'(z^*)$. For any eigenvalue $\lambda$ other than $\lambda^*$ of $f'(z^*)$, if $\Re(\lambda - \lambda^*) < 0$, then $z^*$ is locally stable on $S$.*

---

[1] $\sigma(A)$ denotes the set of eigenvalues of a matrix $A$.

Let us apply the above result to Kendall's model. First, we consider the stability of trivial exponential solutions. The Jacobian matrix of the right-hand side of (4.4) is given by

$$f' = \begin{pmatrix} -\mu_x - \psi_x & -\psi_y & \kappa_x + \mu_y + \sigma \\ -\psi_x & -\mu_y - \psi_y & \kappa_y + \mu_x + \sigma \\ \psi_x & \psi_y & -(\mu_x + \mu_y + \sigma) \end{pmatrix},$$

and its value at $z^* = (1, 0, 0)$ is

$$f'(z^*) = \begin{pmatrix} -\mu_x & -\psi_y(1,0) & \kappa_x + \mu_y + \sigma \\ 0 & -\mu_y - \psi_y(1,0) & \kappa_y + \mu_x + \sigma \\ 0 & \psi_y(1,0) & -(\mu_x + \mu_y + \sigma) \end{pmatrix}.$$

The Jacobian matrix $J$ at $z^* = (1, 0, 0)$ has an eigenvalue $\lambda_1 = -\mu_x$, and the other eigenvalues $\lambda_2$ and $\lambda_3$ are given as

$$\lambda_{2,3} = \frac{-B \pm \sqrt{D}}{2},$$

where

$$\begin{cases} B := \mu_x + 2\mu_y + \sigma + \psi_y(1,0) \\ D := (\mu_x + \sigma + \psi_y(1,0))^2 + 4\kappa_y \psi_y(1,0) \end{cases}.$$

Note that $\lambda_1$ is the exponential growth rate $\lambda^*$. A necessary and sufficient condition for $\lambda_{2,3} - \lambda_1 < 0$ is given by

$$\sqrt{D} < -\mu_x + 2\mu_y + \sigma + \psi_y(1,0),$$

which is equivalent to

$$0 < -\mu_x + 2\mu_y + \sigma + \psi_y(1,0), \tag{4.22}$$

and

$$D < [2(\mu_y - \mu_x) + (\mu_x + \sigma + \psi_y(1,0))]^2.$$

From the latter condition, we have

$$\kappa_y \psi_y(1,0) < (\mu_y - \mu_x)(\mu_y + \sigma + \psi_y(1,0)). \tag{4.23}$$

Because $\psi_y(1,0) \geq 0$, condition (4.23) reduces to

$$\mu_x < \mu_y - \frac{\kappa_y \psi_y(1,0)}{\mu_y + \sigma + \psi_y(1,0)}. \tag{4.24}$$

If this condition holds, then $\mu_x < \mu_y$, and hence, (4.22) is satisfied. Therefore, (4.24) is a necessary and sufficient condition for the trivial exponential solution $(1, 0, 0)e^{-\mu_x t}$ to be locally asymptotically stable. In this case, $(\mu_x, \mu_y)$ does not satisfy (4.16), and there is no non-trivial exponential solution. Using similar arguments, a necessary and sufficient condition for the trivial exponential solution $(0, 1, 0)e^{-\mu_y t}$ being locally asymptotically stable is given by

$$\mu_y + \frac{\kappa_x \psi_x(0,1)}{\mu_x + \sigma + \psi_x(0,1)} < \mu_x. \tag{4.25}$$

If (4.24) or (4.25) holds, there exist only two trivial stationary states on the boundary of $S$, and there is no periodic solution. Thus, a stable stationary state attracts all solution orbits inside of $S$.

To verify the local stability of non-trivial exponential solutions, which exist in the parameter region where (4.16) holds, let us introduce new variables $(\xi, \eta)$ as

$$\xi = \frac{x}{p}, \quad \eta = \frac{y}{p}.$$

The interior of $S$ can then be injectively mapped into the interior of $\mathbb{R}^2_+$, and (4.4) reduces to the two-dimensional system

$$\begin{aligned} \dot{\xi} &= \kappa_x + (\mu_y + \sigma)(1+\xi) - (1+\xi)\psi(\xi,\eta), \\ \dot{\eta} &= \kappa_y + (\mu_x + \sigma)(1+\eta) - (1+\eta)\psi(\xi,\eta). \end{aligned} \tag{4.26}$$

The Jacobian matrix $J(\xi, \eta)$ of the right-hand side is calculated as

$$J(\xi,\eta) = \begin{pmatrix} \mu_x + \sigma - \psi - (1+\xi)\psi_x & -(1+\xi)\psi_y \\ -(1+\eta)\psi_x & \mu_y + \sigma - \psi - (1+\eta)\psi_y \end{pmatrix},$$

and its value at the stationary point $(\xi^*, \eta^*)$ is

$$J(\xi^*,\eta^*) = \begin{pmatrix} -\frac{\kappa_x}{1+\xi^*} - (1+\xi)\psi_x(\xi^*,\eta^*) & -(1+\xi^*)\psi_y(\xi^*,\eta^*) \\ -(1+\eta^*)\psi_x(\xi^*,\eta^*) & -\frac{\kappa_y}{1+\eta^*} - (1+\eta^*)\psi_y(\xi^*,\eta^*) \end{pmatrix}.$$

Because the trace of $J(\xi^*, \eta^*)$ is negative and $\det J(\xi^*, \eta^*) > 0$, all eigenvalues are negative and the stationary point is locally asymptotically stable. Moreover, all non-diagonal entries of $J(\xi, \eta)$ are negative, which means the system is *competitive* in the sense of Hirsch [9], and hence, all bounded trajectories converge to a stationary point. If we consider trajectories on $S$, trivial solution orbits on the boundary $x = 0$ or $y = 0$ are unstable and any other orbits stay in the interior of $S$. These correspond to bounded trajectories in (4.26), so they converge to the stationary point $(\xi^*, \eta^*) > 0$.

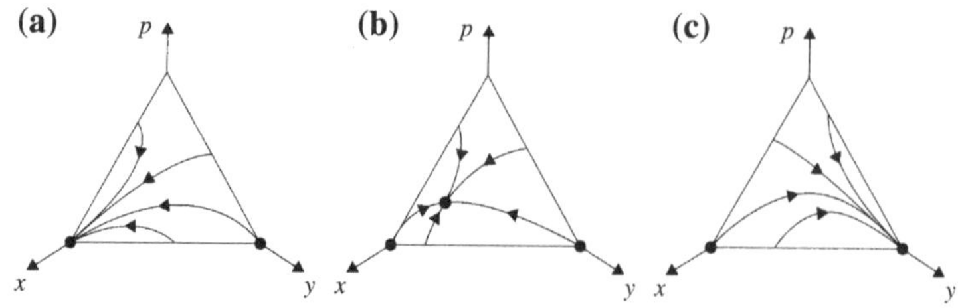


**Fig. 4.2** Phase portraits of a dynamical system on the set $S$ [5]

That is, under the parameter condition (4.16), the positive stationary point on $S$ attracts all trajectories in the interior of $S$. We can therefore conclude the following:

**Proposition 4.3** ([5]) *If* (4.24) *holds,* $(1, 0, 0)$ *is stable and* $(0, 1, 0)$ *is unstable on S. If* (4.16) *holds, there exists a unique positive stationary state that is stable, whereas* $(1, 0, 0)$ *and* $(0, 1, 0)$ *are unstable. If* (4.25) *holds,* $(1, 0, 0)$ *is unstable and* $(0, 1, 0)$ *is stable.*

Roughly speaking, if the difference between the male and female death rates is not overly large, there exists a stable balanced growth orbit. However, for example, if the male death rate becomes too high, the positive stationary point approaches the trivial stationary point $(1, 0, 0)$. The orbit then moves away from $S$, and an exchange of stability occurs. That is, the exponential solution corresponding to the extinction of the female one-sex population becomes stable and any other solution orbits are attracted to this exponential solution (Fig. 4.2).

In summary, Kendall's marriage model shows that the Malthusian growth of a population can be realized under the monogamous persistent union formation rule and that regardless of the high fertility within unions, population decline will occur if the difference between male and female death rates is sufficiently large. This latter point is a characteristic phenomenon of pair formation models.

However, Kendall's model neglects the age structure, making it unrealistic as a human population model in which pair formation (and reproduction) is possible even when male and female singles are infants. By introducing a maturation period, Kendall's model has been extended to a delay-differential equation model [8]. For the delay-differential equation model, we can again obtain conditions for the existence of exponential solutions and their stability, but the existence of a maturation period causes the range of the difference between male and female death rates, such that non-trivial exponential solutions can exist, to become narrower. Another unrealistic aspect of Kendall's model is that there is no growth bound. Castillo-Chavez and Huang [3] examined a pair formation model that takes into account the logistic effect on the birth rate and the divorce rate, which makes the solution bounded.

## 4.3 Pair Formation Models with Age Structure

In 1971, two decades after Kendall's marriage model appeared, Fredrickson [4] formulated a monogamous marriage model with an age structure. Let $x(t,a)$ be the age-density function of female singles at time $t$ and age $a$, $y(t,a)$ be the age-density function of male singles at time $t$, and age $a$ and $p(t,a,b)$ be the age-density function of couples in which the female's age is $a$ and the male's age is $b$ (we refer to these as an $(a,b)$ couple) at time $t$. Let $\mu_x(a)$ and $\mu_y(a)$ be the female and male death rates at age $a$, $\sigma(a,b)$ be the divorce rate of $(a,b)$ couples, $\beta(a,b)$ be the marital fertility rate of $(a,b)$ couples, $\gamma$ be the ratio of female newborns at birth, and $\rho(t,a,b)$ be the density of newly produced $(a,b)$ couples per unit time at time $t$. *Fredrickson's marriage model* is then formulated as follows:

$$
\begin{aligned}
\frac{\partial x(t,a)}{\partial t} + \frac{\partial x(t,a)}{\partial a} &= -\mu_x(a)x(t,a) - \int_0^\infty \rho(t,a,z)dz \\
&\quad + \int_0^\infty p(t,a,z)[\sigma(a,z)+\mu_y(z)]dz, \\
\frac{\partial y(t,a)}{\partial t} + \frac{\partial y(t,a)}{\partial a} &= -\mu_y(a)y(t,a) - \int_0^\infty \rho(t,z,a)dz \\
&\quad + \int_0^\infty p(t,z,a)[\sigma(z,a)+\mu_x(z)]dz, \\
\frac{\partial p(t,a,b)}{\partial t} + \frac{\partial p(t,a,b)}{\partial a} + \frac{\partial p(t,a,b)}{\partial b} & \\
&= \rho(t,a,b) - (\sigma(a,b)+\mu_x(a)+\mu_y(b))p(t,a,b), \\
x(t,0) &= \gamma \int_0^\infty \int_0^\infty \beta(a,b)p(t,a,b)dadb, \\
y(t,0) &= (1-\gamma)\int_0^\infty \int_0^\infty \beta(a,b)p(t,a,b)dadb, \\
p(t,0,b) &= p(t,a,0) = 0,
\end{aligned}
\tag{4.27}
$$

where $\rho(t,a,b)$ is given by a *marriage function* $\Psi(u,v)(a,b)$ as

$$\rho(t,a,b) = \Psi(x(t,\cdot), y(t,\cdot))(a,b),$$

which is a nonlinear function of the density of singles, $x$ and $y$.

Staroverov [28] and Hadeler [6] later extended the Fredrickson model to recognize the duration of pairs. This is a reasonable realistic extension, because the divorce rate and the marital fertility rate are heavily dependent on the duration of pairs. Let $c$ be the duration of pairs and $p(t,a,b,c)$ be the age–duration-density function of couples. Then, (4.27) can be extended as follows (*Staroverov–Hadeler model*):

$$\begin{aligned}
\frac{\partial x(t,a)}{\partial t} + \frac{\partial x(t,a)}{\partial a} &= -\mu_x(a)x(t,a) - \int_0^\infty \rho(t,a,z)dz \\
&\quad + \int_0^\infty \int_0^\infty p(t,a,\zeta,\eta)[\sigma(a,\zeta,\eta) + \mu_y(\zeta)]d\zeta d\eta, \\
\frac{\partial y(t,a)}{\partial t} + \frac{\partial y(t,a)}{\partial a} &= -\mu_y(a)y(t,a) - \int_0^\infty \rho(t,z,a)dz \\
&\quad + \int_0^\infty \int_0^\infty p(t,\zeta,a,\eta)[\sigma(\zeta,a,\eta) + \mu_x(\zeta)]d\zeta d\eta, \\
\frac{\partial p(t,a,b,c)}{\partial t} + \frac{\partial p(t,a,b,c)}{\partial a} &+ \frac{\partial p(t,a,b,c)}{\partial b} + \frac{\partial p(t,a,b,c)}{\partial c} \\
&= -(\sigma(a,b,c) + \mu_x(a) + \mu_y(b))p(t,a,b,c), \\
x(t,0) &= \gamma \int_0^\infty \int_0^\infty \int_0^\infty \beta(a,b,c)p(t,a,b,c)dadbdc, \\
y(t,0) &= (1-\gamma) \int_0^\infty \int_0^\infty \int_0^\infty \beta(a,b,c)p(t,a,b,c)dadbdc, \\
p(t,a,b,0) &= \rho(t,a,b),
\end{aligned} \tag{4.28}$$

where we adopt the convention that $p(t,a,b,c) = 0$ for $c \geq \min(a,b)$.

Inaba [13, 14] also introduced the marital duration variable into the Fredrickson model. However, instead of treating age and duration equivalently, Inaba introduced a marriage cohort equation using the marital duration by age at marriage. Let $p(t,c;a,b)$ be the density of couples at duration $c$ whose bride's age is $a$ and groom's age is $b$ at the point of couple formation. We call this the couple density function of *type* $(a,b)$. Let $\sigma(c;a,b)$ be the divorce rate of type $(a,b)$ couples and $\beta(c;a,b)$ be the marital fertility rate of type $(a,b)$ couples. The monogamous marriage model can then be reformulated as

$$\begin{aligned}
\frac{\partial x(t,a)}{\partial t} + \frac{\partial x(t,a)}{\partial a} &= -\mu_x(a)x(t,a) - \int_0^\infty \rho(t,a,\eta)d\eta \\
&\quad + \int_0^a \int_0^\infty [\mu_y(\tau+\eta) + \delta(\tau;a-\tau,\eta)]p(t,\tau;a-\tau,\eta)d\eta d\tau, \\
\frac{\partial y(t,a)}{\partial t} + \frac{\partial y(t,a)}{\partial a} &= -\mu_y(a)y(t,a) - \int_0^\infty \rho(t,\zeta,a)d\zeta \\
&\quad + \int_0^a \int_0^\infty [\mu_x(\tau+\zeta) + \delta(\tau;\zeta,a-\tau)]p(t,\tau;\zeta,a-\tau)d\zeta d\tau, \\
\frac{\partial p(t,c;a,b)}{\partial t} + \frac{\partial p(t,c;a,b)}{\partial c} &= -[\mu_x(a+c) + \mu_y(b+c) + \delta(c;a,b)]p(t,c;a,b), \\
x(t,0) &= \gamma \int_0^\infty \int_0^\infty \int_0^\infty \beta(c;a,b)p(t,c;a,b)dadbdc, \\
y(t,0) &= (1-\gamma) \int_0^\infty \int_0^\infty \int_0^\infty \beta(c;a,b)p(t,c;a,b)dadbdc, \\
p(t,0;a,b) &= \rho(t,a,b) = \Psi(x(t,\cdot),y(t,\cdot))(a,b),
\end{aligned} \tag{4.29}$$

where the equation for $p$ describes the dynamics of a marriage cohort of type $(a, b)$ couples. As we see in the next section, a major advantage of this cohort equation approach is that the equation for the marriage cohort is easily integrated along the characteristic lines. Hence, if we insert the expression for $p$ into the equations for singles, we can easily eliminate the couple density from the basic system. Unfortunately, this introduces time delay terms into the equations for singles.

The marriage function is a nonlinear operator $\Psi(u, v)$ that maps the female single density $u$ and the male single density $v$ to the density of newly produced couples. It is reasonable to assume that the marriage function satisfies the following axioms:

1. If $(u, v) \geq 0$, then $\Psi(u, v) \geq 0$,
2. For all $(u, v) \geq 0$, $\Psi(u, 0) = \Psi(0, v) = 0$,
3. If $(u, v) \leq (u', v')$, then

$$\int_0^\infty \int_0^\infty \Psi(u, v)(a, b) da db \leq \int_0^\infty \int_0^\infty \Psi(u', v')(a, b) da db,$$

4. If $\alpha \geq 0$, then $\Psi(\alpha u, \alpha v) = \alpha \Psi(u, v)$,
5. For $a \neq x, b \neq x$, it holds that

$$\frac{\partial \Psi(u, v)(a, b)}{\partial u(x)} \leq 0, \quad \frac{\partial \Psi(u, v)(a, b)}{\partial v(x)} \leq 0.$$

Although axioms (1), (2), and (3) are trivial, axioms (4) and (5) are not. The homogeneity condition (axiom (4)) reflects the fact that the number of encounters among individuals per unit time will saturate and become almost constant in a large-scale population, and so will not hold if the population density is not sufficiently high. Axiom (5) implies that there is competition among age classes. The marriage probability of one age class will decrease, or at least not increase, if the population of another age class increases.

As concrete examples, we can consider the following marriage functions for age-structured populations:

$$\Psi(u, v)(a, b) = 2\rho(u, v)(a, b) \frac{\xi(a)u(a)\eta(b)v(b)}{\int_0^\infty \xi(a)u(a)da + \int_0^\infty \eta(b)v(b)db}, \tag{4.30}$$

$$\Psi(u, v)(a, b) = \left[\beta(h(a, b)u(a))^\alpha + (1 - \beta)(g(a, b)v(b))^\alpha\right]^{\frac{1}{\alpha}}. \tag{4.31}$$

Function (4.30) is called the *proportionate mixing* assumption and (4.31) is the generalized mean function, where $0 < \beta < 1, -\infty < \alpha < 1$. If $\alpha = -1$, the generalized mean becomes the weighted harmonic mean

$$\frac{h(a, b)g(a, b)u(a)v(b)}{\beta g(a, b)v(b) + (1 - \beta)h(a, b)u(a)},$$

and if $\alpha \to 0$, it becomes the weighted geometric mean

$$[h(a,b)u(a)]^{\beta}[g(a,b)v(b)]^{1-\beta}.$$

If $\alpha \to \infty$, we have a minimum value function given by

$$\min\{h(a,b)u(a), g(a,b)v(b)\}.$$

All of the above functions are homogeneous of degree one, but only the proportionate mixing function satisfies the axiom for age competition.

There are few studies on what kind of marriage function is most appropriate for real data. Keyfitz [17] reported that the geometric mean slightly weighted toward females fits well with US marital statistics from the 1960s. Martcheva and Milner [21] found that the marriage function estimated from the 1970s US census is an almost linear function mainly weighted on the male population, whereas the age-specific fitness of the estimated marital population using the marriage function is not so good. Readers are referred to [12] for more detailed arguments.

In a real population, it would be difficult to test the specific nonlinearity of marriage functions, because the ratio of males to females is almost one to one. Indeed, a theoretically favorable function satisfying all axioms such as the proportionate mixing function may not necessarily give a good estimate. As the mating process between males and females is very complex, we should examine its structure in depth to elicit further understanding. It would be rather surprising if a marriage function that can be applied to a wide range of sex ratios could be expressed as a simple mathematical function.

## 4.4 Malthusian Growth via Pair Formation

If we observe a population for a limited period, the stable population model often describes the local population dynamics well. Even in such a case, individuals must be reproduced under some kind of marriage rule. Therefore, a natural question is whether the nonlinear marriage model can have a stable exponential solution. Though several authors have given sufficient conditions for age-structured homogeneous marriage models to have exponential solutions [13, 14, 20, 26, 31], there are presently no results on the stability and uniqueness of exponential solutions. Here, we only sketch the existence problem of exponential solutions in age-structured pair formation models. Because the general argument is so difficult and complex, we first consider special examples that enable the existence of exponential solution to be easily proved.

### 4.4.1 Intra-cohort Marriage Models

First, we consider a special case of the Fredrickson model (4.27) in which marriage only occurs between males and females of the same age. In other words, marriage only occurs within the same cohort (the *intra-cohort marriage model*). Moreover, for simplicity, we neglect the difference between male and female death rates and assume that the sex ratio at birth is unity. In this special case, the density of pairs is described by one age and time. Let $p(t,a)$ be the age-density function of couples at age $a$ (which is common for male and female partners), and let $\sigma(a)$, $\rho(t,a)$, and $\beta(a)$ be the divorce rate, the density of newly produced couples, and the age-specific (one-sex) fertility rate, respectively. The intra-cohort marriage model is then formulated as follows:

$$\begin{aligned}
\frac{\partial x(t,a)}{\partial t}+\frac{\partial x(t,a)}{\partial a} &= -\mu(a)x(t,a)+p(t,a)[\sigma(a)+\mu(a)]-\rho(t,a),\\
\frac{\partial y(t,a)}{\partial t}+\frac{\partial y(t,a)}{\partial a} &= -\mu(a)y(t,a)+p(t,a)[\sigma(a)+\mu(a)]-\rho(t,a),\\
\frac{\partial p(t,a)}{\partial t}+\frac{\partial p(t,a)}{\partial a} &= -(\sigma(a)+2\mu(a))p(t,a)+\rho(t,a),\\
x(t,0)=y(t,0) &= \int_0^\infty \beta(a)p(t,a)da,\\
p(t,0) &= 0.
\end{aligned} \tag{4.32}$$

If we define $v(t,a) := x(t,a)-y(t,a)$, then $v$ satisfies the McKendrick equation with the zero boundary condition. Hence, it follows that

$$v(t,a)=\begin{cases}0, & t>a,\\ v(0,a-t)\frac{\ell(a)}{\ell(a-t)}, & a>t,\end{cases}$$

where $\ell(a) := \exp(-\int_0^a \mu(z)dz)$ is the survival probability. Because $v(t,a)\to 0$ as $t\to\infty$, the trivial solution $v\equiv 0$ is globally stable. Therefore, without loss of generality, we can assume that $x(t,a)=y(t,a)$ for all $t\geq 0$ in advance. Then, $\rho$ is given by

$$\rho(t,a)=\Psi(x(t,\cdot),x(t,\cdot))(a).$$

Although $\Psi(x,x)$ is not generally linear with respect to $x$, if we adopt the generalized mean function as the marriage function, we can write

$$\rho(t,a)=\Psi(1,1)(a)x(t,a).$$

Under the above assumption, (4.32) reduces to the multistate stable population model

$$
\begin{aligned}
\frac{\partial x(t,a)}{\partial t} + \frac{\partial x(t,a)}{\partial a} &= -(\mu(a)+\theta(a))x(t,a) + (\sigma(a)+\mu(a))p(t,a), \\
\frac{\partial p(t,a)}{\partial t} + \frac{\partial p(t,a)}{\partial a} &= \theta(a)x(t,a) - (\sigma(a)+2\mu(a))p(t,a), \\
x(t,0) &= \int_0^\infty \beta(a)p(t,a)da, \\
p(t,0) &= 0,
\end{aligned}
\tag{4.33}
$$

where $\theta(a) := \Psi(1,1)(a)$ gives the force of marriage. As we can interpret $p(t,a)$ as the age-density function of the married female population, the above model is equivalent to the one-sex multistate stable population model in which the female population is divided into single and married states and reproduction only occurs in the married state.

Let $\ell_1(a)$ [$\ell_2(a)$] be the probability that a female individual is single [married] at age $a$. Then, it holds that

$$
\frac{d}{da}\begin{pmatrix}\ell_1(a)\\ \ell_2(a)\end{pmatrix} = \begin{pmatrix}-\mu(a)-\theta(a) & \mu(a)+\sigma(a)\\ \theta(a) & -2\mu(a)-\sigma(a)\end{pmatrix}\begin{pmatrix}\ell_1(a)\\ \ell_2(a)\end{pmatrix}.
$$

By solving the above ODE system, we have

$$
\begin{aligned}
\ell_1(a) &= \ell(a) - \ell_2(a), \\
\ell_2(a) &= \int_0^a e^{-\int_s^a (2\mu(z)+\sigma(z)+\theta(z))dz}\theta(s)\ell(s)ds.
\end{aligned}
$$

Thus, we obtain the renewal equation

$$
x(t,0) = \int_0^\infty \beta(a)\ell_2(a)x(t-a,0)da,
$$

where $x(t,0)$ denotes the number of newborns per unit time at time $t$. The characteristic equation for determining the Malthusian parameter $\lambda$ is then given by

$$
\int_0^\infty e^{-\lambda a}\beta(a)\ell_2(a)da = 1,
$$

and the basic reproduction number is

$$
R_0 = \int_0^\infty \beta(a)\int_0^a e^{-\int_s^a (2\mu(z)+\sigma(z)+\theta(z))dz}\theta(s)\ell(s)dsda.
$$

In summary, if there is no difference in the male and female vital rates and marriage occurs only within a cohort, the asymptotic behavior of the pair formation model is described by the stable population model.

Let us now consider the marital duration-dependent model (4.29). Let $\Gamma(c; a, b)$ be the survival probability of a type $(a, b)$ couple given by

$$\Gamma(c; a, b) = \frac{\ell_x(a+c)}{\ell_x(a)} \frac{\ell_y(b+c)}{\ell_y(b)} \exp\left(-\int_0^c \delta(\tau; a, b) d\tau\right),$$

where $\ell_x(a)$ and $\ell_y(a)$ are the female and male survival rates, defined by

$$\ell_x(a) = \exp\left(-\int_0^a \mu_x(\zeta) d\zeta\right), \quad \ell_y(a) = \exp\left(-\int_0^a \mu_y(\eta) d\eta\right).$$

Using these survival functions, the equation for couples can be integrated along the characteristic lines as

$$p(t, c; a, b) = \begin{cases} \Gamma(c; a, b) \Psi(x(t-c, \cdot), y(t-c, \cdot))(a, b), & t - c > 0, \\ p(0, c-t; a, b) \frac{\Gamma(c; a, b)}{\Gamma(c-t; a, b)}, & c - t > 0. \end{cases}$$

Inserting the above expression into the equations for singles, we obtain the following closed system:

$$\begin{aligned}
\frac{\partial x(t,a)}{\partial t} + \frac{\partial x(t,a)}{\partial a} &= -\mu_x(a) x(t,a) - \int_0^\infty \rho(t, a, \eta) d\eta \\
&\quad + \int_0^a \int_0^\infty [\mu_y(\tau + \eta) + \delta(\tau; a - \tau, \eta)] \Gamma(\tau; a - \tau, \eta) \rho(t - \tau, a - \tau, \eta) d\eta d\tau, \\
\frac{\partial y(t,a)}{\partial t} + \frac{\partial y(t,a)}{\partial a} &= -\mu_y(a) y(t,a) - \int_0^\infty \rho(t, \eta, a) d\eta \\
&\quad + \int_0^a \int_0^\infty [\mu_x(\tau + \eta) + \delta(\tau; \eta, a - \tau)] \Gamma(\tau; \eta, a - \tau) \rho(t - \tau, \eta, a - \tau) d\eta d\tau, \\
x(t, 0) &= \gamma \int_0^\infty \int_0^\infty \int_0^\infty \beta(\tau; \zeta, \eta) \Gamma(\tau; \zeta, \eta) \rho(t - \tau, \zeta, \eta) d\zeta d\eta d\tau, \\
y(t, 0) &= (1 - \gamma) \int_0^\infty \int_0^\infty \int_0^\infty \beta(\tau; \zeta, \eta) \Gamma(\tau; \zeta, \eta) \rho(t - \tau, \zeta, \eta) d\zeta d\eta d\tau,
\end{aligned} \tag{4.34}$$

where $\rho(t, a, b) = \Psi(x(t, \cdot), y(t, \cdot))(a, b)$.

Again, let us consider a special case in which there is no difference between male and female vital rates and marriage occurs only in the same cohort. In this case, the male and female age distributions are the same, and we can obtain a simplified one-sex system for the female population (intra-cohort marriage model):

$$\frac{\partial x(t,a)}{\partial t}+\frac{\partial x(t,a)}{\partial a}=-\mu(a)x(t,a)-\rho(a)x(t,a)$$
$$+\int_0^a[\mu(a)+\delta(\tau;a-\tau)]\Gamma(\tau;a-\tau)\rho(a-\tau)x(t-\tau,a-\tau)d\tau,$$
$$x(t,0)=\int_0^\infty\int_0^\infty\beta(\tau;\zeta)\Gamma(\tau;\zeta)\rho(\zeta)x(t-\tau,\zeta)d\zeta d d\tau, \tag{4.35}$$

where

$$\Psi(x(t,\cdot),x(t,\cdot))(a,a)=\rho(a)x(t,a),$$

$$\Gamma(\tau;\zeta):=\exp\left(-\int_0^\tau[\mu(\zeta+\sigma)+\delta(\sigma;\zeta)]d\sigma\right),$$

$\beta(\tau;\zeta)$ is the marital birth rate for female newborns and $\delta(\tau;\zeta)$, $\Gamma(\tau;\zeta)$ are the divorce rate and the survival rate of couples, respectively, at marital duration $\tau$ according to the age at marriage $\zeta$.

System (4.35) can again be considered as a multistate stable population model. Let $\ell_1(a)$ be the probability that a woman is in the single state at age $a$ and $\ell_2(\tau;\zeta)$ be the probability that a woman married at age $\zeta$ remains in the marital status after duration $\tau$. We then have

$$\frac{d\ell_1(a)}{da}=-(\mu(a)+\rho(a))\ell_1(a)+\int_0^a(\mu(a)+\delta(\tau;a-\tau))\ell_2(\tau;a-\tau)d\tau,$$
$$\frac{\partial\ell_2(\tau;\zeta)}{\partial\tau}=-(\mu(\tau+\zeta)+\delta(\tau;\zeta))\ell_2(\tau;\zeta),$$
$$\ell_2(0;\zeta)=\rho(\zeta)\ell_1(\zeta). \tag{4.36}$$

From the above system, we can obtain an integral–differential equation for $\ell_1(a)$:

$$\frac{d\ell_1(a)}{da}=-(\mu(a)+\rho(a))\ell_1(a)$$
$$+\int_0^a(\mu(a)+\delta(\tau;a-\tau))\Gamma(\tau;a-\tau)\rho(a-\tau)\ell_1(a-\tau)d\tau. \tag{4.37}$$

Using the variation-of-constants formula and changing the order of the integrals, (4.37) can be reduced to the integral equation

$$\ell_1(a)=g(a)+\int_0^a R(a,x)\ell_1(x)dx, \tag{4.38}$$

where

$$g(a):=\exp\left(-\int_0^a[\mu(\sigma)+\rho(\sigma)]d\sigma\right),$$

$$R(a,x) := \int_x^a (\mu(z) + \delta(z-x;x))\Gamma(z-x;x)\frac{g(a)}{g(z)}dz.$$

Because (4.38) is a Volterra integral equation, it has a unique nonnegative solution. Using the solution $\ell_1(a)$ and $x(t,a) = B(t-a)\ell_1(a)$ with $x(t,0) =: B(t)$, the following limiting renewal equation is obtained from the boundary condition of (4.35):

$$\begin{aligned}B(t) &= \int_0^\infty \int_0^\infty \beta(\tau;\zeta)\Gamma(\tau;\zeta)\rho(\zeta)\ell_1(\zeta)B(t-\tau-\zeta)d\zeta d\tau \\ &= \int_0^\infty \int_0^a \beta(\tau;a-\tau)\Gamma(\tau;a-\tau)\rho(a-\tau)\ell_1(a-\tau)d\tau B(t-a)da.\end{aligned}$$

Therefore, the birth rate $B(t)$ is determined by Lotka's integral equation with a net reproduction function $\Psi(a)$ given by

$$\Psi(a) = \int_0^a \beta(\tau;a-\tau)\Gamma(\tau;a-\tau)\rho(a-\tau)\ell_1(a-\tau)d\tau,$$

and the basic reproduction number is calculated as

$$R_0 = \int_0^\infty \int_0^\infty \beta(\tau;\xi)\Gamma(\tau;\xi)\rho(\xi)\ell_1(\xi)d\xi d\tau.$$

Again, we conclude that the standard Lotka theory can be applied to this intra-cohort marriage model structured by the marriage duration.

### 4.4.2 Inter-cohort Marriage Models

Finally, we sketch a general approach to the age–duration-dependent model (4.34). To determine exponential solutions, we insert the functions

$$(x,y) = (e^{\lambda t}u(a), e^{\lambda t}v(a)),$$

into (4.34) to obtain

$$\begin{aligned}\frac{du(a)}{da} &= -(\mu_m(a)+\lambda)u(a) - \int_0^\infty \Psi(u,v)(a,\eta)d\eta \\ &+ \int_0^a \int_0^\infty [\mu_f(\tau+\eta) + \delta(\tau;a-\tau,\eta)]e^{-\lambda\tau}\Gamma(\tau;a-\tau,\eta)\Psi(u,v)(a-\tau,\eta)d\eta d\tau, \\ \frac{dv(a)}{da} &= -(\mu_f(a)+\lambda)v(a) - \int_0^\infty \Psi(u,v)(a,\eta)d\eta \\ &+ \int_0^a \int_0^\infty [\mu_m(\tau+\eta) + \delta(\tau;\eta,a-\tau)]e^{-\lambda\tau}\Gamma(\tau;\eta,a-\tau)\Psi(u,v)(\eta,a-\tau)d\eta d\tau,\end{aligned}$$

$$u(0) = (1-\gamma)\int_0^\infty\int_0^\infty\int_0^\infty e^{-\lambda\tau}\beta(\tau;\zeta,\eta)\Gamma(\tau;\zeta,\eta)\Psi(u,v)(\zeta,\eta)d\zeta d\eta d\tau,$$
$$v(0) = \gamma\int_0^\infty\int_0^\infty\int_0^\infty e^{-\lambda\tau}\beta(\tau;\zeta,\eta)\Gamma(\tau;\zeta,\eta)\Psi(u,v)(\zeta,\eta)d\zeta d\eta d\tau. \tag{4.39}$$

The existence of a solution $(u, v, \lambda)$ satisfying the above relation implies that there exists an exponential solution. To seek such a solution, we first fix $\lambda$ and consider a solution of (4.39) with the constant boundary condition

$$u(0) = 1-\gamma, \quad v(0) = \gamma.$$

If there exists such a solution, we denote it as $(u_\lambda, v_\lambda)$. For some $\lambda$, if

$$\int_0^\infty\int_0^\infty\int_0^\infty e^{-\lambda\tau}\beta(\tau;\zeta,\eta)\Gamma(\tau;\zeta,\eta)\Psi(u_\lambda,v_\lambda)(\zeta,\eta)d\zeta d\eta d\tau = 1,$$

then $(e^{\lambda t}u_\lambda(a), e^{\lambda t}v_\lambda(a))$ gives an exponential solution. It can be proved that there exists at least one such solution [14].

If we consider the case $\lambda = 0$, the fixed point theorem can be used to prove that the corresponding solution $(u_0, v_0)$ of (4.39) exists. Then, $(u_0, v_0)$ gives the survival probability for singles with radix $(1-\gamma, \gamma)$. If the condition

$$\mathscr{R} := \int_0^\infty\int_0^\infty\int_0^\infty \beta(\tau;\zeta,\eta)\Gamma(\tau;\zeta,\eta)\Psi(u_0,v_0)(\zeta,\eta)d\zeta d\eta d\tau = 1 \tag{4.40}$$

holds, $(u_0, v_0)$ is the persistent solution with $\lambda = 0$, that is, it is a stationary solution. We can now conjecture that there exists an exponential solution with positive growth rate $\lambda > 0$ if $\mathscr{R} > 1$, while there is no exponential solution with nonnegative growth rate if $\mathscr{R} < 1$. This conjecture was proved by Zacher [31] for the Staroverov–Hadeler model in (4.28).

## 4.5 Semigroup Approach

Mathematical well-posedness for pair formation models, that is, the existence and uniqueness of positive solution, and its continuous dependence on the initial data can be shown by using the semigroup method or by the integral equation approach [10, 12, 14, 19, 27, 29]. Here, we show a semigroup approach to the Fredrickson model [15]. Readers are referred to Chap. 10 for the basics of semigroup theory. Here, we assume that the age space is $[0, \infty)$ and adopt the following technical assumption:

**Assumption 4.4** (1) $\mu_x, \mu_y \in L^\infty(0,\infty)$, $\beta, \sigma \in L^\infty(\mathbb{R}_+ \times \mathbb{R}_+)$,

(2) The marriage function $\Psi$ is a bounded operator from $Y_+ := L^1_+ \times L^1_+$ to $L^1_+(\Omega)$ with $\Omega = \mathbb{R}_+ \times \mathbb{R}_+$, and it is locally Lipschitz continuous in $Y_+$ with $L^1$ norm;

there exists an increasing function $L(r)$ such that $|\Psi(f) - \Psi(g)|_{L^1} \le L(r)|f - g|_Y$ for all $f, g \in \{f \in Y_+ : |f|_Y \le r\}$.

(3) There exists a number $\eta > 0$ such that for any $(u, v) \in Y_+$

$$\int_0^\infty \Psi(u, v)(a, b)db \le \eta u(a), \quad \int_0^\infty \Psi(u, v)(a, b)da \le \eta v(b).$$

The assumption (3) is natural, because it implies that the force of marriage applied to single individuals is uniformly bounded.

Let us consider a real Banach lattice $X = L^1(0, \infty) \times L^1(0, \infty) \times L^1(\Omega)$ with norm $|\cdot|_X$ given by

$$|f|_X := \int_0^\infty |f_1(a)|da + \int_0^\infty |f_2(b)|db + 2\int_0^\infty \int_0^\infty |f_3(a, b)|dadb.$$

Then, the state space of the population is the positive cone $X_+$ and the norm $|f|_X$ gives the total size of the population.

First define the population operator $B : \mathscr{D}(B) \subset X \to X$ as follows:

$$Bf := \left(-\frac{df_1(a)}{da}, -\frac{df_2(b)}{db}, -Df_3(a, b)\right)^{\mathrm{T}}, \tag{4.41}$$

where T denotes the transpose of the vector and the operator $D$ denotes the directional derivative along the vector $(1, 1)$ defined by

$$Df(a, b) = \lim_{h \to 0} \frac{f(a + h, b + h) - f(a, b)}{h}, \tag{4.42}$$

and the domain $\mathscr{D}(B)$ is given by

$$\begin{aligned}
\mathscr{D}(B) :=\{&f \in X : f_1, f_2 \text{ are absolutely continuous}, f_1', f_2' \in L^1(0, \infty),\\
&f_3 \text{ is absolutely continuous along the direction } (1, 1) \text{ for almost every } (a, b),\\
&Df_3 \in L^1(\Omega), f_1(0) = \gamma \int_0^\infty \int_0^\infty \beta(a, b) f_3(a, b)dadb,\\
&f_2(0) = (1 - \gamma) \int_0^\infty \int_0^\infty \beta(a, b) f_3(a, b)dadb, f_3(a, 0) = f_3(0, b) = 0\}.
\end{aligned}$$

The perturbation terms $C : X_+ \to X$ and $F : X_+ \to X$ are defined by

$$Cf = \begin{pmatrix} -\mu_x(a) f_1(a) + \int_0^\infty f_3(a, b)[\sigma(a, b) + \mu_y(b)]db \\ -\mu_y(b) f_2(b) + \int_0^\infty f_3(a, b)[\sigma(a, b) + \mu_x(a)]da \\ -(\mu_x(a) + \mu_y(b) + \sigma(a, b)) f_3(a, b) \end{pmatrix}$$

$$Ff = \begin{pmatrix} -\int_0^\infty \Psi(f_1, f_2)(a, b)db \\ -\int_0^\infty \Psi(f_1, f_2)(a, b)da \\ \Psi(f_1, f_2)(a, b) \end{pmatrix}.$$

Let $A := B + C$. Then, $\mathscr{D}(A) = \mathscr{D}(B)$ and the Fredrickson two-sex population model is formulated as a semilinear Cauchy problem in $X$ as

$$\frac{df(t)}{dt} = Af(t) + Ff(t), \quad f(0) = f_0, \tag{4.43}$$

where $f_0 \in X_+$ is the initial data. Let $\underline{\mu} := \inf\{\mu_x, \mu_y\}$ and $\bar{\beta} := \sup \beta$. Then, we have

**Lemma 4.1** *Let $\Lambda := \{\lambda \in \mathbb{C} : \Re\lambda > -\underline{\mu}\}$. Then, $\Lambda \subset \rho(A)$, where $\rho(A)$ denotes the resolvent set of $A$. Moreover, let $\alpha$ be a number such that $\alpha := (\bar{\beta}/2) - \underline{\mu}$. Then, the Hille–Yosida estimate holds:*

$$\|(\lambda - A))^{-1}\| \le \frac{1}{\lambda - \alpha}, \quad \text{for } \lambda > \alpha. \tag{4.44}$$

*Proof* Let us consider the resolvent equation

$$(\lambda - A)f = \phi, \quad f \in \mathscr{D}(A), \quad \phi \in X. \tag{4.45}$$

Then, we can write

$$\begin{aligned} &\lambda f_1(a) + f_1'(a) + \mu_x(a) f_1(a) - \int_0^\infty f_3(a,b)[\sigma(a,b) + \mu_y(b)]db = \phi_1(a), \\ &\lambda f_2(b) + f_2'(b) + \mu_y(b) f_2(b) - \int_0^\infty f_3(a,b)[\sigma(a,b) + \mu_x(a)]da = \phi_2(b), \\ &\lambda f_3(a,b) + Df_3(a,b) + (\mu_x(a) + \mu_y(b) + \sigma(a,b)) f_3(a,b) = \phi_3(a,b). \end{aligned} \tag{4.46}$$

By formal integration, we obtain the following expression:

$$\begin{aligned} f_1(a) &= f_1(0)e^{-\lambda a}\ell_x(a) \\ &\quad + \int_0^a e^{-\lambda(a-s)}\frac{\ell_x(a)}{\ell_x(s)}\left[\phi_1(s) + \int_0^\infty f_3(s,b)[\sigma(s,b) + \mu_y(b)]db\right]ds, \\ f_2(b) &= f_2(0)e^{-\lambda b}\ell_y(b) \\ &\quad + \int_0^b e^{-\lambda(b-s)}\frac{\ell_y(b)}{\ell_y(s)}\left[\phi_2(s) + \int_0^\infty f_3(a,s)[\sigma(a,s) + \mu_x(a)]da\right]ds, \\ f_3(a,b) &= \begin{cases} \int_0^b \phi_3(a-b+s,s)e^{-\lambda(b-s)}\pi(a-b+s,s;a,b)ds, & (a > b), \\ \int_0^a \phi_3(s,b-a+s)e^{-\lambda(a-s)}\pi(s,b-a+s;a,b)ds, & (a < b), \end{cases} \end{aligned} \tag{4.47}$$

where $\pi$ is the survival function for pairs defined by

$$\pi(a,b;a+h,b+h) := \frac{\ell_x(a+h)}{\ell_x(a)}\frac{\ell_y(b+h)}{\ell_y(b)}\exp\left(-\int_0^h \sigma(a+s,b+s)ds\right),$$

$\ell_x(a) := \exp(-\int_0^a \mu_x(\sigma)d\sigma)$ and $\ell_y(a) := \exp(-\int_0^a \mu_y(\sigma)d\sigma)$. For $\lambda \in \Lambda$, the right-hand side of (4.47) defines a bounded linear operator from $X$ to $\mathscr{D}(A)$ and it is the resolvent operator $(\lambda - A)^{-1}$. Moreover, we know that for $\lambda \in \mathbb{R} \cap \Lambda$, the resolvent operator $(\lambda - A)^{-1}$ is a positive operator, so first we estimate $\|(\lambda - A)^{-1}\|$ by using the positivity on the cone $X_+$. Let us assume that $\phi \in X_+$ and $f = (\lambda - A)^{-1}\phi \in X_+$. Then, by integrating system (4.46) and using the positivity of $\phi$ and $f$, it follows that

$$\begin{aligned}
&\lambda|f_1|_1 + \int_0^\infty f_1'(a)da + \int_0^\infty \mu_x(a)f_1(a)da \\
&\quad - \int_0^\infty da \int_0^\infty f_3(a,b)[\sigma(a,b) + \mu_y(b)]db = |\phi_1|_1, \\
&\lambda|f_2|_1 + \int_0^\infty f_2'(b)db + \int_0^\infty \mu_y(b)f_2(b)db \\
&\quad - \int_0^\infty db \int_0^\infty f_3(a,b)[\sigma(a,b) + \mu_x(a)]da = |\phi_2|_1, \\
&\lambda|f_3|_1 + \int_0^\infty \int_0^\infty Df_3(a,b)dadb \\
&\quad + \int_0^\infty \int_0^\infty (\mu_x(a) + \mu_y(b) + \sigma(a,b))f_3(a,b)dadb = |\phi_3|_1,
\end{aligned} \tag{4.48}$$

where $|\cdot|_1$ denotes the $L^1$ norm. Note that for $\lambda \in \mathbb{R} \cap \Lambda$, $f_1(\infty) = f_2(\infty) = f_3(\infty, b) = f_3(a, \infty) = 0$. By adding term to term in (4.48), we obtain

$$\begin{aligned}
\lambda(|f_1|_1 + |f_2|_1 + 2|f_3|_1) = |\phi_1|_1 + |\phi_2|_1 + 2|\phi_3|_1 \\
- \int_0^\infty \mu_x(a)f_1(a)da - \int_0^\infty \mu_y(b)f_2(b)db \\
+ \int_0^\infty \int_0^\infty (\beta(a,b) - \mu_x(a) - \mu_y(b))f_3(a,b)dadb.
\end{aligned}$$

Let us define a number $\alpha$ such that $\alpha := (\bar{\beta}/2) - \underline{\mu}$. Then, it follows immediately that

$$\lambda(|f_1|_1 + |f_2|_1 + 2|f_3|_1) \le |\phi_1|_1 + |\phi_2|_1 + 2|\phi_3|_1 + \alpha(|f_1|_1 + |f_2|_1 + 2|f_3|_1).$$

Therefore, we conclude that if $\lambda > \alpha$ and $\phi$ is positive, then the estimate (4.44) holds. Next, consider the case that $\phi$ is real but not necessarily positive. $\phi$ can be decomposed as $\phi = \phi_+ + \phi_-$ where $\phi_+ = \max(\phi, 0)$ and $\phi_- = \max(-\phi, 0)$, and we can write $|\phi|_X = |\phi_+|_X + |\phi_-|_X$. Moreover, we obtain

$$f = (\lambda - A)^{-1}\phi = f_+ - f_-,$$

where

$$f_+ := (\lambda - A)^{-1}\phi_+ \in X_+, \quad f_- := (\lambda - A)^{-1}\phi_- \in X_+.$$

From the above argument, we know that for $\lambda > \alpha$

$$|f_+|_X \le \frac{|\phi_+|_X}{\lambda - \alpha}, \quad |f_-|_X \le \frac{|\phi_-|_X}{\lambda - \alpha}.$$

On the other hand, it is easily seen that $\max(f, 0) \le f_+$ and $\max(-f, 0) \le f_-$. Thus, we conclude that

$$\begin{aligned}|f|_X &= |\max(f, 0)|_X + |\max(-f, 0)|_X \le |f_+|_X + |f_-|_X, \\ &\le \frac{1}{\lambda - \alpha}(|\phi_+|_X + |\phi_-|_X) = \frac{|\phi|_X}{\lambda - \alpha}.\end{aligned}$$

Then, we again reach to the estimate (4.44). □

**Lemma 4.2** *The population operator A is a densely defined closed linear operator.*

*Proof* First, it is easy to see that $A$ is a closed linear operator, though we omit the proof. Next, we show that $\mathscr{D}(A)$ is dense. For $\lambda > \alpha$ and any $\phi \in X$, we define $f_\lambda = \lambda(\lambda - A)^{-1}\phi$. Then, if we can show that $\lim_{\lambda \to \infty} f_\lambda = \phi$, our proof completes since $f_\lambda \in \mathscr{D}(A)$. To this end, we write the resolvent as follows:

$$(\lambda - A)^{-1}\phi = (\lambda - A_0)^{-1}\phi + M(\lambda)\phi, \ \lambda > \alpha, \ \phi \in X,$$

where $A_0$ is the operator corresponding to the special case of the operator $A$ with zero boundary condition (i.e., $\beta = 0$) and $M(\lambda)$ is the linear operator defined by the difference between $(\lambda - A)^{-1}$ and $(\lambda - A_0)^{-1}$. Since it is easily seen that $A_0$ is a generator of $C_0$-semigroup, it follows that

$$\lim_{\lambda \to \infty} \lambda(\lambda - A_0)^{-1}\phi = \phi.$$

Hence, it is sufficient to show that $\lim_{\lambda \to \infty} M(\lambda)\phi = 0$ in order to complete our proof. Note that if we write $(g_1(\lambda), g_2(\lambda), g_3(\lambda))^{\mathrm{T}} = M(\lambda)\phi$, then we have $g_3(\lambda) = 0$ and

$$\begin{aligned}g_1(\lambda) &= e^{-\lambda a}\ell_x(a)\gamma \int\int_\Omega \beta(a, b)U(\phi_3)(a, b)dadb, \\ g_2(\lambda) &= e^{-\lambda b}\ell_y(b)(1 - \gamma)\int\int_\Omega \beta(a, b)U(\phi_3)(a, b)dadb,\end{aligned}$$

where $U(\phi_3)$ is given by

$$U(\phi_3)(a, b) = \begin{cases}\int_0^b \phi_3(a - b + s, s)e^{-\lambda(b-s)}\pi(a - b + s, s; a, b)ds, & (a > b), \\ \int_0^a \phi_3(s, b - a + s)e^{-\lambda(a-s)}\pi(s, b - a + s; a, b)ds, & (a < b).\end{cases} \tag{4.49}$$

Let

$$J_i = \int\int_{\Omega_i} |U(\phi_3)(a,b)|dadb, \quad (i=1,2)$$

where $\Omega_1 = \{(a,b) : b \geq a,\ 0 \leq a,\ 0 \leq b\}$, $\Omega_2 = \{(a,b) : b \leq a,\ 0 \leq a,\ 0 \leq b\}$. By changing of variables, we can obtain

$$J_1 = \int\int_{\Delta} |U(\phi_3)(x, x+y)|dxdy$$

where $\Delta := \{(x,y) : 0 \leq x, 0 \leq y\}$. From (4.49), we have

$$\begin{aligned}
J_1 &\leq \int\int_{\Delta} dxdy \int_0^x |\phi_3(s, s+y)| e^{-\lambda(x-s)} \pi(s, s+y; x, x+y) ds \\
&\leq \int\int_{\Delta} dxdy \int_0^x |\phi_3(s, s+y)| e^{-(\lambda+\underline{\sigma})(x-s)} ds \\
&= \int_0^\infty dy \int_0^\infty dx \int_0^x |\phi_3(s, s+y)| e^{-(\lambda+\underline{\sigma})(x-s)} ds \\
&= \int_0^\infty dy \int_0^\infty ds |\phi_3(s, s+y)| \int_s^\infty e^{-(\lambda+\underline{\sigma})(x-s)} dx,
\end{aligned}$$

where $\underline{\sigma} := \inf \sigma$. From the above inequality, it is not difficult to see that

$$J_1 \leq \frac{1}{\lambda + \underline{\sigma}} \int_0^\infty \int_0^\infty |\phi_3(s, s+y)| dsdy.$$

Applying the same kind of argument to $J_2$, we can arrive at the following estimate:

$$|U(\phi_3)|_{L^1(\Omega)} = J_1 + J_2 \leq \frac{1}{\lambda + \underline{\sigma}} |\phi_3|_{L^1(\Omega)}.$$

From the concrete expression of $M(\lambda)$, we can observe that

$$\begin{aligned}
|M(\lambda)\phi|_X &\leq |f_1(0)| \int_0^\infty e^{-\lambda a} da + |f_2(0)| \int_0^\infty e^{-\lambda a} da \\
&\leq \frac{\gamma \bar{\beta}}{\lambda} |U(\phi_3)|_{L^1(\Omega)} + \frac{(1-\gamma)\bar{\beta}}{\lambda} |U(\phi_3)|_{L^1(\Omega)} \leq \frac{\bar{\beta}}{\lambda(\lambda + \underline{\sigma})} |\phi_3|_{L^1(\Omega)}.
\end{aligned}$$

Then, it follows immediately that $\lim_{\lambda \to \infty} \lambda |M(\lambda)\phi|_{\tilde{X}} = 0$. Thus, we can conclude that the operator $A$ is densely defined. □

**Proposition 4.5** *The operator $A$ is an infinitesimal generator of a strongly continuous positive semigroup $T(t)$ such that $\|T(t)\| \leq e^{\alpha t}$.*

*Proof* From Lemmas 4.1 and 4.2, we know that $A$ satisfies the Hille–Yosida condition, so it generates a strongly continuous semigroup $T(t) = e^{tA)}$ on $X$, and from

Lemma 4.1, we can obtain the estimate $\|T(t)\| \leq e^{\alpha t}$. From Hille's formula, we obtain that

$$T(t) = \lim_{n\to\infty}\left(I - \frac{t}{n}A\right)^{-n} = \lim_{n\to\infty}\left(\frac{n}{t}\right)\left(\frac{n}{t} - A\right)^{-1},$$

where lim denotes strong convergence. Since the resolvent operator is a positive operator, we can conclude that $T(t)$ is also positive. □

**Lemma 4.3** *Under the Assumption 4.4, there exists a constant $\varepsilon > 0$ such that $(I + \varepsilon F)(X_+) \subset X_+$, where $I$ denotes the identity operator.*

*Proof* Under the Assumption 4.4, if we choose a number $\varepsilon$ such that $\varepsilon < 1/\eta$, then for $u \in X_+$ we have

$$(I + \varepsilon F)u = \begin{pmatrix} u_1(a) - \varepsilon \int_0^\infty \Psi(u_1, u_2)(a, b)db \\ u_2(b) - \varepsilon \int_0^\infty \Psi(u_1, u_2)(a, b)da \\ u_3(a, b) + \varepsilon\Psi(u_1, u_2) \end{pmatrix}$$
$$\geq \begin{pmatrix} u_1(a) - \frac{1}{\eta} \int_0^\infty \Psi(u_1, u_2)(a, b)db \\ u_2(b) - \frac{1}{\eta} \int_0^\infty \Psi(u_1, u_2)(a, b)da \\ u_3(a, b) + \varepsilon\Psi(u_1, u_2) \end{pmatrix} \geq 0.$$

This completes our proof. □

**Proposition 4.6** *Let $f_0 \in X_+$. Then, the Cauchy problem* (4.43) *has a unique mild solution in $X_+$, which defines a semiflow $S(t)f_0$ such that $S(t)(X_+) \subset X_+$.*

*Proof* Instead of the original Eq. (4.43), let us consider the following equivalent equation:

$$\frac{df(t)}{dt} = \left(A - \frac{1}{\varepsilon}\right)f + \frac{1}{\varepsilon}(I + \varepsilon F)f. \tag{4.50}$$

It is well known that the mild solution of (4.50) is given as a solution of the following integral equation:

$$f(t) = e^{-\frac{1}{\varepsilon}t}e^{At}f_0 + \frac{1}{\varepsilon}\int_0^t e^{-\frac{1}{\varepsilon}(t-s)}e^{A(t-s)}[f(s) + \varepsilon Ff(s)]ds, \tag{4.51}$$

where the constant $\varepsilon$ is chosen as $\varepsilon < 1/\eta$. Since the nonlinear perturbation is assumed to be locally Lipschitz continuous, the local solution of (4.51) is constructed by the positive iteration:

$$z_0(t) = f_0,$$
$$z_{n+1}(t) = e^{-\frac{1}{\delta}t}e^{At}f_0 + \frac{1}{\varepsilon}\int_0^t e^{-\frac{1}{\varepsilon}(t-s)}e^{A(t-s)}[z_n(s) + \varepsilon Fz_n(s)]ds,$$

Thanks to the positivity of $e^{At}$ and $I + \varepsilon F$, we can prove $z_n \in X_+$ iteratively. Indeed, if $z_0,\ z_n \in X_+$, the right-hand side of the iteration equation is a convex linear combination of those positive terms. Since the operator $F$ is locally Lipschitz continuous, the sequence $z_n(t)$ converges to a positive, mild local solution $f(t) \in X_+$. From Proposition 4.5, we have the following estimate:

$$|f(t)| \leq e^{(\alpha-\frac{1}{\varepsilon})t}|f_0| + \frac{K}{\varepsilon}\int_0^t e^{(\alpha-\frac{1}{\varepsilon})(t-s)}|f(s)|ds,$$

where $K := \|I + \varepsilon F\|$. Then, it is easily seen that the following estimate holds:

$$|f(t)| \leq |f_0|e^{(\alpha-\frac{1-K}{\varepsilon})t}.$$

Since the norm of the local solution grows at most exponentially as time evolves, it can be extended to a global solution. Hence, we can define a semiflow $S(t)$ by $S(t)f_0 = f(t)$. □

Let $\phi : X \to \mathbb{R}$ is a bounded linear functional on $X$ defined by

$$\langle \phi, f \rangle := \int_0^\infty f_1(a)da + \int_0^\infty f_2(b)db + 2\int_0^\infty \int_0^\infty f_3(a,b)dadb,$$

where $f = (f_1, f_2, f_3) \in X$. Then, $\langle \phi, f \rangle = |f|_X$ if $f \in X_+$. Moreover, it follows from (4.51) that $\langle \phi, f(t) \rangle > 0$ for the solution $f(t)$ of (4.50) with $f(0) > 0$.

Introducing a new variable $w(t) := f(t)/\langle \phi, f(t) \rangle$, (4.43) is formulated as the normalized system on the hyperplane $\mathscr{P} := \{z \in X_+ : \langle \phi, z \rangle = 1\}$ [11]:

$$\frac{dw(t)}{dt} = (A+F)w(t) - \langle \phi, (A+F)w(t) \rangle w(t), \tag{4.52}$$

with $w(0) = f(0)/\langle \phi, f(0) \rangle \in \mathscr{P}$. In fact, it follows from (4.52) that

$$\frac{d}{dt}\langle \phi, w \rangle = \langle \phi, (A+F)w(t) \rangle (1 - \langle \phi, w(t) \rangle). \tag{4.53}$$

Therefore, we have

$$\langle \phi, w(t) \rangle = 1 - (1 - \langle \phi, w(0) \rangle)e^{-\int_0^t \langle \phi, (A+F)w(s) \rangle ds},$$

from which we conclude that $w(t) \in \mathscr{P}$ if $w(0) \in \mathscr{P}$. If we use the solution $w(t)$ of the normalized system (4.52), the solution $f(t)$ of the original system is given by

$$f(t) = w(t)\exp\left(\int_0^t \langle \phi, (A+F)w(s) \rangle ds\right)\langle \phi, f(0) \rangle. \tag{4.54}$$

Let us consider the existence problem of exponentially growing persistent solutions. Let $w^* \in \mathscr{P}$ be a steady state of the normalized system. Then, we have the nonlinear eigenvalue problem:

$$\lambda^* w^* = Aw^* + Fw^*, \quad w^* \in \mathscr{D}(A) \cap \mathscr{P}, \tag{4.55}$$

where $\lambda^* = \langle \phi, (A+F)w^* \rangle$. Then, it is clear that $e^{\lambda^* t} w^*$ becomes a persistent solution of the original system. Conversely, if there exists a persistent solution $e^{\lambda^* t} f^*$ for the original homogeneous system, $\lambda^*$ and $w^* = f^*/\langle \phi, f^* \rangle$ must satisfy (4.55).

Let $\psi$ be a functional $X_+ \to \mathbb{R}$ given by

$$\begin{aligned} \langle \psi, f \rangle = & \int_0^\infty \int_0^\infty (\beta(a,b) - \mu_x(a) - \mu_y(b)) f_3(t,a,b) da db \\ & - \int_0^\infty \mu_x(a) f_1(t,a) da - \int_0^\infty \mu_y(b) f_2(t,b) db, \end{aligned} \tag{4.56}$$

for $f = (f_1, f_2, f_3) \in X_+$.

**Lemma 4.4** *Let $\bar{\mu} := \sup \mu$. For the functional $\psi$, the following estimate holds:*

$$-\bar{\mu} \langle \phi, f \rangle \le \langle \psi, f \rangle \le \alpha \langle \phi, f \rangle, \quad \forall f \in X_+, \tag{4.57}$$

*and it holds that*

$$\langle \psi, f \rangle = \langle \phi, Af + Ff \rangle, \quad \forall f \in \mathscr{D}(A). \tag{4.58}$$

**Exercise 4.3** Prove the above Lemma 4.4 and

$$\frac{d\langle \phi, z \rangle}{dt} = \langle \psi, w \rangle \langle \phi, z \rangle, \quad w(t) = \frac{z(t)}{\langle \phi, z \rangle},$$

which implies that $\langle \psi, w \rangle$ gives the growth rate of the population $z(t)$.

From the above observations, in order to show the existence of persistent solutions for the original system, it is sufficient to show that Eq. (4.55) has a positive solution $w^*$ with $\lambda^* = \langle \psi, w^* \rangle$. Let us consider the operator $\Phi_\varepsilon$ defined by

$$\Phi_\varepsilon := \frac{1}{\varepsilon} \left( \frac{1}{\varepsilon} - A \right)^{-1} (I + \varepsilon F + \varepsilon H), \tag{4.59}$$

where $\varepsilon$ is a positive number and $Hw = -\langle \psi, w \rangle w$ for $w \in \mathscr{P}$. It is clear that if $\Phi_\varepsilon$ has a positive fixed point $w^* \in \mathscr{P}$, $w^*$ is a positive eigenvector of $A+F$ associated with the eigenvalue $\lambda^* = \langle \psi, w^* \rangle$.

According to [18, Theorem 5.5], we observe that

**Lemma 4.5** *Suppose that $\varepsilon$ is a small number such that $1 - \varepsilon\alpha > 0$. Then, $\Phi_\varepsilon(\mathscr{P}) \subset X_+ \cap \mathscr{D}(A)$. Moreover, it follows that*

$$\inf_{w \in \mathscr{P}} |\Phi_\varepsilon(w)|_X > 0. \tag{4.60}$$

*Proof* Using the Assumption 4.4, it is easy to see that $\Phi_\varepsilon(\mathscr{P}) \subset X_+ \cap \mathscr{D}(A)$. Let us consider the equation $f = \Phi_\varepsilon(w),\ w \in \mathscr{P}$, and $f \in \mathscr{D}(A) \cap X_+$. Then, we can write as follows:

$$\left(\frac{1}{\varepsilon} - A\right) f = \frac{1}{\varepsilon}(I + \varepsilon F + \varepsilon H)w.$$

Integrating both sides with $[0, \infty)$ and $\Omega$ and adding term to term, we obtain the following relation:

$$\frac{1}{\varepsilon}|f|_X - \langle \psi, f\rangle = \frac{1}{\varepsilon}(1 - \varepsilon\langle \psi, w\rangle). \tag{4.61}$$

It follows from (4.57) that

$$\frac{1}{\varepsilon}|f|_X \geq -\bar{\mu}|f|_X + \frac{1}{\varepsilon} - \langle \psi, w\rangle \geq -\bar{\mu}|f|_X + \frac{1}{\varepsilon} - \alpha.$$

Therefore, we can conclude that

$$\inf_{w\in\mathscr{P}} |\Phi_\varepsilon(w)|_X \geq \frac{1 - \varepsilon\alpha}{1 + \varepsilon\bar{\mu}} > 0.$$

This completes our proof. □

Under the assumption of Lemma 4.5, we can conjecture that $\Phi_\varepsilon$ has at least one positive fixed point in $\mathscr{P}$. For example, if we define a operator $\tilde{\Phi}_\varepsilon$ from $\mathscr{P}$ into itself by $\tilde{\Phi}_\varepsilon w := \Phi_\varepsilon(w)/|\Phi_\varepsilon(w)|_X$ and it is a compact operator, there exists a fixed point $w^* \in \mathscr{P}$ such that $\tilde{\Phi}_\varepsilon w^* = w^* \in \mathscr{P}$. From the definition of $\tilde{\Phi}_\varepsilon$, we have $\Phi_\varepsilon(w^*) = cw^*$ with $c := |\Phi_\varepsilon(w^*)|_X > 0$. Using (4.61) with $f = cw^*$, we have

$$\frac{1}{\varepsilon}c - c\langle \psi, w^*\rangle = \frac{1}{\varepsilon}(1 - \varepsilon\langle \psi, w^*\rangle),$$

which implies

$$\left(\frac{1}{\varepsilon} - \langle \psi, w^*\rangle\right)(c - 1) = 0.$$

Since $\varepsilon$ is chosen such that $\frac{1}{\varepsilon} - \langle \psi, w^*\rangle \geq \frac{1}{\varepsilon} - \alpha > 0$, we can conclude that $c = 1$. That is, $w^*$ is a fixed point of $\Phi_\varepsilon$ on $\mathscr{P}$.

In order to see the linearized stability of the exponential solution, the *Euler formula* for the homogeneous system is crucial [30]:

**Proposition 4.7** *If $F$ is Fréchet differentiable at $x \in X_+$, the Euler formula $F'(x)x = F(x)$ holds. Moreover, $F'(cx)x = F'(x)x$ for $c > 0$.*

*Proof* Suppose that $F$ is Fréchet differentiable at $x \in X_+$, for $h \in X_+$ and $c > 0$, define $G_c(h) := F(cx + h) - F(cx) - F'(x)h$. Since $F$ is Fréchet differentiable at $x$, there exists a continuous increasing function $L : \mathbb{R}_+ \to \mathbb{R}_+$ such that $L(0) = 0$

and $|G_1(h)|_X \le L(|h|_X)|h|_X$ for $h \in X_+$. From the homogeneity, we have $G_c(h) = cG_1(h/c)$ for $h \in X_+$ and $c > 0$. Hence, if we let $h = x$, we have

$$F(cx + x) - F(cx) - F'(x)x = F(x) - F'(x)x = G_c(x) = cG_1(x/c).$$

Then, we have $|F(x) - F'(x)x|_X \le L(\frac{|x|_X}{c})|x|_X$, which implies $F'(x)x = F(x)$ because we can take arbitrarily large $c$. For $c > 0$, $F(cx) = F'(cx)cx = cF(x)$, so we have $F'(cx)x = F(x) = F'(x)x$. □

Therefore, if $F$ is Fréchet differentiable at a steady state $w^*$ of the normalized system such that $(A + F)w^* = \lambda^* w^*$, then $w^*$ is the positive eigenvector for the linearized operator $A + F'(w^*)$ associated with eigenvalue $\lambda^*$. Let $\zeta(t) := w(t) - w^*$. Then, the linearized equation for the normalized system is given by

$$\frac{d\zeta(t)}{dt} = B\zeta(t) - \langle \phi, B\zeta(t)\rangle w^* - \lambda^* \zeta(t), \tag{4.62}$$

where $B := A + F'(w^*)$. The exponential solution of the original system (4.43) is called *stable* if the corresponding steady state of the normalized system is locally stable in classical sense. Roughly speaking, as is observed in Proposition 4.2 for the finite-dimensional case, if $\lambda^*$ is the dominant eigenvalue of $B$ with algebraic multiplicity one, the exponential solution corresponding to $w^*$ is stable [11].

If $B$ is the infinitesimal generator of a strongly continuous semigroup of bounded linear operators $T(t), t \ge 0$ in $X$, and it is *asynchronous exponential growth* (see Chap. 10), that is, there exists a nonzero rank one projection $P$ in $X$ such that $\lim_{t\to\infty} e^{-\lambda^* t} T(t) = P$, there exists $\delta > 0$ such that if $x \in U := \{x \in X_+ \setminus \{0\} : |(I - P)x|_X / |Px|_X < \delta\}$, then $Qx := \lim_{t\to\infty} e^{-\lambda^* t} f(t)$ with $f(0) = x$ exists, $Qx \in Range(P)$ and $Qx \ne 0$ [30], which gives another (essentially equivalent) definition of the orbital stability.

As argued above, the Fredrickson pair formation modelhas an exponential persistent solution, but it is not yet known whether it is unique and stable. Four decades have passed since the age-structured pair formation model was first proposed. However, many of its mathematical characteristics are still unknown. Even though the status of marriage in the legal sense is decreasing in developed countries, it is still true that humans need stable pair formation to reproduce future generations. Moreover, after the beginning of the HIV/AIDS pandemic, mathematical epidemiologists became concerned with the human pair formation phenomenon as a transmission mechanism for infectious agents [2]. In this sense, pair formation (marriage) models will continue to be a central challenge in mathematical population dynamics.

## References


1. Bartlett, M.S.: Age distributions. Biometrics **26**, 377–386 (1970)
2. Castillo-Chavez, C. (ed.): Mathematical and Statistical Approaches to AIDS Epidemiology. Lecture Notes in Biomathematics, vol. 83. Springer, Berlin (1989)
3. Castillo-Chavez, C., Huang, W.: The logistic equation revisited: the two-sex case. Math. Biosci. **128**, 299–316 (1995)
4. Fredrickson, A.G.: A mathematical theory of age structure in sexual populations: random mating and monogamous marriage models. Math. Biosci. **10**, 117–143 (1971)
5. Hadeler, K.P., Waldstätter, R., Wörz-Busekros, A.: Models for pair formation in bisexual populations. J. Math. Biol. **26**, 635–649 (1988)
6. Hadeler, K.P.: Pair formation in age-structured populations. Acta. Applic. Math. **14**, 91–102 (1989)
7. Hadeler, K.P.: Periodic solutions of homogeneous equations. J. Diff. Equ. **95**, 183–202 (1992)
8. Hadeler, K.P.: Pair formation models with maturation period. J. Math. Biol. **32**, 1–15 (1993)
9. Hirsch, M.W.: Systems of differential equations which are competitive or cooperative. I: limit sets. SIAM J. Math. Anal. **13**(2), 167–179 (1982)
10. Iannelli, M., Martcheva, M.: A semigroup approach to the well posedness of an age-structured two-sex population model. Dynamic Syst. Appl. **6**, 353–370 (1997)
11. Iannelli, M., Martcheva, M.: Homogeneous dynamical systems and the age-structured SIR model with proportionate mixing incidence. In: Iannelli, M., Lumer, G. (eds.) Evolution Equations: Applications to Physics, Industry, Life Sciences and Economics, Progress in Nonlinear Differential Equations and Their Applications, vol. 55, pp. 227–251. Birkhäuser, Basel (2003)
12. Iannelli, M., Martcheva, M., Milner, F.A.: Gender-Structured Population Modeling, Mathematical Methods, Numerics, and Simulation. Society for Industrial and Applied Mathematics, Philadelphia (2005)
13. Inaba, H.: An age-structured two–sex model for human population reproduction by first marriage, Working Paper Series 15, Institute of Population Problems, Tokyo (1993)
14. Inaba, H.: Persistent age distributions for an age-structured two-sex population model. Math. Popul. Studies **7**(4), 365–398 (2000)
15. Inaba, H.: Semigroup approach to a pair formation model in human demography, RIMS Kokyuroku 1264, Mathematical Economics, Research Institute for Mathematical Sciences, Kyoto University, pp. 27–41 (2002)
16. Kendall, D.G.: Stochastic processes and population growth. J. Roy. Stat. Soc. B **11**, 230–264 (1949)
17. Keyfitz, N.: The mathematics of sex and marriage. In: Proceedings of the Sixth Berkeley Symposium on Mathematical Statistics and Probability, vol. 4, Biology and Health, pp. 89–108. University of California Press, Berkeley (1972)
18. Krasnoselskii, M.A.: Positive Solutions of Operator Equations. Noordhoff, Groningen (1964)
19. Martcheva, M., Milner, F.A.: A two-sex age-structured population model: well-posedness. Math. Popul. Studies **7**(2), 111–129 (1999)
20. Martcheva, M.: Exponential growth in an age-structured two-sex populations. Math. Biosci. **157**, 1–22 (1999)
21. Martcheva, M., Milner, M.: The mathematics of sex and marriage, revisited. Math. Popul. Studies **9**(2), 123–141 (2001)
22. Pollak, R.A.: Two-sex demographic models. J. Polit. Econ. **98**(2), 399–420 (1990)
23. Pollak, R.A.: Two-sex population models and classical stable population theory. In: Adams, J., Hermalin, A., Lam, D., Smouse, P. (eds.) Convergent Issues in Genetics and Demography, pp. 317–333. Oxford University Press, Oxford (1990)
24. Pollard, A.H.: The measurement of reproductivity. J. Inst. Actuaries **74**, 288–318 (1948)
25. Pollard, J.H.: Mathematical Models for the Growth of Human Populations. Cambridge University Press, Cambridge (1973)
26. Prüss, J., Schappacher, W.: Persistent age-distributions for a pair-formation model. J. Math. Biol. **33**, 17–33 (1994)

27. Prüss, J., Schappacher, W.: Semigroup methods for age-structured population dynamics. In: Chatterji, S.D., Fuchssteiner, B., Kulisch, U., Liedl, R. (eds.) Jahrbuch Überblicke Mathematik 1994, pp. 74–90. Viewveg, Braunschweig (1994)
28. Staroverov, O.V.: Reproduction of the structure of the population and marriage. [Russian] Ekonomika i matematiceskije metody **13**, 72–82 (1977)
29. Waldstätter, R.: Models for pair formation with applications to demography and epidemiology, Ph.D. Thesis, University of Tübingen (1990)
30. Webb, G.F.: Asynchronous exponential growth in differential equations with homogeneous nonlinearities. In: Dore, G., Favini, A., Obrecht, E., Venni, A. (eds.) Differential Equations in Banach Spaces. Lecture Notes in Pure and Applied Mathematics, vol. 148, pp. 225–233. Dekker, New York (1993)
31. Zacher, R.: Persistent solutions for age-dependent pair-formation models. J. Math. Biol. **42**, 507–531 (2001)

# Chapter 5
# Basic Ideas in Epidemic Modeling

**Abstract** In this chapter, we start by studying the early Kermack–McKendrick epidemic model, and introduce the basic ideas and ingredients of epidemic modeling. A crucial point is that we cannot precisely interpret the basic ideas and indices of infectious disease epidemiology without knowing the underlying nonlinear population dynamics. The early Kermack–McKendrick model is an infection-age-dependent outbreak model, and its extensions in the late 1970 s opened the door to the recent developments in mathematical epidemiology. The key idea of analyzing epidemic models is the *basic reproduction number* $R_0$ and its well-known *threshold principle*: if $R_0 > 1$, the final size of the epidemic is positive no matter how small the initial infected population, whereas if $R_0 < 1$, the final size becomes zero as the initial number of infected individuals goes to zero. We demonstrate the threshold principle based on the original definition of the final size given by Kermack and McKendrick, although there are slightly different definition for the final size. We then extend the original model to account for the heterogeneity of individuals and derive the pandemic threshold theorem. Subsequently, we introduce the demography of the host population and prove the endemic threshold theorem: if $R_0 > 1$, there exists at least one endemic steady state, whereas if $R_0 < 1$, there is no endemic steady state. This principle, however, does not hold under certain conditions. We provide examples in which subcritical endemic steady states exist even when $R_0 < 1$ because of the reinfection of recovered individuals.

## 5.1 The Early Kermack–McKendrick Model

The model provided by Kermack and McKendrick in their most famous paper [66], the first of a series on epidemics, described an outbreak of infectious disease in a local closed population [67–70].[1] For a many years, this paper had not been examined closely, and it took almost half a century for its deep mathematical implications to be uncovered by Kendall, Reddingius, Thieme, Diekmann, and Metz [19, 64, 80, 91, 97].

[1] Reader are referred to [28] for the origin of epidemic models.



H. Inaba, *Age-Structured Population Dynamics in Demography and Epidemiology*, DOI 10.1007/978-981-10-0188-8_5

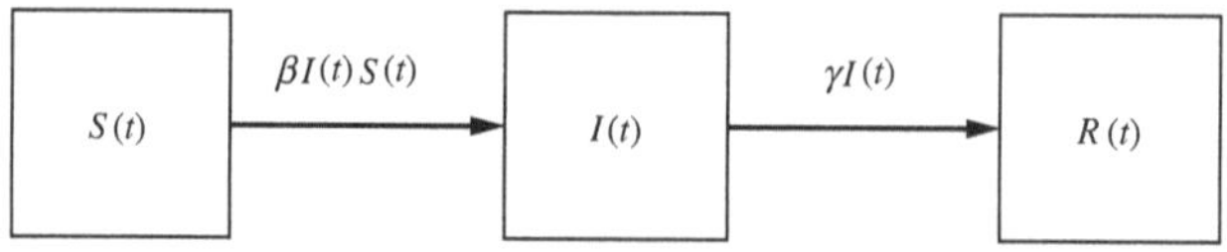


**Fig. 5.1** Flowchart of a SIR model

### *5.1.1 Basic Model*

First, let us consider a closed population without age structure and assume that the epidemic timescale is so short that the demographic factors (birth and death) in the host population can be neglected. We divide the host population into three parts: the susceptible population $S(t)$, the infective population $I(t)$, and the removed population $R(t)$. Hence, this type of compartment model is called the *SIR model.* Note that the infected host is assumed to be infectious, that is, we neglect the *latent period.*[2] The removed population refers to individuals who have recovered from the disease via immunity, been quarantined, or died. We here count the fatalities as part of the removed population. The simplest version of the Kermack–McKendrick model is then formulated as the following system of ordinary differential equations:

$$\begin{aligned} \frac{dS(t)}{dt} &= -\beta S(t)I(t), \\ \frac{dI(t)}{dt} &= \beta S(t)I(t) - \gamma I(t), \\ \frac{dR(t)}{dt} &= \gamma I(t), \end{aligned} \tag{5.1}$$

where $\beta$ is the infection rate and $\gamma$ is the removal rate, that is, the *force of infection* (the rate of infection per unit time), denoted by $\lambda(t)$, is given by $\beta I(t)$. If we divide the infected class into the exposed class (or the latent period) and the infectious class, the above model becomes an SEIR model; if removed individuals can again become susceptible via a loss of immunity, it is called the SIRS model and so on (Fig. 5.1).

The first task in the analysis of epidemic models is to clarify the invasion condition under which an epidemic outbreak can occur if a very small infected population appears in a totally susceptible host population. In the basic model of (5.1), the total population size $S(t) + I(t) + R(t)$ is constant. Let $N$ be this total population size. Then, there are stationary solutions $(N, 0, 0)$, $(S_0, 0, R_0)$, and $(0, 0, N)$, where $S_0$ and $R_0$ are constants such that $S_0 + R_0 = N$. If the host population is in the stationary state $(N, 0, 0)$, that is, all of the host population are susceptible, the dynamics of

[2]The time between the receipt of infection and when the infected individual becomes infective is called the *latent period.* The time between the receipt of infection and the appearance of symptoms is called the *incubation period.* The period during which infectious organisms are discharged is called the *infectious period* [6]. Readers are referred to [13] for more complicated compartment assumptions.

a small infected population invading a totally susceptible host population can be described by the linearized equation at the disease-free steady state $(N, 0, 0)$ as

$$\frac{dy(t)}{dt} = (\beta N - \gamma)y(t),$$

where $y(t)$ denotes the density of the infected population. Therefore, we know that in the initial invasion phase, the infected population dynamics are given by the Malthusian law as $y(t) = y(0)e^{(\beta N - \gamma)t}$. The invasion of the infectious disease occurs if the Malthusian parameter is positive, $\beta N - \gamma > 0$. This condition can be written as

$$R_0 = \frac{\beta N}{\gamma} > 1,$$

whereas disease invasion does not occur if $R_0 < 1$. In other words, there exists a critical community size $N_{cr} := \gamma/\beta$ such that the outbreak does not occur if the (susceptible) community density is less than the critical size ($N < N_{cr}$).

The parameter $R_0$ is called the *basic reproduction number (ratio)*. Because $\beta N$ is the average number of secondary cases produced by an infected individual per unit time in a totally susceptible host population with total size $N$, and $1/\gamma$ is the average duration in the infectious state, $R_0$ is the average number of secondary cases produced by an infected individual during their entire infectious period in a *completely* susceptible host population [24].

If we linearize system (5.1) at $(S_0, 0, R_0)$ with $S_0 + R_0 = N$, that is, the initial (stationary) population is composed of susceptible and recovered individuals, the invasion condition is given by

$$R_e = \frac{\beta S_0}{\gamma} > 1,$$

where $R_e$ is called the *effective reproduction number* or the *initial replacement ratio/number* [102, p. 297]. $R_e$ gives the total number of secondary cases that an average infective individual can produce if the number of susceptible individuals remains at its initial size $S_0$.

Note that if we define $\ell(\tau)$ as the survival probability in the infective state at *infection-age* $\tau$ (the time elapsed since infection), our assumption implies that

$$\frac{d\ell(\tau)}{d\tau} = -\gamma\ell(\tau).$$

Then, $\gamma\ell(\tau) = \gamma e^{-\gamma\tau}$ is the probability density function of recovery occurring, and the expected sojourn time in the infected state is calculated as

$$\int_0^\infty \tau\gamma\ell(\tau)d\tau = \frac{1}{\gamma}.$$

Conversely, the probability density function of secondary infection occurring at infection-age $\tau$ is given by

$$\frac{\beta N \ell(\tau)}{\int_0^\infty \beta N \ell(\tau) d\tau} = \gamma \ell(\tau).$$

Then, the *generation interval* (or *generation time*), which is the duration between the time of infection of a secondary infectee and the time of infection of their primary infector, is also given by $1/\gamma$. This equality, of course, does not hold if $\gamma$ and $\beta$ are not constant.

Using the basic reproduction number, we can expect the disease to invade the susceptible population and the number of infected individuals to increase exponentially in the initial phase if $R_0 > 1$, whereas the disease will be naturally eradicated (if there is no subcritical endemic steady state) if $R_0 < 1$. Such qualitative changes in the solution corresponding to changes in the parameters are called *threshold phenomena*.

For more complex epidemic models, it is not necessarily easy to calculate $R_0$. However, one of the primary motivating factors behind research into epidemic models is to examine effective ways to prevent the epidemic from spreading. Clearly, $R_0$ is a powerful index in the sense that the outbreak of an epidemic via a small invasion of infective individuals can be controlled if $R_0$ remains less than unity. Moreover, it is often observed that the threshold condition $R_0 > 1$ is not only the condition for disease invasion, but also the condition for the existence and stability of endemic steady states when the susceptible population is supplied by childbearing, immigration, or loss of immunity in the recovered population. Therefore, finding a method of calculating $R_0$ and examining the effects of interventions or other epidemic parameter changes on $R_0$ are the most important tasks in mathematical epidemiology. We investigate the theory of $R_0$ in detail in Chap. 9.

The force of infection $\lambda$ is the rate at which the susceptible population becomes infected per unit time and per capita. In general, this could be given as follows:

$$\lambda(t) = \phi c(P(t)) \left[ \frac{I(t)}{P(t)} \right],$$

where $c(P)$ denotes the number of per capita contacts per unit time depending on $P$, which is the size of the population involved in the contact process. The ratio $I/P$ is the probability that an individual is infective, and $\phi$ is the probability that the infectious agent is successfully transmitted by a contact.

Note that $P$ is generally different from the total population size $N = S + I + R$. If the removed class is composed of immunized individuals and their activity is not different from that of susceptibles, they are sufficiently mixed with susceptibles and infecteds in the entire population. In this case, we expect that $P = N$. However, if the removed population is really quarantined from the rest of the population because of sickness or treatment, it may be reasonable to assume that $P < N$.

Although the number of contacts $c(P)$ depends on how the infectious agent is transmitted, it is reasonable to assume that $c(P)$ is monotone increasing and bounded above. If the population size is so large that the number of per capita contacts per unit time is saturated, we could simply assume that $c(P)$ is constant. However, if the domain of $c(P)$ encompasses a broad range of the population size $P$, or if we take into account more realistic aspects such as the time needed for contact to occur, it may be appropriate to use more complex functions [4, 47].

If $P$ is constant, regardless of the form of $c(P)$, the number of newly infected individuals per unit time is given in the bilinear form, often called the *law of mass action* [16, 48]:

$$\lambda(t)S(t) = \beta I(t)S(t),$$

where

$$\beta = \frac{c(P)\phi}{P} = \text{const.}$$

Even in the case that $P$ is not constant, if we can assume that $c(P) = kP$, we again obtain a linear function of the infected population size $I$ as the force of infection. In the Kermack–McKendrick model, the force of infection is given by the linear form, because we implicitly assume that $c(P) = kP$ or that $P = N$ and $N$ is constant. In this case, we will see that the epidemic will end with the extinction of the infectious population, and some of the susceptible individuals survive the epidemic.

However, if we assume that $c(P) = c$ is constant, we obtain the force of infection as

$$\lambda(t) = c\phi\frac{I}{P},$$

which is homogeneous of degree zero (scale-independent), and the nonlinear incidence term $\lambda(t)S(t)$ becomes homogeneous of degree one. An epidemic system with homogeneous nonlinearity is called a *homogeneous epidemic system* [54]. The scale-independent force of infection reflects the saturation effect of the number of contacts in a sufficiently large-scale population and is therefore usually adopted for modeling sexually transmitted diseases. In the epidemiological literature, the incidence rate given by $\beta SI/N$ is often called the *standard incidence rate* or *Macdonald type*. If $N$ is assumed to be constant, for the purpose of qualitative analysis, we do not need to distinguish the mass-action law and the standard type. However, its distinction is important for interpreting and estimating the transmission coefficient $\beta$. Several authors have examined extensions of the basic SIR model (5.1) with a contact number depending on the host size using a general nonlinear law [11, 12, 27].

*Remark 5.1* Applying the variation-of-constants formula to the basic system (5.1), we obtain

$$S(t) = S(0)e^{-\int_0^t \lambda(\sigma)d\sigma},$$
$$I(t) = I(0)e^{-\gamma t} + \int_0^t e^{-\gamma(t-\sigma)}\lambda(\sigma)S(\sigma)d\sigma,$$

where $\lambda(t) = \beta I(t)$. Then, the Kermack–McKendrick model is formulated as a nonlinear integral equation:

$$\lambda(t) = \beta I(0)e^{-\gamma t} + S(0)\int_0^t e^{-\gamma(t-\sigma)}\lambda(\sigma)e^{-\int_0^\sigma \lambda(\zeta)d\zeta}d\sigma.$$

It is a basic observation that an epidemic model can be formulated by a nonlinear renewal integral equation, which we will see again and again.

### 5.1.2 Threshold Theorem and the Final Size Equation

In this section, let us consider the threshold phenomena of the basic SIR model (5.1). It is clear that the dynamics of (5.1) are completely determined by the following two-dimensional dynamical system:

$$\begin{aligned}\frac{dS(t)}{dt} &= -\beta S(t)I(t),\\ \frac{dI(t)}{dt} &= \beta S(t)I(t) - \gamma I(t).\end{aligned} \tag{5.2}$$

The solution orbit of (5.2) with initial data in the domain $\Omega = \{(S, I) : S \geq 0,\ I \geq 0,\ S + I \leq N\}$ exists for $0 < t < \infty$ and remains in $\Omega$. All points on the horizontal axis are equilibrium points, and there is no other equilibrium point in $\Omega$. Because $S(t)$ is monotone decreasing and nonnegative, it has a nonnegative limit $\lim_{t\to\infty} S(t) = S(\infty)$. $I(t)$ is also decreasing if $S < N_{cr}$ and therefore also has a nonnegative limit $I(\infty)$. As the limit point must be an equilibrium point, we have $\lim_{t\to\infty} I(t) = 0$.

From (5.2), we have

$$\frac{dI}{dS} = -1 + \frac{N_{cr}}{S}.$$

For $t \geq 0$, (5.2) can then be integrated as

$$I(t) = I(0) + S(0) - S(t) + N_{cr}\log\frac{S(t)}{S(0)}. \tag{5.3}$$

As a function of $S$, $I$ attains its maximum at $S = N_{cr} = \gamma/\beta$ as

$$I_{max} = S(0) + I(0) - N_{cr} + N_{cr}\log\frac{N_{cr}}{S(0)}.$$

If $N_{cr} < S(0)$, then $(N_{cr}, I_{max}) \in \Omega$, that is, the size of the infected population increases from the initial size $I(0)$ and the epidemic outbreak attains a peak. If $N_{cr} \geq S(0)$, the size of the infected population decreases from the initial size, and no epidemic outbreak occurs.

From (5.3), if we let $t \to \infty$, we obtain

$$S(\infty) = S(0) + I(0) + N_{cr} \log \frac{S(\infty)}{S(0)}.$$

Therefore, we have

$$S(\infty) = S(0) \exp\left( -\frac{S(0) + I(0) - S(\infty)}{N_{cr}} \right), \tag{5.4}$$

which implies that $S(\infty) > 0$, that is, some susceptible individuals can escape infection, and the epidemic ends not by the extinction of susceptibles, but by the eradication of those infected.

For an epidemic starting from the initial point $(S(0), I(0), 0)$ with $S(0) + I(0) = N$ (5.1), we define the *intensity of epidemic* or the *final size of epidemic* by

$$p := \frac{N - S(\infty)}{N} = \frac{R(\infty)}{N},$$

where $p$ is the proportion of the total number of host individuals that finally contracts the disease, provided that $R(0) = 0$. If we assume $R(0) = 0$, we have

$$S(\infty) + R(\infty) = S(0) + I(0) = N.$$

It follows from (5.4) that

$$1 - p = \left( 1 - \frac{I(0)}{N} \right) e^{-R_0 p}. \tag{5.5}$$

A unique positive root of (5.5) gives the intensity of epidemic starting from the initial data $(S(0), I(0), 0)$ with $S(0) + I(0) = N$. The intensity of epidemic was named by Bailey [6], although it was originally introduced by Kermack and McKendrick [66].

**Proposition 5.1** *For any epidemic with initial data such that $S(0) > 0$, $I(0) > 0$, and $R(0) = 0$, it holds that $p \geq p_\infty$, where $p_\infty$ is the largest nonnegative root of the final size equation*

$$1 - x = e^{-xR_0}. \tag{5.6}$$

*Proof* It follows from (5.5) that $1 - p \leq e^{-R_0 p}$. If $R_0 > 1$, it follows that $p = 0$ or $p \geq p_\infty$, where $p_\infty$ is a unique positive root of (5.6). Because $p > 0$, we have $p \geq p_\infty$. If $R_0 \leq 1$, $p \geq p_\infty$ holds trivially, because $p_\infty = 0$ is the largest nonnegative root of the final size equation.

From the above argument, we know that if $R_0 > 1$, the epidemic is *major*; that is, the intensity of epidemic is not necessarily small and is greater than $p_\infty > 0$ no matter how small the initial number of newly infected individuals.

If we solve (5.2) for $t < 0$, the solution orbit goes to an equilibrium point on the horizontal axis for $t \to -\infty$. We can then construct a solution orbit for $-\infty < t < \infty$ connecting two equilibrium points on the horizontal axis. For this *limiting epidemic* orbit, we can define the intensity of epidemic by

$$p_\infty := \frac{N - S(\infty)}{N} = \frac{S(-\infty) - S(\infty)}{S(-\infty)},$$

because $N = S(-\infty)$. For the limiting epidemic process, integrating $dI/dS$ from $-\infty$ to $\infty$, we have

$$0 = S(-\infty) - S(\infty) + \frac{\gamma}{\beta} \log \frac{S(\infty)}{S(-\infty)},$$

from which we have the final size relation

$$1 - p_\infty = e^{-R_0 p_\infty}.$$

Thus, $p_\infty$ is the final size of the limiting epidemic (Fig. 5.2).

In reality, if the scale of the host population is so large that deterministic epidemic modeling is effective, the initial infected population size would be very small in comparison with the susceptible host population size. Therefore, the limiting epidemic process may correspond to a realistic situation, and it is advantageous that we can estimate the final size of the epidemic from $R_0$ alone, without any knowledge of the initial data. Conversely, if we have some data regarding the intensity of epidemic, $R_0$ can be estimated from the final size equation.

From (5.1), we have

$$\frac{d \log S(t)}{dt} = -\frac{\beta}{\gamma} \frac{dR(t)}{dt}.$$

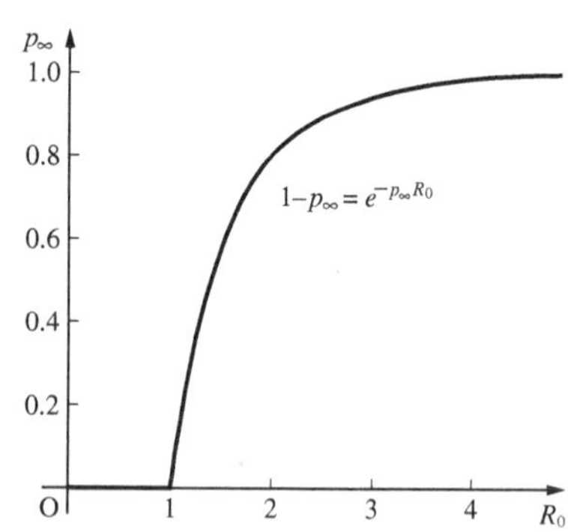


**Fig. 5.2** Final size as a function of $R_0$ in the limiting epidemic

Integrating the above equation from 0 to $t$, we obtain

$$\log \frac{S(t)}{S(0)} = -\frac{\beta}{\gamma}(R(t) - R(0)).$$

Suppose that $R(0) = 0$. Then, we have

$$S(t) = S(0)e^{-\frac{\beta}{\gamma}R(t)}.$$

Therefore, from (5.1) and the conservation of the total population size, we arrive at the initial value problem for $R(t)$:

$$\frac{dR(t)}{dt} = \gamma(N - S(0)e^{-\frac{\beta}{\gamma}R(t)} - R(t)), \quad R(0) = 0, \tag{5.7}$$

where $dR/dt$ is the incidence rate of newly infected individuals. The curve $(t, dR(t)/dt)$ is called the *epidemic curve*.

If $R(t)/N_{cr}$ is sufficiently small, we can apply the approximation

$$e^{-\frac{\beta R(t)}{\gamma}} \approx 1 - \frac{\beta R(t)}{\gamma} + \frac{1}{2}\left(\frac{\beta R(t)}{\gamma}\right)^2$$

to the initial value problem (5.7). This gives the following approximate equation:

$$\frac{dR(t)}{dt} = \gamma\left[N - S(0) + \left(\frac{S(0)}{N_{cr}} - 1\right)R(t) - \frac{S(0)R(t)^2}{2N_{cr}^2}\right].$$

The above equation is the separation of variable type, and it can be solved analytically. In particular, if we consider the limiting case of $I(0) \to 0$, $S(0) \to N$, we have the logistic equation

$$\begin{aligned}\frac{dR(t)}{dt} &= \gamma\left[(R_0 - 1)R(t) - \frac{NR(t)^2}{2N_{cr}^2}\right] \\ &= aR(t)(1 - bR(t)),\end{aligned}$$

where

$$a := \gamma(R_0 - 1), \quad b := \frac{\beta R_0}{2\gamma(R_0 - 1)}.$$

Thus, we obtain the approximate solution

$$R(t) = \frac{R(0)e^{at}}{1 + bR(0)(e^{at} - 1)}.$$

If $a > 0$, that is, the threshold condition $R_0 > 1$ holds, the intensity of epidemic is given by

$$\frac{R(\infty)}{N} = \frac{1}{Nb} = \frac{2}{R_0}\left(1 - \frac{1}{R_0}\right).$$

Let $N = N_{cr} + \rho$ and assume that $\rho$ is sufficiently small in comparison with $N$. Then, we have

$$R(\infty) = 2\rho(1 - \frac{\rho}{N}) \approx 2\rho.$$

This means that $2\rho$ are removed from the initial susceptible population $N_{cr} + \rho$ and $N_{cr} - \rho$ remains susceptible. Conversely, if $N \le N_{cr}$, then $R(\infty) = 0$, that is, none of the population is removed. Thus, we can state the following well-known threshold result given by Kermack and McKendrick:

**Proposition 5.2** *If $N > N_{cr}$ and $\rho = N - N_{cr}$ is sufficiently small in comparison with $N$, the total size of the removed population is approximately $2\rho$, whereas it is zero if $N \le N_{cr}$.*

Readers are referred to [40, 73, 102] for more estimates of the final size of the epidemic.

*Remark 5.2* Note that the basic system (5.1) can be seen as a special case of the well-known *Lotka–Volterra system*. In fact, (5.1) becomes a Lotka–Volterra system if we add a birth term $\alpha S$ with birth rate $\alpha$ to the equation for $S$ in (5.1). The resulting system has periodic solutions if $\alpha > 0$ and is also a Hamiltonian system. In fact, if we let $u = \log S$ and $v = \log I$, (5.1) can be written as

$$\frac{du}{dt} = \frac{\partial H}{\partial v},$$
$$\frac{dv}{dt} = -\frac{\partial H}{\partial u},$$

where $H(u, v) := -\beta(e^u + e^v) + \gamma u$. Thus, it is clear that the Hamiltonian $H(u, v)$ is constant [88].

**Exercise 5.1** If an infectious disease is fatal, we can consider an SIR model with a standard incidence rate as

$$\frac{dS(t)}{dt} = -\frac{\beta S(t)I(t)}{S(t) + I(t)},$$
$$\frac{dI(t)}{dt} = \frac{\beta S(t)I(t)}{S(t) + I(t)} - \gamma I(t),$$
$$\frac{dR(t)}{dt} = \gamma I(t),$$

where $\gamma$ is the removal rate including the death rate and the removed class $R$ is assumed to be excluded from the transmission process. Show that $S(\infty) = 0$ [37].

**Exercise 5.2** Consider a population of size $N$ after one outbreak starting from an infinitesimally small infected population. The total population can then be decomposed into two parts, $S(\infty) = N(1 - p_\infty)$ and $R(\infty) = Np_\infty$, where $p_\infty$ is the positive solution of the final size equation $1 - p_\infty = e^{-R_0 p_\infty}$ with $R_0 > 1$. If a loss of immunity occurs as time passes, the recovered population $R(\infty)$ again becomes susceptible with relative susceptibility $\sigma$, and the effective size of the susceptible population becomes $S(\infty) + \sigma R(\infty) = N((1 - p_\infty) + \sigma p_\infty)$. Show that the effective reproduction number

$$R_e = R_0((1 - p_\infty) + \sigma p_\infty)$$

is greater than unity if $\sigma \geq 1/2$ [63], that is, the host population can be reinvaded by the disease if the relative susceptibility increases to 0.5. This is also the necessary condition that guarantees $R_e > 1$ for any $R_0 > 1$.

**Exercise 5.3** Consider the following integro–differential equation system as the SIR epidemic model for a one-dimensionally distributed population:

$$\begin{aligned}
\frac{\partial u(x,t)}{\partial t} &= -\beta u(x,t) \int_{\mathbb{R}} \lambda(x-y) v(y,t) dy, \\
\frac{\partial v(x,t)}{\partial t} &= -\gamma v(x,t) + \beta u(x,t) \int_{\mathbb{R}} \lambda(x-y) v(y,t) dy, \\
\frac{\partial w(x,t)}{\partial t} &= \gamma v(x,t),
\end{aligned}$$

where $x \in \mathbb{R}$, $t > 0$, $\beta > 0$ is the transmission rate, $\gamma > 0$ is the removal rate, $u(x,t)$ denotes the (normalized) density of susceptible individuals at position $x$ and time $t$, $v(x,t)$ is the density of infected individuals, and $w(x,t)$ is the density of removed individuals. $\lambda(z)$ is the contact rate between susceptibles and infecteds whose distance is $|z|$ and is assumed to be a nonnegative, bounded continuous function such that $\int_{\mathbb{R}} \lambda(z) dz = 1$ and $\lambda(z) = \lambda(-z)$. The initial condition is $u(x,0) = u_0(x)$, $(x,0) = v_0(x)$, and $w(x,0) = 0$, where $u_0, v_0 \in L^1_+(\mathbb{R}) \cap L^\infty_+(\mathbb{R})$ with $u_0(x) + v_0(x) = 1$.

(1) Show that the solution is nonnegative and $u(x,t) + v(x,t) + w(x,t) = 1$ for any $t > 0$, $x \in \mathbb{R}$ and that the basic reproduction number is given by $R_0 = \beta/\gamma$.
(2) Let us define $W(x,t) := \int_{\mathbb{R}} \lambda(x-y) w(y,t) dy$. Show that $w(x,t)$ satisfies the following equation:

$$\frac{1}{\gamma} \frac{\partial w(x,t)}{\partial t} + w(x,t) = 1 - u_0(x) e^{-R_0 W(x,t)}.$$

(3) Show that at each point $x$, $u_\infty(x) = \lim_{t\to\infty} u(x,t)$, $v_\infty(x) = \lim_{t\to\infty} v(x,t)$, and $w_\infty(x) = \lim_{t\to\infty} w(x,t)$ exist and the following holds:

$$u_\infty(x) > 0,\ v_\infty(x) = 0,\ w_\infty(x) > 0.$$

(4) Suppose that $R_0 > 1$. Show that $\inf_{x \in \mathbb{R}} w_\infty(x) \geq p$ holds, where $p$ is the positive root of the final size equation $1 - x = e^{-R_0 x}$. This is a one-dimensional version of Kendall's *Pandemic Threshold Theorem* [65, 107, 108]. Readers are referred to discussion in [89, Sect. 3.3].

## 5.2 Three Applications

Although the SIR model (5.1) is simple, it has many useful extensions and applications to transmission phenomena in the real world. Here, we give three applications of the basic SIR model (5.1).

### 5.2.1 *Transmission by Environmental Contamination*

First, we consider the case in which the infective agent is transmitted by environmental contamination. Let $V(t)$ be the density of the free infective agent in the environment. It is assumed that the force of infection is proportional to this density. The free infective agent is produced by the infected host with production rate $\varepsilon$ and decays with rate $\alpha$ in the environment. The basic system without demography is then formulated as

$$\begin{aligned}\frac{dS(t)}{dt} &= -\beta S(t)V(t),\\ \frac{dI(t)}{dt} &= \beta S(t)V(t) - \gamma I(t),\\ \frac{dV(t)}{dt} &= -\alpha V(t) + \varepsilon I(t),\end{aligned} \tag{5.8}$$

where $\gamma$ denotes the clearance rate of infected hosts.

For example, the model in (5.8) was used by Hahn and Furniss [36, 46] to simulate an anthrax epizootic. In their model, $S(t)$ denotes the density of vulnerable animals at time $t$, $I(t)$ is the density of carcasses (of animals that have died of anthrax), and $V(t)$ is the environmental contamination, defined as the density of anthrax spores. If we consider $S$ to be the density of susceptible (target) cells, $I$ to be the density of infected cells, and $V$ to be the density of free virus in the bloodstream, then (5.8) becomes a model of virus dynamics without cell demography [86]. As a model for virus dynamics, the leading part is $V$, with $S$ and $I$ forming an "environment" in which the virus can reproduce. Therefore, the basic reproduction number is calculated by the renewal equation for the density of newly produced virus.

Applying the variation-of-constants formula to $I$ and $V$, we obtain

$$\begin{aligned} I(t) &= I(0)e^{-\gamma t} + \int_0^t e^{-\gamma(t-\sigma)}\beta S(\sigma)V(\sigma)d\sigma, \\ V(t) &= V(0)e^{-\alpha t} + \int_0^t e^{-\alpha(t-s)}\varepsilon I(s)ds. \end{aligned} \tag{5.9}$$

Note that if initial data are given for $t \in (-\infty, 0]$, (5.8) and (5.9) can be written as a delay-differential equation model:

$$\begin{aligned} \frac{dS(t)}{dt} &= -\beta S(t)V(t), \\ \frac{dI(t)}{dt} &= \beta S(t) \int_0^\infty e^{-\alpha\sigma}\varepsilon I(t-\sigma)d\sigma - \gamma I(t), \end{aligned} \tag{5.10}$$

so (5.8) is a special case of the infection-age-structured epidemic model.

Inserting the expression for $I$ into the equation for $V$, we arrive at a nonlinear renewal equation for the virus density:

$$V(t) = g(t) + \int_0^t \Psi(\sigma)S(t-\sigma)V(t-\sigma)d\sigma, \tag{5.11}$$

where

$$g(t) := V(0)e^{-\alpha t} + \int_0^t e^{-\alpha(t-s)}\varepsilon I(0)e^{-\gamma s}ds,$$

$$\Psi(\sigma) := \begin{cases} \beta\varepsilon \frac{e^{-\gamma\sigma} - e^{-\alpha\sigma}}{\alpha - \gamma} & (\alpha \neq \gamma), \\ \beta\varepsilon\sigma e^{-\alpha\sigma} & (\alpha = \gamma). \end{cases}$$

The linearized equation about the infection-free steady state $(S(0), 0, 0)$ is then given by

$$V(t) = g(t) + S(0)\int_0^t \Psi(\sigma)V(t-\sigma)d\sigma. \tag{5.12}$$

Let $b(t) := \beta S(0)V(t)$. Then, $b(t)$ denotes the density of newly infected hosts in the invasion phase and satisfies

$$b(t) = \beta S(0)g(t) + S(0)\int_0^t \Psi(\sigma)b(t-\sigma)d\sigma. \tag{5.13}$$

Therefore, the basic reproduction number of the infected host population in the infection-free steady state $(S(0), 0, 0)$ is calculated as

$$R_0 = S(0)\int_0^\infty \Psi(\sigma)d\sigma = \frac{\varepsilon\beta S(0)}{\gamma\alpha}, \tag{5.14}$$

where $\varepsilon/\gamma$ is the *burst size*, that is, the total number of infective agents produced by an infected host during its total lifetime. In virus dynamics, the basic reproduction number for the newly produced virus is also given by (5.14), because the density of newly produced virus, given by $\varepsilon I(t)$, satisfies a renewal equation with the same integral kernel as (5.13).

Integrating the renewal equation (5.9) from 0 to $\infty$, we have

$$\int_0^\infty I(t)dt = \frac{I(0)}{\gamma} + \frac{\beta}{\gamma}\int_0^\infty S(\sigma)V(\sigma)d\sigma,$$

and it follows from the differential equation for $S$ that

$$S(\infty) - S(0) = -\beta\int_0^\infty S(\sigma)V(\sigma)d\sigma.$$

Therefore, it holds that

$$\int_0^\infty I(t)dt = \frac{I(0)}{\gamma} + \frac{1}{\gamma}(S(0) - S(\infty)).$$

Next, observe that

$$\int_0^\infty V(t)dt = \frac{v(0)}{\alpha} + \frac{\varepsilon}{\alpha}\int_0^\infty I(s)ds,$$

from which we have

$$\int_0^\infty V(t)dt = \frac{v(0)}{\alpha} + \frac{\varepsilon}{\alpha}\left[\frac{I(0)}{\gamma} + \frac{1}{\gamma}(S(0) - S(\infty))\right].$$

Integrating the relation $S'/S = -\beta V$, we obtain

$$\log\frac{S(\infty)}{S(0)} = -\beta\int_0^\infty V(t)dt = -\frac{\beta v(0)}{\alpha} - \frac{\beta\varepsilon}{\alpha}\left[\frac{I(0)}{\gamma} + \frac{1}{\gamma}(S(0) - S(\infty))\right],$$

from which we arrive at the final size relation

$$\log(1 - p) = -a - R_0 p, \tag{5.15}$$

where

$$p := \frac{S(0) - S(\infty)}{S(0)}, \quad a := \frac{\beta v(0)}{\alpha} + \frac{\beta\varepsilon I(0)}{\alpha\gamma}.$$

If the initial infected host density and the initial virus density go to zero, we again obtain the well-known final size equation

$$1 - p = \exp(-R_0 p). \tag{5.16}$$

That is, the existence of a delay effect in the transmission process does not affect the final size equation.

**Exercise 5.4** If we introduce the recovered class in (5.8), we obtain SIR+V model. Suppose that $N = S + I + R$ is constant. Calculate the basic reproduction number and derive the final size equation [14].

### 5.2.2 *Virus Dynamics*

In cell biology, it has recently been observed that virus transmission can occur between susceptible cells and infected cells by direct contact [61]. Let $S$ be the density of susceptible cells, $I$ be the density of infected cells, and $V$ be the density of free virus in the bloodstream. If we take *cell-to-cell transmission* into account, we can modify the basic model of (5.8) as follows:

$$\begin{aligned}\frac{dS(t)}{dt} &= -(\beta_1 I(t) + \beta_2 V(t))S(t),\\ \frac{dI(t)}{dt} &= (\beta_1 I(t) + \beta_2 V(t))S(t) - \gamma I(t),\\ \frac{dV(t)}{dt} &= -\alpha V(t) + \varepsilon I(t),\end{aligned} \tag{5.17}$$

where $\beta_1$ denotes the transmission coefficient for cell-to-cell transmission and $\beta_2$ denotes the transmission coefficient between susceptible cells and the free virus. Readers are referred to [14, 90] for a recent mathematical analysis of the cell-to-cell transmission model.

Let $S(0)$ be the initial population size of totally susceptible cells. The linearized system about the virus-free steady state $(S(0), 0, 0)$ is then

$$\begin{aligned}\frac{dI(t)}{dt} &= b(t) - \gamma I(t),\\ \frac{dV(t)}{dt} &= \varepsilon I(t) - \alpha V(t),\end{aligned}$$

where $b(t) := S(0)(\beta_1 I(t) + \beta_2 V(t))$ is the density of newly infected cells in the initial invasion phase. Using the variation-of-constants formula in the linearized equation, we obtain the following expressions:

$$\begin{aligned}V(t) &= V(0)e^{-\alpha t} + \int_0^t e^{-\alpha(t-s)}\varepsilon I(s)ds,\\ I(t) &= I(0)e^{-\gamma t} + \int_0^t e^{-\gamma(t-z)}b(z)dz.\end{aligned} \tag{5.18}$$

Inserting the equation for $I(t)$ into the equation for $V(t)$, we have

$$V(t) = g(t) + \varepsilon \int_0^t \int_0^x e^{-\alpha(x-\theta)-\gamma\theta} d\theta b(t-x)dx,$$

where

$$g(t) := V(0)e^{-\alpha t} + \int_0^t e^{-\alpha(t-s)} \varepsilon I(0) e^{-\gamma s} ds.$$

Therefore, we arrive at the following renewal equation for the newly produced cells $b(t)$:

$$b(t) = h(t) + \int_0^t \Psi(x) b(t-x) dx, \tag{5.19}$$

where $h(t)$ (the density of newly infected cells produced by the initial infected cells and the virus) is defined by

$$h(t) := (\beta_2 g(t) + \beta_1 I(0)e^{-\gamma t})S(0),$$

the kernel $\Psi$ is given by

$$\begin{aligned}\Psi(x) &= \beta_2 S(0)\varepsilon \int_0^x e^{-\gamma\theta-\alpha(x-\theta)} d\theta + \beta_1 S(0) e^{-\gamma x} \\ &= \frac{\beta_2 S(0)\varepsilon}{\gamma\alpha} (\phi_1 * \phi_2)(x) + \frac{\beta_1 S(0)}{\gamma} \phi_1(x)\end{aligned}$$

and $\phi_1$ and $\phi_2$ are the probability density functions for the clearance of the infected cells and the clearance of free virus, respectively, which are defined by

$$\phi_1(x) = \gamma e^{-\gamma x}, \quad \phi_2(x) = \alpha e^{-\alpha x}.$$

The basic reproduction number is then calculated as

$$R_0 = \int_0^\infty \Psi(x) dx = \frac{\beta_2 S(0)\varepsilon}{\alpha\gamma} + \frac{\beta_1 S(0)}{\gamma}. \tag{5.20}$$

Moreover, the generation time $G$ is

$$G = \int_0^\infty t \frac{\Psi(t)}{R_0} dt = \frac{R_2}{R_0} G_2 + \frac{R_1}{R_0} G_1 \leq G_2, \tag{5.21}$$

where

$$R_2 := \frac{\beta_2 S(0)\varepsilon}{\alpha\gamma}, \quad R_1 := \frac{\beta_1 S(0)}{\gamma}, \quad G_2 := \frac{1}{\gamma} + \frac{1}{\alpha}, \quad G_1 := \frac{1}{\gamma}.$$

Therefore, $R_2$ denotes the reproduction number for transmission mediated by free virus, and $G_2$ is its generation time; $R_1$ is the reproduction number of the cell-to-cell transmission, and $G_1$ is its generation time. The existence of cell-to-cell transmission shortens the generation time from $G_2$ to $G$.

**Exercise 5.5** Show that the final size equation for the model in (5.17) is again given by (5.16).

**Exercise 5.6** Let $w(t) := \varepsilon I(t)$ be the density of newly produced virus at time $t$. Formulate the renewal equation for $w(t)$ and confirm that its integral kernel is again given by $\Psi$.

### 5.2.3 *Asymptomatic Transmission Model*

We now give another interpretation of model (5.17). In fact, the virus dynamics model (5.17) can be interpreted as the *asymptomatic transmission* model for infectious diseases [55, 56]. Let $S(t)$ be the susceptible population, and let $I(t)$ be the infected population in the *incubation period* (the time between the receipt of infection and the appearance of symptoms). The incubation period is a key index for epidemiology [78, 84, 85]. Let $V(t)$ be the density of infected individuals suffering from the onset of the disease. In this case, we assume that $\gamma = \varepsilon$ (if there is no isolation), which denotes the rate of onset, $\alpha$ is the recovery rate, $\beta_1$ is the asymptomatic transmission coefficient, and $\beta_2$ is the symptomatic transmission coefficient.

From a practical point of view, we are concerned with the dynamics of the infected population with onset $V$, because they are observable and represent the target of the isolation policy. Thus, we calculate the renewal equation for the newly produced infected individuals with onset, denoted by $v(t) := \varepsilon I(t)$. For simplicity, we omit the initial data:

$$\begin{aligned} v(t) &= \varepsilon \int_0^t e^{-\gamma(t-z)} b(z)dz = (\phi_1 * b)(t), \\ b(t) &= R_1 \int_0^t \phi_1(z) b(t-z)dz + R_2 \int_0^t \phi_2(z) v(t-z)dz \\ &= R_1(\phi_1 * b)(t) + R_2(\phi_2 * v)(t). \end{aligned} \tag{5.22}$$

Let $\mathscr{R}(t)$ be the resolvent kernel corresponding to the integral kernel $R_1\phi_1$ defined by the solution of the resolvent equation (see Chap. 10):

$$\mathscr{R} = R_1\phi_1 + R_1\phi_1 * \mathscr{R}. \tag{5.23}$$

Note that the resolvent $\mathscr{R}$ is integrable if and only if $R_1 < 1$ and is calculated as

$$\int_0^\infty \mathscr{R}(t)dt = \frac{R_1}{1 - R_1}.$$

We can solve (5.22) for $b$ as follows:

$$b = f + \mathscr{R} * f, \tag{5.24}$$

where $f := R_2\phi_2 * v$. Inserting (5.24) into the equation for $v$ in (5.22), we arrive at the renewal equation for $v(t)$:

$$v = R_2[\phi_1 * \phi_2 + \phi_1 * (\mathscr{R} * \phi_2)] * v = \Phi * v, \tag{5.25}$$

where $\Phi$ is the reproduction kernel for the newly produced infected individuals with onset $v$.

If $R_1 < 1$, the *state-reproduction number* $T_v$ of $v(t)$ is calculated as

$$T_v := \int_0^\infty \Phi(z)dz = \frac{R_2}{1 - R_1}, \tag{5.26}$$

whereas $T_v = \infty$ if $R_1 \geq 1$. Under the assumption that $R_1 < 1$, that is, the reproduction number of the infected population in the incubation period is subcritical, it follows that $T_v > 1$ if $R_0 > 1$ and $T_v < 1$ if $R_0 < 1$. We can therefore use $T_v$ as a control index.

The idea of the state-reproduction number is similar to the *type-reproduction number* $T$ (Chap. 9), but can be distinguished by considering the "type" to indicate a specific state of the "newly infected" population (*state-at-infection* or *state-at-birth*), whereas "state" indicates any epidemic state [55, 56, 58].

Suppose that we can immediately isolate infected individuals who exhibit the onset of infection. Let $k \in [0, 1]$ be the proportion of isolation (the efficacy of isolation) from new cases with the onset of infection. In this case, $\varepsilon$ is replaced by $(1 - k)\gamma$, the number of secondary cases with onset becomes $(1 - k)R_2$, and so the state-reproduction number becomes $(1 - k)T_v$. Hence, the eradication condition $R_0 < 1$ is attained if

$$k > 1 - \frac{1}{T_v}. \tag{5.27}$$

If we let $\theta$ be the proportion of transmission prior to symptoms given by

$$\theta := \frac{R_1}{R_1 + R_2} = \frac{R_1}{R_0}, \tag{5.28}$$

then the eradication condition can be written as

$$R_1 + (1 - k)R_2 = R_0(1 - k(1 - \theta)) < 1. \tag{5.29}$$

Of course, if $\theta \to 1$, it is more difficult to prevent the disease by isolation. A similar expression to (5.29) is given in [35].

The average interval between the primary onset and the secondary onset, denoted by $L_v$, is calculated as

$$L_v := \frac{1}{T_v}\int_0^\infty z\Phi(z)dz = \frac{1}{\gamma}\frac{1}{1-R_1} + \frac{1}{\alpha}. \tag{5.30}$$

Traditionally, the *serial interval* is defined as the period from the observation of symptoms in one case to the observation of symptoms in a second case "directly" infected from the first [6, p. 21]. Conversely, $L_v$ can be written as

$$L_v = \frac{1}{\alpha} + \frac{1}{\gamma}\sum_{n=0}^{\infty} R_1^n > \frac{1}{\alpha} + \frac{1}{\gamma} = G_2, \tag{5.31}$$

where $G_2$ is the serial interval in the traditional sense and $\sum_{n=1}^{\infty} R_1^n$ reflects the indirect reproduction of secondary cases mediated by asymptomatic transmission. Hence, $L_v$ is the mean interval from the primary onset case to the secondary onset case taking into account all possible reproduction paths. From (5.21) and (5.31), we conclude that $L_v > G_2 > G$, that is, the real serial interval is longer than the serial interval in the traditional sense and the generation time.

The Euler–Lotka equation for the renewal process (5.19) with $\varepsilon = \gamma$ and $S_0 = S(0)$ of newly infected individuals $b$ is given as

$$\int_0^\infty e^{-\lambda x}\Psi(x)dx = \frac{\beta_2 S_0}{\alpha}\hat{\phi}_2(\lambda)\hat{\phi}_1(\lambda) + \frac{\beta_1 S_0}{\gamma}\hat{\phi}_1(\lambda) = 1, \tag{5.32}$$

where $\hat{f}$ denotes the Laplace transformation of $f$. It is easy to see that

$$\hat{\phi}_1(\lambda) = \frac{\gamma}{\gamma+\lambda}, \quad \hat{\phi}_2(\lambda) = \frac{\alpha}{\alpha+\lambda},$$

and so the Euler–Lotka characteristic equation becomes a quadratic:

$$\lambda^2 + \alpha\gamma(G_2(1-R_1) + G_1 R_1)\lambda + \alpha\gamma(1-R_1) = \alpha\gamma R_2. \tag{5.33}$$

If $R_0 > 1$ and $R_1 < 1$, (5.33) has a unique positive root, so it elicits a Malthusian parameter $\lambda$. Dividing both sides of (5.33) by $\alpha\gamma(1-R_1)$, we obtain the approximation

$$T_v = 1 + L_v\lambda + O(\lambda^2). \tag{5.34}$$

We can then estimate the state-reproduction number of infected individuals with onset from the observed Malthusian parameter $\lambda$ and $L_v$.

## 5.3 Infection-Age-Dependent Model

Although (5.1) is usually referred to as *the* Kermack–McKendrick model, it is a special case of the general infection-age-dependent SIR model studied by Kermack and McKendrick in their 1927 paper [66], which was the first of their series of papers dealing with compartmental SIR epidemic models. In this section, we introduce the early model of 1927; we will examine the endemic (reinfection) model of 1933 in Chap. 8.

Let $i(t,\tau)$ be the density of infected individuals at *infection-age* (duration since infection) $\tau$ and time $t$, $\beta(\tau)$ be the infection rate at infection-age $\tau$, and $\gamma(\tau)$ be the removal rate. Instead of (5.1), we can then formulate the following partial differential equation model:

$$\begin{aligned}\frac{dS(t)}{dt} &= -\lambda(t)S(t),\\ \frac{\partial i(t,\tau)}{\partial t}+\frac{\partial i(t,\tau)}{\partial \tau} &= -\gamma(\tau)i(t,\tau),\\ i(t,0) &= \lambda(t)S(t),\\ \frac{dR(t)}{dt} &= \int_0^\infty \gamma(\tau)i(t,\tau)d\tau,\end{aligned} \tag{5.35}$$

where the force of infection $\lambda(t)$ is given by

$$\lambda(t)=\int_0^\infty \beta(\tau)i(t,\tau)d\tau.$$

Kermack and McKendrick did not use the "McKendrick" partial differential equation to formulate the infection-age-dependent SIR model in their 1927 paper, although it had already been introduced by McKendrick in 1926 [79]. Further details of this 1926 paper are given by Dietz [29].

Again, we do not take into account demographic factors such as natural mortality, fertility, and migration of the host population, so the total size of the host population $N$ is constant:

$$N=S(t)+\int_0^\infty i(t,\tau)d\tau+R(t),$$

where we assume that $i(t,\cdot)\in L^1_+(0,\infty)$ and $i(t,\infty)=0$. This boundary condition is satisfied if we assume that $\partial i/\partial\tau$ is integrable.

**Exercise 5.7** Prove that the force of infection $\lambda$ satisfies a "scalar" nonlinear renewal equation

$$\lambda(t)=g(t)+S(0)\int_0^t \beta(\tau)\Gamma(\tau)\lambda(t-\tau)e^{-\int_0^{t-\tau}\lambda(\sigma)d\sigma}d\tau,$$

where $g$ is a force of infection given by the initial infected individuals:

$$g(t) := \int_t^{\infty} \beta(\tau) \frac{\Gamma(\tau)}{\Gamma(\tau - t)} i(0, \tau - t) d\tau,$$

and $\Gamma(\tau)$ is defined below.

## 5.3.1 Linear Invasion Phase and $R_0$

System (5.35) has a *disease-free steady state* (abbreviated as DFSS) without recovered individuals $(S, i, R) = (N, 0, 0)$. The linearized system at the DFSS is given by

$$\begin{aligned} \frac{\partial y(t,\tau)}{\partial t} + \frac{\partial y(t,\tau)}{\partial \tau} &= -\gamma(\tau) y(t,\tau), \\ y(t,0) &= N \int_0^{\infty} \beta(\tau) y(t,\tau) d\tau, \end{aligned} \tag{5.36}$$

where $y(t, \tau)$ denotes the deviation from DFSS (we omit the equations describing the deviation of $S$ and $R$ from their steady states). If a very small number of infected individuals invade a totally susceptible population with size $N$, the initial dynamics of the infected population are described by the linearized equation (5.36). Readers may recall that (5.36) is a stable population model, as discussed in Chap. 1 of this text.

By integrating the McKendrick equation in (5.36) along its characteristic line, we obtain the expression

$$y(t,\tau) = \begin{cases} \Gamma(\tau) v(t-\tau), & t - \tau > 0, \\ \frac{\Gamma(\tau)}{\Gamma(\tau - t)} y_0(\tau - t), & \tau - t > 0, \end{cases} \tag{5.37}$$

where $y_0(\tau) = y(0, \tau)$, $v(t) := y(t, 0)$, and $\Gamma(\tau)$ is the survival probability in the infected state given by

$$\Gamma(\tau) := \exp\left( -\int_0^{\tau} \gamma(\sigma) d\sigma \right).$$

Inserting (5.37) into the boundary condition of (5.36), we obtain the following linear integral equation for the number of newly infected individuals per unit time $v(t)$ in the initial invasion phase:

$$v(t) = N(G y_0)(t) + N \int_0^t \Psi(\tau) v(t - \tau) d\tau, \tag{5.38}$$

where

$$\Psi(\tau) := \beta(\tau)\Gamma(\tau), \quad (Gy_0)(t) := \int_0^\infty \beta(\tau+t)\frac{\Gamma(\tau+t)}{\Gamma(\tau)}y_0(\tau)d\tau.$$

Here, we assume that $G : L^1_+(\mathbb{R}_+) \to L^1_+(\mathbb{R}_+)$ is a positive operator, $\Psi \in L^1_+(0,\infty)$, and $\beta(\tau)$ and $\gamma(\tau)$ are bounded nonnegative measurable functions. Moreover, we assume that

$$\int_A^\infty \gamma(\sigma)d\sigma = +\infty$$

for any $A > 0$, which implies that infected individuals will recover after some time.

Note that the normalized distribution

$$\frac{\Psi(\tau)}{\int_0^\infty \Psi(\tau)d\tau}$$

gives the probability density function of the generation interval (*generation time*), so it is different from the distribution of sojourn time in the infected state given by $\gamma(\tau)\Gamma(\tau)$. The expected sojourn time in the infected state is given by

$$\int_0^\infty \tau\gamma(\tau)\Gamma(\tau)d\tau = \int_0^\infty \Gamma(\tau)d\tau.$$

Because (5.38) is the renewal integral equation (Lotka's integral equation), we know that the basic reproduction number $R_0$ is given by

$$R_0 = N\int_0^\infty \Psi(\tau)d\tau. \tag{5.39}$$

As shown in Chap. 1, we can prove that there exists a constant $q_0 > 0$ such that

$$\lim_{t\to\infty} e^{-\lambda_0 t}v(t) = q_0,$$

where the intrinsic rate of natural increase $\lambda_0$ is given by a real root of the characteristic equation

$$N\int_0^\infty e^{-\lambda_0\tau}\Psi(\tau)d\tau = 1,$$

and the following sign relation holds:

$$\text{sign}(\lambda_0) = \text{sign}(R_0 - 1).$$

Then, if $R_0 > 1$, the disease can successfully invade the host susceptible population, whereas if $R_0 < 1$, it cannot. In the terminology of dynamical systems, the DFSS $(N, 0, 0)$ is locally asymptotically stable if $R_0 < 1$ and is unstable if $R_0 > 1$.

### 5.3.2 Asymptotic Behavior

Let us now consider the asymptotic behavior of the nonlinear system (5.35) with initial data $(S(0), i(0, \tau), R(0)) = (S_0, i_0(\tau), 0)$, that is, we assume that

$$N = S_0 + \int_0^\infty i_0(\tau)d\tau.$$

**Lemma 5.1** *There exists a positive number* $S(\infty)$ *such that*

$$\lim_{t\to\infty} S(t) =: S(\infty).$$

*It follows that*

$$\lim_{t\to\infty} B(t) = \lim_{t\to\infty} I(t) = 0,$$

*where* $B(t) := i(t, 0)$ *and* $I(t) := \int_0^\infty i(t, \tau)d\tau$.

*Proof* First, note that $S(t)$ is monotone decreasing and nonnegative and has a nonnegative limit $\lim_{t\to\infty} S(t) =: S(\infty) \geq 0$. From the relations $\dot{S}(t) = -B(t) = -S(t)\lambda(t)$, we obtain a nonlinear integral equation for $S$:

$$\frac{\dot{S}(t)}{S(t)} = -\lambda(t) = \int_0^t \Psi(\tau)\dot{S}(t-\tau)d\tau - (Gi_0)(t). \tag{5.40}$$

Integrating both sides of (5.40) with respect to $t$ from 0 to a large positive number $T$, we obtain

$$\log\frac{S(T)}{S(0)} = \int_0^T d\sigma \int_0^\sigma \Psi(\tau)\dot{S}(\sigma-\tau)d\tau - \int_0^T (Gi_0)(t)dt.$$

Observe that

$$\begin{aligned}\int_0^T d\sigma \int_0^\sigma \Psi(\tau)\dot{S}(\sigma-\tau)d\tau &= \int_0^T \Psi(\tau)d\tau \int_\tau^T \dot{S}(\sigma-\tau)d\sigma \\ &= \int_0^T \Psi(\tau)(S(T-\tau) - S(0))d\tau.\end{aligned}$$

If we let $T \to \infty$, $S(\infty)$ must satisfy

$$\log\frac{S(\infty)}{S(0)} = (S(\infty) - S(0))\frac{R_0}{N} - \int_0^\infty (Gi_0)(t)dt, \tag{5.41}$$

which shows that $S(\infty)$ is positive.

Next, using our assumption and the Lebesgue dominant convergence theorem, it follows that $\lim_{t\to\infty}(Gi_0)(t) = 0$. Observe that

$$S(\infty) = S(0) - \int_0^\infty B(\sigma)d\sigma > 0,$$

from which we know that $B \in L^1_+(0, \infty)$. Therefore, for any $\varepsilon > 0$, there exists a large $T > 0$ such that $\int_T^\infty B(\tau)d\tau < \varepsilon/|\Psi|_{L^\infty}$. Observe that, for $t > T$,

$$\int_0^t \Psi(t-\tau)B(\tau)d\tau = \int_0^T \Psi(t-\tau)B(\tau)d\tau + \int_T^t \Psi(t-\tau)B(\tau)d\tau =: J_1 + J_2.$$

Thus, we have

$$J_2 := \int_T^t \Psi(t-\tau)B(\tau)d\tau \le \int_T^\infty B(\tau)d\tau|\Psi|_{L^\infty} < \varepsilon.$$

Because $\Psi(t-\tau)B(\tau)$ is a bounded integrable function for $\tau \in [0, T]$, it follows from the Lebesgue dominant convergence theorem that $\lim_{t\to\infty} J_1 = 0$. Hence, $\limsup_{t\to\infty}(J_1 + J_2) \le \varepsilon$, which implies that $\lim_{t\to\infty}(J_1 + J_2) = 0$. Because $S(t) \le N$ and

$$i(t,\tau) = \begin{cases} \Gamma(\tau)B(t-\tau), & t-\tau > 0, \\ \frac{\Gamma(\tau)}{\Gamma(\tau-t)} i_0(\tau-t), & \tau - t > 0, \end{cases}$$

we have

$$B(t) = S(t)\lambda(t) \le N(Gi_0)(t) + N\int_0^t \Psi(\tau)B(t-\tau)d\tau.$$

Therefore, it follows that $\lim_{t\to\infty} B(t) = 0$. By applying a similar argument to the following relation:

$$I(t) = \int_0^t \Gamma(t-\tau)B(\tau)d\tau + \int_0^\infty \frac{\Gamma(\tau+t)}{\Gamma(\tau)} i_0(\tau)d\tau,$$

we have that $\lim_{t\to\infty} I(t) = 0$. □

### 5.3.3 The Intensity of Epidemic and Its Lower Bound

From our assumption that $N = S(0) + I(0)$ and $R(0) = 0$, the definition $(1 - p)N = S(\infty)$, and (5.41), we obtain the final size relation

$$1 - p = \left(1 - \frac{I(0)}{N}\right) e^{-R_0 p + \zeta}, \tag{5.42}$$

where the parameter $\zeta$ is given by

$$\zeta := \frac{R_0 I(0)}{N} - \int_0^\infty (G i_0)(t) dt.$$

We should remark that, in general, the sign of $\zeta$ is not definite. In fact, we can observe that

$$\int_0^\infty (G i_0)(t) dt = \frac{1}{N} \int_0^\infty \nu(\tau) i_0(\tau) d\tau,$$

where $\nu(\tau)$ is the expected number of secondary cases produced by an infected individual with infection-age $\tau$ given by

$$\nu(\tau) := N \int_\tau^\infty \beta(x) \frac{\Gamma(x)}{\Gamma(\tau)} dx,$$

and this is not necessarily monotone decreasing. Let $V(i_0)$ be the average number of secondary cases produced by an initial infected individual, defined by

$$V(i_0) := \int_0^\infty \nu(\tau) \left( \frac{i_0(\tau)}{I(0)} \right) d\tau.$$

Then, we obtain

$$\zeta = \frac{I(0)}{N} (R_0 - V(i_0)).$$

In particular, if we consider the special initial condition $i_0(\tau) = I(0)\delta(\tau)$, where $\delta(\tau)$ is Dirac's delta function, the epidemic spreads from a newly infected population with infection-age zero, and we have $\zeta = 0$ because $V(i_0) = \nu(0) = R_0$. Thus, we again obtain the same intensity equation as in (5.5). More generally, if $\zeta \le 0$, it follows that $1 - p \le e^{-R_0 p}$ and we can repeat the proof of Proposition 5.1, that is, the intensity of epidemic has a lower bound $p_\infty$ given by the largest nonnegative root of the final size Eq. (5.6):

**Proposition 5.3** *Consider system* (5.35) *with initial data* $(S_0, i_0(\cdot), 0)$. *Suppose that* $R_0 \le V(i_0)$. *Then, it holds that* $p \ge p_\infty$, *where* $p_\infty$ *is the largest nonnegative root of the final size equation* $1 - x = e^{-x R_0}$.

Note that $V(i_0) < R_0$ if $\nu(\tau) < \nu(0)$ for all $\tau > 0$. Thus, if $\nu$ is monotone decreasing, the condition that $R_0 \le V(i_0)$ does not hold. However, we can show that $p \ge p_\infty$ holds as long as the initial number of infected individuals is sufficiently small.

Let us introduce the *cumulative force of infection* by

$$\Lambda(t) := \int_0^t \lambda(\sigma) d\sigma = -\log \frac{S(t)}{S(0)}. \tag{5.43}$$

Integrating (5.43) from 0 to $t$, we have

$$-\Lambda(t) = \log \frac{S(t)}{S(0)} = \int_0^t d\sigma \int_0^\sigma \Psi(\tau)\dot{S}(\sigma - \tau)d\tau - \int_0^t (Gi_0)(\sigma)d\sigma$$
$$= \int_0^t \Psi(\tau)S(0)(e^{-\Lambda(t-\tau)} - 1)d\tau - \int_0^t (Gi_0)(\sigma)d\sigma.$$

Therefore, we arrive at a nonlinear renewal integral equation for $\Lambda(t)$:

$$\Lambda(t) = g(t) + R_e \int_0^t \psi(\tau) f(\Lambda(t-\tau))d\tau, \tag{5.44}$$

where

$$g(t) := \int_0^t (Gi_0)(\sigma)d\sigma, \quad \psi(\tau) := \frac{\Psi(\tau)}{\int_0^\infty \Psi(\xi)d\xi}, \quad f(x) := 1 - e^{-x}$$

and

$$R_e := S(0) \int_0^\infty \Psi(\tau)d\tau = \frac{S(0)}{N} R_0$$

is the effective reproduction number.

If $i_0 \neq 0$, $\Lambda(t)$ is monotone increasing and bounded above. It has a positive limit $\Lambda(\infty) = \lim_{t\to\infty} \Lambda(t)$ and satisfies

$$\Lambda(\infty) = g(\infty) + R_e f(\Lambda(\infty)) > 0.$$

Then, we have

$$\Lambda(\infty) \geq R_e(1 - e^{-\Lambda(\infty)}). \tag{5.45}$$

If $R_e > 1$, it follows from (5.45) that $\Lambda(\infty) \geq q$, where $q$ is a unique positive solution of

$$x = R_e(1 - e^{-x}).$$

Conversely, if $R_e \leq 1, x \geq R_e(1 - e^{-x})$ holds for all $x \geq 0$ and $\Lambda(\infty) \geq q$ trivially holds for $q = 0$.

Using $S(t) = S(0)e^{-\Lambda(t)}$, we have

$$p = 1 - \frac{S(\infty)}{N} = 1 - \frac{S(0)}{N} e^{-\Lambda(\infty)} \geq 1 - e^{-q} = \frac{q}{R_e},$$

which implies that $p \geq p_* := q/R_e$. Thus, we arrive at the following conclusion:

**Lemma 5.2** *For any epidemic with initial data $S(0) > 0$, $i_0 \neq 0$, and $R(0) = 0$, it holds that $p \geq p_*$, where $p_*$ is the largest nonnegative root of the intensity equation $1 - x = e^{-R_e x}$.*

The lower bound $p_*$ is not especially useful, because we cannot normally observe the initial data $i_0(\tau)$ and hence do not have $R_e$. However, if the size of the initial infected population is very small, we can again show that the final size $p_\infty$ of the limiting epidemic gives a lower bound of the final size:

**Proposition 5.4** *For any epidemic with initial data $S(0) > 0$, $i_0 \neq 0$, and $R(0) = 0$, it holds that*

$$\liminf_{\varepsilon \downarrow 0} p \geq p_\infty, \tag{5.46}$$

*where $\varepsilon$ is the size of the initial infected population and $p_\infty$ is the largest nonnegative root of the final size equation $1 - x = e^{-R_0 x}$.*

*Proof* Let $\varepsilon := \int_0^\infty i_0(a)da \in (0, N)$ be the size of the initial infected population and $u_0(a)$ be the normalized initial distribution given by $u_0(a) := i_0(a)/\int_0^\infty i_0(a)da$. If we define

$$g_0(t) := \int_0^t d\sigma \int_0^\infty \beta(a+\sigma)\frac{\Gamma(a+\sigma)}{\Gamma(a)}u_0(a)da,$$

we have $g(t) = \varepsilon g_0(t)$. We assume that $g_0(\infty) < \infty$. Let $\Lambda(t;\varepsilon)$ be the solution of the renewal equation

$$\Lambda(t;\varepsilon) = \varepsilon g_0(t) + R_e \int_0^t \psi(\tau) f(\Lambda(t-\tau;\varepsilon))d\tau.$$

Let $\Lambda(\infty;\varepsilon) := \lim_{t\to\infty} \Lambda(t;\varepsilon)$. This is a positive root of the limiting equation

$$\Lambda(\infty;\varepsilon) = \varepsilon g_0(\infty) + \left(1 - \frac{\varepsilon}{N}\right) R_0 f(\Lambda(\infty;\varepsilon)),$$

where we have used $N = S(0) + \varepsilon$. Therefore, we have

$$\Lambda(\infty;\varepsilon) \geq \left(1 - \frac{\varepsilon}{N}\right) R_0 f(\Lambda(\infty;\varepsilon)).$$

Suppose that $R_0 > 1$. We can choose a small $\varepsilon$ such that $\left(1 - \frac{\varepsilon}{N}\right) R_0 > 1$, and so it holds that $\Lambda(\infty;\varepsilon) \geq q(\varepsilon)$, where $q(\varepsilon)$ is a unique positive root of the equation

$$x = \left(1 - \frac{\varepsilon}{N}\right) R_0(1 - e^{-x}).$$

If $R_0 > 1$, the equation $x = R_0(1 - e^{-x})$ has a unique positive root $q(0) > 0$ and $q(\varepsilon)$ converges to $q(0)$ as $\varepsilon \to 0$. Observe that if $\varepsilon \downarrow 0$, we obtain

$$p = 1 - \frac{S(\infty)}{N} = 1 - \frac{S(0)}{N}e^{-\Lambda(\infty;\varepsilon)} \geq 1 - e^{-\Lambda(\infty;\varepsilon)} \geq 1 - e^{-q(\varepsilon)}.$$

Therefore, we have

$$\liminf_{\varepsilon \downarrow 0} p \geq \lim_{\varepsilon \downarrow 0}(1 - e^{-q(\varepsilon)}) = \frac{q(0)}{R_0} = p_\infty,$$

where $p_\infty$ is the final size equation $1 - x = e^{-R_0 x}$ of the limiting epidemic. Conversely, $p \geq p_\infty = 0$ is trivial when $R_0 \leq 1$. □

In the real world, a new epidemic in a large-scale susceptible population may begin from very few cases, which corresponds to the limiting epidemic starting from an infinitesimally small infected population at some indefinite time far in the past. Such a limiting epidemic orbit can be investigated as follows: Fix a number $\zeta \in (0, \Lambda(\infty; \varepsilon))$, and let $t_\zeta(\varepsilon)$ be the time at which $\Lambda(t; \varepsilon)$ first arrives at $\zeta$, that is, $t_\zeta(\varepsilon) := \inf\{t > 0 : \Lambda(t; \varepsilon) = \zeta\}$. This time increases as $\varepsilon$ goes to zero. Shifting the time origin to introduce a new time $h$ as $t = t_\zeta(\varepsilon) + h$, we define $\hat{\Lambda}(h)$ for $h \in (-\infty, +\infty)$ as follows:

$$\hat{\Lambda}(h; \varepsilon) = \begin{cases} \Lambda(t_\zeta(\varepsilon) + h; \varepsilon), & h \geq -t_\zeta(\varepsilon) \\ 0, & h < -t_\zeta(\varepsilon). \end{cases}$$

We can rewrite (5.44) as

$$\hat{\Lambda}(h; \varepsilon) = \varepsilon \hat{g}_0(h) + R_e \int_{-t_\zeta(\varepsilon)}^{h} \psi(h - \tau) f(\hat{\Lambda}(\tau; \varepsilon)) d\tau,$$

where

$$\hat{g}_0(h) := \int_0^{t_\zeta(\varepsilon)+h} d\sigma \int_0^\infty \beta(a + \sigma) \frac{\ell(a + \sigma)}{\ell(a)} u_0(a) da.$$

If we then let $\varepsilon \to 0$, we arrive at the limiting equation for $v(h) := \lim_{\varepsilon \to 0} \hat{\Lambda}(h; \varepsilon)$:

$$v(h) = R_0 \int_{-\infty}^{h} \psi(h - \tau) f(v(\tau)) d\tau = R_0 \int_0^\infty \psi(\tau) f(v(h - \tau)) d\tau. \tag{5.47}$$

We now expect that if $\varepsilon \to 0$, $\hat{\Lambda}(h; \varepsilon)$ will uniformly converge to a bounded continuous function $v(h) \in C(\mathbb{R})$ satisfying the following properties:

(1) $v(0) = \zeta$,
(2) $v(h)$ is monotone increasing and $v(-\infty) = 0 \leq v(h) < v(\infty)$,
(3) $v(h)$ satisfies the limiting equation (5.47).

It follows from $S(t_\zeta(\varepsilon) + h) = S(0)e^{-\hat{\Lambda}(h;\varepsilon)}$ that $\hat{S}(h) = (N - \varepsilon)e^{-\hat{\Lambda}(h;\varepsilon)}$, where $\hat{S}(h) := S(t_\zeta(\varepsilon) + h)$. If we define the limiting orbit as $x(h) := \lim_{\varepsilon \downarrow 0} \hat{S}(h)$, then we obtain $x(h) = Ne^{-v(h)}$. Let us define the final size of the limiting epidemic as

$$p_\infty = 1 - \frac{x(\infty)}{N} = 1 - e^{-v(\infty)}.$$

It then follows from (5.47) that $v(\infty) = R_0(1 - e^{-v(\infty)})$. Hence, we again obtain the final size equation $1 - p_\infty = e^{-R_0 p_\infty}$.

The above conjecture was first stated by Metz [80] and proved by Diekmann [19]. Roughly speaking, we can say that the limiting epidemic is growing exponentially at $t = -\infty$. This kind of phenomenon is called the *nonlinear renewal theorem* [21, 43, 44, 81, 100]. Similar results have also been studied using perturbation–theoretic arguments [40, 73].

## 5.4 Pandemic Threshold Theorem

In 1957, Kendall proposed a spatial extension of the early Kermack–McKendrick epidemic model [66] and stated his *pandemic threshold theorem* (PTT) for the whole space $\mathbb{R}^2$ in a discussion of Bartlett's paper [65]. In the PTT, the term *pandemic* originally referred to the idea that the proportion of individuals contracting the disease is ultimately at least some positive number, given as a root of the final size equation. This remains true however far we move away from the initial focus of infection, provided that the epidemic parameters satisfy a kind of threshold condition [6, p. 176].

Interest in such threshold results for spatial epidemic models was revived by Diekmann [20], Thieme [96, 97], de Mottoni et al. [17], and Webb [107, 108] in the late 1970s, and Kendall's PTT was extended to the infection-age-structured Kermack–McKendrick model with continuous individual heterogeneity. About two decades later, a comprehensive study of the spatial Kermack–McKendrick model with an additional finite-dimensional trait space was given by Rass and Radcliff [89], who considered continuous spatial variables and a transmission kernel that depends on the geographical distance and (finite-dimensional) traits between the susceptible and infected populations.

More recently, reaction–diffusion epidemic models have been examined in light of modern $R_0$ theory [1, 87, 106]. Readers should note that modern $R_0$ theory began with the epoch-defining paper by Diekmann et al. in 1990 [24]. Hence, epidemic models developed prior to 1990 could not make use of this $R_0$ theory. Therefore, we here focus on the role of $R_0$ in a spatial extension of the early Kermack–McKendrick model with nonlocal transmission, where an infectious agent can be transmitted between individuals located in two different states, and the "state" variable representing individual heterogeneity can be interpreted as not only geographical position, but any biological or social "trait" of the individuals. We use the term "pandemic" if an epidemic may occur everywhere in the domain of the heterogeneity variable, regardless of whether the domain is bounded or not, although, as mentioned above, it originally referred to the phenomenon in an unbounded domain (see Exercise 5.3).

### 5.4.1 Basic Model and $R_0$

First, we extend the early Kermack–McKendrick model (5.35) to take into account individual heterogeneity expressed by continuous variables. Let $\xi$ be a (scalar or vector) parameter with domain $\Omega \subset \mathbb{R}^n$ that expresses any biological, epidemiological *state* of individuals. That is, $\xi$ may indicate spatial distribution, genetic, physiological, or behavioral characteristics (e.g., the degree of infection risk) and so on. We consider the case in which individuals do not migrate within the trait space. Readers are referred to [18, 30, 31] for spatial SIR models with a diffusion term and infection-age structure. Early attempts to examine the Kermack–McKendrick model with diffusion can be found in [17, 107, 108].

Let $S(t,\xi)$, $i(t,\tau,\xi)$, and $R(t,\xi)$ be the susceptible, infected, and recovered population densities with state $\xi$ at time $t$, respectively. The early Kermack–McKendrick model (5.35) can then be extended as follows:

$$
\begin{aligned}
\frac{\partial S(t,\xi)}{\partial t} &= -\lambda(t,\xi)S(t,\xi), \\
\frac{\partial i(t,\tau,\xi)}{\partial t} + \frac{\partial i(t,\tau,\xi)}{\partial \tau} &= -\gamma(\tau,\xi)i(t,\tau,\xi), \\
i(t,0,\xi) &= \lambda(t,\xi)S(t,\xi), \\
\frac{\partial R(t,\xi)}{\partial t} &= \int_0^\infty \gamma(\tau,\xi)i(t,\tau,\xi)d\tau,
\end{aligned}
\tag{5.48}
$$

where the force of infection $\lambda(t,\xi)$ is given by

$$
\lambda(t,\xi) = \int_0^\infty \int_\Omega \beta(\tau,\xi,\eta)i(t,\tau,\eta)d\eta d\tau
$$

and $\beta(\tau,\xi,\eta)$ denotes the transmission coefficient between infected individuals with infection-age $\tau$ and state $\eta$ and susceptible individuals with state $\xi$.

Let $S(0,\xi) = S_0(\xi)$ and $i(0,\tau,\xi) = i_0(\tau,\eta)$ be the initial data, and let $N(\xi)$ be the (time-independent) density of the total population at state $\xi$:

$$
N(\xi) := S(t,\xi) + \int_0^\infty i(t,\tau,\xi)d\tau + R(t,\xi).
$$

Then, there exists a DFSS without recovered individuals $(N(\xi),0,0)$.

The linearized equation for infected individuals at $(N,0,0)$ is given by

$$
\begin{aligned}
\frac{\partial y(t,\tau,\xi)}{\partial t} + \frac{\partial y(t,\tau,\xi)}{\partial \tau} &= -\gamma(\tau,\xi)y(t,\tau,\xi), \\
y(t,0,\xi) &= N(\xi)\int_0^\infty \int_\Omega \beta(\tau,\xi,\eta)y(t,\tau,\eta)d\eta d\tau.
\end{aligned}
\tag{5.49}
$$

Integrating the McKendrick equation in (5.49) along the characteristic line, we obtain

$$y(t,\tau,\xi) = \begin{cases} b(t-\tau,\xi)\Gamma(\tau,\xi), & t-\tau>0, \\ \frac{\Gamma(\tau,\xi)}{\Gamma(\tau-t,\xi)} y(0,\tau-t,\xi), & \tau-t>0, \end{cases}$$

where $b(t,\xi) := y(t,0,\xi)$ is the density of newly infected individuals in the initial invasion phase and

$$\Gamma(\tau,\xi) := \exp\left(-\int_0^\tau \gamma(x,\xi)dx\right)$$

is the survival probability at state $\xi$.

Inserting the above equation into the boundary condition of (5.49), we have that the newly infected population density $b(t,\xi)$ at the DFSS without recovered individuals satisfies the renewal equation

$$b(t,\xi) = N(\xi)G[y(0)](t,\xi) + N(\xi)\int_0^t \int_\Omega \Psi(\tau,\xi,\eta)b(t-\tau,\eta)d\eta d\tau, \quad (5.50)$$

where

$$\Psi(\tau,\xi,\eta) := \beta(\tau,\xi,\eta)\Gamma(\tau,\eta),$$

$$G[y(0)](t,\xi) := \int_t^\infty \int_\Omega \beta(\tau,\xi,\eta)\frac{\Gamma(\tau,\eta)}{\Gamma(\tau-t,\eta)} y(0,\tau-t,\eta)d\eta d\tau.$$

As discussed in Chap. 9, the basic reproduction number of the renewal system (5.50) is given by the spectral radius of a linear positive operator (next-generation operator) $K$ on $L^1(\Omega)$ defined by

$$(Ku)(\xi) := N(\xi)\int_0^\infty \int_\Omega \Psi(\tau,\xi,\eta)u(\eta)d\eta d\tau, \quad u \in L^1(\Omega).$$

In the following, we assume that $K$ is positive, non-supporting, and compact. Its spectral radius $r(K)$ is then the dominant positive eigenvalue of $K$ associated with a positive eigenfunction. We can therefore state that the Malthusian parameter

$$\lambda_0 := \lim_{t\to\infty} \frac{\log |b(t,\cdot)|_{L^1}}{t}$$

exists and the sign relation $\text{sign}(\lambda_0) = \text{sign}(R_0 - 1)$ holds [57].

### 5.4.2 The Initial Value Problem

First, let us consider an epidemic starting from time $t = 0$. Integrating the McKendrick equation in (5.48) along the characteristic line, we obtain

$$i(t,\tau,\xi)=\begin{cases} B(t-\tau,\xi)\Gamma(\tau,\xi), & t-\tau>0,\\ \frac{\Gamma(\tau,\xi)}{\Gamma(\tau-t,\xi)}i(0,\tau-t,\xi), & \tau-t>0,\end{cases}$$

where $B(t,\xi):=i(t,0,\xi)=-\dot{S}(t,\xi)$ is the density of newly infected individuals. It follows that

$$\frac{\dot{S}(t,\xi)}{S(t,\xi)}=-\lambda(t,\xi)=\int_0^t\int_\Omega \Psi(\tau,\xi,\eta)\dot{S}(t-\tau,\eta)d\eta d\tau-G[i_0](t,\xi), \tag{5.51}$$

where $i_0:=i(0,\tau,\xi)$.

Define the *cumulative force of infection* as

$$\Lambda(t,\xi):=\int_0^t\lambda(x,\xi)dx=-\log\frac{S(t,\xi)}{S_0(\xi)},$$

where we assume that $S_0(\xi)=S(0,\xi)>0$. By integrating both sides of (5.51) with respect to $t$ from 0 to $t$, we obtain the nonlinear renewal equation

$$\Lambda(t,\xi)=g(t,\xi)+\int_0^t\int_\Omega \Psi(\tau,\xi,\eta)S_0(\eta)f(\Lambda(t-\tau,\eta))d\eta d\tau, \tag{5.52}$$

where

$$f(x):=1-e^{-x},\quad g(t,\xi):=\int_0^t G[i_0](\sigma,\xi)d\sigma.$$

As shown by Diekmann [20], a convenient framework for the study of (5.52) is provided by the Banach space $C_T=C([0,T];BC(\Omega))$ of continuous functions on $[0,T]$ with values in $BC(\Omega)$, which is the set of bounded continuous functions, equipped with the norm $|\Lambda|_{C_T}=\sup_{0\le t\le T}|\Lambda(t,\cdot)|_{BC(\Omega)}$. Readers are referred to Diekmann [20] for the precise assumptions that ensure the existence and uniqueness of solutions to (5.52).

Here, we assume that

$$\sup_{\xi\in\Omega}\int_\Omega\int_0^\infty \Psi(\tau,\xi,\eta)d\tau N(\eta)d\eta<\infty. \tag{5.53}$$

Then, $\Lambda(t,\xi)$ is uniformly bounded, continuous, and monotone increasing with respect to time $t$, so $\Lambda(\infty,\xi):=\lim_{t\to\infty}\Lambda(t,\xi)$ exists in the sense of uniform convergence in compact sets of $\Omega$ and becomes the solution of the following limiting equation:

$$\Lambda(\infty,\xi)=g(\infty,\xi)+\int_\Omega\int_0^\infty \Psi(\tau,\xi,\eta)d\tau S_0(\eta)f(\Lambda(\infty,\eta))d\eta. \tag{5.54}$$

Let us now define the state-specific *final size* by

$$p(\xi) := 1 - \frac{S(\infty, \xi)}{N(\xi)}$$

with the condition $R(0, \xi) = 0$.

Note that some authors have used a different definition of the final size. Diekmann [20] called $\Lambda(\infty, \xi)$ the final size. Rass and Radcliffe [89] considered a model of *epidemics initiated from outside*, in which an epidemic in a totally susceptible host population is triggered by infected individuals introduced from outside who do not comprise the total host population $N(\xi)$. This leads to a final size that is the proportion of the total number of "initial susceptibles" that finally contract the disease. That is, Rass and Radcliffe assume that $S_0(\xi) = N(\xi)$ and its final size is $1 - S(\infty, \xi)/S_0(\xi)$.

Based on the above limiting equation, Kendall's pandemic threshold theorem [65] has been extended to the infection-age-structured Kermack–McKendrick model by Diekmann [20]. Here, we give a formulation of Diekmann's pandemic threshold result. A more general result has been proved by Thieme [99].

**Proposition 5.5** *Suppose that $\Omega$ is compact and connected and that for each $\xi \in \Omega$ there exists $\delta = \delta(\xi) > 0$ such that the set $\{x : |x - \xi| \le \delta\} \cap \Omega$ is contained in the support of $\int_0^\infty \Psi(\tau, \xi, \cdot)d\tau$. Let*

$$R_* := \inf_{\xi \in \Omega} \int_\Omega \int_0^\infty \Psi(\tau, \xi, \eta) d\tau S_0(\eta) d\eta.$$

*Then, if $R_* > 1$, it follows that*

$$\Lambda_* \ge q > 0, \quad p(\xi) \ge p_*, \tag{5.55}$$

*where $\Lambda_* := \inf_{\xi \in \Omega} \Lambda(\infty, \xi)$, $q = R_* p_*$, and $p_*$ is a unique positive root of the intensity equation*

$$1 - x = e^{-R_* x}. \tag{5.56}$$

*Proof* If $R_* > 1$, it is easy to see that $x = R_*(1 - e^{-x})$ has a unique positive root $q > 0$ and $x < R_*(1 - e^{-x})$ for $0 < x < q$. However, it follows from (5.54) and the monotonicity of $1 - e^{-x}$ that $\Lambda_* \ge R_*(1 - e^{-\Lambda_*})$. Therefore, we know that $\Lambda_* = 0$ or $\Lambda_* \ge q$. Suppose that $\Lambda_* = 0$. Because $\Omega$ is compact, there exists some $\xi_0 \in \Omega$ such that $\Lambda(\infty, \xi_0) = 0$. In (5.54), each term is nonnegative, so the assumption for $\Psi$ implies that $\Lambda(\infty, x) = 0$ in the set $\{x : |x - \xi_0| \le \delta\} \cap \Omega$. Repeating this argument, we can find a maximal subset $X$ of $\Omega$ where $\Lambda(\infty, x) = 0$. Because $\partial X \subset \partial \Omega$, we have $X = \Omega$. That is, we conclude that $\Lambda(\infty, \xi) = 0$ for any $\xi \in \Omega$, which contradicts (5.54). Thus, if $R_* > 1$, we have $\Lambda_* \ge q > 0$. Finally, observe that

$$p(\xi) = 1 - \frac{S(0, \xi)}{N(\xi)} e^{-\Lambda(\infty, \xi)} \ge 1 - e^{-q} = \frac{q}{R_*}.$$

Let $p_* = q/R_*$. We can then conclude that $p_*$ is a positive root of the intensity equation (5.56) that gives a lower bound of the intensity $p(\xi)$. This completes the proof. □

The above result tells us that if $R_* > 1$, the epidemic outbreak ultimately occurs everywhere in $\Omega$, no matter how small the initial infected population (the *hair-trigger effect*). Although we omit the argument, a more interesting problem is to give a lower bound for the epidemic in a non-compact domain. Diekmann [20] showed that the same kind of threshold result holds for the non-compact domain $\Omega = \mathbb{R}$ or $\Omega = \mathbb{R}^2$ when the transmission coefficient $\beta$ is given by a separable function such as $\beta(\tau, \xi, \eta) = h(\tau)v(\xi - \eta)$.

Note that Diekmann's pandemic threshold result was not based on the basic reproduction number $R_0$, an aspect that was later improved [27]. Let us check the difference between the threshold number $R_*$ and $R_0$. Let $f^*$ be the adjoint positive eigenfunctional of $K$ associated with the eigenvalue $R_0 = r(K)$. From the definition of $R_*$, we have $KS_0 \geq R_* N$. Then, we have

$$\langle f^*, KS_0 \rangle = \langle K^* f^*, S_0 \rangle = R_0 \langle f^*, S_0 \rangle \geq R_* \langle f^*, N \rangle,$$

from which we obtain

$$R_0 \geq \frac{\langle f^*, N \rangle}{\langle f^*, S_0 \rangle} R_* \geq R_*.$$

We know that $R_* > 1$ is a stronger condition than the invasion condition $R_0 > 1$. When $R_0 > 1 > R_*$, we have $p_* = 0$, which does not work as a lower bound of the intensity of epidemic, although an epidemic outbreak occurs.

### 5.4.3 The Final Size Equation of the Limiting Epidemic

To examine the relation between $R_0$ and the final size, let us consider a limiting epidemic starting from a completely susceptible steady state at $t = -\infty$. Integrating the McKendrick equation in (5.48) along the characteristic line, we obtain

$$i(t, \tau, \xi) = B(t - \tau, \xi)\Gamma(\tau, \xi),$$

where

$$B(t, \xi) := i(t, 0, \xi) = \lambda(t, \xi)S(t, \xi) = -\dot{S}(t, \xi)$$

is the density of newly infected individuals. We then obtain that

$$\frac{\dot{S}(t, \xi)}{S(t, \xi)} = -\lambda(t, \xi) = \int_0^\infty \int_\Omega \Psi(\tau, \xi, \eta)\dot{S}(t - \tau, \eta)d\eta d\tau. \tag{5.57}$$

Define the cumulative force of infection as

$$\Lambda(t,\xi) := \int_{-\infty}^{t} \lambda(x,\xi)dx = -\log\frac{S(t,\xi)}{N(\xi)},$$

where we have assumed that $S(-\infty,\xi) = N(\xi) > 0$ for all $\xi \in \Omega$.

Integrating both sides of (5.57) with respect to $t$ from $-\infty$ to $t$, we obtain the nonlinear renewal equation

$$\Lambda(t,\xi) = \int_0^\infty \int_\Omega \Psi(\tau,\xi,\eta)N(\eta)f(\Lambda(t-\tau,\eta))d\eta d\tau, \qquad (5.58)$$

where $f(x) := 1 - e^{-x}$. Then, $\Lambda(t,\xi)$ is uniformly bounded and monotone increasing with respect to time $t$, so $\Lambda(\infty,\xi) := \lim_{t\to\infty}\Lambda(t,\xi)$ exists and becomes the solution of the following limiting equation:

$$\Lambda(\infty,\xi) = \int_\Omega \int_0^\infty \Psi(\tau,\xi,\eta)d\tau N(\eta)f(\Lambda(\infty,\eta))d\eta. \qquad (5.59)$$

Let us define the intensity of epidemic at state $\xi$ by

$$p(\xi) = 1 - \frac{S(\infty,\xi)}{N(\xi)} = 1 - \frac{S(\infty,\xi)}{S(-\infty,\xi)}.$$

Then, $p(\xi)$ gives the ultimate proportion of recovered individuals with trait $\xi$, which is the final size of the limiting epidemic at $\xi$.

Using $p(\xi)$ and the fact that $S(\infty,\xi) = N(\xi)e^{-\Lambda(\infty,\xi)}$, (5.59) can be written as

$$-\log(1-p(\xi)) = \int_\Omega \int_0^\infty \Psi(\tau,\xi,\eta)d\tau N(\eta)p(\eta)d\eta. \qquad (5.60)$$

Therefore, we can introduce a *final size operator equation* as follows:

$$1 - \phi(\xi) = \exp(-(U^{-1}KU\phi)(\xi)), \quad \phi \in L^\infty(\Omega), \qquad (5.61)$$

where $U : L^\infty \to L^1$ is a multiplication operator defined by

$$(U\phi)(\xi) := N(\xi)\phi(\xi).$$

If the final size operator equation (5.61) has a unique positive solution, it gives the final size distribution $p_\infty$ of the limiting epidemic.

We can now conjecture that under appropriate conditions, the final size operator equation (5.61) has a unique positive solution if $R_0 > 1$, whereas it has no positive solution if $R_0 \le 1$ and that for the initial value problem of (5.48) with $R(0,\xi) = 0$ and initial infected population size $\varepsilon > 0$, $\lim_{\varepsilon\downarrow 0} p(\xi) \ge p_\infty(\xi)$, where $p_\infty(\xi)$ denotes the largest nonnegative solution of the final size operator equation (5.61). Although we omit the proof here, Inaba [59] gave a treatment of the above conjecture. Note that we can show the pandemic threshold result without assuming the compactness of the domain of heterogeneity of the state variables if we assume the compactness and nonsupporting properties of the next-generation operator.

### 5.4.4 Traveling Wave Solutions

Let us now assume that $\gamma$ is independent of the individual character $\xi$ and that the transmission rate can be decomposed into the effect of individual distance in the state space and the effect of infection-age as follows:

$$\beta(\tau, \xi, \eta) = h(\tau)\nu(\xi - \eta),$$

where we have assumed that $\nu(x) = \nu(-x)$ and $\Omega = \mathbb{R}$.

If $N(\xi) = N = const.$, we can use a similar argument to that applied in the previous section to induce a homogeneous problem corresponding to the limiting epidemic orbit:

$$\Lambda(t, \xi) = R_0 \int_0^\infty H(\tau) \int_{-\infty}^\infty V(\xi - \eta) f(\Lambda(t - \tau, \eta)) d\eta d\tau, \tag{5.62}$$

where

$$\Lambda(t, \xi) = -\log \frac{S(t, \xi)}{N},$$

$$H(\tau) := \frac{h(\tau)\Gamma(\tau)}{\int_0^\infty h(\tau)\Gamma(\tau)d\tau}, \quad V(x) := \frac{\nu(x)}{\int_{-\infty}^\infty \nu(x)dx},$$

and the basic reproduction number is calculated as

$$R_0 = N \int_0^\infty h(\tau)\Gamma(\tau)d\tau \int_{-\infty}^\infty \nu(x)dx.$$

In the homogeneous problem (5.62), our interest focuses on whether there is a *traveling wave solution* given by $\Lambda(t, \xi) = w(\xi + ct)$ and how we can estimate its speed $c$. In the following, we mainly consider the case in which $c > 0$, as the case for $c < 0$ can be discussed in the same manner.

Inserting $\Lambda(t, \xi) = w(\xi + ct)$ into (5.62), we have

$$w(x) = R_0 \int_{-\infty}^\infty V_c(x - \zeta) f(w(\zeta)) d\zeta, \tag{5.63}$$

where

$$V_c(z) := \int_0^{\infty} H(\tau)V(z - c\tau)d\tau.$$

The constant solution of (5.63) is given by the roots of the equation $y = R_0 f(y)$, and it has a positive root $y = p > 0$ other than the zero solution if $R_0 > 1$.

Let us consider the linearized equation of (5.63) about $w = 0$:

$$v(x) = R_0 \int_{-\infty}^{\infty} V_c(x - \zeta)v(\zeta)d\zeta. \tag{5.64}$$

As shown in Sect. 2.5, if $R_0 > 1$ and under appropriate conditions, there exists a number $c_0 > 0$ such that (5.64) has a positive solution for any $c > c_0$. In such a case, we can construct a solution of (5.63) such that

$$0 < w(x) < p, \quad \lim_{x \to -\infty} w(x) = 0, \quad \lim_{x \to +\infty} w(x) = p.$$

That is, we can state the following ([20], see also [98, 99] and [102] for a more general result):

**Proposition 5.6** *If $R_0 > 1$, there exists a number $c_0 > 0$ such that for all $c > c_0$, there is a traveling wave solution. The lower bound of the speed of the traveling wave solution $c_0$ is given by $c_0 = \inf\{c|$ there exists $\lambda$ such that $L_c(\lambda) < 1\}$, where $L_c(\lambda) := R_0 \int_{-\infty}^{\infty} e^{-\lambda z} V_c(z)dz$.*

The lower bound of the speed of the traveling wave solution also gives the asymptotic speed of propagation of a perturbation. That is, for any $c_1$ and $c_2$ such that $0 < c_1 < c_0 < c_2$ and for sufficiently large $t$, the solution uniformly converges to zero in the domain $|x| \geq c_2 t$ and it uniformly becomes greater than zero in the domain $|x| \leq c_1 t$ [22, 27]. Recent results for the existence of traveling wave solutions for the SIR model with infection-age can be found in [30].

## 5.5 Endemic Threshold Phenomena

The early Kermack–McKendrick model examined so far describes one outbreak of an epidemic and its extinction. That is, there is no *endemic steady state* in which the infected population becomes established in the host population as there is no supply or reproduction of the susceptible population. If the susceptible population is reproduced by childbearing, supplemented by immigration or the recovered individuals lose their immunity, we can usually expect the threshold condition $R_0 > 1$ to give not only the invasion criterion, but also the condition for the existence of endemic steady states. Early attempts to examine endemic models for infectious diseases were reported by Kermack and McKendrick [67, 68] and Soper [93]. We now consider three endemic models.

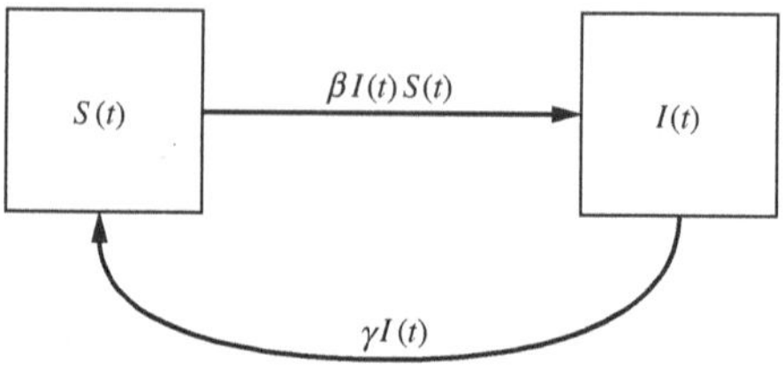


**Fig. 5.3** Flowchart of the SIS model

## 5.5.1 SIS Model Without Demography

First, we consider a very simple model that enables an endemic steady state without demographic factors. If recovered individuals directly return to the susceptible class without any immunity, the basic model (5.1) can be reformulated as the *SIS model* (Fig. 5.3):

$$\begin{aligned} \frac{dS(t)}{dt} &= -\beta S(t)I(t) + \gamma I(t), \\ \frac{dI(t)}{dt} &= \beta S(t)I(t) - \gamma I(t). \end{aligned} \tag{5.65}$$

Because the total size of the host population $N = S(t) + I(t)$ is constant, the SIS model (5.65) reduces to the logistic model

$$\frac{dI(t)}{dt} = (\gamma(R_0 - 1) - \beta I(t))I(t), \tag{5.66}$$

where $R_0 = \beta N/\gamma$. Therefore, it is easy to see that if $R_0 > 1$, the disease can invade the totally susceptible population of size $N$, and the size of the infected population converges to the endemic steady state

$$I^* = \frac{\gamma}{\beta}(R_0 - 1) = N\left(1 - \frac{1}{R_0}\right)$$

as time evolves, whereas it is naturally eradicated if $R_0 \leq 1$.

This simple SIS model without age structure can be extended to the infection-age-dependent model

$$\begin{aligned} \frac{dS(t)}{dt} &= -S(t)\int_0^\infty \beta(\tau)i(t,\tau)d\tau + \int_0^\infty \gamma(\tau)i(t,\tau)d\tau, \quad t > 0, \\ \frac{\partial i(t,\tau)}{\partial t} + \frac{\partial i(t,\tau)}{\partial \tau} &= -\gamma(\tau)i(t,\tau), \quad t > 0, \quad \tau > 0, \\ i(t,0) &= S(t)\int_0^\infty \beta(\tau)i(t,\tau)d\tau, \quad t > 0, \end{aligned} \tag{5.67}$$

where we have assumed that $\beta, \gamma \in L_+^\infty(\mathbb{R}_+)$.

We can interpret (5.67) as an SEIS model or an SEIRS model if there exist some $\omega_1 > 0$ and $\omega_2 > 0$ such that $\beta(\tau) = 0$ for $\tau \notin [\omega_1, \omega_2]$ and $\beta(\tau) > 0$ for $\tau \in [\omega_1, \omega_2]$, because the period $[0, \omega_1)$ can be interpreted as the *latent period*, $[\omega_1, \omega_2]$ is the infective period, and $(\omega_2, \infty)$ is the recovered and immune period. Therefore, the infection-age-dependent model (5.67) allows a wider range of interpretations than the non-structured SIS model (5.65). This "one-clock model" will be discussed again in Chap. 8.

For (5.67), we have neglected the host demographic vital rates, so the total size of the host population $N = S(t) + \int_0^\infty i(t, \tau)d\tau$ remains constant. Therefore, (5.67) can be reduced to a boundary value problem of the infected population $i(t, \tau)$:

$$\begin{aligned} \frac{\partial i(t, \tau)}{\partial t} + \frac{\partial i(t, \tau)}{\partial \tau} &= -\gamma(\tau) i(t, \tau), \\ i(t, 0) &= \left(N - \int_0^\infty i(t, \tau)d\tau\right) \int_0^\infty \beta(\tau) i(t, \tau)d\tau. \end{aligned} \tag{5.68}$$

Let $\Gamma(\tau) := \exp(-\int_0^\tau \gamma(\sigma)d\sigma)$. If we introduce a new distribution $v(t, \tau)$ by $i(t, \tau) = N\Gamma(\tau)v(t, \tau)$, (5.68) can be simplified as

$$\begin{aligned} \frac{\partial v(t, \tau)}{\partial t} + \frac{\partial v(t, \tau)}{\partial \tau} &= 0, \\ v(t, 0) &= R_0 \left(1 - \int_0^\infty \Gamma(\tau) v(t, \tau)d\tau\right) \int_0^\infty \phi(\tau) v(t, \tau)d\tau, \end{aligned} \tag{5.69}$$

where $R_0$ is the basic reproduction number and $\phi(\tau)$ gives the probability distribution of the occurrence of secondary infection:

$$R_0 = N \int_0^\infty \beta(\tau)\Gamma(\tau)d\tau, \quad \phi(\tau) := \frac{\beta(\tau)\Gamma(\tau)}{\int_0^\infty \beta(\tau)\Gamma(\tau)d\tau}.$$

We can observe that equilibrium solutions of (5.69) are constant-valued functions. If we denote an equilibrium solution as $v^*(\tau) = v^*$, it follows from the boundary condition that

$$1 = R_0 \left(1 - v^* \int_0^\infty \Gamma(\tau)d\tau\right).$$

Then, we have

$$v^* = \frac{1}{e_0}\left(1 - \frac{1}{R_0}\right),$$

where $e_0 := \int_0^\infty \Gamma(\tau)d\tau$ is the average duration of the infected state. Although the DFSS $v^* = 0$ always exists, there is no other nonnegative equilibrium solution if $R_0 \leq 1$, whereas there exists a unique endemic equilibrium solution if $R_0 > 1$.

The linearized equation of (5.69) about $v^* = 0$ is given by

$$\frac{\partial u(t,\tau)}{\partial t} + \frac{\partial u(t,\tau)}{\partial \tau} = 0,$$
$$u(t,0) = R_0 \int_0^\infty \phi(\tau) u(t,\tau) d\tau.$$

As shown in Chap. 1, the local stability of the DFSS is then determined by the location of the roots $z$ of Lotka's characteristic equation

$$R_0 \int_0^\infty e^{-z\tau} \phi(\tau) d\tau = 1. \tag{5.70}$$

If $R_0 < 1$, then all real parts of the characteristic roots are negative, so the DFSS is locally asymptotically stable. If $R_0 > 1$, there exists a positive characteristic root and the solution is unstable. Let $v(0,\tau) =: v_0(\tau)$ be the initial data, and let $B(t) := v(t,0)$ be the boundary value of (5.69). We then obtain

$$B(t) \le R_0 \int_0^t \phi(\tau) B(t-\tau) d\tau + g(t),$$

where

$$g(t) := R_0 \int_t^\infty \phi(\tau) v_0(\tau - t) d\tau.$$

If $C(t)$ is the solution of the renewal equation

$$C(t) = R_0 \int_0^t \phi(\tau) C(t-\tau) d\tau + g(t),$$

then we have $B(t) \le C(t)$, and it follows from the renewal theorem that there exists a constant $C_0 > 0$ such that $\lim_{t\to\infty} e^{-\lambda_0 t} C(t) = C_0$, where $\lambda_0$ is the real root of the characteristic equation (5.70). If $R_0 < 1$, then $\lambda_0 < 0$ and we have $v(t,\tau) = B(t-\tau) \le C(t-\tau) \to 0$ as $t \to \infty$, that is, the DFSS is globally asymptotically stable.

**Exercise 5.8** Let $B^\infty := \limsup_{t\to\infty} B(t)$. Show that $B^\infty = 0$ if $R_0 < 1$.

Subsequently, let us consider the stability of the endemic steady state when $R_0 > 1$. Introducing the perturbation $u(t,\tau) := v(t,\tau) - v^*$ to induce the linearized equation at the endemic steady state $v^*$, we have

$$\frac{\partial u(t,\tau)}{\partial t} = -\frac{\partial u(t,\tau)}{\partial \tau},$$
$$u(t,0) = \int_0^\infty \phi(\tau) u(t,\tau) d\tau - \frac{R_0 - 1}{e_0} \int_0^\infty \Gamma(\tau) u(t,\tau) d\tau.$$

Therefore, we obtain the characteristic equation as

$$\int_0^\infty \phi(\tau)e^{-z\tau}d\tau - \frac{R_0 - 1}{e_0}\int_0^\infty \Gamma(\tau)e^{-z\tau}d\tau = 1. \tag{5.71}$$

For a small $\varepsilon$, let $R_0 = 1 + \varepsilon$ and let $z(\varepsilon)$ be the dominant characteristic root. Then, $z(0) = 0$, and it follows from (5.71) that

$$\left.\frac{dz(\varepsilon)}{d\varepsilon}\right|_{\varepsilon=0} = -\frac{1}{\int_0^\infty \tau\phi(\tau)d\tau} < 0.$$

Hence, all real parts of the characteristic roots are negative for sufficiently small $\varepsilon > 0$. That is, for the infection-age-dependent SIS model (5.68), we can prove the following[3]:

**Proposition 5.7** *If $R_0 < 1$, the DFSS uniquely exists and is globally asymptotically stable. If $R_0 > 1$, the DFSS is unstable and a unique endemic steady state exists. The endemic steady state is locally asymptotically stable if $R_0 > 1$ and $|R_0 - 1|$ is sufficiently small.*

If $R_0$ becomes much larger than unity, it has been shown that the endemic steady state can lose its stability. A conjugate pair of complex characteristic roots then cross the imaginary axis into the right half plane with a positive speed, and a periodic solution bifurcates. This type of bifurcation is often observed for epidemics with very long incubation periods [23, 42]. This implies that the mechanism of infection-age dependency could explain the phenomenon of *recurrent outbreak*. Indeed, sustained oscillations can be observed in the chronic-age-structured SIS model [77].

*Remark 5.3* In contrast to (5.67), we can consider the *variable susceptibility model* as follows:

$$\begin{aligned}
&\frac{\partial s(t,\tau)}{\partial t} + \frac{\partial s(t,\tau)}{\partial \tau} = -\beta(\tau)I(t)s(t,\tau),\\
&s(t,0) = \gamma I(t),\\
&\frac{dI(t)}{dt} = -\gamma I(t) + I(t)\int_0^\infty \beta(\tau)s(t,\tau)d\tau,
\end{aligned}$$

where $s(t,\cdot)$ is the duration-density function for the susceptible individuals, $\tau$ denotes the duration in the susceptible status, $\gamma$ is the recovery rate, and $\beta(\tau)$ denotes the susceptibility (or the level of immunity) at duration $\tau$. We study this model in Chap. 8 as the Pease model for influenza epidemic.

---

[3] For more rigorous treatment to characteristic equation (5.71), readers are referred to Remark 3.1 or [50, Chap. IV].

### 5.5.2 SIR Model with Demography

Second, we consider the scenario in which demographic replacement supplements the susceptible population. In this case, we modify the Kermack–McKendrick model (5.1) to take into account the demography of the host population. Let $b$ be the birth rate and $\mu$ the natural (per capita) death rate of the host population. The basic model (5.1) can then be reformulated as follows:

$$\begin{aligned}\frac{dS(t)}{dt} &= b - \mu S(t) - \beta S(t)I(t),\\ \frac{dI(t)}{dt} &= \beta S(t)I(t) - (\mu + \gamma)I(t),\\ \frac{dR(t)}{dt} &= -\mu R(t) + \gamma I(t),\end{aligned} \tag{5.72}$$

where we assume that $R(t)$ denotes the recovered population with lifelong immunity and that there is no vertical transmission from mother to newborns and no disease-induced extra death rate.

For this basic model, the total population size $N(t) = S(t) + I(t) + R(t)$ has the stable equilibrium $b/\mu$, and we can assume without loss of generality that the total initial size is given by its equilibrium value $N := b/\mu$. The basic system can then be reduced to a two-dimensional dynamical system as

$$\begin{aligned}\frac{dS(t)}{dt} &= \mu N - \mu S(t) - \beta S(t)I(t),\\ \frac{dI(t)}{dt} &= \beta S(t)I(t) - (\mu + \gamma)I(t).\end{aligned} \tag{5.73}$$

It is clear that the closed bounded set $\Omega := \{(S, I) : S \geq 0, I \geq 0, S + I \leq N\}$ is positively invariant with respect to the flow defined by system (5.73). That is, a solution orbit starting from an initial point in $\Omega$ is contained in the compact set $\Omega$ for all positive time.

In the early invasion phase, we have $S \approx N$ and the linearized equation is given by

$$\frac{dI(t)}{dt} = (\gamma + \mu)\left[\frac{\beta N}{\gamma + \mu} - 1\right] I(t). \tag{5.74}$$

Therefore, the basic reproduction number is calculated as

$$R_0 = \frac{\beta N}{\gamma + \mu}.$$

Moreover, system (5.73) has two possible steady states, $E_1$ and $E_2$, defined by

$$E_1 := \left(\frac{b}{\mu}, 0\right), \quad E_2 := \left(\frac{N}{R_0}, \frac{\mu}{\beta}(R_0 - 1)\right),$$

where $E_1$ always exists and corresponds to the DFSS, whereas $E_2$ is the endemic steady state that is biologically meaningful only if $R_0 > 1$.

If we denote the endemic steady state as $(S^*, I^*)$, it is easily seen that

$$\frac{S^*}{N} = \frac{1}{R_0}, \quad \frac{I^*}{N} = \frac{\mu}{\mu + \gamma}\left(1 - \frac{1}{R_0}\right). \tag{5.75}$$

That is, at the endemic steady state, the ratio of the susceptible population is inversely proportional to the basic reproduction number, and the prevalence is proportional to $1 - 1/R_0$. We can now prove the following endemic threshold result [49]:

**Proposition 5.8** *If $R_0 \leq 1$, the DFSS is globally asymptotically stable, whereas it is unstable if $R_0 > 1$. The endemic steady state is globally asymptotically stable if $R_0 > 1$.*

*Proof* For the solution orbit $(S(t), I(t)) \subset \Omega$, it follows that

$$\begin{aligned}\frac{dI(t)}{dt} &\leq \beta(N - I(t))I(t) - (\mu + \gamma)I(t) \\ &= (\mu + \gamma)(R_0 - 1)I(t) - \beta I^2(t).\end{aligned}$$

Therefore, if $R_0 \leq 1$, we have $\lim_{t\to\infty} I(t) = 0$. That is, $E_1$ attracts all solution orbits in $\Omega$ except for $S = 0$. However, it is clear that $E_1$ is unstable if $R_0 > 1$, because the linearized equation (5.74) has an exponentially growing solution. If $R_0 > 1$, the Jacobian matrix at the endemic steady state $E_2$ is given by

$$\begin{pmatrix} -\mu R_0 & -(\mu + \gamma) \\ \mu(R_0 - 1) & 0 \end{pmatrix},$$

and its eigenvalues are calculated as

$$\frac{1}{2}\left[-\mu R_0 \pm \sqrt{(\mu R_0)^2 - 4\mu(\mu + \gamma)(R_0 - 1)}\right].$$

Therefore, if $R_0 > 1$, the real parts of the eigenvalues are all negative, and the endemic steady state is locally asymptotically stable. For each solution, define a function $V$ as

$$V(S, I) := S^*\left[\frac{S}{S^*} - 1 - \log\frac{S}{S^*}\right] + I^*\left[\frac{I}{I^*} - 1 - \log\frac{I}{I^*}\right],$$

where $(S^*, I^*)$ denotes $E_2$. It is easily verified that $V(S, I)$ is positive for $(S, I) \in \Omega$ and attains its minimum value of zero at $E_2$. Differentiating $V$ along the solution orbit, we have

$$\frac{dV}{dt} = \frac{\partial V}{\partial S}\frac{dS}{dt} + \frac{\partial V}{\partial I}\frac{dI}{dt} = -\frac{\mu R_0}{S}(S - S^*)^2 < 0.$$

That is, $V$ is monotone strictly decreasing along the solution orbit, which means that there is no periodic solution. From the well-known Poincaré–Bendixon theorem, we can conclude that $E_2$ attracts all solution orbits in $\Omega$ except for the boundary solutions $S = 0$ and $I = 0$. The same conclusion can be derived from the *La Salle Invariance Principle* [2, Proposition (18.3)] [109, Proposition 4.6], because the maximal invariant set for the Lyapunov functional $V$ such that $\dot{V} = 0$ is given by a singleton $\{(S^*, I^*)\}$. □

The above endemic threshold result gives a typical example in which the disease invasion condition (instability of the DFSS) also implies the existence and stability of the endemic steady state, which corresponds to the principle of exchange of stability (see Chap. 3).

For common infectious child diseases such as measles, chicken pox, and rubella, it is known that recurrent outbreaks occur every few years [9]. These are not damping oscillations, but sustained oscillations. However, the endemic SIR model (5.72) does not have a periodic solution, so we have to consider another factor to produce recurrent outbreaks. For example, it is known that seasonal variations in infection rates [71, 82, 94, 95, 110] or nonlinear dependence of the force of infection on the infected population size [74] can produce periodic solutions. The host demography is also important. In fact, if newborns are produced by susceptibles as $b = mS$, where $m > 0$ is the birth rate, the basic system (5.72) becomes the Lotka–Volterra system, which has periodic solutions. Moreover, in the real infection process, the heterogeneity of the host population, particularly its age structure or immunity structure (by infection experience), plays an important role, and we can expect structured population models to produce more complex behavior in their solutions. Indeed, it is known that the host age structure [5] and the *cross-immunity* (*drift* effect, [76]) can lead to sustained oscillations. It is also practically important to introduce new epidemic states. An isolation or quarantine period could also contribute to the occurrence of periodic solutions [33, 34].

Finally, we note the more elementary problem of whether deterministic modeling is appropriate. Even if the SIR model with $R_0 > 1$ describes the endemic infectious disease, there exists a possibility that the disease will be eradicated because of demographic or environmental stochasticity when the prevalence becomes very small in the diminution phase of the epidemic [41]. In such situations, the deterministic approach is no longer effective, and we would need to adopt stochastic models. Indeed, the stochastic SIR model has been shown to explain recurrent outbreaks [8, 10].

*Remark 5.4* If we introduce the infection-age in (5.72), we obtain a structured SIR model [75] as

$$\begin{aligned}
\frac{dS(t)}{dt} &= \mu N - \mu S(t) - \lambda(t)S(t),\\
\frac{\partial i(t,\tau)}{\partial t} + \frac{\partial i(t,\tau)}{\partial t} &= -(\mu+\gamma(\tau))i(t,\tau),\\
i(t,0) &= \lambda(t)S(t),\\
\frac{dR(t)}{dt} &= -\mu R(t) + \int_0^\infty \gamma(\tau)i(t,\tau)d\tau,\\
\lambda(t) &= \int_0^\infty \beta(\tau)i(t,\tau)d\tau.
\end{aligned}$$

Integrating the McKendrick equation along the characteristic line, we obtain

$$i(t,\tau) = \begin{cases} e^{-\mu\tau}\Gamma(\tau)\lambda(t-\tau)S(t-\tau), & t-\tau>0,\\ e^{-\mu t}\frac{\Gamma(\tau)}{\Gamma(\tau-t)}i(0,\tau-t), & \tau-t>0,\end{cases}$$

where $\Gamma(\tau) := \exp(-\int_0^\tau \gamma(\sigma)d\sigma)$ and

$$N = S(t) + \int_0^\infty i(t,\tau)d\tau + R(t),$$

is assumed to be constant. Substituting this expression into the definition of $\lambda$, we have

$$\lambda(t) = g_1(t) + \int_0^t \beta(\tau)e^{-\mu\tau}\Gamma(\tau)\lambda(t-\tau)S(t-\tau)d\tau,$$

where

$$g_1(t) := \int_t^\infty \beta(\tau)e^{-\mu t}\frac{\Gamma(\tau)}{\Gamma(\tau-t)}i(0,\tau-t)d\tau.$$

On the other hand, applying the variation-of-constants formula, we obtain

$$S(t) = g_2(t) - \int_0^t e^{-\mu\tau}\lambda(t-\tau)S(t-\tau)d\tau,$$

where $g_2(t) := S(0)e^{-\mu t} + N(1-e^{-\mu t})$. Define a vector $\xi(t) := (S(t),\lambda(t))^{\mathrm{T}}$ and $g(t) := (g_1(t), g_2(t))^{\mathrm{T}}$. Let $f : \mathbb{R}^2 \to \mathbb{R}^2$ be a nonlinear function defined by $f(x) = (x_1x_2, x_1x_2)^{\mathrm{T}}$ for $x = (x_1, x_2) \in \mathbb{R}^2$. And let $\Psi$ be a matrix defined by

$$\Psi(\tau) = \begin{pmatrix} -e^{-\mu\tau} & 0\\ 0 & e^{-\mu\tau}\beta(\tau)\Gamma(\tau)\end{pmatrix}.$$

Then, we obtain a nonlinear renewal equation for $\xi$:

$$\xi(t) = g(t) + \int_0^t \Psi(\tau) f(\xi(t-\tau))d\tau.$$

This type of the nonlinear renewal equation has been well studied by many authors. Readers are referred to [81]. We can again confirm a fact that an epidemic model can be formulated by a renewal integral equation.

**Exercise 5.9** From the above argument, derive a scalar integral equation for $\lambda(t)$.

**Exercise 5.10** If the total population size is varying, the basic system (5.72) can be replaced by the homogeneous system:

$$\begin{aligned}
\frac{dS(t)}{dt} &= bN(t) - \mu S(t) - \beta \frac{S(t)I(t)}{N(t)}, \\
\frac{dI(t)}{dt} &= \frac{\beta S(t)I(t)}{N(t)} - (\mu + \gamma)I(t), \\
\frac{dR(t)}{dt} &= -\mu R(t) + \gamma I(t).
\end{aligned}$$

Setting $u = S/N, v = I/N, w = R/N$ and derive the normalized system for $(u, v, w)$. What is your conclusion? Next, consider the case that there exists a disease-induced death rate.

**Exercise 5.11** Consider an epidemic model for a fatal disease with demography given by

$$\begin{aligned}
\frac{dS(t)}{dt} &= b(S(t) + I(t)) - \mu S(t) - \frac{\beta S(t)I(t)}{S(t) + I(t)}, \\
\frac{dI(t)}{dt} &= -(\mu + \delta)I(t) + \frac{\beta S(t)I(t)}{S(t) + I(t)},
\end{aligned}$$

where $\beta > 0$ is the transmission coefficient, $b > 0$ is the birth rate, $\mu > 0$ is the death rate, and $\delta > 0$ is the extra death rate caused by the disease. Suppose that $S(0) > 0$, $I(0) > 0$, and $\beta > b + \delta$, and define the incidence $y(t)$ as

$$y(t) := \frac{I(t)}{S(t) + I(t)}.$$

(1) Show that $y(t)$ satisfies

$$\frac{dy}{dt} = (\beta - b - \delta)y - (\beta - \delta)y^2,$$

and solve this differential equation.

(2) Calculate $I(t)$ and prove that

$$\lim_{t\to\infty} e^{-rt} I(t) = I(0) \left( \frac{\beta - b - \delta}{S(0)(\beta - \delta)} \right)^{\frac{\beta}{\beta-\delta}},$$

where

$$r = (\mu + \delta) \left( \frac{R_0 b}{\beta - \delta} - 1 \right)$$

and $R_0 = \frac{\beta}{\mu+\delta}$ is the basic reproduction number of this disease. Hence, $R_0 > 1$ is not sufficient for the disease to be endemic.

**Exercise 5.12** As in Sect. 5.4, let us introduce a continuous variable describing individual heterogeneity. For example, $S(t, \zeta)$ denotes the density of susceptibles with heterogeneity (trait) variable $\zeta \in \Omega \subset \mathbb{R}^k$, etc. Then, the basic system is given as

$$\begin{aligned}
\frac{\partial S(t,\zeta)}{\partial t} &= b(\zeta) - \mu(\zeta)S(t,\zeta) - \lambda(t,\zeta)S(t,\zeta), \\
\frac{\partial I(t,\zeta)}{\partial t} &= \lambda(t,\zeta)S(t,\zeta) - (\mu(\zeta) + \gamma(\zeta))I(t,\zeta), \\
\frac{\partial R(t,\zeta)}{\partial t} &= \gamma(\zeta)I(t,\zeta) - \mu(\zeta)R(t,\zeta),
\end{aligned}$$

where the force of infection $\lambda$ is given by

$$\lambda(t,\zeta) = \int_\Omega \sigma(\zeta,\eta) I(t,\eta) d\eta,$$

with the transmission coefficient $\sigma \in L^\infty_+(\Omega \times \Omega)$. Suppose that the total density is time-independent and given by $N(\zeta) = b(\zeta)/\mu(\zeta)$. Linearize the basic system at the disease-free steady state and derive the renewal equation for the incidence $B(t,\zeta) := \lambda(t,\zeta)N(\zeta)$. Show that the next-generation operator $K$ on $L^1(\Omega)$ is given by

$$(K\phi)(x) = N(x) \int_\Omega \frac{\sigma(x,z)}{\mu(z)+\gamma(z)} \phi(z) dz, \quad \phi \in L^1(\Omega).$$

Next, suppose that the separable mixing assumption holds, that is, there are two functions $\sigma_1$ and $\sigma_2$ such that $\sigma(\zeta,\eta) = \sigma_1(\zeta)\sigma_2(\eta)$. Show that the basic reproduction number (the positive eigenvalue of $K$) is given by

$$R_0 = \int_\Omega \frac{\sigma(z,z)N(z)}{\mu(z)+\gamma(z)} dz.$$

Finally, examine the endemic threshold and stability results.[4]

[4] Kuniya and Wang [72] studied the above model with spatial diffusion terms.

### 5.5.3 Vaccination and Reinfection Model

Using the endemic SIR model (5.72), let us calculate the *vaccination* proportion needed to eradicate the SIR disease. If we can neglect the possible immunological difference between the recovered population and the vaccinated population, the SIR model (5.72) can be extended to take into account the vaccination of newborns as follows:

$$\begin{aligned}\frac{dS(t)}{dt} &= (1-e)b - \mu S(t) - \beta S(t)I(t),\\ \frac{dI(t)}{dt} &= \beta S(t)I(t) - (\mu+\gamma)I(t),\\ \frac{dR(t)}{dt} &= eb - \mu R(t) + \gamma I(t),\end{aligned} \tag{5.76}$$

where $e$ denotes the proportion of infants who are vaccinated. The DFSS is then given by

$$(S^*, I^*, R^*) = \left(\frac{(1-e)b}{\mu}, 0, \frac{eb}{\mu}\right).$$

Therefore, the effective reproduction number [45] for the partially immunized population can be defined as

$$R_e = \frac{\beta}{\mu+\gamma}\frac{(1-e)b}{\mu} = R_0(1-e).$$

(Some authors refer to $R_e$ as the *control reproduction number*.) It is easy to see that the DFSS is globally asymptotically stable if $R_e < 1$. This eradication criterion can be formulated as the following well-known *control relation*:

$$1 - \frac{1}{R_0} < e, \tag{5.77}$$

where $1 - 1/R_0$ gives the *critical proportion of immunization* for newborns. For example, if $R_0 = 10$, over 90 percent of infants must be immunized to eradicate the disease if the vaccination efficacy is 100 percent. In realistic situations, we have to calculate the critical vaccination proportion for a given age class, so we need an age-structured SIR model (see Chap. 9).

If recovery from the infected state does not necessarily lead to complete (permanent) immunity, or the immunity reduces with time, the recovered individuals can be reinfected. The SIR model can then be modified as the *reinfection model* [38, 39, 92]:

$$\begin{aligned}\frac{dS(t)}{dt} &= (1-e)b - \mu S(t) - \beta S(t)I(t),\\ \frac{dI(t)}{dt} &= \beta(S(t)+\sigma R(t))I(t) - (\mu+\gamma)I(t),\\ \frac{dR(t)}{dt} &= eb - \mu R(t) - \sigma\beta R(t)I(t) + \gamma I(t),\end{aligned} \tag{5.78}$$

where $\sigma\beta$ is the reinfection rate of the recovered population, $\sigma \geq 0$ denotes the relative susceptibility of the recovered individuals $R$ to the never infected individuals $S$, and $e$ is the vaccination coverage of newborns. If $\sigma = 0$, (5.78) becomes an SIR model, whereas if $\sigma = 1$, $S$ and $R$ are the same, so we have an SIS model. The origin of the reinfection model can be traced back to the Kermack–McKendrick reinfection model, which is investigated in Chap. 8.

Note that the basic reproduction number of the reinfection model (5.78) with $e = 0$ is the same as $R_0$ of the original SIR model (5.72). However, (5.78) has a DFSS

$$(S, I, R) = \left(\frac{(1-e)b}{\mu}, 0, \frac{be}{\mu}\right),$$

so the effective reproduction number depending on the vaccination proportion $e$ is given by

$$R_e := \frac{\beta b}{\mu(\mu+\gamma)}(1-(1-\sigma)e).$$

If we assume $\sigma < 1$, $R_e$ is monotone decreasing with respect to $e$. Then, $R_e > 1$ for all $e \in [0, 1]$ if $R_1 = \sigma R_0 > 1$. That is, if $R_0 > \sigma^{-1}$, the disease cannot be controlled by the mass vaccination of infants. In fact, if we assume complete vaccination coverage by setting $e = 1$, the susceptible population disappears as time evolves, and system (5.78) reduces to the limiting IR system:

$$\begin{aligned}\frac{dI(t)}{dt} &= \beta\sigma R(t)I(t) - (\mu+\gamma)I(t),\\ \frac{dR(t)}{dt} &= b - \mu R(t) - \sigma\beta R(t)I(t) + \gamma I(t),\end{aligned} \tag{5.79}$$

where the $R$-class has the partial susceptibility to which the reduced infection rate $\sigma\beta$ is applied. The effective reproduction number of the limiting system (5.79) is then $R_1 = \sigma R_0$, and the prevalence at the endemic steady state is given by

$$\frac{I^*}{N} = 1 - \frac{1}{\sigma R_0}, \tag{5.80}$$

where $N = b/\mu$ is the total size of the host stationary population.

As seen in (5.75), for the SIR model with permanent immunity, the prevalence cannot become larger than $\mu/(\mu+\gamma)$, no matter how large is $R_0$, and the upper limit is normally a small number, because $\gamma$ is much larger than $\mu$. Conversely, (5.80) indicates that if reinfection is possible and $R_1 = \sigma R_0 > 1$, the disease cannot be controlled by vaccination, and the prevalence goes to unity if $R_0 \to \infty$, even under complete vaccination coverage.

Because the qualitative change in the epidemiological implication occurs for the prevalence and controllability at $R_0 = 1/\sigma$, Gomes et al. [38, 39] called $\sigma^{-1}$ the *reinfection threshold* of $R_0$. As seen above, the reinfection threshold value of $R_0$ corresponds to the fact that $\sigma R_0$ is the effective reproduction number of the limiting system (5.79), that is, $R_0 = 1/\sigma$ does not imply a bifurcation point of the basic system (5.78), but implies the threshold value of the limiting system (5.79) [15, 39].

In the reinfection model (5.78), let us now consider the case of no vaccination (so $e = 0$). We can assume that without loss of generality, the total size of the host population is constant; $S + I + R = N = b/\mu$. Let $I^*$ be the infected population density of the endemic steady state. It is easy to obtain the following characteristic equation for $I^*$:

$$F(I^*) := \frac{\beta(S^* + \sigma R^*)}{\mu+\gamma} = \frac{\beta}{\mu+\gamma}\left[\frac{b}{\mu+\beta I^*} + \frac{\sigma\gamma I^*}{\mu+\sigma\beta I^*}\right] = 1. \tag{5.81}$$

It is easy to see that $F(0) = R_0$ and

$$F(N) < \frac{\beta}{\mu+\gamma}\left[\frac{b}{\beta N} + \frac{\gamma}{\beta}\right] = 1.$$

Therefore, $F(I^*) = 1$ has at least one positive root in $(0, N)$, so there exists one endemic steady state if $R_0 > 1$.

**Exercise 5.13** Show the following statements:

1. If $R_0 = 1$ and $\sigma > 1 + \mu/\gamma$, there exists at least one endemic steady state.
2. If $R_0 \le 1$ and $\sigma \le 1$, it holds that $\lim_{t\to\infty} I(t) = 0$.

Then, we can show that the infected population $I$ is *uniformly weakly persistent*:

**Proposition 5.9** *If $R_0 > 1$, there exists a number $\varepsilon > 0$ such that for any nonnegative solution with $I(0) > 0$, it holds that* $\limsup_{t\to\infty} I(t) > \varepsilon$.

*Proof* Suppose that for any $\varepsilon > 0$, there exists a solution $I(t)$ with $I(0) > 0$ such that $\limsup_{t\to\infty} I(t) < \varepsilon$. Then, there exists some $t_0 > 0$ such that $I(t) \le \varepsilon$ for all $t \ge t_0$. From $R' \le -\mu R + \gamma I$, we have for $t \ge t_0$,

$$R(t) \le R(t_0)e^{-\mu(t-t_0)} + \frac{\gamma\varepsilon}{\mu}(1 - e^{-\mu(t-t_0)}).$$

If we choose a large $t_1 > t_0$, we can obtain $R(t) < \frac{2\gamma\varepsilon}{\mu}$ for $t > t_1$. Observe that for $t > t_1$,

$$S = N - I - R \geq N - \varepsilon - \frac{2\gamma\varepsilon}{\mu}.$$

and

$$\begin{aligned}\frac{I'}{I} &= -(\mu+\gamma) + \beta(S + \sigma R) \geq -(\mu+\gamma) + \beta\left(N - \varepsilon - \frac{2\gamma\varepsilon}{\mu}\right) \\ &= (\mu+\gamma)\left(-1 + R_0\left(1 - \frac{\varepsilon}{N} - \frac{2\gamma\varepsilon}{N\mu}\right)\right)\end{aligned}.$$

Therefore, if we choose $\varepsilon$ in advance such that

$$\varepsilon < \frac{N(R_0 - 1)}{R_0(1 + \frac{2\gamma}{\mu})},$$

it follows that $I'/I > 0$ for $t > t_1$ and $I \to \infty$, which is a contradiction. □

Next, define a function $F(\varepsilon, x)$ as follows:

$$F(\varepsilon, x) := \frac{\varepsilon\beta}{\mu+\gamma}\left[\frac{b}{\mu + \varepsilon\beta x} + \frac{\sigma\gamma x}{\mu + \sigma\varepsilon\beta x}\right] - 1,$$

where $\varepsilon$ is a bifurcation parameter and other parameters are chosen as

$$R_0 = \frac{\beta b}{\mu(\mu+\gamma)} = 1,$$

so $F(1, 0) = 0$. Then, $\varepsilon$ represents the effective reproduction number of the parametrized system with infection rate $\varepsilon\beta$, and a positive solution $x$ of $F(\varepsilon, x) = 0$ gives the size of the infected population at the endemic steady state. It follows that

$$F_x(1, 0) = \frac{\beta\gamma}{\mu(\mu+\gamma)}\left(\sigma - 1 - \frac{\mu}{\gamma}\right),$$

where we have used the relation $\frac{\beta b}{\mu(\mu+\gamma)} = 1$. Then, if $\sigma \neq 1 + (\mu/\gamma)$, it follows from the implicit function theorem that $F(\varepsilon, x) = 0$ can be locally solved as $x = x(\varepsilon)$ with $x(1) = 0$. Moreover, it follows that

$$x'(1) = -\frac{F_\varepsilon(1, 0)}{F_x(1, 0)}.$$

Because $F_\varepsilon(1, 0) = 1 > 0$, we know that if $\sigma > 1 + (\mu/\gamma)$, then $x'(1) < 0$, so there exists a small number $\eta > 0$ such that the positive solution $x(\varepsilon) > 0$ exists for $\varepsilon \in (1 - \eta, 1)$, which implies that endemic steady states are backwardly bifurcated at $\varepsilon = 1$. However, if $\sigma < 1 + (\mu/\gamma)$, then $x'(1) > 0$, and so there exists a small number $\eta > 0$ such that the positive solution $x(\varepsilon) > 0$ exists for $\varepsilon \in (1, 1 + \eta)$, and there is

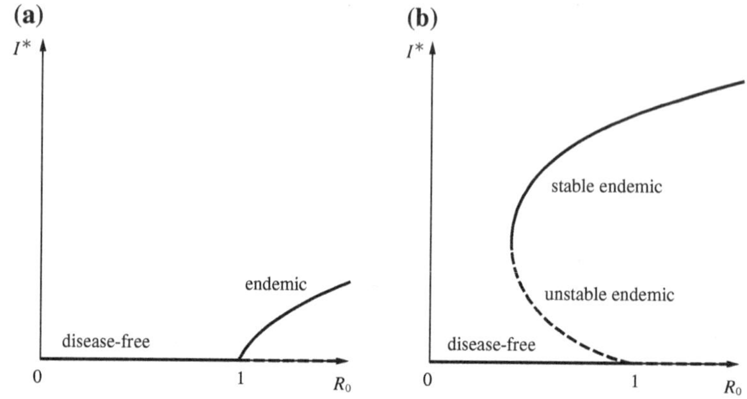


**Fig. 5.4** Schematic representation of forward (**a**) and backward (**b**) bifurcation diagrams for endemic steady states

no positive solution for $\varepsilon \in (1-\eta, 1)$, which means that endemic steady states are forwardly bifurcated at $\varepsilon = 1$ (see Fig. 5.4).

We can conclude that if susceptibility is enhanced through infection experience, a subcritical endemic steady state could exist. In such a *backward bifurcation case*, reducing $R_0$ below unity is sufficient to prevent the initial invasion, but is not sufficient to control the endemic disease. Although the enhancement of susceptibility through infection experience may be an unacceptable assumption in many cases, it may not necessarily be unrealistic if the secondary infection is asymptomatic and does not lead to prevention behavior (see Sect. 8.3). Readers are referred to [103] for mathematical conditions of the existence of the subthreshold equilibria in epidemic models.

**Exercise 5.14** If we neglect demographic factors, that is, $b = \mu = 0$, the concept of a "reinfection threshold" is clarified, because it gives a sharp threshold result without vaccination [62, 101]. Consider the basic system as

$$
\begin{aligned}
\frac{dS(t)}{dt} &= -\beta S(t)I(t),\\
\frac{dI(t)}{dt} &= \beta(S(t)+\sigma R(t))I(t) - \gamma I(t),\\
\frac{dR(t)}{dt} &= -\sigma\beta R(t)I(t) + \gamma I(t).
\end{aligned}
$$

Let $\Omega$ be a set of $\mathbb{R}^3$ such that $\Omega = \{(S, I, R) \in \mathbb{R}^3_+ : S + I + R = 1\}$, and let $\Omega_0$ be a subset of $\Omega$ such that $\Omega_0 = \{(S, I, R) \in \mathbb{R}^3_+ : S = 0, I + R = 1\}$. Prove the following statements[5]:

---

[5]The Pease model studied in Chap. 8 can be seen as an age-dependent extension of this reinfection model on the subset $\Omega_0$.

(1) $\Omega$ and $\Omega_0$ are positively invariant with respect to the solution flow.

(2) Let $R_0 = \beta/\gamma$. There exists a unique endemic steady state if and only if $\sigma R_0 > 1$. Confirm that the endemic threshold $\sigma R_0 = 1$ is also the invasion threshold in the disease-free (partially immunized) steady state $(S, I, R) = (0, 0, 1) \in \Omega_0$ and that this is different from the invasion threshold $R_0 = 1$ in the DFSS with full susceptibility $(1, 0, 0) \in \Omega \setminus \Omega_0$.

(3) Suppose that $S(0) \neq 0$. There is a constant of motion

$$V(S, R) = \frac{1}{S^\sigma}\left(R - \frac{1}{\sigma R_0}\right).$$

(Hint: Use the equation $\frac{dS}{dR} = \frac{-\beta S}{\gamma - \sigma\beta R}$.)

(4) Suppose that $I(0) \neq 0$. It follows that

$$\lim_{t\to\infty} I(t) = \begin{cases} 1 - \frac{1}{\sigma R_0}, & (\sigma R_0 > 1) \\ 0, & (\sigma R_0 \le 1) \end{cases}.$$

**Exercise 5.15** Let us consider another mechanism to create a backward bifurcation. An epidemiological model of hepatitis C with a chronic stage and variable population size is formulated as follows [77]:

$$\begin{aligned}
\frac{dS(t)}{dt} &= bN - \left(\mu + \gamma\frac{I(t)}{N(t)} + \lambda(t)\right)S(t) + \int_0^\infty \alpha(\tau)v(t, \tau)d\tau, \\
\frac{dI(t)}{dt} &= \left(\gamma\frac{I(t)}{N(t)} + \lambda(t)\right)S(t) - (\mu + k)I(t), \\
&\left(\frac{\partial}{\partial t} + \frac{\partial}{\partial \tau}\right)v(t, \tau) = -(\nu + \alpha(\tau))v(t, \tau), \\
v(t, 0) &= kI(t), \\
\lambda(t) &= \frac{1}{N(t)}\int_0^\infty \delta(\tau)v(t, \tau)d\tau,
\end{aligned}$$

where $S$ is susceptibles, $I$ is the infecteds with acute hepatitis C and $v(t, \tau)$ is the infecteds with chronic hepatitis C with class age $\tau$, and $N$ is the total population size given by $N(t) = S(t) + I(t) + \int_0^\infty v(t, \tau)d\tau$. And $b$ is the birth rate, $\mu$ is the natural death rate, $\alpha(\tau)$ is the recovery/treatment rate for the chronic state, $k$ is the rate of progression to chronic stage, $\gamma$ and $\delta(\tau)$ are the transmission coefficients, and $\nu$ is the death rate in the chronic stage.

1. First, we consider the case that $b = \mu = \nu$ and assume that the total population is in a demographic steady state. Let $b(t) := \gamma I(t) + \int_0^\infty \delta(\tau)v(t, \tau)d\tau$. Then, $b(t)$ denotes the density of newly infecteds in the disease-free steady state, and the

linearized equation is given by $\frac{dI(t)}{dt} = b(t) - (\mu + k)I(t)$. Derive the renewal equation for $b(t)$ starting from $t = -\infty$ and show that the basic reproduction number $R_0$ is given by

$$R_0 = \frac{\gamma}{\mu + k} + \frac{k}{\mu + k} \int_0^\infty \delta(\tau) e^{-\mu\tau - \int_0^\tau \alpha(z)dz} d\tau.$$

Next, discuss the endemic threshold results and stability of steady states.

2. For the general case ($b \neq \mu$, $\nu > \mu$), introduce the normalized densities as $s = S/N$, $i = I/N$, and $u = \nu/N$. Derive the normalized system.
3. For the normalized system, define a threshold parameter $R$ by

$$R = \frac{\gamma}{b + k} + \frac{k}{b + k} \int_0^\infty \delta(\tau) e^{-(b+\nu-\mu)\tau - \int_0^\tau \alpha(z)dz} d\tau.$$

Show that there exists a unique endemic steady state for the normalized system (i.e., endemic persistent solution for the homogeneous system) if and only if $R > 1$ provided that $\mu = \nu$ [77]. Next, derive a condition to produce a backward bifurcation at $R = 1$ when $\nu > \mu$ [60].

**Exercise 5.16** Suppose that recovered individuals are protected from infection, but some recovered individuals can lose their immunity completely and return to the full susceptible class. Then, we can formulate the following SIRS model:

$$\begin{aligned}
\frac{dS(t)}{dt} &= (1 - e)b - \mu S(t) - \beta S(t)I(t) + \int_0^\infty \theta(\tau) r(t, \tau) d\tau \\
\frac{dI(t)}{dt} &= \beta S(t)I(t) - (\mu + \gamma)I(t), \\
\left(\frac{\partial}{\partial t} + \frac{\partial}{\partial \tau}\right) & r(t, \tau) = -(\mu + \theta(\tau)) r(t, \tau), \\
r(t, 0) &= eb + \gamma I(t),
\end{aligned}$$

where $e$ is the vaccination coverage for newborns and $\theta(\tau)$ is the recovery-age-dependent reversion rate. Show that the critical proportion of immunization is given by

$$e^* = \frac{1}{k}\left(1 - \frac{1}{R_0}\right),$$

where

$$R_0 = \frac{\beta b}{\mu(\mu + \gamma)}, \quad k = \int_0^\infty \mu e^{-\mu\tau - \int_0^\tau \theta(\zeta)d\zeta} d\tau,$$

and give the biological interpretation for $k$ and explain that the disease cannot be controlled by the vaccination if $k < (1 - 1/R_0)$.

## 5.6 Vector-Transmitted Diseases

In this section, we present another epidemic mechanism leading to the backward bifurcation of endemic steady states. We consider a simple epidemic model for *vector-transmitted disease* [51]. The basic model considered here can be seen as a simplified version of those for Chagas disease developed by Velasco-Hernández [104, 105]. Chagas disease (or *American trypanosomiasis*) is caused by the protozoan parasite *Trypanosoma cruzi* and transmitted by blood-feeding triatomine bugs. It is a chronic, frequently fatal infection that is common in Latin America. Although Chagas disease can also be transmitted by blood transfusion, organ transplants, and the placenta (vertical transmission from mother to babies), we consider only vector transmission and neglect the variable infectivity. Readers are referred to the work of Inaba and Sekine [52] for an infection-age-dependent model.

### 5.6.1 *Basic Model and Invasion Threshold*

We divide the host population into two groups: $S(t)$ denotes the density of the uninfected, but susceptible, host population and $I(t)$ denotes the density of infected hosts, where $t$ denotes time. The infected population is removed by the extra death rate due to disease $\gamma > 0$. The total size of the host population is given by $T(t) := S(t) + I(t)$. Let $M(t)$ denote the density of susceptible vectors at time $t$, $V(t)$ be the density of infected vectors at time $t$, and $U(t) := M(t) + V(t)$ be the total density of vectors. Let $b_1$ $[b_2]$ be the birth rate of the host population [vectors] and $\mu_1$ $[\mu_2]$ be the natural death rate of the host population [vectors]. For the vector population, we assume that there is no extra death rate due to infection.

To model a vector-transmitted disease, the most important ingredient is the formula of the force of transmission. Here, we adopt a simple assumption that is very common in traditional models of vector-transmitted diseases such as malaria. Readers are referred to Esteva and Matias [32] for another type of assumption regarding the force of infection.

Let $a$ be the number of bites per vector per unit time and $c_1$ be the proportion of infected bites that give rise to infection from an infected vector to a susceptible host. For simplicity, we assume that $a$ is a constant. The vector $V$ makes $ac_1V$ infectious bites, of which a fraction $S/T$ are on susceptible hosts. That is, the number of newly infected hosts per unit time by vector transmission is given by $ac_1V(S/T)$.

Hence, the force of infection (the probability per unit time of a susceptible becoming infected) for the host population, denoted by $\lambda_1(t)$, is given by

$$\lambda_1(t) = \alpha\frac{V(t)}{T(t)}, \tag{5.82}$$

where $\alpha := ac_1$. Using a similar argument, the force of infection for the vector population, denoted by $\lambda_2(t)$, can be written as

$$\lambda_2(t) = \beta \frac{I(t)}{T(t)}, \tag{5.83}$$

where $\beta := ac_2$ and $c_2$ denotes the proportion of bites that give rise to infection from an infected host to a susceptible vector.

Under the above assumption, our basic model for a vector-transmitted disease can be formulated as follows:

$$\begin{aligned}
\frac{dS(t)}{dt} &= b_1 - (\mu_1 + \lambda_1(t))S(t), \\
\frac{dI(t)}{dt} &= \lambda_1(t)S(t) - (\mu_1 + \gamma)I(t), \\
\frac{dM(t)}{dt} &= b_2 - (\mu_2 + \lambda_2(t))M(t), \\
\frac{dV(t)}{dt} &= \lambda_2(t)M(t) - \mu_2 V(t),
\end{aligned} \tag{5.84}$$

If we add two equations for the vector population, we obtain a single equation for the total vector population $U = M + V$ as $dU(t)/dt = b_2 - \mu_2 U(t)$. It is easily seen that $U^* := b_2/\mu_2$ is a globally stable steady state. In the following, we assume in advance that $M(t) + V(t) = b_2/\mu_2$ for all $t \geq 0$. Then, it is easy to see that the solution of (5.84) starting from the region

$$\Omega := \{(S, I, M, V) \in \mathbb{R}^4_+ : 0 \leq S + I \leq b_1/\mu_1, M + V = b_2/\mu_2\}$$

remains in $\Omega$ for all $t > 0$, that is, $\Omega$ is positively invariant with respect to the flow generated by the basic system (5.84).

System (5.84) has the DFSS

$$(S^*, I^*, M^*, V^*) = \left(\frac{b_1}{\mu_1}, 0, \frac{b_2}{\mu_2}, 0\right).$$

The dynamics of the initial invasion phase are described by the linearized system at the DFSS, which can be written as

$$\begin{pmatrix} \dot{I} \\ \dot{V} \end{pmatrix} = \begin{pmatrix} -(\mu_1 + \gamma) & \alpha \\ \frac{\beta b_2 \mu_1}{b_1 \mu_2} & -\mu_2 \end{pmatrix} \begin{pmatrix} I \\ V \end{pmatrix}, \tag{5.85}$$

from which we have a system of renewal equations

$$\begin{aligned}
I(t) &= \int_0^\infty e^{-(\mu_1+\gamma)s} \alpha V(t-s) ds, \\
V(t) &= \frac{\beta b_2 \mu_1}{b_1 \mu_2} \int_0^\infty e^{-\mu_2 s} I(t-s) ds,
\end{aligned}$$

where we set the initial time $t = -\infty$. Therefore, we obtain a limiting renewal equation for the newly infected host population density as

$$B(t) := \alpha V(t) = \int_0^\infty \Psi(s)B(t-s)ds, \tag{5.86}$$

where

$$\Psi(s) := \frac{\alpha\beta b_2\mu_1}{b_1\mu_2}\int_0^s e^{-\mu_2 x}e^{-(\mu_1+\gamma)(s-x)}dx.$$

The basic reproduction number for the host population is then calculated as

$$R_0 = \int_0^\infty \Psi(s)ds = \frac{\alpha\beta b_2\mu_1}{b_1\mu_2^2(\mu_1+\gamma)}, \tag{5.87}$$

which is the square of the spectral radius of the next-generation matrix induced from the coefficient matrix of (5.85).

**Exercise 5.17** Let

$$A := \begin{pmatrix} -(\mu_1+\gamma) & \alpha \\ \frac{\beta b_2\mu_1}{b_1\mu_2} & -\mu_2 \end{pmatrix} = Q + M,$$

where

$$Q := \begin{pmatrix} -(\mu_1+\gamma) & 0 \\ 0 & -\mu_2 \end{pmatrix}, \quad M := \begin{pmatrix} 0 & \alpha \\ \frac{\beta b_2\mu_1}{b_1\mu_2} & 0 \end{pmatrix}.$$

Then, the next-generation matrix is given by $K := M(-Q)^{-1}$. Calculate its spectral radius.

Using the basic reproduction number, we can formulate the threshold condition for this vector-transmitted disease as follows:

**Proposition 5.10** *If $R_0 < 1$, then the DFSS is locally asymptotically stable, and if $R_0 > 1$, it is unstable. Moreover, if*

$$R_0 < \frac{\mu_1}{\mu_1+\gamma}, \tag{5.88}$$

*then the DFSS is globally asymptotically stable.*

*Proof* The eigenvalues of the coefficient matrix of (5.85) are the solutions of the quadratic equation

$$\lambda^2 + (\mu_1+\mu_2+\gamma)\lambda + (\mu_1+\gamma)\mu_2(1-R_0) = 0.$$

Then, if $R_0 < 1$, all eigenvalues have negative real parts, whereas if $R_0 > 1$, one of the eigenvalues is positive. Therefore, the first half of this statement is obvious. Let us prove the latter half. It follows from (5.84) that

$$\frac{dT(t)}{dt} = b_1 - \mu_1 T(t) - \gamma I(t) \geq b_1 - (\mu_1 + \gamma)T(t).$$

Then, we obtain

$$T(t) \geq \frac{b_1}{\mu_1 + \gamma} + \left(T(0) - \frac{b_1}{\mu_1 + \gamma}\right) e^{-(\mu_1+\gamma)t}.$$

Hence, if $T(0) \geq b_1/(\mu_1 + \gamma)$, it follows that $T(t) \geq b_1/(\mu_1 + \gamma)$ for all $t \geq 0$. Even if $T(0) < b_1/(\mu_1 + \gamma)$, the condition $T(t) \geq b_1/(\mu_1 + \gamma)$ will be satisfied as time goes to infinity, so without loss of generality, we can assume that $T(0) \geq b_1/(\mu_1 + \gamma)$ holds in advance. Hence, we can assume that $1/T(t) \leq (\mu_1 + \gamma)/b_1$. Next, let us define functions $B(t)$ and $C(t)$ as

$$B(t) := \lambda_1(t)S(t), \quad C(t) =: \lambda_2(t)M(t),$$

that is, $B(t)$ denotes the number of newly infected hosts at time $t$ and $C(t)$ is the number of newly infected vectors at time $t$. From (5.84), we have the following expressions:

$$\begin{aligned}
S(t) &= \frac{b_1}{\mu_1} + (S(0) - \frac{b_1}{\mu_1})e^{-\mu_1 t} - \int_0^t e^{-\mu_1(t-s)} B(s)ds, \\
I(t) &= I(0)e^{-(\mu_1+\gamma)t} + \int_0^t e^{-(\mu_1+\gamma)(t-s)} B(s)ds, \\
V(t) &= V(0)e^{-\mu_2 t} + \int_0^t e^{-\mu_2(t-s)} C(s)ds.
\end{aligned} \tag{5.89}$$

Thus, we obtain a system of Volterra integral equations:

$$\begin{aligned}
B(t) &= \alpha \frac{S(t)}{T(t)} \left( V(0)e^{-\mu_2 t} + \int_0^t e^{-\mu_2(t-s)} C(s)ds \right), \\
C(t) &= \beta \frac{M(t)}{T(t)} \left( I(0)e^{-(\mu_1+\gamma)t} + \int_0^t e^{-(\mu_1+\gamma)(t-s)} B(s)ds \right).
\end{aligned} \tag{5.90}$$

Using the inequalities $S/T \leq 1$ and $M/T \leq (\mu_1 + \gamma)b_2/(b_1\mu_2)$ and inserting the expression for $C(t)$ into the expression for $B(t)$, it follows that

$$B(t) \leq G(t) + \int_0^t K(s)B(t-s)ds,$$

where

$$K(s) := \frac{\alpha\beta(\mu_1 + \gamma)b_2}{b_1\mu_2} \int_0^s e^{-\mu_2(s-\sigma)} e^{-(\mu_1+\gamma)\sigma} d\sigma,$$

$$G(t) := \alpha V(0)e^{-\mu_2 t} + \frac{\alpha\beta(\mu_1+\gamma)b_2}{b_1\mu_2} I(0) \int_0^t e^{-\mu_2(t-s)} e^{-(\mu_1+\gamma)s} ds.$$

Therefore, we conclude that $\lim_{t\to\infty} B(t) = 0$ if

$$\int_0^\infty K(s)ds = \frac{\alpha\beta b_2}{b_1\mu_2^2} = R_0 \frac{\mu_1+\gamma}{\mu_1} < 1.$$

Hence, (5.88) is a sufficient condition for the global stability of the DFSS. □

### 5.6.2 *Backward Bifurcation of Endemic Steady States*

Let us denote $S^*$, $I^*$, $M^*$, and $V^*$ as the stationary states of $S(t)$, $I(t)$, $M(t)$, and $V(t)$, respectively, and let $\lambda_j^*$, $(j = 1, 2)$ be the force of infection corresponding to the stationary state. Then, we have

$$\begin{aligned} &b_1 - \lambda_1^* S^* - \mu_1 S^* = 0, \\ &\lambda_1^* S^* - (\mu_1+\gamma)I^* = 0, \\ &b_2 - \lambda_2^* M^* - \mu_2 M^* = 0, \\ &\lambda_2^* M^* - \mu_2 V^* = 0. \end{aligned} \tag{5.91}$$

It is easy to obtain the expression for the steady states as follows:

$$(S^*, I^*, M^*, V^*) = \left( \frac{b_1}{\mu_1+\lambda_1^*}, \frac{b_1\lambda_1^*}{(\mu_1+\lambda_1^*)(\mu_1+\gamma)}, \frac{b_2}{\mu_2+\lambda_2^*}, \frac{\lambda_2^* b_2}{\mu_2(\mu_2+\lambda_2^*)} \right),$$

From the relations $\lambda_1^* = \alpha V^*/(S^*+I^*)$ and $\lambda_2^* = \beta I^*/(S^*+I^*)$, we obtain

$$\lambda_1^* = \frac{\alpha b_2}{\mu_2 b_1} \frac{(\mu_1+\lambda_1^*)(\mu_1+\gamma)}{\lambda_1^*+\mu_1+\gamma} \frac{\lambda_2^*}{\mu_2+\lambda_2^*}, \tag{5.92}$$

$$\lambda_2^* = \frac{\lambda_1^*\beta}{\mu_1+\gamma+\lambda_1^*}. \tag{5.93}$$

From (5.92) and (5.93), we can derive the following equation that is satisfied by $\lambda_1^*$:

$$\lambda_1^* = \frac{\alpha\beta b_2}{\mu_2 b_1} \frac{(\mu_1+\lambda_1^*)(\mu_1+\gamma)}{\lambda_1^*+\mu_1+\gamma} \frac{\lambda_1^*}{\mu_2(\mu_1+\gamma)+\lambda_1^*(\mu_2+\beta)} = \lambda_1^* R(\lambda_1^*), \tag{5.94}$$

where $R(\lambda)$ is a function defined by

$$R(\lambda) := \frac{\alpha\beta b_2}{\mu_2 b_1} \frac{(\mu_1+\lambda)(\mu_1+\gamma)}{\lambda+\mu_1+\gamma} \frac{1}{\mu_2(\mu_1+\gamma)+\lambda(\mu_2+\beta)}.$$

Using the expression for the steady states, we can decompose $R(\lambda_1^*)$ as

$$R(\lambda_1^*) = \left[\alpha \cdot \frac{1}{\mu_2} \cdot \frac{S^*}{T^*}\right] \cdot \left[\frac{1}{\mu_1 + \gamma} \cdot \beta \cdot \frac{U^*}{T^*} \cdot \frac{M^*}{U^*}\right], \tag{5.95}$$

which shows that $R(\lambda)$ is the product of the average number of infected hosts produced by an infected vector during its infective period and the average number of infected vectors produced by an infected host during its infective period at the endemic steady state. That is, $R(\lambda)$ is the reproduction number at the endemic steady state with the force of host infection $\lambda$. It is then clear that $R(\lambda) = 1$ is the condition of stationarity, and it has a unique positive solution if $R(\lambda)$ is monotone decreasing and $R(0) = R_0 > 1$. However, if $R'(0) > 0$, we can expect a backward bifurcation to occur at $R_0 = 1$.

In the SIR model with the mass-action-type force of infection, the susceptible population decreases as the force of infection increases. However, in the homogeneous model (5.84), the proportion of susceptibles in a host population increases with respect to the force of infection $\lambda_1$, so the average number of infected hosts produced by an infected vector during its infective period is an increasing function of $\lambda_1$, and so we could have $R'(0) > 0$. In fact, we can prove the following bifurcation result for the endemic steady states:

**Proposition 5.11** *Let us define $D$ and $G$ as*

$$D := \mu_1^2 - (\mu_1 + \gamma)G, \quad G := \mu_1 - \frac{\gamma\mu_2}{\mu_2 + \beta}.$$

*If $G \geq 0$, then there exists a unique endemic steady state if and only if $R_0 > 1$. If $G < 0$, then the following hold:*

*(1) If $R_0 \geq 1$, there exists only one endemic steady state.*
*(2) If $R_0 < 1$ and $R(-\mu_1 + \sqrt{D}) > 1$, then there exist two endemic steady states.*
*(3) If $R_0 < 1$ and $R(-\mu_1 + \sqrt{D}) = 1$, then there exists only one endemic steady state.*
*(4) If $R_0 < 1$ and $R(-\mu_1 + \sqrt{D}) < 1$, then there is no endemic steady state.*

*Proof* Observe that

$$R(\lambda) = \frac{\alpha\beta b_2(\mu_1 + \gamma)}{\mu_2 b_1}\left(1 - \frac{\gamma}{\lambda + \mu_1 + \gamma}\right)\frac{1}{\mu_2(\mu_1 + \gamma) + \lambda(\mu_2 + \beta)}.$$

Then, we obtain

$$R'(\lambda) = \frac{\alpha\beta b_2(\mu_1 + \gamma)}{\mu_2 b_1}\frac{-f(\lambda)}{(\mu_1 + \gamma + \lambda)^2(\mu_2(\mu_1 + \gamma) + \lambda(\mu_2 + \beta))^2},$$

where

$$f(\lambda) = (\mu_2 + \beta)(\lambda^2 + 2\mu_1\lambda + (\mu_1 + \gamma)G).$$

If $G \geq 0$, then $R'(\lambda) \leq 0$ for all $\lambda > 0$, so $R(\lambda)$ is monotone decreasing for $\lambda > 0$. Because

$$R(0) = R_0 = \frac{\alpha\beta b_2 \mu_1}{b_1 \mu_2^2(\mu_1 + \gamma)}, \quad \lim_{\lambda\to\infty} R(\lambda) = 0,$$

we know that the characteristic equation $R(\lambda) = 1$ has one positive root if and only if $R_0 > 1$. In contrast, if $G < 0$, then we have $D > 0$ and

$$R'(\lambda) = -\frac{\alpha\beta b_2(\mu_1 + \gamma)}{\mu_2 b_1} \frac{(\lambda + \mu_1 + \sqrt{D})(\lambda + \mu_1 - \sqrt{D})}{(\mu + \gamma + \lambda)^2((\mu_2(\mu_1 + \gamma) + \lambda(\mu_2 + \beta))^2}.$$

Hence, we know that $R(\lambda)$ has a unimodal pattern and attains its maximum at $\lambda = -\mu_1 + \sqrt{D} > 0$. Thus, the above statements follow. □

**Corollary 5.1** *The backward bifurcation does not occur if the extra death rate $\gamma$ is zero.*

From the above proposition, we know that a backward bifurcation of endemic steady states could occur at $R_0 = 1$ under the condition $G < 0$, but we do not yet know what kind of parameter set satisfies the condition $R(-\mu_1 + \sqrt{D}) \geq 1$. However, observe that

$$R'(0) = -\frac{\alpha\beta b_2(\mu_2 + \beta)G}{\mu_2^3 b_1(\mu_1 + \gamma)^2}.$$

If $G < 0$, then $R'(0) > 0$, and there exist two endemic steady states if $R_0 < 1$ and $|R_0 - 1|$ is sufficiently small.

Note that $R(\lambda) = 1$ can be reduced to a quadratic equation

$$g(\lambda) := (\mu_2 + \beta)\lambda^2 + H\lambda + (\mu_1 + \gamma)^2\mu_2(1 - R_0) = 0, \tag{5.96}$$

where $H$ is given by

$$H := (\mu_1 + \gamma)\left(2\mu_2 + \beta - \frac{\alpha\beta b_2}{\mu_2 b_1}\right).$$

Because $g'(0) = H$ and $g(0) = (\mu_1 + \gamma)^2\mu_2(1 - R_0)$, we can easily obtain the following formulation using graphical considerations:

**Proposition 5.12** *Let us define the discriminant $E$ of* (5.96) *as*

$$E := H^2 - 4(\mu_1 + \gamma)^2(\mu_2 + \beta)\mu_2(1 - R_0).$$

*Then, the following hold:*

*(1) If $R_0 > 1$, there is only one endemic steady state,*
*(2) If $R_0 = 1$, there is only one endemic steady state when $H < 0$, and there is no endemic steady state when $H \geq 0$,*

(3) *If* $R_0 < 1$, *there is no endemic steady state when* $H \geq 0$. *If* $H < 0$, *there are two endemic steady states when* $E > 0$, *only one endemic steady state when* $E = 0$, *and no endemic steady state when* $E < 0$.

**Corollary 5.2** *Suppose that*

$$2 + \frac{\beta}{\mu_2} < \frac{\alpha\beta b_2}{\mu_2^2 b_1} < 2 + \frac{\beta}{\mu_2} + 2\sqrt{1 + \frac{\beta}{\mu_2}}. \tag{5.97}$$

*Then, the backward bifurcation occurs at* $R_0 = 1$.

*Proof* First, observe that under condition (5.97), we have

$$H = \mu_2(\mu_1 + \gamma)\left(2 + \frac{\beta}{\mu_2} - \frac{\alpha\beta b_2}{\mu_2^2 b_1}\right) < 0. \tag{5.98}$$

Let us choose $R_0$ as a bifurcation parameter. Consider $\gamma$ as a function of $R_0$ and fix the other parameters. Observe that

$$\gamma = \frac{\alpha\beta b_2 \mu_1}{b_1 \mu_2^2}\frac{1}{R_0} - \mu_1. \tag{5.99}$$

Then, $R_0$ moves from zero to $\alpha\beta b_2/(b_1\mu_2^2)$, and $\gamma$ decreases monotonically from $\infty$ to zero. Moreover, note that $E = 0$ when $R_0 = R_0^*$, where

$$R_0^* := 1 - \frac{\mu_2}{4(\mu_2 + \beta)}\left(2 + \frac{\beta}{\mu_2} - \frac{\alpha\beta b_2}{\mu_2^2 b_1}\right)^2,$$

which is positive under condition (5.97). Thus, we obtain Table 5.1 and it shows that for $R_0^* < R_0 < 1$, there exist two endemic steady states, and the backward bifurcation occurs at $R_0 = 1$. □

We have developed a mathematical model for vector-transmitted diseases with fatality and have calculated the basic reproduction number $R_0$ to show that the disease can invade the susceptible population and that a unique endemic steady state exists if $R_0 > 1$, whereas the disease dies out if $R_0 < \frac{\mu_1}{\mu_1+\gamma}$. We have also proved that a backward bifurcation of the endemic steady state can occur. Thus, even if $R_0 < 1$, there could exist endemic steady states. In our modeling, the disease-induced death rate plays a crucial role in the existence of backward bifurcation.

The presence of a backward bifurcation has practically important consequences for the control of infectious diseases. If the endemic state at $R_0 = 1$ exhibits a forward bifurcation, the size of the infected population will be approximately proportional to the difference $|R_0 - 1|$. In a system with a backward bifurcation, the endemic

**Table 5.1** Changes of $\gamma$ and the sign of $E$

| $R_0$ | 0 | $R_0^*$ | 1 | $\alpha\beta b_2/(\mu_2^2 b_1)$ |
|---|---|---|---|---|
| $\gamma$ | $\infty$ | | $\mu_1(\alpha\beta b_2/(\mu_2^2 b_1) - 1)$ | 0 |
| $E$ | $E < 0$ | 0 | $E > 0$ | |

steady state that exists when $R_0$ is slightly greater than unity could have a large infectious population, so the result of $R_0$ rising above unity is a drastic change in the number of infected individuals. Conversely, reducing $R_0$ back below unity would not necessarily eradicate the disease if the change in $R_0$ is small. That is, if the disease is already endemic, eradication involves reducing the basic reproduction number so far that it enters the region in which the DFSS is globally asymptotically stable and there is no endemic steady state.

In this chapter, we have only discussed the existence of backward bifurcations of endemic steady states. From the general stability principle of subcritically bifurcated steady-state solutions, if $R_0 < 1$ and there are two endemic steady states, we could expect the steady state corresponding to the smaller force of infection to be unstable and the other steady state to be locally stable near the bifurcation point. We will examine such stability arguments again in Chap. 7 (see also [52, 53]).

**Exercise 5.18** Consider an outbreak of a vector-transmitted disease. Let $x$ be the density of the susceptible host population, $y$ be the density of infected hosts, $u$ be the density of susceptible vectors, and $v$ be the density of infected vectors. We neglect the host demography, but take into account the demography of the vector population, because the life cycle of vectors is much faster than that of hosts. The basic system is then given as

$$\begin{aligned}\frac{dx(t)}{dt} &= -\beta x(t)v(t),\\ \frac{dy(t)}{dt} &= \beta x(t)v(t) - \gamma y(t),\\ \frac{du(t)}{dt} &= b - (\mu + \delta y(t))u(t),\\ \frac{dv(t)}{dt} &= \delta y(t)u(t) - \mu v(t),\end{aligned}$$

where $b$ is the birth rate of the vector population, $\mu$ is the death rate of vectors, $\beta, \delta$ are transmission coefficients, and $\gamma$ is the removal rate. Suppose that $u(0) + v(0) = b/\mu$.

(1) Show that the basic reproduction number at the DFSS $(x_0, 0, b/\mu, 0)$ is given by

$$R_0 = \frac{\beta x_0 b\delta}{\mu^2\gamma}.$$

(2) Show that $x_\infty := \lim_{t\to\infty} x(t)$ exists and

$$\int_0^\infty y(t)dt = \frac{1}{\gamma}(y(0) + x(0) - x_\infty).$$

(3) Show that $x_\infty > 0$.

# References


1. Allen, L.J.S., Bolker, B.M., Lou, Y., Nevai, A.L.: Asymptotic profiles of the steady states for an SIS epidemic reaction-diffusion model. Disc. Cont. Dyn. Syst. **21**(1), 1–20 (2008)
2. Amann, H.: Ordinary Differential Equations: An Introduction to Nonlinear Analysis. Walter de Gruyter, Berlin (1990)
3. Anderson, R.M., May, R.M.: Infectious Diseases of Humans: Dynamics and Control. Oxford UP, Oxford (1991)
4. Andreason, V.: Disease regulation of age-structured host populations. Theor. Popul. Biol. **36**, 214–239 (1989)
5. Andreasen, V.: Instability in an SIR-model with age-dependent susceptibility. In: Arino, O., Axelrod, D., Kimmel, M., Langlais, M. (eds.) Mathematical Population Dynamics. Theory of Epidemics, vol. 1, pp. 3–14. Wuerz Pub, Winnipeg (1995)
6. Bailey, N.T.J.: The Mathematical Theory of Infectious Diseases and its Applications, 2nd edn. Charles Griffin, London (1975)
7. Bailey, N.T.J.: The Biomathematics of Malaria. Charles Griffin, London (1982)
8. Bartlett, M.S.: Deterministic and stochastic models for recurrent epidemics. In: Neyman, J. (ed.) Proceedings of the Third Berkeley Symposium on Mathematical Statistics and Probability, vol. IV, pp. 81–109. University of California Press, California (1956)
9. Bartlett, M.S.: Measles periodicity and community size. J. Roy. Stat. Soc. A **120**, 48–70 (1957)
10. Bartlett, M.S.: Stochastic Population Models in Ecology and Epidemiology, Methuen and Co. Ltd., London, Wiley Inc., New York (1960)
11. Brauer, F.: The Kermack and McKendrick epidemic model revisited. Math. Biosci. **198**, 119–131 (2005)
12. Brauer, F., van den Driessche, P., Wu, J. (eds.): Mathematical Epidemiology, Mathematical Biosciences Subseries. Lecture Notes in Mathematics, vol. 1945. Springer, Berlin (2008)
13. Brauer, F., Castillo-Chávez, C.: Mathematical Models in Population Biology and Epidemiology. Texts in Applied Mathematics 40, 2nd edn. Springer, Berlin (2012)
14. Brauer, F.: A new epidemic model with indirect transmission. J. Biol. Dyn. (2016). doi:10.1080/17513758.2016.1207813
15. Breban, R., Blower, S.: Letter to Editor: The reinfection threshold does not exist. J. Theor. Biol. **235**, 151–152 (2005)
16. de Jong, M.C.M., Diekmann, O., Heesterbeek, H.: How does transmission of infection depend on population size. In: Mollison, D. (ed.) Epidemic Models: Their Structure and Relation to Data, pp. 84–94. Cambridge U. P., Cambridge (1995)
17. de Mottoni, P., Orlandi, E., Tesei, A.: Asymptotic behavior for a system describing epidemics with migration and spatial spread of infection. Nonl. Anal. Theory, Meth. Appl. **3**(5), 663–675 (1979)
18. Di Blasio, G.: Mathematical analysis for an epidemic model with spatial and age structure. J. Evol. Equ. **10**, 929–953 (2010)
19. Diekmann, O.: Limiting behaviour in an epidemic model. Nonl. Anal. Theory Meth. Appl. **1**, 459–470 (1977)

20. Diekmann, O.: Thresholds and travelling waves for the geographical spread of infection. J. Math. Biol. **6**, 109–130 (1978)
21. Diekmann, O., Kaper, H.G.: On the bounded solutions of a nonlinear convolution equation. Nonl. Anal. Theory Meth. Appl. **2**(6), 721–737 (1978)
22. Diekmann, O.: Run for your life. A note on the asymptotic speed of propagation of an epidemic. J. Diff. Equ. **33**, 58–73 (1979)
23. Diekmann, O., Montijn, R.: Prelude to Hopf bifurcation in an epidemic model: analysis of a characteristic equation associated with a nonlinear Volterra integral equation. J. Math. Biol. **14**, 117–127 (1982)
24. Diekmann, O., Heesterbeak, J.A.P., Metz, J.A.J.: On the definition and the computation of the basic reproduction ratio $R_0$ in models for infectious diseases in heterogeneous populations. J. Math. Biol. **28**, 365–382 (1990)
25. Diekmann, O., Heesterbeek, H., Metz, J.A.J.: In: Mollison, D. (ed.) The Legacy of Kermack and McKendrick, in Epidemic Models: Their Structure and Relation to Data, pp. 95–115. Cambridge University Press, Cambridge (1995)
26. Diekmann, O., de Koeijer, A.A., Metz, J.A.J.: On the final size of epidemic within herds. Canad. Appl. Math. Q. **4**(1), 21–30 (1996)
27. Diekmann, O., Heesterbeek, J.A.P., Britton, T.: Mathematical Tools for Understanding Infectious Disease Dynamics. Princeton University Press, Princeton and Oxford (2013)
28. Dietz, K.: The first epidemic model: A historical note on P.D. En'ko. Austral. J. Statist. **30A**, 56–65 (1988)
29. Dietz, K.: Introduction to McKendrick (1926) Applications of Mathematics to Medical Sciences. In: Kotz, S., Johnson, N.L. (eds.) Breakthroughs in Statistics, pp. 17–26. Springer, New York (1997)
30. Ducrot, A., Magal, P.: Travelling wave solutions for an infection-age structured model with diffusion. Proc. Roy. Soc. Edinb. **139A**, 459–482 (2009)
31. Ducrot, A., Magal, P., Ruan, S.: Travelling wave solutions in multigroup age-structured epidemic models. Arch. Rat. Mech. Anal. **195**, 311–331 (2010)
32. Esteva, L., Matias, M.: A model for vector transmitted diseases with saturation incidence. J. Biol. Sys. **9**(4), 235–245 (2001)
33. Feng, Z., Thieme, H.R.: Recurrent outbreaks of childhood diseases revisited: the impact of isolation. Math. Biosci. **128**, 93–130 (1995)
34. Feng, Z., Thieme, H.R.: Endemic models with arbitrarily distributed periods of infection II. Fast disease dynamics and permanent recovery. SIAM J. Appl. Math. **61**, 983–1012 (2000)
35. Fraser, C., Riley, S., Anderson, R.M., Ferguson, N.M.: Factors that make an infectious disease outbreak controllable. Proc. Natl. Aca. Sci. **101**(16), 6146–6151 (2004)
36. Friedman, A., Yakubu, A.A.: Anthrax epizootic and migration: persistence or extinction. Math. Biosci. **241**, 137–144 (2013)
37. Gleißner, W.: The spread of epidemics. Appl. Math. Comp. **27**, 167–171 (1988)
38. Gomes, M.G., White, L.J., Medley, G.F.: Infection, reinfection, and vaccination under suboptimal immune protection: epidemiological perspectives. J. Theor, Biol. **228**, 539–549 (2004)
39. Gomes, M.G., White, L.J., Medley, G.F.: The reinfection threshold. J. Theor, Biol. **236**, 111–113 (2005)
40. Grasman, J., Matkowsky, B.J.: Singular perturbations of epidemic models involving a threshold. In: Dold, A., Eckmann, B. (eds.) Asymptotic Analysis II, vol. LNM 985, pp. 400–412. Springer, Berlin (1983)
41. Grasman, J., van Herwaarden, O.A.: Asymptotic Methods for the Fokker–Planck Equation and the Exit Problem in Applications. Springer, Berlin (1999)
42. Gripenberg, G.: On some epidemic models. Quart. Appl. Math. **39**, 317–327 (1981)
43. Gripenberg, G.: On a nonlinear integral equation modelling an epidemic in an age-structured population. J. Reine. Angew. Math. **341**, 54–67 (1983)
44. Gripenberg, G.: An estimate for the solution of a Volterra equation describing an epidemic. Nonl. Anal. Theory Meth. Appl. **7**(2), 161–165 (1983)

45. Gumel, A.B., Ruan, S., Day, T., Watmough, J., Brauer, F., van den Driessche, P., Gabrielson, D., Bowman, C., Alexander, M.E., Ardral, S., Wu, J., Sahai, B.M.: Modelling strategies for controlling SARS outbreak. Proc. R. Soc. Lond. B **271**, 2223–2232 (2004)
46. Hahn, B.D., Furniss, P.R.: A deterministic model of an anthrax epizootic: threshold results. Ecol. Model. **20**, 233–241 (1983)
47. Heesterbeek, J.A.P., Metz, J.A.J.: The saturating contact rate in marriage and epidemic models. J. Math. Biol. **31**, 529–539 (1993)
48. Heesterbeek, J.A.P.: The law of mass-action in epidemiology: a historical perspective. In: Cuddington, K., Beisner, B.E. (eds.) Ecological Paradigms Lost, pp. 81–105. Elsevier, Amsterdam (2005)
49. Hethcote, H.W.: Asymptotic behaviour and stability in epidemic models. In: van den Driessche, P. (ed.) Mathematical Problems in Biology. Lecture notes biomath, vol. 2, pp. 83–92. Springer, Berlin (1974)
50. Iannelli, M.: Mathematical Theory of Age-Structured Population Dynamics. Giardini Editori e Stampatori in Pisa (1995)
51. Inaba, H.: Backward bifurcation in a model for vector transmission disease. In: Sekimura, T., Noji, S., Ueno, N., Maini, P.K. (eds.) Morphogenesis and Pattern Formation in Biological Systems. pp. 271–279. Springer, Berlin (2003)
52. Inaba, H., Sekine, H.: A mathematical model for Chagas disease with infection-age-dependent infectivity. Math. Biosci. **190**, 39–69 (2004)
53. Inaba, H.: Endemic threshold results in an age-duration-structured population model for HIV infection. Math. Biosci. **201**, 15–47 (2006)
54. Inaba, H.: Age-structured homogeneous epidemic systems with application to the MSEIR epidemic model. J. Math. Biol. **54**, 101–146 (2007)
55. Inaba, H., Nishiura, H.: The state-reproduction number for a multistate class age structured epidemic system and its application to the asymptomatic transmission model. Math. Biosci. **216**, 77–89 (2008)
56. Inaba, H., Nishiura, H.: The type-reproduction number, the serial interval and the intrinsic growth rate: The basic epidemiological indices for asymptomatic transmission, RIMS Kokyuroku 1597, Theory of Biomathematics and its Applications IV, Research Institute for Mathematical Sciences, Kyoto University, 173–180 (2008)
57. Inaba, H.: On a new perspective of the basic reproduction number in heterogeneous environments. J. Math. Biol. **65**, 309–348 (2012)
58. Inaba, H.: On the definition and the computation of the type-reproduction number $T$ for structured populations in heterogeneous environments. J. Math. Biol. **66**, 1065–1097 (2013)
59. Inaba, H.: On a pandemic threshold theorem of the early Kermack-McKendrick model woth individual heterogeneity. Math. Poul. Stud. **21**, 95–111 (2014)
60. Isono, S.: Mathematical Analysis of a Hepatitis C Model, MA thesis, Graduate School of Mathematical Sciences, University of Tokyo, [Japanese] (2007)
61. Iwami, S. et al.: Cell-to-cell infection by HIV contributes over half of virus infection, eLIFE, 2015;4:e0850. doi:10.7554/eLife.08150 (2015)
62. Katriel, G.: Epidemics with partial immunity to reinfection. Math. Biosci. **228**, 153–159 (2010)
63. Katriel, G.: The size of epidemics in populations with heterogeneous susceptibility. J. Math. Biol. **65**, 237–262 (2012)
64. Kendall, D.G.: Deterministic and stochastic epidemics in closed populations, In Proceedings of the Third Berkeley Symposium on Mathematical Statistics and Probability, Neyman, J. (ed.), Vol. IV, University of Calfornia Press, 149–165 (1956)
65. Kendall, D.G.: Discussion of Measles periodicity and community size by M.S. Bartlett. J. Roy. Statist. Soc. **A120**, 48–70 (1957)
66. Kermack, W.O., McKendrick, A.G.: Contributions to the mathematical theory of epidemics I. In: Proceedings of the Royal Society 115A, 700–721 (1927): reprinted in Bulletin of Mathematical Biology 53(1/2), 33–55 (1991)

67. Kermack, W.O., McKendrick, A.G.: Contributions to the mathematical theory of epidemics II. The problem of endemicity, Proceedings of the Royal Society 138A, 55–83 (1932): reprinted in Bulletin of Mathematical Biology 53(1/2), 57–87 (1991)
68. Kermack, W.O., McKendrick, A.G.: Contributions to the mathematical theory of epidemics III. Further studies of the problem of endemicity, Proceedings of the Royal Society 141A, 94–122 (1933): reprinted in Bulletin of Mathematical Biology 53(1/2), 89–118 (1991)
69. Kermack, W.O., McKendrick, A.G.: Contributions to the mathematical theory of epidemics IV. Analysis of experimental epidemics of the virus disease mouse ectromelia, Journal of Hygiene, Cambridge **37**, 172–187 (1937)
70. Kermack, W.O., McKendrick, A.G.: Contributions to the mathematical theory of epidemics V. Analysis of experimental epidemics of mouse typhoid; A bacterial disease conferring incomplete immunity, Journal of Hygiene, **39**, 271–288. Cambridge (1939)
71. Kuznetsov, Y.A., Piccardi, C.: Bifurcation analysis of periodic SEIR and SIR epidemic models. J. Math. Biol. **32**, 109–121 (1994)
72. Kuniya, T., Wang, J.: Lyapunov functions and global stability for a spatially diffusive SIR epidemic model, Applicable Analysis, http://dx.doi.org/10.1080/00036811.2016.1199796 (2016)
73. Lauwerier, H.A.: Mathematical Models of Epidemics, 2nd printing, Mathematical Centre Tracts 138. Mathematisch Centrum, Amsterdam (1984)
74. Liu, W.M., Levin, S.A., Iwasa, Y.: Influence of nonlinear incidence rates upon the behavior of SIRS epidemiological models. J. Math. Biol. **23**, 187–204 (1986)
75. Magal, P., McCluskey, C.C., Webb, G.F.: Lyapunov functional and global asymptotic stability for an infection-age model. Appl. Anal. **89**(7), 1109–1140 (2010)
76. Magal, P., Ruan, S.: Sustained oscillations in an evolutionary epidemiological model of influenza A drift. Proc. Roy. Soc. A **466**, 965–992 (2010)
77. Martcheva, M., Castillo-Chavez, C.: Diseases with chronic stage in a population with varying size. Math. Biosci. **182**, 1–25 (2003)
78. McKendrick, A.G., Morison, M.J.: The determination of incubation periods from maritime statistics, with particular reference to the incubation period of influenza. Ind. J. Med. Res. **7**, 364–371 (1919)
79. McKendrick, A.G.: Application of mathematics to medical problems. Proc. Edinburgh. Math. Soc. **44**, 98–130 (1926)
80. Metz, J.A.J.: The epidemic in a closed population with all susceptibles equally vulnerable; some results for large susceptible populations and small initial infections. Acta Biotheoretica **27**(1/2), 75–123 (1978)
81. Metz, J.A.J., Diekmann, O.: The Dynamics of Physiologically Structured Populations. Lecture Notes in Biomathematics, vol. 68. Springer, Berlin (1986)
82. Nakata, Y., Kuniya, T.: Global dynamics of a class of SEIR epidemic models in a periodic environment. J. Math. Anal. Appl. **363**, 230–237 (2010)
83. Nakata, Y., et al.: Stability of epidemic models with waning immunity. SUT J. Math. **50**(2), 205–245 (2014)
84. Nishiura, H.: Early efforts in modellng the incubation period of infectious diseases with an acute course of illness. Emerg. Themes Epidmiol. **4**, 2 (2007)
85. Nishiura, H., Inaba, H.: Estimation of the incubation period of influenza A (H1N1-2009) among imported cases: addressing censoring using outbreak data at the origin of importation. J. Theor. Biol. **272**, 123–130 (2011)
86. Nowak, M.A., May, R.M.: Virus Dynamics: Mathematical Principles of Immunology and Virology. Oxford University Press, Oxford (2000)
87. Peng, R., Zhao, X.Q.: A reaction-diffusion SIS epidemic model in a time-periodic environment. Nonlinearity **25**, 1451–1471 (2012)
88. Perthame, B.: Transport Equations in Biology. Birkhäuser, Basel (2007)
89. Rass, L., Radcliffe, J.: Spatial Deterministic Epidemics, American Mathematical Society (2003)

90. Pourbashash, H., Pilyugin, S.S., De Leenheer, P.: Global analysis of within host virus models with cell-to-cell viral transmission. Disc. Conti. Dyn. Syst. B **19**(10), 3341–3357 (2014)
91. Reddingius, J.: Notes on the mathematical theory of epidemics. Acta Biotheoretica **20**, 125–157 (1971)
92. Safan, M., Heesterbeek, H., Dietz, K.: The minimum effort required to eradicate infections in models with backward bifurcation. J. Math. Biol. **53**, 703–718 (2006)
93. Soper, H.E.: The interpretation of periodicity in disease prevalence. J. Roy. Stat. Soc. **92**, 34–73 (1929)
94. Smith, H.L.: Suharmonic bifurcation in an S-I-R epidemic model. J. Math. Biol. **17**, 163–177 (1983)
95. Smith, H.L.: Multiple stable subharmonics for a periodic epidemic model. J. Math. Biol. **17**, 179–190 (1983)
96. Thieme, H.R.: A model for the spatial spread of an epidemic. J. Math. Biol. **4**, 337–351 (1977)
97. Thieme, H.R.: The asymptotic behaviour of solutions of nonlinear integral equations. Math. Z. **157**, 141–154 (1977)
98. Thieme, H.R.: Asymptotic estimate of the solutions of nonlinear integral equations and asymptotic speeds for the spread of populations. J. Reine Angew. Math. **306**, 94–121 (1979)
99. Thieme, H.R.: On the boundedness and the asymptotic behaviour of the non-negative solutions to Volterra-Hammerstein integral equations. Manuscr. math. **31**, 379–412 (1980)
100. Thieme, H.R.: Renewal theorems for some mathematical models in epidemiology. J. Inte. Equ. **8**, 185–216 (1985)
101. Thieme, H.R., Yang, J.: An endemic model with variable re-infection rate and applications to influenza. Math. Biosci. **180**, 207–235 (2002)
102. Thieme, H.R.: Mathematics in Population Biology. Princeton University Press, Princeton (2003)
103. van den Driessche, P., Watmough, J.: Reproduction numbers and sub-threshold endemic equilibria for compartmental models of disease transmission. Math. Biosci. **180**, 29–48 (2002)
104. Velasco-Hernández, J.X.: An epidemiological model for the dynamics of Chagas disease. Biosystem **26**, 127–134 (1991)
105. Velasco-Hernández, J.X.: A model for Chagas disease involving transmission by vectors and blood transfusion. Theor. Popul. Biol. **46**, 1–31 (1994)
106. Wang, W., Zhao, X.Q.: A nonlocal and time-delayed reaction-diffusion model of dengue transmission. SIAM J. Appl. Math. **71**(1), 147–168 (2011)
107. Webb, G.F.: An age-dependent epidemic model with spatial diffusion. Arch. Rat. Mech. Anal. **75**, 91–102 (1980)
108. Webb, G.F.: A reaction-diffusion model for a deterministic diffusive epidemic. J. Math. Anal. Appl. **84**, 150–161 (1981)
109. Webb, G.F.: Theory of Nonlinear Age-Dependent Population Dynamics. Marcel Dekker, New York and Basel (1985)
110. Xinli, H.: Threshold dynamics for SIR epidemic model in periodic environments, 2010 International Conference on Computer Application and System Modeling, V7, 41–45 (2010)

# Chapter 6
# Age-Structured SIR Epidemic Model

**Abstract** The transmission process of infectious agents is largely affected by biological factors, social behavior, and the activity of hosts. These components are essentially (chronological) age dependent; in addition, any mass vaccination policy targets individuals in specific age classes. Therefore, it is crucial to take into account the host age structure to enable realistic modeling for disease prevention policy. In this chapter, we investigate the age-structured susceptible–infected–recovered (SIR) epidemic model, which is very useful when applied to real age-dependent data of common childhood infectious diseases. For simplicity, we assume that the host demography is not affected by the presence of diseases, and that the host population has already attained a stable age distribution. This is not the case if we cannot disregard the disease-induced mortality, as in the case of HIV/AIDS, but is appropriate for many non-lethal diseases. We again establish the endemic threshold principle based on the basic reproduction number $R_0$, which is defined by the spectral radius of the next-generation operator. Here, we sketch a functional analytic approach to the age-dependent model, because it is a natural means of dealing with infinite-dimensional problems. Although the infection-age-dependent model was already introduced by Kermack and McKendrick in the late 1920s, it was not until the 1970s that the chronological-age-dependent epidemic model was considered. The basic endemic threshold results for the age-structured SIR model have been established over the past three decades. However, even now, our understanding of the global dynamics of the age-structured SIR epidemic model remains limited.

## 6.1 SIR Epidemic Model with Age Structure

### *6.1.1 Basic Model*

We consider an infectious disease in a closed age-structured host population, and assume that the disease confers permanent immunity on the recovered individuals. For simplicity, we neglect the disease-induced death rate, the effect of infection on fertility, the latent period, and the infection-age dependency of the parameters. Let $S(t, a)$ be the age density of susceptible individuals at time $t$, $I(t, a)$ be the age



H. Inaba, *Age-Structured Population Dynamics in Demography and Epidemiology*, DOI 10.1007/978-981-10-0188-8_6

density of infected individuals at time $t$, and $R(t, a)$ be the age density of recovered individuals at time $t$. The basic age-structured SIR model is then formulated by the following system of McKendrick equations[1]:

$$\begin{aligned}
&\frac{\partial S(t,a)}{\partial t} + \frac{\partial S(t,a)}{\partial a} = -(\mu(a) + \theta(a) + \lambda(t,a))S(t,a),\\
&\frac{\partial I(t,a)}{\partial t} + \frac{\partial I(t,a)}{\partial a} = \lambda(t,a)S(t,a) - (\mu(a) + \gamma(a))I(t,a),\\
&\frac{\partial R(t,a)}{\partial t} + \frac{\partial R(t,a)}{\partial a} = \theta(a)S(t,a) + \gamma(a)I(t,a) - \mu(a)R(t,a),\\
&S(t,0) = \int_0^\omega m(a)(S(t,a) + (1-q)I(t,a) + R(t,a))da,\\
&I(t,0) = q\int_0^\omega m(a)I(t,a)da,\\
&R(t,0) = 0,\\
&S(0,a) = S_0(a), \quad I(0,a) = I_0(a), \quad R(0,a) = R_0(a),
\end{aligned} \tag{6.1}$$

where $\omega$ is the maximum attainable chronological age of the host population, $m(a)$ denotes the age-specific birth rate, $\mu(a)$ is the age-specific death rate, $\gamma(a)$ is the recovery rate, $q \in [0, 1]$ is the proportion of vertically transmitted newborns, and $\theta(a)$ is the vaccination rate. We assume that the force of infection $\lambda(t, a)$ is given by the *true mass-action model*:

$$\lambda(t,a) := \frac{1}{N(t)}\int_0^\omega \beta(a,\sigma)I(t,\sigma)d\sigma, \tag{6.2}$$

where

$$N(t) := \int_0^\omega P(t,a)da$$

is the total size of the host population, $P(t, a) := S(t, a) + I(t, a) + R(t, a)$ denotes the age density of the host population, $\beta(a, \sigma)$ is the transmission coefficient between susceptible individuals at age $a$ and infected individuals at age $\sigma$, and $(S_0, I_0, R_0)$ are the initial data. For the definition of the true mass-action model, readers are referred to [13].

The force of infection by the true mass-action model is traditionally called the *standard incidence* or *Macdonald type*. In the *pseudo-mass-action model* [13], the force of infection is given by

[1]For early studies of the age-structured SIR epidemic model, readers are referred to [1–4, 7, 9, 20–23, 26, 27, 30, 53]. Effects of the disease-induced death rate in the age-structured SIR model [7] were discussed in [2, 45, 48].

$$\lambda(t,a) = \int_0^\omega \beta(a,\sigma) I(t,\sigma) d\sigma. \tag{6.3}$$

However, as long as the total size $N$ is constant, both models are qualitatively equivalent in their mathematical nature, and the scale factor $1/N$ may be omitted.

It should also be remarked that the true mass-action model for the force of infection is homogeneous of degree one, which implies that the force of infection is independent of the scale of the host population. Hence, we call the above epidemic model the *homogeneous SIR epidemic system.*

By adding the three PDEs in (6.1), we know that the age density of the host population $P(t,a)$ satisfies the stable population model[2]:

$$\begin{aligned}
&\frac{\partial P(t,a)}{\partial t} + \frac{\partial P(t,a)}{\partial a} = -\mu(a) P(t,a),\\
&P(t,0) = \int_0^\omega m(a) P(t,a) da,\\
&P(0,a) = P_0(a),
\end{aligned} \tag{6.4}$$

where $P_0(a) := S_0(a) + I_0(a) + R_0(a)$. The dynamics of the host population can then be determined independently from the epidemic. If we define new variables $(s,i,r)$ by

$$S(t,a) = P(t,a)s(t,a), \quad I(t,a) = P(t,a)i(t,a), \quad R(t,a) = P(t,a)r(t,a),$$

we obtain the following non-autonomous system for $(s,i,r)$:

$$\begin{aligned}
&\frac{\partial s(t,a)}{\partial t} + \frac{\partial s(t,a)}{\partial a} = -(\theta(a) + \lambda(t,a)) s(t,a),\\
&\frac{\partial i(t,a)}{\partial t} + \frac{\partial i(t,a)}{\partial a} = \lambda(t,a) s(t,a) - \gamma(a) i(t,a),\\
&\frac{\partial r(t,a)}{\partial t} + \frac{\partial r(t,a)}{\partial a} = \theta(a) s(t,a) + \gamma(a) i(t,a),\\
&s(t,0) = 1 - q \int_0^\omega \phi(t,a) i(t,a) da,\\
&i(t,0) = q \int_0^\omega \phi(t,a) i(t,a) da,\\
&r(t,0) = 0,\\
&\lambda(t,a) = \int_0^\omega \beta(a,\sigma) \psi(t,\sigma) i(t,\sigma) d\sigma,
\end{aligned} \tag{6.5}$$

---

[2]Here, the gender of the host population is not considered, and the age-specific vital rates are genderless rates, which are usually not used in demography.

where

$$\phi(t,a) := \frac{m(a)P(t,a)}{\int_0^\omega m(a)P(t,a)da}, \quad \psi(t,a) := \frac{P(t,a)}{\int_0^\omega P(t,a)da}.$$

Although the new system includes time-dependent parameters, the behavior of $P(t,a)$ is determined independently from $(s,i,r)$, and hence, we can see that $\phi$ and $\psi$ are known functions. Moreover, it follows from $s(t,a)+i(t,a)+r(t,a)=1$ that the new system is essentially determined by $(s,i)$ or $(i,r)$.

From the stable population theory, we know that $\phi$ and $\psi$ uniformly converge to time-independent normalized age distributions given by

$$\lim_{t\to\infty}\phi(t,a) = c_1(a) := \frac{e^{-r_0 a}m(a)\ell(a)}{\int_0^\omega e^{-\lambda_0 a}m(a)\ell(a)da} = e^{-r_0 a}m(a)\ell(a),$$

$$\lim_{t\to\infty}\psi(t,a) = c_2(a) := \frac{e^{-r_0 a}\ell(a)}{\int_0^\omega e^{-r_0 a}\ell(a)da} = b_0 e^{-r_0 a}\ell(a),$$

where $r_0$ is the intrinsic growth rate of the host population given as a unique real root of the characteristic equation

$$\int_0^\omega e^{-r_0 a}m(a)\ell(a)da = 1,$$

$\ell(a)$ is the survival probability defined by $\ell(a) := \exp(-\int_0^a \mu(\sigma)d\sigma)$, $c_1(a)$ is the probability density of age at childbearing under the stable population, $c_2(a)$ is the stable age profile, and $b_0$ is the crude birth rate in the stable population given by

$$b_0 = \frac{1}{\int_0^\omega e^{-r_0 a}\ell(a)da}.$$

Because the new system (6.5) is asymptotically autonomous, we can examine its asymptotic behavior by investigating the limiting autonomous system, that is, we can assume that the host population has already attained the stable age distribution. A mathematical justification of this assumption is discussed in [32, 33]. Readers may also refer to [5, 28] for a different treatment of homogeneous age-structured epidemic models.

The limiting autonomous system, which we call the *normalized epidemic system*, is given as follows:

$$\begin{aligned}
&\frac{\partial s(t,a)}{\partial t} + \frac{\partial s(t,a)}{\partial a} = -(\theta(a) + \lambda(t,a))s(t,a), \\
&\frac{\partial i(t,a)}{\partial t} + \frac{\partial i(t,a)}{\partial a} = \lambda(t,a)s(t,a) - \gamma(a)i(t,a), \\
&\frac{\partial r(t,a)}{\partial t} + \frac{\partial r(t,a)}{\partial a} = \theta(a)s(t,a) + \gamma(a)i(t,a), \\
&s(t,0) = 1 - q\int_0^\omega c_1(a)i(t,a)da, \\
&i(t,0) = q\int_0^\omega c_1(a)i(t,a)da, \\
&r(t,0) = 0, \\
&\lambda(t,a) = \int_0^\omega \beta(a,\sigma)c_2(\sigma)i(t,\sigma)d\sigma,
\end{aligned} \tag{6.6}$$

where the state space is given by

$$\Omega = \{(s,i,r) \in L^1_+(0,\omega) \times L^1_+(0,\omega) \times L^1_+(0,\omega) : s + i + r = 1\}.$$

In the following, we mainly investigate the normalized system to consider the epidemic in a stable population. Note that the steady state solution of (6.6) corresponds to a persistent (exponentially growing) solution of the original homogeneous epidemic system (6.1). That is, if $(s^*(a), i^*(a), r^*(a))$ is a steady state of (6.6),

$$(e^{r_0(t-a)}\ell(a)s^*(a), e^{r_0(t-a)}\ell(a)i^*(a), e^{r_0(t-a)}\ell(a)r^*(a))$$

becomes an exponential solution of (6.1).

**Exercise 6.1** We can consider another type of normalized system by introducing new distributions $u$, $v$, and $w$ as

$$S(t,a) = N(t)u(t,a), \quad I(t,a) = N(t)v(t,a), \quad R(t,a) = N(t)w(t,a).$$

Write down the normalized system for $(u, v, w)$ (see Sect. 4.5 and [28]). This normalization is especially effective if the host demography is affected by the disease (Chap. 7).

### *6.1.2 Abstract Approach to the Well-Posedness*

We now present a semigroup proof for the existence and uniqueness of nonnegative solutions of the normalized system (6.6). In fact, to deal with this problem, the semigroup approach is more powerful than the classical integral equation approach (see Chap. 10).

Because $s(t,a) + i(t,a) + r(t,a) = 1$, for the normalized system (6.6), it is sufficient to consider the $(i,r)$ system as follows:

$$\begin{aligned}
&\frac{\partial i(t,a)}{\partial t} + \frac{\partial i(t,a)}{\partial a} = \lambda(t,a)(1 - i(t,a) - r(t,a)) - \gamma(a)i(t,a),\\
&\frac{\partial r(t,a)}{\partial t} + \frac{\partial r(t,a)}{\partial a} = \theta(a)(1 - i(t,a) - r(t,a)) + \gamma(a)i(t,a),\\
&i(t,0) = q\int_0^\omega c_1(a)i(t,a)da,\\
&r(t,0) = 0,\\
&\lambda(t,a) = \int_0^\omega \beta(a,\sigma)c_2(\sigma)i(t,\sigma)d\sigma,
\end{aligned} \tag{6.7}$$

where the state space of $(i,r)$ is defined by

$$\Omega := \{(i,r) \in L^1_+(0,\omega) \times L^1_+(0,\omega) : 0 \le i + r \le 1\}.$$

We choose the $(i,r)$ system instead of the $(s,i)$ system because the boundary condition of the $(i,r)$ system is homogeneous and linear. The operators $A$, with domain $\mathscr{D}(A)$, and $F$ acting on the state space $E := L^1(0,\omega) \times L^1(0,\omega)$ are defined as

$$(A\phi)(a) = \begin{pmatrix} -\frac{d}{da} & 0 \\ 0 & -\frac{d}{da} \end{pmatrix} \begin{pmatrix} \phi_1(a) \\ \phi_2(a) \end{pmatrix},$$

$$\mathscr{D}(A) = \left\{ \phi = \begin{pmatrix} \phi_1 \\ \phi_2 \end{pmatrix} \in E : \phi_j \in W^{1,1}(0,\omega), \begin{pmatrix} \phi_1(0) \\ \phi_2(0) \end{pmatrix} = \begin{pmatrix} q\int_0^\omega c_1(a)\phi_1(a)da \\ 0 \end{pmatrix} \right\},$$

$$F(\phi)(a) = \begin{pmatrix} \lambda[a \mid \phi_1](1 - \phi_1(a) - \phi_2(a)) - \gamma\phi_1 \\ \theta(a)(1 - \phi_1(a) - \phi_2(a)) + \gamma\phi_1 \end{pmatrix},$$

where $\mathscr{D}(A)$ denotes the domain of the population operator $A$ and $W^{1,1}$ is the set of absolutely continuous functions.

Let $u = (i,r)^{\mathrm{T}}$ be the state vector of the $(i,r)$ system.[3] The $(i,r)$-system can then be formulated as a semilinear Cauchy problem on the Banach space $E$:

$$\frac{du(t)}{dt} = Au(t) + F(u(t)), \quad u(0) = u_0. \tag{6.8}$$

It is easy to see that the operator $A$ generates a strongly continuous semigroup $e^{tA}$. In fact, the semigroup $e^{tA}$ can be expressed as follows:

$$e^{tA}\begin{pmatrix} \phi_1(a) \\ \phi_2(a) \end{pmatrix} = \begin{pmatrix} (e^{tA_1}\phi_1)(a) \\ (e^{tA_2}\phi_2)(a) \end{pmatrix},$$

[3] T denotes the transpose of the vector.

where $e^{tA_1}$ and $e^{tA_2}$ are semigroups on $L^1(0,\omega)$, which are defined for $u \in L^1(0,\omega)$ as

$$(e^{tA_1}u)(a) = \begin{cases} B(t-a), & t-a>0, \\ u(a-t), & a-t>0, \end{cases}$$

$$(e^{tA_2}u)(a) = \begin{cases} 0, & t-a>0, \\ u(a-t), & a-t>0. \end{cases}$$

Here, $B(t)$ denotes the solution of the renewal integral equation given by

$$B(t) = G(t) + q\int_0^t c_1(a)B(t-a)da,$$

where

$$G(t) := q\int_{t\wedge\omega}^{\omega} c_1(a)u(a-t)da$$

and $t \wedge \omega := \min\{t,\omega\}$. It can easily be verified that the state space $\Omega$ is positively invariant with respect to the semiflow $e^{tA}$: $e^{tA}(\Omega) \subset \Omega$.

In the following, we assume that $\beta, \theta$, and $\gamma$ are essentially bounded nonnegative measurable functions. Then, we can conclude that

**Lemma 6.1** *The mapping $F : \Omega \to E$ is Lipschitz continuous and there exists a number $\alpha \in (0,1)$ such that*

$$(I + \alpha F)(\Omega) \subset \Omega. \tag{6.9}$$

*Proof* Because the Lipschitz continuity is clear, let us show (6.9). If we define the vector $v_j$ as

$$(I+\alpha F)(u_1,u_2)^{\mathrm{T}} = (v_1,v_2)^{\mathrm{T}},$$

then we have

$$v_1(a) + v_2(a) = \alpha(\lambda[a|u_1] + \theta(a))(1 - u_1(a) - u_2(a)) + (u_1(a) + u_2(a)).$$

On the other hand, $\sup\lambda \le \sup\beta$ for $u = (u_1,u_2)^{\mathrm{T}} \in \Omega$. If we define $\lambda^+ := \sup\lambda$ and $\theta^+ := \sup\theta$, we obtain $v_1(a) + v_2(a) \le 1$ if we choose $\alpha$ such that $\alpha \le (\lambda^+ + \theta^+)^{-1}$. Moreover, if we let $\gamma^+ := \sup\gamma$, then $v_1 \ge 0$ if $\alpha\gamma^+ \le 1$. Therefore, if we choose

$$0 < \alpha < \min\left\{\frac{1}{\lambda^+ + \theta^+}, \frac{1}{\gamma^+}\right\},$$

then we know that (6.9) holds. □

Following the method of Busenberg et al. [6], we can rewrite the Cauchy problem (6.8) as follows:

$$\frac{du(t)}{dt} = \left(A - \frac{1}{\alpha}\right)u(t) + \frac{1}{\alpha}(I + \alpha F)u(t), \quad u(0) = u_0, \tag{6.10}$$

where $\alpha$ is chosen such that (6.9) holds. The mild solution of this problem is then given by the variation-of-constants formula [47, Chap. 6]:

$$u(t) = e^{-\frac{1}{\alpha}t}e^{tA}u_0 + \int_0^t e^{-\frac{1}{\alpha}(t-s)}e^{(t-s)A}[u(s) + \alpha F(u(s))]ds. \tag{6.11}$$

The mild solution defines the semiflow $S(t)u_0$, $t > 0$ by $S(t)u_0 = u(t)$. Define an iterative sequence by

$$u^0(t) = u_0, \quad u^{n+1}(t) \quad = e^{-\frac{1}{\alpha}t}e^{tA}u_0 + \int_0^t e^{-\frac{1}{\alpha}(t-s)}e^{(t-s)A}[u^n(s) + \alpha F(u^n(s))]ds.$$

If $u^n \in \Omega$, it follows that $e^{tA}u_0, e^{(t-s)A}[u^n(s)+\alpha F(u^n(s))] \in \Omega$. Hence, $u^{n+1} \in \Omega$, as it is the convex sum of two elements of the convex set $\Omega$. It follows from the Lipschitz continuity that the iterative sequence $u_n$ uniformly converges to the mild solution $S(t)u_0 \in \Omega$. Thus, we can conclude the existence and uniqueness result:

**Proposition 6.1** *The Cauchy problem* (6.8) *has a unique mild solution* $S(t)u_0$ *and* $\Omega$ *is positively invariant with respect to the semiflow* $S(t)$. *If the initial data are included in* $\mathscr{D}(A)$, *the mild solution becomes a classical solution.*

*Remark 6.1* Note that the abstract Eq. (6.10) is also useful for showing the existence of an endemic steady state. Let $u^*$ denote a stationary solution of (6.10). Then, we have

$$\left(A - \frac{1}{\alpha}\right)u^* + \frac{1}{\alpha}(I + \alpha F)u^* = 0.$$

Let $v^* := -(A - \frac{1}{\alpha})u^*$. Because $-(A - \frac{1}{\alpha})$ is positively invertible, we have $u^* = (-(A - \frac{1}{\alpha}))^{-1}v^*$. Thus, we obtain a fixed point equation for $v^*$:

$$v^* = \frac{1}{\alpha}(I + \alpha F)(-(A - \frac{1}{\alpha}))^{-1}v^* = \Psi_\alpha(v^*),$$

where

$$\Psi_\alpha := (I + \alpha F)(I - \alpha A)^{-1}$$

is a positive nonlinear operator that was introduced in [6].

If $\Psi_\alpha$ has a positive fixed point $v^*$, $u^* = \left(-\left(A - \frac{1}{\alpha}\right)\right)^{-1}v^*$ gives a positive stationary solution of the $(i, r)$ system (6.7). Define the linearized operator at the origin as follows:

$$\Psi'_\alpha[0] := K_\alpha = (I + \alpha F'[0])(I - \alpha A)^{-1},$$

where $F'[0]$ is the Fréchet derivative of the operator $F$ at the origin. If we can check the conditions of Krasnoselskii's theorem (Proposition 10.32, [37]), the fixed point equation has at least one positive solution if $\Psi_\alpha$ and $\Psi'_\alpha[0]$ are compact positive operators and $r(\Psi'_\alpha[0]) > 1$. However, there is no endemic steady state if $r(K_\alpha) \le 1$, because $\Psi_\alpha \le K_\alpha$ and we can apply Proposition 10.34.

Note that the spectral radius $r(K_\alpha)$ is a surrogate index of the basic reproduction number $R_0$ (for the definition, see Chap. 9) such that

$$\mathrm{sign}(R_0 - 1) = \mathrm{sign}(r(K_\alpha) - 1). \tag{6.12}$$

In fact, as shown in Sect. 6.4, the linearized operator $A + F'[0]$ has the dominant eigenvalue $r_0$ such that $\mathrm{sign}(r_0) = \mathrm{sign}(r(K) - 1)$, where $K = F'[0](-A)^{-1}$ is the next-generation operator (NGO) and $R_0 = r(K)$. Corresponding to the decomposition $A + F'[0] = (A - 1/\alpha) + (1/\alpha)(I + \alpha F'[0])$, its NGO is calculated as follows:

$$\frac{1}{\alpha}(I + \alpha F'[0])\left(-\left(A - \frac{1}{\alpha}\right)\right)^{-1} = (I + \alpha F'[0])(I - \alpha A)^{-1} = K_\alpha.$$

From the theory of $R_0$, we have that $\mathrm{sign}(r_0) = \mathrm{sign}(r(K_\alpha) - 1)$, which shows that (6.12) holds.

Therefore, the endemic criterion $r(K_\alpha) > 1$ is independent of the choice of $\alpha$, so we can conclude that there exists at least one endemic steady state if $R_0 > 1$, whereas there is no endemic steady state if $R_0 \le 1$. This kind of argument can be applied to a wide range of semilinear problems appearing in age-dependent population dynamics.

## 6.2 Epidemic in a Demographic Steady State

### *6.2.1 Horizontal Transmission and its $R_0$*

In this section, we examine a simple case in which the host population is in a demographic steady state and there is no vertical transmission or vaccination.[4] That is, we assume that $q = 0$, $\theta(a) \equiv 0$, and

$$\int_0^\omega m(a)\ell(a)da = 1,$$

which implies that the (demographic) net reproduction rate of the host population is unity and $r_0 = 0$. We can then assume that the host population is a demographic stationary population given by

[4] A detailed analysis of this model was given in [30].

$$P(t,a) = P_0(a) := B\ell(a), \quad \forall t > 0,$$

where $B$ is the birth rate (number of newborns per unit time). Hence, $c_2(a) = b_0\ell(a)$, where $b_0 = 1/\int_0^\omega \ell(a)da$ denotes the crude birth rate in the stationary population. In the following, we mainly consider the $(s,i)$ system:

$$\begin{aligned}
&\frac{\partial s(t,a)}{\partial t} + \frac{\partial s(t,a)}{\partial a} = -\lambda(t,a)s(t,a),\\
&\frac{\partial i(t,a)}{\partial t} + \frac{\partial i(t,a)}{\partial a} = \lambda(t,a)s(t,a) - \gamma(a)i(t,a),\\
&\lambda(t,a) = \int_0^\omega \beta(a,\sigma)c_2(\sigma)i(t,\sigma)d\sigma,\\
&s(t,0) = 1, \quad i(t,0) = 0.
\end{aligned} \tag{6.13}$$

The state space of this system is $E_+ = L^1_+(0,\omega) \times L^1_+(0,\omega)$, and it is assumed that $\gamma \in L^\infty_+(0,\omega)$ and $\beta \in L^\infty_+((0,\omega)\times(0,\omega))$.

First, let us consider the steady state of (6.13). Let $s^*(a)$ and $i^*(a)$ be the steady state solution and let $\lambda^*(a)$ be the corresponding force of infection. Thus, we have

$$\begin{aligned}
&\frac{ds^*(a)}{da} = -\lambda^*(a)s^*(a),\\
&\frac{di^*(a)}{da} = -\gamma(a)i^*(a) + \lambda^*(a)i^*(a),\\
&\lambda^*(a) = \int_0^\omega \beta(a,\sigma)c_2(\sigma)i^*(\sigma)d\sigma,\\
&s^*(0) = 1, \quad i^*(0) = 0.
\end{aligned} \tag{6.14}$$

By solving (6.14), we obtain the following expressions:

$$\begin{aligned}
s^*(a) &= \exp\left(-\int_0^a \lambda^*(\sigma)d\sigma\right),\\
i^*(a) &= \int_0^a \frac{\Gamma(a)}{\Gamma(\zeta)}\lambda^*(\zeta)e^{-\int_0^\zeta \lambda^*(z)dz}d\zeta,
\end{aligned} \tag{6.15}$$

where

$$\Gamma(a) := \exp\left(-\int_0^a \gamma(\sigma)d\sigma\right).$$

Inserting (6.15) into the expression for $\lambda^*$ in (6.14) and changing the order of integrals, we arrive at the nonlinear integral equation for the unknown $\lambda^*(a)$:

$$\lambda^*(a) = \int_0^\omega \phi(a,\zeta)\lambda^*(\zeta)e^{-\int_0^\zeta \lambda^*(z)dz}d\zeta, \tag{6.16}$$

where

$$\phi(a,\zeta) := \int_\zeta^\omega \beta(a,\sigma)c_2(\sigma)\frac{\Gamma(\sigma)}{\Gamma(\zeta)}d\sigma. \tag{6.17}$$

Then, there exists an endemic steady state, expressed as (6.15), if and only if the nonlinear Eq. (6.16) has a positive solution $\lambda^*$. It is also clear that $\lambda^*(a) \equiv 0$ is a trivial solution of (6.16) that corresponds to the disease-free steady state $(s^*, i^*) = (1, 0)$.

As we deal with the general transmission coefficient in Sect. 6.4, here, we treat a simple case in which there exist functions $\beta_1(a)$ and $\beta_2(\zeta)$ such that $\beta(a,\zeta)$ can be factorized as

$$\beta(a,\zeta) = \beta_1(a)\beta_2(\zeta). \tag{6.18}$$

Assumption (6.18) is called the *separable mixing assumption*, which implies that there is no correlation between the age of the infected individuals and that of the susceptible individuals, with $\beta_1(a)$ reflecting the susceptibility and $\beta_2(\zeta)$ indicating the infectivity. In particular, if $\beta_1 = \beta_2$, we have the *proportionate mixing* assumption. (Traditionally, these assumptions are often called the proportionate mixing assumption without distinguishing between separable and proportional [14, 24].)

Under the separable mixing assumption, (6.16) can be written as

$$\lambda^*(a) = \beta_1(a)\int_0^\omega \pi(\zeta)\lambda^*(\zeta)e^{-\int_0^\zeta \lambda^*(z)dz}d\zeta, \tag{6.19}$$

where

$$\pi(\zeta) := \int_\zeta^\omega \beta_2(\sigma)c_2(\sigma)\frac{\Gamma(\sigma)}{\Gamma(\zeta)}d\sigma.$$

Then, (6.19) becomes a one-dimensional equation and its solution must be written as $\lambda^*(a) = c\beta_1(a)$, where $c$ is an unknown constant. Inserting this expression into (6.19), we arrive at an equation for the unknown number $c$:

$$1 = \int_0^\omega \phi(\zeta,\zeta)e^{-c\int_0^\zeta \beta_1(z)dz}d\zeta. \tag{6.20}$$

The right-hand side of (6.20) is a strictly decreasing function of $c$ that goes to zero as $c \to \infty$. Therefore, if the condition

$$\int_0^\omega \phi(\zeta,\zeta)d\zeta = \int_0^\omega \beta_1(\zeta)\pi(\zeta)d\zeta > 1 \tag{6.21}$$

holds, the characteristic Eq. (6.20) has a unique positive solution.

As we discuss in Sect. 6.4, the left-hand side of the threshold condition (6.21) gives the *basic reproduction number* $R_0$ of the system (6.13). Hence, we can summarize the above argument as follows:

**Proposition 6.2** *For the normalized epidemic system* (6.13) *with the separable mixing assumption, there exists a unique endemic steady state if and only if the basic reproduction number is greater than unity:*

$$R_0 = \int_0^\omega \phi(\zeta, \zeta)d\zeta > 1. \tag{6.22}$$

From the above expression, $R_0$ for the separable mixing case is given by

$$\begin{aligned} R_0 &= \int_0^\omega \beta_1(\zeta) \int_\zeta^\omega \beta_2(\sigma)c_2(\sigma)\frac{\Gamma(\sigma)}{\Gamma(\zeta)}d\sigma d\zeta \\ &= \int_0^\omega \beta_1(\zeta)c_2(\zeta) \int_\zeta^\omega \beta_2(\sigma)\frac{\ell(\sigma)\Gamma(\sigma)}{\ell(\zeta)\Gamma(\zeta)}d\sigma d\zeta. \end{aligned}$$

If the transmission coefficient $\beta$ and the recovery rate $\gamma$ are constant, we have

$$R_0 = \frac{\beta}{\gamma}\int_0^\omega c_2(\sigma)(1 - e^{-\gamma\sigma})d\sigma, \tag{6.23}$$

where $c_2(a) = \ell(a)/\int_0^\omega \ell(a)da$ is the age profile of the stationary host population. Thus, $R_0$ becomes larger as the host population ages under an improvement in the death rate. Moreover, if $\gamma = 0$, that is, infected individuals become lifelong carriers, we have the simple relation

$$R_0 = \beta A_0,$$

where $A_0$ denotes the average age of the host stationary population:

$$A_0 := \int_0^\omega ac_2(a)da.$$

Note that if $\beta$ is constant, the force of infection at the endemic steady state $\lambda^*$ is constant and the probability density of infection is given by

$$p(a) := \frac{e^{-\lambda^* a}\ell(a)}{\int_0^\omega e^{-\lambda^* a}\ell(a)da}.$$

Let $A$ be the average age of infection given by

$$A := \int_0^\omega ap(a)da.$$

If we adopt the Type I survival rate, that is, $\ell(a) = 1$ for $a \in [0, \omega)$, we have the well-known relation $A = 1/\lambda^*$.

### 6.2.2 Local Stability of Endemic Steady State

Let us now consider the stability problem of the steady states. For illustrative purposes, we adopt the simplified assumption that the transmission coefficient $\beta(a, \zeta)$ and the recovery rate $\gamma(a)$ are the constants $\beta$ and $\gamma$, respectively. That is, the simplified system is formulated as

$$\begin{aligned}
&\frac{\partial s(t,a)}{\partial t} + \frac{\partial s(t,a)}{\partial a} = -\lambda(t,a)s(t,a), \\
&\frac{\partial i(t,a)}{\partial t} + \frac{\partial i(t,a)}{\partial a} = -\gamma i(t,a) + \lambda(a)s(t,a), \\
&\lambda(t,a) = \beta \int_0^\omega c_2(\sigma) i(t,\sigma) d\sigma, \\
&s(t,0) = 1, \\
&i(t,0) = 0.
\end{aligned} \tag{6.24}$$

Let $(s^*(a), i^*(a))$ be the steady state solution and let $(u(t,a), v(t,a))$ be the perturbation from the steady state:

$$\begin{aligned}
s(t,a) &= s^*(a) + u(t,a), \\
i(t,a) &= i^*(a) + v(t,a),
\end{aligned} \tag{6.25}$$

where

$$\begin{aligned}
s^*(a) &= e^{-\lambda^* a}, \\
i^*(a) &= \lambda^* \int_0^a e^{-\lambda^* \sigma - \gamma(a-\sigma)} d\sigma.
\end{aligned}$$

Inserting (6.25) into (6.24) and neglecting the second-order terms of $(u, v)$, we obtain the linearized equation

$$\begin{aligned}
&\frac{\partial u(t,a)}{\partial t} + \frac{\partial u(t,a)}{\partial a} = -\lambda^* u(t,a) - s^*(a)\beta \int_0^\omega c_2(\sigma) v(t,\sigma) d\sigma, \\
&\frac{\partial v(t,a)}{\partial t} + \frac{\partial v(t,a)}{\partial a} = -\gamma v(t,a) + \lambda^* u(t,a) + s^*(a)\beta \int_0^\omega c_2(\sigma) v(t,\sigma) d\sigma, \\
&\lambda^* = \beta \int_0^\omega c_2(\sigma) i^*(\sigma) d\sigma, \\
&u(t,0) = v(t,0) = 0,
\end{aligned} \tag{6.26}$$

where we have used the fact that $(s^*(a), i^*(a))$ satisfies (6.14).

If we assume separation of variable solutions such as

$$u(t,a) = e^{\alpha t} g(a), \quad v(t,a) = e^{\alpha t} h(a),$$

then we have

$$\begin{aligned}
\frac{dg(a)}{da} &= -(\alpha+\lambda^*)g(a) - s^*(a)\beta\int_0^\omega c_2(\sigma)h(\sigma)d\sigma,\\
\frac{dh(a)}{da} &= -(\alpha+\gamma)h(a) + \lambda^* g(a) + s^*(a)\beta\int_0^\omega c_2(\sigma)h(\sigma)d\sigma, \qquad (6.27)\\
g(0) &= h(0) = 0.
\end{aligned}$$

By solving the equation for $g$, we obtain

$$g(a) = -\beta\int_0^a e^{-(\alpha+\lambda^*)(a-z)-\lambda^* z}dz\int_0^\omega c_2(\sigma)h(\sigma)d\sigma.$$

Inserting the above expression into the equation for $h(a)$, it follows that

$$\frac{dh(a)}{da} = -(\alpha+\gamma)h(a) + \beta e^{-\lambda^* a}\left[1-\lambda^*\int_0^a e^{-\alpha(a-z)}dz\right]\int_0^\omega c_2(\sigma)h(\sigma)d\sigma.$$

Therefore, we have

$$h(a) = \int_0^a e^{-(\gamma+\alpha)(a-\zeta)}\beta e^{-\lambda^*\zeta}\left[1-\lambda^*\int_0^\zeta e^{-\alpha(\zeta-z)}dz\right]d\zeta\int_0^\omega c_2(\sigma)h(\sigma)d\sigma.$$

Multiplying both sides of the above expression by $c_2(a)$, integrating from 0 to $\omega$, and dividing the result by the common factor $\int_0^\omega c_2(\sigma)h(\sigma)d\sigma$, we arrive at the following characteristic equation:

$$1 = \int_0^\omega c_2(a)\int_0^a e^{-(\gamma+\alpha)(a-\zeta)}\beta e^{-\lambda^*\zeta}\left[1-\lambda^*\int_0^\zeta e^{-\alpha(\zeta-z)}dz\right]d\zeta da, \qquad (6.28)$$

which determines the growth rate $\alpha$ of the perturbation term. By changing the variables and the order of integrals, (6.28) can be written as the Laplace transform

$$\hat{J}(\alpha,\lambda^*) := \int_0^\omega e^{-\alpha x}J(x,\lambda^*)dx = 1,$$

where $\hat{J}$ denotes the Laplace transform of a function $J$ defined by

$$\begin{aligned}
J(x,\lambda^*) &:= G(x,\lambda^*) - \lambda^* H(x,\lambda^*),\\
G(x,\lambda^*) &:= \beta e^{-\gamma x}\int_x^\omega c_2(a)e^{-\lambda^*(a-x)}da,\\
H(x,\lambda^*) &:= \beta\int_x^\omega c_2(a)\int_{a-x}^a e^{-\gamma(a-\zeta)-\lambda^*\zeta}d\zeta da.
\end{aligned}$$

Because $\lambda^*$ is the force of infection at the steady state, it must satisfy

$$\lambda^* = \beta \int_0^\omega c_2(\sigma)\lambda^* \int_0^\sigma e^{-\lambda^*\zeta-\gamma(\sigma-\zeta)} d\zeta d\sigma.$$

If $\lambda^* > 0$, by changing the order of integrals, the above condition can be rewritten as follows:

$$\hat{G}(0, \lambda^*) = 1. \tag{6.29}$$

Therefore, if $\alpha \geq 0$ and $\lambda^* > 0$, it follows from $J(x, \lambda^*) < G(x, \lambda^*)$ that

$$\hat{J}(\alpha, \lambda^*) < \hat{G}(\alpha, \lambda^*) \leq \hat{G}(0, \lambda^*) = 1,$$

which shows that $\alpha$ is not a characteristic root. That is, there is no nonnegative characteristic root at the endemic steady state. However, there remains a possibility that conjugate pairs of complex roots with nonnegative real parts exist.

Observe that $J(x, \lambda^*)$ can be rewritten as

$$J(x, \lambda^*) = \beta \int_x^\omega c_2(a) m(a, x, \lambda^*) da,$$

where

$$m(a, x, \lambda^*) := e^{-\gamma x - \lambda^*(a-x)} - \lambda^* \int_0^x e^{-\gamma\sigma - \lambda^*(a-\sigma)} d\sigma.$$

$J(x, \lambda^*)$ is defined for $0 \leq x \leq a \leq \omega$. We can see that

$$\frac{\partial m}{\partial x} = -\gamma e^{-\gamma x - \lambda^*(a-x)} < 0.$$

Thus, we have that

$$m(a, x, \lambda) \geq m(a, a, \lambda) = e^{-\gamma a} - \lambda^* \int_0^a e^{-\gamma\sigma - \lambda^*(a-\sigma)} d\sigma.$$

By partial integration, we obtain

$$m(a, a, \lambda) = e^{-\lambda^* a} \left(1 - \gamma \int_0^a e^{(\lambda^* - \gamma)\sigma} d\sigma\right).$$

Therefore, if $\lambda^*$ satisfies the condition

$$1 > \gamma \int_0^\omega e^{(\lambda^* - \gamma)\sigma} d\sigma, \tag{6.30}$$

then $m(a, a, \lambda)$, and hence $J(x, \lambda)$, is positive almost everywhere. Condition (6.30) is satisfied if $\lambda^*$ is sufficiently small. In such a case, the distribution of roots of the characteristic equation $\hat{J}(\alpha, \lambda^*) = 1$ is the same as for Lotka's characteristic equation, so it has no root in the right half-plane because

$$\hat{J}(0, \lambda^*) = \hat{G}(0, \lambda^*) - \lambda^* \hat{H}(0, \lambda^*) < 1.$$

That is, as long as $\lambda^*$ is sufficiently small, the corresponding endemic steady state is locally asymptotically stable.

From (6.29), we know that $\lambda^*$ is small if

$$\hat{G}(0, 0) = \frac{\beta}{\gamma} \int_0^{\omega} c_2(a)(1 - e^{-\gamma a}) da$$

is larger than (but very close to) unity. As seen from (6.23), the right-hand side of the above expression is the basic reproduction number $R_0$ for system (6.24).

If $\lambda^* = 0$, the characteristic equation is given by

$$\hat{G}(\alpha, 0) = \int_0^{\omega} e^{-\alpha x} G(x, 0) dx = 1.$$

Because $G(x, 0) > 0$, it follows that all characteristic roots have negative real parts if $\hat{G}(0, 0) = \int_0^{\omega} G(x, 0)dx < 1$, whereas there exists a positive characteristic root if $\hat{G}(0, 0) > 1$. In summary, we can state the following threshold properties:

**Proposition 6.3** *Suppose that $\beta(a, \zeta)$ and $\gamma(a)$ are constant. Let*

$$R_0 = \hat{G}(0, 0) = \beta \int_0^{\omega} e^{-\gamma x} \int_x^{\omega} c_2(a) da dx.$$

*Then, the disease-free steady state is locally asymptotically stable if $R_0 < 1$, whereas it is unstable if $R_0 > 1$. The endemic steady state uniquely exists if and only if $R_0 > 1$, and it is locally asymptotically stable if $|R_0 - 1|$ is sufficiently small.*

*Remark 6.2* In order to show the above conclusion rigorously, we need a functional analytic formulation (see Chap. 10). The linearized system (6.26) is formulated as an abstract Cauchy problem on $E := L^1(0, \omega) \times L^1(0, \omega)$:

$$\frac{dx(t)}{dt} = (A + B)x(t),$$

where $x(t) = (u(t), v(t))^{\mathrm{T}}$, operators $A$ with domain $\mathscr{D}(A)$ and $B$ are defined as

$$(A\phi)(a) = \begin{pmatrix} -\frac{d}{da} & 0 \\ 0 & -\frac{d}{da} - \gamma \end{pmatrix} \begin{pmatrix} \phi_1(a) \\ \phi_2(a) \end{pmatrix},$$

$$\mathcal{D}(A) = \left\{\phi = (\phi_1, \phi_2) \in E : \phi_j \in W^{1,1}(0, \omega), \phi_1(0) = \phi_2(0) = 0\right\},$$

$$(B\phi)(a) = \begin{pmatrix} -\lambda^*\phi_1(a) - s^*(a)\beta \int_0^\omega c_2(\sigma)\phi_2(\sigma)d\sigma \\ \lambda^*\phi_1(a) + s^*(a)\beta \int_0^\omega c_2(\sigma)\phi_2(\sigma)d\sigma \end{pmatrix}.$$

Characteristic roots $\alpha$ are eigenvalues of the linearized operator $A + B$ at the steady state. Since $A + B$ has a compact resolvent, the spectrum of $A + B$ is the point spectrum. Because $e^{t(A+B)}$ is an eventually compact semigroup, the spectral mapping theorem holds and the spectral bound of $A + B$ gives the growth bound of the linearized semigroup $e^{(A+B)t}$. Using the principle of linearized stability, we can conclude that the local stability of the steady state is determined by the sign of the real part of the dominant eigenvalue of $A + B$ [30, 54].

*Remark 6.3* Although we have shown the local stability [54] of the disease-free steady state when $R_0 < 1$, our argument also implies the global stability of the disease-free steady state. To see this, let us consider the linearized system at the disease-free steady state for the infected population:

$$\begin{aligned} &\frac{\partial v(t,a)}{\partial t} + \frac{\partial v(t,a)}{\partial a} = -\gamma v(t,a) + \beta \int_0^\omega c_2(\sigma) v(t,\sigma) d\sigma, \\ &v(t,0) = 0. \end{aligned} \tag{6.31}$$

As we have seen above, the local stability of the disease-free steady state implies that $v(t, a)$ converges to zero if $R_0 < 1$. However, using the comparative argument, we know that $i(t, a) \le v(t, a)$ because $s \le 1$. Thus, we can conclude that the local stability of the disease-free steady state implies its global stability.

## 6.3 Epidemic in a Stable Population

In this section, we return to the normalized problem (6.6). That is, we take into account vertical transmission and the vaccination term, and remove the assumption of demographic stationarity for the host population. Instead, for simplicity, we assume that all epidemic parameters ($\beta$, $\gamma$, and $\theta$) are constant. Readers who are interested in the general age-dependent case for this model may consult [31].

### *6.3.1 Threshold Condition for Invasion and Endemicity*

First, let us calculate the steady state solutions and the *threshold number*. Originally, the definition of the basic reproduction number assumed that the host population was in a demographic steady state [15, 16], and so we here use the term "threshold number" instead of the basic reproduction number to formulate the invasion condition in an exponentially growing (or shrinking) host population, although we can define the basic reproduction number in a non-stationary population ([35], Chap. 9). If we denote the steady state solutions as $(s^*, i^*)$, then it follows that

$$\begin{aligned} \frac{ds^*(a)}{da} &= -(\theta + \lambda^*)s^*(a), \\ \frac{di^*(a)}{da} &= \lambda^* s^*(a) - \gamma i^*(a), \\ s^*(0) &= 1 - q\int_0^\omega c_1(a)i^*(a)da, \\ i^*(0) &= q\int_0^\omega c_1(a)i^*(a)da, \\ \lambda^* &= \beta\int_0^\omega c_2(\sigma)i^*(\sigma)d\sigma. \end{aligned} \tag{6.32}$$

By solving the above equations, we obtain

$$\begin{aligned} s^*(a) &= s^*(0)e^{-(\theta+\lambda^*)a}, \\ i^*(a) &= i^*(0)e^{-\gamma a} + s^*(0)\lambda^*\int_0^a e^{-\gamma(a-s)-(\theta+\lambda^*)s}ds. \end{aligned}$$

Inserting the above expressions into the boundary condition and using the relation $s^*(0) + i^*(0) = 1$, we have

$$i^*(0) = qAi^*(0) + (1 - i^*(0))qB(\lambda^*),$$

where $A$ and $B(\lambda^*)$ are defined by

$$A := \int_0^\omega c_1(a)e^{-\gamma a}da, \quad B(\lambda^*) := \int_0^\omega c_1(a)\int_0^a e^{-\gamma(a-s)-(\theta+\lambda^*)s}dsda.$$

Therefore, we obtain

$$i^*(0) = \frac{q\lambda^* B(\lambda^*)}{1 - qA + q\lambda^* B(\lambda^*)}.$$

Inserting the above equation into the expression for $\lambda^*$, we have

$$\lambda^* = \beta\lambda^* D(\lambda^*) + \frac{q\lambda^* B(\lambda^*)(\beta C - \beta\lambda^* D(\lambda^*)}{1 - qA + q\lambda^* B(\lambda^*)}, \tag{6.33}$$

where

$$C := \int_0^\omega c_2(a)e^{-\gamma a}da,$$
$$D(\lambda^*) := \int_0^\omega c_2(a)\int_0^a e^{-\gamma(a-s)-(\theta+\lambda^*)s}dsda.$$

From the above argument, we know that there exists a steady state solution if and only if (6.33) has a nonnegative solution $\lambda^*$. Clearly, $\lambda^* = 0$ is the trivial solution corresponding to the disease-free steady state $(s^*, i^*) = (e^{-\theta a}, 0)$. To find a positive solution $\lambda^* > 0$, let us consider the characteristic equation

$$F(\lambda^*) = 1,$$

where

$$F(x) := \beta D(x) + \frac{q\beta B(x)(C - xD(x))}{1 - qA + qxB(x)}.$$

That is, the positive solution of (6.33) is given by a positive root of the equation $F(x) = 1$. Because it is easy to see that $\lim_{x\to\infty} F(x) = 0$, we know that there exists at least one positive root if

$$F(0) = \beta D(0) + \frac{q\beta CB(0)}{1 - qA} > 1.$$

In particular, if there is no vertical vaccination ($q = 0$), $F(x)$ becomes a monotone decreasing function, and the endemic steady state exists uniquely under the condition $\beta D(0) > 1$. Thus, $F(0)$ becomes a threshold number for endemicity.

### 6.3.2 *Local Stability of Steady States*

Let us now examine the stability of the steady states. Observe that the linearized equation at the disease-free steady state is given as

$$\begin{aligned}
\frac{\partial i(t,a)}{\partial t} + \frac{\partial i(t,a)}{\partial a} &= \beta e^{-\theta a}\int_0^\omega c_2(\sigma)i(t,\sigma)d\sigma - \gamma i(t,a), \\
i(t,0) &= q\int_0^\omega c_1(a)i(t,a)da.
\end{aligned} \tag{6.34}$$

If we assume a separation of variables solution such as $i(t,a) = e^{zt}y(a)$, we obtain

$$\begin{aligned}
\frac{dy(a)}{da} &= -(z+\gamma)y(a) + \beta e^{-\theta a}\int_0^\omega c_2(\sigma)y(\sigma)d\sigma, \\
y(0) &= q\int_0^\omega c_1(a)y(a)da.
\end{aligned} \tag{6.35}$$

By solving the above equation, we have

$$y(a) = qe^{-(z+\gamma)a}\langle c_1, y\rangle + \beta \int_0^a e^{-(z+\gamma)(a-s)-\theta s} ds \langle c_2, y\rangle, \tag{6.36}$$

where we have used the notation $\langle f, g\rangle := \int_0^\omega f(a)g(a)da$. Multiplying both sides of (6.36) by $c_1(a)$ or $c_2(a)$ and integrating from 0 to $\omega$, it follows that

$$\begin{pmatrix} \langle c_1, y\rangle \\ \langle c_2, y\rangle \end{pmatrix} = \begin{pmatrix} qJ(z) & \beta K(z) \\ qL(z) & \beta M(z) \end{pmatrix} \begin{pmatrix} \langle c_1, y\rangle \\ \langle c_2, y\rangle \end{pmatrix}, \tag{6.37}$$

where

$$\begin{aligned} J(z) &:= \int_0^\omega c_1(a)e^{-(z+\gamma)a}da, \\ K(z) &:= \int_0^\omega c_1(a) \int_0^a e^{-(z+\gamma)(a-s)-\theta s} dsda, \\ L(z) &:= \int_0^\omega c_2(a)e^{-(z+\gamma)a}da, \\ M(z) &:= \int_0^\omega c_2(a) \int_0^a e^{-(z+\gamma)(a-s)-\theta s} dsda. \end{aligned}$$

The simultaneous Eq. (6.37) for the unknown numbers $\langle c_1, y\rangle$ and $\langle c_2, y\rangle$ has non-trivial solutions if and only if

$$(1 - qJ(z))(1 - \beta M(z)) - q\beta K(z)L(z) = 0, \tag{6.38}$$

which is the characteristic equation for determining the growth rate of the infected population. If $\Re z \geq 0$, (6.38) can be rewritten as

$$1 = G(z) := \beta M(z) + \frac{q\beta K(z)L(z)}{1 - qJ(z)}.$$

Because $G(+\infty) = 0$, there exists at least one positive root if $G(0) > 1$. Hence, the disease-free steady state is unstable if $G(0) = F(0) > 1$. However, if $F(0) = G(0) < 1$, there is no root in the right half-plane. In fact, if $z$ is a root of (6.38) with $\Re z \geq 0$, it follows that

$$\begin{aligned} 0 =& |(1 - qJ(z))(1 - \beta M(z)) - q\beta K(z)L(z)| \\ &\geq (1 - q|J(z)|)(1 - \beta|M(z)|) - q\beta|K(z)||L(z)| \\ &\geq (1 - qJ(0))(1 - \beta M(0)) - q\beta K(0)L(0) = (1 - qJ(0))(1 - G(0)) > 0, \end{aligned}$$

which is a contradiction. We can conclude that the disease-free steady state is locally asymptotically stable if $F(0) = G(0) < 1$. We can summarize the above argument as follows:

**Proposition 6.4** *Suppose that $\beta$, $\gamma$, and $\theta$ are constant. If $F(0) > 1$, then there exists at least one endemic steady state and the disease-free steady state is unstable, whereas it is locally asymptotically stable if $F(0) < 1$. If there is no vertical transmission, the endemic steady state uniquely exists if $F(0) > 1$.*

To observe how the epidemic depends on the host demography, it would be interesting to examine the sensitivity of $F(0)$ with respect to the Malthusian parameter $r_0$. Let $R_1(r_0)$ be the threshold number related to direct horizontal transmission, which is a function of the stable growth rate $r_0$. Then, we have

$$R_1(r_0) := \beta D(0) = \beta \int_0^\omega c_2(a, r_0) \int_0^a e^{-\gamma(a-s)-\theta s} ds da = \beta \int_0^\omega c_2(a, r_0) f(a) da,$$

where

$$f(a) := \int_0^a e^{-\gamma(a-s)-\theta s} ds = \frac{e^{-\theta a} - e^{-\gamma a}}{\gamma - \theta}$$

and $c_2(a, r_0)$ denotes the stable age profile with the stable growth rate $r_0$:

$$c_2(a, r_0) := \frac{e^{-r_0 a}\ell(a)}{\int_0^\omega e^{-r_0 a}\ell(a) da}.$$

By a simple calculation, we can observe that

$$\frac{\partial R_1(r_0)}{\partial r_0} = \beta \int_0^\omega f(a) c_2(a, r_0)(\bar{a}_2 - a) da,$$

where

$$\bar{a}_2 := \int_0^\omega a c_2(a, r_0) da$$

is the average age of the host stable age distribution. If we consider the case in which there is no vaccination ($\theta = 0$), $f(a)$ is a monotone increasing function. We can then observe that

$$\begin{aligned}
&\int_0^\omega f(a) c_2(a, r_0)(\bar{a}_2 - a) da \\
&= \int_0^{\bar{a}_2} f(a) c_2(a, r_0)(\bar{a}_2 - a) da + \int_{\bar{a}_2}^\omega f(a) c_2(a, r_0)(\bar{a}_2 - a) da \\
&\le f(\bar{a}_2) \int_0^{\bar{a}_2} c_2(a, r_0)(\bar{a}_2 - a) da + f(\bar{a}_2) \int_{\bar{a}_2}^\omega c_2(a, r_0)(\bar{a}_2 - a) da = 0.
\end{aligned}$$

Therefore, $R_1(r_0)$ is a monotone decreasing function of $r_0$. Because the stable age structure $c_2(a, r_0)$ is monotone decreasing with respect to the intrinsic growth rate $r_0$, $R_1$ becomes larger if $r_0$ is smaller; that is, the age structure is aging. However, if the growth rate $r_0$ is larger, meaning that the age structure is younger, the threshold number related to horizontal transmission is smaller. This fact supports an observation made by McLean [45]. However, if the vaccination term is nonzero, $f(a)$ has a unimodal pattern, and population aging does not necessarily lead to an increase in the threshold number.

Finally, let us consider the threshold number related to horizontal transmission mediated by vertical transmission defined by

$$R_2(r_0) := \frac{V(r_0)W(r_0)}{1 - qA(r_0)},$$

where

$$A(r_0) := \int_0^\omega c_1(a, r_0)e^{-\gamma a}da,$$
$$V(r_0) := \beta \int_0^\omega c_2(a, r_0)e^{-\gamma a}da,$$
$$W(r_0) := q \int_0^\omega c_1(a, r_0) \int_0^a e^{-\gamma(a-s)-\theta s}dsda.$$

Using a similar argument to that for the horizontal transmission case, if $\theta = 0$, we obtain

$$\frac{\partial A(r_0)}{\partial r_0} > 0, \quad \frac{\partial V(r_0)}{\partial r_0} > 0, \quad \frac{\partial W(r_0)}{\partial r_0} < 0.$$

Therefore, even in the case of no vaccination, $\partial R_2(r_0)/\partial r_0$ does not necessarily have a constant sign, and the response of $R_2$ to the intrinsic growth rate is complex.

## 6.4 Threshold Principle and $R_0$

Although we have so far considered the basic model (6.1) under some simplified conditions for the transmission coefficient, such as constant parameters or the separable mixing assumption, we can remove these restrictions to study the invasion threshold condition under full age dependency. Our purpose here is to introduce a general framework for calculating $R_0$, the NGO, and the intrinsic growth rate of the infected population in the invasion phase for age-structured populations. Our argument is rather formal; a biologically reasonable interpretation for the definition of $R_0$ and its general theory are discussed in Chap. 9.

We assume that the host population is in a demographic steady state and there is no vaccination term, because originally the "basic" reproduction number was defined for a completely susceptible population in a demographic steady state. Next, we will treat the case in which the host population is a stable population with a nonzero intrinsic growth rate. In cases where the host population is not in a demographic steady state, we can distinguish relative eradication and absolute eradication of the disease, because the proportion of infected individuals in the total population can decrease, even when the number of infected individuals increases. Therefore, in general, the invasion threshold condition does not necessarily imply the endemic threshold [34, 39].

The disease invasion process at the disease-free (totally susceptible) steady state is described by the following linearized equation:

$$\frac{\partial I(t,a)}{\partial t} + \frac{\partial I(t,a)}{\partial a} = -(\mu(a)+\gamma(a))I(t,a) + c_2(a)\int_0^{\omega} \beta(a,\sigma)I(t,\sigma)d\sigma,$$
$$I(t,0) = q\int_0^{\omega} m(a)I(t,a)da, \tag{6.39}$$

where $c_2(a) = \ell(a)/\int_0^{\omega} \ell(x)dx$ is the age profile of the completely susceptible host population at a demographic steady state. In the following, we outline the computation of $R_0$ based on (6.39).

### 6.4.1 *Horizontal Transmission*

First, we consider the purely horizontal transmission model with $q = 0$. The assumption $I(t,0) = 0$ implies that newborns of age zero are not horizontally infected by the existing infected population. This is a biologically reasonable assumption because of the existence of maternal antibodies.

There are two perspectives for describing the asymptotic dynamics governed by system (6.39). The *period (cross-sectional) approach* observes how the age distribution at each time $t$ evolves with time. This approach is well described by formulating (6.39) as a dynamical system (ordinary differential equations in a Banach space):

$$\frac{dI(t)}{dt} = (A+F)I(t), \quad I(0) = I_0, \tag{6.40}$$

where $I(t)$ is a function that takes values in the state space $L^1(0,\omega)$ defined by $I(t)(a) := I(t,a)$, and the linear operator $A$ (the population operator) and $F$ are defined by

$$(A\phi)(a) := -\frac{d\phi(a)}{da} - (\mu(a)+\gamma(a))\phi(a),$$
$$(F\phi)(a) := c_2(a)\int_0^{\omega} \beta(a,\sigma)\phi(\sigma)d\sigma,$$

where the domain of $A$ is given by

$$\mathscr{D}(A) = \{\phi \in L^1(0,\omega) : \phi(0) = 0,\ A\phi \in L^1(0,\omega)\}.$$

Let $\hat{I}(z,a) := \int_0^{\infty} e^{-zt} I(t,a)dt$, $z \in \mathbb{C}$. Taking the Laplace transformation on both sides of (6.40), we obtain

$$\hat{I}(z,a) - I_0(a) = (A+F)\hat{I}(z,\cdot)(a),$$

then by the inverse Laplace transformation, it follows that

$$I(t,a) = \frac{1}{2\pi i}\int_{\Gamma} e^{zt}(zI - (A+F))^{-1}I(0,\cdot)(a)dz,$$

where $\Gamma$ is an integral path in the resolvent set of $A+F$. As is seen in Chap. 3, we can expect that $I(t,a)$ admits an asymptotic eigenfunction expansion as

$$I(t,a) \approx \sum_{k=0}^{\infty} \alpha_k e^{z_k t}\phi_k(a),$$

where $\phi_k$ is the eigenvector of $A+F$ associated with its eigenvalue $z_k$. Observe that the eigenvalue equation $(A+F)\phi = z\phi$ can be transformed into a fixed point equation as $\psi = F(z-A)^{-1}\psi$, where $\psi := (z-A)\phi$. Then, it is not difficult to see

$$\phi(a) = ((z-A)^{-1}\psi)(a) = \int_0^a e^{-z(a-s)}\frac{\ell(a)\Gamma(a)}{\ell(s)\Gamma(s)}\psi(s)ds.$$

Therefore, we have

$$\psi = F(z-A)^{-1}\psi =: \hat{\Pi}(z)\psi,$$

where $\hat{\Pi}(z)$ is an integral operator acting on $L^1(0,\omega)$ defined by

$$\begin{aligned} \hat{\Pi}(z) &:= \int_0^{\omega} e^{-z\tau}\Pi(\tau)d\tau, \\ (\Pi(\tau)f)(a) &:= c_2(a)\int_0^{\omega}\beta(a,\tau+\zeta)\frac{\ell(\tau+\zeta)\Gamma(\tau+\zeta)}{\ell(\zeta)\Gamma(\zeta)}f(\zeta)d\zeta, \end{aligned} \tag{6.41}$$

where we adopt a convention that $\beta(a,\sigma) = 0$ for $\sigma > \omega$ and $\ell(a) = \Gamma(a) = 0$ for $a > \omega$. Then, $z \in \mathbb{C}$ is the eigenvalue of $A+F$ if and only if $\hat{\Pi}(z)$ has an eigenvalue unity. Since $\hat{\Pi}(z)$, $z \in \mathbb{R}$ is a positive operator, and its spectral radius, denoted by $r(\hat{\Pi}(z))$, is the dominant positive eigenvalue whose associated eigenvector is a unique eigenvector in the positive cone if $\hat{\Pi}(z)$, $z \in \mathbb{R}$ is compact and non-supporting.[5] Moreover, $r(\hat{\Pi}(z))$ decreases from $+\infty$ to 0 as $z \in \mathbb{R}$ moves from $-\infty$ to $+\infty$, and there exists a unique positive root $r_0$ of equation $r(\hat{\Pi}(z)) = 1$ such that the *sign relation* holds: $\text{sign}(r_0) = \text{sign}(\hat{\Pi}(0) - 1)$. Since we can prove that $r_0$ is the dominant eigenvalue of $A+F$ (see [30] and Chap. 10), $r_0$ is the asymptotic growth rate (the intrinsic growth rate or the Malthusian parameter) of the infected population in the initial invasion phase.

In order to see that $\hat{\Pi}(0)$ is the next-generation operator, let us consider the *cohort approach*. Let $J(t,\tau;a)$ be the density of infected population at infection-age $\tau$

[5] For the definition of the non-supporting operator, see Chap. 10

whose age at infection is $a$. That is, $J(t, \tau; a)$ denotes the density of the *infection cohort* with the age at infection $a$. Then, we have

$$I(t, a) = \int_0^a J(t, \tau; a - \tau) d\tau,$$

and we can rewrite (6.39) as the cohort equation for $J$:

$$\begin{aligned} \left( \frac{\partial}{\partial t} + \frac{\partial}{\partial \tau} \right) J(t, \tau; a) &= -(\mu(a + \tau) + \gamma(a + \tau)) J(t, \tau; a), \\ J(t, 0; a) &= c_2(a) \int_0^\omega \beta(a, \sigma) \int_0^\sigma J(t, \tau; \sigma - \tau) d\tau d\sigma, \end{aligned} \tag{6.42}$$

where $J(t, 0; a)$ denotes the incidence rate of the newly infected population. From the above McKendrick equation, if we define $v(t, a) := J(t, 0; a)$, it follows that

$$J(t, \tau; a) = \begin{cases} \frac{\ell(a+\tau)\Gamma(a+\tau)}{\ell(a)\Gamma(a)} v(t - \tau, a), & t - \tau > 0, \\ \frac{\ell(a+\tau)\Gamma(a+\tau)}{\ell(a+\tau-t)\Gamma(a+\tau-t)} J(0, \tau - t; a), & \tau - t > 0. \end{cases}$$

Inserting the above expression into the boundary condition and changing the order of integrals, we have

$$v(t, a) = g(t, a) + c_2(a) \int_0^t d\tau \int_\tau^\omega \beta(a, \sigma) \frac{\ell(\sigma)\Gamma(\sigma)}{\ell(\sigma - \tau)\Gamma(\sigma - \tau)} v(t - \tau, \sigma - \tau) d\tau d\sigma,$$

where

$$g(t, a) := c_2(a) \int_{t \wedge \omega}^\omega d\tau \int_\tau^\omega \beta(a, \sigma) \frac{\ell(\sigma)\Gamma(\sigma)}{\ell(\sigma - t)\Gamma(\sigma - t)} J(0, \tau - t; \sigma - t) d\sigma.$$

The above equation can be seen as an abstract renewal integral equation:

$$v(t) = g(t) + \int_0^t \Pi(\tau) v(t - \tau) d\tau, \tag{6.43}$$

where $g$ and $v$ are functions from $\mathbb{R}_+$ to $L^1(0, \omega)$ defined by $g(t)(a) := g(t, a)$ and $v(t)(a) := v(t, a)$, and $\Pi(\tau)$ is the operator on $L^1$ defined above.

Let $v_n(t)$, called the $n$-th generation distribution, be the density of newly infecteds produced per unit time at time $t$. Let $g(t)$ be the zero-th generation (secondary cases) directly transmitted from the initial infected population (primary cases) $J(0, \tau; a)$ and so we can define the successive infective generations as follows:

$$v_0(t) = g(t), \quad v_{n+1}(t) = \int_0^t \Pi(\tau) v_n(t - \tau) d\tau, \quad (n = 0, 1, 2, \ldots),$$

where the size of $n$-th generation distribution is given by its norm:

$$|v_n|_Y = \int_0^\infty \int_0^\omega |v_n(t,a)|dadt,$$

where $Y := L^1(\mathbb{R}_+ \times (0,\omega))$. The positive operator $K_Y$ on $Y_+$ defined by

$$K_Y : f \to \int_0^t \Pi(\tau) f(t-\tau)d\tau$$

is called the *generation evolution operator* (GEO) (see Chap. 9). Then, the $n$-th generation distribution is given by $v_n = K_Y^n g$. Integrating both sides of the iterative relation from 0 to $\infty$, we have

$$\int_0^\infty v_{n+1}(t)dt = \int_0^\infty \int_0^t \Pi(\tau)v_n(t-\tau)d\tau dt = \int_0^\infty \Pi(\tau)d\tau \int_0^\infty v_n(t)dt.$$

Therefore, the "time-aggregated" $n$-th generation density of the newly infected population, denoted by $w_n := \int_0^\infty v_n(t)dt$, is calculated as $w_n = K^n w_0$ where $K := \hat{\Pi}(0) = \int_0^\infty \Pi(\tau)d\tau$. That is, the biological generation evolution process $v_n = K_Y^n g$ on $Y_+$ is reduced to an iteration process $w_n = K^n w_0$, evolved by the NGO, on the space of time-aggregated generation distributions and $|w_n|_{L^1} = |v_n|_Y$. Since the spectral radius of the next-generation operator $K$ is defined by $r(K) = \lim_{n\to\infty} \sqrt[n]{\|K^n\|_{\mathscr{L}(E)}}$, where $\|\cdot\|_{\mathscr{L}(E)}$ denotes the operator norm for bounded linear operators on $E = L^1(0,\omega)$, it follows from $|w_n|_{L^1} \le |K^n|_{\mathscr{L}(E)}|w_0|_{L^1}$ that $\lim_{n\to\infty} \sqrt[n]{|v_n|_Y} = \lim_{n\to\infty} \sqrt[n]{|w_n|_{L^1}} \le r(K)$, where the left-hand side is, if the limit exists, the asymptotic per-generation growth factor. If $K$ is a non-supporting compact operator, $r(K)$ becomes the positive dominant eigenvalue of $K$, that is, $|\mu| < r(K)$ for any eigenvalue $\mu$ of $K$. In such a case, we can expect that the $n$-th infective generation $w_n$ will grow geometrically with the growth factor $r(K)$. More precisely speaking, there exists a positive functional $F_0$ associated with the eigenvalue $r(K)$ such that $\lim_{n\to\infty} r(K)^{-n}K^n w_0 = \frac{\langle F_0, w_0\rangle}{\langle F_0, \psi_0\rangle}\psi_0$, where $\psi_0$ is the positive eigenvector of $K$ corresponding to the eigenvalue $r(K)$ (Proposition 10.11). Therefore, it follows that $\lim_{n\to\infty} \sqrt[n]{|v_n|_Y} = \lim_{n\to\infty} \sqrt[n]{|w_n|_{L^1}} = r(K)$, which shows that the spectral radius $r(K)$ gives the asymptotic per-generation growth factor with respect to the total size of generation. From the above argument, $r(K)$ satisfies not only the sign relation but also the generational interpretation, and we can define the basic reproduction number $R_0$ by $R_0 = r(K)$ (see Chap. 9).

*Remark 6.4* Applying the variation-of-constants formula to (6.40), we have

$$I(t) = e^{tA}I(0) + \int_0^t e^{\sigma A}FI(t-\sigma)d\sigma,$$

where $e^{tA}$ is the $C_0$-semigroup generated by $A$. Then, if we define $v(t) := FI(t)$, it follows that

$$v(t) = Fe^{tA}I(0) + \int_0^t Fe^{\sigma A}v(t-\sigma)d\sigma.$$

Therefore, as seen in Sect. 2.5, the NGO for (6.40) is calculated as

$$\int_0^\infty Fe^{\sigma A}d\sigma = F(-A)^{-1},$$

where we have used the fact that $A$ has no spectrum in the half-plane $\Re z \geq 0$ and the well-known relation $(z-A)^{-1} = \int_0^\infty e^{-zt}S(t)dt$ for $z \in \rho(A)$ and $S(t) = e^{tA}$. The sign relation $\text{sign}(r_0) = \text{sign}(R_0 - 1)$ is clear, because $r_0$ is given as a real number such that the positive operator

$$\int_0^\infty e^{-z\sigma}Fe^{\sigma A}d\sigma = F(z-A)^{-1} = \hat{\Pi}(z), \quad z > s(A),$$

has a spectral radius of unity at $z = r_0$. Then, there exists a positive eigenfunction $\phi$ such that $F(r_0 - A)^{-1}\phi = \phi$, which implies $(A+F)v = r_0 v$ with $v := (r_0 - A)^{-1}\phi$. Hence, $r_0$ is the real (indeed, dominant) eigenvalue of the linearized generator $A+F$, and it is positive if $R_0 = r(F(-A)^{-1}) > 1$ and negative if $R_0 < 1$.

We can summarize the above argument as the following general principle:

**Proposition 6.5** (Invasion Threshold Principle) *Let $R_0$ be the basic reproduction number given by the dominant positive eigenvalue (the spectral radius) of the NGO $K = F(-A)^{-1}$. The disease can invade the completely susceptible host population if $R_0 > 1$, whereas it cannot if $R_0 < 1$. $R_0$ gives the asymptotic per-generation growth factor of the size of each generation of infected individuals, the intrinsic growth rate $r_0$ of the infected population is given by the dominant real eigenvalue of the linearized generator $A + F$, and the sign relation* $\text{sign}(r_0) = \text{sign}(R_0 - 1)$ *holds.*

Note that we assume the disease is carried by a small population. As seen in the previous chapter, the invasion threshold principle concerns the local stability of the disease-free steady state, so $R_0 < 1$ does not necessarily exclude the existence of an endemic steady state.

An important special case for which we can compute $R_0$ explicitly is the separable mixing case, where the transmission coefficient is written in the separated form $\beta(a,\sigma) = \beta_1(a)\beta_2(\sigma)$. If we adopt the separable mixing assumption, the next-generation operator becomes a one-dimensional operator:

$$(Kf)(a) = c_2(a)\beta_1(a)\int_0^\omega\int_0^\omega \beta_2(\tau+\zeta)\frac{\ell(\tau+\zeta)\Gamma(\tau+\zeta)}{\ell(\zeta)\Gamma(\zeta)}d\tau f(\zeta)d\zeta.$$

Then, $c_2\beta_1$ is the positive eigenfunction of $K$ associated with $R_0 = r(K)$, and so we have

$$R_0 = \int_0^\omega\int_0^\omega \beta_2(\tau+\zeta)\frac{\ell(\tau+\zeta)\Gamma(\tau+\zeta)}{\ell(\zeta)\Gamma(\zeta)}d\tau c_2(\zeta)\beta_1(\zeta)d\zeta, \tag{6.44}$$

which is already given in (6.22).

*Remark 6.5* If we use the integral equation approach, we can reach the next-generation operator by a classical calculation. Let

$$B(t,a) := c_2(a)\int_0^\omega \beta(a,\sigma)I(t,\sigma)d\sigma,$$

be the density of newly infected individuals. Then, we have

$$\frac{\partial I(t,a)}{\partial t} + \frac{\partial I(t,a)}{\partial a} = -(\mu(a)+\gamma(a))I(t,a) + B(t,a),$$

with the boundary condition $I(t,0) = 0$. Using the method of characteristic lines, we obtain

$$I(t,a) = \int_0^a \frac{\ell(a)\Gamma(a)}{\ell(s)\Gamma(s)}B(t-a+s,s)ds,$$

for $t-a>0$. Inserting this expression into the definition of $B$, we have for $t>\omega$

$$\begin{aligned}B(t,a) &= c_2(a)\int_0^\omega \beta(a,\sigma)\int_0^\sigma \frac{\ell(\sigma)\Gamma(\sigma)}{\ell(s)\Gamma(s)}B(t-\sigma+s,s)ds \\ &= c_2(a)\int_0^\omega d\tau \int_0^\omega \beta(a,\tau+\zeta)\frac{\ell(\tau+\zeta)\Gamma(\tau+\zeta)}{\ell(\zeta)\Gamma(\zeta)}B(t-\tau,\zeta)d\zeta \\ &= \int_0^\omega (\Pi(\tau)B(t-\tau,\cdot))(a)d\tau.\end{aligned}$$

Then, we can again arrive at the next-generation operator $K = \int_0^\infty \Pi(\tau)d\tau$.

### 6.4.2 Vertical Transmission

The above invasion threshold result also holds for a disease with vertical transmission, although we omit the proof here. Readers are referred to Inaba [31] for the basic theory of age-structured SIR models with vertical transmission. Instead, we only give a formal calculation of the NGO for the case $q>0$. For this purpose, we formally adopt the extended state space method used by Thieme (see Chap. 10).

Let $X := L^1(0,\omega)$ be the state space of age-density vectors, let $Z := \mathbb{R}\times X$ be the extended state space for the age-density function and its boundary value, and let $Z_0 := \{0\}\times X$ be its closed subspace. Define an operator $\mathscr{A}$ on $Z$ by

$$\mathscr{A}(0,\psi)^{\mathrm{T}} := \left(-\psi(0), -\frac{d\psi(a)}{da} - (\mu(a)+\gamma(a))\psi(a)\right)^{\mathrm{T}}, \quad (0,\psi)\in\mathscr{D}(\mathscr{A})$$

and a bounded linear perturbation operator $\mathscr{B} : Z_0 \to Z$ by

$$\mathscr{B}(0, \psi)^{\mathrm{T}} = \left( q \int_0^{\omega} m(a)\psi(a)da, c_2(a) \int_0^{\omega} \beta(a, \sigma)\psi(\sigma)d\sigma \right)^{\mathrm{T}}, \quad (0, \psi) \in Z_0.$$

The linearized equation for the infected population can then be written as a linear Cauchy problem in the extended state space $Z$:

$$\frac{du(t)}{dt} = \mathscr{A}u(t) + \mathscr{B}u(t), \quad u(0) = (0, \phi) \in Z_0, \tag{6.45}$$

where $u(t) = (0, I(t, \cdot)) \in Z_0$.

From the general framework for defining the NGO, we calculate the following operator from $Z$ into itself: $\mathscr{K} := \mathscr{B}(-\mathscr{A})^{-1}$. Let $(-\mathscr{A})^{-1}(z_1, z_2)^{\mathrm{T}} = (0, \psi)^{\mathrm{T}}$. Then, we have the expression:

$$\psi(a) = \ell(a)\Gamma(a)z_1 + \int_0^a \frac{\ell(a)\Gamma(a)}{\ell(s)\Gamma(s)} z_2(s)ds.$$

It follows that

$$\mathscr{B}(-\mathscr{A})^{-1} \begin{pmatrix} z_1 \\ z_2 \end{pmatrix} = \begin{pmatrix} q \int_0^{\omega} m(a) \left[ \ell(a)\Gamma(a)z_1 + \int_0^a \frac{\ell(a)\Gamma(a)}{\ell(\sigma)\Gamma(\sigma)} z_2(\sigma)d\sigma \right] da \\ c_2(a) \int_0^{\omega} \beta(a, \zeta) \left[ \ell(\zeta)\Gamma(\zeta)z_1 + \int_0^{\zeta} \frac{\ell(\zeta)\Gamma(\zeta)}{\ell(\sigma)\Gamma(\sigma)} z_2(\sigma)d\sigma \right] d\zeta \end{pmatrix}.$$

Thus, the spectral radius $r(\mathscr{K})$ gives the basic reproduction number.

As an example, let us consider the separable mixing case. Suppose that $\beta(a, \sigma)$ can be separated as $\beta(a, \sigma) = \beta_1(a)\beta_2(\sigma)$. Then, we have

$$\mathscr{K} \begin{pmatrix} z_1 \\ z_2 \end{pmatrix} = \begin{pmatrix} q \int_0^{\omega} m(a) \left[ \ell(a)\Gamma(a)z_1 + \int_0^a \frac{\ell(a)\Gamma(a)}{\ell(\sigma)\Gamma(\sigma)} z_2(\sigma)d\sigma \right] da \\ c_2(a)\beta_1(a) \int_0^{\omega} \beta_2(\zeta) \left[ \ell(\zeta)\Gamma(\zeta)z_1 + \int_0^{\zeta} \frac{\ell(\zeta)\Gamma(\zeta)}{\ell(\sigma)\Gamma(\sigma)} z_2(\sigma)d\sigma \right] d\zeta \end{pmatrix}.$$

Therefore, the positive eigenvector of $\mathscr{K}$ associated with $R_0 = r(\mathscr{K})$ is expressed as $(\alpha_1, c_2(a)\beta_1(a)\alpha_2)^{\mathrm{T}}$, where $\alpha_j > 0$. It follows that

$$R_0 \begin{pmatrix} \alpha_1 \\ \alpha_2 \end{pmatrix} = A \begin{pmatrix} \alpha_1 \\ \alpha_2 \end{pmatrix},$$

where $A$ is a $2 \times 2$ positive matrix defined by

$$A := \begin{pmatrix} q \int_0^{\omega} m(a)\ell(a)\Gamma(a)da & q \int_0^{\omega} m(a) \int_0^a \frac{\ell(a)\Gamma(a)}{\ell(\sigma)\Gamma(\sigma)} c_2(\sigma)\beta_1(\sigma)d\sigma da \\ \int_0^{\omega} \beta_2(\zeta)\ell(\zeta)\Gamma(\zeta)d\zeta & \int_0^{\omega} \beta_2(\zeta) \int_0^{\zeta} \frac{\ell(\zeta)\Gamma(\zeta)}{\ell(\sigma)\Gamma(\sigma)} c_2(\sigma)\beta_1(\sigma)d\sigma d\zeta \end{pmatrix}.$$

We can calculate $R_0$ as the positive eigenvalue of the two-dimensional matrix $A$, that is, $R_0 = r(A)$.

To support the above formal calculation, let us introduce the renewal equation for horizontally and vertically transmitted diseases. Let $I_v(t,a)$ be the age density of vertically transmitted infected individuals and let $b(t) = I_v(t,0)$ be the number of vertically transmitted newborns at time $t$. Then, we have

$$I(t,a) = \int_0^a J(t,\tau;a-\tau)d\tau + I_v(t,a),$$

where $J(t,\tau;a)$, $a > 0$ denotes the density of horizontally transmitted infected individuals with age at infection $a$ and infection-age $\tau$ (i.e., the chronological age is $a+\tau$). Let $v(t,a) := J(t,0;a)$ be the age density of newly horizontally transmitted infected individuals. Then, we obtain

$$\begin{aligned} v(t,a) &= c_2(a)\int_0^\omega \beta(a,\sigma)I(t,\sigma)d\sigma \\ &= c_2(a)\int_0^\omega \beta(a,\sigma)\left\{\int_0^\sigma J(t,\tau;\sigma-\tau)d\tau + I_v(t,\sigma)\right\}d\sigma, \\ b(t) &= q\int_0^\omega m(a)I(t,a)da = q\int_0^\omega m(a)\left\{\int_0^a J(t,\tau;a-\tau)d\tau + I_v(t,a)\right\}da. \end{aligned} \tag{6.46}$$

Inserting the expressions

$$\begin{aligned} J(t,\tau;a) &= \begin{cases} \frac{\ell(a+\tau)\Gamma(a+\tau)}{\ell(a)\Gamma(a)}v(t-\tau,a), & (t-\tau>0), \\ \frac{\ell(a+\tau)\Gamma(a+\tau)}{\ell(a+\tau-t)\Gamma(a+\tau-t)}J(0,\tau-t;a), & (\tau-t>0), \end{cases} \\ I_v(t,a) &= \begin{cases} \ell(a)\Gamma(a)b(t-a), & (t-a>0), \\ \frac{\ell(a)\Gamma(a)}{\ell(a-t)\Gamma(a-t)}I_v(0,a-t), & (a-t>0) \end{cases} \end{aligned} \tag{6.47}$$

into the above integral equations, we obtain a system of renewal integral equations:

$$\begin{aligned} v(t,a) &= c_2(a)\int_0^t d\tau \int_\tau^\omega \beta(a,\sigma)\frac{\ell(\sigma)\Gamma(\sigma)}{\ell(\sigma-\tau)\Gamma(\sigma-\tau)}v(t-\tau,\sigma-\tau)d\sigma \\ &\quad + c_2(a)\int_0^t \beta(a,\sigma)\ell(\sigma)\Gamma(\sigma)b(t-\sigma)d\sigma + g(t,a), \\ b(t) &= q\int_0^t d\tau \int_\tau^\omega m(a)\frac{\ell(a)\Gamma(a)}{\ell(a-\tau)\Gamma(a-\tau)}v(t-\tau,a-\tau)da \\ &\quad + q\int_0^t m(a)\ell(a)\Gamma(a)b(t-a)da + h(t), \end{aligned} \tag{6.48}$$

where $g$ and $h$ are the starting functions. Define linear operators $\pi_{ij}(\tau)$ as follows:

$$\pi_{11}(\tau)z_1 := qm(\tau)\ell(\tau)\Gamma(\tau)z_1,$$
$$\pi_{12}(\tau)z_2 := q\int_\tau^\omega m(a)\frac{\ell(a)\Gamma(a)}{\ell(a-\tau)\Gamma(a-\tau)}z_2(a-\tau)da,$$
$$(\pi_{21}(\tau)z_1)(a) := c_2(a)\beta(a,\tau)\ell(\tau)\Gamma(\tau)z_1,$$
$$(\pi_{22}(\tau)z_2)(a) := \int_\tau^\omega \beta(a,\sigma)\frac{\ell(\sigma)\Gamma(\sigma)}{\ell(\sigma-\tau)\Gamma(\sigma-\tau)}z_2(\sigma-\tau)d\sigma,$$

where $\phi \in L^1(0,\omega)$. That is, $\pi_{11}$ and $\pi_{21}$ are operators from $\mathbb{R}$ into $L^1$, $\pi_{12}$ is an operator from $L^1$ into $\mathbb{R}$, and $\pi_{22}$ is an operator from $L^1$ into itself. Let $\Pi(\tau) := (\pi_{ij}(\tau))_{1\le i,j\le 2}$ be a positive operator from $L^1_+ \times \mathbb{R}_+$ to itself. The renewal system (6.48) can then be written as

$$\begin{pmatrix} b(t) \\ v(t) \end{pmatrix} = \int_0^t \Pi(\tau)\begin{pmatrix} b(t-\tau) \\ v(t-\tau) \end{pmatrix}d\tau + \begin{pmatrix} g(t) \\ h(t) \end{pmatrix}.$$

Therefore, the NGO is calculated as

$$K = \int_0^\omega \Pi(\tau)d\tau,$$

which is identical to $\mathscr{K} = \mathscr{B}(-\mathscr{A})^{-1}$.

Finally, let us calculate the type-reproduction operator for horizontal transmission, which maps a horizontally transmitted primary case distribution to a secondary horizontally transmitted case distribution intermediated by vertical transmission. For the definition of the type-reproduction number, see Chap. 9. Using the type-reproduction number for horizontal transmission, we can obtain a control relation for disease prevention based only on the control of horizontal transmission.

Define a splitting of the NGO as

$$\mathscr{K} = \mathscr{K}_1 + \mathscr{K}_2,$$

where $\mathscr{K}_j := \mathscr{B}_j(-\mathscr{A})^{-1}$ and the operators $\mathscr{B}_j$ are defined by

$$\mathscr{B}_1(0,\psi) := \left(q\int_0^\omega m(a)\psi(a)da, 0\right), \quad \mathscr{B}_2(0,\psi) := \left(0, c_2(a)\int_0^\omega \beta(a,\sigma)\psi(\sigma)d\sigma\right),$$

for $(0,\psi) \in Z_0$.

Note that a demographic steady state with $q < 1$ has $r(\mathscr{K}_1) < 1$, because the infected population cannot be maintained by vertical transmission alone. The type-reproduction operator for horizontal transmission $\mathscr{M}_2$ is then given by $\mathscr{M}_2 := \mathscr{K}_2(I - \mathscr{K}_1)^{-1}$. Observe that

$$\begin{aligned}\mathscr{M}_2 &= \mathscr{B}_2(-\mathscr{A})^{-1}(I - \mathscr{B}_1(-\mathscr{A})^{-1})^{-1} \\ &= \mathscr{B}_2((I - \mathscr{B}_1(-\mathscr{A})^{-1})(-\mathscr{A}))^{-1} = \mathscr{B}_2(-(\mathscr{A} + \mathscr{B}_1))^{-1}.\end{aligned}$$

Let $(-(\mathscr{A}+\mathscr{B}_1))^{-1}(z_1, z_2)^{\mathrm{T}} = (0, \psi)^{\mathrm{T}}$. Then, we have

$$\psi(a) = (z_1 + q\langle m, \psi\rangle)\ell(a)\Gamma(a) + \int_0^a \frac{\ell(a)\Gamma(a)}{\ell(\sigma)\Gamma(\sigma)} z_2(\sigma)d\sigma,$$

where

$$\langle m, \psi\rangle = \frac{1}{1-\langle m, \ell\Gamma\rangle}\left[\langle m, \ell\Gamma\rangle z_1 + \int_0^\omega m(a)\int_0^a \frac{\ell(a)\Gamma(a)}{\ell(\sigma)\Gamma(\sigma)} z_2(\sigma)d\sigma da\right].$$

Using the above expression for $\psi$, we obtain

$$\mathscr{M}_2(z_1, z_2)^{\mathrm{T}} = \left(0, c_2(a)\int_0^\omega \beta(a,\sigma)\psi(\sigma)d\sigma\right)^{\mathrm{T}},$$

and the type-reproduction number for horizontal transmission, denoted by $T_2$, is given by its spectral radius: $T_2 = r(\mathscr{M}_2)$.

Because the range of $\mathscr{M}_2$ is included in $Z_0$, its positive eigenfunction associated with the type-reproduction number $T_2 = r(\mathscr{M}_2)$ should be an element of $Z_0$. Let $(0, z_2)^{\mathrm{T}}$ be the positive eigenfunction associated with $T_2 = r(\mathscr{M}_2)$. From the above expression, we have $r(\mathscr{M}_2)z_2 = K_2 z_2$, where $K_2$ is a positive operator on $L^1$ given by

$$\begin{aligned}(K_2 f)(a) &= c_2(a)\int_0^\omega \beta(a,\zeta)\int_0^\zeta \frac{\ell(\zeta)\Gamma(\zeta)}{\ell(\sigma)\Gamma(\sigma)} f(\sigma)d\sigma d\zeta \\ &+ \frac{qc_2(a)\int_0^\omega \beta(a,\zeta)\ell(\zeta)\Gamma(\zeta)d\zeta}{1-q\int_0^\omega m(x)\ell(x)\Gamma(x)dx}\int_0^\omega m(a)\int_0^a \frac{\ell(a)\Gamma(a)}{\ell(\sigma)\Gamma(\sigma)} f(\sigma)d\sigma da,\end{aligned} \tag{6.49}$$

where it is noted that $1 - q\int_0^\omega m(x)\ell(x)\Gamma(x)dx > 0$ if $q < 1$.

The first part of $K_2$ corresponds to direct horizontal transmission, and the second part is the NGO of indirect horizontal transmission by which the horizontally transmitted secondary cases are intermediated by vertically transmitted infected individuals. Thus, $r(\mathscr{M}_2) = r(K_2)$ and $R_0 > 1$ if and only if $r(K_2) > 1$, so $K_2$ is also a kind of type-reproduction operator for horizontal transmission.

### 6.4.3 *Threshold Number in the Normalized System*

As noted above, the basic reproduction number $R_0$ for the infected population was originally defined for a host population in a demographic steady state. In Sect. 6.3, we introduced a *threshold number* to formulate the invasion threshold condition for the normalized system, where the original population system is *not* in a demographic steady state. We now formally extend the idea of $R_0$ to the case where the host is a stable population and discuss the relation between $R_0$ and the threshold number of the normalized system.

In general, the threshold number for the normalized system is defined by the same procedure as the basic reproduction number of the original system. For simplicity, we assume that $q = 0$ and $\theta = 0$. The linearized system for the infected population in the normalized system (6.6) is then given by

$$\begin{aligned} &\frac{\partial i(t,a)}{\partial t} + \frac{\partial i(t,a)}{\partial a} = -\gamma(a)i(t,a) + \int_0^\omega \beta(a,\sigma)c_2(\sigma;r_0)i(t,\sigma)d\sigma, \\ &i(t,0) = 0, \end{aligned} \tag{6.50}$$

where

$$c_2(a;r_0) := \frac{e^{-r_0 a}\ell(a)}{\int_0^\infty e^{-r_0 z}\ell(z)dz}$$

is the age profile of the completely susceptible host stable population with the Malthusian parameter $r_0$.

Let $\tilde{K}$ be the next-generation(-like) operator for the normalized system. Similar to expression $K = \hat{\Pi}(0)$, we can calculate $\tilde{K}$ as

$$(\tilde{K}f)(a) := \int_0^\omega \tilde{\Phi}(a,\zeta)f(\zeta)d\zeta, \quad f \in L^1(0,\omega),$$

where

$$\tilde{\Phi}(a,\zeta) := \int_0^\omega \beta(a,\tau+\zeta)c_2(\tau+\zeta;r_0)\frac{\Gamma(\tau+\zeta)}{\Gamma(\zeta)}d\tau$$

and the threshold number is given by the spectral radius $r(\tilde{K})$. As for the case of $r_0 = 0$, it can then be proved that the disease-free steady state of the normalized system is globally stable if $r(\tilde{K}) < 1$ and there exists an endemic steady state if $r(\tilde{K}) > 1$ [30].

Observe that the above integral kernel can be rewritten as follows:

$$\tilde{\Phi}(a,\zeta) = \int_0^\omega e^{-r_0\tau}\beta(a,\tau+\zeta)\frac{\ell(\tau+\zeta)\Gamma(\tau+\zeta)}{\ell(\zeta)\Gamma(\zeta)}d\tau c_2(\zeta;r_0),$$

and let $L : L^1(0,\omega) \to L^1(0,\omega)$ be a positive operator defined by

$$(Lf)(a) := c_2(a;r_0)f(a).$$

If the host population is a stable population with the intrinsic growth rate $r_0$, the NGO of the original system (6.1) is obtained from $K = \hat{\Pi}(0)$ by replacing $c_2(a)$ with $c_2(a;r_0)$:

$$(K(r_0)f)(a) := c_2(a;r_0)\int_0^\omega\int_0^\omega \beta(a,\tau+\zeta)\frac{\ell(\tau+\zeta)\Gamma(\tau+\zeta)}{\ell(\zeta)\Gamma(\zeta)}d\tau f(\zeta)d\zeta,$$

where $K(0) = K$. Then, we can observe that if $r_0 > 0$, $K(r_0)L \geq L\tilde{K}$, whereas if $r_0 < 0$, it holds that $K(r_0)L \leq L\tilde{K}$. If we assume that $K(r_0)$ and $\tilde{K}$ are non-supporting operators, the comparison theorem holds (Chap. 10, [44]). Then, we have $r(K(r_0)) = r(L^{-1}K(r_0)L) > r(\tilde{K})$ if $r_0 > 0$ and $r(K(r_0)) < r(\tilde{K})$ if $r_0 < 0$.

Therefore, if we define $r(K(r_0))$ as the basic reproduction number $R_0$ for the disease in a stable population, we conclude that if the host population grows ($r_0 > 0$), there is a possibility that $R_0 > 1 > r(\tilde{K})$; that is, the infected population grows exponentially, but the disease is relatively eradicated from the host population and does not become endemic. Conversely, if the host population is decreasing ($r_0 < 0$) and $R_0 < 1 < r(\tilde{K})$, the infected population is also shrinking exponentially, but the prevalence of the disease in the host population expands and the disease becomes endemic. That is, $R_0$ is the threshold value for disease invasion, whereas the threshold number $r(\tilde{K})$ of the normalized system is the index that determines whether the disease becomes endemic or not in the host population.

### 6.4.4 Endemic Threshold Condition

As seen in previous sections, we can expect the invasion threshold condition to give the endemic threshold condition when the host population is in a demographic steady state. It is a crucial observation that the endemicity is also characterized by the basic reproduction number $R_0$. To formulate the endemic threshold results, let us return to the original model (6.1) without normalization and, for simplicity, again assume that $q = 0$, $\theta = 0$ and the host population is in a demographic steady state.

Let $S^*$ and $I^*$ be age-density functions of the susceptible and infected populations, respectively. From (6.1), we obtain the following expressions:

$$\begin{aligned} S^*(a) &= B\ell(a)\exp\left(-\int_0^a \lambda^*(\sigma)d\sigma\right), \\ I^*(a) &= \int_0^a \frac{\ell(a)\Gamma(a)}{\ell(\zeta)\Gamma(\zeta)}\lambda^*(\zeta)S^*(\zeta)d\zeta. \end{aligned} \tag{6.51}$$

Inserting the above into the definition of the force of infection and changing the order of integrals, we have

$$\begin{aligned} \lambda^*(a) &= \frac{1}{N}\int_0^\omega d\zeta\int_\zeta^\omega \beta(a,\sigma)\frac{\ell(\sigma)\Gamma(\sigma)}{\ell(\zeta)\Gamma(\zeta)}d\sigma\lambda^*(\zeta)S^*(\zeta) \\ &= \int_0^\omega \Phi(a,\zeta)\lambda^*(\zeta)e^{-\int_0^\zeta \lambda^*(x)dx}d\zeta, \end{aligned} \tag{6.52}$$

where $\Phi(a,\zeta)$ is given by $\Phi(a,\zeta) := \int_\zeta^\omega \beta(a,\sigma)\frac{\ell(\sigma)\Gamma(\sigma)}{\ell(\zeta)\Gamma(\zeta)}d\sigma c_2(\zeta)$ and $N = \int_0^\omega B\ell(a)da$ is the total size of the host stationary population.

The above equation can be considered as a fixed point equation for the force of infection at the endemic steady state

$$\lambda^* = \Psi(\lambda^*), \tag{6.53}$$

where $\Psi$ is a nonlinear positive operator given by

$$(\Psi f)(a) := \int_0^\omega \Phi(a,\zeta)f(\zeta)e^{-\int_0^\zeta f(\sigma)d\sigma}d\zeta, \quad f \in L^1(0,\omega). \tag{6.54}$$

As $\Psi$ is not necessarily monotonic and $\Psi(0) = 0$, it is not generally easy to determine the number of positive fixed points. However, we can use Krasnoselskii's theorem (Proposition 10.32) to prove that $\Psi$ has at least one positive fixed point if $\Psi'[0]$ (the Fréchet derivative of $\Psi(x)$ at $x = 0$) is a non-supporting compact operator and $r(\Psi'[0]) > 1$, because $\Psi(L^1_+)$ is a bounded set in the positive cone $L^1_+$. By a simple calculation, it can be seen that $r(\Psi'[0]) = r(K)$, so we know that the invasion threshold condition $R_0 = r(K) > 1$ implies the existence of an endemic steady state. However, as shown in Sect. 6.1, if $R_0 = r(K) < 1$, there is no endemic steady state, and the local stability of the disease-free steady state implies its global stability. We can summarize the above steady state analysis as follows [30]:

**Proposition 6.6** *Suppose that $q = 0$, $\theta \equiv 0$, and the NGO $K$ is compact and non-supporting. The disease-free steady state is globally asymptotically stable if $R_0 = r(K) < 1$, whereas it is unstable and there exists at least one endemic steady state if $R_0 = r(K) > 1$.*

From the above argument, we know that as the basic reproduction number $R_0$ crosses unity from below, the disease-free steady state loses its stability and a locally stable endemic steady state bifurcates, that is, an exchange of stability occurs. If $R_0$ increases further, there is a possibility that the bifurcated endemic steady state will become unstable and a periodic solution may bifurcate under appropriate conditions [3, 12, 50], although sufficient universal conditions for the existence of periodic solutions are not yet known.

Although the uniqueness and stability of an endemic steady state for the age-structured SIR model with general transmission coefficients remains an open problem, we can show that the endemic steady state uniquely exists if $\Psi$ is monotone and has appropriate concavity (Proposition 10.31, [30]), and it is locally asymptotically stable if the force of infection is sufficiently small. Other examples of unique endemic steady states are given in [10]. Note that the same kind of endemic threshold results also hold for vertically transmitted diseases [11, 31]. Recently, Franceschetti et al. [19] showed that the age-structured SIR epidemic model can have multiple endemic steady states and complex dynamical behavior. Conversely, Kuniya [38] found that a discretization of the age-structured SIR model leads to a system of equations for the multigroup epidemic model, and that this system has a globally stable endemic steady state if and only if $R_0 > 1$. If we assume that the transmission coefficient is independent of the age of infected individuals, we can obtain results for the global stability [41, 46].

If recovery from infection does not confer any immunity, then instead of the SIR model, we can formulate the SIS model. For the basic theory on the age-structured SIS model, readers are referred to [6, 8, 17]. If the transmission coefficient is comparable with a separable mixing function, it has been proved that the age-structured SIS epidemic model has a globally stable endemic steady state if and only if $R_0 > 1$. Even when the parameters are time periodic, we can prove that there exists a time-periodic endemic solution if $R_0 > 1$ [39, 40].

We can extend the age-dependent SIR model to reflect more realistic aspects of disease dynamics. For example, if we take into account the incubation (exposed) class, the basic model becomes the SEIR model [42], and further, considering the effect of maternal antibodies gives the MSEIR model [32], where $M$ denotes the class of passive immunity due to the maternal antibodies. If the host population is not closed, but allows infected immigration, there is no disease-free steady state. However, it is known that threshold-like results can be obtained for the age-structured SIR model with immigration [18, 29].

## 6.5 Infection-Age Dependency

It is a most important extension for the age-dependent SIR model to introduce the *infection-age* dependency. In fact, as seen in the previous chapter, the natural history of infection is well described by infection-age, so the infection-age is a basic key parameter in describing the reproduction of infective individuals, while the chronological age can be seen as an additional trait parameter. It is quite natural to assume that the infectivity parameter and the recovery rate are functions of infection-age.

A homogeneous age-structured SIR model that considers the infection-age has been proposed by Dietz and Schenzle [14], and studied by Inaba and Nishiura [34] (see Chap. 9). For the special case in which the chronological age structure can be neglected, the stability of the endemic steady state has been proved [25, 43], whereas the stability of the endemic steady state in the full age-structured model remains undetermined. Here, we sketch a method of calculating $R_0$ and the basic renewal equation.

For simplicity, suppose that $q = 0$ and $\theta = 0$. Then, if we introduce the infection-age $\tau$ into the basic model of (6.1), the basic system becomes

$$
\begin{aligned}
&\frac{\partial S(t,a)}{\partial t}+\frac{\partial S(t,a)}{\partial a}=-(\mu(a)+\lambda(t,a))S(t,a),\\
&S(t,0)=\int_0^{\omega} m(a)P(t,a)da,\\
&\frac{\partial I(t,\tau,a)}{\partial t}+\frac{\partial I(t,\tau,a)}{\partial s}+\frac{\partial I(t,\tau,a)}{\partial a}=-(\mu(a)+\gamma(\tau))I(t,\tau,a),\\
&I(t,0,a)=\lambda(t,a)S(t,a),\\
&\frac{\partial R(t,a)}{\partial t}+\frac{\partial R(t,a)}{\partial a}=-\mu(a)R(t,a)+\int_0^{a}\gamma(\tau)I(t,\tau,a)d\tau,\\
&R(t,0)=0,
\end{aligned}
\tag{6.55}
$$

where $\tau$ is the infection-age, $\gamma(\tau)$ is the infection-age-dependent recovery rate, and $\lambda(t,a)$ is given by

$$
\lambda(t,a)=\frac{1}{N(t)}\int_0^{\omega}\int_0^{\sigma}\beta(a,\tau,\sigma)I(t,\tau,\sigma)d\tau d\sigma. \tag{6.56}
$$

Here, $\beta(a,\tau,\sigma)$ is the transmission coefficient between susceptible individuals with age $a$ and infected individuals with chronological age $\sigma$ and infection-age $\tau$. The density of infectives $I(t,\cdot,\cdot)$ is a function from $\mathbb{R}_+$ to $L^1(\Omega)$ with $\Omega=\{(\tau,a)\in\mathbb{R}_+^2 : 0\le\tau\le a\le\omega\}$.

### 6.5.1 *The Basic Reproduction Number*

In the invasion phase, the linearized equation for the infective population is formulated as the following boundary value problem:

$$
\begin{aligned}
&\frac{\partial I(t,\tau,a)}{\partial t}+\frac{\partial I(t,\tau,a)}{\partial \tau}+\frac{\partial I(t,\tau,a)}{\partial a}=-(\mu(a)+\gamma(\tau))I(t,\tau,a),\\
&I(t,0,a)=c_2(a)\int_0^{\omega}\int_0^{\sigma}\beta(a,\tau,\sigma)I(t,\tau,\sigma)d\tau d\sigma,
\end{aligned}
\tag{6.57}
$$

where $c_2(a)$ is the stable age profile. Integrating (6.57) along the characteristic line, we obtain for $a\ge\tau$

$$
I(t,\tau,a)=\begin{cases}\Gamma(\tau)\frac{\ell(a)}{\ell(a-\tau)}B(t-\tau,a-\tau), & t-\tau>0,\\ \frac{\Gamma(\tau)\ell(a)}{\Gamma(\tau-t)\ell(a-t)}I_0(\tau-t,a-t), & \tau-t>0,\end{cases} \tag{6.58}
$$

where $\Gamma(\tau):=\exp(-\int_0^{\tau}\gamma(z)dz)$, $I_0(\tau,a)=I(0,\tau,a)$ is the initial data, and $B(t,a):=I(t,0,a)$ is the density of newly infected individuals.

Inserting expression (6.58) into the definition

$$B(t,a) = c_2(a) \int_0^{\omega} d\tau \int_{\tau}^{\omega} \beta(a,\tau,\sigma) I(t,\tau,\sigma) d\sigma, \tag{6.59}$$

we arrive at an abstract renewal equation in $L^1(0,\omega)$:

$$B(t) = G(t) + \int_0^t \Psi(\tau) B(t-\tau) d\tau, \tag{6.60}$$

where $B(t) = B(t,\cdot) \in L^1(0,\omega)$, $\Psi(\tau)$ is a positive linear operator on $L^1(0,\omega)$ defined by

$$(\Psi(\tau)\phi)(a) = c_2(a) \int_{\tau}^{\omega} \beta(a,\tau,\sigma) \Gamma(\tau) \frac{\ell(\sigma)}{\ell(\sigma-\tau)} \phi(\sigma-\tau) d\sigma, \quad \phi \in L^1(0,\omega) \tag{6.61}$$

and

$$G(t,a) := c_2(a) \int_t^{\omega} d\tau \int_{\tau}^{\omega} \beta(a,\tau,\sigma) \frac{\Gamma(\tau)\ell(\sigma)}{\Gamma(\tau-t)\ell(\sigma-t)} I_0(\tau-t,\sigma-t) d\sigma,$$

with the convention that $\Psi = 0$ for $\tau > \omega$ and $G = 0$ for $t > \omega$.

Therefore, the NGO $K$ on $L^1(0,\omega)$ is given as follows:

$$\begin{aligned} (K\phi)(a) &= \int_0^{\omega} (\Psi(\tau)\phi)(a) d\tau \\ &= c_2(a) \int_0^{\omega} d\tau \int_{\tau}^{\omega} \beta(a,\tau,\sigma) \Gamma(\tau) \frac{\ell(\sigma)}{\ell(\sigma-\tau)} \phi(\sigma-\tau) d\sigma, \end{aligned} \tag{6.62}$$

for $\phi \in L^1(0,\omega)$. Thus, the spectral radius $r(K)$ is the basic reproduction number for the age–duration-dependent SIR epidemic model. We will use this model in Chap. 9 to estimate $R_0$ based on endemic data.

**Exercise 6.2** Suppose that there exist functions $\beta_1$ and $\beta_2$ such that $\beta(a,\tau,\sigma) = \beta_1(a)\beta_2(\tau,\sigma)$ (separable mixing assumption). Prove that

$$R_0 = \int_0^{\omega} d\tau \int_{\tau}^{\omega} e^{r_0 \tau} \beta_2(\tau,\sigma) \Gamma(\tau) c_2(\sigma) \beta_1(\sigma-\tau) d\sigma,$$

where $r_0$ is the stable growth rate of the host population.

### 6.5.2 *Integral Equation Approach*

For the basic system (6.55), the normalization method is useful again. Let us introduce new variables $s(t,a)$, $i(t,s,a)$, and $r(t,a)$ by

$$S(t,a)=s(t,a)P(t,a), \quad I(t,\tau,a)=i(t,\tau,a)P(t,a), \quad R(t,a)=r(t,a)P(t,a),$$

where $P(t,a)$ is assumed to be a persistent solution of the host stable population. Then, we obtain

$$s(t,a)+\int_0^a i(t,\tau,a)d\tau + r(t,a) = 1.$$

Then, the basic system is reduced to a normalized S-I system:

$$\begin{aligned}
&\frac{\partial s(t,a)}{\partial t}+\frac{\partial s(t,a)}{\partial a} = -b(t,a),\\
&\frac{\partial i(t,\tau,a)}{\partial t}+\frac{\partial i(t,\tau,a)}{\partial a}+\frac{\partial i(t,\tau,a)}{\partial s} = -\gamma(\tau)i(t,\tau,a),\\
&s(t,0)=1,\\
&i(t,0,a)=b(t,a),\\
&b(t,a)=s(t,a)\int_0^\omega\int_0^\sigma \beta(a,\tau,\sigma)c_2(\sigma)i(t,\tau,\sigma)d\tau d\sigma,
\end{aligned} \tag{6.63}$$

where $b(t,\cdot)$ is the age-specific incidence rate of new infecteds. Then, we have

$$i(t,\tau,a)=\begin{cases}\Gamma(\tau)b(t-\tau,a-\tau), & t-\tau>0,\\ \frac{\Gamma(\tau)}{\Gamma(\tau-t)}i_0(\tau-t,a-t), & \tau-t>0.\end{cases}$$

Substituting this expression into the definition of $b$, we obtain an abstract integral equation

$$b(t)=s(t)\left[G_0(t)+\int_0^t \Psi_0(\tau)b(t-\tau)d\tau\right], \tag{6.64}$$

where $\Psi_0(\tau)$ is a positive linear operator on $L^1(0,\omega)$ defined by

$$(\Psi_0(\tau)\phi)(a)=\int_\tau^\omega \beta(a,\tau,\sigma)c_2(\sigma)\Gamma(\tau)\phi(\sigma-\tau)d\sigma, \quad \phi\in L^1(0,\omega)$$

and

$$G_0(t,a):=\int_t^{\omega\vee t}d\tau\int_\tau^\omega \beta(a,\tau,\sigma)\frac{\Gamma(\tau)}{\Gamma(\tau-t)}i_0(\tau-t,\sigma-t)d\sigma.$$

And it is easy to see that

$$s(t,a)=\begin{cases}1-\int_0^a b(t-a+\sigma,\sigma)d\sigma, & t-a>0,\\ s(0,a-t)-\int_0^t b(\sigma,a-t+\sigma)d\sigma, & a-t>0.\end{cases} \tag{6.65}$$

Therefore, the basic system can be reduced to a nonlinear integral equation of $b$, and we can apply the standard fixed point argument to prove the existence and uniqueness for the solution. Define the next-generation operator for the normalized system by

$$K_0 = \int_0^\infty \Psi_0(s)ds.$$

Based on the comparison argument, it follows from (6.64) that $\lim_{t\to\infty} b(t) = 0$ if $r(K_0) < 1$, which implies that the infected population is relatively eradicated from the host stable population if $r(K_0) < 1$.

**Exercise 6.3** We can repeat the argument in Sect. 6.4.3. Define a parametrized operator $K(r) = \int_0^\infty \Psi_r(s)ds$ by the kernel

$$(\Psi_r(s)\phi)(a) = c_2(a) \int_s^\omega \beta(a, \tau, \sigma)\Gamma(\tau)e^{-r\tau} \frac{\ell(\sigma)}{\ell(\sigma - \tau)} \phi(\sigma - \tau)d\sigma,$$

and define a multiplication operator $L$ by $(L\phi)(a) = c_2(a)\phi(a)$. Then, $K(0) = K$. Show that $K_0 = L^{-1}K(r_0)L$, hence $r(K_0) = r(K(r_0))$, where $r_0$ is the stable growth rate of the host population. Since $r(K(r))$ is a decreasing function of $r \in \mathbb{R}$, it follows that $r(K_0) < R_0$ if $r_0 > 0$, $r(K_0) = R_0$ if $r_0 = 0$, and $r(K_0) > R_0$ if $r_0 < 0$.

Another formulation by an integral equation based on the force of infection is also interesting. Let $\lambda_0$ be the force of infection for the normalized system:

$$\lambda_0(t, a) = \int_0^\omega \int_0^\sigma \beta(a, \tau, \sigma)c_2(\sigma)i(t, \tau, \sigma)d\tau d\sigma.$$

Since $b(t, a) = \lambda_0(t, a)s(t, a)$, we have

$$s(t, a) = \begin{cases} e^{-\int_0^a \lambda_0(t-a+\sigma,\sigma)d\sigma}, & t - a > 0, \\ s_0(a - t)e^{-\int_0^t \lambda_0(\sigma,a-t+\sigma)d\sigma}, & a - t > 0, \end{cases} \tag{6.66}$$

where $s_0(a) := s(0, a)$. Inserting the expression (6.66) into the definition of $\lambda_0$, we obtain an integral equation

$$\lambda_0(t, a) = h(t, a) + \int_0^t d\sigma \int_0^\sigma \beta(a, \tau, \sigma)c_2(\sigma)\Gamma(\tau)\lambda_0(t-\tau, \sigma-\tau)e^{-\int_\tau^\sigma \lambda_0(t-\zeta,\sigma-\zeta)d\zeta} d\tau, \tag{6.67}$$

where

$$h(t, a) := \int_t^{\omega\vee t} d\sigma \int_0^\sigma \beta(a, \tau, \sigma)c_2(\sigma)\frac{\Gamma(\tau)}{\Gamma(\tau - t)} i_0(\tau - t, \sigma - t)d\tau.$$

Then, we can again observe that the age-dependent SIR model can be formulated by a "scalar" nonlinear renewal equation.

**Exercise 6.4** Show that in the demographic steady state assumed in Sect. 6.2 ($r_0 = 0$), the force of infection at the endemic steady states of the age–duration (infection-age)-dependent SIR model satisfies Eq. (6.16) with an integral kernel given by

$$\phi(a, \zeta) = \int_{\zeta}^{\omega} \beta(a, \sigma - \zeta, \sigma) c_2(\sigma) \Gamma(\sigma - \zeta) d\sigma.$$

Using Krasnoselskii's fixed point theorem (Proposition 10.32), prove that there exists at least one endemic steady state if $R_0 > 1$. In particular, the endemic steady state uniquely exists if we adopt the separable mixing assumption (see Exercise 6.2).

For the separable mixing case, it is easy to see that there exists a unique endemic steady state if $R_0 > 1$, but its global stability is an open problem. However, we can show *uniform weak persistence* of the infected population:

**Proposition 6.7** *Suppose that* $r_0 = 0$, *the separable mixing assumption holds and* $R_0 > 1$. *Define*

$$\phi(t) := \int_0^{\omega} d\sigma \int_0^{\sigma} \beta_2(\tau, \sigma) c_2(\sigma) i(t, \tau, \sigma) d\tau.$$

*For any initial data* $(s_0, i_0)$ *such that* $\phi(t) > 0$ *for all* $t \geq 0$, *there exists a number* $\varepsilon > 0$ *such that* $\limsup_{t \to \infty} \phi(t) > \varepsilon$.

*Proof* Observe that the force of infection is given by $\lambda_0(t, a) = \beta_1(a)\phi(t)$. For $t > \omega$, we have

$$\begin{aligned} i(t, \tau, a) &= \Gamma(\tau)\lambda_0(t - \tau, a - \tau)s(t - \tau, a - \tau) \\ &= \Gamma(\tau)\beta_1(a - \tau)\phi(t - \tau)s(t - \tau, a - \tau), \end{aligned}$$

Then, it follows that for $t > \omega$

$$\begin{aligned} \phi(t) &= \int_0^{\omega} d\sigma \int_0^{\sigma} \beta_2(\tau, \sigma) c_2(\sigma) \Gamma(\tau) \beta_1(\sigma - \tau) \phi(t - \tau) s(t - \tau, \sigma - \tau) d\tau \\ &= \int_0^t \phi(t - \tau) \int_{\tau}^t \beta_2(\tau, \sigma) c_2(\sigma) \Gamma(\tau) \beta_1(\sigma - \tau) s(t - \tau, \sigma - \tau) d\sigma d\tau, \end{aligned}$$

where we adopt a convention that $\beta_2(\tau, a) = 0$ for $\tau > a$ and $a > \omega$. Suppose that for any $\varepsilon > 0$, there exists an initial data $(s_0, i_0)$ and a large number $T_0 > 0$ such that $\phi(t) \leq \varepsilon$ for all $t \geq T_0$ and $\phi(t) > 0$. From

$$s(t, a) = e^{-\int_0^a \lambda(t - a + \sigma, \sigma) d\sigma},$$

for $t > \omega$, we obtain an estimate

$$s(t, a) \geq e^{-\varepsilon \int_0^a \beta_1(\sigma) d\sigma},$$

for $t > T_0 + \omega$. Let $\phi_d(t) := \phi(t + d)$ for $d > 0$. If we take large $d$ and $T > T_0$ such that $T + \omega < d$, it follows that

$$\begin{aligned}\phi_d(t) &= \int_0^{t+d} \phi(t+d-\tau) \int_\tau^{t+d} \beta_2(\tau,\sigma)c_2(\sigma)\Gamma(\tau)\beta_1(\sigma-\tau)s(t+d-\tau,\sigma-\tau)d\sigma d\tau \\ &\geq \int_0^t \phi_d(t-\tau) \int_\tau^{\tau+T} \beta_2(\tau,\sigma)c_2(\sigma)\Gamma(\tau)\beta_1(\sigma-s)s(t+d-\tau,\sigma-\tau)d\sigma d\tau, \\ &\geq \int_0^t \phi_d(t-\tau) \int_\tau^{\tau+T} \beta_2(\tau,\sigma)c_2(\sigma)\Gamma(\tau)\beta_1(\sigma-\tau)e^{-\varepsilon \int_0^{\sigma-\tau} \beta_1(\zeta)d\zeta} d\sigma d\tau, \\ &= \int_0^t \phi_d(t-\tau)\psi_\varepsilon(\tau)d\tau,\end{aligned}$$

where

$$\psi_\varepsilon(\tau) := \int_\tau^{\tau+T} \beta_2(\tau,\sigma)c_2(\sigma)\Gamma(\tau)\beta_1(\sigma-\tau)e^{-\varepsilon \int_0^{\sigma-\tau} \beta_1(\zeta)d\zeta} d\sigma.$$

Taking the Laplace transform on both sides, we have

$$\hat{\phi}_d(z) \geq \hat{\phi}_d(z)\hat{\psi}_\varepsilon(z),$$

where $\hat{\phi}_d(z)$ exists for all $z > 0$ because $\phi_d$ is bounded and $\hat{\psi}_\varepsilon(z)$ exists for all $z \geq 0$ and $\varepsilon \geq 0$. Moreover, we can observe that

$$\lim_{T\to\infty} \hat{\psi}_0(0) = \int_0^\infty d\tau \int_\tau^\infty \beta_2(\tau,\sigma)c_2(\sigma)\Gamma(\tau)\beta_1(\sigma-\tau)d\sigma = R_0 > 1.$$

Since $\hat{\psi}_\varepsilon(z)$ is continuous with respect to $\varepsilon$, $T$, and $z$, then we can choose small $\varepsilon > 0$, $z > 0$, and a large $T > 0$ in advance such that $\hat{\psi}_\varepsilon(z) > 1$. Then, we have $\hat{\phi}_d(z) = 0$, so $\phi_d(t) = \phi(t + d) = 0$ for almost all $t \geq 0$. This contradicts our assumption. □

Although we omit the proof, under appropriate additional conditions, uniform weak persistence implies uniform strong persistence, that is, for any initial data $(s_0, i_0)$ such that $\phi(t) > 0$ for all $t \geq 0$, there exists a number $\varepsilon > 0$ such that $\liminf_{t\to\infty} \phi(t) > \varepsilon$ [36, 49, 51, 52].

## References

1. Anderson, R., May, R.M.: Infectious Diseases of Humans: Dynamics and Control. Oxford UP, Oxford (1991)
2. Andreason, V.: Disease regulation of age-structured host populations. Theor. Popul. Biol. **36**, 214–239 (1989)

3. Andreasen, V.: Instability in an SIR-model with age-dependent susceptibility. In: Arino, O., Axelrod, D., Kimmel, M., Langlais, M. (eds.) Mathematical Population Dynamics. Theory of Epidemics, vol. 1, pp. 3–14. Wuerz Publishing, Winnipeg (1995)
4. Busenberg, S., Cooke, K., Iannelli, M.: Endemic threshold and stability in a class of age-structured epidemic. SIAM J. Appl. Math. **48**, 1379–1395 (1988)
5. Busenberg, S., Hadeler, K.P.: Demography and epidemics. Math. Biosci. **101**, 63–74 (1990)
6. Busenberg, S., Iannelli, M., Thieme, H.: Global behaviour of an age-structured S-I-S epidemic model. SIAM J. Math. Anal. **22**, 1065–1080 (1991)
7. Busenberg, S., Cooke, K.: Vertically Transmitted Diseases: Models and Dynamics. Biomathematics, vol. 23. Springer, Berlin (1993)
8. Busenberg, S., Iannelli, M., Thieme, H.: Dynamics of an age-structured epidemic model. In: Liao Shan-Tao, Ye Yan-Qian, Ding Tong-Ren (eds.) Dynamical Systems. Nankai Series in Pure, Applied Mathematics and Theoretical Physics, vol. 4, pp. 1–19. World Scientific, Singapore (1993)
9. Capasso, V.: Mathematical Structures of Epidemic System. Lecture Notes in Biomathematics, vol. 97. Springer, Berlin (1993)
10. Cha, Y., Iannelli, M., Milner, F.A.: Are multiple endemic equilibria possible? In: Arino, O., Axelrod, D., Kimmel, M. (eds.) Advances in Mathematical Population Dynamics-Molecules, Cells and Man, pp. 779–788. World Scientific, Singapore (1997)
11. Cha, Y., Iannelli, M., Milner, F.A.: Existence and uniqueness of endemic states for the age-structured S-I-R epidemic model. Math. Biosci. **150**, 177–190 (1998)
12. Cha, Y., Iannelli, M., Milner, F.A.: Stability change of an epidemic model. Dynamic Syst. Appl. **9**, 361–376 (2000)
13. De Jong, M.C.M., Diekmann, O., Heesterbeek, H.: How does transmission of infection depend on population size? In: Mollison, D. (ed.) Epidemic Models: Their Structure and Relation to Data, pp. 84–94. Cambridge U.P, Cambridge (1995)
14. Dietz, K., Schenzle, D.: Proportionate mixing models for age-dependent infection transmission. J. Math. Biol. **22**, 117–120 (1985)
15. Diekmann, O., Heesterbeek, J.A.P., Metz, J.A.J.: On the definition and the computation of the basic reproduction ratio $R_0$ in models for infectious diseases in heterogeneous populations. J. Math. Biol. **28**, 365–382 (1990)
16. Diekmann, O., Heesterbeek, J.A.P., Britton, T.: Mathematical Tools for Understanding Infectious Disease Dynamics. Princeton University Press, Princeton (2013)
17. Feng, Z., Huang, W., Castillo-Chavez, C.: Global behavior of a multi-group SIS epidemic model with age structure. J. Diff. Equ. **218**, 292–324 (2005)
18. Franceschetti, A., Pugliese, A.: Threshold behaviour of a SIR epidemic model with age structure and immigration. J. Math. Biol. **57**(1), 1–27 (2008)
19. Franceschetti, A., Pugliese, A., Breda, D.: Multiple endemic states age-structured SIR epidemic models. Math. Biosci. Eng. **9**(3), 577–599 (2012)
20. Greenhalgh, D.: Analytical results on the stability of age-structured recurrent epidemic models. IMA J. Math. Appl. Med. Biol. **4**, 109–144 (1997)
21. Greenhalgh, D.: Analytical threshold and stability results on age-structured epidemic models with vaccination. Theor. Popul. Biol. **33**, 266–290 (1988)
22. Greenhalgh, D.: Threshold and stability results for an epidemic model with an age-structured meeting rate. IMA J. Math. Appl. Med. Biol. **5**, 81–100 (1988)
23. Greenhalgh, D.: Existence, threshold and stability results for an age-structured epidemic model with vaccination and a non-separable transmission coefficient. Int. J. Sys. Sci. **24**(4), 641–668 (1993)
24. Greenhalgh, D., Dietz, K.: Some bounds on estimates for reproductive ratios derived from the age-specific force of infection. Math. Biosci. **124**, 9–57 (1994)
25. Hethcote, H.W., Tudor, D.W.: Integral equation models for endemic infectious diseases. J. Math. Biol. **9**, 37–47 (1980)
26. Hethcote, H.W.: The mathematics of infectious diseases. SIAM Rev. **42**(4), 599–653 (2000)
27. Hoppensteadt, F.: An age dependent epidemic model. J. Franklin Inst. **297**(5), 325–333 (1974)

28. Iannelli, M., Martcheva, M.: Homogeneous dynamical systems and the age-structured SIR model with proportionate mixing incidence. In: Iannelli, M., Lumer, G. (eds.) Evolution Equations: Applications to Physics, Industry, Life Sciences and Economics (Progress in Nonlinear Differential Equations and Their Applications), vol. 55, pp. 227–251. Birkhäuser, Basel (2003)
29. Iannelli, M., Manfredi, P.: Demographic changes and immigration in age-structured epidemic models. Math. Popul. Studies **14**(3), 169–191 (2007)
30. Inaba, H.: Threshold and stability results for an age-structured epidemic model. J. Math. Biol. **28**, 411–434 (1990)
31. Inaba, H.: Mathematical analysis of an age-structured SIR epidemic model with vertical transmission. Disc. Conti. Dyna. Sys. Series B **6**(1), 69–96 (2006)
32. Inaba, H.: Age-structured homogeneous epidemic systems with application to the MSEIR epidemic model. J. Math. Biol. **54**, 101–146 (2007)
33. Inaba, H.: Homogeneous epidemic systems in the stable population, In Mathematical Economics, RIMS Kokyuroku 1557, Research Institute for Mathematical Sciences, Kyoto University, Kyoto, pp. 28–44 (2007)
34. Inaba, H., Nishiura, H.: The basic reproduction number of an infectious disease in a stable population: the impact of population growth rate on the eradication threshold. Math. Model. Nat. Phenom. **3**(7), 194–228 (2008)
35. Inaba, H.: On a new perspective of the basic reproduction number in heterogeneous environments. J. Math. Biol. **65**, 309–348 (2012)
36. Kawachi, K.: A note on persistence about structured population models. J. Biol. Dyn. **2**(4), 449–464 (2008)
37. Krasnoselskii, M.A.: Positive Solutions of Operator Equations. Noordhoff, Groningen (1964)
38. Kuniya, T.: Global stability analysis with a discretization approach for an age-structured multi-group SIR epidemic model. Nonlinear Analysis RWA **12**, 2640–2655 (2011)
39. Kuniya, T., Inaba, H.: Endemic threshold results for age-structured SIS epidemic model with periodic parameters. J. Math. Anal. Appls. **402**, 477–492 (2013)
40. Kuniya, T., Iannelli, M.: $R_0$ and the global behavior of an age-structured SIS epidemic model with periodicity and vertical transmission. Math. Biosci. Eng. **11**(4), 929–945 (2014)
41. Kuniya, T., Wang, J., Inaba, H.: A multi-group SIR epidemic model with age structure. Disc. Cont. Dyn. Sys. B **21**(10), 3515–3550 (2016)
42. Li, X.Z., Gupur, G., Zhu, G.T.: Threshold and stability results for an age-structured SEIR epidemic model. Comp. Math. Appls. **42**, 883–907 (2001)
43. Magal, P., McCluskey, C.C., Webb, G.F.: Lyapunov functional and global asymptotic stability for an infection-age model. Appl. Anal. **89**(7), 1109–1140 (2010)
44. Marek, I.: Frobenius theory of positive operators: comparison theorems and applications. SIAM J. Appl. Math. **19**, 607–628 (1970)
45. McLean, A.: Dynamics of childhood infections in high birthrate countries. In: Hoffmann, G.W., Hraba, T. (eds.) Immunology and Epidemiology. Lecture Notes in Biomathematics, vol. 65, pp. 171–197. Springer, Berlin (1986)
46. Melnik, A.V., Korobeinikov, A.: Lyapunov functions and global stability for SIR and SEIR models with age-dependent susceptibility. Math. Biosci. Eng. **10**(2), 369–378 (2013)
47. Pazy, A.: Semigroups of Linear Operators and Applications to Partial Differential Equations. Springer, Berlin (1983)
48. Safan, M.: Spread of infectious diseases: impact on demography, and the eradication effort in models with backward bifurcation, Ph.D. Thesis, Faculty of Mathematics and Phyiscs, Eberhard-Karls University of Tuebingen (2006)
49. Smith, H.L. and Thieme, H.R.: Dynamical Systems and Population Persistence, Graduate Studies in Mathematics, American Mathematical Society, vol. 118. Providence, Rhode Island (2011)
50. Thieme, H.R.: Stability change for the endemic equilibrium in age-structured models for the spread of S-I-R type infectious diseases. Differential Equation Models in Biology, Epidemiology and Ecology. Lecture Notes in Biomathematics, vol. 92, pp. 139–158. Springer, Berlin (1991)

51. Thieme, H.R.: Disease extinction and disease persistence in age structured epidemic models. Nonl. Anal. **47**, 6181–6194 (2001)
52. Thieme, H.R.: Mathematics in Population Biology. Princeton University Press, Princeton (2003)
53. Tudor, D.W.: An age-dependent epidemic model with applications to measles. Math. Biosci. **73**, 131–147 (1985)
54. Webb, G.F.: Theory of Nonlinear Age-Dependent Population Dynamics. Marcel Dekker, New York (1985)

# Chapter 7
# Epidemic Models for HIV Infection

**Abstract** Since the 1980s, the HIV/AIDS epidemic has become a global pandemic and is one of the foremost public health problems in the world today. The HIV/AIDS epidemic also had a huge impact on the development of mathematical epidemiology for infectious diseases, because it was the first worldwide outbreak of a newly emerging virus after mathematical epidemiology had established itself against classical endemic diseases, such as measles, rubella, influenza, and malaria. Compared with these infectious diseases, HIV/AIDS has very different features that made it necessary to refine the basic ideas and extend the existing epidemic models. In this chapter, we first summarize the epidemiological features of HIV/AIDS and examine mathematical models to describe the initial invasion phase of the epidemic. Next, we consider the age-structured HIV epidemic model at the population level, taking into account variable infectivity, the host age structure, and the pair formation phenomenon. Finally, we consider an age-structured model of HIV infection at the cell population level.

## 7.1 Modeling the Invasion Phase

The etiological agent for acquired immunodeficiency syndrome (AIDS) is human immunodeficiency virus (HIV). Those infected with HIV have a high probability of developing AIDS if they do not receive effective medical treatment. The major possible infection routes of HIV are limited to sexual contact, needle sharing, blood transfusion, and vertical transmission.

The HIV epidemic has various characteristics that distinguish it from traditional infectious diseases. First, it is well known that HIV has a long incubation and infectious period, estimated at 8–10 years on average. Secondly, there exists some evidence that during the long incubation period, the infectivity of infected people varies depending on the time since infection (*infection age*). In fact, a short initial (acute) phase and a longer (symptomatic) terminal phase of high infectivity are separated by a long (asymptomatic) phase during which infectivity is very low. It appears that infectivity is proportional to the density of virus in the bloodstream. Accordingly, the timescale of HIV transmission is so long that demographic or compositional


H. Inaba, *Age-Structured Population Dynamics in Demography and Epidemiology*, DOI 10.1007/978-981-10-0188-8_7

changes in the host population could affect the transmission process. Conversely, the HIV/AIDS epidemic increases the death rate of infected individuals and hence generally affects the growth and structure of the host population [45].

From the above characteristics, when modeling HIV infection at the population level, it is important to consider the infection-age structure and the variable infectivity and to formulate the appropriate transmission submodels according to the infection routes. For example, the pattern of the epidemic given by considering heterosexual transmission would be different from the spread of HIV via needle sharing among drug users.

### 7.1.1 Malthus Model and Prevalence

During the long incubation period of HIV infection, infected individuals would not know they were infected without taking an antibody test. Thus, it seems that when full-blown AIDS cases are being observed, the peak of the epidemic in a cohort has passed, and the major part of the host population would already have been infected. Although full-blown AIDS cases can be accurately reported, it is not normally possible to determine the size of the HIV-infected population directly without a large-scale sample survey. Therefore, the first mathematical problem in HIV epidemiology is to estimate the unknown number of HIV-infected individuals from the data of AIDS cases.

In the following, the infected population refers to the asymptomatic HIV-infected individuals who have not yet developed AIDS. Let $i(t,\tau)$ be the density of HIV-infected individuals at time $t$ and infection age (the time since infection) $\tau$. First, we neglect the host chronological age structure and the natural death rate. Hence, the number of infected individuals at time $t$ is given by

$$I(t) = \int_0^\infty i(t,\tau)d\tau.$$

Let $\gamma(\tau)$ be the rate at which AIDS develops at infection-age $\tau$, and let $\Gamma(\tau)$ be the corresponding survival probability defined by

$$\Gamma(\tau) := \exp\left(-\int_0^\tau \gamma(\sigma)d\sigma\right),$$

where we assume that $\Gamma(\infty) = 0$, that is, the survival probability $\Gamma(\tau)$ is applied to infected individuals who will ultimately develop AIDS. The boundary value $i(t,0) =: B(t)$ denotes the number of newly infected individuals at time $t$. The differences between individuals' immune systems are significant, so some infected individuals will not necessarily develop AIDS (the long-term survivors). Let $\alpha$ be the proportion of infected people who will finally develop AIDS. Under these assumptions, it follows that

$$i(t,\tau) = (\alpha\Gamma(\tau) + (1-\alpha))B(t-\tau) = \Gamma^*(\tau)B(t-\tau),$$

where $\Gamma^*(\tau) := \alpha\Gamma(\tau) + (1-\alpha)$ is the survival probability for an infected cohort. The incubation period distribution is then given by

$$F(\tau) := \alpha(1-\Gamma(\tau)) = 1 - \Gamma^*(\tau).$$

If we let $A(t)$ be the cumulative number of AIDS cases until time $t$, then we have

$$A(t) = \int_0^t ds \int_0^\infty \gamma(\tau)\Gamma(\tau)\alpha B(s-\tau)d\tau.$$

Taking the partial integral with the assumption $B(t) = 0$ for $t < 0$, it is easy to see that

$$A(t) = \int_0^t F(t-\tau)B(\tau)d\tau. \tag{7.1}$$

In general, there are reliable data for $A(t)$, but the number of HIV-infected individuals $I(t)$ cannot be measured directly, and the survival probability $\Gamma(\tau)$ is estimated by cohort observations of HIV-infected individuals. If $B(t)$ is obtained from $A(t)$ and $F(t)$ by the deconvolution of (7.1), we can estimate the density distribution $i(t,\tau)$ or the total size $I(t)$ of the infected population. This approach is known as *backcalculation* [5, 13]. Backcalculation is a method for numerically solving the first kind of Volterra integral equation (7.1), which is known as an *ill-posed* problem in the sense that it is difficult to obtain a statistically robust result. In fact, the backcalculation solution is unstable and very sensitive to the choice of incubation distribution $F(\tau)$. Because $F(0) = 0$ and $F(\tau)$ takes very small values around $\tau = 0$, and as $A(t)$ is censored in that it is given for a time interval $t \in [0, t_0]$, the number of infected individuals $B(t)$ around $t = t_0$ will scarcely affect $A(t)$. Therefore, if the reliability of the data for $A(t)$ is not so high, it would be difficult to estimate $B(t)$ uniquely. Hence, backcalculation normally requires some additional assumption (constraint condition), such that $B(t)$ has a given form or appropriate smoothness.

*Remark 7.1* If $A(t)$ and $F(t)$ are given by mathematical functions, it is not difficult to solve (7.1) formally. In fact, if we define $\hat{f}(\lambda)$ as the Laplace transform of a function $f(t)$, it follows from the convolution theorem for the Laplace transformation that $\hat{A}(\lambda) = \hat{F}(\lambda)\hat{B}(\lambda)$, where $\hat{A}(\lambda) = \int_0^\infty e^{-\lambda t}A(t)dt$, etc. Therefore, it follows from the inversion theorem that

$$B(t) = \frac{1}{2\pi i}\int_\Theta e^{\lambda t}\frac{\hat{A}(\lambda)}{\hat{F}(\lambda)}d\lambda,$$

where $\Theta$ is an integral path included in the half plane of convergence. However, $A(t)$ and $F(t)$ must be known appropriate functions in order to calculate the inverse formula analytically, and such functions would not necessarily fit the real data [13].

*Remark 7.2* Consider a closed HIV-infected population with the initial infection-age-density function $i_0(\tau)$. Suppose that $\alpha = 1$. Let $C(t)$ be the number of new *onsets* at time $t$. Then, it is easy to see that

$$C(t) = \int_0^\infty f(t+\tau)\frac{i_0(\tau)}{\Gamma(\tau)}d\tau,$$

where $f(\tau) := \gamma(\tau)\Gamma(\tau)$ is the probability density function that the onset occurs at infection-age $\tau$. Since $\int_0^\infty C(t)dt = \int_0^\infty i_0(s)ds$, the time distribution for the onset is given by

$$c(t) := \frac{C(t)}{\int_0^\infty C(t)dt} = \int_0^\infty \frac{f(t+\tau)}{\Gamma(\tau)}\frac{i_0(\tau)}{\int_0^\infty i_0(s)ds}d\tau.$$

If we assume that the initial distribution is in a steady state, there exists a positive number $B$ such that $i_0(\tau) = B\Gamma(\tau)$ and we obtain

$$c(t) = \frac{\Gamma(t)}{\int_0^\infty \Gamma(s)ds}.$$

Therefore, we can estimate the distribution of the incubation period from the observable data $c(t)$ [30, 35].

As the simplest assumption, let us assume the Malthusian law:

$$B(t) = B_0 e^{\lambda_0 t}, \tag{7.2}$$

whereby the infected population has already passed its transient phase and is now in the stable growth orbit. Because the initial data are unknown, we assume that (7.2) holds for $t < 0$. The cumulative number of AIDS cases is then calculated as

$$A(t) = \int_0^\infty F(\tau)B(t-\tau)d\tau = B_0 e^{\lambda_0 t}\int_0^\infty e^{-\lambda_0 \tau}F(\tau)d\tau,$$

and the distribution of infected individuals is given by

$$i(t,\tau) = (1-F(\tau))B(t-\tau) = B_0 e^{\lambda_0(t-\tau)}(1-F(\tau)).$$

Therefore, the infection-age profile of infected individuals is also time-independent:

$$\frac{i(t,\tau)}{I(t)} = be^{-\lambda_0\tau}(1-F(\tau)),$$

where

$$b := \frac{i(t,0)}{I(t)} = \frac{1}{\int_0^\infty e^{-\lambda_0\tau}(1-F(\tau))d\tau} \tag{7.3}$$

is the crude rate at which secondary cases are produced per unit time and per infected individual. Similarly, we can define the *crude rate of developing AIDS* as follows:

$$d := \frac{1}{I(t)}\frac{dA(t)}{dt} = \frac{\int_0^\infty e^{-\lambda_0 \tau}\Gamma^*(\tau)\gamma(\tau)d\tau}{\int_0^\infty e^{-\lambda_0 \tau}(1-F(\tau))d\tau}. \tag{7.4}$$

Let $\kappa$ be the crude rate of developing AIDS for the infection cohort leading to full-blown AIDS:

$$\kappa := \frac{\int_0^\infty e^{-\lambda_0 \tau}\Gamma(\tau)\gamma(\tau)d\tau}{\int_0^\infty e^{-\lambda_0 \tau}\Gamma(\tau)d\tau}. \tag{7.5}$$

By partially integrating the numerator, we have

$$\kappa = \frac{1}{\int_0^\infty e^{-\lambda_0 \tau}\Gamma(\tau)d\tau} - \lambda_0.$$

From these definitions, it is clear that

$$b = \frac{\lambda_0(\lambda_0+\kappa)}{\lambda_0+(1-\alpha)\kappa}, \quad d = \frac{\alpha\lambda_0\kappa}{\lambda_0+(1-\alpha)\kappa}, \quad \lambda_0 = b-d.$$

Taking the ratio of the size of the infected population to the cumulative number of AIDS cases, we obtain

$$\frac{I(t)}{A(t)} = \frac{\int_0^\infty e^{-\lambda_0 \tau}(1-F(\tau))d\tau}{\int_0^\infty e^{-\lambda_0 \tau}F(\tau)d\tau} = \frac{1}{\lambda_0\int_0^\infty e^{-\lambda_0 \tau}F(\tau)d\tau} - 1.$$

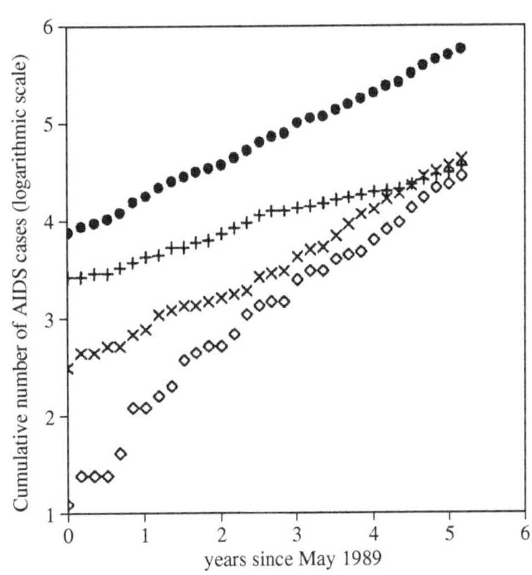


**Fig. 7.1** Cumulative number of AIDS cases for the exponential invasion phase in Japan. •: total, ×: heterosexual, +: homosexual, ⋄: unknown

By a simple calculation, we also have

$$I(t) = \left[\frac{1}{\alpha}\left(\frac{\lambda_0}{\kappa}+1\right)-1\right]A(t). \tag{7.6}$$

Because it is $A(t)$ that can be observed in the real data, if the cumulative number of AIDS cases is growing exponentially, we can determine the ratio $I(t)/A(t)$ using the Malthusian parameter $\lambda_0$ and the survival probability by assuming that the invasion process can be described by the Malthusian population model.

For example, let us apply the above argument to real data. From 1989 to 1994, the HIV epidemic in Japan was in the invasion phase, as the cumulative AIDS cases exhibited exponential growth with $\lambda_0 = 0.359$ (Fig. 7.1). If we use the well-known Weibull distribution and the log-logistic distribution to describe the incubation [5] and assume that $\alpha = 0.9$, then we obtain the results presented in Table 7.1.

In Table 7.1, $e_0 := \int_0^\infty \Gamma(\tau)d\tau$ denotes the average sojourn time of HIV-infected individuals in the incubation period. The World Health Organization suggests that the cumulative rate of developing AIDS in one infection cohort is about 50 percent over 10 years, which almost corresponds to model III in Table 7.2 (Weibull distribution) (Fig. 7.2). Therefore, if we assume that all AIDS cases are reported correctly and $\alpha = 0.9$, it is estimated that the size of the HIV-infected population in the early 1990s was 10–17 times higher than the cumulative number of AIDS cases at any point in time. The ratio $I/A$ is larger because the average sojourn time in the incubation period is larger and $\alpha$ is smaller. As the Malthusian parameter $\lambda_0$ grows, the ratio $I/A$ increases rapidly and will exceed 100 if $\lambda_0 = 0.9$, as was observed in the gay community of San Francisco.

*Remark 7.3* In the estimation formula (7.6), it is clear that the crude rate of developing AIDS $\kappa$ is a key parameter. Here, let us introduce an approximation formula

**Table 7.1** Stable population parameters for $\lambda_0 = 0.359$ and $\alpha = 0.9$ per year

| Models | $e_0$ (years) | $\kappa$ | $d$ | $b$ | $A/I$ |
|---|---|---|---|---|---|
| I | 8.5 | 0.03895 | 0.03468 | 0.39368 | 0.09660 |
| II | 9 | 0.03151 | 0.02811 | 0.38711 | 0.07831 |
| III | 10 | 0.02592 | 0.02316 | 0.38216 | 0.06453 |
| IV | 11 | 0.02423 | 0.02166 | 0.38066 | 0.06033 |

**Table 7.2** Examples of survival probability ($\tau$ in years)

| Model | Survival function |
|---|---|
| I | $\Gamma(\tau) = \exp(-0.004\tau^{2.438})$ |
| II | $\Gamma(\tau) = \exp(-0.0021\tau^{2.65})$ |
| III | $\Gamma(\tau) = \exp(-0.0021\tau^{2.516})$ |
| IV | $\Gamma(\tau) = \frac{1}{1+(0.1\tau)^{3.08}}$ |

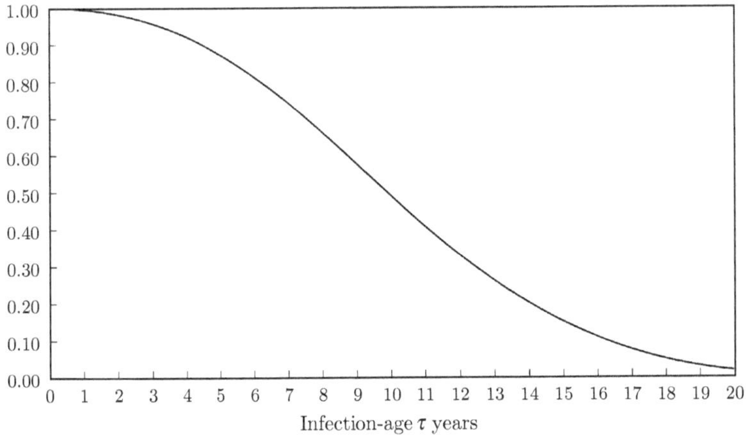


**Fig. 7.2** An example of survival probability (model III)

for $\kappa$ [19]. Let

$$J := \int_0^\infty e^{-\lambda_0 \tau} \Gamma(\tau) \gamma(\tau) d\tau,$$

and let the probability density function $\gamma(\tau)\Gamma(\tau)$ be a distribution concentrated about the average sojourn time $e_0$ in the asymptomatic phase. Observe that

$$J = -e^{-\lambda_0 e_0} \int_0^\infty e^{-\lambda_0(\tau - e_0)} \frac{d\Gamma(\tau)}{d\tau} d\tau,$$

where it follows from our assumption that

$$\int_0^\infty e^{-\lambda_0(\tau-e_0)} \frac{d\Gamma(\tau)}{d\tau} d\tau \approx \int_0^\infty \left[ 1 - \lambda_0(\tau - e_0) + \frac{\lambda_0^2(\tau - e_0)^2}{2} \right] \frac{d\Gamma(\tau)}{d\tau} d\tau.$$

Therefore, we have

$$J \approx e^{-\lambda_0 e_0} \left[ 1 + \frac{\lambda_0^2 \sigma^2}{2} \right],$$

where

$$\sigma^2 = -\int_0^\infty (\tau - e_0)^2 \frac{d\Gamma(\tau)}{d\tau} d\tau$$

denotes the dispersion of the probability density function $\gamma(\tau)\Gamma(\tau)$. However, it follows from partial integration that

$$J = 1 - \lambda_0 \int_0^\infty e^{-\lambda_0 \tau} \Gamma(\tau) d\tau,$$

so we have

$$\kappa = \frac{\lambda_0 J}{1 - J}.$$

Using the above approximation for $J$, we arrive at an approximation formula for $\kappa$:

$$\kappa \approx \frac{\lambda_0(1 + \frac{1}{2}\lambda_0^2\sigma^2)}{e^{\lambda_0 e_0} - (1 + \frac{1}{2}\lambda_0^2\sigma^2)}. \tag{7.7}$$

The crude rate $\kappa$ is then an increasing function of the dispersion $\sigma^2$. For example, in the extreme case of $\sigma = 0$, $\kappa \approx 0.01$ when $\lambda_0 = 0.35$ (observed growth rate of AIDS cases in Japan around the 1990s) and $e_0 = 10$, so the number of unobserved HIV-infected individuals is $35 \sim 40$ times the number of confirmed AIDS cases.

**Exercise 7.1** The infection-age structure of the reported (diagnosed) HIV-infected individuals is usually very different from that of (not necessarily reported) total HIV-infected individuals. Therefore, $\kappa$ is also different from the (observable) crude rate of developing AIDS in the reported HIV-infected individuals, even when $B(t)$ is an exponential function [20, 34, 36]. Let us check this fact in a simple situation. Suppose that $\alpha = 1$. Let $h$ be a time-independent rate of diagnosis, and let $i_d(t, \tau)$ be the infection-age density of diagnosed HIV infecteds at time $t$. Show that $i_d(t, \tau) = (1 - e^{-h\tau})\Gamma(\tau)B(t - \tau)$. Define the capture ratio $C(t)$, the crude rate of developing AIDS $\kappa(t)$, and the crude rate of developing AIDS for the diagnosed HIV-infected individuals $\kappa_d(t)$ as

$$C(t) = \frac{\int_0^\infty i_d(t, \tau)d\tau}{\int_0^\infty i(t, \tau)d\tau}, \quad \kappa(t) = \frac{\int_0^\infty \gamma(\tau)\Gamma(\tau)B(t - \tau)d\tau}{\int_0^\infty \Gamma(\tau)B(t - \tau)d\tau},$$

$$\kappa_d(t) = \frac{\int_0^\infty (1 - e^{-h\tau})\gamma(\tau)\Gamma(\tau)B(t - \tau)d\tau}{\int_0^\infty (1 - e^{-h\tau})\Gamma(\tau)B(t - \tau)d\tau}.$$

Show that $C(t)\kappa_d(t)/\kappa(t)$ gives the ratio of AIDS cases produced from previously diagnosed HIV infecteds to total AIDS cases, so it is not equal to the capture ratio if $\kappa \neq \kappa_d$.

*Remark 7.4* If we introduce unobserved susceptible population and the transmission parameter, we can extend the above Malthus model to a stable population model. Due to the individual difference of the immune system, the survival probability and the infectivity of infected individuals have several variations, called the *profile*. Let $j = 0, 1, 2, \ldots$ be the number indicating the profile of survival probability and infectivity, and let $i_j(t, \tau)$ be the infection-age density of infecteds with $j$th profile [28]. We assume that this profile is not inherited and the probability that a new infection leads the $j$th profile, denoted by $\zeta_j > 0$, is constant. Let $S_0$ be the size of susceptible population, $\Lambda(S_0)$ the number of susceptibles with which an infected individual has contacts per unit time, $\beta_j(\tau)$ the transmission rate of the $j$th profile, $\Gamma_j(\tau)$ be the survival probability of the $j$th profile, and $B(t)$ the number of newly infecteds at time $t$. Biologically, the hazard rate $\gamma_j$ of the $j$th profile would have a positive correlation with the infectivity $\beta_j(\tau)$, since they are related to the density of virus in bloodstream. Then, we obtain

$$B(t) = \Lambda(S_0) \int_0^\infty \left[ \sum_j \beta_j(\tau) \Gamma_j(\tau) \zeta_j \right] B(t-\tau) d\tau.$$

Since we assume that $S_0$ is a constant, the above equation is the Lotka-type renewal integral equation with the net reproduction function

$$\Psi(\tau) := \Lambda(S_0) \left[ \sum_j \beta_j(\tau) \Gamma_j(\tau) \zeta_j \right].$$

That is, the infected population dynamics in the early stage of the HIV epidemic can be described by the stable population model. Then, the basic reproduction number is calculated as $R_0 = \int_0^\infty \Psi(\tau) d\tau$. Moreover, the intrinsic growth rate of the invading infecteds is given by the real root $\lambda_0$ of the Lotka's characteristic equation:

$$\int_0^\infty e^{-\lambda_0 \tau} \Psi(\tau) d\tau = 1.$$

However, the net reproduction function $\Psi$ is usually not observed, and $\lambda_0$ must be estimated from the observed data for AIDS cases. Conversely, if we have any information about the normalized pattern $\psi(\tau) := \Psi(\tau)/R_0$, the basic reproduction number can be estimated as $R_0 = \hat{\psi}(\lambda_0)^{-1}$ [47].

**Exercise 7.2** Suppose that the transmission coefficient $\beta$ is constant. Show that if the incidence rate is given by the Malthusian law (7.2) and $\alpha = 1$, it holds that $R_0 = be_0 = (\lambda_0 + \kappa)e_0$.

## 7.1.2 *Risk-Based Model*

We now consider a population structured by the risk of HIV infection (*risk-based model*), where the risk is measured by the average number of sexual contacts per unit time [1, 3]. Let $\zeta$ be a continuous risk variable with state space $\Omega = \mathbb{R}_+$, let $S(\zeta)$ be the density of susceptible individuals with risk $\zeta$ in the initial invasion phase, and let $I(t, \zeta)$ be the infected population density at time $t$ and risk $\zeta$. The linearized equation for the infected population is then given by[1]:

[1] The full nonlinear model is considered in [6, 17, 23, 24, 44].

$$\frac{\partial I(t,\zeta)}{\partial t} = \lambda(t)\zeta S(\zeta) - (\mu+\gamma)I(t,\zeta), \tag{7.8}$$

where $\mu$ is the natural death rate, $\gamma$ is the removal rate, $\lambda(t)$ is the force of infection given by

$$\lambda(t) = \beta\frac{\int_0^\infty \zeta I(t,\zeta)d\zeta}{\int_0^\infty \zeta S(\zeta)d\zeta} \tag{7.9}$$

and $\beta$ is the transmission rate per contact. The fraction in the composition of $\lambda$ denotes the probability that a partner is infected. Tentatively, we assume that the state space of the risk distribution is given by

$$E := \left\{\phi \in L^1(\mathbb{R}_+) : \int_0^\infty x^n|\phi(x)|dx < \infty, \quad n = 1, 2, \ldots\right\},$$

and assume that $S \in E_+$ and $I(0,\cdot) \in E_+$.

Applying the variation-of-constants formula to (7.8), we obtain

$$\begin{aligned} I(t,\zeta) &= g(t,\zeta) + \int_0^t e^{-(\mu+\gamma)(t-\tau)}\lambda(\tau)\zeta S(\zeta)d\tau \\ &= g(t,\zeta) + \Lambda(S)(\zeta)\int_0^t e^{-(\mu+\gamma)\tau}\beta\int_0^\infty \eta I(t-\tau,\eta)d\eta d\tau, \end{aligned} \tag{7.10}$$

where

$$\Lambda(S)(\zeta) := \frac{\zeta S(\zeta)}{\int_0^\infty \zeta S(\zeta)d\zeta}, \quad g(t,\zeta) := I(0,\zeta)e^{-(\mu+\gamma)t}.$$

Let $B(t,\zeta) := \lambda(t)\zeta S(\zeta)$ be the density of newly produced infected individuals. Then, we have an abstract renewal equation:

$$B(t,\zeta) = G(t,\zeta) + \int_0^t (\Psi(\tau)B(t-\tau,\cdot))(\zeta)d\tau, \tag{7.11}$$

where $B(t,\cdot)$ is an $E$-valued function, $\Psi(\tau)$ is a bounded linear operator on $E$ defined by

$$(\Psi(\tau)\phi)(\zeta) := \Lambda(S)(\zeta)\beta e^{-(\mu+\gamma)\tau}\int_0^\infty \eta\phi(\eta)d\eta, \quad \phi \in E \tag{7.12}$$

and

$$G(t,\zeta) := \beta\Lambda(S)(\zeta)\int_0^\infty \eta I(0,\eta)d\eta e^{-(\mu+\gamma)t}.$$

Therefore, the next-generation operator (NGO) $K$ on $E$ is given by

$$(K\phi)(\zeta) = \int_0^\infty \Psi(\tau)\phi d\tau = \Lambda(S)(\zeta)\frac{\beta}{\mu+\gamma}\int_0^\infty \eta\phi(\eta)d\eta, \quad \phi \in E. \tag{7.13}$$

Because $K$ is a one-dimensional operator, its spectral radius is easily calculated as

$$R_0 = \frac{\beta}{\mu+\gamma}\int_0^\infty \eta \Lambda(S)(\eta)d\eta, \tag{7.14}$$

and $R_0$ gives the asymptotic per-generation growth factor of the infected population (Chap. 9).

Let $m$ be the average number of partners, and let $\sigma^2$ be its dispersion:

$$m := \int_0^\infty \zeta \frac{S(\zeta)}{\int_0^\infty S(\eta)d\eta} d\zeta,$$

$$\sigma^2 = \int_0^\infty (\zeta - m)^2 \frac{S(\zeta)}{\int_0^\infty S(\eta)d\eta} d\zeta = \int_0^\infty \zeta^2 \frac{S(\zeta)}{\int_0^\infty S(\eta)d\eta} d\zeta - m^2.$$

On the other hand, observe that

$$\int_0^\infty \zeta \Lambda(S)(\zeta)d\zeta = \frac{\int_0^\infty \zeta^2 S(\zeta)d\zeta}{\int_0^\infty \zeta S(\zeta)d\zeta} = \frac{(m^2+\sigma^2)\int_0^\infty S(\zeta)d\zeta}{\int_0^\infty \zeta S(\zeta)d\zeta}. \tag{7.15}$$

Therefore, it follows that

$$R_0 = \frac{\beta}{\mu+\gamma}\frac{\langle\zeta^2\rangle}{\langle\zeta\rangle} = \frac{\beta m}{\mu+\gamma}\left[1+\left(\frac{\sigma}{m}\right)^2\right], \tag{7.16}$$

where

$$\langle\zeta^n\rangle := \frac{\int_0^\infty \zeta^n S(\zeta)d\zeta}{\int_0^\infty S(\zeta)d\zeta}.$$

The result in (7.16) was derived in [1], which demonstrated a serious effect of individual heterogeneity on the basic reproduction number. If the dispersion is zero (the host population is homogeneous), the basic reproduction number attains its minimum value $R_0 = \beta m/(\mu+\gamma)$. However, if we assume the individual heterogeneity obeys a *power law* (as observed in scale-free networks, [29]), the dispersion can become very large. In this case, $R_0$ may become huge, resulting in the host susceptible population being easily invaded by HIV. This type of vulnerability was observed in the HIV epidemic in the USA, where $S(\zeta)$ was proportional to $\zeta^{-3}$ [7]. Then, $\langle\zeta^2\rangle = \log(b/a)$ if $S$ has a support $[a, b]$ with $b > a > 0$, so $R_0$ goes to infinity if $b \to \infty$. That is, the long tail of the power law implies a large $R_0$. Therefore, it is dangerous to focus only on the average behavior of host individuals when formulating an epidemic model.

### 7.1.3 Pair Formation Model

As discussed above, HIV has various possible infection routes, such as sexual contact, blood transfusions, contaminated needle sharing, and vertical transmission. The transmission process through heterosexual contact, however, should receive the most attention regarding the future course of the epidemic, because everyone engaging in sexual activity forms the largest risk group with respect to this infection route. If we formulate a model for HIV infection by heterosexual contact, it is essential to take into account the *pair formation* phenomenon between both sexes. In fact, the major proportion of sexual contact occurs within more or less stable unions, and HIV can be transmitted between males and females not only by promiscuous sexual relations, but also by persistent unions. As was pointed out by several authors, the persistence of unions would affect the HIV transmission process. For example, two susceptible individuals who form a pair without sexual contact outside the pair are immune to sexually transmitted diseases as long as they do not separate. In this case, the basic reproduction number of HIV infection depends on the pair formation and separation rates. However, owing to its complexity, it is very difficult to analyze the pair formation model for HIV infection taking into account the persistence of unions, the variable infectivity (infection-age dependency), and the host (chronological) age structure. For the formulation of complex HIV infection models, readers are referred to the work of Hadeler [15] and Inaba [21].

We focus on the HIV invasion process by pair formation in bisexual populations. We consider only the linear process and do not discuss the full nonlinear dynamics. However, it should be noted that the following linear model can be obtained by linearizing the basic nonlinear model, which is too complex to analyze here but describes the whole dynamics of the HIV epidemic.

Suppose that the host population is structured by pair duration and infection-age, but for simplicity, we neglect the chronological age, because one of our main purposes here is to understand the effects of variable infectivity and union persistence on $R_0$. Moreover, we assume serial monogamy—that is, individuals repeat the pair formation and dissolution processes, but sexual contact occurs only within (heterosexual) pairs. According to the convention of Dietz and Hadeler [11], we assume that a pair is formed after one sexual contact, so we disregard the existence of pairs without sexual contact and a courtship period between the formation of the pair and the first sexual contact [9].

First, let us calculate the (multistate) survival probability for an infection cohort, where state one denotes the infected state coupled with an infected partner, state two denotes the infected single state, and state three denotes the infected state coupled with a susceptible partner. As infection can only occur within a pair, the infection cohort starts from state one. Let $\ell_j^m(\tau)$ [$\ell_j^f(\tau)$] be the male [female] survival probability that individuals in state one at $\tau = 0$ are alive in state $j$ and have infection-age $\tau$. Let $\rho_m$ [$\rho_f$] be the pair formation rate for infected males [females]. These are assumed to be constant, though they are originally functions of the density of susceptible and infected individuals and would vary as the infection process progresses.

Let $\gamma$ be the rate of developing AIDS, $\sigma$ be the rate of pair dissolution, $\eta$ be the number of sexual contacts per unit time in a pair, and $\mu_m$ [$\mu_f$] be the male [female] natural death rate. These are also assumed to be constant.

We assume that a pair can be dissolved not only by spontaneous pair dissolution, but also by one partner dying or developing AIDS and that there is no other reason for the interruption of sexual activity. Let $\beta_{mf}(\tau)$ [$\beta_{fm}(\tau)$] be the HIV infection rate per sexual contact from males [females] at infection-age $\tau$ to his [her] female [male] partner.

We now introduce vector notation. Let

$$\ell^m(\tau) = (\ell_1^m(\tau), \ell_2^m(\tau), \ell_3^m(\tau))^{\mathrm{T}}, \quad \ell^f(\tau) = (\ell_1^f(\tau), \ell_2^f(\tau), \ell_3^f(\tau))^{\mathrm{T}}$$

be the male and female survival probability vectors, where T denotes the transpose of the vector. Let $A^m$ and $A^f$ be $3 \times 3$ matrices defined by

$$A^m(\tau) := \begin{pmatrix} -2\gamma - \sigma - \mu_m - \mu_f & \beta_{10}(\tau)\rho_m & \eta\beta_{10}(\tau) \\ \sigma + \gamma + \mu_f & -\gamma - \rho_m - \mu_m & \sigma + \mu_f \\ 0 & (1 - \beta_{10}(\tau))\rho_m & -\eta\beta_{10}(\tau) - \gamma - \sigma - \mu_m - \mu_f \end{pmatrix},$$

$$A^f(\tau) := \begin{pmatrix} -2\gamma - \sigma - \mu_m - \mu_f & \beta_{01}(\tau)\rho_f & \eta\beta_{01}(\tau) \\ \sigma + \gamma + \mu_m & -\gamma - \rho_f - \mu_f & \sigma + \mu_m \\ 0 & (1 - \beta_{01}(\tau))\rho_f & -\eta\beta_{01}(\tau) - \gamma - \sigma - \mu_m - \mu_f \end{pmatrix}.$$

From our assumptions, we know that $\ell^m(\tau)$ and $\ell^f(\tau)$ satisfy the following differential equations:

$$\begin{aligned} \frac{d\ell^m(\tau)}{d\tau} &= A^m(\tau)\ell^m(\tau), \quad \ell^m(0) = (1,0,0)^{\mathrm{T}}, \\ \frac{d\ell^f(\tau)}{d\tau} &= A^f(\tau)\ell^f(\tau), \quad \ell^f(0) = (1,0,0)^{\mathrm{T}}. \end{aligned} \tag{7.17}$$

Let $B_m(t)$ and $B_f(t)$ be the number of newly HIV-infected males and females produced per unit time. It follows that

$$\begin{aligned} B_m(t) &= \int_0^\infty [\rho_f\beta_{fm}(\tau)\ell_2^f(\tau) + \eta\beta_{fm}(\tau)\ell_3^f(\tau)]B_f(t-\tau)d\tau, \\ B_f(t) &= \int_0^\infty [\rho_m\beta_{mf}(\tau)\ell_2^m(\tau) + \eta\beta_{mf}(\tau)\ell_3^m(\tau)]B_m(t-\tau)d\tau. \end{aligned} \tag{7.18}$$

The above system of integral equations is the same as Pollard's linear two-sex model (Chap. 4), so its basic reproduction number can be calculated as follows:

**Proposition 7.1** *The basic reproduction number for HIV infection via pair formation is given by* $R_0 = \sqrt{R_m R_f}$, *where* $R_m$ [$R_f$] *is the reproduction number whereby male [female] infected individuals produce female [male] infected individuals given by*

$$R_m := \int_0^\infty \beta_{mf}(\tau)[\rho_m \ell_2^m(\tau) + \eta \ell_3^m(\tau)]d\tau,$$
$$R_f := \int_0^\infty \beta_{fm}(\tau)[\rho_f \ell_2^f(\tau) + \eta \ell_3^f(\tau)]d\tau.$$

To simplify the model and obtain a concrete expression for $R_0$, let us neglect the parameter differences between the sexes. That is, we assume that $\beta(\tau) := \beta_{mf}(\tau) = \beta_{fm}(\tau)$, $\rho := \rho_m = \rho_f$, and $\mu := \mu_m = \mu_f$, so we omit the subscripts indicating male and female. It is easy to see that the basic reproduction number for this symmetric case is given by

$$R_0 = \int_0^\infty \beta(\tau)[\rho \ell_2(\tau) + \eta \ell_3(\tau)]d\tau. \tag{7.19}$$

Note that the common survival rate vector $\ell(\tau) := (\ell_1(\tau), \ell_2(\tau), \ell_3(\tau))^{\mathrm{T}}$ satisfies the following system of ordinary differential equations:

$$\frac{d\ell(\tau)}{d\tau} = A(\tau)\ell(\tau), \quad \ell(0) = (1, 0, 0)^{\mathrm{T}}, \tag{7.20}$$

where the matrix $A$ is given by

$$A(\tau) := \begin{pmatrix} -2\gamma - \sigma - 2\mu & \beta(\tau)\rho & \eta\beta(\tau) \\ \sigma + \gamma + \mu & -\gamma - \rho - \mu & \sigma + \mu \\ 0 & (1 - \beta(\tau))\rho & -\eta\beta(\tau) - \gamma - \sigma - 2\mu \end{pmatrix}.$$

Although we cannot give a general analytical solution for (7.20), we can solve it for the special case in which we neglect the rate of developing AIDS, that is, $\gamma = 0$. Though this assumption is not realistic for the long-run behavior, it is reasonable for the initial invasion phase of HIV. Under this assumption, (7.20) has two integrals:

$$\ell_1(\tau) + \ell_2(\tau) + \ell_3(\tau) = e^{-\mu\tau},$$
$$\ell_1(\tau) + \ell_3(\tau) = e^{-(\sigma+\rho+2\mu)\tau} + \rho \int_0^\tau e^{-(\sigma+\rho+2\mu)(\tau-s)} e^{-\mu s} ds.$$

From the above integrals, we can obtain the analytic solution

$$\ell_2(\tau) = \frac{\sigma + \mu}{\sigma + \rho + \mu} e^{-\mu\tau}(1 - e^{-(\sigma+\rho+\mu)\tau}),$$
$$\ell_3(\tau) = \frac{\rho(\sigma + \mu)}{\sigma + \rho + \mu} \int_0^\tau e^{-\int_s^\tau (\eta\beta(\zeta)+\sigma+2\mu)d\zeta}(1 - \beta(s))(e^{-\mu s} - e^{-(\sigma+\rho+2\mu)s})ds. \tag{7.21}$$

If we define $\phi(\tau) := \rho\ell_2(\tau)$, then $\phi(\tau)$ gives the probability that pair formation occurs at infection-age $\tau$:

$$\phi(\tau) = \frac{\rho(\sigma+\mu)}{\sigma+\rho+\mu} e^{-\mu\tau}(1 - e^{-(\sigma+\rho+\mu)\tau}). \tag{7.22}$$

Therefore, the total average number of partners during the sexually active life is calculated as

$$\int_0^\infty \phi(\tau) d\tau = \frac{\rho(\sigma+\mu)}{\mu(\sigma+\rho+2\mu)}. \tag{7.23}$$

From (7.19), (7.21)–(7.23), we can state the following:

**Proposition 7.2** *For the symmetrical case, if $\gamma = 0$, the basic reproduction number is calculated as follows:*

$$R_0 = \int_0^\infty S(\tau)\phi(\tau) d\tau, \tag{7.24}$$

*where*

$$S(\tau) := T(\tau)(1 - \beta(\tau)) + \beta(\tau),$$
$$T(\tau) := \int_\tau^\infty \eta\beta(z) e^{-\int_\tau^z (\beta(\zeta)\eta+\sigma+2\mu)d\zeta} dz.$$

The function $S(\tau)$ gives the infection probability that an infected individual who forms a pair with a susceptible partner at infection-age $\tau$ gives rise to an infection per pair formation. This is composed of the probability for infection by persistent contact occurring in a pair and the probability that the infection occurs at the moment of pair formation. If the infection probability $\beta$ per contact is small, the former part will play a more important role in $R_0$. $T(\tau)$ is the probability that an infected individual with a susceptible partner will ever infect the partner.

It is important to examine the sensitivity of the basic reproduction number to the relevant parameters, because the disease prevention strategy must be planned so as to control $R_0$ at below unity. It is easy to see that the basic reproduction number given by (7.24) is a monotone increasing function of the pair formation rate $\rho$. Therefore, it is clear that lowering the pair formation rate will effectively decrease $R_0$. However, the response of $R_0$ to the pair dissolution rate $\sigma$ is more complex.

For simplicity, let us consider the case where $\beta$ is constant. In this case, $R_0$ is calculated as

$$R_0 = \frac{\rho(\sigma+\mu)}{\mu(\sigma+\rho+2\mu)} \cdot \frac{\beta(\eta+\sigma+2\mu)}{\beta\eta+\sigma+2\mu}.$$

Differentiating $R_0$ with respect to $\sigma$, we have

$$\frac{\partial R_0}{\partial \sigma} = \frac{\beta\rho(A\sigma^2 + 2B\sigma + C)}{\mu(\beta\eta+\sigma+2\mu)^2(\sigma+\rho+2\mu)^2},$$

where $A$, $B$, and $C$ are given by

$$\begin{aligned} A &:= \rho + \mu - \eta(1-\beta), \\ B &:= \rho\beta\eta + 2\mu(\rho + \beta\eta + \mu) - \mu\eta, \\ C &:= 4\mu^3 + 4(\rho + \beta\eta)\mu^2 + (\beta\eta^2 + 3\rho\beta\eta + \rho\eta)\mu + \beta\eta^2\rho. \end{aligned}$$

If we define $D := B^2 - AC$, we can conclude that if $D < 0$, then the denominator of $\partial R_0/\partial\sigma$ is positive definite, so $R_0$ becomes a monotone increasing function of $\sigma$. However, if $D > 0$, there exist some numbers $\alpha_1 < \alpha_2$ such that

$$\frac{\partial R_0}{\partial \sigma} = \frac{\beta\rho A(\sigma - \alpha_1)(\sigma - \alpha_2)}{\mu(\beta\eta + \sigma + 2\mu)^2(\sigma + \rho + 2\mu)^2}.$$

Because we can observe that

$$\left[\frac{\partial R_0}{\partial \sigma}\right]_{\sigma=0} = \frac{\beta\rho C}{\mu(\beta\eta + 2\mu)^2(\rho + 2\mu)^2} > 0,$$

if $A > 0$, $R_0$ becomes a monotone increasing function of $\sigma$ in the case that $\alpha_1 < \alpha_2 < 0$. If $0 < \alpha_1 < \alpha_2$, $R_0$ attains a local maximum at $\sigma = \alpha_1$ and a local minimum at $\sigma = \alpha_2$. However, if $A < 0$, we have $\alpha_1 < 0 < \alpha_2$, so $R_0$ reaches a maximum at $\sigma = \alpha_2$.

For realistic parameter values, we can expect that $A < 0$, and hence, $D > 0$. In fact, we can rewrite $A$ as

$$A = (\rho + \mu)\eta\left(\frac{1}{\eta} - \frac{1-\beta}{\rho + \mu}\right),$$

where $1/\eta$ is the average interval between sexual contact in a pair and $1/(\rho+\mu)$ is the average duration in the single state. Therefore, for realistic situations, we assume that $1/\eta \ll 1/(\rho+\mu)$ and $1-\beta$ is close to unity and hence expect that $A < 0$.

Thus, if the infection rate $\beta$ is constant, we can conclude that realistic parameter values give a basic reproduction number that exhibits a unimodal pattern as a function of the pair dissolution rate $\sigma$. Therefore, if the pair dissolution rate is already small and the pairs are stable, $R_0$ will decrease as $\sigma$ becomes smaller. However, in the region where $\sigma$ is large, $R_0$ decreases as $\sigma$ increases because the pair is not stable long enough for infection to occur. In such parameter regions, reducing the pair dissolution rate would paradoxically lead to a larger value of $R_0$.

Subsequently, we consider the relation between $R_0$ and the infection probability $\beta(a)$, which is not necessarily constant. We consider the infection rate per partner $S(\tau)$ and $T(\tau)$ to be functionals of the infection probability $\beta(\tau)$, so we write them as $S(\beta)(\tau)$ and $T(\beta)(\tau)$. Integrating partially, we have

$$T(\beta)(\tau) = 1 - (\sigma + 2\mu) \int_\tau^\infty e^{-\int_\tau^z (\beta(\zeta)\eta + \sigma + 2\mu) d\zeta} dz,$$
$$S(\beta)(\tau) = 1 - (1 - \beta(\tau))(\sigma + 2\mu) \int_\tau^\infty e^{-\int_\tau^z (\beta(\zeta)\eta + \sigma + 2\mu) d\zeta} dz, \tag{7.25}$$

where $T(\beta)$ and $S(\beta)$ are monotone increasing with respect to $\beta$. Let $R_0(\beta)$ be the basic reproduction ratio corresponding to $\beta$. We can write

$$R_0(\beta) = \int_0^\infty S(\beta)(a)\phi(a)da.$$

Thus, the monotonicity holds, that is, $\beta_1 \le \beta_2$ implies $R_0(\beta_1) \le R_0(\beta_2)$. However, we have no monotonic relation if $\beta_1$ and $\beta_2$ are not comparable.

In the case of HIV infection, there is evidence for a short initial phase and a longer terminal phase of high infectivity, separated by a long phase during which infectivity is very low [2, 4, 5, 41]. Note that constant infectivity and variable infectivity are not comparable with each other if they have the same average infectivity. As an important consequence of variable infectivity, several authors have pointed out that the non-constant infectivity pattern of HIV decreases $R_0$ compared with constant infectivity if they have the same average infectivity [12, 15, 16].

To examine the above observation in our modeling approach, we follow Hadeler [15] by defining the *average infectivity* $\bar{\beta}$ and the normalized infectivity pattern $h(\tau)$ as

$$\bar{\beta} := \mu \int_0^\infty e^{-\mu\tau} \beta(\tau) d\tau, \quad h(\tau) := \frac{\beta(\tau)}{\bar{\beta}}. \tag{7.26}$$

Note that the standard schedule of constant infectivity is given by $h_0(a) \equiv 1$.

First, consider the special case in which there are no persistent unions ($\sigma = \infty$). For this special case, it follows from (7.22) and (7.24) that

$$R_0(\beta) = \rho \int_0^\infty e^{-\mu a} \beta(a) da,$$

because $S(\beta)(\tau) \to \beta(\tau)$ and $\phi(\tau) \to \rho \exp(-\mu\tau)$ if $\sigma \to \infty$. However, using the definition of the average infectivity (7.26), we obtain

$$R_0(\bar{\beta} h_0) = \frac{\rho\bar{\beta}}{\mu} = R_0(\beta).$$

For the case of non-persistent union, the basic reproduction number of the non-constant infectivity is the same as $R_0$ with constant infectivity if the average infectivity of both is the same. Therefore, the possible dependence of $R_0$ on the infectivity pattern under the same average infectivity must be caused by the persistence of unions.

For the case of persistent unions ($\sigma < \infty$), $R_0$ could vary depending on the pattern of the variable infectivity. It is difficult to prove that a variable infectivity pattern always leads to a lower $R_0$ than constant infectivity when there is the same average infectivity. However, this conjecture has been confirmed for many cases by numerical calculations [21]. Based on (7.25), we can conclude that the main reason is that $S(\beta)$ exhibits a similar pattern to $\beta$ when $\beta$ is variable. From (7.22), the probability density of pair formation $\phi$ has a unimodal pattern, so the initial and terminal phases of $S(\beta)$ play relatively minor roles in the calculation of $R_0$. If $S(\beta)$ has a variable pattern, we could expect that $R_0(\beta) < R_0(\bar{\beta}h_0)$, because the probability density of pair formation $\phi$ is relatively large in the long intermediate phase during which $S(\beta) < S(\bar{\beta}h_0)$.

*Remark 7.5* Since the late 1980s, pair formation models for sexually transmitted diseases have been developed by many authors. Epidemic pair formation models without structural variables (ordinary differential equation models) have been investigated in detail by Dietz and Hadeler [11], Dietz [10], and Waldstätter [46]. Castillo-Chavez [6] formulated a pair formation model structured by infection-age. Diekmann et al. [8, 9], Heesterbeek [16], and Dietz et al. [12] described the transmission dynamics of sexually transmitted diseases as a discrete process on the generation of infected individuals and established algorithms for calculating $R_0$ that account for an arbitrary but finite number of disease states and partnership states. An important observation from their numerical calculations is that $R_0$ decreases for variable infectivity if the average infectivity remains constant. Knolle [26] considered the transmission of HIV as a Markov process to calculate its basic reproduction number. Mode and Dietz [32] pointed out that many results derived from deterministic pair formation models for sexually transmitted diseases could be induced by some stochastic models for individual processes. Hadeler's model [14, 15] is closest to the basic model described in this section, as both are deterministic structured population models that consider pair formation and infection-age-dependent infectivity. Though we do not consider the chronological age, Hadeler's formula is more convenient when incorporating other structural variables such as chronological age. In contrast, our modeling strategy constructs a structured population model along the individual event history, effectively bridging between Markovian models for the individual processes developed to calculate $R_0$ in the initial invasion phase and dynamical population-level models of the whole epidemic dynamics. In other words, we could embed the stochastic view of the individual transmission process into nonlinear deterministic dynamical systems. As a result, we could induce some new formulas for $R_0$ and clarify their relation to the assumptions about individual behavior.

Our basic assumptions about human sexual behavior are extremely simple and can therefore be criticized in many aspects. First, though we disregard the existence of pairs without sexual contact, in reality such pairs exist, just as there may be a courtship period between pair formation and the first sexual contact or a period between an unstable union with discrete sexual contact and the formation of a stable union with continuous sexual contact. Second, we neglect the possibility of sexual contact outside of the union and the existence of commercial sex workers. Third, we do not incorporate different levels of individual sexual activity. More realistic pair formation models that account for these phenomena have been studied; readers are referred to early studies by Watts and May [49] and Kretzschmar et al. [27] for further details.

From a biological standpoint, it is also interesting to consider why HIV has evolved such a variable infectivity pattern, because $R_0$ could be interpreted as a measure of its fitness. Unlike our simple model, if the rate of developing AIDS is correlated with the transmission probability, the pattern of $\beta$ could be interpreted as an optimum schedule for the growth of the pathogen strain [39]. In fact, the process of developing AIDS can be understood as an interaction between HIV and the host immune system.

## 7.2 Age-Structured Model for HIV Infection in a Homosexual Community

### *7.2.1 Basic Model*

After the (linear) invasion phase, the nonlinear interaction between susceptible and infected individuals determines the spread of an infectious disease. To consider the long-term dynamics of HIV infection, we now develop an epidemic model for HIV infection in a homosexual population. Though we do not take into account the effect of persistent unions and assume random contact, we introduce the chronological age and the infection-age (duration) structure. In fact, the chronological age is one of the most important parameters in determining individual sexual behavior and strongly affects the transmission process. If sexual contact occurs only within a cohort, infectious disease without vertical transmission will be automatically eradicated when the cohort dies out. Therefore, it is natural to think that the basic reproduction number or the severity of horizontally transmitted diseases will be heavily dependent on the age differences among individual contacts [26].

Let us formulate the age–duration-structured HIV epidemic model for homosexual individuals with a constant birth rate [22]. For simplicity, individuals are assumed to be homogeneous with respect to their sexual activity, though the following argument could easily be extended to the risk-based model without any essential modifications. Individuals engage in sexual contact with one another at random, and the duration of an exclusive partnership is negligibly short. We divide the homosexual population into three groups: $S$ (uninfected but susceptible), $I$ (incubation class; HIV-infected asymptomatically with infectivity), and $A$ (fully developed AIDS symptoms). We do not introduce an explicit latent class, because the latent period of AIDS is negligibly short in comparison with its long incubation period and its existence can be expressed by the infection-age-dependent infectivity. Individuals in state $A$ are assumed to be sexually inactive and thus not involved in the transmission process.

Let $S(t,a)$ be the age density of the susceptible population at time $t$ and age $a$, and let $B$ be the birth rate of the susceptible population. Let $I(t,\tau;a)$ be the density of the infected population at time $t$ and infection-age (duration since infection) $\tau$ with the (chronological) age at infection $a$, that is, the chronological age is $a+\tau$. Let $A(t,\tau;a)$ be the density of the AIDS population at time $t$ and *disease age*[2] $\tau$ for individuals who have developed AIDS at age $a$. Let $\mu(a)$ be the age-specific natural death rate, $\gamma(\tau;a)$ be the rate of developing AIDS at infection age $\tau$ for individuals whose chronological age at infection is $a$, $\delta(\tau;a)$ be the death rate at disease age $\tau$ due to AIDS for individuals who have developed AIDS at age $a$, and $\lambda(t,a)$ be the force of infection at age $a$ and time $t$. The dynamics of the population are then governed by the following system:

$$\begin{aligned}
&\left(\frac{\partial}{\partial t}+\frac{\partial}{\partial a}\right)S(t,a)=-(\mu(a)+\lambda(t,a))S(t,a),\\
&\left(\frac{\partial}{\partial t}+\frac{\partial}{\partial \tau}\right)I(t,\tau;a)=-(\mu(a+\tau)+\gamma(\tau;a))I(t,\tau;a),\\
&\left(\frac{\partial}{\partial t}+\frac{\partial}{\partial \tau}\right)A(t,\tau;a)=-(\mu(a+\tau)+\delta(\tau;a))A(t,\tau;a),\\
&S(t,0)=B,\\
&I(t,0;a)=\lambda(t,a)S(t,a),\\
&A(t,0;a)=\int_0^a\gamma(\tau;a-\tau)I(t,\tau;a-\tau)d\tau.
\end{aligned}\tag{7.27}$$

The force of infection $\lambda(t,a)$ is assumed to have the following expression:

$$\lambda(t,a)=\int_0^{\omega}\int_0^{b}\beta(a,b,\tau)\xi(a,b,N(t,\cdot))\frac{I(t,\tau;b-\tau)}{N(t,b)}d\tau db,$$

where $N(t,a)$ is the age density of the sexually active population at time $t$ given by

$$N(t,a)=S(t,a)+\int_0^a I(t,\tau;a-\tau)d\tau,$$

$\beta(a,b,\tau)$ is the transmission probability for a susceptible person of age $a$ being infected by sexual contact with an infected partner of age $b$ and infection age $\tau$, and $\omega$ denotes the upper bound of age in the sexually active population.

The *mating function* $\xi(a,b,N(t,\cdot))$ depending on the population density $N(t,\cdot)$ denotes the average number of sexual partners of age $b$ an individual aged $a$ has per unit time at time $t$. From its physical meaning, the mating function must satisfy the following condition:

$$N(t,a)\xi(a,b,N(t,\cdot))=N(t,b)\xi(b,a,N(t,\cdot)).\tag{7.28}$$

[2] The time elapsed since the appearance of symptoms is called the *disease age*.

In the following, we assume that the mating function can be expressed as

$$\xi(a, b, N(t, \cdot)) = C(P(t))\frac{N(t, b)}{P(t)}, \tag{7.29}$$

where $P(t)$ is the total size of the sexually active population given by

$$P(t) = \int_0^{\omega} N(t, \sigma)d\sigma$$

and $C(P)$ denotes the mean number of sexual partners an average individual has per unit time when the population size is $P$. It is easy to see that the mating function (7.29) satisfies condition (7.28). Under the above assumptions, the force of infection is given by

$$\lambda(t, a) = \frac{C(P(t))}{P(t)} \int_0^{\omega} \int_0^{b} \beta(a, b, \tau) I(t, \tau; b - \tau) d\tau db. \tag{7.30}$$

From its biological meaning, it is reasonable to assume that the function $C(P) : \mathbb{R}_+ \to \mathbb{R}_+$ is monotone increasing and upper-bounded. Typical examples of $C(P)$ are as follows:

$$(i) \quad C(P) = \alpha_0 P, \quad (ii) \quad C(P) = \frac{\alpha_0 \alpha_\infty P}{\alpha_0 P + \alpha_\infty}, \quad (iii) \quad C(P) = \alpha_\infty,$$

where $\alpha_0$ and $\alpha_\infty$ are given positive numbers. Note that the saturating contact law (ii) approaches the mass-action-type contact law (i) when $P \to 0$, whereas it becomes the homogeneous of degree one (scale-independent) contact law (iii) if $P \to \infty$. In the following, we assume Lipschitz continuity as follows:

**Assumption 7.3** $C(x)/x$ is a monotone decreasing function for $x \geq 0$. There exists a constant $L > 0$ for any $x, y \geq 0$ such that $|C(x)/x - C(y)/y| \leq L|x - y|$.

The mathematically well-posed nature of the time evolution problem (7.27) can be proved by several approaches. Though we skip the details, the semigroup solution can be constructed using the perturbation method of non-densely defined operators [22]. The semigroup approach is most advantageous for establishing the principle of linearized stability.

*Remark 7.6* When the host demographic age structure is not so relevant and could be neglected, we can formulate the simplified model

$$
\begin{aligned}
&\frac{dS(t)}{dt} = b - (\mu + \lambda(t))S(t), \\
&\left(\frac{\partial}{\partial t} + \frac{\partial}{\partial \tau}\right) I(t, \tau) = -(\mu + \gamma(\tau))I(t, \tau), \\
&I(t, 0) = \lambda(t)S(t), \\
&\frac{dR(t)}{dt} = -(\mu + \delta)R(t) + \int_0^\infty \gamma(\tau)I(t, \tau)d\tau, \\
&\lambda(t) = \frac{C(P(t))}{P(t)} \int_0^\infty \beta(\tau)I(t, \tau)d\tau, \\
&P(t) = S(t) + \int_0^\infty I(t, \tau)d\tau.
\end{aligned}
\tag{7.31}
$$

The above simplified model has shown its usefulness in predicting the HIV epidemic in drug users in Italy [18], and it has been mathematically analyzed in [6, 42].

### 7.2.2 Criterion for HIV Invasion

Let us now consider the invasion problem to establish the basic reproduction number for HIV infection. Note that the basic system (7.27) has a disease-free steady state $(S_0, I_0) := (B\ell(a), 0)$, where $\ell(a)$ is the survival function defined by $\ell(a) := \exp\left(-\int_0^a \mu(\sigma)d\sigma\right)$. By linearizing the basic system about the disease-free steady state, we know that in the early stages of the epidemic, the dynamics of the infected population are described by the following equation:

$$
\begin{aligned}
&\left(\frac{\partial}{\partial t} + \frac{\partial}{\partial \tau}\right) I(t, \tau; a) = -(\mu(a + \tau) + \gamma(\tau; a))I(t, \tau; a), \\
&I(t, 0; a) = S_0(a)\frac{C(P_0)}{P_0} \int_0^\omega \int_0^b \beta(a, b, \tau)I(t, \tau; b - \tau)d\tau db,
\end{aligned}
\tag{7.32}
$$

where $S_0(a) = B\ell(a)$ and $P_0 := \int_0^\omega S_0(a)da$.

By integrating (7.32) along the characteristic line, we obtain the expression

$$
I(t, \tau; a) = \begin{cases} \Gamma(\tau; a)\frac{\ell(a+\tau)}{\ell(a)} I(t - \tau, 0; a), & (t - \tau > 0) \\ \frac{\Gamma(\tau; a)\ell(a+\tau)}{\Gamma(\tau-t; a)\ell(a+\tau-t)} I(0, \tau - t; a), & (\tau - t > 0) \end{cases}
\tag{7.33}
$$

where $I(0, \tau; a)$ is a given initial data point and $\Gamma(\tau; a)$ is the survival function defined by

$$
\Gamma(\tau; a) := \exp\left(-\int_0^\tau \gamma(\sigma; a)d\sigma\right).
$$

Inserting (7.33) into the boundary condition in (7.32), we obtain the following renewal integral equation for the density of the newly infected population $B(t,a) := I(t,0;a)$:

$$B(t,a) = G(t,a) + S_0(a)\frac{C(P_0)}{P_0}\int_0^t \int_\tau^\omega K_0(a,b,\tau)B(t-\tau,b-\tau)dbd\tau, \quad (7.34)$$

where $G$ and $K_0$ are given by

$$G(t,a) := S_0(a)\frac{C(P_0)}{P_0}\int_t^\omega \int_\tau^\omega \beta(a,b,\tau)\frac{\Gamma(\tau;b-\tau)\ell(b)}{\Gamma(\tau-t;b-\tau)\ell(b-t)}I(0,\tau-t;b-\tau)dbd\tau,$$

$$K_0(a,b,\tau) := \beta(a,b,\tau)\Gamma(\tau;b-\tau)\frac{\ell(b)}{\ell(b-\tau)},$$

and we adopt the convention that $G = 0$ for $t > \omega$.

For technical reasons, it is convenient to adopt the following assumption from this point on:

**Assumption 7.4** The age-dependent functions $\ell(a)$, $K_0(a,b,\tau)$ and $\Gamma(\tau;a)$ are extended as zero-valued functions outside the age interval $[0,\omega]$ and for $b < \tau$. Moreover, $\beta$ is a uniformly bounded function, $\inf_{a\geq 0}\mu(a) =: \underline{\mu} > 0$ and $\inf_{a\geq 0}\gamma(\sigma) =: \underline{\gamma} > 0$.

Let us consider $G(t,a)$ and $B(t,a)$ as $L^1$-valued functions of $t > 0$, and let $\Pi(\tau)$ be a linear positive operator from $L^1_+(0,\omega)$ to itself defined by

$$(\Pi(\tau)\psi)(a) := S_0(a)\frac{C(P_0)}{P_0}\int_\tau^\omega K_0(a,b,\tau)\psi(b-\tau)db$$

for $\tau \in [0,\omega]$ and $\Pi(\tau) = 0$ for $\tau > \omega$. We can then consider (7.34) as an abstract renewal integral equation in $L^1$:

$$B(t) = G(t) + \int_0^t \Pi(\tau)B(t-\tau)d\tau, \quad t > 0. \quad (7.35)$$

As we have seen in Chaps. 2 and 6, if the Laplace transform

$$\hat{\Pi}(z) := \int_0^\infty e^{-z\tau}\Pi(\tau)d\tau$$

is compact and non-supporting on the real axis, its spectral radius (the Frobenius root of the positive operator $\hat{\Pi}(z)$ for real $z$), denoted by $r(\hat{\Pi}(z))$, is monotone decreasing from $+\infty$ to zero, and there exists a unique real root $r_0$ such that $r(\hat{\Pi}(r_0)) = 1$ and $\text{sign}(r_0) = \text{sign}(r(\hat{\Pi}(0)) - 1)$. Moreover, we can prove the renewal theorem for the existence of a nonnegative constant $c$ such that $\lim_{t\to\infty} e^{-r_0 t}B(t) = c\psi_0$, where $\psi_0$ is the positive eigenvector corresponding to the Frobenius eigenvalue one of $\hat{\Pi}(r_0)$.

From the general theory for $R_0$ (see Chap. 9), we can define $R_0$ for HIV infection by $R_0 = r(\hat{\Pi}(0))$, where $\hat{\Pi}(0)$ is the NGO. We can then state the criterion for HIV invasion: The disease can invade the completely susceptible population if $R_0 > 1$, whereas it cannot if $R_0 < 1$. Formally, this can be written as follows:

**Proposition 7.5** *The disease-free steady state is locally asymptotically stable if* $R_0 = r(\hat{\Pi}(0)) < 1$, *whereas it is unstable if* $R_0 > 1$.

If we adopt the separable mixing assumption, that is, assume that the transmission rate $\beta$ can be decomposed as $\beta(a, b, \tau) = \beta_1(a)\beta_2(b, \tau)$, we have

$$
\begin{aligned}
(\hat{\Pi}(0)\psi)(a) &= S_0(a)\frac{C(P_0)}{P_0}\beta_1(a) \\
&\quad \times \int_0^{\omega} d\tau \int_z^{\omega} \beta_2(b, b-z)\Gamma(\tau; b-\tau)\frac{S_0(b)}{S_0(b-\tau)}\psi(b-\tau)db.
\end{aligned}
$$

That is, in this special case, the range of the NGO is a one-dimensional space spanned by $\beta_1(a)S_0(a)$, which is none other than the Frobenius eigenvector of the NGO. Therefore, we obtain the eigenvalue equation $\hat{\Pi}(0)\beta_1 S_0 = R_0\beta_1 S_0$ from which we can explicitly calculate the basic reproduction number as

$$
R_0 = \frac{C(P_0)}{P_0}\int_0^{\infty}\int_{\tau}^{\infty}\beta_2(b, \tau)\Gamma(\tau; b-\tau)S_0(b)\beta_1(b-\tau)dbd\tau. \tag{7.36}
$$

### 7.2.3 *Bifurcation of Endemic Equilibria*

Let us now consider the existence problem of endemic steady states. Let $(S^*(a), I^*(\tau; a))$ be the steady state for system (7.27), and let $\lambda^*(a)$ be the force of infection in the steady state. It follows that

$$
\begin{aligned}
S^*(a) &= S_0(a)e^{-\int_0^a \lambda^*(\xi)d\xi}, \\
I^*(\tau; a) &= \Gamma(\tau; a)\frac{\ell(a+\tau)}{\ell(a)}B^*(a),
\end{aligned} \tag{7.37}
$$

where $B^*(a) := \lambda^*(a)S^*(a)$ is the density of the newly infected population.

It follows from (7.30) and (7.37) that $B^*$ satisfies the following renewal integral equation:

$$
B^*(a) = S^*(a)\frac{C(P[\lambda^*])}{P[\lambda^*]}\int_0^{\omega}\int_{\tau}^{\omega} K_0(a, b, \tau)B^*(b-\tau)dbd\tau, \tag{7.38}
$$

where $P[\lambda^*]$ denotes the size of the steady state population with the force of infection $\lambda^*$ given by

$$P[\lambda^*] := \int_0^\omega S_0(a)\left[e^{-\int_0^a \lambda^*(\xi)d\xi} + \int_0^a \Gamma(a-\tau;\tau)\lambda^*(\tau)e^{-\int_0^\tau \lambda^*(\xi)d\xi}d\tau\right]da.$$

Inserting the relation

$$B^*(a) = \lambda^*(a)S_0(a)e^{-\int_0^a \lambda^*(\xi)d\xi}$$

into (7.38), we know that $\lambda^*$ must satisfy the following nonlinear integral equation:

$$\lambda^*(a) = \frac{C(P[\lambda^*])}{P[\lambda^*]}\int_0^\omega db \int_0^b K(a,b,\tau)\lambda^*(b-\tau)e^{-\int_0^{b-\tau}\lambda^*(\xi)d\xi}d\tau, \qquad (7.39)$$

where

$$K(a,b,\tau) := K_0(a,b,\tau)S_0(b-\tau) = \beta(a,b,\tau)S_0(b)\Gamma(\tau;b-\tau).$$

Conversely, the nonnegative solution of (7.39) determines the steady state through (7.37). It is clear that $\lambda^* = 0$ is a trivial solution of (7.39) that corresponds to the disease-free steady state.

We now give an elementary proof for the existence of an endemic steady state, that is, we show that the nonlinear equation (7.39) has a positive solution under additional conditions.

**Proposition 7.6** *Suppose that $C(P)/P$ is non-increasing with respect to $P$, $K(a,b,\tau)$ is differentiable with respect to $b$, $\Gamma(a;\tau)$ is differentiable with respect to $\tau$ and*

$$\frac{\partial K(a,b,\tau)}{\partial b} \le 0, \quad \frac{\partial \Gamma(a-\tau;\tau)}{\partial \tau} \ge 0. \qquad (7.40)$$

*Then, if $R_0 > 1$, there exists at least one endemic steady state.*

*Proof* First, we define a nonlinear positive operator $F$ on $L^1(0,\omega)$ as follows:

$$F(\lambda)(a) := \frac{C(P[\lambda])}{P[\lambda]}\int_0^\omega db \int_0^b d\tau K(a,b,\tau)\lambda(b-\tau)e^{-\int_0^{b-\tau}\lambda(\xi)d\xi}.$$

The problem, then, is to seek a positive fixed point of $F$. Under our assumptions, we can prove that the operator $F$ becomes monotone. In fact, by changing the order of integrals, we have

$$\begin{aligned}
&\int_0^\omega db \int_0^b d\tau K(a,b,\tau)\lambda(b-\tau)e^{-\int_0^{b-\tau}\lambda(\xi)d\xi} \\
&= \int_0^\omega d\tau \int_\tau^\omega db K(a,b,\tau)\left[-\frac{\partial}{\partial b}e^{-\int_0^{b-\tau}\lambda(\xi)d\xi}\right] \\
&= \int_0^\omega \left[K(a,\tau,\tau) + \int_\tau^\omega \frac{\partial K(a,b,\tau)}{\partial b}e^{-\int_0^{b-\tau}\lambda(\xi)d\xi}db\right]d\tau.
\end{aligned}$$

Therefore, under our assumption in (7.40), the double integral part becomes monotone increasing with respect to $\lambda$. By exchanging the order of integrals in the definition of $P[\lambda]$, it follows that

$$P[\lambda] = \int_0^\omega B\ell(a)\Gamma(a;0)da + \int_0^\omega e^{-\int_0^\tau \lambda(\xi)d\xi} d\tau \int_\tau^\omega B\ell(a)\frac{\partial \Gamma(a-\tau;\tau)}{\partial \tau}da.$$

Again, under our assumption in (7.40), $P[\lambda]$ is monotone decreasing with respect to $\lambda$, so $C(P[\lambda])/P[\lambda]$ becomes a non-decreasing function of $\lambda$. Then, $F$ becomes monotone (non-decreasing) with respect to $\lambda$. Moreover, we can observe that for $\lambda \in L^1_+$,

$$\begin{aligned} F(\lambda)(a) &\geq \frac{C(P[0])}{P[0]} e^{-|\lambda|_{L^1}} \int_0^\omega d\tau \int_\tau^\omega K(a,b,\tau)\lambda(b-\tau)db \\ &= e^{-|\lambda|_{L^1}} S_0^{-1}(a) \int_0^\omega (\Pi(\tau)S_0\lambda)(a)d\tau \\ &= e^{-|\lambda|_{L^1}} (U^{-1}\hat{\Pi}(0)U\lambda)(a), \end{aligned} \tag{7.41}$$

where $U$ is an operator defined by $(U\psi)(a) := S_0(a)\psi(a)$. Now, let $x_0$ be the positive eigenvector of the similar NGO $U^{-1}\hat{\Pi}(0)U$ associated with the eigenvalue $R_0 > 1$, and let $y_0$ be a positive vector defined by

$$y_0 := \frac{\log R_0}{|x_0|_{L^1}} x_0 \in L^1_+.$$

It follows from (7.41) that

$$F(y_0) \geq e^{-|y_0|_{L^1}} (U^{-1}\hat{\Pi}(0)U)y_0 = y_0.$$

Define a sequence of functions $\{y_n\}_{n=0,1,2,\ldots}$ by $y_n = F(y_{n-1})$. From the monotonicity of $F$, we have $y_0 \leq y_1 \leq \cdots \leq y_n \leq y_{n+1} \leq \ldots$. As $\{y_n\}$ is a monotone sequence and is bounded above, it follows from Levi's theorem that there exists a limit $y_\infty \in L^1_+$ such that $\lim_{n\to\infty} y_n = y_\infty$. It is clear that this limit function is a positive fixed point of $F$. □

Although it seems that condition (7.40) guaranteeing the monotonicity of $F$ is restrictive, it is satisfied if $\beta$ is constant and if $\Gamma(a;\tau)$ is independent of the age at infection $\tau$. For the general case without such restrictive conditions, we can again use Krasnoselskii's theorem (Proposition 10.32) to prove the existence of a positive fixed point of $F$. In fact, if we define $F'[0]$ as the Fréchet derivative of $F$ at the origin, then $F'[0] = U^{-1}\hat{\Pi}(0)U$ and $F(L^1_+)$ is a bounded set. Therefore, if the NGO $\hat{\Pi}(0)$ is compact and non-supporting, there exists a positive eigenvector associated with the positive eigenvalue $R_0 = r(F'[0]) = r(\hat{\Pi}(0)) > 1$ and it is a unique eigenvector in the positive cone. In such a case, the assumptions of Krasnoselskii's theorem are satisfied, and we know that $F$ has at least one positive fixed point. That is, the threshold condition $R_0 > 1$ is not only the disease invasion condition, but also the condition for the existence of an endemic steady state:

**Proposition 7.7** *Suppose that the next-generation operator $\hat{\Pi}(0)$ is compact and non-supporting. If $R_0 = r(\hat{\Pi}(0)) > 1$, there exists at least one endemic steady state.*

To characterize the endemicity of the HIV epidemic, let us now introduce the *NGO under a given force of infection* $\lambda^*$, denoted by $\Psi[\lambda^*]$, as follows:

$$(\Psi[\lambda^*]\psi)(a) := S^*(a)\frac{C(P[\lambda^*])}{P[\lambda^*]}\int_0^\omega\int_\tau^\omega K(a,b,\tau)\psi(b-\tau)dbd\tau. \tag{7.42}$$

Using the NGO at the endemic steady state, the renewal process (7.38) can be written as $B^* = \Psi[\lambda^*]B^*$. This means that $B^*$ must be simply reproduced at the endemic steady state, and hence, $B^*$ is the Frobenius eigenvector corresponding to the Frobenius eigenvalue one of $\Psi[\lambda^*]$. Then, if we define its spectral radius (the Frobenius eigenvalue) $r(\Psi[\lambda^*])$ as the *effective reproduction number* for the force of infection $\lambda^*$, denoted by $R[\lambda^*]$, that is, $R[\lambda^*] := r(\Psi[\lambda^*])$, the criterion for endemicity is formulated as

$$R[\lambda^*] = 1. \tag{7.43}$$

In other words, $\lambda^*$ gives the force of infection corresponding to the endemic steady state if and only if $\lambda^*$ satisfies condition (7.43).

The formal endemicity condition (7.43) provides an intuitive understanding of why backward bifurcation and multiple endemic equilibria could occur. In fact, if $R[\lambda^*]$ is monotone with respect to the size of $\lambda^*$, there exists at most one endemic equilibrium and the bifurcation will become a forward one, as $R[0] = R_0$ and $R[\infty] = 0$. However, if $R'[0] > 0$ and $R[0] = 1$, endemic equilibria bifurcate backward when $R_0$ crosses unity.

To calculate $R[\lambda^*]$ explicitly and prove the above intuition, we again use the separable mixing assumption. That is, we assume that the transmission rate $\beta$ can be decomposed as $\beta(a,b,\tau) = \beta_1(a)\beta_2(b,\tau)$. In this case, from (7.39), we can write $\lambda^*(a) = c\beta_1(a)$ for some positive constant $c > 0$. The range of $\Psi[\lambda^*]$ is then spanned by $\beta_1(a)\exp(-c\int_0^a \beta_1(\sigma)d\sigma)S_0(a)$, so it follows from the renewal relation

$$\Psi[c\beta_1]\cdot\beta_1 e^{-c\int_0^a \beta_1(\sigma)d\sigma}S_0 = R[c\beta_1]\cdot\beta_1 e^{-c\int_0^a \beta_1(\sigma)d\sigma}S_0$$

that we have

$$R[c\beta_1] = \frac{C(P[c\beta_1])}{P[c\beta_1]}\int_0^\omega\int_\tau^\omega k_2(b,\tau)\beta_1(b-\tau)e^{-c\int_0^{b-\tau}\beta_1(\sigma)d\sigma}dbd\tau, \tag{7.44}$$

where $k_2(b,\tau) := \beta_2(b,\tau)\Gamma(\tau;b-\tau)S_0(b)$.

Thus, we conclude that if the characteristic equation $R[c\beta_1] = 1$ has a positive root $c > 0$, $c\beta_1$ gives the force of infection of an endemic steady state. If $R[c\beta_1]$ is monotone with respect to $c \geq 0$, there exists at most one endemic steady state, and we have a forward bifurcation. This situation is realized if $C(P)$ is proportional to $P$. In contrast, we can consider a scenario in which the existence of multiple endemic steady states produced by backward bifurcations is possible. Let us introduce the bifurcation parameter $\varepsilon$ as

$$R[c\beta_1] = \varepsilon \frac{C(P[c\beta_1])}{P[c\beta_1]} \int_0^{\omega} \int_{\tau}^{\omega} k_2(b,\tau)\beta_1(b-\tau)e^{-c\int_0^{b-\tau}\beta_1(\sigma)d\sigma} dbd\tau, \quad (7.45)$$

where we assume that the other parameters are normalized as

$$\frac{C(P[0])}{P[0]} \int_0^{\omega} \int_{\tau}^{\omega} k_2(b,\tau)\beta_1(b-\tau)dbd\tau = 1. \quad (7.46)$$

Thus, $\varepsilon = R[0]$ gives the basic reproduction number for the parameterized system.

We now define a function $F$ as $F(c,\varepsilon) := R[c\beta_1] - 1$, such that $F(0,1) = 0$. Suppose that

$$\frac{\partial F}{\partial c}(0,1) > 0. \quad (7.47)$$

It then follows from the implicit function theorem that $F(c,\varepsilon) = 0$ can be solved as $c = c(\varepsilon)$ in the neighborhood of $(c,\varepsilon) = (0,1)$, and

$$\left.\frac{dc}{d\varepsilon}\right|_{\varepsilon=1} = -\frac{F_\varepsilon(0,1)}{F_c(0,1)} = -\frac{1}{F_c(0,1)} < 0.$$

Because $c(1) = 0$, for small $\eta > 0$, we have $c(\varepsilon) > 0$ for $\varepsilon \in (1-\eta, 1)$. Let us fix such an $\varepsilon \in (1-\eta, 1)$ and consider $F(c,\varepsilon)$ as a function of $c$. Then, we know that $F(0,\varepsilon) = \varepsilon - 1 < 0$, $F(c(\varepsilon),\varepsilon) = 0$, and $F(\infty,\varepsilon) = -1$. Moreover, from (7.47), $\partial F/\partial c$ is positive at $c = c(\varepsilon)$ if $\varepsilon$ is sufficiently small. Therefore, we conclude that if $\varepsilon$ is less than one but very near to unity, then there exist at least two endemic steady states that are given as positive roots of $F(c,\varepsilon) = 0$. Furthermore, if $\varepsilon = 1$, then there exists at least one endemic steady state under condition (7.47), which means that the endemic equilibria bifurcate backward from the disease-free steady state when $\varepsilon$ crosses unity.

If we assume $\partial F(0,1)/\partial c < 0$, we have $dc/d\varepsilon|_{\varepsilon=1} > 0$. In this case, for small $\eta$, there is no $c(\varepsilon) > 0$ such that $F(c(\varepsilon),\varepsilon) = 0$, $\varepsilon \in (1-\eta, 1)$. The possible bifurcation is then supercritical. If we calculate condition (7.47) explicitly, we can conclude the following:

**Proposition 7.8** *Under the bifurcation scenario given by* (7.45) *and* (7.46), *the endemic equilibria bifurcate backward from the disease-free steady state when* $\varepsilon$ *crosses unity if the following holds:*

$$\left(1-\frac{C'(P[0])P[0]}{C(P[0])}\right)\int_0^\omega \frac{\ell(a)}{e_0}\int_0^a (1-\Gamma(a-\tau;\tau))\beta_1(\tau)d\tau da$$
$$> \frac{\int_0^\omega \int_\tau^\omega k_2(b,\tau)\beta_1(b-\tau)\int_0^{b-\tau}\beta_1(\sigma)d\sigma dbd\tau}{\int_0^\infty \int_\tau^\infty k_2(b,\tau)\beta_1(b-\tau)dbd\tau}, \tag{7.48}$$

*where* $e_0 := \int_0^\omega \ell(a)da$ *is the life span of susceptible individuals. If the opposite inequality holds, we have a forward bifurcation at* $\varepsilon = 1$.

From (7.48), we know that if $C(P)$ is constant (i.e., $\lambda$ is homogeneous of degree one with respect to the scale of the population size), a backward bifurcation at $\varepsilon = 1$ is possible, whereas if $C(P) = \alpha_0 P$ for some constant $\alpha_0 > 0$ (i.e., $\lambda$ is given by the law of mass action), then the bifurcation at $\varepsilon = 1$ is always supercritical. As is usually observed, we can prove that as long as the force of infection is sufficiently small, the forward-bifurcating endemic steady state is locally asymptotically stable, whereas the endemic steady state bifurcating backward from the disease-free steady state is unstable [22, Proposition 6.3].

Although the stability of the endemic steady state is inconclusive even under the separable mixing assumption, Kawachi [25] proved that system (7.27) is *uniformly strongly persistent* in a sense that if $R_0 > 1$, there exists some $\varepsilon > 0$ such that

$$\liminf_{t\to\infty}\int_0^\omega \int_0^b \beta_2(b,\tau)I(t,\tau;b-\tau)d\tau db > \varepsilon,$$

for all solutions satisfying $\int_0^\omega \int_0^b \beta_2(b,\tau)I(t,\tau;b-\tau)d\tau db > 0$. Roughly speaking, except for the trivial initial data, the disease does not go to extinction if $R_0 > 1$. Readers are referred to [40, 43] for definitions and results of persistence theory.

If the basic reproduction number is sufficiently small, we can prove that there is no endemic steady state and the disease-free steady state becomes globally asymptotically stable. For example, we can state the following:

**Proposition 7.9** *Suppose that the NGO* $\hat{\Pi}(0)$ *is compact and non-supporting, and there exists a number* $\alpha > 0$ *such that, for any* $\phi \in L^1_+$,

$$F'[0]\phi - \alpha F(\phi) \in L^1_+ \setminus \{0\}.$$

*Then, if* $R_0 = r(\hat{\Pi}(0)) = r(F'[0]) \le \alpha$, *F has no nonzero fixed point in the positive cone* $L^1_+$.

*Proof* Assume that $R_0 \le \alpha$. If $F$ has a nonzero fixed point in $E := L^1_+$, denoted by $\phi \in E \setminus \{0\}$, we have

$$\alpha\phi = \alpha F(\phi) \le F'[0]\phi.$$

Let $\xi \in E^* \setminus \{0\}$ be the adjoint eigenvector of $F'[0]$ associated with the eigenvalue $R_0$, which is a strictly positive linear functional in the adjoint cone $E^*$. If we write the value of $\xi \in E^*$ at $x \in E$ as $\langle \xi, x\rangle$, it follows from our assumption that

$$\langle \xi, F'[0]\phi - \alpha F(\phi)\rangle = (R_0 - \alpha)\langle \xi, \phi\rangle > 0.$$

Then, we have $R_0 > \alpha$, which is a contradiction. This completes the proof. □

It is easy to prove the following corollary if we use a comparison argument [22]:

**Corollary 7.1** *Assume that* $\sup_{x\geq 0} C(x)/x < \infty$, *and define* $\alpha \in (0, 1)$ *such that* $\alpha := [\sup_{x\geq 0} C(x)/x]^{-1} C(P_0)/P_0$. *Then, if* $R_0 < \alpha$, *there is no endemic steady state and the disease-free steady state is globally asymptotically stable.*

## 7.3 Age-Structured Model of In Vivo HIV Infection

To understand the dynamical process of developing AIDS in vivo and to develop effective treatments, we can use an age-structured model for HIV infection at the cell population level. Here, we introduce an age–duration-structured HIV model with reverse transcriptase (RT) inhibitor and protease inhibitor. This model was introduced in [37] as an extension of the model examined in [33, 38]. An important point of our extension is to introduce local time structures in infected T cells. This enables the model to handle general probability density functions of the sojourn time from infection to completeness of reverse transcription and from completeness of reverse transcription to viral reproduction.

In vivo, HIV infection begins by the attachment of a virus to a $CD4^+$ cell. Inside the cell, the HIV-1 enzyme RT makes a DNA copy of the virus's RNA genome. During this process, if an RT inhibitor is present, then the viral genome will not be copied into DNA, and the host cell will therefore not produce a new virus. When the virus replicates, its DNA is read out to produce viral proteins. A large polyprotein is made, and a viral protease is needed to cut the long polypeptide chains into the individual components necessary to produce infectious virus particles. If the HIV-1 protease is inhibited, the newly produced virus will be non-infectious.

### 7.3.1 *Basic Model and Parameters*

Based on the above observation of the HIV life cycle and the role of inhibitors, the sojourn time distributions of T cells in each infected status are important in studying the HIV dynamics under the influence of antiretroviral treatments. To account for the time course of reverse transcription in infected T cells, we propose the following age–duration-structured model of HIV infection with drug treatment[3]:

[3]In this model, we do not take into account the cell-to-cell transmission (see Sect. 5.2.2).

$$
\begin{aligned}
&\frac{dT(t)}{dt} = b - \mu T(t) - (1-\kappa)kV(t)T(t) + \xi \int_0^\infty \gamma(a)T_1(t,a)da, \\
&\left(\frac{\partial}{\partial t} + \frac{\partial}{\partial a}\right) T_1(t,a) = -(\delta_1(a) + \gamma(a))T_1(t,a), \\
&T_1(t,0) = (1-\kappa)kV(t)T(t), \\
&\left(\frac{\partial}{\partial t} + \frac{\partial}{\partial \tau}\right) T_2(t,\tau;a) = -\delta_2(\tau;a)T_2(t,\tau;a), \\
&T_2(t,0;a) = (1-\xi)\gamma(a)T_1(t,a), \\
&\frac{dV(t)}{dt} = (1-\eta)\int_0^\infty \int_0^\infty m(\tau;a)T_2(t,\tau;a)dad\tau - cV(t),
\end{aligned}
\tag{7.49}
$$

where $T(t)$ denotes the concentration of uninfected target T cells at time $t$, $T_1(t,a)$ denotes the concentration of infected T cells of infection age $a$ (i.e., the time since an HIV virion penetrated the cell) at time $t$ in which RT has not been completed (called "preRT cells"), $T_2(t,\sigma;a)$ denotes the concentration of infected T cells of infection age $a+\sigma$ (i.e., $\sigma$ denotes the duration in the postRT stage) at time $t$ that has completed RT at infection age $a$ (called "postRT cells"), and $V(t)$ denotes the concentration of *infectious* virus at time $t$. $b$ is the recruitment rate of uninfected T cells, $\mu > 0$ is the per capita natural death rate of uninfected T cells, $\delta_1(a)$ is the per capita death rate of infected preRT cells at infection age $a$, $\delta_2(\tau;a)$ is the per capita death rate of infected postRT cells at infection age $a+\tau$, whose infection age at the completion of RT is $a$, $\gamma(a)$ is the age-dependent rate of reverse transcription, $c$ is the *clearance rate* of virions, $k$ is the rate at which uninfected cells become infected by the virus, $m(\tau;a)$ is the virion reproduction rate of infected cells with duration $\tau$ and age of RT $a$ (i.e., RT completes at infection age $a$) in the postRT phase, and $\kappa$, $\xi$, and $\eta$ denote the efficacy of the therapy with entry inhibitors, RT inhibitors, and protease inhibitors, respectively ($\kappa, \xi, \eta \in (0,1)$). We assume that the preRT cells revert to the uninfected state after the failure of reverse transcription.

**Assumption 7.10** As the basic parameters, we assume that $0 < \mu \le \delta_1(a) \le \sup \delta_1 < \infty$, $0 < \mu \le \delta_2(\tau;a) \le \sup \delta_2 < \infty$, $\gamma, m \in L_+^\infty(\mathbb{R}_+)$, and $\inf \gamma > 0$.

Let us introduce the basic renewal equations associated with system (7.49). Integrating the equation for $T_1$ along the characteristic line, we have

$$
T_1(t,a) = \begin{cases} (1-\kappa)\Delta_1(a)\Gamma(a)B(t-a), & (t-a>0), \\ \frac{\Delta_1(a)\Gamma(a)}{\Delta_1(a-t)\Gamma(a-t)} T_1(0,a-t), & (a-t>0), \end{cases} \tag{7.50}
$$

where $\Gamma(a)$ and $\Delta_1(a)$ denote the survival functions given by

$$
\Gamma(a) := \exp\left(-\int_0^a \gamma(\sigma)d\sigma\right), \quad \Delta_1(a) := \exp\left(-\int_0^a \delta_1(\sigma)d\sigma\right),
$$

and

$$B(t) := T_1(t,0) = kV(t)T(t),$$

denotes the incidence rate of newly infected T cells.

The life expectancy of the preRT cells is then given by

$$L_1 := \int_0^\infty \Gamma(a)\Delta_1(a)da.$$

Let $\Psi_1(a) := \gamma(a)\Gamma(a)\Delta_1(a)$ be the probability that RT occurs at infection age $a$. The total probability of occurrence of RT, denoted by $P_{RT}$, is given by

$$P_{RT} = \int_0^\infty \Psi_1(a)da, \tag{7.51}$$

and the probability density function of occurrence of RT is given by $\psi_1(a) := \Psi_1(a)/P_{RT}$.

Integrating the equation for $T_2$ along the characteristic line, we have

$$T_2(t,\tau;a) = \begin{cases} (1-\xi)\gamma(a)T_1(t-\tau,a)\Delta_2(\tau;a), & (t-\tau>0), \\ \frac{\Delta_2(\tau;a)}{\Delta_2(\tau-t;a)}T_2(0,\tau-t;a), & (\tau-t>0), \end{cases} \tag{7.52}$$

where $\Delta_2$ is the survival probability of postRT-infected T cells with respect to the infection age of RT:

$$\Delta_2(\tau;a) := \exp\left(-\int_0^\tau \delta_2(\sigma;a)d\sigma\right).$$

Then, $\Psi_2(\tau;a) := m(\tau;a)\Delta_2(\tau;a)$ denotes the *net viral reproduction rate* at duration $\tau$ since the completion of RT. Note that the life expectancy of postRT cells that complete RT at infection age $a$ is given by

$$L_2(a) := \int_0^\infty \Delta_2(\tau;a)d\tau,$$

and the average life expectancy of postRT cells is given by

$$L_2 := \int_0^\infty L_2(a)\psi_1(a)da.$$

Using $\Psi_1$ and $\Psi_2$, we can define the *infection-age-specific net virion reproduction rate* as follows:

$$\begin{aligned}\Psi(a) :&= \int_0^a \Psi_2(a-\tau;\tau)\Psi_1(\tau)d\tau,\\ &= \int_0^a m(a-\tau;\tau)\Delta_2(a-\tau;\tau)\gamma(\tau)\Gamma(\tau)\Delta_1(\tau)d\tau.\end{aligned}$$

Moreover, the *infection-age-specific virion reproduction rate*, denoted by $p(a)$, is given by

$$p(a) := \int_0^a m(a-\tau;\tau)\gamma(\tau)\Gamma(\tau)d\tau,$$

which is a virion reproduction rate at infection age $a$ under the condition of no death. If the death rate of infected T cells is not affected by RT, we can write

$$\Delta_2(\tau;a) = \frac{\Delta_1(a+\tau)}{\Delta_1(a)},$$

so we obtain $\Delta_2(a-\tau;\tau)\Delta_1(\tau) = \Delta_1(a)$ and $\Psi(a) = p(a)\Delta_1(a)$, which have been used in previous studies [33].

The *burst size* (*total viral reproduction rate*) of infected T cells, denoted by $N$, is the total number of viral particles produced by an infected cell in its life span, which is given by $N := \int_0^\infty \Psi(a)da$. The burst size can also be expressed as follows:

$$N = \int_0^\infty N(a)\Psi_1(a)da = \int_0^\infty N(a)\psi_1(a)da \times P_{RT},$$

where

$$N(a) := \int_0^\infty \Psi_2(\tau;a)d\tau.$$

Then, $N(a)$ gives the total number of viral particles produced by an infected cell that completes RT at infection age $a$ and $\int_0^\infty N(a)\psi_1(a)da$ gives the *average burst size* of a postRT-infected T cell. If we can assume that the virion production process is described by a time-independent Poisson process, then $m(\tau;a)$ is a constant $m$ that describes the (age-independent) force of virion production. Hence, we have $\psi_2(a) = mL_2(a)$, and

$$N = m\int_0^\infty L_2(a)\psi_1(a)da = mL_2.$$

*Remark 7.7* Let $\beta(t,a)$ be the proportion of infected cells that have not completed RT, that is, the age-specific proportion of preRT cells among the infected T cells. This is given by

$$\beta(t,a) = \frac{T_1(t,a)}{T_1(t,a) + \int_0^a T_2(t,\tau;a-\tau)d\tau}.$$

Then, for $t > a$, it becomes a time-independent ratio given by

$$\frac{\Delta_1(a)\Gamma(a)}{\Delta_1(a)\Gamma(a) + (1-\xi)\int_0^a \gamma(a-\tau)\Delta_1(a-\tau)\Gamma(a-\tau)\Delta_2(\tau; a-\tau)d\tau}.$$

Moreover, if we can assume that the death rate of infected T cells is not affected by RT, we obtain

$$\frac{T_1(t,a)}{T_1(t,a) + \int_0^a T_2(t,\tau; a-\tau)d\tau} = \frac{\Gamma(a)}{\Gamma(a) + (1-\xi)(1-\Gamma(a))},$$

which shows that $\beta$ escapes from time dependency after losing the effect of the initial data, but remains affected by $\xi$. Thus, under the existence of RT inhibitor, it is not appropriate to assume the $\xi$-independence of $\beta$, as in [38].

Inserting expression (7.50) into the equation for the uninfected T cells in (7.49), we obtain an integro–differential equation:

$$\frac{dT(t)}{dt} = b - \mu T(t) - (1-\kappa)B(t) + \xi \int_0^t \Psi_1(a)B(t-a)da + G_1(t),$$

where

$$G_1(t) := \xi \int_t^\infty \gamma(a)\frac{\Delta_1(a)\Gamma(a)}{\Delta_1(a-t)\Gamma(a-t)} T_1(0, a-t)da.$$

Applying the variation-of-constants formula, we have

$$\begin{aligned} T(t) &= e^{-\mu t}T(0) \\ &+ \int_0^t e^{-\mu(t-\sigma)}\left[b - (1-\kappa)B(\sigma) + \xi\int_0^\sigma \Psi_1(a)B(\sigma-a)da + G_1(\sigma)\right]d\sigma. \end{aligned}$$

Changing the order of integrals, we arrive at a convolution equation:

$$T(t) = F_1(t) - \int_0^t K_1(a)B(t-a)da, \tag{7.53}$$

where

$$\begin{aligned} K_1(a) &:= (1-\kappa)\left(e^{-\mu a} - \xi\int_0^a e^{-\mu(a-x)}\Psi_1(x)dx\right), \\ F_1(t) &:= e^{-\mu t}T(0) + \int_0^t e^{-\mu(t-\sigma)}(b + G_1(\sigma))d\sigma. \end{aligned}$$

Inserting expression (7.52) into the equation for $V(t)$ in (7.49), we obtain a second integro–differential equation:

$$\frac{dV(t)}{dt} = -cV(t) + (1-\eta)\int_0^t \int_0^\infty (1-\xi)\gamma(a)T_1(t-\tau, a)\Psi_2(\tau; a)dad\tau$$
$$+ (1-\eta)\int_t^\infty \int_0^\infty m(\tau; a)\frac{\Delta_2(\tau; a)}{\Delta_2(\tau - t; a)}T_2(0, \tau - t; a)dad\tau.$$

Moreover, if we replace $T_1$ by the expression in (7.50), we have

$$\frac{dV(t)}{dt} = G_2(t) - cV(t)$$
$$+ (1-\kappa)(1-\eta)(1-\xi)\int_0^t \int_0^{t-\tau} \Psi_1(a)\Psi_2(\tau; a)B(t-\tau-a)dad\tau,$$

where

$$G_2(t) := (1-\eta)(1-\xi)\int_0^t \int_{t-\tau}^\infty T_1(0, a-t+\tau)$$
$$\times \frac{\gamma(a)\Gamma(a)\Delta_1(a)}{\Gamma(a-t+\tau)\Delta_1(a-t+\tau)}\Psi_2(\tau; a)dad\tau$$
$$+ (1-\eta)\int_t^\infty \int_0^\infty m(\tau; a)\frac{\Delta_2(\tau; a)}{\Delta_2(\tau-t; a)}T_2(0, \tau - t; a)dad\tau.$$

Applying the variation-of-constants formula again, we have

$$V(t) = e^{-ct}V(0) + \int_0^t e^{-c(t-\sigma)}G_2(\sigma)d\sigma$$
$$+ (1-\kappa)(1-\eta)(1-\xi)\int_0^t e^{-c(t-\sigma)}$$
$$\times \int_0^\sigma \int_0^{\sigma-\tau} \Psi_1(a)\Psi_2(\tau; a)B(\sigma-\tau-a)dad\tau d\sigma.$$

Changing the order of integrals, we arrive at another convolution equation:

$$V(t) = F_2(t) + \int_0^t K_2(a)B(t-a)da, \tag{7.54}$$

where

$$K_2(a) := (1-\kappa)(1-\eta)(1-\xi)\int_0^a e^{-c(a-y)}\Psi(y)dy,$$
$$F_2(t) := e^{-ct}V(0) + \int_0^t e^{-c(t-\sigma)}G_2(\sigma)d\sigma.$$

Let $x(t) := (T(t), V(t))^{\mathrm{T}}$. Define a nonlinear function $g : \mathbb{R}^2 \to \mathbb{R}^2$ such that $g(x) = (kx_1x_2, kx_1x_2)^{\mathrm{T}}$, where $x = (x_1, x_2)^{\mathrm{T}} \in \mathbb{R}^2$. Let

$$K(a) := \begin{pmatrix} -K_1(a) & 0 \\ 0 & K_2(a) \end{pmatrix}, \quad f(t) := \begin{pmatrix} F_1(t) \\ F_2(t) \end{pmatrix}.$$

It then follows from the above two convolution equations that

$$x(t) = f(t) + \int_0^t K(a)g(x(t-a))da. \tag{7.55}$$

For the nonlinear convolution equation (7.55), the existence and uniqueness of a positive solution have already been established [31], so we omit the argument about the well-posedness of the basic system in (7.49).

### 7.3.2 The Basic Reproduction Number $R_0$

We now calculate the basic reproduction number $R_0$ and related indices for the HIV in vivo infection model. Let $(x, y)$ be a perturbation from the infection-free steady state $(T, V) = (b/\mu, 0)$. Then, $B(t) = k(b/\mu + x)y \sim (kb/\mu)y$, so we define $b(t) := (kb/\mu)y(t)$ as the density of newly infected cells in the invasion phase. Let $u_1(t, a)$ and $u_2(t, \tau; a)$ be the density of infected T cells with infection age $a$ in the preRT stage and the density of infected T cells of infection age $a+\tau$ whose infection age at RT is $a$, respectively. Then,

$$u_1(t,a) = \begin{cases} (1-\kappa)\Delta_1(a)\Gamma(a)b(t-a), & (t-a>0), \\ \frac{\Delta_1(a)\Gamma(a)}{\Delta_1(a-t)\Gamma(a-t)}T_1(0, a-t), & (a-t>0), \end{cases} \tag{7.56}$$

$$u_2(t,\tau;a) = \begin{cases} (1-\xi)\gamma(a)u_1(t-\tau, a)\Delta_2(\tau;a), & (t-\tau>0), \\ \frac{\Delta_2(\tau;a)}{\Delta_2(\tau-t;a)}T_2(0, \tau-t;a), & (\tau-t>0). \end{cases} \tag{7.57}$$

Moreover, $y(t)$ satisfies

$$\frac{dy(t)}{dt} = (1-\eta)\int_0^\infty \int_0^\infty m(\tau;a)u_2(t,\tau;a)da d\tau - cy(t). \tag{7.58}$$

Using the same kind of calculation as above, we arrive at the linearized equation for the density of newly infected cells $b(t)$ as

$$\begin{aligned} b(t) &= \frac{kb}{\mu}\left(F_2(t) + \int_0^t K_2(a)b(t-a)da\right) \\ &= \frac{kb}{\mu}F_2(t) + (1-\kappa)(1-\xi)(1-\eta)\int_0^t K_0(a)b(t-a)da, \end{aligned} \tag{7.59}$$

where

$$K_0(a) := \frac{kb}{\mu} \int_0^a e^{-c(a-\sigma)} \Psi(\sigma) d\sigma$$

is the net reproduction function of the infected T cells in the invasion phase. From (7.59), $\kappa$, $\xi$, and $\eta$ have the same effect regarding the sensitivity of the Malthusian parameter for the density of newly infected T cells.

The effective reproduction number of the infected cells under the *highly active antiretroviral therapy* (HAART) parameter $(\kappa, \xi, \eta)$ is calculated as

$$R_e = \frac{kb}{\mu} \int_0^\infty K_2(a) da = (1-\kappa)(1-\eta)(1-\xi) R_0, \tag{7.60}$$

where $R_0$ is the basic reproduction number for infected cells given by

$$R_0 = \int_0^\infty K_0(a) da = \frac{kb}{\mu c} \int_0^\infty \Psi(a) da = \frac{kbN}{\mu c}. \tag{7.61}$$

The basic reproduction number $R_0$ gives the expected number of infected T cells produced by an infected T cell during its entire life span in totally uninfected T cells.

Let $\Omega$ be a subset of the positive cone in $\mathbb{R}^3$ defined by

$$\Omega := \left\{ (\kappa, \xi, \eta) \in [0,1]^3 | \quad (1-\kappa)(1-\xi)(1-\eta) < \frac{1}{R_0} \right\}.$$

As shown in the next subsection, the infected T cells are eradicated if the efficacy parameter set $(\kappa, \xi, \eta)$ is in $\Omega$.

Let $\phi(a)$ be the probability density function defined by

$$\phi(a) := \frac{K_0(a)}{R_0} = \frac{c}{N} \int_0^a e^{-c(a-\sigma)} \Psi(\sigma) d\sigma.$$

Then, $\phi(a)$ is the probability that secondary infected T cells are produced after duration $a$ since the time of infection of the primary infected T cells. Define

$$\psi(a) := \frac{\Psi(a)}{N}, \quad \psi_3(a) := ce^{-ca},$$

where $\psi(a)$ is the probability density of free virus production at infection age $a$ of the infected T cells and $\psi_3(a)$ is the probability density of entry into the T cells after duration $a$ since the time of "birth" of the free virus. Then, $\phi = \psi * \psi_3$, where $*$ denotes the convolution of functions. The average interval (*generation time*) from the primary infection (by entry of virus) to the secondary infection of T cells is given by

$$\int_0^\infty a\phi(a) da = A_0 + \frac{1}{c},$$

where

$$A_0 := \int_0^\infty a\psi(a)da$$

is the average infection age of virus production.

In the early stage of the reproduction cycle of infected T cells, we can expect the density of newly infected T cells to grow exponentially with a Malthusian parameter $\lambda$; therefore, Lotka's well-known characteristic equation holds:

$$\int_0^\infty e^{-\lambda a} K_0(a)da = 1, \tag{7.62}$$

from which we have

$$\frac{1}{R_0} = \int_0^\infty e^{-\lambda a}\phi(a)da = \frac{c}{\lambda + c}\int_0^\infty e^{-\lambda a}\psi(a)da. \tag{7.63}$$

If we can observe the Malthusian parameter $\lambda$ and the probability density $\psi$, we can estimate $R_0$ from (7.63).

### 7.3.3 Extinction and Persistence of Infected T Cells

Finally, we prove that the effective reproduction number $R_e$ gives a threshold value for the extinction or persistence of infected T cells, which is the mathematical basis for HAART.

**Proposition 7.11** *There exists a unique infected steady state if and only if* $R_e > 1$.

*Proof* Let $T^*$, $T_1^*(a)$, $T_2^*(\tau, a)$, and $V^*$ be a steady state of the basic system (7.49). Then, it is easy to show that

$$T^* = \frac{c}{(1-\xi)(1-\eta)Nk} = \frac{W_0}{R_e},$$

where $W_0 := b/\mu$ is the size of the T cell population at the infection-free steady state, and

$$\begin{aligned} T_1^*(a) &= \Delta_1(a)\Gamma(a)kV^*T^*, \\ T_2^*(\tau; a) &= (1-\xi)\Delta_2(\tau; a)\gamma(a)T_1^*(a), \\ V^* &= \frac{b - \mu T^*}{(1-\kappa)kT^*(1-\xi P_{RT})} = \frac{\mu(R_e - 1)}{(1-\kappa)k(1-\xi P_{RT})}. \end{aligned} \tag{7.64}$$

Because $\xi P_{RT} < 1$, we have $V^* > 0$ if and only if $R_e > 1$. □

From expression (7.64), we have

$$V^* = \frac{\mu[(1-\kappa)(1-\xi)(1-\eta)R_0 - 1]}{(1-\kappa)k(1-\xi P_{RT})}. \tag{7.65}$$

From this, it is easy to see that the density of infection-free virus is a linearly decreasing function of $\eta$, which represents the efficacy of the protease inhibitor, but the dependence on $\xi$ and $\kappa$ is not self-evident. However, it is easy to see that $V^*$ is a decreasing function of $\kappa, \xi$, and $\eta$, and the sensitivity of $V^*$ to a change in efficacy of the entry inhibitor $\kappa$ does not depend on the burst size $N$. Additionally, the prevalence of infected T cells is a decreasing function of the efficacy parameters $\kappa$, $\xi$, and $\eta$.

**Proposition 7.12** *If $R_e < 1$, then $\lim_{t\to\infty} B(t) = 0$, and the infection-free steady state is globally asymptotically stable. If $R_e > 1$, this steady state is unstable.*

*Proof* Let the total size of the infected T cell population at time $t$ be

$$W(t) := T(t) + \int_0^\infty T_1(t,a)da + \int_0^\infty \int_0^\infty T_2(t,\tau;a)dad\tau.$$

Then, we have

$$\begin{aligned}\frac{dW(t)}{dt} &= b - \mu W(t) - \int_0^\infty (\delta_1(a)-\mu)T_1(t,a) - \int_0^\infty \int_0^\infty (\delta_2(a)-\mu)T_2(t,\tau;a)dad\tau \\ &\le b - \mu W(t),\end{aligned}$$

which shows that

$$W(t) \le W(0)e^{-\mu t} + \frac{b}{\mu}(1-e^{-\mu t}).$$

Hence, we obtain

$$T^\infty := \limsup_{t\to\infty} T(t) \le \limsup_{t\to\infty} W(t) \le b/\mu. \tag{7.66}$$

Moreover, we have

$$\frac{dV(t)}{dt} \le (1-\eta)\bar{m}\bar{W} - cV(t),$$

where $\bar{m} := \sup_{\tau\ge0,a\ge0} m(\tau;a)$ and $\bar{W} := \sup_{t\ge0} W(t) \le W(0) + b/\mu < \infty$. Then, we know that $V^\infty := \limsup_{t\to\infty} V(t) < \infty$. Since $B(t) = kV(t)T(t)$, we obtain

$$B^\infty := \limsup_{t\to\infty} B(t) \le kV^\infty T^\infty < \infty.$$

There exists a large $t_0 > 0$ such that $B(t) \le kV^\infty T^\infty$ for all $t > t_0$. Observe that, for $t > 2t_0$,

$$\int_0^t K_2(a)B(t-a)da = \int_0^{t_0} K_2(a)B(t-a)da + \int_{t_0}^t K_2(a)B(t-a)da,$$

where

$$\int_0^{t_0} K_2(a)B(t-a)da \le kV^\infty T^\infty \int_0^{t_0} K_2(a)da.$$

For any $\varepsilon > 0$, we can choose some $t_0$ such that

$$\int_{t_0}^{t} K_2(a)B(t-a)da < \varepsilon,$$

because $K_2$ is integrable on $\mathbb{R}_+$ and $B$ is uniformly bounded above. Therefore, we have

$$\int_0^{t} K_2(a)B(t-a)da \le kV^\infty T^\infty \int_0^{\infty} K_2(a)da + \varepsilon,$$

which shows that

$$\limsup_{t\to\infty} \int_0^{t} K_2(a)B(t-a)da \le kV^\infty T^\infty \int_0^{\infty} K_2(a)da.$$

We then arrive at

$$V^\infty \le \limsup_{t\to\infty} \int_0^{t} K_2(a)B(t-a)da \le kV^\infty T^\infty \int_0^{\infty} K_2(a)da \le R_e V^\infty,$$

which shows that $V^\infty = 0$ if $R_e < 1$. This gives $\lim_{t\to\infty} B(t) = 0$, which implies the global stability of the infection-free steady state. The density of newly infected cells in the linearized equation is given by (7.59). Under Assumption 7.10, we have

$$\begin{aligned} G_2(t) &\le (1-\xi)(1-\eta)|T_1(0,\cdot)|_{L^1} \\ &\times \int_0^t e^{-\mu(t-\tau)-\mu\tau} d\tau + (1-\eta)\bar{m}e^{-\mu t} \int_0^\infty \int_0^\infty T_2(0,x;a)dadx. \end{aligned}$$

Therefore, $G_2 \in L^1_+(\mathbb{R}_+) \cap L^\infty_+(\mathbb{R}_+)$. If $R_e > 1$, the density of newly infected cells $b(t)$ has a positive Malthusian parameter $\lambda_0$ such that

$$\frac{kb}{\mu} \int_0^\infty e^{-\lambda_0 a} K_2(a)da = 1,$$

and it follows from the well-known renewal theorem that

$$\lim_{t\to\infty} e^{-\lambda_0 t} b(t) = \frac{\int_0^\infty e^{-\lambda_0 t} G(t)dt}{(kb/\mu)\int_0^\infty aK_2(a)da}.$$

Therefore, the infection-free steady state is unstable if $R_e > 1$. □

**Proposition 7.13** *If $R_e > 1$, then the infected T cells persist uniformly weakly in the sense that there exists some $\varepsilon > 0$, which does not depend on the initial data, such that* $\limsup_{t\to\infty} B(t) > \varepsilon$.

*Proof* Suppose that for any $\varepsilon > 0$, there exists a solution $B(t)$ such that

$$\limsup_{t\to\infty} B(t) \leq \varepsilon.$$

Then, there exists some $t_0 > 0$ such that $B(t) \leq \varepsilon$ for all $t \geq t_0$. Observe that

$$T(t+t_0) = F_1(t+t_0) - \int_0^{t+t_0} K_1(a)B(t+t_0-a)da,$$

where

$$\int_0^{t+t_0} K_1(t+t_0-a)B(a)da \leq \int_0^{t_0} K_1(t+t_0-a)B(a)da + \int_{t_0}^{t+t_0} K_1(t+t_0-a)B(a)da.$$

Then, it is easy to see that

$$\int_0^{t_0} K_1(t+t_0-a)B(a)da \leq \frac{e^{-\mu t}}{\mu} \sup_{0\leq t\leq t_0} B(t),$$

$$\int_{t_0}^{t+t_0} K_1(t+t_0-a)B(a)da \leq \frac{\varepsilon}{\mu}.$$

Therefore, we have

$$T(t+t_0) \geq F_1(t+t_0) - \frac{e^{-\mu t}}{\mu} \sup_{0\leq t\leq t_0} B(t) - \frac{\varepsilon}{\mu}. \tag{7.67}$$

Next, observe that

$$G_1(t) \leq \xi e^{-(\mu+\underline{\gamma})t}\bar{\gamma}|T_1(0,\cdot)|_{L^1},$$

where $\underline{\gamma} := \inf \gamma$ and $\bar{\gamma} := \sup \gamma$. It follows that

$$\int_0^t e^{-\mu(t-\sigma)}G_1(\sigma)d\sigma \leq \xi \left[\frac{\sup \gamma}{\inf \gamma}\right] |T_1(0,\cdot)|e^{-\mu t} \to 0, \quad t\to\infty.$$

Therefore, we obtain

$$\lim_{t\to\infty} F_1(t) = \frac{b}{\mu},$$

and hence, for any $\varepsilon > 0$, if we choose a large $t_0 > 0$ in advance, it follows that

$$F_1(t+t_0) \geq \frac{b}{\mu} - \varepsilon$$

for all $t > 0$. If we choose a large $t_1 > 0$ such that, for all $t > t_1$,

$$\frac{e^{-\mu t}}{\mu} \sup_{0 \leq t \leq t_0} B(t) \leq \varepsilon,$$

then for $t > t_1$, we have

$$T(t+t_0) \geq \frac{b}{\mu} - \varepsilon', \tag{7.68}$$

where $\varepsilon' := (2 + 1/\mu)\varepsilon$. Let $t_2 = t_0 + t_1$. Observe that

$$\begin{aligned} V(t+t_2) &= F_2(t+t_2) + k \int_0^{t+t_2} K_2(t+t_2-a)T(a)V(a)da \\ &= F_2(t+t_2) + k \int_0^{t_2} K_2(t+t_2-a)T(a)V(a)da \\ &\quad + k \int_{t_2}^{t+t_2} K_2(t+t_2-a)T(a)V(a)da, \end{aligned}$$

where it follows that

$$k \int_{t_2}^{t+t_2} K_2(t+t_2-a)T(a)V(a)da \geq k \left( \frac{b}{\mu} - \varepsilon' \right) \int_0^t K_2(a)V(t+t_2-a)da.$$

Defining $C(t) := V(t+t_2)$, we obtain the integral inequality

$$C(t) \geq F_2(t+t_2) + k \int_0^{t_2} K_2(t+t_2-a)T(a)V(a)da + \int_0^t \Pi(a)C(t-a)da, \tag{7.69}$$

where

$$\Pi(a) := k \left( \frac{b}{\mu} - \varepsilon' \right) K_2(a).$$

Suppose that $\sup_{t \geq 0} C(t) < \infty$. Then, $C(t)$ formally satisfies a renewal equation as

$$C(t) = H(t) + \int_0^t \Pi(a)C(t-a), \tag{7.70}$$

where

$$\begin{aligned} H(t) := C(t) - \int_0^t \Pi(a)C(t-a)da \\ \geq F_2(t+t_2) + k\int_0^{t_2} K_2(t+t_2-a)T(a)V(a)da > 0. \end{aligned}$$

From our assumption that $\sup_{t\geq 0} C(t) < \infty$, we know that $\Pi \in L^1_+(\mathbb{R}_+)$ and $H \in L^\infty_+(\mathbb{R}_+)$. Moreover, we can choose $\varepsilon'$ in advance to be sufficiently small that

$$\int_0^\infty \Pi(a)da = \left(1 - \frac{\mu\varepsilon'}{b}\right) R_e > 1.$$

Then, there exists a unique real positive number $\lambda$ such that

$$\int_0^\infty e^{-\lambda a}\Pi(a)da = 1,$$

and by the well-known renewal theorem, it follows that

$$\lim_{t\to\infty} e^{-\lambda t}C(t) = \frac{\int_0^\infty e^{-\lambda t}H(t)dt}{\int_0^\infty a\Pi(a)da} > 0,$$

which contradicts our assumption of $\sup_{t\geq 0} C(t) < \infty$. Hence, we have $\sup_{t\geq 0} C(t) = \infty$. Because

$$B(t+t_2) = kC(t)T(t+t_0) \geq kC(t)\left(\frac{b}{\mu} - \varepsilon'\right),$$

we have $\sup_{t\geq 0} B(t+t_0) = \infty$, which contradicts our assumption. We therefore conclude that there exists some $\varepsilon > 0$, which does not depend on the initial data, such that $\limsup_{t\to\infty} B(t) > \varepsilon$. □

Based on the persistence theory [40], we can conjecture that $B(t)$ is persistent in the strong sense; that is, there exists some $\varepsilon > 0$, which does not depend on the initial data, such that $\liminf_{t\to\infty} B(t) > \varepsilon$. Readers are referred to [48] for a global stability result for the infected steady state under a slightly different setting.

## References


1. Anderson, R.M., Medley, G.F., May, R.M., Johnson, A.M.: A preliminary study of the transmission dynamics of the human immunodeficiency virus (HIV), the causative agent of AIDS. IMA J. Math. Appl. Med. Biol. **3**, 229–263 (1986)
2. Anderson, R.M.: The epidemiology of HIV infection: variable incubation plus infectious period and heterogeneity in sexual activity. J. R. Stat. Soc. A **151**(Part 1), 66–93 (1988)
3. Anderson, R.M., May, R.M.: Infectious Diseases of Humans: Dynamics and Control. Oxford University Press, Oxford (1991)

4. Blythe, S.P., Anderson, R.M.: Variable infectiousness in HIV transmission models. IMA J. Math. Appl. Med. Biol. **5**, 181–200 (1988)
5. Brookmeyer, R., Gail, M.H.: AIDS Epidemiology: A Quantitative Approach. Oxford University Press, Oxford (1994)
6. Castillo-Chavez, C. (ed.): Mathematical and Statistical Approaches to AIDS Epidemiology. Lecture Notes in Biomathematics, vol. 83. Springer, Berlin (1989)
7. Colgate, S.A., Stanley, E.A., Hyman, J.M., Layne, S.P., Qualls, C.: Risk behavior-based model of the cubic growth of acquired immunodeficiency syndrome in the United States. Proc. Natl. Acad. Sci **86**, 4793–4797 (1989)
8. Diekmann, O., Heesterbeak, J.A.P., Metz, J.A.J.: On the definition and the computation of the basic reproduction ratio $R_0$ in models for infectious diseases in heterogeneous populations. J. Math. Biol. **28**, 365–382 (1990)
9. Diekmann, O., Dietz, K., Heesterbeek, J.A.P.: The basic reproduction ratio for sexually transmitted diseases I. Theoretical considerations. Math. Biosci. **107**, 325–339 (1991)
10. Dietz, K.: On the transmission dynamics of HIV. Math. Biosci. **90**, 397–414 (1988)
11. Dietz, K., Hadeler, K.P.: Epidemiological models for sexually transmitted diseases. J. Math. Biol. **26**, 1–25 (1988)
12. Dietz, K., Heesterbeek, J.A.P., Tudor, D.W.: The basic reproduction ratio for sexually transmitted diseases Part II. Effects of variable HIV infectivity. Math. Biosci. **117**, 35–47 (1993)
13. Freund, H.P., Book, B.L.: Determination of the spread of HIV from the AIDS incidence history. Math. Biosci. **98**, 227–241 (1990)
14. Hadeler, K.P.: Modeling AIDS in structured populations. In: Proceedings of I.S.I. 47th Session, Paris, pp. 83–99 (1989)
15. Hadeler, K.P.: Structured population models for HIV infection pair formation and non-constant infectivity. In: Jewell, N.P., Dietz, K., Farewell, V.T. (eds.), AIDS Epidemiology: Methodological Issues, Birkhäuser, Boston, pp. 156–173 (1992)
16. Heesterbeek, J.A.P.: $R_0$, Ph.D. thesis, Centrum voor Wiskunde en Informatica, Amsterdam (1992)
17. Hyman, J.M., Ann Stanley, E.: Using mathematical models to understand the AIDS epidemic. Math. Biosci. **90**, 415–473 (1988)
18. Iannelli, M., Loro, R., Milner, F., Pugliese, A., Rabbiolo, G.: An AIDS model with distributed incubation and variable infectiousness: applications to IV drug users in Latium, Italy. Eur. J. Epidemiol. **8**(4), 585–593 (1992)
19. Inaba, H.: Estimation of the number of HIV infecteds in the early stage of epidemic and control strategy. J. Popul. Probl. **49**(4), 23–33 (1994) [in Japanese]
20. Inaba, H.: On trends of AIDS and an estimate for the HIV infecteds in Japan. J. Popul. Probl. **50**(4), 31–44 (1995) [in Japanese]
21. Inaba, H.: Calculating $R_0$ for HIV infection via pair formation. In: Arino, O., Axelrod, D., Kimmel, M. (eds.), Advances in Mathematical Population Dynamics -Molecules, Cells and Man, pp. 355–382. World Scientific, Singapore (1997)
22. Inaba, H.: Endemic threshold results in an age-duration-structured population model for HIV infection. Math. Biosci. **201**, 15–47 (2006)
23. Iwami, S., et al.: Pandemic HIV-1 Vpu overcomes intrinsic herd immunity mediated by tetherin. Sci. Rep. **5**, 12256 (2015). doi:10.1038/srep12256
24. Kaplan, E.H., Brandeau, M.L. (eds.): Modeling the AIDS Epidemic: Planning, Policy, and Prediction. Raven Press, New York (1994)
25. Kawachi, K.: A note on persistence about structured population models. J. Biol. Dyn. **2**(4), 449–464 (2008)
26. Knolle, H.: Age preference in sexual choice and the basic reproduction number of HIV/AIDS. Biom. J. **32**(2), 243–256 (1990)
27. Kretzschmar, M., Jager, J.C., Reinking, D.P., van Zessen, G., Brouwers, H.: The basic reproduction ratio $R_0$ for a sexually transmitted disease in a pair formation model with two types of pairs. Math. Biosci. **124**, 181–205 (1994)

28. Levin, B.R., Bull, J.J., Stewart, F.M.: The intrinsic rate of increase of HIV/AIDS: epidemiological and evolutionary implications. Math. Biosci. **132**, 69–96 (1996)
29. May, R.M., Lloyd, A.L.: Infection dynamics on scale-free network. Phys. Rev. E **64**, 066112 (2001)
30. McKendrick, A.G., Morison, M.J.: The determination of incubation periods from maritime statistics, with particular reference to the incubation period of influenza. Ind. J. Med. Res. **7**, 364–371 (1919)
31. Metz, J.A.J., Diekmann, O.: The Dynamics of Physiologically Structured Populations. Lecture Notes in Biomathematics, vol. 68. Springer, Berlin (1986)
32. Mode, C.J., Dietz, K.: On some formulas in a partnership model from the perspective of a semi-Markov process. J. Math. Biol. **32**, 161–169 (1994)
33. Nelson, P.W., Gilchrist, M.A., Coombs, D., Hyman, J.M., Preleson, A.S.: An age-structured model of HIV infection that allows for variations in the production rate of viral particles and the death rate of productively infected cells. Math. Biosci. Eng. **1**(2), 267–288 (2004)
34. Nishiura, H.: Lessons from previous predictions of HIV/AIDS in the United States and Japan: epidemiologic models and policy formulation. Epidemiol. Perspect. Innov. **4**(3), 1–7 (2007)
35. Nishiura, H., Inaba, H.: Estimation of the incubation period of influenza A (H1N1-2009) among imported cases: addressing censoring using outbreak data at the origin of importation. J. Theor. Biol. **272**, 123–130 (2011)
36. Nishiura, H., Mizumoto, K., Miyamatsu, Y., Kinoshita, R.: Policy Application of Mathematical Models to Infectious Disease Control in Asia. Springer, Tokyo (2016)
37. Ochiai, K.: Mathematical Analysis for a HIV-1 Model in Vivo, Master Thesis in the Graduate School of Mathematical Sciences, University of Tokyo (2009)
38. Rong, L., Feng, Z., Perelson, A.S.: Mathematical analysis of age-structured HIV-1 dynamics with combination antiretroviral therapy. SIAM J. Appl. Math. **67**(3), 731–756 (2007)
39. Sasaki, A., Iwasa, Y.: Optimal growth schedule of pathogens within a host: switching between lytic and latent cycles. Theor. Popul. Biol. **39**, 201–239 (1991)
40. Smith, H.L., Thieme, H.R.: Dynamical Systems and Population Persistence. Graduate Studies in Mathematics, vol. 118. American Mathematical Society, Providence (2011)
41. Thieme, H.R., Castillo-Chavez, C.: On the role of variable infectivity in the dynamics of the human immunodeficiency virus epidemic. In: Castillo-Chavez, C. (ed.), Mathematical and Statistical Approaches to AIDS Epidemiology. Lecture Notes in Biomathematics, vol. 83, pp. 157–176. Springer, Berlin (1989)
42. Thieme, H.R., Castillo-Chavez, C.: How may infection-age-dependent infectivity affect the dynamics of HIV/AIDS? SIAM J. Appl. Math. **53**(5), 1447–1479 (1993)
43. Thieme, H.R.: Mathematics in Population Biology. Princeton University Press, Princeton (2003)
44. Thieme, H.R.: Distributed susceptibility: a challenge to persistence theory in infectious disease models. Discet. Contin. Dyn. Syst. B **12**(4), 865–882 (2009)
45. United Nations and World Health Organization (ed.), The AIDS epidemic and its demographic consequences. In: Proceedings of the United Nations/World Health Organization Workshop on Modelling the Demographic Impact of the AIDS Epidemic in Pattern II Countries: Progress to Date and Policies for the Future, New York, 13–15 December 1989, United Nations Publication (1991)
46. Waldstätter, R.: Pair formation in sexually-transmitted diseases. In: Castillo-Chavez, C. (ed.), Mathematical and Statistical Approaches to AIDS Epidemiology. Lecture Notes in Biomathematics, vol. 83, pp. 260–274. Springer, Berlin (1989)
47. Wallinga, J., Lipsitch, M.: How generation intervals shape the relationship between growth rates and reproductive numbers. Proc. R. Soc. B **274**, 599–604 (2007)
48. Wang, J., Huang, G., Takeuchi, Y.: Global asymptotic stability for HIV-1 dynamics with two distributed delays. Math. Med. Biol. **29**, 283–300 (2012)
49. Watts, C.H., May, R.M.: The influence of concurrent partnerships on the dynamics of HIV/AIDS. Math. Biosci. **108**, 89–104 (1992)

# Chapter 8
# Variable Susceptibility, Reinfection, and Immunity

**Abstract** It is widely recognized that dynamic changes in host susceptibility resulting from the evolution of infectious agents and changes in the host immunity distribution play important roles in the spread of infectious diseases. Even for common childhood diseases such as measles, the natural decay of host immunity occurs if the environmental virus disappears and the booster effect is lost. In this chapter, we first consider the Pease model for type A influenza epidemics, which was an early attempt to take into account the decay of host immunity due to antigenic changes in the virus population. A remarkable feature of this model is that for realistic parameter values, the antigenic drift of a dominant virus is a possible mechanism for recurrent outbreaks. Next, we formulate the Kermack–McKendrick reinfection model using the standard age-dependent population dynamics equations and examine its basic properties. The potential importance of the Kermack–McKendrick reinfection model is that it can take into account variable susceptibility and reinfection, and will thus be a useful starting point in considering the epidemiological life history of individuals. The Pease influenza model can be seen as a special case of this reinfection model and has a reinfection threshold. Moreover, subcritical endemic steady states may be created by a backward bifurcation. We show some realistic examples of the reproductivity enhancement that can create the backward bifurcation. Finally, we introduce Aron's malaria model, which can be interpreted as a model for acquired immunity boosted by exposure to infection. It was originally developed to understand the functional relation between the force of infection and the reversion rate and the age prevalence curve of malaria. However, our purpose here is not to give a realistic account of the malaria epidemic, but to show that the age-dependent model is a useful tool for describing the boosting mechanism. Although our analysis is far from complete, we can show that the functional relationship in the endemic steady state is naturally induced from the basic structured population dynamics, and the age-dependent model is essential in formulating epidemiological indices.



H. Inaba, *Age-Structured Population Dynamics in Demography and Epidemiology*, DOI 10.1007/978-981-10-0188-8_8

## 8.1 Pease Model for Type A Influenza Epidemics

### 8.1.1 Basic Model

Most traditional epidemic models do not take genetic or evolutionary changes in the virus into account. In fact, for common childhood diseases such as measles, mumps, and rubella, it is a reasonable assumption that the genetic characteristics of the virus that could affect epidemics do not change and that recovery from these diseases can lead to lifelong immunity. In such a case, recurrent outbreaks of the epidemic would be caused by demographic replacement of the susceptible population.[1]

However, with epidemics such as type A influenza, genetic changes in the virus are thought to play an important role in causing recurrent epidemics. In the case of type A influenza, the virus changes genetically, and hence immunologically, from one epidemic to the next. Therefore, a descendant virus strain can infect hosts who are immune to the progenitor strain of the disease, and reinvade communities that recently suffered an epidemic of the progenitor strain. It has also been observed that the more a virus has changed genetically from its progenitor, the more easily it will be able to reinfect a host that is immune to its progenitor. In the influenza virus, the terms *drift* and *shift* are used to distinguish two alternative mechanisms of evolution. Drift occurs by point mutation, and possibly by short deletions and insertions, and causes continual, gradual changes in the influenza antigen. Shift occurs irregularly and causes abrupt and large changes in the influenza antigens.

Pease [35] first proposed a mathematical model that accounted for the drift effect, and suggested that this evolutionary mechanism could lead to a dampening epidemic oscillation. Several authors later tried to develop evolutionary epidemic models [2, 3, 18–20, 30, 44]. In the following, we mainly discuss the mathematical aspects of Pease's evolutionary epidemic model.

Following Pease, we make four major biological assumptions:

[i] The probability of reinfection is a monotone increasing function of the number of amino acid substitutions between the immunizing and challenge virus strains.
[ii] Only one virus strain circulates in a human community at any one time.
[iii] Random drift, rather than the host's frequency-dependent selection, causes amino acid substitutions to occur in the influenza virus. Random drift occurs continually and causes gradual changes in the virus antigens; thereby, genetic changes in the pathogen from epidemic to epidemic cause previously immune hosts to become susceptible.
[iv] The demography of the host population and the existence of a never-infected population are not taken into account.

---

[1] Even in these classical childhood diseases, however, the vaccinated population could give rise to more complex epidemiologic consequences. For example, see [33] and Sect. 8.3.

Note that in Pease's original assumptions, the probability in Assumption [i] is proportional to the number of amino acid substitutions. However, we assume that the infection rate is upper bounded as the arbitrarily large susceptibility seems to be unrealistic, as Pease pointed out. As long as we consider a short timescale in comparison with the demographic timescale, Assumption [iv] is reasonable.

Let $I(t)$ be the number of infected hosts at time $t$ and $S(t,a)$ be the density of uninfected hosts at time $t$ who have $a$ amino acid substitutions. Thus, $\int_{a_0}^{a_1} S(t,a)da$ is the number of uninfected hosts that were last infected by a virus that differed by more than $a_0$ and less than $a_1$ amino acid substitutions from the virus strain prevailing at time $t$. Note that, in the next section, we use the notation $R(t,a)$ instead of $S(t,a)$, because all susceptible individuals in the Pease model are assumed to be recovered individuals with variable immunity. For now, we adopt the notation used by Pease.

We assume that the number of amino acid substitutions is a continuous variable and that it causes an antigenic drift in the virus strain. The host population size $N(t)$ at time $t$ is given by

$$N(t) = \int_0^\infty S(t,a)da + I(t).$$

Under these basic assumptions, the *Pease model* is formulated as follows:

$$\begin{aligned}
&\frac{\partial S(t,a)}{\partial t} + k\frac{\partial S(t,a)}{\partial a} = -\gamma(a)S(t,a)I(t), \\
&\frac{dI(t)}{dt} = -\nu I(t) + I(t)\int_0^\infty \gamma(a)S(t,a)da, \\
&kS(t,0) = \nu I(t),
\end{aligned} \tag{8.1}$$

where $\nu$ is the rate at which infected hosts recover, $k$ is the (constant) rate at which amino acid substitutions occur in the virus population, and $\gamma(a)$ specifies how amino acid substitutions affect the probability of reinfection, which is assumed to be *monotone increasing* under Assumption [i].

The most important feature of the Pease model is that the antigenic drift of the virus is expressed by the relative change in host susceptibility. Because $k$ is assumed to be constant, the amino acid substitution $a$ essentially plays the same role as the duration in the susceptible class.

Suppose that $S(t,\infty) = 0$, which is satisfied if $S(t,\cdot) \in W^{1,1}(\mathbb{R}_+)$. Then, we know that the total population size $N$ is constant:

$$N = \int_0^\infty S(t,a)da + I(t).$$

Therefore, system (8.1) can be reduced to an initial–boundary value problem of a single equation for $S(t,a)$:

$$
\begin{aligned}
&\frac{\partial S(t,a)}{\partial t} + k\frac{\partial S(t,a)}{\partial a} = -\gamma(a)S(t,a)\left[N - \int_0^\infty S(t,a)da\right], \\
&kS(t,0) = \nu\left[N - \int_0^\infty S(t,a)da\right], \\
&S(0,a) = S_0(a),
\end{aligned}
\tag{8.2}
$$

where $S_0(a)$ denotes the initial data.

By the standard method of integrating along the characteristic line, the McKendrick equation (8.2) can be solved as follows:

$$
S(t,a) = \begin{cases} S(t - \frac{a}{k}, 0)e^{-\int_0^{\frac{a}{k}} \gamma(k\sigma)I(t-\frac{a}{k}+\sigma)d\sigma}, & t - \frac{a}{k} > 0, \\ S(0, a - kt)e^{-\int_0^t \gamma(a-kt+k\sigma)I(\sigma)d\sigma}, & t - \frac{a}{k} < 0. \end{cases} \tag{8.3}
$$

From the boundary condition, we have

$$
S\left(t - \frac{a}{k}, 0\right) = \frac{\nu}{k} I\left(t - \frac{a}{k}\right).
$$

Inserting expression (8.3) into the boundary condition in (8.2), we arrive at the following nonlinear integral equation for $I(t)$:

$$
\begin{aligned}
I(t) = N &- \frac{\nu}{k}\int_0^{kt} I\left(t - \frac{a}{k}\right) e^{-\int_0^{\frac{a}{k}} \gamma(k\sigma)I(t-\frac{a}{k}+\sigma)d\sigma} da \\
&- \int_{kt}^\infty S_0(a - kt)e^{-\int_0^t \gamma(a-kt+k\sigma)I(\sigma)d\sigma} da.
\end{aligned}
\tag{8.4}
$$

We can prove that the nonlinear integral equation (8.4) has a unique nonnegative global solution if $S_0(a) \in L^1_+(0,\infty)$ and $\int_0^\infty S_0(a)da < N$. Therefore, $S(t,a)$ is determined by (8.3) and gives the solution for (8.2) [18].

If the condition that $|S_0|_{L^1} < N$ is not satisfied, we have $|S_0|_{L^1} = N$, that is, there is no infected population for all time $t \geq 0$ and

$$
S(t,a) = \begin{cases} 0, & t - \frac{a}{k} > 0, \\ S_0(a - kt), & t - \frac{a}{k} < 0. \end{cases}
$$

Even if there is no infected population, the antigenic drift increases the relative genetic distance between the present virus and the progenitor strains that infected the host population in the past. Hence, the immunity structure of the susceptible host population is "aging" and its power of protection decreases, because there is no replacement by recovery. From a mathematical point of view, system (8.1) has no disease-free steady state. This important and interesting feature of the Pease model is due to the neglect of the never-infected population class.

For any function $f \in L^1_+(\mathbb{R}_+)$ satisfying $|f|_{L^1} = N$, we can construct the *trivial solution* for system (8.1) as follows:

$$U_f(t,a) := \begin{cases} 0, & t - \frac{a}{k} > 0, \\ f(a-kt), & t - \frac{a}{k} < 0. \end{cases}$$

If we now fix a trivial solution $U_f$ and introduce a new variable $x(t,a)$ as

$$x(t,a) = U_f(t,a) - S(t,a),$$

then we know that $x(t,a)$ satisfies the following initial value problem with homogeneous boundary conditions:

$$\begin{aligned} &\frac{\partial x(t,a)}{\partial t} + k\frac{\partial x(t,a)}{\partial a} = \gamma(a)(U_f(t,a) - x(t,a)) \int_0^\infty x(t,a)da, \\ &x(t,0) = -\frac{v}{k}\int_0^\infty x(t,a)da, \\ &x(0,a) = f(a) - S_0(a). \end{aligned} \tag{8.5}$$

System (8.5) is a non-autonomous semilinear Cauchy problem, so we can construct its evolution operator by the classical variation-of-constants formula. The solution $S(t,a) = U_f(t,a) - x(t,a)$ is nonnegative if the initial data are nonnegative. That is, the Pease model forms a mathematically well-posed problem as a dynamical system [20].

### 8.1.2 Threshold Condition and Persistence

In this section, we first consider the existence and uniqueness of steady states of the Pease model. Let $(S^*(a), I^*)$ be the steady state solution of system (8.1). Then, we have

$$\frac{dS^*(a)}{da} = -\frac{\gamma(a)I^*}{k}S^*(a), \quad kS^*(0) = vI^*.$$

Therefore, we obtain the following expression:

$$S^*(a) = \frac{vI^*}{k}e^{-\frac{I^*}{k}\int_0^a \gamma(\sigma)d\sigma}. \tag{8.6}$$

Because the total population size $N$ is constant, we have an equation for the unknown value $I^*$:

$$N = I^* + \int_0^\infty S^*(a)da = I^*\left(1 + \frac{v}{k}\int_0^\infty e^{-\frac{I^*}{k}\int_0^a \gamma(\sigma)d\sigma}da\right). \tag{8.7}$$

Therefore, we can conclude that there exists a steady state if the above equation has a positive solution in the interval $[0, N]$.

Because there is no disease-free steady state, it seems that we cannot define the basic reproduction number for the Pease model. However, we could expect never-infected individuals to have the maximum susceptibility, so it would be reasonable to assume that their infection rate is given by

$$\gamma(\infty) := \sup_{a \geq 0} \gamma(a) = \lim_{a \to \infty} \gamma(a).$$

Therefore, in the following, we assume that the infection rate of never-infected individuals is given by $\gamma(\infty)$. Then, the basic reproduction number for the never-infected population of size $N$ is given by

$$R_0 = \frac{N\gamma(\infty)}{\nu}. \tag{8.8}$$

If never-infected individuals have a larger susceptibility than $\sup \gamma(\cdot)$, the basic reproduction number is bigger than the above $R_0$, a case that was treated in [24, 44].

Using the basic reproduction number $R_0$, we can prove the following endemic threshold result:

**Proposition 8.1** *If $R_0 \leq 1$, the epidemic is naturally eradicated, whereas if $R_0 > 1$, there exists a unique endemic steady state.*

*Proof* From (8.1), it follows that, for $t > 0$,

$$I'(t) \leq -\nu I(t) + I(t)\gamma(\infty)(N - I(t)) = \nu(R_0 - 1)I(t) - \gamma(\infty)I^2(t).$$

It is then easy to see that $\lim_{t\to\infty} I(t) = 0$ if $R_0 \leq 1$; that is, the disease is eradicated as time passes. Next, let us assume that $R_0 > 1$. Define a function $F(x)$, $x \in (0, N]$, as follows:

$$F(x) := x\left(1 + \frac{\nu}{k}\int_0^\infty e^{-\frac{x}{k}\int_0^a \gamma(\sigma)d\sigma} da\right).$$

From (8.7), we know that an endemic steady state exists if and only if the equation $F(x) = N$ has a positive root in the interval $(0, N]$. By changing the variables, we have

$$F(x) = x + \int_0^\infty e^{-\int_0^{z/\nu} \gamma(ky/x)dy} dz,$$

which shows that $F$ is a monotone non-decreasing function. Moreover, it is easy to see that $\lim_{x\downarrow 0} F(x) = \nu/\gamma(\infty) = N/R_0$ and $F(N) > N$. Thus, we know that $F(x) = N$ has a unique root in $(0, N)$ if and only $R_0 > 1$. □

If $R_0 > 1$, we can define the *prevalence*, denoted by $\pi$, in the steady state:

$$\pi := \frac{I^*}{N} = \frac{I^*}{I^* + \int_0^\infty S^*(a)da}.$$

Then, there is a simple relation between the prevalence and the basic reproduction number that can be written as

$$\pi \le 1 - \frac{1}{R_0}. \tag{8.9}$$

In fact, we can observe that

$$\begin{aligned}\int_0^\infty S^*(a)da &= \frac{\nu I^*}{k}\int_0^\infty e^{-\frac{I^*}{k}\int_0^a \gamma(\sigma)d\sigma}da \\ &\ge \frac{\nu I^*}{k}\int_0^\infty e^{-\frac{I^*}{k}\gamma(\infty)a}da = \frac{\nu}{\gamma(\infty)} = \frac{N}{R_0}.\end{aligned}$$

It then follows that

$$\pi = \frac{I^*}{I^* + \int_0^\infty S^*(a)da} \le \frac{I^*}{I^* + \frac{N}{R_0}},$$

from which we immediately obtain (8.9).

More important information about the prevalence of the influenza epidemic can be derived from the timescale argument. In the endemic steady state, the ratio

$$\frac{S^*(a)}{S^*(0)} = e^{-\frac{I^*}{k}\int_0^a \gamma(\sigma)d\sigma}$$

can be interpreted as a survival ratio in a susceptible cohort, that is, the rate that a newly recovered host will be uninfected at duration $a/k$. Therefore, the average number of amino acid substitutions, denoted by $A_0$, which occurs before a host individual can become reinfected, is calculated as

$$A_0 = \int_0^\infty a\frac{I^*\gamma(a)}{k}e^{-\frac{I^*}{k}\int_0^a \gamma(\sigma)d\sigma}da = \int_0^\infty e^{-\frac{I^*}{k}\int_0^a \gamma(\sigma)d\sigma}da.$$

Hence, the length of time required for $A_0$ amino acid substitutions to occur is $A_0/k$. In addition, $1/\nu$ is the mean length of time that a host is infected and infectious. Because $A_0/k$ is typically measured in years while $1/\nu$ is measured in days, it would be reasonable to assume that $A_0/k \gg 1/\nu$ in real influenza epidemics. Thus, the ratio of the long to short timescales $q$ is given by

$$q := \frac{A_0\nu}{k} = \frac{\nu}{k}\int_0^\infty e^{-\frac{I^*}{k}\int_0^a \gamma(\sigma)d\sigma}da = \frac{1}{I^*}\int_0^\infty S^*(a)da = \frac{1-\pi}{\pi}.$$

It then follows that

$$\pi = \frac{1}{1+q}.$$

For example, if we assume that $q$ may be greater than 50 in reality, the prevalence would be at most several percent under the evolutionary mechanism.

We have so far discussed the basic reproduction number as the endemic threshold, because there is no disease-free steady state. However, the reproduction number can be used to formulate a condition for disease invasion. For the Pease model, a time-dependent effective reproduction number can be defined as

$$R_t := \frac{1}{\nu}\int_0^\infty \gamma(a)S(t,a)da.$$

Using the effective reproduction number, the growth rate of the infected population is

$$\frac{1}{I(t)}\frac{dI(t)}{dt} = \nu(R_t - 1).$$

Therefore, the invasion condition is given by $R_t > 1$. If the initial population is composed of entirely susceptible (but once infected) individuals, it follows that

$$R_t = \frac{1}{\nu}\int_{kt}^\infty \gamma(a)S_0(a-kt)da = \frac{1}{\nu}\int_0^\infty \gamma(a+kt)S_0(a)da,$$

from which we have $\lim_{t\to\infty} R_t = R_0$. Thus, if $R_0 > 1$, the susceptible host population will be ultimately invaded by the disease as time passes, even though $R_{t=0} < 1$ at the initial moment.

Moreover, if $R_0 > 1$ and the virus has initially invaded the community, we can state that the disease is not naturally eradicated in the following sense:

**Proposition 8.2** *If $R_0 > 1$ and $I(0) > 0$, $I(t)$ is uniformly strongly persistent—that is, there exists a number $\varepsilon > 0$ that is independent of the initial data and for which* $\liminf_{t\to\infty} I(t) \geq \varepsilon$.

Although omitted here, there are several proofs of the above proposition [44, 47]. The concept of persistence in a dynamical system has also been discussed at length [39, 43, 45]. We instead prove a weaker result formulated by Yang [47]:

**Proposition 8.3** *If $R_0 > 1$ and $I(0) > 0$, then $I(t)$ is uniformly weakly persistent; that is, there exists a number $\varepsilon > 0$ that is independent of the initial data and for which* $\limsup_{t\to\infty} I(t) \geq \varepsilon$.

*Proof* Suppose that $I(t)$ is not uniformly weakly persistent. Then, for any $\varepsilon > 0$, there exists a solution $I(t)$ such that $I(0) > 0$ and $\limsup_{t\to\infty} I(t) < \varepsilon$. Therefore, there exists some $t_\varepsilon > 0$ such that $I(t) \leq \varepsilon$ for $t \geq t_\varepsilon$. For any $A > 0$, if $t - A/k \geq t_\varepsilon$, it follows from expression (8.3) that

$$\int_0^A S(t,a)da \leq \frac{\nu\varepsilon A}{k}.$$

Conversely, it follows from the monotonicity of $\gamma$ that

$$\begin{aligned}\int_0^\infty \gamma(a)S(t,a)da &= \int_0^A \gamma(a)S(t,a)da + \int_A^\infty \gamma(a)S(t,a)da\\ &\geq \int_0^A \gamma(a)S(t,a)da + \gamma(A)\int_A^\infty S(t,a)da\\ &= \gamma(A)\int_0^\infty S(t,a)da - \int_0^A (\gamma(A)-\gamma(a))S(t,a)da\\ &\geq \gamma(A)(N-I(t)) - \gamma(A)\int_0^A S(t,a)da.\end{aligned}$$

Applying the above relation to the equation for $I(t)$, we have

$$\begin{aligned}\frac{I'(t)}{I(t)} &\geq -v+\gamma(A)(N-I(t)) - \gamma(A)\int_0^A S(t,a)da\\ &\geq -v+\gamma(A)N - \left[\gamma(A)I(t) + \gamma(A)\int_0^A S(t,a)da\right]\\ &\geq -v+\gamma(A)N - \left[\gamma(A)\varepsilon + \gamma(A)\frac{v\varepsilon A}{k}\right]\\ &\geq v\left(-1+\frac{\gamma(A)N}{v}\right) - \varepsilon\gamma(\infty)\left(1+\frac{vA}{k}\right)\end{aligned}$$

for $t > t_\varepsilon + A/k$. From our assumption that $R_0 > 1$, there exists a large $A > 0$ such that $\gamma(A)N/v > 1$. Fixing such a large $A$, we choose $\varepsilon > 0$ such that

$$0 < \varepsilon < \frac{v(-1+\frac{\gamma(A)N}{v})}{\gamma(\infty)(1+\frac{vA}{k})}.$$

Then, there exists a solution $I(t)$ and a number $t_\varepsilon > 0$ such that $I(t) \leq \varepsilon$ for $t \geq t_\varepsilon$. From the choice of $\varepsilon$, we know that if $t - A/k \geq t_\varepsilon$,

$$\liminf_{t\to\infty} \frac{I'(t)}{I(t)} > 0.$$

As $I(t)$ is always positive, we have $\lim_{t\to\infty} I(t) \to \infty$, which contradicts the boundedness of $I(t)$. Thus, we can conclude that $I(t)$ is uniformly weakly persistent. □

### 8.1.3 Stability of the Endemic Steady State

In the following, we assume that $R_0 > 1$; as such, there exists an endemic steady state (ESS). We now consider the stability problem of the ESS. Let us denote $\zeta(t,a)$ as the perturbation from the ESS:

$$S(t,a) = S^*(a) + \zeta(t,a),$$

where $S^*(a)$ is the susceptible population at the ESS. The linearized equation of (8.2) at the ESS is given by

$$\begin{aligned} \frac{\partial \zeta(t,a)}{\partial t} + k\frac{\partial \zeta(t,a)}{\partial a} &= -\gamma(a)I^*\zeta(t,a) + \gamma(a)S^*(a)\int_0^\infty \zeta(t,a)da, \\ k\zeta(t,0) &= -v\int_0^\infty \zeta(t,a)da. \end{aligned} \tag{8.10}$$

Let $\zeta(t,a) = e^{\lambda t}v(a)$. Then, $\lambda$ satisfies the following characteristic equation:

$$\begin{aligned} 1 = \Delta(\lambda) := &-\frac{v}{k}\int_0^\infty e^{-\frac{\lambda}{k}a-\frac{I^*}{k}\int_0^a \gamma(\sigma)d\sigma}da \\ &+\frac{1}{k}\int_0^\infty\int_0^a e^{-\frac{\lambda}{k}(a-s)-\frac{I^*}{k}\int_s^a \gamma(\sigma)d\sigma}\gamma(s)S^*(s)dsda. \end{aligned} \tag{8.11}$$

From the principle of linearized stability, the locations of the roots of the characteristic equation $\Delta(\lambda) = 1$ determine whether the ESS is locally stable. Because $1 - \Delta(\lambda)$ is holomorphic in the half-plane $\Re\lambda > -\gamma(\infty)I^*$, the set of characteristic roots that are zeros of the holomorphic function

$$\Lambda := \{\lambda \in \mathbb{C} : \Delta(\lambda) = 1, \Re\lambda > -\gamma(\infty)I^*\}$$

is composed of discrete points. Moreover, the characteristic equation $\Delta(\lambda) = 1$ can be written as follows:

$$\Delta(\lambda) = -\int_0^\infty e^{-\frac{\lambda}{k}a}\Phi(a)da = 1, \tag{8.12}$$

where

$$\Phi(a) := \frac{v}{k}e^{-\frac{I^*}{k}\int_0^a \gamma(\sigma)d\sigma}\left(1 - \frac{I^*}{k}\int_0^\infty \gamma(s)e^{-\frac{I^*}{k}\int_a^{a+s}\gamma(\sigma)d\sigma}ds\right).$$

In fact, changing the order of the integrals in (8.11), we have

$$\begin{aligned}
&\frac{1}{k}\int_0^\infty \int_0^a e^{-\frac{\lambda}{k}(a-s)-\frac{I^*}{k}\int_s^a \gamma(\sigma)d\sigma}\gamma(s)S^*(s)dsda\\
&=\frac{1}{k}\int_0^\infty \gamma(s)S^*(s)ds\int_s^\infty e^{-\frac{\lambda}{k}(a-s)-\frac{I^*}{k}\int_s^a \gamma(\sigma)d\sigma}da\\
&=\frac{1}{k}\int_0^\infty \gamma(s)S^*(s)ds\int_0^\infty e^{-\frac{\lambda}{k}z-\frac{I^*}{k}\int_s^{s+z} \gamma(\sigma)d\sigma}dz\\
&=\frac{1}{k}\int_0^\infty \gamma(s)ds\int_0^\infty e^{-\frac{\lambda}{k}z}S^*(s+z)dz\\
&=\frac{1}{k}\int_0^\infty e^{-\frac{\lambda}{k}a}\int_0^\infty S^*(s+a)\gamma(s)dsda.
\end{aligned}$$

Therefore, using (8.6) and (8.11), we arrive at the following expression:

$$\Delta(\lambda) = -\int_0^\infty e^{-\frac{\lambda}{k}a}\frac{S^*(a)}{I^*}\left[1-\frac{I^*}{k}\int_0^\infty \gamma(s)\frac{S^*(a+s)}{S^*(a)}ds\right]da.$$

If we again use (8.6), we have (8.12). Subsequently, if we define

$$\phi(a) := 1-\frac{I^*}{k}\int_0^\infty \gamma(s)e^{-\frac{I^*}{k}\int_a^{a+s}\gamma(\sigma)d\sigma}ds,$$

then $\phi(a)$ is non-decreasing because $\gamma(a)$ is monotone increasing. From

$$\phi(0) = 1+\int_0^\infty \frac{d}{ds}\left[e^{-\frac{I^*}{k}\int_0^s \gamma(\sigma)d\sigma}\right]ds = 0,$$

we know that for all $a \ge 0$, $\phi(a) \ge 0$. For any $a \ge 0$, we have $\Phi(a) \ge 0$, and for real $\lambda$, we observe that $\Delta(\lambda) \le 0$. Then, $\Delta(\lambda) = 1$ has no real roots. Moreover, if $\lambda \in \Lambda$, it follows that $\bar{\lambda} \in \Lambda$ and $\Lambda$ is composed of pairs of complex conjugates. As for Lotka's characteristic equation, there exist at most finitely many characteristic roots in the right half-plane. In summary, we have the following result:

**Proposition 8.4** *It follows that $\Lambda \cap \mathbb{R} = \emptyset$ and $\Lambda$ is composed of discrete pairs of complex conjugates. For any $\alpha > -\gamma(\infty)I^*$, there exist at most finitely many elements of $\Lambda$ in the half-plane $\Re\lambda \ge \alpha$.*

Therefore, we can state the principle of linearized stability of (8.2) as follows [20]:

**Proposition 8.5** *For all $\lambda \in \Lambda$, if $\Re\lambda < 0$, the ESS is locally asymptotically stable.*

It follows from the above proposition that the ESS is locally asymptotically stable if the linearized operator has no eigenvalue in the right half-plane. That is, to determine the stability of the ESS, it is sufficient to examine the location of the characteristic roots. Let us rewrite the characteristic equation as follows:

$$\hat{K}(\lambda) = \int_0^\infty e^{-\lambda\tau} K(\tau) d\tau = \frac{1}{\Delta(0)}, \tag{8.13}$$

where the integral kernel $K(\tau)$ is defined as

$$K(\tau) := \frac{k\Phi(k\tau)}{\int_0^\infty \Phi(a)da} = -\frac{k\Phi(k\tau)}{\Delta(0)}.$$

Then, note that $\hat{K}(0) = 1$. We can now show the following:

**Proposition 8.6** *If $\Delta(0) \geq -1$, all characteristic roots have negative real parts, so the ESS is locally asymptotically stable.*

*Proof* Suppose that $\lambda = \alpha + i\beta \neq 0$, $\alpha \geq 0$ is a characteristic root. It follows from the characteristic equation (8.13) that

$$\left| \frac{1}{\Delta(0)} \right| \leq \int_0^\infty e^{-\alpha\tau} K(\tau) |\cos \beta\tau| d\tau < \int_0^\infty K(\tau) d\tau = 1.$$

Thus, we know that $\Delta(0) < -1$ is a necessary condition for the existence of characteristic roots with nonnegative real parts. Therefore, if $\Delta(0) \geq -1$, the ESS is locally asymptotically stable. □

It is an interesting point that the prevalence of a disease is related to the stability of the ESS; that is, we can show the following *50 percent prevalence rule*:

**Corollary 8.1** *If $\pi \geq \frac{1}{2}$, the ESS is locally asymptotically stable.*

*Proof* From Proposition 8.6, it is sufficient to show that $\Delta(0) \geq -1$ provided that $\pi \geq \frac{1}{2}$. From $\pi \geq \frac{1}{2}$, we have

$$\int_0^\infty S^*(a)da \leq I^*.$$

Then, it follows from $0 \leq \phi(a) < 1$ that

$$-\Delta(0) = \frac{\nu}{k} \int_0^\infty e^{-\frac{I^*}{k}\int_0^a \gamma(\sigma)d\sigma} \phi(a)da \leq \frac{1}{I^*} \int_0^\infty S^*(a)da \leq 1,$$

from which we arrive at the conclusion. □

However, because the prevalence in the real world may be small (at most several percent), the 50 percent prevalence rule would not cover the domain of realistic parameter values for type A influenza epidemics. For type A influenza, between pandemics, we can observe recurrent small outbreaks caused by the antigenic drift of a dominant virus. Therefore, it is interesting to consider whether the Pease model could allow sustained periodic solutions for parameter values escaping from the 50 percent prevalence rule.

A possible mechanism for periodic solutions is a Hopf bifurcation of an equilibrium. This occurs if a pair of complex conjugate characteristic roots crosses the imaginary axis transversally from the left half-plane to the right half-plane. In fact, as seen in the Easterlin model of Chap. 3, the characteristic equation $\Delta(0)\hat{K}(\lambda) = 1$ may have characteristic roots with negative real parts provided that $\Delta(0) < -1$ and $|\Delta(0)|$ is increasing. In such a case, it is expected that the destabilization of the ESS will lead to a periodic solution. As $\partial|\Delta(0)|/\partial I^* < 0$, the larger $|\Delta(0)|$ corresponds to smaller prevalence, and we could conjecture that an epidemic with smaller realistic prevalence would have a tendency to oscillate. In fact, Yang [47] showed that a periodic solution can bifurcate if the reinfection rate is a step function. Readers are referred to Magal and Ruan [30] for a more detailed treatment of this bifurcation problem.

### *8.1.4 Effects of Vaccination*

In this section, we consider how vaccination strategies affect the dynamics of the evolutionary epidemic model. In reality, vaccination policies are usually based on age-dependent schedules. However, we cannot deal with age-dependent problems at this point, as the Pease model neglects the chronological-age structure.

Although many vaccination strategies exist, we first assume the simplest case in which uninfected hosts are vaccinated at a constant rate per unit time. Under the evolutionary mechanism, the effect of vaccination is temporal, even if we were to use a vaccine fitted to the present virus. Moreover, the vaccine-induced immunity could be weaker than the immunity of recovered individuals. However, for simplicity, we assume that newly vaccinated individuals can be identified as recovered individuals.

Let $\varepsilon > 0$ be the rate of vaccination. Because the amino acid substitution $a$ of newly vaccinated hosts is reduced to zero, we obtain a new system:

$$\begin{aligned}
&\frac{\partial S(t,a)}{\partial t} + k\frac{\partial S(t,a)}{\partial a} = -\varepsilon S(t,a) - \gamma(a)S(t,a)I(t), \\
&\frac{dI(t)}{dt} = -\nu I(t) + I(t)\int_0^\infty \gamma(a)S(t,a)da, \\
&kS(t,0) = \nu I(t) + \varepsilon\int_0^\infty S(t,a)da.
\end{aligned} \tag{8.14}$$

The above vaccination model is very different from the basic model without vaccination in that it has a disease-free steady state. In fact, even if there are no infected individuals, the host immunity distribution is replaced by the constant vaccination policy. It is easy to see that the disease-free steady state is given by

$$S^*(a) = \frac{\varepsilon N}{k} e^{-\frac{\varepsilon}{k}a}, \quad I^* = 0.$$

Therefore, we can define the effective reproduction number depending on the vaccination rate $\varepsilon$ as

$$\begin{aligned} R_e(\varepsilon) := &\frac{1}{v}\int_0^\infty \gamma(a)S^*(a)da = \frac{\varepsilon N}{kv}\int_0^\infty e^{-\frac{\varepsilon}{k}a}\gamma(a)da \\ = &\frac{N}{v}\int_0^\infty e^{-x}\gamma\left(\frac{kx}{\varepsilon}\right)dx. \end{aligned} \tag{8.15}$$

Then, $R_e(\varepsilon)$ is monotone decreasing with respect to $\varepsilon$ and

$$\frac{\gamma(0)N}{v} \le R_e(\varepsilon) \le R_0 = \frac{\gamma(\infty)N}{v}.$$

If $\gamma(0)N/v < 1$, for sufficiently large $\varepsilon$, we have $R_e(\varepsilon) < 1$. In this case, even if $R_0 > 1$, the entire susceptible population can be protected from the disease by increasing the vaccination rate. In contrast, if $\gamma(0)N/v > 1$, we cannot control the disease.

Let us now consider the ESS $(S^*(a), I^*)$. We then have that

$$\begin{aligned} \frac{dS^*(a)}{da} &= -\frac{\varepsilon}{k}S^*(a) - \frac{\gamma(a)I^*}{k}S^*(a), \\ kS^*(0) &= vI^* + \varepsilon\int_0^\infty S^*(a)da, \end{aligned} \tag{8.16}$$

from which it follows that

$$S^*(a) = S^*(0)e^{-\frac{\varepsilon}{k}a - \frac{I^*}{k}\int_0^a \gamma(\sigma)d\sigma}. \tag{8.17}$$

From (8.16) and (8.17), we obtain

$$\int_0^\infty S^*(a)da = \frac{\frac{vI^*}{k}\int_0^\infty e^{-\frac{\varepsilon}{k}a - \frac{I^*}{k}\int_0^a \gamma(\sigma)d\sigma}da}{1 - \frac{\varepsilon}{k}\int_0^\infty e^{-\frac{\varepsilon}{k}a - \frac{I^*}{k}\int_0^a \gamma(\sigma)d\sigma}da}.$$

Because the total size of the host population is constant, we have

$$N = I^* + \frac{\frac{vI^*}{k}\int_0^\infty e^{-\frac{\varepsilon}{k}a - \frac{I^*}{k}\int_0^a \gamma(\sigma)d\sigma}da}{1 - \frac{\varepsilon}{k}\int_0^\infty e^{-\frac{\varepsilon}{k}a - \frac{I^*}{k}\int_0^a \gamma(\sigma)d\sigma}da}. \tag{8.18}$$

If (8.18) has a positive root $I^* > 0$ in the interval $[0, N]$, there exists an ESS.

**Proposition 8.7** *If $R_e(\varepsilon) > 1$, there exists at least one ESS.*

*Proof* Define a function $\psi(x)$ by

$$\psi(x) = x\left(1 + \frac{v}{\varepsilon}\frac{\phi(x)}{1-\phi(x)}\right),$$

where

$$\phi(x) := \frac{\varepsilon}{k}\int_0^\infty e^{-\frac{\varepsilon}{k}a - \frac{x}{k}\int_0^a \gamma(\sigma)d\sigma} da.$$

It is then sufficient to show that the equation $\psi(x) = N$ has a root $x \in (0, N)$. Note that $\psi(N) > N$ and

$$\lim_{x\to+0} \psi(x) = \frac{v}{\varepsilon}\lim_{x\to+0}\frac{x\phi(x)}{1-\phi(x)} = \frac{v}{\varepsilon}\frac{\phi(0)}{-\phi'(0)} = \frac{N}{R_e(\varepsilon)}.$$

Then, we know that if $R_e(\varepsilon) > 1$, $\psi(+0) < N$, and there exists at least one root in the interval $(0, N)$. □

We can now consider the stability of the ESS. Let $x(t, a)$ be the perturbation from the stationary solution $S^*(a)$. The linearized equation about the steady state can be written as

$$\begin{aligned} \frac{\partial x(t,a)}{\partial t} + k\frac{\partial x(t,a)}{\partial a} &= -(\varepsilon + \gamma(a)I^*)x(t,a) + \gamma(a)S^*(a)\int_0^\infty x(t,a)da, \\ x(t,0) &= \frac{\varepsilon - v}{k}\int_0^\infty x(t,a)da. \end{aligned} \tag{8.19}$$

If we suppose that $x(t, a) = e^{\lambda t}v(a)$, then the characteristic equation satisfied by $\lambda$ is given as follows:

$$\begin{aligned} 1 = \Delta_\varepsilon(\lambda) : &= \frac{\varepsilon - v}{k}\int_0^\infty e^{-\frac{\varepsilon+\lambda}{k}a - \frac{I^*}{k}\int_0^a \gamma(\sigma)d\sigma} da \\ &+ \frac{1}{k}\int_0^\infty\int_0^a e^{-\frac{\varepsilon+\lambda}{k}(a-s) - \frac{I^*}{k}\int_s^a \gamma(\sigma)d\sigma}\gamma(s)S^*(s)dsda. \end{aligned}$$

Again, if the characteristic equation $\Delta_\varepsilon(\lambda) = 1$ has no root in the right half-plane $\Re\lambda \geq 0$, the steady state is locally asymptotically stable. Note that $\Delta_\varepsilon(\lambda)$ defines a holomorphic function in the half-plane $\Re\lambda > -\varepsilon - \gamma(\infty)I^*$. If $\varepsilon \geq v$, the integral kernel of the characteristic equation becomes positive, and we can apply the same kind of argument as for the Euler–Lotka characteristic equation. Then, we have the following conclusion:

**Proposition 8.8** *If $\varepsilon \geq v$, the steady state is locally asymptotically stable if and only if $\Delta_\varepsilon(0) < 1$. Then, the disease-free steady state is locally asymptotically stable if $R_e(\varepsilon) < 1$, whereas it is unstable if $R_e(\varepsilon) > 1$.*

*Proof* If $\varepsilon \geq \nu$, for real $\lambda$, $\Delta_\varepsilon(\lambda)$ decreases monotonically from $+\infty$ to zero and there exists a unique real root. This root is dominant and negative if $\Delta_\varepsilon(0) < 1$. For the disease-free steady state, the characteristic equation is given as follows:

$$\Delta_\varepsilon(\lambda) = \frac{1}{\varepsilon+\lambda}\frac{\varepsilon N}{k}\int_0^\infty \gamma(a)e^{-\frac{\varepsilon}{k}a}da + \frac{\varepsilon-\nu}{\varepsilon+\lambda} = 1.$$

The unique root in the half-plane $\Re\lambda > -\varepsilon$ is then given by

$$\lambda_0 = -\nu + \frac{\varepsilon N}{k}\int_0^\infty \gamma(a)e^{-\frac{\varepsilon}{k}a}da = \nu(R_e(\varepsilon)-1) > -\varepsilon.$$

Thus, we conclude that the disease-free steady state is locally asymptotically stable if $R_e(\varepsilon) < 1$, whereas it is unstable if $R_e(\varepsilon) > 1$. □

In the Pease model, the $i$-state (individual state) is characterized by parameter $a$ (the amount of amino acid substitutions), which plays essentially the same role as the duration since the last infection, because the rate of amino acid substitutions is equal to the mutation rate and is assumed to be constant. Hence, the $i$-state of susceptible individuals is locally determined and is independent of the state of the epidemic. However, if amino acid substitutions are caused by frequency-dependent selection and several types of virus strain can circulate at the same time in a community, we have to deal with much more complex evolutionary models in which the $i$-state depends on the global state of the epidemic.

We have studied a modified Pease model with a simple vaccination term, in which a constant vaccination rate is assumed and the vaccinated susceptible individuals are identified along with the individuals who have recovered from real infection. However, in reality, the vaccination phenomena associated with an evolutionary epidemic would be much more complex. For example, if we use the immunizing virus isolated in the past, according to the number of amino acid substitutions that have occurred between the immunizing virus strain and the strain circulating at given time $t$, the induced immunity will be incomplete; that is, the result of vaccination is to make the susceptible host "younger" with respect to the "age" $a$, but it does not necessarily reset the "age" to zero. Therefore, it seems that the possible rate of vaccination $\delta(a)$ will be of the form $\delta(a) = \delta_c(a) + \delta_p(a)$, where $\delta_c(a)$ denotes the rate of complete immunization and $\delta_p(a)$ is the rate of partial immunization. Let $\beta(a, \alpha)$ be the probability per unit amino acid substitution that a partial vaccination in a susceptible host of "age" $\alpha$ leads to a susceptible host with "age" $a$ ($a \leq \alpha$). We assume that $\beta(a, \alpha) = 0$ for $a > \alpha$ and, for all $\alpha$,

$$\int_0^\alpha \beta(a, \alpha)da = 1.$$

If we combine the vaccination term defined above with the basic model of (8.2), we obtain the following model:

$$
\begin{aligned}
\frac{\partial S(t,a)}{\partial t}+k\frac{\partial S(t,a)}{\partial a} &= -\gamma(a)S(t,a)\left[N-\int_0^\infty S(t,a)da\right] \\
&\quad -\delta(a)S(t,a)+\int_a^\infty \beta(a,\alpha)\delta_p(\alpha)S(t,\alpha)d\alpha, \\
kS(t,0) &= v\left[N-\int_0^\infty S(t,a)da\right]+\int_0^\infty \delta_c(a)S(t,a)da.
\end{aligned}
$$

An important difference from the original model is that this vaccination model has a disease-free steady state. It can be shown that the stationary state for the above modified model is determined by solving a Fredholm integral equation in the same manner as for the metapopulation model of Hastings [16]. The analysis of the threshold phenomena of this vaccination model is incomplete.

## 8.2 Kermack–McKendrick Reinfection Model

In a seminal series of papers published in the 1930s, Kermack and McKendrick proposed an infection-age-structured endemic model that takes into account the demography of the host population, the *waning immunity* (variable susceptibility), and the *reinfection* of recovered individuals [26, 27]. This model received less attention than their well-known outbreak model of 1927 [25]. In their reinfection model, the total population is decomposed into three compartments, the never-infected (fully susceptible), infectious, and recovered populations. The host population is structured by a duration variable for each status, and the chronological age is neglected. The susceptibility of recovered individuals depends on the time that has passed since the last recovery, giving the model sufficient flexibility to capture many facets of reinfection phenomena.

The concept of reinfection is becoming increasingly important in understanding emerging and re-emerging infectious diseases because it makes the control of infectious diseases difficult, and a waning immunity is widely observed if there is no (natural or artificial) *boosting*. In fact, recovered or vaccinated individuals can be reinfected as time passes owing to the natural decay of host immunity or a genetic change in the virus. Reinfection often leads to non-clinical infection, and so its occurrence is likely to be overlooked. This causes the basic reproduction number and the critical coverage of immunization to be miscalculated.

As shown in Sect. 5.5 and [14], we can define the *reinfection threshold* of $R_0$ at which a qualitative change in the epidemiological implication occurs for the prevalence and controllability. Moreover, owing to the enhancement of net reproductivity for secondary cases by reinfection, we expect there to be a backward bifurcation of endemic steady states. In such a case, we have bistable endemic steady states, and attaining a subcritical level of $R_0$ is not a complete policy for disease prevention.

In this section, we first reformulate the forgotten Kermack–McKendrick reinfection model as an age-structured population model and examine basic endemic threshold phenomena in order to present a condition in which backward bifurcation occurs [23]. We then extend the basic model to a chronological-age-dependent model and calculate some basic epidemiological indices. Consideration of the chronological-age structure is crucial to real-world applications, because prevention or vaccination policies usually target age classes. Finally, we again extend the basic model to recognize the subclinical infection observed in malaria and measles epidemics and examine the conditions under which subcritical endemic steady states exist, because failure to do so is likely to produce incorrect estimates and interpretations of epidemiological indices. Although we only consider the variable susceptibility described by local time in this chapter, readers interested in the time-independent distributed susceptibility may refer to [7, 17, 41, 46]. The risk-based models for HIV infection (Sect. 7.1) also incorporate variable susceptibility.

### *8.2.1 Basic Model*

First, we formulate the Kermack–McKendrick reinfection model from a modern perspective [23]. Let $s(t,\tau)$ be the density of the susceptible population who have never been infected (*virgin population* in the terminology of Kermack and McKendrick) at time $t$ and duration (the time elapsed since entry into the $s$-state) $\tau$. Let $i(t,\tau)$ be the density of the infected and infectious population at time $t$ and *infection-age* (the time elapsed since infection) $\tau$, and let $r(t,\tau)$ be the density of the recovered population (partially susceptible population) at time $t$ and *recovery-age* $\tau$ (the time elapsed since the last recovery). Let $m$ and $\mu$ denote the birth (or immigration) rate and the death rate, respectively, and $\gamma(\tau)$ denote the recovery rate at infection-age $\tau$.

We assume that the force of infection applied to the fully susceptible population (virgin population) is given by

$$\lambda(t) := \int_0^\infty \beta(\sigma)i(t,\sigma)d\sigma, \tag{8.20}$$

where $\beta(\tau)$ denotes the infectivity to the virgin population at infection-age $\tau$. The force of (re)infection applied to the recovered population is assumed to be given by $\theta(\tau)\lambda(t)$, where $\theta(\tau)$ is the relative susceptibility schedule of recovered individuals at recovery-age $\tau$. The relative susceptibility is inversely correlated with the waning of immunity. In the following, we assume that $\beta, \gamma, \theta \in L^\infty_+(\mathbb{R}_+)$, and the state space of the age distribution functions $s$, $i$, and $r$ is $L^1_+(\mathbb{R}_+)$. The Kermack–McKendrick reinfection model is then formulated as

$$
\begin{aligned}
&\frac{\partial s(t,\tau)}{\partial t}+\frac{\partial s(t,\tau)}{\partial \tau}=-\mu s(t,\tau)-\lambda(t)s(t,\tau),\\
&\frac{\partial i(t,\tau)}{\partial t}+\frac{\partial i(t,\tau)}{\partial \tau}=-(\mu+\gamma(\tau))i(t,\tau),\\
&\frac{\partial r(t,\tau)}{\partial t}+\frac{\partial r(t,\tau)}{\partial \tau}=-\mu r(t,\tau)-\theta(\tau)\lambda(t)r(t,\tau),\\
&s(t,0)=m\int_0^\infty (s(t,\tau)+i(t,\tau)+r(t,\tau))d\tau,\\
&i(t,0)=\lambda(t)\int_0^\infty (s(t,\tau)+\theta(\tau)r(t,\tau))\,d\tau,\\
&r(t,0)=\int_0^\infty \gamma(\tau)i(t,\tau)d\tau,
\end{aligned}
\tag{8.21}
$$

with initial data

$$s(0,\tau)=s_0(\tau),\quad i(0,\tau)=i_0(\tau),\quad r(0,\tau)=r_0(\tau).$$

If $\theta\equiv 0$, (8.21) becomes the susceptible–infective–recovered (SIR) model with permanent immunity, and if $\theta\equiv 1$, the recovered population can be identified with the virgin population $s$, and model (8.21) reduces to the duration-dependent SIS epidemic model.

Let $N(t)$ be the total size of the host population given by

$$N(t):=\int_0^\infty (s(t,\tau)+i(t,\tau)+r(t,\tau))d\tau.$$

If $m=\mu$, the total size of the host population is constant. In the following, we consider a constant total population size, denoted by $N$, so the boundary condition of $s(t,a)$ is replaced by $s(t,0)=\mu N$. The disease-free steady state is then $(s^*,i^*,r^*)=(\mu N e^{-\mu\tau},0,0)$, and the renewal equation in the linear invasion phase is

$$b(t)=N\lambda(t)=N\int_0^\infty \beta(\tau)\Gamma(\tau)e^{-\mu\tau}b(t-\tau)d\tau, \tag{8.22}$$

where $b(t)=i(t,0)$ and $\Gamma(\tau):=\exp(-\int_0^\tau \gamma(x)dx)$ is the probability that an infected individual has not yet recovered at infection-age $\tau$. The basic reproduction number of (8.22) is then given by

$$R_0=N\int_0^\infty e^{-\mu\tau}\beta(\tau)\Gamma(\tau)d\tau. \tag{8.23}$$

By the principle of linearized stability, the stability of the zero solution of (8.22) determines the local stability of the disease-free steady state of system (8.21). Thus, the disease-free steady state is locally asymptotically stable if $R_0<1$, whereas it is unstable if $R_0 > 1$. The role of the basic reproduction number in population dynamics is discussed in [11, 12, 22] and Chap. 9.

Model (8.21) can be rewritten as the Gurtin–MacCamy model (see Chap. 3) for an age-dependent population. The mathematical well-posedness of this model has been established in [19], and we can use the integrated semigroup formulation to give the solution semiflow [29]. We therefore skip the mathematical well-posedness problem here. The ODE version of (8.21) is studied in Sect. 5.5.3.

### 8.2.2 *Integral Equations*

We now sketch an alternative formulation of the reinfection model using integral equations [8]. For simplicity, instead of the initial value problem, we assume that the epidemic starts at $t = -\infty$. Integrating the partial differential equations in (8.21) along the characteristic line, we have the following set of equations:

$$\begin{aligned} s(t,\tau) &= \mu N e^{-\mu\tau - \int_0^\tau \lambda(t-\tau+\sigma)d\sigma}, \\ i(t,\tau) &= b_1(t-\tau)e^{-\mu\tau}\Gamma(\tau), \\ r(t,\tau) &= b_2(t-\tau)e^{-\mu\tau - \int_0^\tau \lambda(t-\tau+\sigma)\theta(\sigma)d\sigma}, \end{aligned} \tag{8.24}$$

where $b_1(t) := i(t,0)$ and $b_2(t) := r(t,0)$. Inserting (8.24) into the boundary conditions in (8.21), we obtain a set of integral equations:

$$\begin{aligned} b_1(t) &= \lambda(t)\left[\int_0^\infty \mu N e^{-\mu\tau - \int_0^\tau \lambda(t-\tau+\sigma)d\sigma} d\tau \right. \\ &\qquad \left. + \int_0^\infty \theta(\tau) b_2(t-\tau) e^{-\mu\tau - \int_0^\tau \lambda(t-\tau+\sigma)\theta(\sigma)d\sigma} d\tau \right], \\ b_2(t) &= \int_0^\infty b_1(t-\tau)e^{-\mu\tau}\gamma(\tau)\Gamma(\tau)d\tau, \end{aligned} \tag{8.25}$$

where

$$\lambda(t) = \int_0^\infty e^{-\mu\tau}\beta(\tau)\Gamma(\tau)b_1(t-\tau)d\tau. \tag{8.26}$$

Inserting the expression for $b_2$ into the equation for $b_1$ in (8.25) and changing the order of integrals, we obtain

$$b_1(t) = \lambda(t)\int_0^\infty S(t,\tau)d\tau, \tag{8.27}$$

$$
\begin{aligned}
S(t,\tau) :&= s(t,\tau) + \theta(\tau) r(t,\tau) \\
&= \mu N e^{-\mu\tau - \int_0^\tau \lambda(t-\tau+\sigma) d\sigma} \\
&\quad + b_1(t-\tau) e^{-\mu\tau} \int_0^\tau \theta(\sigma) e^{-\int_0^\sigma \theta(\zeta)\lambda(t-\sigma+\zeta) d\zeta} \gamma(\tau-\sigma)\Gamma(\tau-\sigma) d\sigma.
\end{aligned}
\tag{8.28}
$$

where $\int_0^\infty S(t,\tau)d\tau$ is the *effective size of the susceptible population*. Expression (8.27) implies a simple fact that the new incidence at time $t$ is given by the force of infection times the effective size of the susceptible population. The first part $\lambda(t)\int_0^\infty s(t,\tau)d\tau$ gives the incidence of the primary infection, and the second part $\lambda(t)\int_0^\infty \theta(\tau) r(t,\tau)d\tau$ gives the incidence of reinfection, which is not monotone with respect to the force of infection.

From (8.27) and (8.28), we obtain a linear renewal equation for $b_1$ if we consider the force of infection $\lambda$ as a given function. Thus, by solving the linear renewal equation formally, we have an expression for $b_1$ containing the unknown $\lambda$. Inserting this solution into (8.26), we arrive at a nonlinear "scalar" renewal equation for $\lambda$. Alternatively, eliminating $\lambda$ from (8.26)–(8.28), we again obtain a nonlinear scalar integral equation for $b_1$. We can then establish the well-posedness of the Kermack–McKendrick endemic model (8.21) based on the well-known method of nonlinear integral equations.

If $\theta \equiv 0$, (8.21) becomes the SIR model with permanent immunity. This has a unique ESS that is globally stable if and only if $R_0 > 1$ [29, 39]. If $\theta \equiv 1$, the recovered population can be identified with the virgin population, so (8.21) reduces to the infection-age-dependent SIS epidemic model (which can be reduced to a nonlinear renewal equation). The ESS of this model is unique, but when $R_0 > 1$, it may lose its stability and Hopf bifurcations can occur [9, 10, 32]. Under the assumption that $\theta$ is monotone increasing and less than unity, it is concluded that if $R_0 > 1$, there exists a unique ESS that is locally asymptotically stable as long as $|R_0 - 1|$ is sufficiently small [19].

If $\sup\theta > 1$, we can conjecture that the subcritical condition $R_0 < 1$ does not necessarily guarantee the eradication of the disease. In fact, from (8.27), we can formally define a time-dependent (period) reproduction number as

$$
\mathscr{R}(t) := \tilde{S}(t) \int_0^\infty \beta(\tau)\Gamma(\tau)e^{-\mu\tau} d\tau,
$$

where $\tilde{S}(t) := \int_0^\infty S(t,\tau)d\tau$ is the effective size of susceptibility. Because $\tilde{S}(t)$ can be larger than the total population size $N$, $\mathscr{R}(t)$ can also be greater than $R_0$, in which case $R_0 < 1$ would not be a sufficient condition for the eradication of the disease.

Let $\alpha := \max\{1, \sup_{\tau\geq 0}\theta(\tau)\}$. Then, $\tilde{S} \leq \alpha N$, and it follows from (8.27) that

$$
b_1(t) \leq \alpha N \int_0^\infty \beta(\tau)\Gamma(\tau)e^{-\mu\tau} b_1(t-\tau) d\tau.
$$

Using the comparison argument, we know that $\lim_{t\to\infty} b_1(t) = 0$ if $\alpha R_0 < 1$. Thus, we have a simple criterion for the global stability of the disease-free steady state:

**Proposition 8.9** *If* $R_0 < 1/\alpha$, *the disease-free steady state of* (8.21) *is globally asymptotically stable.*

Note that another extreme scenario to recover susceptibility is to assume that recovered individuals are completely immune, but return to the susceptible class at recovery-age $\tau$ with the force of reversion $\delta(\tau)$. In this case, instead of (8.21), we obtain

$$\begin{aligned}
&\frac{\partial r(t,\tau)}{\partial t} + \frac{\partial r(t,\tau)}{\partial \tau} = -\mu r(t,\tau) - \delta(\tau) r(t,\tau),\\
&s(t,0) = \mu N + \int_0^\infty \delta(\tau) r(t,\tau) d\tau,\\
&i(t,0) = \lambda(t) \int_0^\infty s(t,\tau) d\tau,\\
&r(t,0) = \int_0^\infty \gamma(\tau) i(t,\tau) d\tau,
\end{aligned} \tag{8.29}$$

where we omit the McKendrick equations for $s$ and $i$, because they are the same as the equations in (8.21). That is, we obtain an SIRS model with infection-age, as was studied by Nakata et al. [34] in the case where $\delta$ is constant. In this model, the density of the susceptible population is given by

$$s(t,\tau) := e^{-\mu\tau - \int_0^\tau \lambda(t-\tau+\sigma)d\sigma} \left[ \mu N + \int_0^\infty b_1(t-\tau) e^{-\mu\tau} P(\tau) d\tau \right], \tag{8.30}$$

where

$$P(\tau) = \int_0^\tau \delta(x) e^{-\int_0^x \delta(z)dz} \gamma(\tau - x) \Gamma(\tau - x) dx \tag{8.31}$$

denotes the probability density of an infected individual returning to the susceptible state at time $\tau$ since the last infection.

### 8.2.3 Bifurcation of Endemic Steady States

Let us now examine the bifurcation of endemic steady states of (8.21). Let $s^*(\tau)$, $i^*(\tau)$ and $r^*(\tau)$ be the steady state solution. Then, it holds that

$$\begin{aligned}
s^*(\tau) &= \mu N e^{-(\mu+\lambda^*)\tau},\\
i^*(\tau) &= i^*(0) e^{-\mu\tau} \Gamma(\tau),\\
r^*(\tau) &= r^*(0) e^{-\mu\tau - \lambda^* \int_0^\tau \theta(\sigma)d\sigma},
\end{aligned}$$

where

$$\begin{aligned} i^*(0) &= \lambda^* \int_0^\infty (s^*(\tau) + \theta(\tau) r^*(\tau)) d\tau, \\ r^*(0) &= \int_0^\infty \gamma(\tau) i^*(\tau) d\tau \end{aligned} \tag{8.32}$$

and $\lambda^*$ is the force of infection at the steady state given by

$$\lambda^* = \int_0^\infty \beta(\tau) i^*(\tau) d\tau = b^* \langle \beta, \Gamma \rangle. \tag{8.33}$$

In expression (8.33), $b^* := i^*(0)$ is the density of newly infecteds at the steady state and we have used the notation

$$\langle \beta, \Gamma \rangle := \int_0^\infty \beta(\tau) \Gamma(\tau) e^{-\mu\tau} d\tau.$$

Inserting (8.33) into the first equation of (8.32), we obtain

$$b^* = b^* \langle \beta, \Gamma \rangle \int_0^\infty \left( \mu N e^{-(\mu+\lambda^*)\tau} + r^*(0) \theta(\tau) e^{-\mu\tau - \lambda^* \int_0^\tau \theta(\sigma) d\sigma} \right) d\tau,$$

which gives a renewal relation in a steady state with force of infection $\lambda^*$. Because $\langle \beta, \Gamma \rangle = R_0/N$ and $r^*(0) = b^* \langle \gamma, \Gamma \rangle$, we arrive at an equation for the unknown $\lambda^*$:

$$\begin{aligned} R(\lambda^*) : &= \frac{\mu R_0}{\mu + \lambda^*} + \langle \gamma, \Gamma \rangle \lambda^* \int_0^\infty \theta(\tau) e^{-\mu\tau - \lambda^* \int_0^\tau \theta(\sigma) d\sigma} d\tau \\ &= \frac{\mu R_0}{\mu + \lambda^*} + \langle \gamma, \Gamma \rangle \left( 1 - \int_0^\infty \mu e^{-\mu\tau - \lambda^* \int_0^\tau \theta(\sigma) d\sigma} d\tau \right) = 1, \end{aligned} \tag{8.34}$$

where we have used the notation

$$\langle \gamma, \Gamma \rangle := \int_0^\infty \gamma(\tau) \Gamma(\tau) e^{-\mu\tau} d\tau.$$

The first part of $R(\lambda^*)$ gives the reproduction number for primary infection, and the second part is the reproduction number by reinfection. Equation (8.34) implies that the effective reproduction number, given by $R(\lambda^*)$, must be unity in a steady state. Indeed, it follows from (8.27) that we obtain a formal renewal relation at the endemic steady state

$$b_1^* = \lambda^* \int_0^\infty \left[ \mu N e^{-\mu\tau - \lambda^* \tau} d\tau + e^{-\mu\tau} \int_0^\tau \theta(\sigma) e^{-\int_0^\sigma \theta(\zeta) \lambda^* d\zeta} \gamma(\tau - \sigma) \Gamma(\tau - \sigma) d\sigma \right] d\tau,$$

and $\lambda^* = \langle \beta, \Gamma \rangle b_1^*$, from which we obtain (8.34).

It follows from (8.34) that there exists at least one ESS if $R_0 > 1$, because $R(0) = R_0 > 1$ and $\lim_{\lambda\to\infty} R(\lambda) = \langle\gamma, \Gamma\rangle < 1$. Because $R(\lambda^*)$ is not monotone decreasing, there is a possibility that multiple endemic steady states exist.

**Proposition 8.10** *If the following inequality holds:*

$$\langle\gamma, \Gamma\rangle\theta^* > 1, \tag{8.35}$$

*where*

$$\theta^* := \int_0^\infty \theta(\tau)\mu e^{-\mu\tau} d\tau, \tag{8.36}$$

*then endemic steady states backwardly bifurcate from the disease-free steady state when $R_0$ crosses unity. That is, multiple endemic steady states exist if $R_0 < 1$ and $|R_0 - 1|$ is sufficiently small.*

*Proof* Define a function $f(\lambda, R_0) := R(\lambda) - 1$, where $R_0$ is seen as a bifurcation parameter and $f(0, 1) = 0$. Observe that

$$\left.\frac{\partial f}{\partial \lambda}\right|_{(\lambda,R_0)=(0,1)} = \frac{1}{\mu}(\theta^*\langle\gamma, \Gamma\rangle - 1), \quad \left.\frac{\partial f}{\partial R_0}\right|_{(\lambda,R_0)=(0,1)} = 1.$$

Therefore, if condition (8.35) holds, then $f = 0$ can be solved as $\lambda = \lambda(R_0)$ with $\lambda(1) = 0$ in the neighborhood of $(\lambda, R_0) = (0, 1)$. Because $d\lambda(1)/dR_0 < 0$, we have $\lambda(R_0) > 0$ for $R_0 \in (1 - \eta, 1)$ for sufficiently small $\eta > 0$. For each $R_0 \in (1 - \eta, 1)$, we have $f(0, R_0) < 1$, $f(\lambda(R_0), R_0) = 0$, and $\lim_{\lambda\to\infty} f(\lambda, R_0) = \langle\gamma, \Gamma\rangle - 1 < 0$. Thus, there exist at least two endemic steady states. □

Condition (8.35) was first given in [37] using an ordinary differential equation form of (8.21) (see Sect. 5.5.3). It is easily seen that condition (8.35) does not hold if there is no enhancement of susceptibility; that is, if $\theta(\tau) \le 1$ for all $\tau \ge 0$.

*Remark 8.1* If the epidemic timescale is much shorter than the timescale of the host demography, we can neglect the birth and death rates—that is, $\mu = 0$. Moreover, if $\beta$ and $\gamma$ are assumed to be constant, the Kermack–McKendrick model (8.21) can be formulated as follows:

$$\begin{aligned}
\frac{dU(t)}{dt} &= -\beta I(t)U(t),\\
\frac{dI(t)}{dt} &= -\gamma I(t) + \beta I(t)\left(U(t) + \int_0^\infty \theta(\tau)r(t,\tau)d\tau\right),\\
\frac{\partial r(t,\tau)}{\partial t} + \frac{\partial r(t,\tau)}{\partial \tau} &= -\beta\theta(\tau)r(t,\tau)I(t),\\
r(t,0) &= \gamma I(t),
\end{aligned} \tag{8.37}$$

where $U(t) := \int_0^\infty s(t,\tau)d\tau$ and $I(t) := \int_0^\infty i(t,\tau)d\tau$. In this case, it is easy to see that $R_0 = \beta N/\gamma$, and the following endemic threshold property holds:

**Proposition 8.11** *Suppose that $\theta(\tau)$ is a monotone non-decreasing function and there exists a limit $\theta(\infty) = \lim_{\tau\to\infty}\theta(\tau)$. If $R_0\theta(\infty) \le 1$, there is no ESS. If $R_0\theta(\infty) > 1$, there exists a unique ESS.*

*Proof* Let $(U^*, I^*, r^*(\tau))$ be an ESS. Then, we have $U^* = 0$ and

$$N = I^* + \int_0^\infty r^*(\tau)d\tau = I^* + \gamma I^* \int_0^\infty e^{-\beta I^* \int_0^\tau \theta(x)dx} d\tau. \tag{8.38}$$

By changing the variables, we obtain

$$I^* \int_0^\infty e^{-\beta I^* \int_0^\tau \theta(x)dx} d\tau = \int_0^\infty e^{-\beta \int_0^\tau \theta(\frac{x}{I^*})dx} d\tau.$$

Therefore, the right-hand side of (8.38) is a monotone increasing function of $I^* \in [0, N]$. If $I^* \to 0$, the right-hand side of (8.38) goes to $\gamma/(\beta\theta(\infty))$, so it has a unique positive root $I^*$ if and only if $R_0\theta(\infty) > 1$. □

Therefore, if $\theta(\infty) < 1$ and $1 < R_0 < \theta(\infty)^{-1}$, the disease can invade the completely susceptible host population (i.e., an outbreak occurs), but the disease will be naturally eradicated and there is no ESS. Hence, the invasion threshold is not the same as the endemic threshold. These phenomena have been observed by Thieme and Yang [44] and Katriel [24]. Model (8.37) incorporating the host demography has also been studied [31, 44]. Note that the subset $\Omega_0 := \{0\} \times \mathbb{R}_+ \times L^1(\mathbb{R}_+)$ of the state space of (8.37) is positively invariant, and system (8.37) on $\Omega_0$ is described by the following $IR$ system on $\mathbb{R}_+ \times L^1(\mathbb{R}_+)$:

$$\begin{aligned}
\frac{dI(t)}{dt} &= -\gamma I(t) + \beta I(t) \int_0^\infty \theta(\tau) r(t,\tau)d\tau,\\
\frac{\partial r(t,\tau)}{\partial t} + \frac{\partial r(t,\tau)}{\partial \tau} &= -\beta\theta(\tau)I(t)r(t,\tau),\\
r(t,0) &= \gamma I(t).
\end{aligned} \tag{8.39}$$

This is none other than the Pease model for type A influenza studied in the previous section.

### 8.2.4 Vaccination

We now introduce a mass vaccination (host immunization) effect into the basic model (8.21). It is intuitively clear that reinfection phenomena would make disease control more difficult and complex, and we need an index to capture this difficulty. An important effect of vaccination policies is to reduce the effective size of the susceptible population. In the reinfection model, there is a possibility that a disease can invade a fully vaccinated population, and we are naturally led to the idea of the *reinfection threshold*.

Suppose that newborns or immigrants in the virgin population are mass vaccinated with coverage $\varepsilon \in [0, 1]$, and for simplicity, assume that the immunological status of newly vaccinated individuals is identical to that of the newly recovered individuals. This assumption will be relaxed in Sect. 8.3. The boundary condition in the basic system (8.21) is then replaced by:

$$\begin{aligned} s(t,0) &= (1-\varepsilon)\mu N, \\ i(t,0) &= \lambda(t)\int_0^\infty (s(t,\tau)+\theta(\tau)r(t,\tau))\,d\tau, \\ r(t,0) &= \varepsilon\mu N + \int_0^\infty \gamma(\tau)i(t,\tau)d\tau. \end{aligned} \tag{8.40}$$

The disease-free steady state can then be written as

$$(s^*, i^*, r^*) = ((1-\varepsilon)\mu N e^{-\mu\tau}, 0, \varepsilon\mu N e^{-\mu\tau}),$$

so the linearized renewal equation for the initial invasion phase is

$$\xi(t) = \big((1-\varepsilon)N + \varepsilon N\theta^*\big)\int_0^\infty e^{-\mu\tau}\beta(\tau)\Gamma(\tau)\xi(t-\tau)d\tau,$$

where $\xi(t) := \zeta(t,0)$ denotes a small perturbation in the infected population density.

Therefore, the effective reproduction number, denoted by $\mathscr{R}(\varepsilon)$, in the partially immunized disease-free steady state is given by

$$\mathscr{R}(\varepsilon) = (1-\varepsilon)R_0 + \varepsilon R_1 = (1-\varepsilon(1-\theta^*))R_0, \tag{8.41}$$

where $R_1 := \theta^* R_0$. Then, if $\mathscr{R}(\varepsilon) < 1$, the disease-free steady state is locally asymptotically stable, whereas it is unstable if $\mathscr{R}(\varepsilon) > 1$. However, it is unclear whether the disease-free steady state becomes globally asymptotically stable when $\mathscr{R}(\varepsilon) < 1$.

Note that $R_1$ is the effective reproduction number for the fully vaccinated system. In fact, if $\varepsilon = 1$, the virgin population is eradicated, and we obtain the limiting recovered–infected–recovered (RIR) system as

$$\begin{aligned} \frac{\partial i(t,\tau)}{\partial t} + \frac{\partial i(t,\tau)}{\partial \tau} &= -(\mu+\gamma(\tau))i(t,\tau), \\ \frac{\partial r(t,\tau)}{\partial t} + \frac{\partial r(t,\tau)}{\partial \tau} &= -\mu r(t,\tau) - \theta(\tau)\lambda(t)r(t,\tau), \\ i(t,0) &= \lambda(t)\int_0^\infty \theta(\tau)r(t,\tau)d\tau, \\ r(t,0) &= \mu N + \int_0^\infty \gamma(\tau)i(t,\tau)d\tau. \end{aligned} \tag{8.42}$$

If we consider the recovered class as a new susceptible class, system (8.42) can be considered as a duration-dependent SIS model with vaccination. Then, (8.42) has a disease-free steady state $(i^*, r^*) = (0, \mu N e^{-\mu\tau})$, and the linearized system at the disease-free steady state is given as

$$\frac{\partial \zeta(t,\tau)}{\partial t} + \frac{\partial \zeta(t,\tau)}{\partial \tau} = -(\mu + \gamma(\tau))\zeta(t,\tau),$$
$$\zeta(t,0) = \theta^* N \int_0^\infty \beta(\tau)\zeta(t,\tau)d\tau.$$

Therefore, the effective reproduction number for the limiting system (8.42) is given by $R_1 = \theta^* R_0$.

Suppose that $R_0 > 1$. From (8.41), the critical coverage of immunization $\varepsilon^*$ such that $\mathscr{R}(\varepsilon^*) = 1$ is given by

$$\varepsilon^* = \left(1 - \frac{1}{R_0}\right)\frac{1}{1-\theta^*}, \tag{8.43}$$

but this is only meaningful when $\theta^* < 1$; the disease cannot be controlled by vaccination if $\theta^* \geq 1$. Moreover, if $R_1 = \theta^* R_0 > 1$, we have $\mathscr{R}(\varepsilon) > 1$ for all $\varepsilon \in [0, 1]$, and the disease is once again uncontrollable by vaccination, because the fully vaccinated population can be invaded by the disease.

Because there is a qualitative change in the epidemiological implication for the prevalence and controllability at $R_1 = \theta^* R_0 = 1$, following Gomes et al. [14, 15], we call $1/\theta^*$ the *reinfection threshold* of $R_0$. As seen above, the reinfection threshold corresponds to the fact that $R_0 = 1/\theta^*$ does not imply a bifurcation point of the basic system (8.21), but is the threshold condition $R_1 = 1$ of the fully vaccinated system (8.42).

Let $(s^*, i^*, r^*)$ be the steady state of the basic system (8.21) with the boundary condition given by (8.40). Then, we have

$$s^*(\tau) = (1-\varepsilon)\mu N e^{-\mu\tau - \lambda^*\tau},$$
$$i^*(\tau) = i^*(0)e^{-\mu\tau}\Gamma(\tau),$$
$$r^*(\tau) = r^*(0)e^{-\mu\tau - \lambda^* \int_0^\tau \theta(x)dx},$$

where

$$\lambda^* = i^*(0)\langle \beta, \Gamma \rangle,$$
$$i^*(0) = \lambda^* \int_0^\infty (s^*(\tau) + \theta(\tau) r^*(\tau))d\tau,$$
$$r^*(0) = \varepsilon\mu N + i^*(0)\langle \gamma, \Gamma \rangle.$$

From the above equations, we can calculate $i^*(0)$ as

$$\begin{aligned} i^*(0) &= \lambda^* \int_0^\infty (s^*(\tau) + \theta(\tau) r^*(\tau)) d\tau \\ &= \lambda^* \frac{(1-\varepsilon)\mu N}{\mu + \lambda^*} + \lambda^* r^*(0) \int_0^\infty \theta(\tau) e^{-\mu\tau - \lambda^* \int_0^\tau \theta(x) dx} d\tau \\ &= \lambda^* \frac{(1-\varepsilon)\mu N}{\mu + \lambda^*} + \lambda^* \left( \varepsilon \mu N + i^*(0) \langle \gamma, \Gamma \rangle \right) \int_0^\infty \theta(\tau) e^{-\mu\tau - \lambda^* \int_0^\tau \theta(x) dx} d\tau. \end{aligned}$$

We then have the expression

$$i^*(0) = \frac{\lambda^* \frac{(1-\varepsilon)\mu N}{\mu+\lambda^*} + \varepsilon \mu N \lambda^* \int_0^\infty \theta(\tau) e^{-\mu\tau - \lambda^* \int_0^\tau \theta(x) dx} d\tau}{1 - \lambda^* \langle \gamma, \Gamma \rangle \int_0^\infty \theta(\tau) e^{-\mu\tau - \lambda^* \int_0^\tau \theta(x) dx} d\tau}. \tag{8.44}$$

From (8.44) and the relation

$$\lambda^* = \frac{R_0}{N} i^*(0),$$

we know that a positive root $\lambda^* > 0$ must satisfy

$$1 = R_0 \frac{v(\lambda^*)}{u(\lambda^*)}, \tag{8.45}$$

where

$$\begin{aligned} v(\lambda) &:= \frac{(1-\varepsilon)\mu}{\mu + \lambda} + \varepsilon \mu \int_0^\infty \theta(\tau) e^{-\mu\tau - \lambda \int_0^\tau \theta(x) dx} d\tau, \\ u(\lambda) &:= 1 - \langle \gamma, \Gamma \rangle \phi(\lambda). \end{aligned}$$

Here, we have used the notation

$$\phi(\lambda) := \lambda \int_0^\infty \theta(\tau) e^{-\mu\tau - \lambda \int_0^\tau \theta(x) dx} d\tau. \tag{8.46}$$

Observe that

$$\lambda \int_0^\infty \theta(\tau) e^{-\mu\tau - \lambda \int_0^\tau \theta(x) dx} d\tau = 1 - \int_0^\infty \mu e^{-\mu\tau - \lambda \int_0^\tau \theta(x) dx} d\tau.$$

Thus, $\phi$ is an increasing function and $u(\lambda)$ is a decreasing function. We can now conclude the following.

**Proposition 8.12** *If $\mathscr{R}(\varepsilon) > 1$, there exists at least one ESS. Suppose that the following condition holds:*

$$\theta^* \langle \gamma, \Gamma \rangle > \frac{1 - \varepsilon(1 - \theta^{**})}{1 - \varepsilon(1 - \theta^*)}, \tag{8.47}$$

*where*

$$\theta^{**} := \mu^2 \int_0^\infty e^{-\mu\tau}\theta(\tau)\int_0^\tau \theta(x)dxd\tau. \tag{8.48}$$

*Hence, endemic steady states backwardly bifurcate from the disease-free steady state when $\mathscr{R}(\varepsilon)$ crosses unity. That is, multiple endemic steady states exist if $\mathscr{R}(\varepsilon) < 1$ and $|\mathscr{R}(\varepsilon) - 1|$ is sufficiently small.*

*Proof* The relation in (8.45) implies that the effective reproduction number at the ESS with the force of infection $\lambda^*$ is given by

$$R(\lambda^*) = R_0\frac{v(\lambda^*)}{u(\lambda^*)} = \frac{\mathscr{R}(\varepsilon)}{v(0)}\frac{v(\lambda^*)}{u(\lambda^*)}.$$

Then, $R(0) = \mathscr{R}(\varepsilon)$ and $R(\infty) = 0$, so $R(\lambda^*) = 1$ has at least one positive root if $\mathscr{R}(\varepsilon) > 1$, which implies that there exists one ESS. If $R(0) = \mathscr{R}(\varepsilon) = R_0 v(0) = 1$ and condition (8.47) holds, $R'(0) = v(0)^{-1}v'(0) - u'(0) > 0$. Then, $R(\lambda^*) = 1$ has at least one positive root. Moreover, it has at least two positive roots if $R(0) = \mathscr{R}(\varepsilon) < 1$ and $|\mathscr{R}(\varepsilon) - 1|$ is sufficiently small. To clarify this fact, let us again define a function $f(\lambda, R_0) := R(\lambda) - 1$. Then, $f(0, v(0)^{-1}) = 0$ and

$$\left.\frac{\partial f}{\partial R_0}\right|_{(\lambda, R_0)=(0,v(0)^{-1})} = 1, \quad \left.\frac{\partial f}{\partial \lambda}\right|_{(\lambda, R_0)=(0,v(0)^{-1})} = v(0)^{-1}v'(0) - u'(0),$$

where

$$v'(0) = -\frac{1}{\mu}(1 - \varepsilon(1 - \theta^{**})), \quad u'(0) = -\frac{1}{\mu}\langle\gamma, \Gamma\rangle\theta^*.$$

If condition (8.47) holds, $f = 0$ can be solved as $\lambda = \lambda(R_0)$ satisfying $\lambda(v(0)^{-1}) = 0$ and $d\lambda(v(0)^{-1})/dR_0 < 0$ in the neighborhood of $(\lambda, R_0) = (0, v(0)^{-1})$. If $R_0 v(0) < 1$ and $|R_0 v(0) - 1|$ is sufficiently small, for each $R_0$, there exist multiple positive roots such that $f(\lambda, R_0) = 0$, because $f(0, R_0) < 1$, $f(\lambda(R_0), R_0) = 0$ and $f(\infty, R_0) = -1 < 0$. □

Proposition 8.12 tells us that the subcritical condition $\mathscr{R}(\varepsilon) < 1$ is not sufficient to eradicate the disease if condition (8.47) holds. Note that if $\varepsilon = 1$ in (8.47), we know that a backward bifurcation occurs even in the RIR model if $(\theta^*)^2 > \theta^{**}$, though this condition does not hold when $\theta$ is constant.

**Exercise 8.1** Suppose that some recovered individuals can lose their immunity completely and return to the full susceptible class. For simplicity, assume that $\beta$ and $\gamma$ are constants. Then, the basic model (8.21) with vaccination can be formulated as the following SIRS model [38]:

$$\frac{dS(t)}{dt} = (1-e)\mu N - \mu S(t) - \beta S(t)I(t) + \int_0^\infty \delta(\tau)r(t,\tau)d\tau,$$
$$\frac{dI(t)}{dt} = \lambda(t)\left(S(t) + \int_0^\infty \theta(\tau)r(t,\tau)d\tau\right) - (\mu+\gamma)I(t),$$
$$\left(\frac{\partial}{\partial t} + \frac{\partial}{\partial \tau}\right)r(t,\tau) = -(\mu + \delta(\tau) + \lambda(t)\theta(\tau))r(t,\tau),$$
$$r(t,0) = e\mu N + \gamma I(t),$$

where $S(t) := \int_0^\infty s(t,\tau)d\tau$, $I(t) := \int_0^\infty i(t,\tau)d\tau$, $\delta(\tau)$ is the recovery-age dependent reversion rate, and $\lambda(t) = \beta I(t)$. Calculate the effective reproduction number $R_e$ and prove that the disease-free steady state is locally asymptotically stable if $R_e < 1$, while it is unstable and there exists at least one endemic steady state if $R_e > 1$. Is there a possibility that a backward bifurcation of endemic steady states occurs at $R_e = 1$ ?

### 8.2.5 *One Clock or Two Clocks?*

We now present another formulation of the reinfection model known as an age-structured susceptible–infected (S-I) model, that is, a two-compartment model. We divide the host population into two subpopulations of susceptible individuals $s(t,\tau)$ and infected individuals $i(t,\tau)$, where the susceptible population comprises the never-infected individuals as before, but the infected population comprises individuals who have been infected at least once, regardless of whether they have recovered. We can rewrite the basic model of (8.21) as follows:

$$\begin{aligned}
\frac{\partial s(t,\tau)}{\partial t} + \frac{\partial s(t,\tau)}{\partial \tau} &= -\mu s(t,\tau) - \lambda(t)s(t,\tau),\\
\frac{\partial i(t,\tau)}{\partial t} + \frac{\partial i(t,\tau)}{\partial \tau} &= -(\mu + \lambda(t)\theta(\tau))i(t,\tau),\\
s(t,0) &= \mu N,\\
i(t,0) &= \lambda(t)\int_0^\infty (s(t,\tau) + \theta(\tau)i(t,\tau))\,d\tau,
\end{aligned} \tag{8.49}$$

where the force of infection is given by (8.20).

The key idea of the S-I formulation (8.49) is the assumption that the infectivity and the susceptibility of once-infected individuals can be expressed by $\beta(\tau)$ and $\theta(\tau)$, which are functions of the infection-age (the duration since the last infection) $\tau$. The S-I system is referred to as a "one-clock" model, because the epidemiological status of once infected individuals is parametrized only by infection-age. In the original Kermack–McKendrick reinfection model, we used two "clocks": one for the infected and one for the recovered individuals. The reduction in susceptibility was then a function of the time since recovery as given by the second clock. The one-clock model has a simpler structure, although the epidemiological schedules $\beta$ and $\theta$ are more complex and are not monotonic. It is a challenge to develop submodels that can induce $\theta$-function as a result of in vivo dynamics.

In the two-compartment model, recovery is expressed as the loss of infectivity and acquired immunity, whereas waning immunity is expressed as an increase in susceptibility. Thus, it is reasonable to assume that there exist numbers $0 < \tau_1 \le \tau_2$ such that $\beta(\tau) = 0$ for $\tau > \tau_1$ and $\theta(\tau) = 0$ for $\tau < \tau_2$, and so the interval $[\tau_1, \tau_2]$ is the complete immune period. The basic reproduction number is then given by

$$R_0 = N \int_0^\infty \beta(\tau) e^{-\mu\tau} d\tau. \tag{8.50}$$

If we omit the initial data (by assuming that the initial time is $t = -\infty$), the model in (8.49) reduces to a system of renewal equations:

$$\begin{aligned} \lambda(t) &= \int_0^\infty \beta(\tau) e^{-\mu\tau - \int_0^\tau \lambda(t-\tau+\zeta)\theta(\zeta)d\zeta} b(t-\tau) d\tau, \\ b(t) &= \lambda(t) \int_0^\infty \left[ \mu N e^{-\mu\tau - \int_0^\tau \lambda(t-\tau+\zeta)d\zeta} + \theta(\tau) e^{-\mu\tau - \int_0^\tau \lambda(t-\tau+\zeta)\theta(\zeta)d\zeta} b(t-\tau) \right] d\tau, \end{aligned} \tag{8.51}$$

where $b(t) := i(t, 0)$ is the density of newly infected individuals. Again, we obtain a scalar nonlinear renewal equation for $\lambda$ by inserting the expression for $b(t)$ in (8.51) into the first equation for $\lambda(t)$. This point was first stressed by Breda et al. [8].

One problem with the S-I model concerns how a vaccinated population should be introduced. A simple solution is to introduce a vaccinated class $v(t, \tau)$, where $\tau$ denotes the time elapsed since vaccination, and its relative susceptibility schedule $\tilde{\theta}(\tau)$, which may be different from $\theta$. The limiting system (the fully vaccinated model) is then formulated as

$$\begin{aligned} \frac{\partial v(t,\tau)}{\partial t} + \frac{\partial v(t,\tau)}{\partial \tau} &= -(\mu + \lambda(t)\tilde{\theta}(\tau)) v(t,\tau), \\ \frac{\partial i(t,\tau)}{\partial t} + \frac{\partial i(t,\tau)}{\partial \tau} &= -(\mu + \lambda(t)\theta(\tau)) i(t,\tau), \\ v(t,0) &= \mu N, \\ i(t,0) &= \lambda(t) \int_0^\infty \left( \tilde{\theta}(\tau) v(t,\tau) + \theta(\tau) i(t,\tau) \right) d\tau. \end{aligned} \tag{8.52}$$

We thus obtain a model with variable susceptibility and infectivity.

**Exercise 8.2** How can we modify model (8.52) to incorporate the boosting effect?

It is clear that the effective reproduction number is given by

$$R_e = R_0 \int_0^\infty \tilde{\theta}(\tau) \mu e^{-\mu\tau} d\tau,$$

so the reinfection threshold is $1/\tilde{\theta}^* := \left(\int_0^\infty \tilde{\theta}(\tau)\mu e^{-\mu\tau}d\tau\right)^{-1}$. If $R_e = \tilde{\theta}^* R_0 > 1$, the disease can invade the fully vaccinated host:

**Proposition 8.13** *If $R_e > 1$, there exists at least one ESS for* (8.52).

*Proof* Let $\lambda^*$ be the force of infection at the steady state. Then, the steady state is calculated as

$$v^*(\tau) = \mu N e^{-\mu\tau - \lambda^* \int_0^\tau \tilde{\theta}(\sigma)d\sigma}, \quad i^*(\tau) = i^*(0)e^{-\mu\tau - \lambda^* \int_0^\tau \theta(\sigma)d\sigma}.$$

Inserting the above expressions into the boundary condition and using the relation

$$\lambda^* = i^*(0)\int_0^\infty \beta(\tau)e^{-\mu\tau - \lambda^* \int_0^\tau \theta(\sigma)d\sigma}d\tau,$$

we have

$$\begin{aligned} 1 = \lambda^* &\int_0^\infty \theta(\tau)e^{-\mu\tau - \lambda^* \int_0^\tau \theta(\sigma)d\sigma}d\tau \\ &+ \mu N \int_0^\infty \tilde{\theta}(\tau)e^{-\mu\tau - \lambda^* \int_0^\tau \tilde{\theta}(\sigma)d\sigma}d\tau \int_0^\infty \beta(\tau)e^{-\mu\tau - \lambda^* \int_0^\tau \theta(\sigma)d\sigma}d\tau =: f(\lambda^*). \end{aligned} \tag{8.53}$$

Then, $f(0) = R_e$ and $f(\infty) = 0$, so $f(\lambda^*) = 1$ has at least one positive root if $R_e > 1$, and this root gives the force of infection at the ESS. □

Equation (8.53) suggests that a backward bifurcation of endemic steady states could occur, at least if $R_e > R_0$; that is, if there is some enhancement of reproductivity after recovery from the primary infection.

Finally, let us extend the S-I model (8.49) to a chronological-age-structured model, which is essential for considering real-world applications.

$$\begin{aligned} &\frac{\partial s(t,a)}{\partial t} + \frac{\partial s(t,a)}{\partial \tau} = -(\mu(a) + \lambda(t))s(t,a), \\ &\frac{\partial i(t,\tau;a)}{\partial t} + \frac{\partial i(t,\tau;a)}{\partial \tau} = -\mu(a+\tau)i(t,\tau;a) - \theta(\tau)\lambda(t)i(t,\tau;a), \\ &s(t,0) = \int_0^\infty m(a)\left(s(t,a) + \int_0^a i(t,\tau;a-\tau)d\tau\right)da, \\ &i(t,0;a) = \lambda(t)\left(s(t,a) + \int_0^a \theta(\tau)i(t,\tau;a-\tau)d\tau\right), \end{aligned} \tag{8.54}$$

where $m(\cdot)$ is the age-specific birth rate, the force of infection is given by

$$\lambda(t) = \int_0^\infty \int_0^\infty \beta(\tau)i(t,\tau;a)d\tau da,$$

the variable $a$ denotes the chronological age at the time of infection, and $i(t,\tau;a)$ denotes the density of infected individuals with infection-age $\tau$ who are infected at age $a$; that is, the chronological age is $a+\tau$ and the equation for $i$ is an infection-cohort equation.

**Exercise 8.3** Using the joint distribution $i(t,\tau,a)$ (the density of infected individuals with infection-age $\tau$ and age $a$ at time $t$), rewrite the equation (8.54).

If we assume that the host population is in a demographic steady state, the boundary condition of $s(t,0)$ is replaced by $s(t,0)=b=1/\int_0^\infty \ell(x)dx$, where $b$ is the crude birth rate and $\ell(a)=\exp(-\int_0^a \mu(\sigma)d\sigma)$ is the demographic survival probability. By integrating along the characteristic line, we have

$$\begin{aligned} s(t,a) &= b\ell(a)e^{-\int_0^a \lambda(t-a+\sigma)d\sigma}, \\ i(t,\tau;a) &= i(t-\tau,0;a)\frac{\ell(a+\tau)}{\ell(a)}e^{-\int_0^\tau \theta(\sigma)\lambda(t-\tau+\sigma)d\sigma}. \end{aligned}$$

Let $B(t,a):=i(t,0;a)$ be the density of newly infected individuals. Then, we have

$$\begin{aligned} \lambda(t) &= \int_0^\infty\int_0^\infty \beta(\tau)B(t-\tau,a)\frac{\ell(a+\tau)}{\ell(a)}e^{-\int_0^\tau \theta(\sigma)\lambda(t-\tau+\sigma)d\sigma}d\tau da, \\ B(t,a) &= \lambda(t)\Big(b\ell(a)e^{-\int_0^a \lambda(t-a+\sigma)d\sigma} \\ &\quad + \int_0^a \theta(\tau)e^{-\int_0^\tau \theta(\sigma)\lambda(t-\tau+\sigma)d\sigma}\frac{\ell(a)}{\ell(a-\tau)}B(t-\tau,a-\tau)d\tau\Big), \end{aligned}$$

from which we can induce a nonlinear scalar renewal equation for $B$ or $\lambda$.

In the invasion phase, the age density of newly infected individuals satisfies the renewal equation

$$B(t,a)=N(a)\int_0^\infty\int_0^\infty \beta(\tau)\frac{\ell(\tau+\eta)}{\ell(\eta)}B(t-\tau,\eta)d\eta d\tau,$$

where $N(a):=b\ell(a)$ is the host steady state population. Thus, the next-generation operator $K$ is given by

$$(Kf)(a)=N(a)\int_0^\infty\int_0^\infty \beta(\tau)\frac{\ell(\tau+\eta)}{\ell(\eta)}f(\eta)d\eta d\tau, \quad f\in L^1(\mathbb{R}_+),$$

and the basic reproduction number is calculated as follows:

$$R_0=r(K)=\int_0^\infty\int_0^\infty \beta(\tau)\frac{\ell(\eta+\tau)}{\ell(\eta)}N(\eta)d\eta d\tau.$$

Although the detailed analysis of this age-dependent model remains an open problem, it suggests that incorporating the individual epidemiological history into the host population dynamics is an important means of developing more realistic epidemic models. In fact, we could consider the functions $\beta$ and $\theta$ to reflect the in vivo dynamics of a virus (or parasite), in which case the S-I formulation naturally expresses the continuous process of the development of infectivity and immunity. In the above setting, however, the reinfection resets the local time to zero, and so the memory of infectivity and immunity is lost. If the development of immunity after the secondary infection is different from that after the primary infection, our model should be extended to recognize the number of infections.

## 8.3 Reproductivity Enhancement: Examples

As we have shown above, subcritical endemic steady states are likely to exist if the enhancement of susceptibility is induced by reinfection. However, it would be more interesting to know whether there are realistic reinfection mechanisms that can lead to backward bifurcations without the direct susceptibility enhancement, because the recovered individuals are usually less susceptible than the virgin population.

### *8.3.1 Malaria*

In the original Kermack–McKendrick endemic model, reinfected individuals are not distinguished from primary infected individuals. Thus, a natural extension is to consider the epidemiological parameters for reinfected individuals to be different from those infected individuals who are produced from completely susceptible individuals by primary infection. In fact, Águas et al. [1] developed an age-structured population model for the dynamics of malaria transmission and observed that stable endemic steady states subcritically coexist with stable disease-free steady states. In their model, infections of completely susceptible individuals are clinical malaria cases, and recovery from clinical cases confers protection against clinical manifestation of the disease, but not against infection *per se*. A recovered individual can therefore be reinfected and develop a non-clinical form of malaria, which we refer to as *asymptomatic infection*.

According to the above consideration, we can extend the basic model of (8.21) to an asymptomatic transmission model. For simplicity, we again neglect the chronological-age structure. Let $i_1(t, \tau)$ be the density of infected individuals resulting from the infection of completely susceptible individuals, and let $\beta_1$, $\gamma_1$ be the associated transmission coefficient and recovery rate. Let $i_2(t, \tau)$ be the density of reinfected individuals, $\beta_2$ be the transmission coefficient of reinfected individuals, and $\gamma_2$ be the recovery rate of reinfected individuals. We can then rewrite the basic model in (8.21) as follows:

$$\begin{aligned}
\frac{\partial s(t,\tau)}{\partial t} + \frac{\partial s(t,\tau)}{\partial \tau} &= -\mu s(t,\tau) - \lambda(t)s(t,\tau),\\
\frac{\partial i_1(t,\tau)}{\partial t} + \frac{\partial i_1(t,\tau)}{\partial \tau} &= -(\mu + \gamma_1(\tau))i_1(t,\tau),\\
\frac{\partial i_2(t,\tau)}{\partial t} + \frac{\partial i_2(t,\tau)}{\partial \tau} &= -(\mu + \gamma_2(\tau))i_2(t,\tau),\\
\frac{\partial r(t,\tau)}{\partial t} + \frac{\partial r(t,\tau)}{\partial \tau} &= -\mu r(t,\tau) - \theta(\tau)\lambda(t)r(t,\tau),
\end{aligned} \tag{8.55}$$

where the boundary condition is given as

$$\begin{aligned}
s(t,0) &= \mu N,\\
i_1(t,0) &= \lambda(t)\int_0^\infty s(t,\tau)d\tau,\\
i_2(t,0) &= \lambda(t)\int_0^\infty \theta(\tau)r(t,\tau)d\tau,\\
r(t,0) &= \int_0^\infty \gamma_1(\tau)i_1(t,\tau)d\tau + \int_0^\infty \gamma_2(\tau)i_2(t,\tau)d\tau,\\
\lambda(t) &= \int_0^\infty \beta_1(\tau)i_1(t,\tau)d\tau + \int_0^\infty \beta_2(\tau)i_2(t,\tau)d\tau.
\end{aligned}$$

In the model (8.55), we disregard a return path from $r$-class to $s$-class by waning immunity, although it is a popular assumption in malaria models. However, if $\theta \to 1$ as $\tau \to \infty$, the asymptotic recovery of the complete susceptibility can be included in our modeling.

Again, we define

$$\langle \gamma_j, \Gamma_j \rangle := \int_0^\infty e^{-\mu\tau}\gamma_j(\tau)\Gamma_j(\tau)d\tau, \quad \langle \beta_j, \Gamma_j \rangle := \int_0^\infty e^{-\mu\tau}\beta_j(\tau)\Gamma_j(\tau)d\tau$$

for $j = 1, 2$. The basic reproduction number $R_0$ and the effective reproduction number $R_1$ of the fully immunized system[2] are then given by

$$R_0 = N\langle \beta_1, \Gamma_1 \rangle, \quad R_1 = \theta^* N\langle \beta_2, \Gamma_2 \rangle. \tag{8.56}$$

Let $\lambda^*$ be the force of infection at an ESS. Then, we have

$$i_2^*(0) = \frac{\mu N\lambda^*}{\mu + \lambda^*}\frac{\langle \gamma_1, \Gamma_1 \rangle \phi(\lambda^*)}{1 - \langle \gamma_2, \Gamma_2 \rangle \phi(\lambda^*)},$$

[2]The fully immunized system is the disease-free steady state of the system where all newborns are immunized, so it is composed of the recovered population $r^*(\tau) = \mu N e^{-\mu\tau}$, its effective size of susceptibility is $N\theta^*$ and the infection process is described by a renewal equation $i_2(t,0) = \theta^* N \int_0^\infty \beta_2(\tau)i_2(t,\tau)d\tau$.

where $\phi$ is a function defined by (8.46). Because

$$\lambda^* = \langle \beta_1, \Gamma_1 \rangle \frac{\mu N \lambda^*}{\mu + \lambda^*} + \langle \beta_2, \Gamma_2 \rangle i_2^*(0),$$

we have an equation satisfied by the force of infection at the ESS:

$$R(\lambda^*) = \frac{\mu}{\mu + \lambda^*} \left\{ R_0 + \frac{R_1}{\theta^*} \frac{\langle \gamma_1, \Gamma_1 \rangle \phi(\lambda^*)}{1 - \langle \gamma_2, \Gamma_2 \rangle \phi(\lambda^*)} \right\} = 1, \tag{8.57}$$

where $R(0) = R_0$. Using a similar argument as in the previous section, we have

**Proposition 8.14** *Endemic steady states backwardly bifurcate at $R_0 = 1$ if the following condition holds:*

$$\theta^* \frac{\langle \beta_2, \Gamma_2 \rangle}{\langle \beta_1, \Gamma_1 \rangle} \langle \gamma_1, \Gamma_1 \rangle > 1. \tag{8.58}$$

If the *net reproductivity* of asymptomatic cases given by $\langle \beta_2, \Gamma_2 \rangle$ is larger than that of clinical cases, it is possible to satisfy condition (8.58), even when $\theta^* \leq 1$ (no enhancement of susceptibility). This situation could occur if the duration of infection of asymptomatic cases is sufficiently longer than that of clinical cases. This is not unrealistic, because it is likely that asymptomatic infection will persist without clinical treatment.

Let us now calculate the effective reproduction number of the extended model (8.55). Suppose that $\varepsilon$ denotes the proportion of immunization at the disease-free steady state. Let $\zeta_j(t) := i_j(t, 0)$ be the density of newly infected individuals. Then, we obtain a system of renewal equations describing the disease invasion at the disease-free steady state:

$$\begin{aligned} \zeta_1(t) &= (1 - \varepsilon) N (\psi_1 * \zeta_1)(t) + (1 - \varepsilon) N (\psi_2 * \zeta_2), \\ \zeta_2(t) &= \varepsilon N \theta^* (\psi_1 * \zeta_1)(t) + \varepsilon N \theta^* (\psi_2 * \zeta_2), \end{aligned}$$

where $\psi_j(\tau) := e^{-\mu\tau} \beta_j(\tau) \Gamma_j(\tau)$. Therefore, the next generation matrix is given by

$$K = \begin{pmatrix} (1 - \varepsilon) N \langle \beta_1, \Gamma_1 \rangle & (1 - \varepsilon) N \langle \beta_2, \Gamma_2 \rangle \\ \varepsilon N \theta^* \langle \beta_1, \Gamma_1 \rangle & \varepsilon N \theta^* \langle \beta_2, \Gamma_2 \rangle \end{pmatrix},$$

and the effective (vaccine) reproduction number is given by its spectral radius:

$$\begin{aligned} \mathscr{R}(\varepsilon) &= (1 - \varepsilon) N \langle \beta_1, \Gamma_1 \rangle + \varepsilon N \theta^* \langle \beta_2, \Gamma_2 \rangle \\ &= (1 - \varepsilon) R_0 + \varepsilon R_1 \\ &= (1 - (1 - \sigma)\varepsilon) R_0, \end{aligned}$$

where $\sigma := R_1 / R_0$ and $1/\sigma$ is the reinfection threshold of $R_0$. Thus, the disease is uncontrollable if $R_0 > 1/\sigma$, and condition (8.58) becomes $\sigma \langle \gamma_1, \Gamma_1 \rangle > 1$. If $\sigma >$

1, which is necessary to satisfy condition (8.58), we have $\mathscr{R}(\varepsilon) > R_0$. Then, we conclude that vaccination can increase the reproduction number, although it decreases the reproduction number of clinical cases, if the reproductivity enhancement occurs.

### 8.3.2 Measles in a Vaccinated Population

Next, let us consider an epidemic model for measles with a fluctuating immunity level for vaccinees. In this model, we again assume that there are two infectious states. The host population is divided into five subpopulations: the completely susceptible population $s$, vaccinated population $v$, recovered population with complete immunity $r$, classical infectious population for measles $i_1$, and subclinical infectious population for measles $i_2$. Unlike the Kermack–McKendrick reinfection model, the recovered individuals are now assumed to have complete immunity and no susceptibility, and the vaccinated individuals are considered to have partial susceptibility (by waning of immunity) depending on the duration since vaccination. Following infection, a portion of the vaccinated individuals develop *subclinical infection*, and the immunity level of the remaining part is boosted to the level of newly vaccinated individuals. The vaccinated population is structured by the duration since vaccination. We can formulate the vaccine-induced subclinical infection model for measles as follows:

$$\begin{aligned}
\frac{\partial s(t,\tau)}{\partial t} + \frac{\partial s(t,\tau)}{\partial \tau} &= -\mu s(t,\tau) - \lambda(t) s(t,\tau), \\
\frac{\partial i_1(t,\tau)}{\partial t} + \frac{\partial i_1(t,\tau)}{\partial \tau} &= -(\mu + \gamma_1(\tau)) i_1(t,\tau), \\
\frac{\partial i_2(t,\tau)}{\partial t} + \frac{\partial i_2(t,\tau)}{\partial \tau} &= -(\mu + \gamma_2(\tau)) i_2(t,\tau), \\
\frac{\partial r(t,\tau)}{\partial t} + \frac{\partial r(t,\tau)}{\partial \tau} &= -\mu r(t,\tau), \\
\frac{\partial v(t,\tau)}{\partial t} + \frac{\partial v(t,\tau)}{\partial \tau} &= -\mu v(t,\tau) - \lambda(t)\theta(\tau) v(t,\tau),
\end{aligned} \tag{8.59}$$

where the boundary condition is given as

$$\begin{aligned}
s(t,0) &= (1-\varepsilon)\mu N, \\
i_1(t,0) &= \lambda(t) \int_0^\infty s(t,\tau) d\tau, \\
i_2(t,0) &= (1-\kappa)\lambda(t) \int_0^\infty \theta(\tau) v(t,\tau) d\tau, \\
r(t,0) &= \varepsilon p \mu N + \int_0^\infty \gamma_1(\tau) i_1(t,\tau) d\tau + \int_0^\infty \gamma_2(\tau) i_2(t,\tau) d\tau, \\
v(t,0) &= \varepsilon \mu N (1-p) + \kappa \lambda(t) \int_0^\infty \theta(\tau) v(t,\tau) d\tau,
\end{aligned}$$

the force of infection is given by

$$\lambda(t) = \int_0^\infty \beta_1(\tau) i_1(t,\tau) d\tau + \int_0^\infty \beta_2(\tau) i_2(t,\tau) d\tau,$$

$\varepsilon$ is the vaccination coverage of newborns, $p$ is the probability that vaccinated newborns develop complete immunity and $\kappa$ is the probability that the immunity level of vaccinated individuals is boosted by infection. Thus, $1-\kappa$ gives the probability that the infection of vaccinated individuals leads to subclinical infection. The boosting effect is expressed by "resetting" the local time to zero. Kishida [28] investigated a special case of model (8.59) in which $\gamma_j$ and $\beta_j$ are constants, and gave numerical examples in which multiple endemic steady states can exist.

Let $\lambda^*$ be the force of infection at an ESS. Again, it is easy to derive the following characteristic relation:

$$\begin{aligned} R(\lambda^*) :=& \frac{\mu N(1-\varepsilon)}{\mu+\lambda^*} \langle \beta_1, \Gamma_1 \rangle \\ &+ \frac{(1-\kappa)(1-p)\varepsilon\mu N \langle \beta_2, \Gamma_2 \rangle}{1-\kappa\phi(\lambda^*)} \int_0^\infty \theta(\tau) e^{-\mu\tau - \lambda^* \int_0^\tau \theta(\zeta) d\zeta} d\tau = 1. \end{aligned}$$

The effective reproduction number $R(0)$ is then:

$$R(0) = R_e = (1-\varepsilon) R_0 + \varepsilon R_1,$$

where $R_0$ is the basic reproduction number and $R_1$ is the effective reproduction number of the fully vaccinated system:

$$R_0 = N\langle \beta_1, \Gamma_1 \rangle, \quad R_1 = (1-\kappa)(1-p) N \theta^* \langle \beta_2, \Gamma_2 \rangle,$$

where $\theta^*$ is defined by (8.36).

If $p=1$ or $\kappa=1$, we have $R_1=0$ and the critical coverage of immunization is given by $\varepsilon^* = 1 - 1/R_0$. However, if $R_1 > 0$ and the reinfection threshold $\sigma = R_1/R_0$ is less than unity, the critical coverage of immunization is given by

$$\varepsilon^* = \frac{1}{1-\sigma}\left(1 - \frac{1}{R_0}\right) > 1 - \frac{1}{R_0},$$

which suggests that if we take into account subclinical infection, the coverage of immunization to eradicate the disease must be larger than the critical proportion of immunization calculated from the standard SIR model neglecting subclinical cases.

Because $\lim_{\lambda\to\infty} R(\lambda) = 0$, there exists at least one ESS if $R_e > 1$, although it is unclear whether this is unique. Observe that

$$R'(0) = -\frac{1}{\mu}(1-\varepsilon) R_0 + \varepsilon D R_1,$$

where the parameter $D$ is defined by

$$D = -\frac{1}{\theta^*}\int_0^\infty \mu\theta(\tau)e^{-\mu\tau}\int_0^\tau \theta(\zeta)d\zeta d\tau + \kappa\frac{\theta^*}{\mu}.$$

Because $D$ is independent of $\beta_j$ and could become positive, it is possible that $R'(0) > 0$ when $R(0) = R_e = 1$. In such a case, a backward bifurcation occurs at $R_e = 1$, and there exist subcritical endemic steady states.

**Proposition 8.15** *Endemic steady states backwardly bifurcate at $R_e = 1$ if and only if the following condition holds: $1 < (1 + \mu D\sigma)\varepsilon$.*

If $\theta$ is constant, $D = (\kappa - 1)\theta/\mu < 0$, and thus, the bifurcation at $R(0) = 1$ is supercritical. Moreover, observe that

$$\sigma = \frac{R_1}{R_0} = (1-\kappa)(1-p)\theta^*\frac{\langle\beta_2, \Gamma_2\rangle}{\langle\beta_1, \Gamma_1\rangle}.$$

Then, $D\sigma$ is a quadratic function of $\kappa$ and the inequality $1 < (1 + \mu D\sigma)\varepsilon$ does not hold if $\kappa = 1$ or $\kappa = 0$. Therefore, the existence of imperfect boosting is necessary to lead the backward bifurcation. The introduction of imperfect vaccination would make it difficult to eradicate measles, although it could decrease the number of clinical cases.

### 8.3.3 Tuberculosis

Following primary tuberculosis (TB) infection, only a small proportion of individuals develop active TB, and most people remain latent and are at risk of developing active TB by *exogenous reinfection* or by *endogenous reinfection* of latent bacilli. That is, for TB infection, it is not the recovered class but the exposed class to have the partial susceptibility and to be reinfected. Feng, Castillo-Chavez and Capurro [13] have formulated an ODE model for TB infection that incorporates exogenous reinfection and showed that a subcritical bifurcation can be induced by this reinfection mechanism.

Here, we extend this TB model to take into account the class-age structure. To focus on the effect of exogenous reinfection, instead of using the original model as a base, we formulate a class-age structured version of Roberts' simplified model [36] that the treated class and the disease-induced death rate are omitted and the linear force of infection is adopted:

$$\begin{aligned}
\frac{\partial s(t,\tau)}{\partial t}+\frac{\partial s(t,\tau)}{\partial \tau}&=-(\mu+\lambda(t))s(t,\tau),\\
\frac{\partial e(t,\tau)}{\partial t}+\frac{\partial e(t,\tau)}{\partial \tau}&=-(\mu+\kappa(\tau)+\theta(\tau)\lambda(t))e(t,\tau),\\
\frac{\partial i(t,\tau)}{\partial t}+\frac{\partial i(t,\tau)}{\partial \tau}&=-(\mu+\gamma(\tau))i(t,\tau),
\end{aligned} \tag{8.60}$$

where the boundary condition is given as

$$\begin{aligned}
s(t,0)&=\mu N,\\
e(t,0)&=\lambda(t)\int_0^\infty s(t,\tau)d\tau,\\
i(t,0)&=\int_0^\infty (\lambda(t)\theta(\tau)+\kappa(\tau))e(t,\tau)d\tau,\\
\lambda(t)&=\int_0^\infty \beta(\tau)i(t,\tau)d\tau,
\end{aligned}$$

where $e(t,\cdot)$ denotes the class-age distribution of the exposed (latent) individuals, $\kappa(\tau)$ is the endogenous reactivation rate of latent bacilli, $\gamma(\tau)$ is the removal rate (by treatment or by recovery), and $\theta(\tau)$ is the relative susceptibility of the exposed class such that $\lambda(t)\theta(\tau)$ gives the force of exogenous reactivation (reinfection).

Let $b(t) := e(t,0)$ be the density of newly infecteds. From (8.60), we have

$$b(t)=S(t)\int_0^\infty \beta(\tau)\Gamma(\tau)e^{-\mu\tau}i(t-\tau,0)d\tau,$$

where

$$S(t):=\int_0^\infty s(t,\tau)d\tau=\mu N\int_0^\infty e^{-\mu\tau-\int_0^\tau \lambda(\eta)d\eta}d\tau.$$

Observe that

$$i(t,0)=\int_0^\infty (\lambda(t)\theta(\tau)+\kappa(\tau))e^{-\mu\tau-\int_0^\tau[\lambda(\zeta)\theta(\zeta)+\kappa(\zeta)]d\zeta}b(t-\tau)d\tau.$$

Therefore, we obtain a renewal equation for $b$:

$$b(t)=S(t)\int_0^\infty dx\int_0^x \beta(\tau)\Gamma(\tau)e^{-\mu\tau}\pi(t-\tau,x-\tau)e^{-\mu(x-\tau)-\int_0^{x-\tau}\pi(t-x+\zeta,\zeta)d\zeta}d\tau b(t-x),$$

where $\pi(t,\tau) := \lambda(t)\theta(\tau)+\kappa(\tau)$.

Then, it is easy to see that the effective reproduction number corresponding to the force of infection $\lambda^*$ at the endemic steady state is given by

$$R(\lambda^*)=\frac{\mu N\langle\beta,\Gamma\rangle}{\mu+\lambda^*}\int_0^\infty (\kappa(\tau)+\lambda^*\theta(\tau))e^{-\mu\tau-\int_0^\tau(\kappa(x)+\lambda^*\theta(x))dx}d\tau,$$

where $R(0) = R_0$ is the basic reproduction number.

**Proposition 8.16** *A subcritical bifurcation at* $R_0 = 1$ *occurs if*

$$\int_0^\infty \left(\theta(\tau) - \kappa(\tau)\int_0^\tau \theta(\sigma)d\sigma\right) e^{-\mu\tau - \int_0^\tau \kappa(x)dx} d\tau > \int_0^\infty \kappa(\tau) e^{-\mu\tau - \int_0^\tau \kappa(x)dx} d\tau.$$

**Exercise 8.4** Prove the above proposition.

## 8.4 Chronological-Age-Dependent Reinfection Model

### *8.4.1 Basic Model*

We now extend the Kermack–McKendrick reinfection model to take into account the chronological-age structure of host individuals. Let $S(t, a)$ be the density of the susceptible population who have never been infected at time $t$ and chronological-age $a$. Let $i(t, \tau, a)$ be the density of the infected and infectious population at time $t$, age $a$ and infection-age $\tau$ and let $r(t, \tau, a)$ be the density of the recovered population at time $t$, age $a$, and recovery-age $\tau$ (the time elapsed since the last recovery). Let $m(a)$ and $\mu(a)$ denote the age-dependent fertility rate and the force of mortality at age $a$, respectively, and $\gamma(\tau)$ be the recovery rate at infection-age $\tau$.

We assume that the force of infection applied to the fully susceptible population (virgin population) is given by

$$\lambda(t) = \frac{1}{N(t)} \int_0^\infty \int_0^a \beta(\tau) i(t, \tau, a) d\tau da,$$

where $N(t)$ is the total host population size. The Kermack–McKendrick reinfection model can then be extended to a demographic age-structured model as follows:

$$\begin{aligned}
&\frac{\partial S(t,a)}{\partial t} + \frac{\partial S(t,a)}{\partial a} = -\mu(a)S(t,a) - \lambda(t)S(t,a),\\
&\frac{\partial i(t,\tau,a)}{\partial t} + \frac{\partial i(t,\tau,a)}{\partial \tau} + \frac{\partial i(t,\tau,a)}{\partial a} = -(\mu(a) + \gamma(\tau))i(t,\tau,a),\\
&\frac{\partial r(t,\tau,a)}{\partial t} + \frac{\partial r(t,\tau,a)}{\partial \tau} + \frac{\partial r(t,\tau,a)}{\partial a} = -\mu(a)r(t,\tau,a) - \theta(\tau)\lambda(t)r(t,\tau,a),\\
&S(t,0) = \int_0^\infty m(a)(S(t,a) + I(t,a) + R(t,a))da,\\
&i(t,0,a) = \lambda(t)\left[S(t,a) + \int_0^a \theta(\tau)r(t,\tau,a)d\tau\right],\\
&r(t,0,a) = \int_0^a \gamma(\tau)i(t,\tau,a)d\tau,
\end{aligned} \tag{8.61}$$

with initial data

$$S(0,a) = S_0(a), \quad i(0,\tau,a) = i_0(\tau,a), \quad r(0,\tau,a) = r_0(\tau,a),$$

where $I$ and $R$ denote the age-density functions of the infected and recovered populations aggregated with respect to a duration variable:

$$I(t,a) := \int_0^a i(t,\tau,a)d\tau, \quad R(t,a) := \int_0^a r(t,\tau,a)d\tau.$$

Let $P(t,a)$ be the age-density function of the host population given by $P(t,a) := S(t,a) + I(t,a) + R(t,a)$, so $N(t) = \int_0^\infty P(t,a)da$. In the following, we assume that the host population is in a demographic steady state, so the following condition holds:

$$\int_0^\infty m(a)\ell(a)da = 1,$$

where

$$\ell(a) = \exp\left(-\int_0^a \mu(\sigma)d\sigma\right)$$

is the demographic survival probability. Thus, we assume that there exists a constant $B > 0$ such that $P(t,a) = P^*(a) = B\ell(a)$ for all $t \geq 0$ and $N(t) = N^* = \int_0^\infty P^*(a)da = Be_0$, where $e_0 = \int_0^\infty \ell(a)da$ is the average life expectancy of host individuals.

Although we do not discuss the well-posedness of (8.61), it is notable that the basic system (8.61) can be reduced to a system of integral equations. In fact, the partial differential equations in (8.61) can be integrated along the characteristic lines:

$$\begin{aligned}
S(t,a) &= B_1(t-a)\ell(a)e^{-\int_0^a \lambda(t-a+\sigma)d\sigma},\\
i(t,\tau,a) &= B_2(t-\tau,a-\tau)\frac{\ell(a)}{\ell(a-\tau)}\Gamma(\tau),\\
r(t,\tau,a) &= B_3(t-\tau)\frac{\ell(a)}{\ell(a-\tau)}e^{-\int_0^\tau \lambda(t-\tau+\sigma)\theta(\sigma)d\sigma},
\end{aligned}$$

where $B_1(t) := S(t,0)$, $B_2(t,a) := i(t,0,a)$, and $B_3(t,a) := r(t,0,a)$. Note that the force of infection and the density of the newly recovered population are determined by the age-specific incidence rate $B_2$ as

$$\begin{aligned}
\lambda(t) &= \frac{1}{N(t)}\int_0^\infty \int_0^a \beta(\tau)\Gamma(\tau)\frac{\ell(a)}{\ell(a-\tau)}B_2(t-\tau,a-\tau)d\tau da,\\
B_3(t,a) &= \int_0^a \gamma(\tau)\Gamma(\tau)\frac{\ell(a)}{\ell(a-\tau)}B_2(t-\tau,a-\tau)d\tau.
\end{aligned}$$

Therefore, using the boundary conditions for $S(t, 0)$ and $i(t, 0, a)$ in (8.61), we can obtain a nonlinear system of renewal integral equations for $B_1$ and $B_2$, for which we can adopt a classical fixed point method to show the existence and uniqueness of a local solution. For a semigroup approach to (8.61), we can adopt the method described by Thieme [42] (see Chap. 10).

### 8.4.2 Invasion Problem and $R_0$

For the basic model (8.61), there exists a disease-free steady state $(S, i, r) = (P^*, 0, 0)$ in which the linearized equation for the infected population is given by

$$\frac{\partial \zeta(t,\tau,a)}{\partial t} + \frac{\partial \zeta(t,\tau,a)}{\partial \tau} + \frac{\partial \zeta(t,\tau,a)}{\partial a} = -(\mu(a) + \gamma(\tau))\zeta(t,\tau,a),$$
$$\zeta(t,0,a) = w^*(a)\int_0^\infty \int_0^a \beta(\tau)\zeta(t,\tau,a)d\tau da,$$

where $\zeta$ is the density of infected individuals in the initial invasion phase and $w^*(a) = \ell(a)/e_0$ is the age profile of the host population at the demographic steady state.

The above linearized equation for the infective population can be formulated as an abstract boundary value problem as follows:

$$\begin{aligned}&\frac{\partial \zeta(t,\tau,\cdot)}{\partial t} + \frac{\partial \zeta(t,\tau,\cdot)}{\partial \tau} = A\zeta(t,\tau,\cdot) - \gamma(\tau)\zeta(t,\tau,\cdot),\\ &\zeta(t,0,a) = w^*(a)\int_0^\infty \int_0^a \beta(\tau)\zeta(t,\tau,a)d\tau da,\end{aligned} \tag{8.62}$$

where $A$ is a linear operator defined by

$$(A\phi)(a) = -\frac{d\phi(a)}{da} - \mu(a)\phi(a),$$
$$\phi \in \mathscr{D}(A) = \{\phi \in L^1(0,\infty) : A\phi \in L^1, \phi(0) = 0\}.$$

Let $T(t) = e^{tA}$ be the strongly continuous semigroup generated by $A$. Then, we have an explicit expression

$$(T(t)\phi)(a) = \begin{cases} 0, & t - a > 0, \\ \frac{\ell(a)}{\ell(a-t)}\phi(a-t), & a - t > 0. \end{cases} \tag{8.63}$$

Integrating (8.62) along the characteristic line, it follows that

$$\zeta(t,\tau,a)=\begin{cases}\Gamma(\tau)(T(\tau)\zeta(t-\tau,0,\cdot))(a), & t-\tau>0,\\ \frac{\Gamma(\tau)}{\Gamma(\tau-t)}(T(t)\zeta_0(\tau-t,\cdot))(a), & \tau-t>0,\end{cases}$$

where $\zeta_0(\tau,a)=\zeta(0,\tau,a)$ is the initial data.

Let $E:=L^1((0,\infty)\times(0,\infty))$. Note that a biologically meaningful state space for the age–duration distributions is the subset of $E$ such that $E_0:=\{\psi\in E:\psi(\tau,a)=0,\text{ a.e. if }\tau\geq a\}$. From (8.63), it is evident that $\zeta(t,\cdot,\cdot)\in E_0$ if $\zeta_0\in E_0$.

Let $\xi(t,a):=\zeta(t,0,a)$. Inserting expression (8.63) into the definition

$$\xi(t,a)=w^*(a)\int_0^\infty d\tau\int_\tau^\infty\beta(\tau)\zeta(t,\tau,a)da,$$

we arrive at an abstract renewal equation in $L^1(0,\omega)$:

$$\xi(t)=G(t)+\int_0^t\Pi(\tau)\xi(t-\tau)ds,$$

where $\xi(t)=\xi(t,\cdot)\in L^1(0,\infty)$, $\Pi(\tau)$ is a positive linear operator on $L^1(0,\infty)$ defined by

$$(\Pi(\tau)\phi)(a):=w^*(a)\int_\tau^\infty\beta(\tau)\Gamma(\tau)(T(\tau)\phi)(\sigma)d\sigma,\quad\phi\in L^1(0,\infty)$$

and

$$G(t,a):=w^*(a)\int_t^\infty d\tau\int_\tau^\infty\beta(\tau)\frac{\Gamma(\tau)}{\Gamma(\tau-t)}(T(t)\zeta_0(\tau-t,\cdot))(\sigma)d\sigma.$$

Therefore, the next-generation operator $K$ on $L^1(0,\infty)$ is given as follows [22]:

$$\begin{aligned}(K\phi)(a)&=\int_0^\infty(\Pi(\tau)\phi)(a)d\tau\\&=w^*(a)\int_0^\infty d\tau\int_\tau^\infty\beta(\tau)\Gamma(\tau)(T(\tau)\phi)(\sigma)d\sigma\\&=w^*(a)\int_0^\infty d\tau\int_\tau^\infty\beta(\tau)\Gamma(\tau)\frac{\ell(\sigma)}{\ell(\sigma-\tau)}\phi(\sigma-\tau)d\sigma,\\&=w^*(a)\int_0^\infty\Psi(a)\phi(a)da,\end{aligned}$$

where

$$\Psi(a):=\int_0^\infty\beta(\tau)\Gamma(\tau)\frac{\ell(a+\tau)}{\ell(a)}d\tau,$$

is the total infectivity of infected individuals who are infected at age $a$.

In this case, the next-generation operator is a one-dimensional positive map, and its spectral radius can be easily calculated as

$$r(K) = R_0 = \int_0^\infty \Psi(a) w^*(a) da.$$

Although we omit the proof, the following statement is obvious:

**Proposition 8.17** *If $R_0 < 1$, the disease-free steady state is locally asymptotically stable, whereas if $R_0 > 1$, it is unstable. If $\theta(\tau) \le 1$ for all $\tau \ge 0$, the disease-free steady state is globally asymptotically stable if $R_0 < 1$.*

### 8.4.3 Endemic Steady States

Let $S^*(a), i(\tau, a)$, and $r^*(\tau, a)$ be the age–duration-density functions of susceptible, infected, and recovered individuals, respectively, at the ESS. Then, it is easy to obtain the following expressions:

$$\begin{aligned} S^*(a) &= P^*(a) e^{-\lambda^* a}, \\ i^*(\tau, a) &= \frac{\ell(a)}{\ell(a-\tau)} \Gamma(\tau) i^*(0, a-\tau), \\ r^*(\tau, a) &= \frac{\ell(a)}{\ell(a-\tau)} e^{-\lambda^* \int_0^\sigma \theta(\zeta) d\zeta} r^*(0, a-\tau). \end{aligned} \tag{8.64}$$

Let $b_1^*(a) := i^*(0, a)$ and $b_2^*(a) := r^*(0, a)$. Inserting expression (8.64) into the boundary conditions, we have

$$\begin{aligned} b_1^*(a) &= \lambda^* \left( P^*(a) e^{-\lambda^* a} + \int_0^a \theta(\tau) \frac{\ell(a)}{\ell(a-\tau)} e^{-\lambda^* \int_0^\tau \theta(\zeta) d\zeta} b_2^*(a-\tau) d\tau \right), \\ b_2^*(a) &= \int_0^a \gamma(\tau) \Gamma(\tau) \frac{\ell(a)}{\ell(a-\tau)} b_1^*(a-\tau) d\tau. \end{aligned} \tag{8.65}$$

From (8.65), the age density $b_1^*$ of newly infected individuals at the ESS is the unique solution of the renewal equation

$$b_1^*(a) = g(a, \lambda^*) + \int_0^a H(a, \eta, \lambda^*) b_1^*(a-\eta) d\eta, \tag{8.66}$$

where

$$\begin{aligned} g(a, \lambda^*) &:= P^*(a) \lambda^* e^{-\lambda^* a}, \\ H(a, \eta, \lambda^*) &:= \int_0^\eta \frac{\ell(a)}{\ell(a-\eta)} \lambda^* \theta(\tau) e^{-\lambda^* \int_0^\tau \theta(\zeta) d\zeta} \gamma(\eta-\tau) \Gamma(\eta-\tau) d\tau. \end{aligned}$$

Define the resolvent kernel $\mathscr{R}$ as the solution of the resolvent equation

$$\mathscr{R}(a, \eta, \lambda^*) = H(a, \eta, \lambda^*) + \int_0^{\eta} H(a, x, \lambda^*)\mathscr{R}(a - x, \eta - x, \lambda^*)dx. \quad (8.67)$$

The renewal equation (8.66) can be solved as follows:

$$b_1^*(a) = g(a, \lambda^*) + \int_0^a \mathscr{R}(a, x, \lambda^*)g(a - x, \lambda^*)dx. \quad (8.68)$$

From

$$\lambda^* = \frac{1}{N^*}\int_0^{\infty} \Psi(a)b_1^*(a)da = \frac{1}{P^*(a)}(Kb_1^*)(a), \quad (8.69)$$

we can write (8.68) as

$$b_1^* = (\tilde{w} + \mathscr{R} * \tilde{w})\langle \Psi, b_1^* \rangle = (K_e b_1^*)(a),$$

where

$$\tilde{w}(a) := w^*(a)e^{-\lambda^* a}, \quad \langle \Psi, b_1^* \rangle := \int_0^{\infty} \Psi(a)b_1^*(a)da,$$

and $*$ denotes the convolution operation in which $(f * g)(t) := \int_0^t f(\tau)g(t - \tau)d\tau$. The integral operator $K_e$ is the *effective next-generation operator* at the ESS [21] and $\tilde{w} + \mathscr{R} * \tilde{w}$ is its positive eigenvector. Therefore, the spectral radius of $K_e$ should be unity:

$$\langle \Psi, \tilde{w} + \mathscr{R} * \tilde{w} \rangle = 1,$$

which would imply that

$$\begin{aligned} \mathscr{F}(\lambda^*) : &= \int_0^{\infty} \Psi(a)w^*(a)e^{-\lambda^* a}da \\ &+ \int_0^{\infty} \Psi(a) \int_0^a \mathscr{R}(a, a - x, \lambda^*)w^*(x)e^{-\lambda^* x}dxda = 1. \end{aligned} \quad (8.70)$$

Conversely, if $\lambda^*$ is a positive root of (8.70),

$$b_1^* = \lambda^* N^*(\tilde{w} + \mathscr{R} * \tilde{w})$$

satisfies (8.68) and (8.69). Therefore, to show the existence of endemic steady states, it is sufficient to show that $\mathscr{F}(\lambda^*) = 1$ has a positive root.

**Proposition 8.18** *Suppose that* $\underline{\mu} := \inf_{a \geq 0} \mu(a) > 0$. *If* $R_0 > 1$, *there exists at least one ESS.*

*Proof* Because $\mathscr{F}(0) = R_0$, it is sufficient to show that $\limsup_{\lambda^* \to \infty} \mathscr{F}(\lambda^*) < 1$. Changing the order of integrals in (8.70), we have

$$\begin{aligned}\int_0^{\infty} \Psi(a) \int_0^a \mathscr{R}(a, a-x, \lambda^*) w^*(x) e^{-\lambda^* x} dx da \\ = \int_0^{\infty} dx \int_0^{\infty} \mathscr{R}(x+z, z) dz w^*(x) e^{-\lambda^* x}.\end{aligned} \tag{8.71}$$

From the resolvent equation (8.67), it follows that

$$\mathscr{R}(x+z, z) = H(x+z, z) + \int_0^z H(x+z, \zeta) \mathscr{R}(x+z-\zeta, z-\zeta) d\zeta. \tag{8.72}$$

Observe that we can obtain an estimate as

$$H(a, \eta, \lambda^*) \leq L(\eta) := e^{-\underline{\mu}\eta} (\phi_1 * \phi_2)(\eta), \tag{8.73}$$

where

$$\phi_1(\tau) := \lambda^* \theta(\tau) e^{-\lambda^* \int_0^{\tau} \theta(\zeta) d\zeta}, \quad \phi_2(\tau) := \gamma(\tau) \Gamma(\tau). \tag{8.74}$$

Let $G(z)$ be a solution of the renewal equation

$$G(z) = L(z) + \int_0^z L(\zeta) G(z-\zeta) d\zeta. \tag{8.75}$$

Then, it follows from (8.72) and (8.73) that $\mathscr{R}(x+z, z) \leq G(z)$ and

$$\int_0^{\infty} G(z) dz \leq \frac{u_1 u_2}{1 - u_1 u_2},$$

where $u_j := \int_0^{\infty} e^{-\underline{\mu}\tau} \phi_j(\tau) d\tau < 1$. Therefore,

$$\int_0^{\infty} \mathscr{R}(x+z, z) dz \leq \frac{u_1 u_2}{1 - u_1 u_2},$$

and so it follows from (8.71) that $\lim_{\lambda^* \to \infty} \mathscr{F}(\lambda^*) = 0$. That is, $\mathscr{F}(\lambda^*) = 1$ has at least one positive root if $R_0 > 1$. This completes the proof. □

The conditions under which the chronological-age-dependent reinfection model (8.61) has a unique ESS remains an open problem, as does the existence of multiple endemic steady states.

### 8.4.4 Prevalence and Total Infection Rate

Let $p_1(a)$ be the age-specific incidence rate and $p_2(a)$ be the age-specific recovery rate at the ESS. Then, we have

$$p_1(a) = \frac{b_1^*(a)}{P^*(a)}, \quad p_2(a) = \frac{g(a, \lambda^*)}{P^*(a)}.$$

From (8.65), we obtain the system of equations

$$\begin{aligned} p_1(a) &= q(a) + \int_0^a \phi_1(\tau) p_2(a-\tau) d\tau, \\ p_2(a) &= \int_0^a \phi_2(\tau) p_1(a-\tau) d\tau, \end{aligned} \tag{8.76}$$

where $q(a) := \lambda^* e^{-\lambda^* a}$ and $\phi_j$ are defined by (8.74). Then, $q(\cdot)$ gives the probability density that the first infection occurs, $\phi_1(\tau)$ is the probability density that an infection occurs for recovered individuals at recovery-age $\tau$, and $\phi_2(\tau)$ is the probability density of recovery for infected individuals at infection-age $\tau$. Thus, $p_1$ is the solution of the renewal equation

$$p_1(a) = q(a) + \int_0^a (\phi_1 * \phi_2)(\eta) p_1(a-\eta) d\eta.$$

This individual transition process between two states can be thought of as a semi-Markov process, similar to that which appeared in the marriage model of Sect. 2.3.2. Then, the total expected number of infections that a newborn will suffer during its entire life is given by

$$R(\lambda^*) = \int_0^\infty p_1(a) \ell(a) da.$$

Using the probability $p_1$, we have $i^*(0, a) = P^*(a) p_1(a)$. Then, the *age-specific prevalence* at the ESS is calculated as

$$\frac{I^*(a)}{P^*(a)} = \frac{1}{B\ell(a)} \int_0^\infty \frac{\ell(a)}{\ell(a-\tau)} \Gamma(\tau) i^*(0, a-\tau) d\tau = \int_0^a \Gamma(a) p_1(a-\tau) d\tau.$$

The ratio of recovered individuals at age $a$ is also given by

$$\frac{R^*(a)}{P^*(a)} = \frac{1}{P^*(a)} \int_0^\infty r^*(\tau, a) d\tau = \int_0^a e^{-\lambda^* \int_0^\tau \theta(\zeta) d\zeta} p_2(a-\tau) d\tau.$$

Let $P_1$ be the total number of infections (under no death) and $P_2$ be the total number of recoveries per capita at the ESS. Integrating (8.76), we obtain

$$P_1 = 1 + Q_1 P_2, \quad P_2 = Q_2 P_1,$$

where

$$Q_1 := \int_0^\infty \phi_1(\tau) d\tau = 1 - e^{-\lambda^* \int_0^\infty \theta(x) dx}$$

is the total probability of (re)infection for recovered individuals and

$$Q_2 := \int_0^\infty \phi_2(\tau)d\tau = 1 - \Gamma(\infty)$$

is the total probability of recovery for infected individuals. Therefore, if $Q_1 Q_2 < 1$, we have

$$P_1 = \frac{1}{1 - Q_1 Q_2}, \quad P_2 = \frac{Q_2}{1 - Q_1 Q_2},$$

which are increasing functions of $\lambda^*$.

### 8.4.5 Vaccination

Let $v(a)$ be the force of vaccination at age $a$. If we can identify the vaccinated status with the recovered status, the basic model with vaccination is formulated as follows:

$$\begin{aligned}
&\frac{\partial S(t,a)}{\partial t} + \frac{\partial S(t,a)}{\partial a} = -(v(a) + \mu(a) + \lambda(t))S(t,a), \\
&\frac{\partial i(t,\tau,a)}{\partial t} + \frac{\partial i(t,\tau,a)}{\partial \tau} + \frac{\partial i(t,\tau,a)}{\partial a} = -(\mu(a) + \gamma(\tau))i(t,\tau,a), \\
&\frac{\partial r(t,\tau,a)}{\partial t} + \frac{\partial r(t,\tau,a)}{\partial \tau} + \frac{\partial r(t,\tau,a)}{\partial a} = -\mu(a)r(t,\tau,a) - \theta(\tau)\lambda(t)r(t,\tau,a), \\
&S(t,0) = \int_0^\infty m(a)(S(t,a) + I(t,a) + R(t,a))da, \\
&i(t,0,a) = \lambda(t)\left[S(t,a) + \int_0^a \theta(\tau)r(t,\tau,a)d\tau\right], \\
&r(t,0,a) = v(a)S(t,a) + \int_0^a \gamma(\tau)i(t,\tau,a)d\tau.
\end{aligned} \tag{8.77}$$

Then, there exists a disease-free (immunized) steady state

$$S^*(a) = B\ell(a)V(a), \quad r^*(\tau,a) = P^*(a)W(a-\tau),$$

where $V(a) := e^{-\int_0^a v(\zeta)d\zeta}$ is the proportion of susceptible individuals that are not yet vaccinated at age $a$, and $W(a) := v(a)V(a)$ gives the probability density that vaccination occurs at age $a$.

The effective susceptible population density is then given by

$$S_e(a) = S^*(a) + \int_0^a \theta(\tau)r^*(\tau,a)d\tau = P^*(a)[V(a) + (\theta * W)(a)],$$

and the effective reproduction number is calculated as follows:

$$\mathscr{R}(v) = \int_0^\infty \Psi(a)\frac{S_e(a)}{N^*}da = \int_0^\infty \Psi(a)w^*(a)[V(a) + (\theta * W)(a)]da.$$

Suppose that susceptible individuals are vaccinated at age $a_0$ and the coverage proportion is $\varepsilon \in [0, 1]$. Let $W(a) = \varepsilon\delta(a - a_0)$ and $V(a) = 1 - \varepsilon H(a - a_0)$, where $\delta$ is the Dirac function and $H$ denotes the Heaviside function. We can calculate the effective reproduction number as

$$\begin{aligned}\mathscr{R}(\varepsilon) = &\int_0^{a_0} \Psi(a)w^*(a)da + (1-\varepsilon)\int_{a_0}^\infty \Psi(a)w^*(a)da \\ &+ \varepsilon \int_{a_0}^\infty \Psi(a)w^*(a)\theta(a-a_0)da,\end{aligned}$$

and the critical coverage of immunization $\varepsilon^*$ such that $\mathscr{R}(\varepsilon^*) = 1$ is calculated as

$$\varepsilon^* = \frac{R_0 - 1}{\int_{a_0}^\infty \Psi(a)w^*(a)(1 - \theta(a - a_0))da}.$$

In particular, if $a_0$ goes to zero, that is, newborns are vaccinated, we have

$$\mathscr{R}(\varepsilon) = (1-\varepsilon)R_0 + \varepsilon R_1,$$

where

$$R_1 := \int_0^\infty \Psi(a)w^*(a)\theta(a)da$$

is the effective reproduction number of individuals vaccinated at age zero. In this case, the reinfection threshold condition is given by $R_1 = 1$. Although a detailed analysis of the chronological-age-dependent reinfection model has not yet been conducted, the incorporation of individual epidemiological histories with the host demography is crucial to developing epidemic models that are more realistic.

## 8.5 Acquired Immunity Boosted by Exposure

Using the Kermack–McKendrick reinfection model, we can consider the situation in which the host immunity among recovered individuals decays with time, and reinfection can occur. Conversely, the immunity to some endemic diseases such as malaria appears to be sustained by continued exposure, which is called the *boosting* effect.

In a series of papers [4–6], Aron developed simple mathematical models for malaria by which we can take into account the boosting effect in SIRS epidemics. Aron assumes that the subclinically infected population, which is composed of recovered individuals, can contribute to the force of infection and that the recovered (asymptomatic) status is maintained by the boosting of immunity. In 1983, Aron developed a delay-differential equation model to explain the functional relationship between the rate of reversion from the immune class to the susceptible class and the force of infection at the endemic steady state [4]. This functional relationship shows that the boosting mechanism may explain the age-specific prevalence of acute malaria [6]. Aron also combined the functional relation with an SIRS epidemic model to discuss the existence of endemic steady states and the basic reproduction number [5].

In the following, we first show that Aron's delay-differential equation model can be formulated as an age-structured epidemic model in which the boosting mechanism is expressed by resetting the local time via boundary feedback. The functional relationship between the reversion rate and the force of infection is then naturally induced, and Aron's formula is obtained as a special case in which the probability density function for the loss of immunity is given by a delta function. Subsequently, we present a possible extension of the basic model that takes into account the chronological-age structure of the host population, which is necessary for examining the age-specific prevalence curve.

## 8.5.1 *Basic Model*

We assume that the host population is divided into the susceptible class $S(t)$, clinical (symptomatic) infectious class $I(t)$, and subclinical (asymptomatic) infectious class $r(t,\tau)$, where $\tau$ denotes the time that has elapsed since recovery from the symptomatic infectious class. For simplicity, we assume that there is no disease-induced death rate, so the total population size

$$N = \frac{b}{\mu} = S(t) + I(t) + \int_0^\infty r(t,\tau)d\tau$$

is constant, where $b$ is the crude birth rate and $\mu$ is the natural force of mortality of host individuals.

Let $q$ be the recovery rate from clinical infection, $\gamma(\tau)$ be the rate of loss of immunity at duration $\tau$ in the recovered class, and $\lambda(t)$ be the force of infection given by the standard incidence:

$$\lambda(t) = \frac{1}{N}\left(\beta_1 I(t) + \int_0^\infty \beta_2(\tau) r(t,\tau)d\tau\right), \tag{8.78}$$

where $\beta_1$ and $\beta_2(\tau)$ denote the transmission coefficients. That is, the immune individuals recover with subclinical infectivity.[3] It would be reasonable to assume that $0 \le \beta_2 \le \beta_1$ and $\beta_2$ is a monotone increasing function because of the decay of immunity, although the following arguments do not depend on these specific assumptions.

An age-structured model for the acquired immunity boosted by exposure can then be formulated as

$$\begin{aligned}
\frac{dS(t)}{dt} &= \mu N - \mu S(t) - \lambda(t)S(t) + \int_0^\infty \gamma(\tau)r(t,\tau)d\tau, \\
\frac{dI(t)}{dt} &= \lambda(t)S(t) - (\mu + q)I(t), \\
\frac{\partial r(t,\tau)}{\partial t} + \frac{\partial r(t,\tau)}{\partial \tau} &= -(\lambda(t) + \mu + \gamma(\tau))r(t,\tau) \\
r(t,0) &= qI(t) + \lambda(t)\int_0^\infty r(t,\tau)d\tau,
\end{aligned} \tag{8.79}$$

where the re-exposure of immune individuals, expressed by $\lambda(t)r(t,\tau)$, resets the local time $\tau$ to zero (*boosting effect*). The basic model (8.79) is an age-structured version of the ODE model by Aron [5].

*Remark 8.2* For simplicity, according to Aron's original setting, we assume that re-exposed individuals' immune status is the same as that of individuals who have just recovered from clinical infection, and the susceptibility of the subclinical individuals to the force of infection $\lambda$ is independent of the local time (*immune-age*) $\tau$. If we introduce a susceptibility schedule $\theta(\tau)$ for the subclinical class, the boundary condition in (8.79) is replaced by

$$r(t,0) = qI(t) + \lambda(t)\int_0^\infty \theta(\tau)r(t,\tau)d\tau.$$

### 8.5.2 Basic Reproduction Number

Let $(y(t), z(t,\tau))$ be a perturbation of the infected population from the disease-free steady state $(S, I, r) = (N, 0, 0)$. Then, the initial dynamics of the infected population are described by the linearized equation at the disease-free steady state:

[3] In the malaria model by Aguas et al. (Sect. 8.3.1), reinfected individuals is defined as the (secondary) infected class. In Aron's model, reinfection of recovered individuals is considered as the boosting of the immune status and formulated by the boundary feedback.

$$\begin{aligned}\frac{dy(t)}{dt} &= \lambda(t)N - (\mu+q)y(t),\\ \frac{\partial z(t,\tau)}{\partial t} + \frac{\partial z(t,\tau)}{\partial \tau} &= -(\mu+\gamma(\tau))z(t,\tau)\\ z(t,0) &= qy(t),\end{aligned} \tag{8.80}$$

where

$$\lambda(t) = \frac{1}{N}\left(\beta_1 y(t) + \int_0^\infty \beta_2(\tau)z(t,\tau)d\tau\right).$$

Using the variation-of-constants formula, we obtain

$$y(t) = e^{-(\mu+q)t}y(0) + \int_0^t e^{-(\mu+q)(t-\sigma)}N\lambda(\sigma)d\sigma, \quad t > 0. \tag{8.81}$$

Integrating the McKendrick equation in (8.80) along the characteristic line, we obtain

$$z(t,\tau) = \begin{cases} qy(t-\tau)e^{-\mu\tau}\Gamma(\tau), & t-\tau > 0,\\ z(0,\tau-t)\frac{\Gamma(\tau)}{\Gamma(\tau-t)}e^{-\mu t}, & \tau - t > 0,\end{cases} \tag{8.82}$$

where

$$\Gamma(\tau) := \exp\left(-\int_0^\tau \gamma(x)dx\right).$$

Let $v(t) := N\lambda(t)$ be the density of newly infected individuals. Then, it follows that

$$\begin{aligned} v(t) &= \beta_1 y(t) + \int_0^\infty \beta_2(\tau)z(t,\tau)d\tau\\ &= g(t) + \beta_1\int_0^t e^{-(\mu+q)(t-\sigma)}v(\sigma)d\sigma\\ &\quad + \int_0^t qe^{-\mu\tau}\Gamma(\tau)\beta_2(\tau)\int_0^{t-\tau} e^{-(\mu+q)(t-\tau-\sigma)}v(\sigma)d\sigma d\tau,\end{aligned}$$

where

$$g(t) := e^{-(\mu+q)t}y(0) + \int_t^\infty \beta_2(\tau)\frac{\Gamma(\tau)}{\Gamma(\tau-t)}e^{-\mu t}z(0,\tau-t)d\tau$$

is given by the initial data $(y(0), z(0,\tau))$.

Observe that

$$\begin{aligned}&\int_0^t qe^{-\mu\tau}\Gamma(\tau)\beta_2(\tau)\int_0^{t-\tau} e^{-(\mu+q)(t-\tau-\sigma)}v(\sigma)d\sigma d\tau\\ &\quad = \int_0^t v(t-\tau)d\tau\int_0^\tau qe^{-\mu\sigma}\beta_2(\sigma)\Gamma(\sigma)e^{-(\mu+q)(\tau-\sigma)}d\sigma.\end{aligned}$$

Therefore, we arrive at the renewal integral equation

$$v(t) = g(t) + \int_0^t \Phi(\tau)v(t-\tau)d\tau, \tag{8.83}$$

where

$$\Phi(\tau) := \beta_1 e^{-(\mu+q)\tau} + \int_0^\tau q\beta_2(\sigma)e^{-\mu\sigma}\Gamma(\sigma)e^{-(\mu+q)(\tau-\sigma)}d\sigma.$$

Hence, we can calculate the basic reproduction number as

$$R_0 = \int_0^\infty \Phi(\tau)d\tau = \frac{\beta_1}{\mu+q} + \frac{q}{\mu+q}\int_0^\infty e^{-\mu\sigma}\beta_2(\sigma)\Gamma(\sigma)d\sigma, \tag{8.84}$$

and we conclude the following:

**Proposition 8.19** *If $R_0 < 1$, the disease-free steady state is locally stable, whereas it is unstable if $R_0 > 1$.*

### 8.5.3 Endemic Steady State

Consider the endemic steady states. Let $\lambda^*$ be the force of infection at the steady state and $(S^*, I^*, r^*(\tau))$ be the ESS. Then, it is clear that

$$\begin{aligned} S^* &= \frac{1}{\mu+\lambda^*}(b + \langle\gamma, r^*\rangle), \\ I^* &= \frac{\lambda^*}{(\mu+\lambda^*)(\mu+q)}(b + \langle\gamma, r^*\rangle), \\ r^*(\tau) &= \frac{qI^*e^{-(\mu+\lambda^*)\tau}\Gamma(\tau)}{1-\lambda^*\int_0^\infty e^{-(\mu+\lambda^*)\tau}\Gamma(\tau)d\tau}, \end{aligned} \tag{8.85}$$

where $\langle f, g\rangle := \int_0^\infty f(x)g(x)dx$.

Inserting the above expressions into the definition of $\lambda^*$, we have

$$\begin{aligned} \lambda^* &= \frac{1}{N}(\beta_1 I^* + \langle\beta_2, r^*\rangle) \\ &= \frac{1}{N}I^*F(\lambda^*) \\ &= \frac{\lambda^*}{(\mu+\lambda^*)(\mu+q)}\left[\frac{bF(\lambda^*)}{N} + \frac{F(\lambda^*)}{N}\frac{qI^*\int_0^\infty e^{-(\mu+\lambda^*)\tau}\gamma(\tau)\Gamma(\tau)d\tau}{1-\lambda^*\int_0^\infty e^{-(\mu+\lambda^*)\tau}\Gamma(\tau)d\tau}\right], \\ &= \frac{\lambda^*}{(\mu+q)(\mu+\lambda^*)}\left[\mu F(\lambda^*) + \frac{\lambda^* q\int_0^\infty e^{-(\mu+\lambda^*)\tau}\gamma(\tau)\Gamma(\tau)d\tau}{1-\lambda^*\int_0^\infty e^{-(\mu+\lambda^*)\tau}\Gamma(\tau)d\tau}\right], \end{aligned}$$

where

$$F(\lambda^*) := \beta_1 + \frac{q\int_0^\infty e^{-(\mu+\lambda^*)\tau}\beta_2(\tau)\Gamma(\tau)d\tau}{1-\lambda^*\int_0^\infty e^{-(\mu+\lambda^*)\tau}\Gamma(\tau)d\tau}.$$

Therefore, the force of infection at the ESS is given by positive solutions of the following equation:

$$H(\lambda^*) = 1,$$

where

$$H(\lambda^*) := \frac{1}{(\mu+q)(\mu+\lambda^*)}\left[\mu F(\lambda^*) + \frac{\lambda^* q\int_0^\infty e^{-(\mu+\lambda^*)\tau}\gamma(\tau)\Gamma(\tau)d\tau}{1-\lambda^*\int_0^\infty e^{-(\mu+\lambda^*)\tau}\Gamma(\tau)d\tau}\right].$$

**Proposition 8.20** *If $R_0 > 1$, there exists at least one ESS.*

*Proof* Because $H(0) = R_0$, it is sufficient to show that $\limsup_{x\to\infty} H(x) < 1$. Define $\underline{\gamma} := \inf_{\tau\geq 0}\gamma(\tau)$. Observe that

$$\int_0^\infty e^{-(\mu+\lambda^*)\tau}\Gamma(\tau)d\tau \leq \frac{1}{\mu+\lambda^*+\underline{\gamma}},$$

from which we know that $F$ is upper bounded:

$$F(x) \leq \beta_1 + \frac{q\sup\beta_2}{\mu+\underline{\gamma}}.$$

Let

$$\phi(x) := \frac{\int_0^\infty e^{-(\mu+x)\tau}\gamma(\tau)\Gamma(\tau)d\tau}{1-x\int_0^\infty e^{-(\mu+x)\tau}\Gamma(\tau)d\tau}.$$

Then, it follows that

$$\limsup_{x\to\infty} H(x) = \frac{q}{\mu+q}\limsup_{x\to\infty}\phi(x),$$

and it is sufficient to show $\limsup_{x\to\infty}\phi(x) \leq 1$. From the partial integral, however, it is easy to see that

$$\phi(x) = \frac{1-(\mu+x)\int_0^\infty e^{-(\mu+x)\tau}\Gamma(\tau)d\tau}{1-x\int_0^\infty e^{-(\mu+x)\tau}\Gamma(\tau)d\tau} \leq 1.$$

Therefore, the equation $H(x) = 1$ has at least one positive solution for $x \in [0,\infty)$ if $R_0 > 1$. □

Although it is still unclear whether multiple endemic steady states exist, Yukihiko Nakata[4] proved the unique existence of the endemic steady state for the case that $\beta_1 > \sup \beta_2$. If $\beta_2$ and $\gamma$ are age-independent, the disease-free steady state is globally asymptotically stable if $R_0 < 1$, whereas there exists a unique ESS that is globally stable if $R_0 > 1$ [40].

**Exercise 8.5** If $H(0) = R_0 = 1$ and $H'(0) > 0$, we can expect a backward bifurcation to occur at $R_0 = 1$ and subcritical endemic steady states to exist. Show that this situation does not occur if the following holds:

$$\int_0^\infty e^{-\mu\tau}\Gamma(\tau)d\tau \leq \int_0^\infty \tau \left[ \frac{e^{-\mu\tau}\beta_2(\tau)\Gamma(\tau)}{\int_0^\infty e^{-\mu\zeta}\beta_2(\zeta)\Gamma(\zeta)d\zeta} \right] d\tau.$$

This condition is satisfied if $\beta_2$ and $\gamma$ are constant; however, for more general case, it is difficult to hold biologically, why?

Let $p(\lambda^*)$ be the *reversion rate* at the ESS with force of infection $\lambda^*$. Then, we have

$$p(\lambda^*) = \frac{\int_0^\infty \gamma(\tau) r^*(\tau)d\tau}{\int_0^\infty r^*(\tau)d\tau} = \frac{\int_0^\infty \gamma(\tau)\Gamma(\tau)e^{-(\mu+\lambda^*)\tau}d\tau}{\int_0^\infty \Gamma(\tau)e^{-(\mu+\lambda^*)\tau}d\tau}. \tag{8.86}$$

If we assume that $\gamma(\tau)\Gamma(\tau) = \delta(\tau - \tau_0)$ ($\delta(\cdot)$ is a delta function), it follows that

$$p(\lambda^*) = \frac{(\mu+\lambda^*)e^{-(\mu+\lambda^*)\tau_0}}{1 - e^{-(\mu+\lambda^*)\tau_0}}, \tag{8.87}$$

which is the functional relation between the force of infection and the reversion rate at the ESS given by Aron [4, 5]. The reversion rate (8.87) is a monotone decreasing function of the force of infection, so the immunity level of recovered individuals is maintained by exposure to infection.

Moreover, the prevalence $I^*/N$ is not necessarily monotone with respect to the force of infection $\lambda^*$. In fact, it follows from (8.85) that

$$\frac{I^*}{N} = \frac{\lambda^*(1 - \lambda^* \int_0^\infty e^{-(\mu+\lambda^*)\tau}\Gamma(\tau)d\tau)}{q + \mu + \lambda^*(1 - (\mu+\lambda^*)\int_0^\infty e^{-(\mu+\lambda^*)\tau}\Gamma(\tau)d\tau)}.$$

**Exercise 8.6** Calculate the prevalence $I^*/N$ as a function of $\lambda^*$.

That is, we have a counterintuitive conclusion that a decline in the force of infection does not necessarily imply a smaller prevalence of disease in the ESS.

---

[4]personal communication (2016).

**Exercise 8.7** Suppose that $\gamma(\tau)\Gamma(\tau) = \delta(\tau - \tau_0)$. Let $z(t) := \int_0^\infty r(t, \tau)d\tau$. Show that the following delay-differential equation holds:

$$\frac{dz}{dt} = qI(t) - \mu z(t) - [qI(t - \tau_0) + \lambda(t - \tau_0)z(t - \tau_0)]e^{-\mu\tau_0 - \int_{t-\tau_0}^{t} \lambda(\zeta)d\zeta}.$$

Compare the above equation with equation (2.1) of [4].

### 8.5.4 Age-Dependent Extension

In papers [4, 6], Aron considered a model for the age-specific prevalence at the endemic steady state, but did not formulate the basic dynamical system. A most striking feature of the age-specific pattern of prevalence (proportion symptomatically infected) in cross-sectional surveys is that the prevalence among adults is highest at intermediate force of infection, and hence, partial control could increase adult prevalence. To obtain the age-specific prevalence curve, however, the basic model must be extended to a chronological-age-dependent model.

Let $s(t, a)$ be the age density of susceptible individuals at time $t$, $i(t, a)$ be the age density of clinical infected individuals, and $r(t, \tau, a)$ be the duration-age density of subclinical infected individuals at time $t$ A simple chronological-age-dependent extension of (8.79) is then

$$\begin{aligned}
&\frac{\partial s(t,a)}{\partial t} + \frac{\partial s(t,a)}{\partial a} = -\mu(a)s(t,a) - \lambda(t)s(t,a) + \int_0^a \gamma(\tau)r(t,\tau,a)d\tau,\\
&\frac{\partial i(t,a)}{\partial t} + \frac{\partial i(t,a)}{\partial a} = \lambda(t)s(t,a) - (\mu(a) + q)i(t,a),\\
&\frac{\partial r(t,\tau,a)}{\partial t} + \frac{\partial r(t,\tau,a)}{\partial \tau} + \frac{\partial r(t,\tau,a)}{\partial a} = -(\lambda(t)\theta(\tau) + \gamma(\tau) + \mu(a))r(t,\tau,a)\\
&r(t,0,a) = qi(t,a) + \lambda(t)\int_0^a \theta(\tau)r(t,\tau,a)d\tau,
\end{aligned} \tag{8.88}$$

where $\mu(\cdot)$ is the age-specific mortality rate, $\theta(\tau)$ is the relative susceptibility of the subclinical individuals at immune-age $\tau$, and the initial condition is given as

$$s(t,0) = b, \quad i(t,0) = 0, \quad r(t,\tau,0) = 0, \forall\tau \geq 0$$

and

$$\lambda(t) = \beta_1 \int_0^\infty i(t,a)da + \int_0^\infty da \int_0^a \beta_2(\tau)r(t,\tau,a)d\tau,$$

where we omit the scale factor $1/N$ for simplicity and the direct recovery rate from $i$-class to $s$-class. Of course, if we introduce the chronological-age-dependent parameters $\beta_j$ and $\gamma$, the force of infection depends on the chronological age, so the model can cover more general situations.

The linearized system at the disease-free steady state can then be formulated as follows:

$$\begin{aligned}
&\frac{\partial y(t,a)}{\partial t} + \frac{\partial y(t,a)}{\partial a} = \lambda(t)n(a) - (\mu(a)+q)y(t,a),\\
&\frac{\partial z(t,\tau,a)}{\partial t} + \frac{\partial z(t,\tau,a)}{\partial \tau} + \frac{\partial z(t,\tau,a)}{\partial a} = -(\mu(a)+\gamma(\tau))z(t,\tau,a)\\
&z(t,0,a) = qy(t,a),\\
&\lambda(t) = \beta_1 \int_0^\infty y(t,a)da + \int_0^\infty da \int_0^a \beta_2(\tau)z(t,\tau,a)d\tau,
\end{aligned} \tag{8.89}$$

where $n(a) := b\ell(a)$ denotes the host stationary age distribution.

Let $v(t,a) := \lambda(t)n(a)$ be the age density of newly infected individuals. For simplicity, if we assume that the epidemic starts at $t = -\infty$, it follows that

$$\begin{aligned}
y(t,a) &= \int_0^a \frac{\ell(a)}{\ell(\sigma)} e^{-q(a-\sigma)} v(t-a+\sigma,\sigma)d\sigma,\\
z(t,\tau,a) &= qy(t-\tau,a-\tau)\frac{\ell(a)}{\ell(a-\tau)}\Gamma(\tau).
\end{aligned} \tag{8.90}$$

Therefore, the limiting equation for the initial invasion phase is given by a homogeneous Volterra integral equation as

$$\begin{aligned}
v(t,a) &= n(a)\beta_1 \int_0^\infty d\eta \int_0^\eta \frac{\ell(\eta)}{\ell(\eta-\sigma)} e^{-q\sigma} v(t-\sigma,\eta-\sigma)d\sigma\\
&+ n(a)q \int_0^\infty d\eta \int_0^\eta d\tau \beta_2(\tau)\frac{\ell(\eta)}{\ell(\eta-\tau)}\Gamma(\tau)\int_\tau^\eta \frac{\ell(\eta-\tau)}{\ell(\eta-z)} e^{-q(z-\tau)} v(t-z,\eta-z)dz.
\end{aligned} \tag{8.91}$$

The integral kernel (net reproduction operator) $\Psi(\cdot)$ acting on $L^1$ is then defined as follows:

$$\begin{aligned}
(\Psi(z)\phi)(a) &= n(a)\beta_1 \int_z^\infty \frac{\ell(\eta)}{\ell(\eta-z)} e^{-qz}\phi(\eta-z)d\eta\\
&+ n(a)q \int_z^\infty d\eta \int_0^z d\tau \beta_2(\tau)\Gamma(\tau)\frac{\ell(\eta)}{\ell(\eta-z)} e^{-q(z-\tau)}\phi(\eta-z),
\end{aligned} \tag{8.92}$$

by which (8.91) can be written as an abstract convolution equation in $L^1$:

$$v(t) = \int_0^\infty \Psi(z)v(t-z)dz. \tag{8.93}$$

Because the range of the next-generation operator $K = \int_0^\infty \Psi(\sigma)d\sigma$ is a one-dimensional space spanned by $n$, it is easy to calculate the basic reproduction number as its spectral radius:

$$
\begin{aligned}
R_0 = r(K) = b\beta_1 \int_0^\infty dz \int_z^\infty \ell(\eta)e^{-qz}d\eta \\
+ qb \int_0^\infty dz \int_z^\infty d\eta \int_0^z \beta_2(\tau)\Gamma(\tau)\ell(\eta)e^{-q(z-\tau)}d\tau.
\end{aligned}
\tag{8.94}
$$

Next let us consider the endemic steady state. Let $s^*(a)$, $i^*(a)$ and $r^*(\tau, a)$ be the age density functions at the ESS. It follows that

$$
\begin{aligned}
&\frac{ds^*(a)}{da} = -\mu(a)s^*(a) - \lambda^* s^*(a) + \int_0^a \gamma(\tau)r^*(\tau, a)d\tau, \\
&\frac{di^*(a)}{da} = \lambda^* s^*(a) - (\mu(a) + q)i^*(a), \\
&\frac{\partial r^*(\tau, a)}{\partial \tau} + \frac{\partial r^*(\tau, a)}{\partial a} = -(\lambda^*\theta(\tau) + \mu(a) + \gamma(\tau))r^*(\tau, a) \\
&r^*(0, a) = qi^*(a) + \lambda^* \int_0^a \theta(\tau)r^*(\tau, a)d\tau.
\end{aligned}
\tag{8.95}
$$

Therefore, we obtain

$$
\begin{aligned}
s^*(a) &= b\ell(a)e^{-\lambda^* a} + \int_0^a \frac{\ell(a)}{\ell(\sigma)}e^{-\lambda^*(a-\sigma)} \int_0^\sigma \gamma(\tau)r^*(\tau, \sigma)d\tau d\sigma, \\
i^*(a) &= \int_0^a \frac{\ell(a)}{\ell(\sigma)}e^{-q(a-\sigma)}\lambda^* s^*(\sigma)d\sigma, \\
r^*(\tau, a) &= r^*(0, a-\tau)\frac{\ell(a)}{\ell(a-\tau)}\Gamma(\tau)e^{-\lambda^* \int_0^\tau \theta(\zeta)d\zeta}.
\end{aligned}
\tag{8.96}
$$

We then have an integral equation for the boundary value $r^*(0, a)$:

$$
r^*(0, a) = qi^*(a) + \lambda^* \int_0^a \theta(\tau)r^*(0, a-\tau)\frac{\ell(a)}{\ell(a-\tau)}\Gamma(\tau)e^{-\lambda^* \int_0^\tau \theta(\zeta)d\zeta}d\tau,
\tag{8.97}
$$

and

$$
\lambda^* = \beta_1 \int_0^\infty i^*(a)da + \int_0^\infty da \int_0^a \beta_2(\tau)r^*(\tau, a)d\tau.
\tag{8.98}
$$

From (8.96), (8.97), and (8.98), we arrive at a system of equations for $r^*(0, a)$ and $\lambda^*$. Then, we can prove that:

**Proposition 8.21** *If $R_0 > 1$, there exists at least one endemic steady state.*

*Proof* Let $u(a) := r^*(0, a)$. It follows from (8.97) that

$$
\begin{aligned}
u(a) :&= qi^*(a) + \lambda^* \int_0^a \theta(\tau) u(a-\tau) \frac{\ell(a)}{\ell(a-\tau)} \Gamma(\tau) e^{-\lambda^* \int_0^\tau \theta(\zeta) d\zeta} d\tau \\
&= q \int_0^a \frac{\ell(a)}{\ell(\sigma)} e^{-q(a-\sigma)} \lambda^* \left[ b\ell(\sigma) e^{-\lambda^* \sigma} + \int_0^\sigma \frac{\ell(\sigma)}{\ell(\zeta)} e^{-\lambda^*(\sigma-\zeta)} \right. \\
&\quad \times \left. \int_0^\zeta \gamma(\tau) u(\zeta-\tau) \frac{\ell(\zeta)}{\ell(\zeta-\tau)} \Gamma(\tau) e^{-\lambda^* \int_0^\tau \theta(z) dz} d\tau d\zeta \right] d\sigma \\
&\quad + \lambda^* \int_0^a \theta(\tau) u(a-\tau) \frac{\ell(a)}{\ell(a-\tau)} \Gamma(\tau) e^{-\lambda^* \int_0^\tau \theta(\zeta) d\zeta} d\tau \\
&=: \Phi_1(u, \lambda^*).
\end{aligned} \tag{8.99}
$$

Moreover, from (8.98), we obtain

$$
\begin{aligned}
\lambda^* &= \beta_1 \int_0^\infty i^*(a) da + \int_0^\infty da \int_0^a \beta_2(\tau) r^*(\tau, a) d\tau \\
&= \beta_1 \int_0^\infty \int_0^a \frac{\ell(a)}{\ell(\sigma)} e^{-q(a-\sigma)} \lambda^* s^*(\sigma) d\sigma da \\
&\quad + \int_0^\infty da \int_0^a \beta_2(\tau) u(a-\tau) \frac{\ell(a)}{\ell(a-\tau)} \Gamma(\tau) e^{-\lambda^* \int_0^\tau \theta(\zeta) d\zeta} d\tau \\
&= \beta_1 \int_0^\infty \int_0^a \frac{\ell(a)}{\ell(\sigma)} e^{-q(a-\sigma)} \lambda^* \left[ b\ell(\sigma) e^{-\lambda^* \sigma} + \int_0^\sigma \frac{\ell(\sigma)}{\ell(\zeta)} e^{-\lambda^*(\sigma-\zeta)} \right. \\
&\quad \times \left. \int_0^\zeta \gamma(\tau) u(\zeta-\tau) \frac{\ell(\zeta)}{\ell(\zeta-\tau)} \Gamma(\tau) e^{-\lambda^* \int_0^\tau \theta(z) dz} d\tau d\zeta \right] d\sigma da \\
&\quad + \int_0^\infty da \int_0^a \beta_2(\tau) u(a-\tau) \frac{\ell(a)}{\ell(a-\tau)} \Gamma(\tau) e^{-\lambda^* \int_0^\tau \theta(\zeta) d\zeta} d\tau \\
&=: \Phi_2(u, \lambda^*).
\end{aligned} \tag{8.100}
$$

Then, $u$ and $\lambda^*$ should be given as a fixed point of the nonlinear operator: $\Phi = (\Phi_1, \Phi_2)$ from $L^1_+(0, \infty) \times \mathbb{R}_+$ into itself. Since it is easy to see that $\Phi(0) = 0$, $\Phi$ is a positive compact operator and its derivative at infinity is zero. Let $\Phi'[0]$ be the Fréchet derivative of $\Phi$ at the origin. Then, we have

$$
\Phi'[0] \begin{pmatrix} u \\ \lambda \end{pmatrix} = \begin{pmatrix} \lambda b q \ell(a) \int_0^a e^{-q(a-\sigma)} d\sigma \\ \lambda b \beta_1 \int_0^\infty \ell(a) \int_0^a e^{-q(a-\sigma)} d\sigma da + \int_0^\infty da \int_0^a \beta_2(\tau) u(a-\tau) \frac{\ell(a)}{\ell(a-\tau)} \Gamma(\tau) d\tau \end{pmatrix}.
$$

Let $\rho = r(\Phi'[0])$ be the spectral radius of $\Phi'[0]$. Then, it is the positive eigenvalue corresponding to a positive eigenvector $(u, \lambda)$. Thus we have

$$
\begin{aligned}
\rho u &= \lambda b q \ell(a) \int_0^a e^{-q(a-\sigma)} d\sigma, \\
\rho \lambda &= \lambda b \beta_1 \int_0^\infty \ell(a) \int_0^a e^{-q(a-\sigma)} d\sigma da + \int_0^\infty da \int_0^a \beta_2(\tau) u(a-\tau) \frac{\ell(a)}{\ell(a-\tau)} \Gamma(\tau) d\tau,
\end{aligned}
$$

from which we obtain an equation:

$$\rho = b\beta_1 \int_0^\infty \ell(a) \int_0^a e^{-q(a-\sigma)} d\sigma da + \frac{bq}{\rho} \int_0^\infty da \int_0^a \beta_2(\tau)\ell(a)\Gamma(\tau) \int_\tau^a e^{-q(z-\tau)} dz d\tau,$$

Then, it is easy to see from (8.94) that $\rho > 1$ if and only if $R_0 > 1$. From Krasnoselskii's theorem (Proposition 10.32), we can conclude that there exists at least one endemic steady state if $R_0 > 1$. □

In order to calculate the age-dependent prevalence, let us define the *age-specific incidence rate* $p_1(a)$ and the *age-specific recovery rate* $p_2(a)$ at the endemic steady state:

$$p_1(a) := \frac{\lambda^* s^*(a)}{P^*(a)}, \quad p_2(a) = \frac{r^*(0,a)}{P^*(a)}.$$

From (8.96) and (8.97), it follows that

$$\begin{aligned} p_1(a) &= \phi_0(a) + \int_0^a \phi_0(a-\sigma) \int_0^\sigma \phi_1(\tau) p_2(\sigma - \tau) d\tau d\sigma, \\ p_2(a) &= \int_0^a \phi_3(a-\sigma) p_1(\sigma) d\sigma + \int_0^a \phi_2(\tau) p_2(a-\tau) d\tau, \end{aligned} \tag{8.101}$$

where the probability densities $\phi_j$ $(1 \le j \le 4)$ are given as

$$\begin{aligned} &\phi_0(a) := \lambda^* e^{-\lambda^* a}, \quad \phi_1(\tau) := \gamma(\tau)\Gamma(\tau) e^{-\lambda^* \int_0^\tau \theta(\zeta) d\zeta}, \\ &\phi_2(\tau) := \lambda^* \theta(\tau)\Gamma(\tau) e^{-\lambda^* \int_0^\tau \theta(\zeta) d\zeta}, \quad \phi_3(a) := q e^{-qa}. \end{aligned}$$

Then, we obtain a system of renewal equations:

$$\begin{pmatrix} p_1(a) \\ p_2(a) \end{pmatrix} = \begin{pmatrix} \phi_0(a) \\ 0 \end{pmatrix} + \int_0^a \begin{pmatrix} 0 & (\phi_0 * \phi_1)(\tau) \\ \phi_3(\tau) & \phi_2(\tau) \end{pmatrix} \begin{pmatrix} p_1(a-\tau) \\ p_2(a-\tau) \end{pmatrix} d\tau, \tag{8.102}$$

which can be solved by a standard iteration argument.

Let $P_1$ be the total number of infections and $P_2$ be the total number of recoveries during the entire life without termination by death at the ESS. Integrating (8.102), we obtain

$$P_1 = 1 + Q_1 P_2, \quad P_2 = P_1 + Q_2 P_2,$$

where

$$Q_1 := \int_0^\infty \phi_1(\tau) d\tau$$

is the total probability of reversion for recovered individuals and

$$Q_2 := \int_0^\infty \phi_2(\tau) d\tau$$

is the total probability of boosting for recovered individuals. Therefore, if $Q_1 + Q_2 < 1$, we have

$$P_1 = 1 + \frac{Q_1}{1 - Q_1 - Q_2}, \quad P_2 = \frac{1}{1 - Q_1 - Q_2},$$

which are increasing functions of $\lambda^*$.

Using the incidence rates, we obtain the age-specific prevalence as

$$\begin{aligned} \frac{i^*(a)}{P^*(a)} &= \int_0^a e^{-q(a-\sigma)} p_1(\sigma) d\sigma, \\ \frac{1}{P^*(a)} \int_0^a r^*(\tau, a) d\tau &= \int_0^a p_2(a-\tau) \Gamma(\tau) e^{-\lambda^* \int_0^\tau \theta(\zeta) d\zeta} d\tau. \end{aligned} \tag{8.103}$$

Note that the total expected number of infections that a newborn will suffer during its entire life is given by

$$R(\lambda^*) = \int_0^\infty p_1(a) \ell(a) da, \tag{8.104}$$

and the reversion rate is given by

$$p^*(a) := \frac{\int_0^a \gamma(\tau) r^*(\tau, a) d\tau}{\int_0^a r^*(\tau, a) d\tau} = \frac{\int_0^a \phi_2(\tau) e^{-\lambda^* \int_0^\tau \theta(\zeta) d\zeta} p_2(a-\tau) d\tau}{\int_0^a \Gamma(\tau) e^{-\lambda^* \int_0^\tau \theta(\zeta) d\zeta} p_2(a-\tau) d\tau}. \tag{8.105}$$

which is an age-dependent complex function of $\lambda^*$. It would be interesting to consider the response of $p^*$ and the age-dependent prevalence with respect to $\lambda^*$. We can conjecture that the reversion rate is a non-increasing function of the force of infection, and the prevalence is not monotone with respect to the force of infection, so a decline in the force of infection does not necessarily imply a smaller prevalence of disease in the ESS; hence, partial effective disease control may be harmful because reduction in transmission may lead to increased prevalence of illness. However, a detailed analysis of the age-dependent model in (8.88) is left as a challenge for future study.

## References

1. Águas, R., White, L.J., Snow, R.W., Gomes, M.G.M.: Prospects for malaria eradication in Sub-Saharan Africa. PLos ONE **3**(3), e1767 (2008)
2. Andreasen, V., Levin, S., Lin, J.: A model of influenza A drift evolution. Z. angew. Math. Mech. **76**(S2), 421–424 (1996)
3. Andreasen, V., Lin, J., Levin, S.A.: The dynamics of cocirculating influenza strains conferring partial cross-immunity. J. Math. Biol. **35**, 825–842 (1997)
4. Aron, J.L.: Dynamics of acquired immunity boosted by exposure to infection. Math. Biosci. **64**, 249–259 (1983)

5. Aron, J.L.: Acquired immunity dependent upon exposure in an SIRS epidemic model. Math. Biosci. **88**, 37–47 (1988)
6. Aron, J.L.: Mathematical modelling of immunity of malaria. Math. Biosci. **90**, 385–396 (1988)
7. Bonzi, B., Fall, A.A., Iggidr, A., Sallet, G.: Stability of differential susceptibility and infectivity epidemic models. J. Math. Biol. **62**(1), 39–64 (2011)
8. Breda, D., Diekmann, O., de Graaf, W.F., Pugliese, A., Vermiglio, R.: On the formulation of epidemic models (an appraisal of Kermack and McKendrick). J. Biol. Dyn. **6**(Suppl. 2), 103–117 (2012)
9. Diekmann, O., Montijn, R.: Prelude to Hopf bifurcation in an epidemic model: analysis of a characteristic equation associated with a nonlinear Volterra integral equation. J. Math. Biol. **14**, 117–127 (1982)
10. Diekmann, O., van Gils, S.A.: Invariant manifolds for Volterra integral equations of convolution type. J. Differ. Equ. **54**, 139–180 (1984)
11. Diekmann, O., Heesterbeak, J.A.P., Metz, J.A.J.: On the definition and the computation of the basic reproduction ratio $R_0$ in models for infectious diseases in heterogeneous populations. J. Math. Biol. **28**, 365–382 (1990)
12. Diekmann, O., Heesterbeek, J.A.P., Britton, T.: Mathematical Tools for Understanding Infectious Disease Dynamics. Princeton University Press, Princeton (2013)
13. Feng, Z., Castillo-Chavez, C., Capurro, A.F.: A model for tuberculosis with exogenous reinfection. Theor. Pop. Biol. **57**, 235–247 (2000)
14. Gomes, M.G., White, L.J., Medley, G.F.: Infection, reinfection, and vaccination under suboptimal immune protection: epidemiological perspectives. J. Theor. Biol. **228**, 539–549 (2004)
15. Gomes, M.G., White, L.J., Medley, G.F.: The reinfection threshold. J. Theor. Biol. **236**, 111–113 (2005)
16. Hastings, A.: A metapopulation model with population jumps of varying sizes. Math. Biosci. **128**, 285–298 (1995)
17. Hyman, J.M., Li, J.: Differential susceptibility epidemic models. J. Math. Biol. **50**, 626–644 (2005)
18. Inaba, H.: Mathematical analysis for an evolutionary epidemic model. In: Horn, M.A., Simonett, G., Webb, G.F. (eds.) Mathematical Models in Medical and Health Sciences, pp. 213–236. Vanderbilt University Press, Nashville (1998)
19. Inaba, H.: Kermack and McKendrick revisited: the variable susceptibility model for infectious diseases. Jpn. J. Indust. Appl. Math. **18**(2), 273–292 (2001)
20. Inaba, H.: Endemic threshold and stability in an evolutionary epidemic model. In: Castillo-Chaves, C., et al. (eds.) Mathematical Approaches for Emerging and Reemerging Infectious Diseases. The IMA Volumes in Mathematics and its Applications 126, pp. 337–359. Springer, Heidelberg (2002)
21. Inaba, H., Nishiura, H.: The basic reproduction number of an infectious disease in a stable population: the impact of population growth rate on the eradication threshold. Math. Model. Nat. Phenom. **3**(7), 194–228 (2008)
22. Inaba, H.: On a new perspective of the basic reproduction number in heterogeneous environments. J. Math. Biol. **65**, 309–348 (2012)
23. Inaba, H.: Endemic threshold analysis for the Kermack-McKendrick reinfection model. Josai Math. Monogr. **9**, 105–133 (2016)
24. Katriel, G.: Epidemics with partial immunity to reinfection. Math. Biosci. **228**, 153–159 (2010)
25. Kermack, W.O., McKendrick, A.G.: Contributions to the mathematical theory of epidemics I. In: Proceedings of the Royal Society, vol. 115A, pp. 700–721 (1927). (Reprinted in Bulletin of Mathematical Biology, **53**(1/2), 33–55 (1991))
26. Kermack, W.O., McKendrick, A.G.: Contributions to the mathematical theory of epidemics II. The problem of endemicity. In: Proceedings of the Royal Society, vol. 138A, pp. 55–83 (1932). (Reprinted in Bulletin of Mathematical Biology, **53**(1/2), 57–87 (1991))
27. Kermack, W.O., McKendrick, A.G.: Contributions to the mathematical theory of epidemics III. Further studies of the problem of endemicity. In: Proceedings of the Royal Society, vol. 141A, pp. 94–122 (1933). (Reprinted in Bulletin of Mathematical Biology, **53**(1/2), 89–118 (1991))

28. Kishida, M.: A Mathematical Model for Measles with Waning of Immunity, Boosting and Subclinical Infection, MA thesis, Graduate School of Mathematical Sciences, University of Tokyo (2010). (Japanese)
29. Magal, P., McCluskey, C.C., Webb, G.F.: Lyapunov functional and global asymptotic stability for an infection-age model. Appl. Anal. **89**(7), 1109–1140 (2010)
30. Magal, P., Ruan, S.: Sustained oscillations in an evolutionary epidemiological model of influenza A drift. Proc. Roy. Soc. A **466**, 965–992 (2010)
31. Martcheva, M.: An Introduction to Mathematical Epidemiology. Texts in Applied Mathematics 61. Springer, New York (2015)
32. Metz, J.A.J., Diekmann, O.: The Dynamics of Physiologically Structured Populations. Lecture Notes in Biomathematics, vol. 68. Springer, Berlin (1986)
33. Mossong, J., Nokes, D.J., Edmunds, W.J., Cox, M.J., Rathnam, S., Muller, C.P.: Modeling the impact of subclinical measles transmission in vaccinated populations with waning immunity. Am. J. Epidemiol. **150**(11), 1238–1249 (1999)
34. Nakata, Y., Enatsu, Y., Inaba, H., Kuniya, T., Muroya, Y., Takeuchi, Y.: Stability of epidemic models with waning immunity. SUT J. Math. **50**(2), 205–245 (2014)
35. Pease, C.M.: An evolutionary epidemiological mechanism, with applications to type A influenza. Theor. Poul. Biol. **31**, 422–452 (1987)
36. Roberts, M.G.: The pluses and minuses of $\mathscr{R}_0$. J. R. Soc. Interface **4**, 949–961 (2007)
37. Safan, M., Heesterbeek, H., Dietz, K.: The minimum effort required to eradicate infections in models with backward bifurcation. J. Math. Biol. **53**, 703–718 (2006)
38. Sakamoto, H.: An SIRS epidemic model with vaccination and decay of immunity, MA thesis, Graduate School of Mathematical Sciences, The University of Tokyo (2005)
39. Smith, H.L., Thieme, H.R.: Dynamical Systems and Population Persistence. Graduate Studies in Mathematics 118. American Mathematical Society, Providence (2011)
40. Tapaswi, P.K., Chattopadhyay, J.: Global stability results of a "susceptible–infective–immune–susceptible" (SIRS) epidemic model. Ecol. Model. **87**, 223–226 (1996)
41. Thieme, H.R.: Renewal theorems for some mathematical models in epidemiology. J. Integral Equ. **8**, 185–216 (1985)
42. Thieme, H.R.: Analysis of age-structured population models with additional structure. In: Arino, O., Axelrod, D.E., Kimmel, M. (eds.) Mathematical Population Dynamics, pp. 115–126. Marcel Dekker, New York (1991)
43. Thieme, H.R.: Uniform persistence and permanence for non-autonomous semiflows in population biology. Math. Biosci. **166**, 173–201 (2000)
44. Thieme, H.R., Yang, J.: An endemic model with variable re-infection rate and applications to influenza. Math. Biosci. **180**, 207–235 (2002)
45. Thieme, H.R.: Mathematics in Population Biology. Princeton University Press, Princeton (2003)
46. Thieme, H.R.: Distributed susceptibility; a challenge to persistence theory in infectious disease models. Disc. Cont. Dyn. Sys. Ser. B **12**(4), 865–882 (2009)
47. Yang, J.: An Evolutionary Epidemic Model with Application to Type A Influenza, Ph.D. thesis, Arizona State University (2000)

# Chapter 9
# Basic Reproduction Number $R_0$

**Abstract** The basic reproduction number $R_0$ plays a central role in structured population dynamics. Although some roots of $R_0$ can be traced back to the nineteenth century, the specific concept was introduced to the demography literature in 1925. It took a further half century for this number to mature as a key concept in mathematical epidemiology, and it is only recently that the stable population theory has become a popular tool in the field. However, the progress of mathematical epidemiology over the past two decades has been remarkable, and the basic concept and applications of $R_0$ are now better developed in epidemiology than in demography. In particular, the successful introduction of a general definition of the basic reproduction number for structured populations in the context of epidemic models gave rise to a new epoch in our understanding. Since then, the theory of the basic reproduction number has been developed as a central tenet of both infectious disease epidemiology and general population dynamics. Recently, this basic idea has evolved considerably to allow its application to time-heterogeneous environments. In this chapter, we sketch a general theory of $R_0$. First, we formulate a general definition for the basic reproduction number $R_0$ of structured populations in time-heterogeneous environments. Based on the generation evolution operator, we show that the basic reproduction number can be calculated as the spectral radius of the next-generation operator in a constant environment or in a periodic environment. Subsequently, we define the type-reproduction number in a time-heterogeneous environment and examine some applications in demography and epidemiology. Finally, we discuss some methods to estimate $R_0$ from available data.

## 9.1 Definition of $R_0$ in Heterogeneous Environments

We start our argument with some basic observations about the essential components of the basic reproduction number $R_0$ for structured populations.[1] For a given linear population system, the basic reproduction number is uniquely defined as long as we

[1]For the historical origin of $R_0$, readers are referred to [23, 30, 48, 49, 53, 54].



H. Inaba, *Age-Structured Population Dynamics in Demography and Epidemiology*, DOI 10.1007/978-981-10-0188-8_9

interpret it as the asymptotic per-generation growth factor of newly produced individuals (although there are many surrogate indices which share the threshold property with the basic reproduction number). Here, we state our theory based on the terminology of general structured population dynamics (that is, in a demographic setting). Readers can easily interpret the basic model as an epidemic model by considering childbearing to represent the reproduction of new infection.

Suppose that the host individuals are characterized by a variable $\zeta \in \Omega$, which is called the *h-state variable* ($h$ for heterogeneous). The set $\Omega \subset \mathbb{R}^n$ is the $h$-state space' (or $i$-state space, where $i$ stands for individual). Although, in general, $\Omega$ may be composed of continuous variables and discrete variables, for simplicity, we mainly treat the case in which the $h$-state variable is continuous.

Define $A(t, \tau, \zeta, \eta)$ as the expected number of newborns with $h$-state $\zeta$ produced per unit time at time $t$ by an individual who was born $\tau$ units of time ago at $h$-state $\eta$. If migration exists in the $h$-state, $A$ can be decomposed as

$$A(t, \tau, \zeta, \eta) = \int_{\Omega} \beta(t, \tau, \zeta, \eta') W(t, \tau, \eta' | t - \tau, 0, \eta) d\eta', \qquad (9.1)$$

where $\beta(t, \tau, \zeta, \eta)$ is the expected number of newborns at state $\zeta$ produced by individuals at state $\eta$, age $\tau$, and time $t$, and $W(t+h, a+h, \eta | t, a, \zeta)$ denotes the survival probability (Green's function, see Sect. 2.5) of individuals at time $t$, age $a$, and state $\zeta$ surviving to age $a + h$ and state $\eta$ after $h$ units of time.

Let $b(t, \zeta)$, $\zeta \in \Omega_b$, denote the density of newborns at time $t$, where $\Omega_b \subset \Omega$ is the set of *states-at-birth*, which are the $h$-states in which newborns can be produced. For a detailed argument about the state-at-birth (or state-at-infection), readers are referred to [20]. We assume that $b(t, \cdot) \in E_+ = L^1_+(\Omega_b)$, and we call $E_+$ the *b-state space* and $\Omega \setminus \Omega_b$ the *transient state*. If $\Omega_b$ is a finite discrete set, the corresponding $b$-state space is a finite-dimensional vector space.

The real-time development of newborns is then described by the renewal integral equation

$$b(t, \zeta) = g(t, \zeta) + \int_0^t \int_{\Omega_b} A(t, \tau, \zeta, \eta) b(t - \tau, \eta) d\eta d\tau, \quad t > 0, \qquad (9.2)$$

where $g(t, \zeta)$ is the density of newborns produced by the initial population.

Let $E_+ := L^1_+(\Omega_b)$ be the set of density distributions of newborns. Define a linear positive integral operator $\Psi(t, \tau)$ that leaves the cone $E_+$ invariant by

$$(\Psi(t, \tau) f)(\zeta) := \int_{\Omega_b} A(t, \tau, \zeta, \eta) f(\eta) d\eta, \quad f \in E_+.$$

Then, $\Psi(t, \tau)$, which we call the *net reproduction operator* (NRO), is an operator that maps the density (distribution) of newborns at time $t - \tau$ to the density of their children produced at $\tau$ time later.

**Exercise 9.1** Let us define the fertility operator $M$ and the survival operator $L$ as follows:

$$(M(t,\tau)\phi)(\zeta) := \int_{\Omega} \beta(t,\tau,\zeta,\eta)\phi(\eta)d\eta, \quad \phi \in L^1_+(\Omega),$$

$$(L(t,\tau;t-\tau,0)f)(\zeta) := \int_{\Omega_b} W(t,\tau,\zeta|t-\tau,0,\eta)f(\eta)d\eta, \quad f \in E_+.$$

Show that the NRO can be decomposed as $\Psi(t,\tau) = M(t,\tau)L(t,\tau;t-\tau,0)$ when we assume that (9.1) holds.

If we set $b(t) := b(t,\cdot) \in E$, so that $b(t)$ is interpreted as an $E$-valued function, (9.1) can be written as an abstract renewal equation in $E$:

$$b(t) = g(t) + \int_0^t \Psi(t,\tau)b(t-\tau)d\tau, \quad t > 0. \tag{9.3}$$

It is well known that, for any $t > 0$, the density of newborns at time $t$ is given by the *generation expansion*

$$b(t) = \sum_{m=0}^{\infty} b_m(t), \tag{9.4}$$

where $b_m(t)$ is given by the iterative process

$$b_0(t) = g(t), \quad b_m(t) = \int_0^t \Psi(t,\tau)b_{m-1}(t-\tau)d\tau, \quad m = 1,2,\dots \tag{9.5}$$

Thus, $b_m(t) \in E_+$ gives the density of the $m$th generation of newborns at time $t$, which we call the *generation distribution*.

In biological terms, it is reasonable to adopt an $L^1$-space as the state space of the generation distribution:

$$b_m \in Y_+ := L^1_+(\mathbb{R}_+;E) \cong L^1_+(\mathbb{R}_+ \times \Omega_b),$$

where $Y_+$ is the positive cone of the Banach lattice $Y := L^1(\mathbb{R}_+;E) \cong L^1(\mathbb{R}_+ \times \Omega_b)$ with norm defined by

$$|b_m|_Y := \int_0^\infty |b_m(t)|_E dt = \int_0^\infty \int_{\Omega_b} |b_m(t,\zeta)|d\zeta dt. \tag{9.6}$$

If we see the time variable $t$ as a kind of $h$-state variable, $\mathbb{R}_+ \times \Omega_b$ is the extended $i$-state space of newborns and $Y_+$ is called the *extended b-state space*. Then, $|b_m|_Y$ gives the total size of the $m$th generation (total number of newborns produced as the $m$th generation). In the following, we assume that the initial data is *non-trivial*, that is, $b_0 \in Y_+ \setminus \{0\}$.

The *generation evolution operator* (GEO) associated with the NRO $\Psi(t, \tau)$ is the positive integral operator acting on the extended $b$-state space $Y_+$ defined by

$$(K_Y f)(t) = \int_0^t \Psi(t, \tau) f(t - \tau) d\tau, \quad f \in Y_+. \tag{9.7}$$

The GEO $K_Y$ produces a birth genealogy $\{b_0, b_1, b_2, \ldots\} \subset Y_+$ via the iteration process $b_m = K_Y b_{m-1}$. Therefore, a simple idea is that the spectral radius $r(K_Y)$ of GEO would give the basic reproduction number. As we see below, this is true for constant or periodic environments. Note that $r(K_Y) < 1$ is always a sufficient condition for population extinction, because (9.3) can be solved as $b = (I - K_Y)^{-1} g$ in the extended state space $Y$, which implies that $\lim_{t \to \infty} \int_t^\infty |b(\zeta)|_E d\zeta = 0$.

Here, we assume that

$$\sup_{\tau \geq 0} \int_0^\infty \|\Psi(s + \tau, s)\|_{\mathscr{L}(E)} ds < \infty,$$

where $| \cdot |_{\mathscr{L}(E)}$ denotes the operator norm. Then, $K_Y$ is a bounded linear operator from $Y$ to itself that leaves the cone $Y_+$ invariant. Moreover, $K_Y$ is strictly positive if the *cohort net reproduction operator*

$$K_\tau : \phi \to \int_0^\infty \Psi(s + \tau, s) \phi ds, \quad \phi \in E$$

is strictly positive for almost all $\tau \geq 0$.

**Definition 9.1** For non-trivial initial data $b_0 \in Y_+ \setminus \{0\}$, the *asymptotic per-generation growth factor* for a birth genealogy $\{b_m\}_{m=0,1,2,\ldots}$ produced by the generation evolution operator $K_Y$ is defined by

$$\mathscr{R} := \lim_{m \to \infty} \sqrt[m]{|b_m|_Y} = \lim_{m \to \infty} \sqrt[m]{|K_Y^m b_0|_Y}, \tag{9.8}$$

if the limit exists. For the solution $b(t)$ of (9.3), if the limit

$$\lambda_0 := \lim_{t \to \infty} \frac{\log |b(t)|_E}{t} \tag{9.9}$$

exists, we call $\lambda_0$ the *Malthusian parameter* for $b(t)$. If $\mathscr{R}$ and $\lambda_0$ exist independently from the initial data $b_0$ and the following *sign relation* holds:

$$\text{sign}(\mathscr{R} - 1) = \text{sign}(\lambda_0), \tag{9.10}$$

then $\mathscr{R}$ is called the *basic reproduction number* of the renewal process (9.2) and is denoted by $R_0$.

From the above, we know that as long as $R_0$ exists, then *by definition* the generational interpretation holds; that is, $R_0$ gives the per-generation growth factor (the asymptotic ratio of the size of successive generations). Our problem is then to determine the conditions under which $R_0$ exists in the above sense, the Malthusian parameter $\lambda_0$ exists, and the sign relation holds between $R_0$ and $\lambda_0$.

*Remark 9.1* As shown in [40], if we define a reproduction number

$$\bar{R}_0 := \limsup_{m \to \infty} \sqrt[m]{|b_m|_Y},$$

then it is always defined and $\bar{R}_0 \leq r(K_Y)$. Under appropriate conditions for $\Psi$, we can prove the dichotomy that $\sum_{m=0}^{\infty} |b_m|_Y < +\infty$ and $\bar{\lambda}_0 \leq 0$ if $\bar{R}_0 < 1$ and $\sum_{m=0}^{\infty} |b_m|_Y = +\infty$ and $\bar{\lambda}_0 \geq 0$ if $\bar{R}_0 > 1$, where $\bar{\lambda}_0 := \limsup_{t \to \infty} (\log |b(t)|_E)/t$ is the *Lyapunov order number*.

## 9.2 The Next-Generation Operator

We now show that $R_0$ exists in constant environments and in periodic environments and that it is calculated as the spectral radius of the *next-generation operator*. The method of NGO was first clearly formulated by Diekmann et al. in 1990 [17].

### 9.2.1 Constant Environments

First, suppose that the NRO $\Psi$ is time-independent. The renewal process of newborns is then described by the abstract renewal equation in $E$

$$b(t) = g(t) + \int_0^t \Psi(\tau) b(t - \tau) d\tau, \quad t > 0. \tag{9.11}$$

Let $\hat{\Psi}(\lambda)$ be the Laplace transform of the operator $\Psi$:

$$\hat{\Psi}(\lambda) := \int_0^{\infty} e^{-\lambda \tau} \Psi(\tau) d\tau.$$

If we assume that the age space is a finite interval, there exists some $\tau_{\dagger} > 0$ such that $\Psi = 0$ for $\tau > \tau_{\dagger}$, so the Laplace transform $\hat{\Psi}(\lambda)$ exists for all $\lambda \in \mathbb{C}$.

Using positivity arguments [33, 35], if $\hat{\Psi}(\lambda)$ is a non-supporting compact operator for all $\lambda \in \mathbb{R}$, there exists a real $\lambda_0$ such that $r(\hat{\Psi}(\lambda_0)) = 1$. Furthermore, the well-known renewal theorem (see Sect. 2.5) holds; that is, there is a positive number $q_0$ depending on the initial data $g$ such that

$$\lim_{t\to\infty} e^{-\lambda_0 t} b(t) = q_0 \psi_0, \tag{9.12}$$

where $\psi_0$ is a positive eigenvector of $\hat{\Psi}(\lambda_0)$ associated with the unity eigenvalue. Moreover, it holds that

$$\text{sign}(\lambda_0) = \text{sign}(r(\hat{\Psi}(0)) - 1). \tag{9.13}$$

Therefore, we know that in a constant environment, $\lambda_0$ gives the Malthusian parameter of the renewal process (9.11).

Let $K_Y : Y \to Y$ be the GEO in a constant environment:

$$(K_Y f)(t) := \int_0^t \Psi(\tau) f(t-\tau) d\tau, \quad f \in Y_+. \tag{9.14}$$

**Lemma 9.1** *Suppose that*

$$\int_0^\infty \|\Psi(\tau)\|_{\mathscr{L}(E)} d\tau < \infty. \tag{9.15}$$

*Then, $K_Y$ is a bounded linear operator from $Y$ to itself that leaves the cone $Y_+$ invariant and*

$$|K_Y|_{\mathscr{L}(Y)} \le \int_0^\infty |\Psi(\tau)|_{\mathscr{L}(E)} d\tau. \tag{9.16}$$

*Proof* Observe that, for $f = f(t, \zeta) \in Y$,

$$\begin{aligned}
|K_Y f|_Y &= \int_0^\infty dt \int_{\Omega_b} d\zeta \left| \int_0^t d\tau \int_{\Omega_b} d\eta A(\tau, \zeta, \eta) f(t-\tau, \eta) \right| \\
&\le \int_0^\infty dt \int_{\Omega_b} d\zeta \int_0^t d\tau \int_{\Omega_b} d\eta A(\tau, \zeta, \eta) |f(t-\tau, \eta)| \\
&= \int_{\Omega_b} d\zeta \int_{\Omega_b} d\eta \int_0^\infty d\tau \int_\tau^\infty dt A(\tau, \zeta, \eta) |f(t-\tau, \eta)| \\
&= \int_0^\infty d\tau \int_0^\infty dt \int_{\Omega_b} d\zeta \int_{\Omega_b} d\eta A(\tau, \zeta, \eta) |f(t, \eta)|.
\end{aligned}$$

For $f \in Y$, we define its positive part as $f_+ := \max\{f, 0\}$ and negative part as $f_- := \max\{-f, 0\}$. Then, $f = f_+ - f_-$ and $|f| = f_+ + f_-$. Hence, we can see

$$\begin{aligned}
&\int_{\Omega_b} d\zeta \int_{\Omega_b} d\eta A(\tau, \zeta, \eta)|f(t, \eta)| \\
&= \int_{\Omega_b} d\zeta \int_{\Omega_b} d\eta A(\tau, \zeta, \eta)(f_+(t, \eta) + f_-(t, \zeta)) \\
&= |\Psi(\tau) f_+(t, \cdot)|_E + |\Psi(\tau) f_-(t, \cdot)|_E \\
&\le |\Psi(\tau)|_{\mathscr{L}(E)}(|f_+(t, \cdot)|_E + |f_-(t, \cdot)|_E) = |\Psi(\tau)|_{\mathscr{L}(E)}|f(t, \cdot)|_E.
\end{aligned}$$

Therefore, we obtain that

$$|K_Y f|_Y \le \int_0^\infty d\tau \int_0^\infty dt \|\Psi(\tau)\|_{\mathscr{L}(E)} |f(t, \cdot)|_E = \int_0^\infty |\Psi(\tau)|_{\mathscr{L}(E)} d\tau |f|_Y,$$

which shows that inequality (9.16) holds, and so $K_Y$ is a bounded linear operator from $Y$ to itself. This completes the proof. □

The *next-generation operator* (NGO) is defined by

$$K_E := \hat{\Psi}(0) = \int_0^\infty \Psi(\tau) d\tau, \tag{9.17}$$

and for $f = f(t, \zeta) \in Y$, $(t, \zeta) \in \mathbb{R}_+ \times \Omega_b$, we introduce an aggregation operator $T : Y \to E_+$ by

$$(Tf)(\zeta) := \int_0^\infty |f(t, \zeta)| dt. \tag{9.18}$$

It is easy to see that $T$ is a bounded operator and that the following exchange property holds:

**Lemma 9.2** *It holds that*

$$|f|_Y = |Tf|_E, \tag{9.19}$$

*and so the operator norm of* $T$ *is unity. Moreover, for* $f \in Y_+$, *it follows that*

$$TK_Y f = K_E T f. \tag{9.20}$$

*Proof* Observe that

$$|f|_Y = \int_0^\infty dt \int_{\Omega_b} d\zeta |f(t, \zeta)| = \int_{\Omega_b} d\zeta (Tf)(\zeta) = |Tf|_E.$$

Next, for $f \in Y_+$, we can observe that

$$TK_Y f = \int_0^\infty dt \int_0^t \Psi(s) f(t - s) ds = \int_0^\infty \Psi(s) ds \int_0^\infty f(t) dt = K_E T f.$$

This completes the proof. □

Although newborns are originally identified by their time at birth $t$ and $h$-state variable $\zeta$, in a constant environment newborns with the same $h$-state are identical with respect to their life course, even though they are produced at different times. Therefore, aggregating the generation distribution with respect to the time parameter, we can define the aggregated (timeless) $m$th generation distribution as

$$Tb_m = \int_0^\infty b_m(t)dt \in E_+.$$

It follows from (9.20) that the generation evolution process in $Y_+$ is reduced to the iterative process on $E_+$ given by

$$Tb_m = TK_Y b_{m-1} = K_E T b_{m-1}. \tag{9.21}$$

Because of the positivity of $K_E$ (see Chap. 10), under suitable conditions such as compactness and primitivity (non-supporting property), the spectral radius $r(K_E)$ is the dominant eigenvalue of $K_E$ associated with a positive eigenvector $f_E \in E_+$, and there exists a positive eigenfunctional $F_E \in E_+^*$ such that

$$Tb_m = K_E^m T b_0 \sim \langle F_E, Tb_0 \rangle r(K_E)^m f_E, \quad m \to \infty, \tag{9.22}$$

where $E^*$ denotes the dual space and $\langle F_E, \phi \rangle$ denotes the value of $F_E$ at $\phi \in E$ [51].

From (9.19), we have $|Tb_m|_E = |b_m|_Y$, and so it follows from (9.22) that

$$\lim_{m\to\infty} \sqrt[m]{|b_m|_Y} = \lim_{m\to\infty} \sqrt[m]{|Tb_m|_E} = r(K_E). \tag{9.23}$$

Therefore, the asymptotic per-generation growth factor for the genealogy $\{b_m\}$ exists and is independent of the starting distribution $b_0$. Moreover, the sign relation holds as $K_E = \hat{\Psi}(0)$. We can therefore state the following:

**Proposition 9.1** *In a constant environment, the basic reproduction number $R_0$ exists and is calculated as the spectral radius of the NGO $K_E$.*

Moreover, we can prove that:

**Proposition 9.2** *The spectral radius of the GEO gives $R_0$, that is, it holds that*

$$r(K_Y) = r(K_E) = \lim_{m\to\infty} \sqrt[m]{|b_m|_Y}. \tag{9.24}$$

*Proof* Observe from $|b_m|_Y \le \|K_Y^m\|_{\mathscr{L}(Y)} |b_0|_Y$ that

$$\lim_{m\to\infty} \sqrt[m]{|b_m|_Y} \le \lim_{m\to\infty} \sqrt[m]{\|K_Y^m\|_{\mathscr{L}(Y)}} \lim_{m\to\infty} \sqrt[m]{|b_0|_Y} = r(K_Y).$$

Therefore, we have

$$R_0 = \lim_{m\to\infty} \sqrt[m]{|b_m|_Y} \le r(K_Y). \tag{9.25}$$

Because $R_0 = r(K_E)$, we have $r(K_E) \le r(K_Y)$. It is then sufficient to show that $r(K_E) \ge r(K_Y)$. For $f \in Y_+$, observe that

$$\begin{aligned}
|K_Y f|_Y &= \int_{\Omega_b} d\zeta \int_0^\infty dt \int_0^t \Psi(s) f(t-s) ds \\
&= \int_{\Omega_b} d\zeta \int_0^\infty ds \int_s^\infty \Psi(s) f(t-s) dt = \int_{\Omega_b} d\zeta \int_0^\infty \Psi(s) ds \int_0^\infty f(t) dt \\
&= \int_{\Omega_b} d\zeta K_E T f = |K_E T f|_E.
\end{aligned}$$

Moreover, if $|K_Y^n f|_Y = |K_E^n T f|_E$ for $f \in Y_+$, we have

$$|K_Y^{n+1} f|_Y = |K_Y^n (K_Y f)|_Y = |K_E^n T K_Y f|_E = |K_E^n K_E T f|_E = |K_E^{n+1} T f|_E,$$

where we have used (9.20). By mathematical induction, it holds that

$$|K_Y^n f|_Y = |K_E^n T f|_E, \quad n = 1, 2, \ldots$$

Therefore, for any $f \in Y$, we obtain

$$|K_Y^n f|_Y \le |K_Y^n f_+|_Y + |K_Y^n f_-|_Y = |K_E^n T f_+|_E + |K_E^n T f_-|_E. \tag{9.26}$$

Here, we can observe that

$$|K_E^n T f|_E = |K_E^n T f_+|_E + |K_E^n T f_-|_E, \tag{9.27}$$

because $Tf = Tf_+ + Tf_-$ and $|K_E^n (f+g)|_E = |K_E^n f|_E + |K_E^n g|_E$ if $f, g \in E_+$. From (9.26) and (9.27), we have $|K_Y^n f|_Y \le |K_E^n T f|_E$. Using (9.19), we obtain

$$\frac{|K_Y^n f|_Y}{|f|_Y} \le \frac{|K_E^n T f|_E}{|Tf|_E}$$

for $f \ne 0$. Note that $Tf \ne 0$ if $f \ne 0$. Therefore, it follows that

$$\begin{aligned}
|K_Y^n|_{\mathscr{L}(Y)} &= \sup_{f \in Y \setminus \{0\}} \frac{|K_Y^n f|_Y}{|f|_Y} \le \sup_{f \in Y \setminus \{0\}} \frac{|K_E^n T f|_E}{|Tf|_E} \\
&\le \sup_{\phi \in E \setminus \{0\}} \frac{|K_E^n \phi|_E}{|\phi|_E} = |K_E^n|_{\mathscr{L}(E)},
\end{aligned}$$

which shows that

$$r(K_Y) = \lim_{n\to\infty} \sqrt[n]{|K_Y^n|_{\mathscr{L}(Y)}} \le \lim_{n\to\infty} \sqrt[n]{|K_E^n|_{\mathscr{L}(E)}} = r(K_E).$$

This completes the proof. □

### 9.2.2 Periodic Environments

We now consider the renewal process in a periodic environment:

$$b(t) = g(t) + \int_0^t \Psi(t, \tau)b(t - \tau)d\tau, \quad t > 0, \tag{9.28}$$

where the NRO $\Psi$ has a period $\theta > 0$ as

$$\Psi(t + \theta, \tau) = \Psi(t, \tau), \quad t \in \mathbb{R}, \quad \tau > 0.$$

In the definition given by Bacaër and Guernaoui (the BG definition) [5], the basic reproduction number $R_0$ for the periodic case is given by the spectral radius of the positive integral operator defined by

$$f \to \int_0^\infty \Psi(t, \tau) f(t - \tau)d\tau, \quad f \in C_\theta(\mathbb{R}; E), \tag{9.29}$$

where $C_\theta$ is the set of $\theta$-periodic, continuous $E$-valued functions.[2]

Let $K_\theta(\lambda)$ ($\lambda \in \mathbb{C}$) be the integral operator on $C_\theta$ defined by

$$(K_\theta(\lambda)f)(t) := \int_0^\infty e^{-\lambda\tau}\Psi(t, \tau) f(t - \tau)d\tau, \quad f \in C_\theta(\mathbb{R}; E), \tag{9.30}$$

so the operator (9.29) is given by $K_\theta(0)$. According to the periodic renewal theorem [42, 63], the solution of (9.28) satisfies $b(t) \sim e^{\lambda_0 t}\psi_0(t)$, $t \to \infty$, where $\psi_0 \in C_\theta$ is a positive eigenvector of $K_\theta(\lambda_0)$ associated with the eigenvalue unity, and the asymptotic growth rate $\lambda_0$ is a real number such that $r(K_\theta(\lambda_0)) = 1$. Moreover, it holds that

$$\text{sign}(\lambda_0) = \text{sign}(r(K_\theta(0)) - 1), \tag{9.31}$$

which shows that the BG definition $R_0 = r(K_\theta(0))$ is reasonable from a real-time perspective, because the sign relation holds. In fact, the state space $C_\theta$ can be replaced by the space of locally integrable $\theta$-periodic, $E$-valued functions [39, 65].

Different from the constant environment case, however, it is evident that the state space of $K_\theta(0)$ cannot be interpreted as the space of generation distributions aggregated by the operator $T$, because $K_\theta(0)$ acts on the space of periodic functions. Therefore, in the time-dependent case, we cannot define a next-generation(-like) operator acting between two successive time-aggregated generation distributions. However, another aggregation leads to a reasonable definition of the NGO in a periodic environment.

---

[2] This $R_0$ theory for periodic environments has been developed in a series of papers by Bacaër and his collaborators [5–13]. For a slightly different formulation, see [70]. An early use of the spectral radius of the integral operator as a threshold value for a periodic system is found in [4].

Newborns are identified by their time of birth and $h$-state variable. However, for the periodic case, if $t_1 \equiv t_2 \pmod{\theta}$, time $t_1$ and time $t_2$ play the same role as an (extended) $h$-state variable, because individuals born at $t_1$ and $t_2$ will experience the same life history because of the periodicity of the environment. This observation suggests that the NGO could be defined on the space of $\theta$-periodic functions. The time parameter in the extended $h$-state space would no longer denote real chronological time, but instead be an index to indicate a season (with mod $\theta$) at which childbearing occurs. This kind of idea is well known in the context of matrix population models [9, 16].

To formulate the above perspective under the natural state space $L^1$, we first define a space $Y_\theta$ (the periodic $b$-state space) as the set of locally integrable $\theta$-periodic $E$-valued functions with norm

$$|f|_{Y_\theta} := \int_0^\theta |f(t)|_E dt = \int_0^\theta dt \int_{\Omega_b} |f(t,\zeta)| d\zeta,$$

and, according to the BG definition, define the NGO $K_\theta = K_\theta(0)$ for a periodic environment given by

$$(K_\theta f)(t) := \int_0^\infty \Psi(t,\tau) f(t-\tau) d\tau, \quad f \in Y_\theta.$$

For the time-dependent net reproduction kernel, we can again define the GEO as

$$(K_Y f)(t) := \int_0^t \Psi(t,\tau) f(t-\tau) d\tau, \quad f \in Y_+. \tag{9.32}$$

Hence, the generation evolution process is again expressed as an iterative process in $Y_+$ as $b_m = K_Y b_{m-1}$, where we assume that $K_Y$ defines a bounded linear operator from $Y_+$ into itself.

To aggregate the generation distributions, we introduce a periodization operator $U : Y \to (Y_\theta)_+$ by

$$(Uf)(t) := \sum_{n=-\infty}^{+\infty} |f^*(t+n\theta)|, \quad t \in \mathbb{R},$$

where $f^* \in L^1(\mathbb{R} \times \Omega_b)$ is defined as $f^*(t) = f(t)$ for $t \geq 0$ and $f^*(t) = 0$ for $t < 0$. The periodization operation $U$ can then be seen as an aggregation of generation distributions by identifying $f \in Y_+$ with its $\theta$-shifted distributions $f^*(t+n\theta)$.

**Lemma 9.3** *It holds that*

$$|f|_Y = |Uf|_{Y_\theta}, \tag{9.33}$$

*and the operator norm of U is unity. Moreover, it follows that*

$$UK_Y f = K_\theta U f, \quad f \in Y_+. \tag{9.34}$$

*Proof* If $f \in Y_+$, it follows that

$$\begin{aligned}
|Uf|_{Y_\theta} &= \int_0^\theta dt \int_{\Omega_b} d\zeta \sum_{n=-\infty}^{+\infty} f^*(t+n\theta) = \sum_{n=-\infty}^{+\infty} \int_{n\theta}^{(n+1)\theta} dt \int_{\Omega_b} d\zeta f^*(t) \\
&= \sum_{n=0}^{+\infty} \int_{n\theta}^{(n+1)\theta} dt \int_{\Omega_b} d\zeta f(t) = \int_0^\infty |f(t)|_E dt = |f|_Y.
\end{aligned}$$

For a general $f \in Y$, we have $f = f_+ - f_-$ and

$$|f|_Y = |f_+|_Y + |f_-|_Y = |Uf_+|_{Y_\theta} + |Uf_-|_{Y_\theta}.$$

On other hand, we can observe that

$$\begin{aligned}
(Uf)(t) &= \sum_{n=-\infty}^{+\infty} |f^*(t+n\theta)| \\
&= \sum_{n=-\infty}^{+\infty} f_+^*(t+n\theta) + \sum_{n=-\infty}^{+\infty} f_-^*(t+n\theta) \\
&= (Uf_+)(t) + (Uf_-)(t),
\end{aligned}$$

which shows that

$$|Uf|_{Y_\theta} = |Uf_+|_{Y_\theta} + |Uf_-|_{Y_\theta}.$$

Thus, we obtain (9.33). Next, let us derive (9.34). Observe that, for $f \in Y$,

$$(K_Y f)^*(t) = \int_0^\infty \Psi(t,s) f^*(t-s) ds.$$

Therefore, if $f \in Y_+$, we obtain

$$\begin{aligned}
(UK_Y f)(t) &= \sum_{n=-\infty}^{+\infty} \int_0^\infty \Psi(t+n\theta, s) f^*(t+n\theta-s) ds \\
&= \int_0^\infty \Psi(t,s)(Uf)(t-s) ds,
\end{aligned}$$

which shows that (9.34) holds. □

The generation evolution process in $Y$-space is therefore reduced to an iteration process in $Y_\theta$-space. In fact, if we apply $U$ to $b_m = K_Y b_{m-1}$ and use the exchange property (9.34), we have

$$Ub_m = UK_Y b_{m-1} = K_\theta U b_{m-1}. \tag{9.35}$$

That is, the evolution process $b_m = K_Y b_{m-1}$ in the real generation distribution space $Y$ reduces to the evolution on the periodic $b$-state space, and for this reduction, the generation size is preserved because $|b_m|_Y = |Ub_m|_{Y_\theta}$. From (9.35), it is reasonable to call $K_\theta$ the NGO, because it evolves the time-aggregated generation distributions.

To apply the positive operator theory, let us carry out a second aggregation. Using the periodicity, $K_\theta$ reduces to an integral operator on $Z := L^1([0,\theta]; E)$.[3] Define a positive operator $K_Z : Z \to Z$ as follows:

$$(K_Z\phi)(t) := \int_0^\theta \Pi(t,s)\phi(s)ds, \quad t \in [0,\theta), \quad \phi \in Z, \tag{9.36}$$

where

$$\Pi(t,s) := \begin{cases} \sum_{n=0}^\infty \Psi(t, t-s+n\theta), & t > s, \\ \sum_{n=1}^\infty \Psi(t, t-s+n\theta), & t < s. \end{cases}$$

Let $V : Y_\theta \to Z$ be an operator such that $(Vf)(t) = f(t)$ for $t \in [0,\theta]$. Then, we have the following:

**Lemma 9.4** *If it holds that*

$$|f|_{Y_\theta} = |Vf|_Z, \tag{9.37}$$

*and the operator norm of V is unity. Moreover, it follows that*

$$VK_\theta = K_Z V. \tag{9.38}$$

If we define $V^{-1} : Z \to Y_\theta$ as the operator that maps $\phi \in Z$ to its periodization in $Y_\theta$, then $V$ becomes a bijection from $Y_\theta$ to $Z$. Therefore, we have $K_\theta = V^{-1}K_Z V$, and we can state the following:

**Lemma 9.5** *It holds that*

$$r(K_Z) = r(K_\theta). \tag{9.39}$$

Using (9.38), the iteration process in (9.35) reduces to an iteration process in the function space $Z$:

$$VUb_m = VK_\theta U b_{m-1} = K_Z VU b_{m-1}. \tag{9.40}$$

[3] As the reduction of $K_\theta$ to $K_Z$ is introduced in [7, 40], we omit the proof.

$Z$ is the set of state vectors in which the time parameter does not represent chronological time, but is instead a heterogeneity parameter to indicate the season in which newborns are produced.

Because we can usually expect $K_Z$ to be a compact, positive non-supporting operator, we can again apply the positive operator theory to conclude that

$$VUb_m = K_Z^m VUb_0 \sim \langle F_Z, VUb_0\rangle r(K_Z)^m f_Z, \quad m \to \infty, \tag{9.41}$$

where $F_Z \in Z_+^*$ denotes the dual eigenfunctional with respect to the positive eigenvalue $r(K_Z)$ associated with the positive eigenfunction $f_Z \in Z_+$, and $\langle F_Z, \phi\rangle$ denotes the value of $F_Z$ at $\phi \in Z$. From (9.33), (9.37), (9.39), and (9.41), we have

$$\begin{aligned}\lim_{m\to\infty} \sqrt[m]{|VUb_m|_Z} &= r(K_Z) = r(K_\theta)\\ &= \lim_{m\to\infty} \sqrt[m]{|Ub_m|_{Y_\theta}} = \lim_{m\to\infty} \sqrt[m]{|b_m|_Y}.\end{aligned} \tag{9.42}$$

Therefore, we know that $r(K_\theta)$ is the asymptotic per-generation growth factor. As we can show that the periodic renewal system has a Malthusian parameter $\lambda_0$ under appropriate conditions and the sign relation holds between $\lambda_0$ and $r(K_\theta)$ [39, 63], we can state that:

**Proposition 9.3** *In a periodic environment, the basic reproduction number $R_0$ is calculated as the spectral radius of the NGO $K_\theta$:*

$$R_0 = r(K_\theta) = \lim_{m\to\infty} \sqrt[m]{|b_m|_Y}. \tag{9.43}$$

*Remark 9.2* To induce the NGO $K_\theta$ of the BG definition in the above argument, we adopted two-step aggregation of the space $Y$ ($Y \to Y_\theta \to Z$). However, the one-step aggregation $Y \to Z$ is also possible, as adopted by Bacaër [10]. Let us define a one-sided aggregation operator $U_+ : Y \to Z$ as follows:

$$(U_+ f)(t) := \sum_{n=0}^{+\infty} |f(t+n\theta)|, \quad t \in [0,\theta), \quad f \in Y.$$

It is again easy to see that $|U_+ f|_Z = |f|_Y$, and we can prove that

$$U_+ K_Y f = K_Z U_+ f, \quad f \in Y_+. \tag{9.44}$$

That is, the evolution process $b_m = K_Y b_{m-1}$ can be reduced to the evolution process $U_+ b_m = K_Z U_+ b_{m-1}$ on the space $Z$ of aggregated generation distributions. Thus, we can prove that

$$r(K_Z) = \lim_{m\to\infty} \sqrt[m]{|U_+ b_m|_Z} = \lim_{m\to\infty} \sqrt[m]{|b_m|_Y}.$$

In this case, $K_Z$ is the NGO acting on the extended $b$-state space $Z$ and $R_0 = r(K_Z)$.

Finally, we establish the fact that the basic reproduction number is again given by the spectral radius of the GEO:

**Proposition 9.4** *If the NRO $\Psi$ is $\theta$-periodic with respect to time $t$, it holds that*

$$r(K_Y) = r(K_\theta) = \lim_{m\to\infty} \sqrt[m]{|b_m|_Y}. \tag{9.45}$$

*Proof* It follows from (9.43) that $r(K_\theta) \le r(K_Y)$. Thus, it is sufficient to show that $r(K_\theta) \ge r(K_Y)$. For $f \in Y_+$, observe that

$$\begin{aligned}
|K_Y f|_Y &= \int_{\Omega_b} d\zeta \int_0^\infty dt \int_0^t \Psi(t,s) f(t-s) ds \\
&= \int_{\Omega_b} d\zeta \int_0^\infty dt \int_0^\infty \Psi(t,s) f^*(t-s) ds \\
&= \int_{\Omega_b} d\zeta \sum_{n=-\infty}^{\infty} \int_{n\theta}^{(n+1)\theta} dt \int_0^\infty \Psi(t,s) f^*(t-s) ds \\
&= \int_{\Omega_b} d\zeta \int_0^\theta dx \int_0^\infty \sum_{n=-\infty}^{\infty} \Psi(n\theta + x, s) f^*(n\theta + x - s) ds \\
&= \int_{\Omega_b} d\zeta \int_0^\theta dx \int_0^\infty \Psi(x,s) \sum_{n=-\infty}^{\infty} f^*(n\theta + x - s) ds \\
&= \int_{\Omega_b} d\zeta \int_0^\theta dx \int_0^\infty \Psi(x,s)(Uf)(x-s) ds \\
&= \int_{\Omega_b} d\zeta \int_0^\theta dx K_\theta U f = |K_\theta U f|_{Y_\theta}.
\end{aligned}$$

If we assume that $|K_Y^n f|_Y = |K_\theta^n U f|_{Y_\theta}$ for $f \in Y_+$, we have

$$|K_Y^{n+1} f|_Y = |K_Y^n (K_Y f)|_Y = |K_\theta^n U K_Y f|_{Y_\theta} = |K_\theta^{n+1} U f|_{Y_\theta},$$

where we have used (9.34). By mathematical induction, it holds that

$$|K_Y^n f|_Y = |K_\theta^n U f|_{Y_\theta}, \quad n = 1, 2, \ldots$$

for $f \in Y_+$. Therefore, for any $f \in Y$, we obtain

$$|K_Y^n f|_Y \le |K_Y^n f_+|_Y + |K_Y^n f_-|_Y = |K_\theta^n U f_+|_{Y_\theta} + |K_\theta^n U f_-|_{Y_\theta}. \tag{9.46}$$

On the other hand, we can observe that

$$|K_\theta^n U f|_{Y_\theta} = |K_\theta^n U f_+|_{Y_\theta} + |K_\theta^n U f_-|_{Y_\theta}, \tag{9.47}$$

because $Uf = Uf_+ + Uf_-$ and $|K_\theta^n(f+g)|_{Y_\theta} = |K_\theta^n f|_{Y_\theta} + |K_\theta^n g|_{Y_\theta}$ if $f, g \in E_+$. From (9.46) and (9.47), we have $|K_Y^n f|_Y \le |K_\theta^n Uf|_{Y_\theta}$. Using (9.33), we obtain

$$\frac{|K_Y^n f|_Y}{|f|_Y} \le \frac{|K_\theta^n Uf|_{Y_\theta}}{|Uf|_{Y_\theta}}$$

for $f \ne 0$. Note that $Uf \ne 0$ if $f \ne 0$. Therefore, it follows that

$$\begin{aligned}\|K_Y^n\|_{\mathscr{L}(Y)} &= \sup_{f\in Y\setminus\{0\}} \frac{|K_Y^n f|_Y}{|f|_Y} \le \sup_{f\in Y\setminus\{0\}} \frac{|K_\theta^n Uf|_{Y_\theta}}{|Uf|_{Y_\theta}} \\ &\le \sup_{\phi\in Y_\theta\setminus\{0\}} \frac{|K_\theta^n \phi|_{Y_\theta}}{|\phi|_{Y_\theta}} = \|K_\theta^n\|_{\mathscr{L}(Y_\theta)},\end{aligned}$$

which shows that

$$r(K_Y) = \lim_{n\to\infty} \sqrt[n]{\|K_Y^n\|_{\mathscr{L}(Y)}} \le \lim_{n\to\infty} \sqrt[n]{\|K_\theta^n\|_{\mathscr{L}(Y_\theta)}} = r(K_\theta).$$

This completes our proof. □

As seen above, there exists a Malthusian parameter $\lambda_0$ for constant or periodic environments, and the sign relation holds between $R_0$ and $\lambda_0$. The general conditions under which there exists a Malthusian parameter and $R_0$, and whether the sign relation between them holds, remain an open question. Recently, several authors have shown that the basic reproduction number can be defined for a stochastic environment [13] and an almost periodic environment [72]. Even in these cases, we can conjecture that the spectral radius of the GEO gives $R_0$.

As we have shown in Part II, we can expect that there exists an endemic steady state if $R_0 > 1$ in constant environments. Although we only discuss the role of $R_0$ as the invasion threshold here, $R_0$ for the periodic system also operates as the endemic threshold [43–45, 74].

**Exercise 9.2** Consider the SIS model without demography:

$$\begin{aligned}\frac{dS(t)}{dt} &= -\beta(t)S(t)I(t) + \gamma I(t), \\ \frac{dI(t)}{dt} &= \beta(t)S(t)I(t) - \gamma I(t),\end{aligned} \tag{9.48}$$

where $I(t)$ denotes the density of the infective population, $S(t)$ is the density of susceptibles, $\beta(t)$ is a $\theta$-periodic contact coefficient, and $\gamma$ is the recovery rate (see Sects. 5.1 and 6.1). Show that

$$R_0 = \frac{N}{\gamma\theta}\int_0^\theta \beta(t)dt, \tag{9.49}$$

where $N$ denotes the total size of the host population. Show that $\lim_{t\to\infty} I(t) = 0$ if $R_0 \leq 1$, whereas $I(t)$ converges to a $\theta$-periodic function if $R_0 > 1$ and $I(0) > 0$ [5, 34].

## 9.3 Reconciliation with Differential Equation Models

As seen in the previous sections, the basic reproduction number $R_0$ is naturally defined by the renewal equation formulation, because it is an essential expression of the life course of individuals. However, many traditional population models do not explicitly account for the age structure, and they are formulated by ordinary differential equations (ODEs). Therefore, we now consider the relation between the renewal equation formulation (9.2) and differential equation models to derive a general recipe for calculating $R_0$ in such models.

### 9.3.1 *ODE Case*

Suppose that the state space of individuals, denoted by $\Omega = \{1, 2, \ldots, N\}$, is a finite set and neglects the age structure so that all life cycle parameters are age-independent. Without loss of generality, we can assume that the birth state space is given by $\Omega_b = \{1, 2, \ldots, N_b\}$ with $N_b \leq N$. Let $p_k(t)$, $k \in \Omega$, be the size of the population in the $k$th state and $p(t) = (p_1(t), \ldots, p_N(t))^{\mathrm{T}}$ be the density vector of the multistate population. For the finite-dimensional autonomous case, the basic population evolution equation is given by

$$\frac{dp(t)}{dt} = Ap(t) = (Q + M)p(t), \tag{9.50}$$

where $A = Q + M$, $M$ is an $N \times N$ nonnegative, nonzero matrix called the *reproduction matrix* whose $(j, k)$th entry $m_{jk}$ denotes the number of newborns produced at state $j \in \Omega_b$ per unit time and per individual at state $k \in \Omega$, and $Q$ is an $N \times N$ *transition intensity* matrix whose $(j, k)$th entry $q_{jk} \geq 0$ $(j \neq k)$ denotes the transition intensity from the $k$th to $j$th state. The diagonal elements $q_{kk}(t)$ of $Q$ are given by $q_{kk} = -\mu_k - \sum_{j\neq k} q_{jk}$, where $\mu_k$ denotes the removal rate (by death or other causes) in the $k$th state. We assume that $Q$ is a nonzero *essentially nonnegative matrix*; that is, all off-diagonal entries are nonnegative. The *survival matrix* is given by $L(t) = \exp(Qt)$.

An essentially nonnegative nonzero matrix is also called *quasi-positive* [65]. Here, we write $A > 0$ if all entries $a_{ij}$ of a matrix $A = (a_{ij})$ are positive, whereas we write $A \geq 0$ if $a_{ij} \geq 0$ for all $i$ and $j$. According to the standard definition, a matrix $A$ is called *nonnegative* if $A \geq 0$, whereas $A$ is said to be *positive* if $A > 0$ [68, Definition 2.1]. On the other hand, as is shown in Chap. 10, in the context of positive operator theory, a bounded linear "operator" that leaves a positive cone invariant is called *positive*. Therefore, the linear operator $x \mapsto Ax$ in $\mathbb{R}^n$ is a "positive" operator if the matrix $A$ is nonnegative, whereas the linear operator $x \mapsto Ax$ in $\mathbb{R}^n$ is a "strongly positive" operator if $A$ is positive. For any elements $x$ and $y$ in a positive cone, the notation $x < y$ implies that $y - x \geq 0$ and $y - x \neq 0$. If a square matrix $A$ is irreducible and essentially nonnegative, it is called *essentially positive*. Then, the following holds[4]:

**Lemma 9.6** ([68], Sect. 8.2) *A matrix $A$ is essentially positive if and only if $A + sI_d$ is a nonnegative, irreducible and primitive matrix for all sufficiently large $s > 0$.*

*Proof* Because the "if" part is clear, we prove the "only if" part. If a matrix $A$ is essentially positive, it is clear that $A + sI_d$ is nonnegative and irreducible for all sufficiently large $s > 0$. Then, we can assume that $A + (s-1)I_d$ is a nonnegative and irreducible $n \times n$ matrix for some large $s > 0$. From a well-known result (e.g., see [25], Chapter XIII, Lemma 1; [68], Lemma 2.2), we have $(A + sI_d)^{n-1} = [(A + (s-1)I_d) + I_d]^{n-1} > 0$, which implies that $A + sI_d$ is primitive.[5] □

**Lemma 9.7** ([14], Lemmas 2, 4; [68], Theorem 8.2) *A square matrix $A$ is essentially nonnegative if and only if $e^{At} \geq 0$ for all $t \geq 0$, and $A$ is essentially positive if and only if $e^{At} > 0$ for all $t > 0$.*

Let $s(A)$ be the spectral bound of a matrix $A$, that is, $s(A) := \max_{\lambda \in \sigma(A)} \Re\lambda$, where $\sigma(A)$ is the set of eigenvalues of $A$. It is natural to assume $s(Q) < 0$, because individuals die out as time evolves. Then, the following holds (see [19, Lemma 6.12]):

**Lemma 9.8** *$s(Q) < 0$ if and only if $-Q$ is nonnegatively invertible, that is, $(-Q)^{-1}$ exists and $(-Q)^{-1} \geq 0$.*

Applying the variation-of-constants formula to (9.50), we have

$$p(t) = e^{Qt}p(0) + \int_0^t e^{Q(t-\tau)}Mp(\tau)d\tau.$$

Multiplying $M$ from the left-hand side, we obtain a renewal equation formulation of system (9.50):

$$\begin{aligned} Mp(t) &= Me^{Qt}p(0) + \int_0^t Me^{Q(t-\tau)}Mp(\tau)d\tau, \\ &= ML(t)p(0) + \int_0^t \Psi(\tau)Mp(t-\tau)d\tau, \end{aligned}$$

where $Mp(t)$ gives a vector of newborns at time $t$, and $\Psi(\tau) := ML(\tau)$ is the net reproduction matrix. Let $v(t) := Mp(t)$ be the birth rate vector of newborns. Then, we obtain the renewal equation formulation

---

[4] $I_d$ denotes the identity matrix.

[5] A nonnegative matrix $A$ is primitive if and only if there exists an integer $n$ such that $A^n > 0$.

$$v(t) = g(t) + \int_0^t \Psi(\tau)v(t-\tau)d\tau, \tag{9.51}$$

where $g(t) = ML(t)p(0)$. If every state is the birth state ($\Omega = \Omega_b$), the *next-generation matrix* (NGM) is calculated as follows:

$$K := \int_0^\infty \Psi(\tau)d\tau = \int_0^\infty Me^{Q\tau}d\tau = M(-Q)^{-1}.$$

If $N_b < N$, the renewal Eq. (9.51) is redundant in the sense that it includes trivial relations for non-birth states; that is, $v_j = 0$ and the $j$th row of $K$ is zero for $j > N_b$. If the state space is larger than the birth state space, $K$ is the *next-generation matrix with large domain* [20]. The $(i, j)$th entry of $(-Q)^{-1} = \int_0^\infty e^{Q\tau}d\tau$ denotes the expected sojourn time that an individual born in state $j$ spends in the $i$th state, so $M(-Q)^{-1}$ maps a time-aggregated density vector of newborns of the $m$th generation to a time-aggregated density vector of the next generation of newborns.

Recently, several authors have established the following relation between $r(K)$ and the spectral bound of $A$ (see [20, Theorem A.1], [65, Theorem2.3], [67, Theorem 2]):

**Proposition 9.5** *Suppose that $M$ is a nonnegative matrix, $Q$ is an essentially non-negative matrix with $s(Q) < 0$ and $K = M(-Q)^{-1}$. Let $R_0 = r(K)$. Then, it follows that*

$$\operatorname{sign}(s(A)) = \operatorname{sign}(R_0 - 1). \tag{9.52}$$

**Exercise 9.3** Suppose that $s(Q) < 0$ and $K = M(-Q)^{-1}$ is indecomposable. Let $\Sigma := \{\lambda \in \mathbb{C} : \Re\lambda > s(Q), \lambda \in \sigma(M+Q)\}$ and $\Lambda := \{\lambda \in \mathbb{C} : \Re\lambda > s(Q), \det(I - \hat{\Psi}(\lambda)) = 0\}$, where $\hat{\Psi}(\lambda) = \int_0^\infty e^{-\lambda\sigma}Me^{\sigma Q}d\sigma$.

1. Show that $\Sigma = \Lambda$.
2. Show that if $R_0 = r(K) \geq 1$, there exists a nonnegative dominant eigenvalue $\lambda_0 \in \Sigma$.
3. Show that $\operatorname{sign}(s(M+Q)) = \operatorname{sign}(R_0 - 1)$.
4. Show that $s(R_0^{-1}M + Q) = 0$.

From Proposition 9.5, we know that the stability of the trivial steady state (the steady state corresponding to population extinction) can be formulated by the spectral radius of the NGM $K$. Based on the previous argument, the basic reproduction number $R_0$ for the population model (9.50) is given by $r(K)$. Although there are often surrogate threshold parameters that share the same kind of sign relation with $R_0$ [32], $R_0 = r(K)$ is distinguished from other indices by its generational interpretation.

**Lemma 9.9** *Under the assumption of Proposition 9.5, there exists a nonnegative exponential solution $e^{\lambda_0 t}\phi_0$ for the basic system (9.50), where $\phi_0$ is a nonnegative eigenvector of $A$ associated with eigenvalue $\lambda_0 = s(A)$.*

*Proof* Because $A$ is essentially nonnegative, there exists some $\alpha > 0$ such that $A + \alpha I \geq 0$. From the Perron–Frobenius Theorem, there exists a nonnegative eigenvector $\phi$ such that $(A + \alpha I)\phi = r(A + \alpha I)\phi$. Because $\lambda \in \sigma(A + \alpha I)$ if and only if $\lambda - \alpha \in \sigma(A)$, we have $s(A) = r(A + \alpha I) - \alpha \in \sigma(A)$ and $\phi$ as its corresponding eigenvector. Therefore, we know that (9.50) has a nonnegative exponential solution $e^{s(A)t}\phi$. □

If any solution of the basic Eq. (9.50) is asymptotically proportional to the exponential solution, we call it the *dominant exponential solution*. If a dominant positive exponential solution exists, its exponent $\lambda_0$ is none other than the Malthusian parameter of the renewal system (9.51). To ensure the existence of a positive dominant exponential solution, we need additional conditions:

**Proposition 9.6** ([14]) *If $A$ is essentially positive, it has a unique (up to a constant factor) strictly positive eigenvector $\phi_0$ associated with a real, simple eigenvalue $\lambda_0$ such that $\lambda_0 > \Re\lambda$ for any $\lambda \in \sigma(A) \setminus \{\lambda_0\}$. Then, the basic system (9.50) has a positive dominant exponential solution $e^{\lambda_0 t}\phi_0$.*

We remark that if $e^{Qt} > 0$, that is, an individual born in any state can reach every state, $Q$ is essentially positive. Hence, $A$ is also essentially positive, and there exists a dominant positive exponential solution.

*Example 9.1* (*Multigroup SEIR model*) As a concrete example, consider a two-state susceptible–exposed–infective–recovered (SEIR) model as follows:

$$\begin{aligned}
\frac{dS_i(t)}{dt} &= -S_i(t)\sum_{j=1}^{2}\beta_{ij}I_j(t),\\
\frac{dE_i(t)}{dt} &= S_i(t)\sum_{j=1}^{2}\beta_{ij}I_j(t) - \varepsilon_i E_i(t),\\
\frac{dI_i(t)}{dt} &= \varepsilon_i E_i(t) - \gamma_i I_i(t),\\
\frac{dR_i(t)}{dt} &= \gamma_i I_i(t),
\end{aligned} \tag{9.53}$$

where $i = 1, 2$, $\beta_{ij}$ is the transmission coefficient between susceptible host $i$ and infective host $j$, $\varepsilon_i$ is the transition rate from the exposed class to the infectious class, and $\gamma_i$ is the recovery rate. For example, the first hosts are children and the second hosts are the adult population. Then, the linearized equation for classes $E$ and $I$ at the disease-free steady state can be written as

$$\frac{d}{dt}\begin{pmatrix} E_1 \\ E_2 \\ I_1 \\ I_2 \end{pmatrix} = \begin{pmatrix} -\varepsilon_1 & 0 & \beta_{11}N_1 & \beta_{12}N_1 \\ 0 & -\varepsilon_2 & \beta_{21}N_2 & \beta_{22}N_2 \\ \varepsilon_1 & 0 & -\gamma_1 & 0 \\ 0 & \varepsilon_2 & 0 & -\gamma_2 \end{pmatrix}\begin{pmatrix} E_1 \\ E_2 \\ I_1 \\ I_2 \end{pmatrix}, \tag{9.54}$$

where $N_i$ denotes the size of the $i$th population (which is assumed to be totally susceptible in the initial invasion phase), $E_1$ and $E_2$ are birth states, and $I_1$ and $I_2$ are non-birth states. We can then decompose the coefficient matrix as follows:

$$M = \begin{pmatrix} 0 & 0 & N_1\beta_{11} & N_1\beta_{12} \\ 0 & 0 & N_2\beta_{21} & N_2\beta_{22} \\ 0 & 0 & 0 & 0 \\ 0 & 0 & 0 & 0 \end{pmatrix} = \begin{pmatrix} M_{11} & M_{12} \\ M_{21} & M_{22} \end{pmatrix}, \quad Q = \begin{pmatrix} -\varepsilon_1 & 0 & 0 & 0 \\ 0 & -\varepsilon_2 & 0 & 0 \\ \varepsilon_1 & 0 & -\gamma_1 & 0 \\ 0 & \varepsilon_2 & 0 & -\gamma_2 \end{pmatrix} = \begin{pmatrix} Q_{11} & Q_{12} \\ Q_{21} & Q_{22} \end{pmatrix},$$

where $M_{ij}$ and $Q_{ij}$ are $2 \times 2$ block matrices. Therefore, if we formally apply the well-known recipe, we obtain the NGM with large domain as

$$\tilde{K} = M(-Q)^{-1} = \begin{pmatrix} \frac{N_1\beta_{11}}{\gamma_1} & \frac{N_1\beta_{12}}{\gamma_2} & \frac{N_1\beta_{11}}{\gamma_1} & \frac{N_1\beta_{12}}{\gamma_2} \\ \frac{N_2\beta_{21}}{\gamma_1} & \frac{N_2\beta_{22}}{\gamma_2} & \frac{N_2\beta_{21}}{\gamma_1} & \frac{N_2\beta_{22}}{\gamma_2} \\ 0 & 0 & 0 & 0 \\ 0 & 0 & 0 & 0 \end{pmatrix},$$

and its upper left-hand corner principal minor gives the NGM, as the exposed class is the state-at-birth:

$$K = \begin{pmatrix} k_{11} & k_{12} \\ k_{21} & k_{22} \end{pmatrix} = \begin{pmatrix} \frac{N_1\beta_{11}}{\gamma_1} & \frac{N_1\beta_{12}}{\gamma_2} \\ \frac{N_2\beta_{21}}{\gamma_1} & \frac{N_2\beta_{22}}{\gamma_2} \end{pmatrix}.$$

Thus, we have $R_0 = r(K) = r(\tilde{K})$.

Indeed, if we formulate a renewal equation for the state-at-infection vector, we can calculate $K$ directly. Let $I = (I_1, I_2)^{\mathrm{T}}$. Then, we obtain

$$\begin{aligned} \frac{dE(t)}{dt} &= Q_{11}E(t) + M_{12}I(t), \\ \frac{dI(t)}{dt} &= Q_{21}E(t) + Q_{22}I(t). \end{aligned}$$

Using the variation-of-constants formula, we have

$$\begin{aligned} E(t) &= \int_0^\infty e^{Q_{11}\sigma} M_{12} I(t-\sigma)d\sigma, \\ I(t) &= \int_0^\infty e^{Q_{22}\sigma} Q_{21} E(t-\sigma)d\sigma. \end{aligned}$$

Let $B(t) := M_{12}I(t)$. Then, $B$ is the density of newly infecteds. Eliminating $E$ from the above equations, we obtain a renewal equation for $B$:

$$\begin{aligned} B(t) &= M_{12} \int_0^\infty e^{Q_{22}\sigma} Q_{21} \int_0^\infty e^{Q_{11}\tau} B(t-\sigma-\tau)d\tau \\ &= \int_0^\infty \Psi(z)B(t-z)dz, \end{aligned}$$

where

$$\Psi(z) := M_{12} \int_0^z e^{Q_{22}\sigma} Q_{21} e^{Q_{11}(z-\sigma)} d\sigma.$$

Then, we have $K = \int_0^\infty \Psi(z)dz$.

*Remark 9.3* In many ODE epidemic models, the authors do not clearly distinguish the state-at-birth and the transient states. Hence, the division of the generator $M + Q$ is not uniquely determined, and $M(-Q)^{-1}$ does not necessarily give the NGO. This aspect has been examined by various researchers [20, 37, 67]. For example, we can consider another decomposition of the coefficient matrix of (9.53) as

$$M^* = \begin{pmatrix} 0 & 0 & N_1\beta_{11} & N_1\beta_{12} \\ 0 & 0 & N_2\beta_{21} & N_2\beta_{22} \\ \varepsilon_1 & 0 & 0 & 0 \\ 0 & \varepsilon_2 & 0 & 0 \end{pmatrix}, \quad Q^* = \begin{pmatrix} -\varepsilon_1 & 0 & 0 & 0 \\ 0 & -\varepsilon_2 & 0 & 0 \\ 0 & 0 & -\gamma_1 & 0 \\ 0 & 0 & 0 & -\gamma_2 \end{pmatrix}.$$

Then, we can compute the following NGM-like matrix:

$$\tilde{K}^* = M^*(-Q^*)^{-1} = \begin{pmatrix} 0 & 0 & \frac{N_1\beta_{11}}{\gamma_1} & \frac{N_1\beta_{12}}{\gamma_2} \\ 0 & 0 & \frac{N_2\beta_{21}}{\gamma_1} & \frac{N_2\beta_{22}}{\gamma_2} \\ 1 & 0 & 0 & 0 \\ 0 & 1 & 0 & 0 \end{pmatrix},$$

and it is easy to calculate the spectral radius as follows:

$$r(\tilde{K}^*) = \sqrt{r(\tilde{K})} = \sqrt{r(K)}.$$

Therefore, the decomposition $M^* + Q^*$ does not give $R_0$ but $\sqrt{R_0}$, although it shares the same threshold property as $R_0$. However, as seen above, the decomposition $M^* + Q^*$ does not give the NGM; that is, the spectral radius of $\tilde{K}^*$ is not the asymptotic per-generation growth factor for newly infected individuals. Some authors focus solely on the threshold property (the sign relation) in their interpretation of $R_0$ and thus conclude that $R_0$ is not uniquely determined; however, this is a misunderstanding (e.g., see [47]). Although there are many surrogate indices that share the sign relation with $R_0$, $R_0$ is determined uniquely as the asymptotic per-generation growth factor for newly infected individuals.

**Exercise 9.4** To check that the matrix $K$ defined in the above example is the NGM, write down the renewal integral equation for $E(t)$ and calculate the integral of its kernel. Confirm that we can obtain the same NGM by calculating the integral of the kernel of the renewal integral equation that is induced based on the decomposition $M^* + Q^*$.

We now assume that individuals' reproduction and survival parameters change with time. If we consider the case in which density-dependent effects can be neglected, the population vector $p(t)$ satisfies a non-autonomous linear differential equation system as follows:

$$\frac{dp(t)}{dt} = A(t)p(t) = (Q(t) + M(t))p(t), \quad t \in \mathbb{R}, \tag{9.55}$$

where $p(t) \in \mathbb{R}_+^N$.

Define the survival matrix $L(t)$ as the solution of a matrix differential equation

$$\frac{dL(t)}{dt} = Q(t)L(t), \quad L(0) = I_d, \tag{9.56}$$

where $I_d$ denotes the $N \times N$ identity matrix. In other words, $L(t)$ is the fundamental matrix of the $N$-dimensional non-homogeneous ODE system $dx(t)/dt = Q(t)x(t)$ and is therefore invertible. Let $\ell_{ij}(t)$ be the $(i, j)$th entry of $L(t)$. Then, $\ell_{ij}(t)$ for $j \in \Omega_b$ is the probability that a newborn produced at state $j$ and time zero will survive to reach state $i$ and time $t$. Define a two-parameter system (the transition matrix) $L(t, s) := L(t)L(s)^{-1}$, $t \geq s$. This forms an *evolutionary system* generated by $Q(t)$; that is, the following properties hold:

$$\begin{aligned} &\frac{\partial L(t,s)}{\partial t} = Q(t)L(t,s), \\ &L(t,s)L(s,r) = L(t,r), \quad r \leq s \leq t, \\ &L(s,s) = I_d. \end{aligned} \tag{9.57}$$

The transition matrix is nonnegative; that is, $L(t, s) \geq 0$, because $Q(t)$ is essentially nonnegative. The *exponential growth bound* of the evolutionary system $L(t, s)$ is defined by $\omega(L) := \inf\{\omega :$ there exists some $\bar{L} \geq 1$ such that $|L(\tau + s, s)| \leq \bar{L}e^{\omega\tau}, \forall s \in \mathbb{R}, \tau \geq 0\}$, where $|A|$ denotes the norm of a matrix $A$. We now adopt the following assumption:

**Assumption 9.1** We assume that $Q(t)$ and $M(t)$ are uniformly bounded and continuous on $\mathbb{R}$, so $\bar{Q} := \sup_{t\in\mathbb{R}} |Q(t)| < \infty$ and $\bar{M} := \sup_{t\in\mathbb{R}} |M(t)| < \infty$. For the survival evolutionary system, we assume that $\omega(L) < 0$.

The above assumption of $\omega(L) < 0$ is biologically reasonable, because it implies that a closed population becomes extinct if there is no reproduction. For example, if we assume that $\inf_{j,t} \mu_j(t) =: \underline{\mu} > 0$, then we can observe that $|L(t, \tau)| \leq e^{-\underline{\mu}(t-\tau)}, \quad t \geq \tau$, from which it follows that $\omega(L) \leq -\underline{\mu}$.

Next, let $U(t, s)$, $t \geq s$ be the evolutionary system associated with the generator $A(t)$. Using a fundamental matrix $\Phi(t)$ of (9.55), we have $U(t, s) = \Phi(t)\Phi(s)^{-1}$. The solution of (9.55) is then given by $p(t) = U(t, s)p(s)$, and $\{U(t, s) : t \geq s\}$ again becomes an evolutionary system acting on $\mathbb{R}^N$.

Applying the variation-of-constants formula, we have

$$U(t,s) = L(t,s) + \int_s^t L(t,\sigma)M(\sigma)U(\sigma,s)d\sigma, \quad t \geq s. \tag{9.58}$$

Then, $L(t,s) \geq 0$ implies $U(t,s) \geq 0$, so $\{U(t,s) : t \geq s\}$ is a positive evolutionary system. Moreover, we note that $U(t,s)$ is *strongly positive* if $L(t,s)$ is strongly positive. (see Chap. 10) Based on our convention, if $L(t,s)$ is a positive "matrix," $x \mapsto L(t,s)x$ is a strongly positive "operator" on $\mathbb{R}^n$.

Let us define the birth rate vector as $b(t) := M(t)p(t) = M(t)U(t,s)p(s)$ for $t > s$. Then, we have the following renewal equation system:

$$b(t) = g(t) + \int_0^{t-s} \Psi(t,\sigma)b(t-\sigma)d\sigma, \tag{9.59}$$

where $\Psi(t,\sigma) = M(t)L(t,t-\sigma)$ and $g(t) = M(t)L(t,s)p(s)$.

As a special case, let us consider the time-periodic system. Define the state space $E := C_\theta(\mathbb{R};\mathbb{R}^N)$, which is a Banach space of all $\theta$-periodic continuous functions from $\mathbb{R}$ to $\mathbb{R}^N$ equipped with maximum norm. Let $E_+$ be its positive cone. Suppose that $Q(t)$ and $M(t)$ are $\theta$-periodic (matrix-valued) functions. Then, it follows that

$$L(t+\theta,\tau+\theta) = L(t,\tau), \quad U(t+\theta,\tau+\theta) = U(t,\tau),$$

and $\Psi(t+\theta,\sigma) = \Psi(t,\sigma)$.

If $\lambda \in \mathbb{C}$ and $\phi \in E \setminus \{0\}$ are such that $e^{\lambda t}\phi(t)$ satisfies (9.55) for all $t \in \mathbb{R}$, we call this the *exponential solution* with *exponent* $\lambda$. If the basic system (9.55) has a positive exponential solution and any solution is asymptotically proportional to the exponential solution, we call the exponent of the exponential solution the *Malthusian parameter* (or the intrinsic rate of natural increase). For the weakly ergodic evolutionary system (Chap. 10), if there exists a positive exponential solution, any positive solution is asymptotically proportional to the exponential solution because of the weak ergodicity. Thus, the positive exponential solution necessarily dominates the asymptotic behavior of the evolutionary system, and the exponent of the positive exponential solution becomes the Malthusian parameter.

If the periodic system (9.55) has a positive exponential solution $e^{\lambda_0 t}\phi(t)$, the monodromy matrix $U(s+\theta,s)$ has a positive eigenvector $\phi(s)$ associated with a positive eigenvalue $e^{\lambda_0\theta}$, because $e^{\lambda_0(s+\theta)}\phi(s+\theta) = U(s+\theta,s)e^{\lambda_0 s}\phi(s)$ and $\phi(s+\theta) = \phi(s)$. If $U(s+\theta,s)$ is primitive, it follows from the Perron–Frobenius Theorem that $e^{\lambda_0\theta} = r(U(s+\theta,s))$, so $\lambda_0$ is the dominant Floquet exponent. Thus, the following sign relation holds:

$$\text{sign}(\lambda_0) = \text{sign}(r(U(s+\theta,s)) - 1),$$

by which $r(U(s+\theta,s))$ is a surrogate index for the basic reproduction number [28, 29]. However, the monodromy operator is not the NGO, because $r(U(s+\theta,s))$ does not have a generational interpretation.

As shown in the previous section, based on the renewal equation formulation, the NGO $K_\theta$ for the periodic ODE system is defined as

$$(K_\theta \phi)(t) = \int_0^\infty \Psi(t,\sigma)\phi(t-\sigma)d\sigma, \tag{9.60}$$

where $K_\theta$ acts on $E = C_\theta(\mathbb{R}; \mathbb{R}^N)$, and the basic reproduction number is the spectral radius of the positive operator $K_\theta$. In fact, if $s \to -\infty$, (9.59) becomes the (limiting) homogeneous equation

$$b(t) = \int_0^\infty \Psi(t,\sigma)b(t-\sigma)d\sigma. \tag{9.61}$$

If we insert the exponential solution $e^{\lambda t}\phi(t)$ into the homogeneous renewal equation, it follows that

$$\phi(t) = \int_0^\infty e^{-\lambda\sigma}\Psi(t,\sigma)\phi(t-\sigma)d\sigma = (K_\theta(\lambda)\phi)(t), \tag{9.62}$$

which shows that $\phi$ is the (periodic) eigenvector of $K_\theta(\lambda)$ associated with eigenvalue unity. As shown in the previous section, $r(K_\theta(0))$ becomes a per-generation growth factor, and the Malthusian parameter $\lambda_0 \in \mathbb{R}$ exists such that $r(K_\theta(\lambda)) = 1$ and $r(K_\theta(0)) - 1$ and $\lambda_0$ have the same sign.

*Remark 9.4* As shown above, we can define the NGO even for the periodic case, but the NGO for a general time-heterogeneous environment remains unclear. The spectral radius of the GEO would work as a threshold value for the growth bound of the population evolution family $U$, although there remains a gap between $r(K_Y)$ and $R_0$, because the Malthusian parameter does not necessarily exist. In fact, for a general time-heterogeneous environment, Thieme [65] has shown that the spectral radius of the operator $K_\mathbb{F}$ defined by[6]

$$(K_\mathbb{F}\phi)(t) = \int_0^\infty \Psi(t,\sigma)\phi(t-\sigma)d\sigma, \tag{9.63}$$

for $\phi \in \mathbb{F} = C_0(\mathbb{R}; E)$ or $\phi \in \mathbb{F} = L^p(\mathbb{R}; E)$, becomes a threshold value in the sense that the sign relation between the growth bound of the evolution family $U(t,s)$, $t \geq s$ and $r(K_\mathbb{F})$ holds as

$$\text{sign}(\omega(U)) = \text{sign}(r(K_\mathbb{F}) - 1). \tag{9.64}$$

Then, $r(K_\mathbb{F}) < 1$ is a sufficient condition for the population to become extinct, whereas $r(K_\mathbb{F}) > 1$ would be a sufficient condition for population persistence. Instead of $Y = L^1(\mathbb{R}_+; E)$, if we adopt $\mathbb{F} = L^1(\mathbb{R}; E)$ as a space of generation

[6] $C_0(\mathbb{R}; E)$ is the set of continuous $E$-valued functions $f$ on $\mathbb{R}$ such that $\lim_{t\to\pm\infty} f(t) = 0$.

distributions, $K_{\mathbb{F}}$ can be seen as the *generation evolution operator on the line*, while we can call $K_Y$ the *generation evolution operator on the half line*.

*Remark 9.5* Note that delay-differential equations that serve as population models can also be transformed into renewal equations. Thus, their basic reproduction number is essentially defined by the above arguments. As a simple example, consider a delay-differential equation as

$$\frac{dp(t)}{dt} = L(t, t-\tau)M(t-\tau)p(t-\tau) + Q(t)p(t), \tag{9.65}$$

where the initial data $p(\theta) \in C[-\tau, 0]$ are given, and $M$, $Q$ are defined on $[-\tau, \infty)$. Then, (9.65) can be interpreted as a model for an adult (reproductive) population with maturation period $\tau$. Let $b(t) := M(t)p(t)$ be the density of newborns. It is easily checked that the renewal equation holds as

$$b(t) = g(t) + \int_0^t \Psi(t, \eta)b(t-\eta)d\eta, \quad t > 0, \tag{9.66}$$

where

$$g(t) := M(t)L(t, 0)p(0) + \int_{-\tau}^0 M(t)L(t, \sigma)p(\sigma)d\sigma,$$

$$\Psi(t, \eta) := \begin{cases} M(t)L(t, t-\eta), & \eta > \tau, \\ 0, & \eta < \tau. \end{cases}$$

If $M$ and $Q$ are time-independent, the next-generation matrix is given by $M(-Q)^{-1}$, and if $M(t)$ and $Q(t)$ are $\theta$-periodic, the NGO is given by a map $\phi \to \int_0^\infty \Psi(t, \eta)\phi(t-\eta)d\eta$ on $C_\theta(\mathbb{R}; \mathbb{R}^N)$ [40, 65, 75].

**Exercise 9.5** Prove that the DDE model (9.65) can be derived from an age-structured population model as

$$\begin{aligned} &\frac{\partial u(t,a)}{\partial t} + \frac{\partial u(t,a)}{\partial a} = (Q(t) + \varepsilon(a)I)u(t,a), \\ &\frac{\partial v(t,a,\sigma)}{\partial t} + \frac{\partial v(t,a,\sigma)}{\partial a} + \frac{\partial v(t,a,\sigma)}{\partial \sigma} = Q(t)v(t,a,\sigma), \\ &u(t,0) = M(t)\int_0^\infty \int_0^a v(t,a,\sigma)d\sigma da, \\ &v(t,a,0) = \varepsilon(a)u(t,a), \end{aligned} \tag{9.67}$$

where $a$ is age, $\sigma$ is duration elapsed from entering into the adult state, $I$ is the identity matrix, $\varepsilon(\cdot)$ is a scalar transition intensity (common to all states) from the juvenile population $u$ to the adult (reproductive) population $v$ such that $\varepsilon(a)e^{-\int_0^a \varepsilon(\zeta)d\zeta} = \delta(a-\tau)$, and the total size of the adult population is given by $p(t) = \int_0^\infty \int_0^a v(t,a,\sigma)d\sigma da$.

### 9.3.2 PDE Case

In the formulation of general structured population models, Thieme [64] and Diekmann et al. [18] stated that an age-space-dependent population model, which was traditionally formulated by an age-dependent diffusion equation (a partial differential equation, PDE), could be written as a renewal integral equation. Moreover, Bacaër and Ait Dads [12] pointed out that the basic reproduction number could be calculated for the reaction–diffusion equation model by transforming it to a renewal integral equation. A crucial point is that any population model can, in principle, be formulated as an age-dependent model in as far as it describes the renewal process of individuals, where "age" denotes the time elapsed from birth. Here, we sketch the basic idea that differential equation population models can be transformed into renewal equation models.

Let us first consider the non-autonomous ODE case. If we introduce an "implicit" age-density distribution $u(t,a)$ such that

$$p(t) = \int_0^\infty u(t,a)da,$$

then the ODE model (9.55) can be formulated as the age-dependent model

$$\begin{aligned} \frac{\partial u(t,a)}{\partial t} + \frac{\partial u(t,a)}{\partial a} &= Q(t)u(t,a), \\ u(t,0) := b(t) &= \int_0^\infty M(t)u(t,a)da. \end{aligned} \tag{9.68}$$

Integrating along the characteristic line, we have

$$u(t,a) = \begin{cases} L(t,t-a)b(t-a), & t > a, \\ L(t,0)u(0,a-t), & a > t, \end{cases}$$

where $b(t) = u(t,0)$. Inserting the above expression into the boundary condition, it follows that

$$\begin{aligned} b(t) &= M(t)\int_0^\infty u(t,a)da \\ &= \int_0^t M(t)L(t,t-a)b(t-a)da + \int_t^\infty M(t)L(t,0)u(0,a-t)da, \\ &= \int_0^t \Psi(t,t-a)b(t-a)da + M(t)L(t,0)p(0), \end{aligned} \tag{9.69}$$

which is none other than the renewal Eq. (9.59) with $s = 0$.

For the case of a PDE population model without age structure, let us consider a simple diffusion equation:

$$\frac{\partial p(t,x)}{\partial t} = \Delta p(t,x) + (Q(x) + M(x))p(t,x), \tag{9.70}$$

where $p(t, x)$ denotes the (finite-dimensional) population density at time $t$ and position $x$, $Q(x)$ is the state transition matrix at $x$, $M(x)$ is the birth rate matrix at $x$, and $\Delta$ is the diffusion operator with respect to $x$.

Let us introduce an implicit age-density function $u$ as

$$p(t,x) = \int_0^\infty u(t,a,x)da.$$

Then, the above age-independent model can be written in age-dependent form as

$$\begin{aligned} &\frac{\partial u(t,a,x)}{\partial t} + \frac{\partial u(t,a,x)}{\partial a} = \Delta u(t,a,x) + Q(x)u(t,a,x), \\ &u(t,0,x) = \int_0^\infty M(x)u(t,a,x)da. \end{aligned} \tag{9.71}$$

Integrating the above McKendrick equation along the characteristic line, we obtain

$$u(t,a,x) = \begin{cases} L(a,0)b(t-a,\cdot)(x), & t > a, \\ L(a,a-t)u(0,a-t,\cdot)(x), & a > t, \end{cases} \tag{9.72}$$

where $b(t, x) = u(t, 0, x)$ and $L(a, \sigma)$ is the solution operator on $L^1(\Omega)$ of the autonomous system

$$\frac{\partial L(a,\sigma)}{\partial a} = (\Delta + Q)L(a,\sigma), \quad L(\sigma,\sigma) = I_d, \tag{9.73}$$

that is, there exists a semigroup $T(t)$, $t \geq 0$, such that $L(a, \sigma) = T(a - \sigma)$ and its generator is $\Delta + Q$.

Inserting the above expression into the boundary condition, we have

$$
\begin{aligned}
b(t,x) &= M(x)\int_0^\infty u(t,a,x)da \\
&= \int_0^t M(x)L(a,0)b(t-a,\cdot)(x)da + M(x)\int_t^\infty L(a,a-t)u(0,a-t,\cdot)(x)da, \\
&= M(x)\int_0^t (T(a)b(t-a,\cdot))(x)da + M(x)(T(t)p(0,\cdot))(x),
\end{aligned}
\tag{9.74}
$$

which is a renewal equation whose NGO is given by

$$
K = M\int_0^\infty T(a)da = M(-A)^{-1}, \tag{9.75}
$$

where $M$ denotes the multiplication operator defined by $(Mf)(x) = M(x)f(x)$ on $L^1(\Omega)$, and $A = \Delta + Q$ is the generator of the semigroup $T(t)$. As we saw in Sect. 2.5, under appropriate conditions, the operator $\int_0^\infty T(a)da$ can be rewritten as an integral operator on $L^1(\Omega)$. Several authors have introduced the basic reproduction number for the reaction–diffusion equation based on the idea of the renewal equation formulation [57, 71].

*Remark 9.6* As we have already seen in the ODE case, $R_0$ is also characterized as a number such that $s(R_0^{-1}M + A) = 0$. Then, there exists a positive eigenvector of $\Delta + Q + R_0^{-1}M$ associated with the dominant eigenvalue zero. By using the variational method, $R_0$ can be estimated from $M$ and $Q$ [2, 15].

Finally, let us consider a stable population model with diffusion in an additional trait space:

$$
\begin{aligned}
&\frac{\partial p(t,a,x)}{\partial t} + \frac{\partial p(t,a,x)}{\partial a} = (Ap(t,a,\cdot))(x) - \mu(a)p(t,a,x), \\
&p(t,0,x) = \int_0^\infty \beta(a)p(t,a,x)da, \\
&p(0,a,x) = p_0(a,x),
\end{aligned}
\tag{9.76}
$$

where $p(t,\cdot,\cdot) \in E = L^1(\mathbb{R}_+; X(\Omega))$ and $A$ denotes an infinitesimal generator of a strongly continuous semigroup $T(t) = e^{tA}$ on a Banach space $X(\Omega)$ with the trait variable space $\Omega \subset \mathbb{R}^n$. Using the characteristics method, the solution can be expressed as

$$
p(t,a,\cdot) = \begin{cases} \ell(a)T(a)b(t-a,\cdot), \\ \frac{\ell(a)}{\ell(a-t)}T(t)p_0(a-t,\cdot), \end{cases} \tag{9.77}
$$

where $b(t,x) = p(t,0,x)$ and $\ell(a) = \exp(-\int_0^a \mu(\zeta)d\zeta)$. Inserting expression (9.77) into the boundary condition, we again arrive at the renewal equation on $X(\Omega)$:

$$
b(t.\cdot) = g(t,\cdot) + \int_0^t \Psi(a)b(t-a,\cdot)da, \tag{9.78}
$$

where $\Psi(a) := \beta(a)\ell(a)T(a)$ and

$$g(t, \cdot) := \int_t^\infty \beta(a)\frac{\ell(a)}{\ell(a-t)}T(t)p_0(a-t, \cdot)da.$$

Therefore, the PDE model (9.76) is again reduced to an abstract integral equation, and its basic reproduction number is calculated as the spectral radius of the NGO $\int_0^\infty \beta(a)\ell(a)T(a)da$. Marcati and Serafini [50] and Webb [73] investigated this problem in the case where $A$ is the diffusion operator with Dirichlet or Neumann boundary conditions. Readers can find another treatment for this problem in [4].

## 9.4 Type-Reproduction Number

Although the definition and theoretical implications of $R_0$ in heterogeneous populations have been successfully formulated, it is now understood that we cannot always rely on $R_0$ alone when considering forms of population control. In fact, if an intervention policy can only be applied to a specific host type, $R_0$ for the multistate host population does not offer a simple threshold condition for controlling a specific host type, because $R_0$ for the multistate population is the asymptotic ratio (growth factor) of the vector quantity describing the successive generations of individuals. When any intervention for population planning can only be targeted at a specific subpopulation, it is of practical importance to know whether the population can be controlled by intervention with respect to the specific host type alone.

As an improvement on this issue, in the context of epidemic models, Heesterbeek and Roberts proposed the *type-reproduction number* $T$ [31, 58]. The type-reproduction number for a specific (target) host type can be interpreted as the average number of secondary cases of that type produced by the primary cases of the same host type during the entire course of infection. Here, it must be noted that $T$ takes into account not only the secondary cases *directly* transmitted from the specific host, but also the cases *indirectly* transmitted by way of other types (hosts) who were infected from the primary cases in the specific host with no intermediate cases in the target host. Roberts and Heesterbeek showed that $T$ is a useful measure when a particular single host type is targeted as a disease control effort in a community with various host types because, under appropriate assumptions, the threshold can be formulated as $T < 1$, referring only to the target host type. In particular, the well-known *control relation* $\varepsilon = 1 - 1/T$ (where $\varepsilon$ is the critical coverage of immunization) can be used as the critical condition for eradication by means of a control effort targeted only at a specific type of host in a heterogeneous population. Although the type-reproduction number is defined for the birth state (state-at-infection)—that is, the individual state at which new infection can occur—Inaba and Nishiura [36] extended the idea to non-birth states, known as the *state-reproduction number*. Shuai et al. [61] introduced another extension, called the *target reproduction number*, to allow disease control targeting contacts between types.

We now introduce the type-reproduction number (TRN) based on the continuous state space in the demographic setting. Although we deal with the definition of TRN in a constant environment, readers may also be interested in the definition and applications in time-heterogeneous environments [41].

### 9.4.1 A Demographic Example

Before stating the general theory for the TRN, we first give a demographic example [38]. Let us consider a two-region population model and its urban–rural (two-regional) net reproduction matrix (see Chap. 2) as

$$K = \begin{pmatrix} k_{11} & k_{12} \\ k_{21} & k_{22} \end{pmatrix},$$

where $k_{ij}$ is the expected number of female newborns in region $i$ produced by a woman born in region $j$ during her entire life. For instance, let region one be the urban area and region two be the rural area, and assume that $R_0 = r(K) > 1$ and $k_{22} > 1$. Suppose that we are interested in a population control method whereby a stationary population will be attained.

If all entries are uniformly reduced as much as $\varepsilon$, it is clear that the multiregional net reproduction rate becomes $(1-\varepsilon)r(K)$. In this uniform reduction case, the critical proportion of reduction $\varepsilon^*$ to maintain a multiregional zero-growth population is again given by $1 - 1/R_0$, whereas if the proportion of reduction is not uniform, this is not the case.

Suppose that the fertility of the rural area (region two) decreases by as much as $\varepsilon \in (0, 1)$. Hence, the controlled net reproduction matrix, denoted by $K(\varepsilon)$, becomes

$$K(\varepsilon) = \begin{pmatrix} k_{11} & k_{12} \\ (1-\varepsilon)k_{21} & (1-\varepsilon)k_{22} \end{pmatrix}.$$

Alternatively, if individuals born in the rural area reduce their fertility by as much as $\varepsilon$, then the controlled net reproduction matrix becomes

$$K(\varepsilon) = \begin{pmatrix} k_{11} & (1-\varepsilon)k_{12} \\ k_{21} & (1-\varepsilon)k_{22} \end{pmatrix}.$$

In both cases, the net reproduction rate is calculated as

$$R_0 = r(K(\varepsilon)) = \frac{1}{2}\left[k_{11} + (1-\varepsilon)k_{22} + \sqrt{(k_{11} + (1-\varepsilon)k_{22})^2 - 4(1-\varepsilon)\det K}\right],$$

which is a complex function of the reduction rate $\varepsilon$. Because the critical proportion of fertility reduction $\varepsilon^*$ is determined by $r(K(\varepsilon^*)) = 1$, there is no simple relation between $\varepsilon^*$ and $R_0$. If the dimension of the state space becomes larger, it is almost impossible to give an explicit relationship between $R_0$ and the critical proportion of fertility reduction.

If we assume that $k_{11} < 1$—that is, the urban-born population cannot replace itself without migration from the rural area and the rural-born population is the target—the *type-reproduction matrix* $M_2$ is calculated as

$$M_2 = K_2(I - K_1)^{-1},$$

where

$$K_1 := \begin{pmatrix} k_{11} & k_{12} \\ 0 & 0 \end{pmatrix}, \quad K_2 := \begin{pmatrix} 0 & 0 \\ k_{21} & k_{22} \end{pmatrix}.$$

Then, the TRN of the rural region (target region), denoted by $T_2$, can be calculated as its spectral radius:

$$T_2 = r(M_2) = k_{22} + \frac{k_{21}k_{12}}{1 - k_{11}}.$$

The first term, $k_{22}$, denotes the total number of female newborns in the rural region directly produced by a woman born in the rural region. The second part,

$$\frac{k_{21}k_{12}}{1 - k_{11}} = k_{21}(1 + k_{11} + k_{11}^2 + \ldots)k_{12},$$

denotes the number of rural offspring produced by urban-born individuals who are descendants of a rural-born woman with no intermediate rural-born descendants. In fact, a rural-born woman produces $k_{12}$ female newborns in the urban area, an urban-born woman leaves $1/(1 - k_{11}) = 1 + k_{11} + k_{11}^2 + \ldots$ descendants born in the urban area, and each urban-born woman produces $k_{21}$ offspring in the rural area. As shown above, we can prove that $R_0 > 1$ if $T_2 > 1$, $R_0 = 1$ if $T_2 = 1$, and $R_0 < 1$ if $T_2 < 1$; that is, the threshold condition for population growth is expressed by $T_2 = 1$.

It is easily seen that if we decrease the rural fertility by as much as $\varepsilon$—that is, $k_{22}$ and $k_{21}$ are replaced by $(1 - \varepsilon)k_{22}$ and $(1 - \varepsilon)k_{21}$, respectively—or decrease the fertility of females born in the rural area by as much as $\varepsilon$—that is, $k_{22}$ and $k_{12}$ are replaced by $(1 - \varepsilon)k_{22}$ and $(1 - \varepsilon)k_{12}$, respectively—then $T_2$ becomes $(1 - \varepsilon)T_2$ in both cases. The critical reduction rate can then be calculated as

$$\varepsilon^* = 1 - \frac{1}{T_2}.$$

That is, if we decrease the rural fertility by $\varepsilon^*$, or the native-dependent net reproduction rate of the rural area is reduced by $\varepsilon^*$, $T_2$ and $R_0$ become unity and the population converges to a stationary state.

Let us now consider the case in which two control programs are acting independently. Suppose that the rural fertility decreases by $\varepsilon_1$, and the native-dependent net reproduction rate of the rural area decreases by $\varepsilon_2$. The net reproduction matrix then becomes

$$K(\varepsilon_1, \varepsilon_2) = \begin{pmatrix} k_{11} & (1-\varepsilon_2)k_{12} \\ (1-\varepsilon_1)k_{21} & (1-\varepsilon_1)(1-\varepsilon_2)k_{22} \end{pmatrix}.$$

Thus, the TRN of the rural area is $(1-\varepsilon_1)(1-\varepsilon_2)T_2$, and the critical reduction level, denoted by $(\varepsilon_1^*, \varepsilon_2^*)$, is a set given by the nonlinear relation

$$(1-\varepsilon_1^*)(1-\varepsilon_2^*) = \frac{1}{T_2}. \tag{9.79}$$

In such a case, though we no longer have a simple linear control relation, the control set of parameters given by $(1-\varepsilon_1^*)(1-\varepsilon_2^*) < 1/T_2$ is much simpler than the control set defined by $R_0 = r(K(\varepsilon_1, \varepsilon_2)) < 1$.

As a numerical example, let us consider an urban–rural model given by Rogers [60, p. 128]. Suppose that the NGM is given by

$$K = \begin{pmatrix} k_{11} & k_{12} \\ k_{21} & k_{22} \end{pmatrix} = \begin{pmatrix} \frac{3}{4} & \frac{1}{2} \\ \frac{1}{4} & 1 \end{pmatrix}.$$

Then, the reproductivity of this two-region system is above the replacement level:

$$R_0 = r(K) = \frac{1}{2}\left[k_{11} + k_{22} + \sqrt{(k_{11}-k_{22})^2 + 4k_{12}k_{21}}\right] = \frac{5}{4} > 1.$$

Because $k_{11} < 1$, we can calculate the TRN for the rural area as

$$T_2 = k_{22} + \frac{k_{12}k_{21}}{1-k_{11}} = \frac{3}{2}.$$

Therefore, we know that $\varepsilon^* = 1 - 1/T_2 = 1/3$, and hence, the total population will be controlled to the stationary state if we decrease the rural fertility, or the rural-born individuals' fertility, by as much as 33 percent. In contrast, the total population will reach the stationary state if we decrease everyone's fertility by $(1-1/R_0) \times 100 = 20$ percent.

### 9.4.2 General Theory

Let $\Omega_b \subset \mathbb{R}^n$ be an $i$-state space indicating the birth state (state-at-infection in the epidemiological setting) and $E = L^1(\Omega_b)$ be the $p$-state space (the space of density functions of state-specific populations). We focus on the renewal process of a specific type set of a population. Without loss of generality, we assume that the host-type set can be decomposed as $\Omega_b = \Omega_1 \cup \Omega_2$, $\Omega_1 \cap \Omega_2 = \emptyset$. The host type $\Omega_1$ is the target host for which we would like to calculate the TRN.

Let $\chi_j(\zeta)$ be the definition function of a subset $\Omega_j$ such that

$$\chi_j(\zeta) = \begin{cases} 1, & \zeta \in \Omega_j, \\ 0, & \zeta \in \Omega_b \setminus \Omega_j \end{cases}.$$

The projection operator $P_j$ is defined by

$$(P_j u)(\zeta) = \chi_j(\zeta)u(\zeta), \quad u \in E.$$

Note that $P_1 + P_2 = I_d$. Thus, we can define a projected NGO as $K_j := P_j K$, which produces a distribution of newborns with $i$-state variable belonging to $\Omega_j$.

Let $b_1(t) := P_1 b(t)$ be the density of newborns of a target population and $b_2(t) := P_2 b(t)$ be the density of newborns of a nontarget population. Then, corresponding to the host-type decomposition, the basic Eq. (9.3) is decomposed into a system of renewal equations:

$$b_1(t) = g_1(t) + (\Psi_1 * b_1)(t) + (\Psi_1 * b_2)(t), \tag{9.80}$$

$$b_2(t) = g_2(t) + (\Psi_2 * b_1)(t) + (\Psi_2 * b_2)(t), \tag{9.81}$$

where $\Psi_j := P_j \Psi$, $g_j(t) := P_j g(t)$, and $*$ denotes the convolution of functions.

Let us introduce the resolvent kernel $\Phi_2$ corresponding to the integral kernel $\Psi_2$ as the solution of the resolvent equation

$$\Phi_2 = \Psi_2 + \Psi_2 * \Phi_2 = \Psi_2 + \Phi_2 * \Psi_2. \tag{9.82}$$

Using the resolvent kernel, if we consider $g_2 + \Psi_2 * b_1$ as the initial data in (9.81), we can formally solve for $b_2$ as

$$\begin{aligned} b_2 &= g_2 + \Psi_2 * b_1 + \Phi_2 * (g_2 + \Psi_2 * b_1) \\ &= g_2 + \Phi_2 * g_2 + (\Psi_2 + \Phi_2 * \Psi_2) * b_1 \\ &= g_2 + \Phi_2 * g_2 + \Phi_2 * b_1. \end{aligned}$$

Inserting the above expression for $b_2$ into (9.80), we arrive at a single renewal equation for $b_1$:

$$b_1 = g_1 + \Psi_1 * g_2 + (\Psi_1 * \Phi_2) * g_2 + [\Psi_1 + \Psi_1 * \Phi_2] * b_1, \tag{9.83}$$

which expresses a *hypothetical* renewal process of the target individuals. The *net reproduction operator* for the target population is then given by

$$\Pi_1(\tau) := \Psi_1(\tau) + (\Psi_1 * \Phi_2)(\tau). \tag{9.84}$$

Define the NGO $K$ and its decomposition $K_j$ as

$$K = \int_0^\infty \Psi(\tau)d\tau, \quad K_j = \int_0^\infty \Psi_j(\tau)d\tau.$$

If $r(K_2) < 1$, then the positive operator

$$(I - K_2)^{-1} = \sum_{m=0}^{\infty} K_2^m$$

exists and the resolvent kernel $\Phi_2$ is integrable on $[0, \infty)$ as

$$Q_2 := \int_0^\infty \Phi_2(\tau)d\tau = \sum_{m=1}^{\infty} K_2^m = (I - K_2)^{-1}K_2 = K_2(I - K_2)^{-1}.$$

Hence, $\Pi_1$ is also integrable, and we have

$$M_1 := \int_0^\infty \Pi_1(\tau)d\tau = K_1 + K_1 Q_2, \tag{9.85}$$

where $M_1$ is the NGO for the target hosts, called the *type-reproduction operator* (TRO), and $K_j$ is the type-specific NGO. From (9.85), we obtain

$$M_1 = K_1(I + Q_2) = K_1(I - K_2)^{-1}. \tag{9.86}$$

Therefore, in a constant environment, the TRN $T$ for the target host is computed as the spectral radius of the TRO $M_1$:

$$T = r(M_1) = r(K_1(I - K_2)^{-1}), \tag{9.87}$$

which is a well-known formula for the case of a finite-dimensional state space [58].

We can now apply the following result of Thieme:

**Proposition 9.7** ([65], Theorem 3.10) *Let $X$ be an ordered Banach space with a closed convex cone $X_+$ that is normal and generating. Let $A_j : X \to X$ $(j = 1, 2)$ be positive linear operators with $r(A_2) < 1$. Then, it follows that*

$$\mathrm{sign}(r(A_1 + A_2) - 1) = \mathrm{sign}(r(A_1(I - A_2)^{-1}) - 1). \tag{9.88}$$

Because $R_0 = r(K)$, applying the above general result to the splitting of $K$ as $K = K_1 + K_2$ with $r(K_2) < 1$, we have the following sign relation:

**Proposition 9.8** *Suppose that $r(K_2) < 1$. Then, the sign relation*

$$\mathrm{sign}(R_0 - 1) = \mathrm{sign}(T - 1) \tag{9.89}$$

*holds.*

From (9.89), the TRN is a surrogate index of $R_0$, and it does not give a per-generation growth factor for a real biological renewal process. If we assume a kind of irreducibility of $K$, the sign relation (9.89) is extended to a trichotomy which is an infinite-dimensional extension of Li–Schneider theorem [46]. For finite-dimensional case, this kind of characterization is already given for $R_0$ [20]:

**Proposition 9.9** *Assume that $E$ has a total positive cone $E_+$ and the NGO $K$ is compact and semi-nonsupporting. If $K$ is split as $K = K_1 + K_2$, where $K_1$ and $K_2$ are positive compact operators with $r(K_2) < 1$ and $T = r(M_1) > 0$, then it follows that*

$$r(\frac{K_1}{T} + K_2) = 1, \tag{9.90}$$

*and one of the following holds:*

$$R_0 = T = 1, \quad 1 < R_0 < T, \quad 0 < T < R_0 < 1. \tag{9.91}$$

*Proof* By the Krein–Rutman Theorem, $T$ is an eigenvalue of a compact positive operator $M_1$ corresponding to a nonnegative eigenfunction $\phi \in E$. Then, we have

$$M_1\phi = T\phi = K_1(I - K_2)^{-1}\phi.$$

Let $\psi := (I - K_2)^{-1}\phi$. It follows from $\psi \in E_+ \setminus \{0\}$ and $K_1\psi = T(I - K_2)\psi$ that

$$\left(\frac{K_1}{T_1} + K_2\right)\psi = \psi.$$

Therefore, $\psi$ is a nonnegative eigenfunction associated with the unity eigenvalue of $K_1/T + K_2$. Because $K_1/T + K_2 \geq \min(1, 1/T)K$, $K_1/T + K_2$ is also compact and semi-nonsupporting. Thus, it must hold that $r(\frac{K_1}{T} + K_2) = 1$. If $T = 1$, we have $R_0 = r(K) = r(K_1 + K_2) = 1$. If $T > 1$, we have

$$\frac{K_1}{T} + K_2 < K_1 + K_2 < K_1 + TK_2.$$

As $K = K_1 + K_2$ is compact and semi-nonsupporting, it follows from the comparison theorem [52] that

$$1 = r\left(\frac{K_1}{T} + K_2\right) < r(K) = R_0 < Tr\left(\frac{K_1}{T} + K_2\right) = T.$$

Finally, if $0 < T < 1$,

$$1 = r\left(\frac{K_1}{T} + K_2\right) > r(K) = R_0 > Tr\left(\frac{K_1}{T} + K_2\right) = T.$$

This completes the proof. □

*Remark 9.7* Let us calculate the TRO based on the ODE formulation (9.50):

$$\frac{dp(t)}{dt} = (Q + M)p(t) = Qp(t) + (P_1 M + P_2 M)p(t). \tag{9.92}$$

Then, the type-specific NGO is calculated from the generator as follows:

$$K_j = \int_0^\infty P_j M e^{Q\sigma} d\sigma = P_j M(-Q)^{-1}. \tag{9.93}$$

From (9.86), we obtain a formula for the TRO:

$$M_1 = P_1 M(-Q)^{-1}(I - P_2 M(-Q)^{-1})^{-1} = P_1 M(-(Q + P_2 M))^{-1}. \tag{9.94}$$

In fact, for the resolvent positive operator $Q + P_2 M$, it follows that $\text{sign}(s(Q + P_2 M)) = \text{sign}(r(P_2 M(-Q)^{-1}) - 1)$. Then, $r(K_2) < 1$ implies that $0 \in \rho(Q + P_2 M)$ and $-(Q + P_2 M)$ is nonnegatively invertible. Therefore, the TRN is characterized as a number $T$ such that

$$s(\frac{P_1 M}{T} + P_2 M + Q) = 0. \tag{9.95}$$

Expressions (9.94) and (9.95) show that the TRN $T$ can be interpreted as a basic reproduction number in cases where the birth state (state-at-infection in epidemic models) is limited to the target host and the birth of nontarget hosts is interpreted as a transition to the nontarget state.

## 9.5 Applications in Epidemiology

### *9.5.1 Calculating $R_0$ in a Periodic Environment*

Because the NGO for the periodic case is an infinite-dimensional operator, it is generally difficult to obtain an analytical expression of the basic reproduction number in a periodic environment. However, in simple scalar cases, it is possible to calculate $R_0$ in a periodic environment.

First, let us consider a non-structured SIR epidemic model with time-periodic parameters. The linearized dynamics of the infected population are described by a scalar equation:

$$\frac{dI(t)}{dt} = (\beta(t)S(t) - \mu(t))I(t), \tag{9.96}$$

where $S(t)$ is the susceptible host population density at the disease-free state, $\beta(t)$ denotes the transmission rate, and $\mu(t)$ denotes the removal rate of infected individuals. We assume that $S$, $\beta$, and $\mu$ have some period $\theta > 0$. Using the variation-of-constants formula, we have

$$I(t) = e^{-\int_0^t \mu(x)dx} I(0) + \int_0^t e^{-\int_s^t \mu(x)dx} \beta(s)S(t)I(s)ds.$$

Let $v(t) := \beta(t)S(t)I(t)$ be the incidence of new infection. We then obtain a renewal equation:

$$v(t) = \beta(t)S(t)e^{-\int_0^t \mu(x)dx} I(0) + \beta(t)S(t) \int_0^t e^{-\int_{t-s}^t \mu(x)dx} v(t-s)ds. \quad (9.97)$$

The NGO is defined by

$$(K_\theta \phi)(t) = \int_0^\infty \Psi(t,s)\phi(t-s)ds, \quad (9.98)$$

where

$$\Psi(t,s) := \beta(t)S(t)e^{-\int_{t-s}^t \mu(x)dx}$$

and $\phi \in C_\theta(\mathbb{R})$, which is a set of $\theta$-periodic continuous functions. As shown by Bacaër and Guernaoui [5], we can calculate the basic reproduction number as follows:

$$R_0 = r(K_\theta) = \frac{\int_0^\theta \beta(t)S(t)dt}{\int_0^\theta \mu(t)dt}. \quad (9.99)$$

In fact, $R_0$ is a positive number such that the modified system

$$\frac{dI(t)}{dt} = \left(\frac{\beta(t)S(t)}{R_0} - \mu(t)\right) I(t)$$

has a zero Malthusian parameter. Because

$$\lim_{t\to\infty} \frac{\log I(t)}{t} = \lim_{t\to\infty} \frac{1}{t} \int_0^t \left(\frac{\beta(\sigma)S(\sigma)}{R_0} - \mu(\sigma)\right) d\sigma$$
$$= \frac{1}{\theta} \int_0^\theta \left(\frac{\beta(\sigma)S(\sigma)}{R_0} - \mu(\sigma)\right) d\sigma = 0,$$

then we have expression (9.99). Traditionally, (9.99) has been used as the basic reproduction number for a seasonal homogeneous SIR model [26].

An interesting application of (9.99) is the calculation of the effective reproduction number for a *pulse vaccination* strategy, which is a powerful tool for disease eradication [1, 55, 56, 66]. Under the pulse vaccination scheme, a fraction $\varepsilon$ of the entire susceptible population are vaccinated in a single pulse, which is applied every $\theta$ years [62]. That is, under a pulse vaccination strategy, the environment (the susceptible host population) of infected individuals becomes artificially periodic.

First, we consider a simple case in which the environment is constant and the susceptible (non-structured) population is described by

$$\frac{dS(t)}{dt} = b - \mu S(t), \quad t \in (n\theta, (n+1)\theta)$$

with the pulse condition

$$S(n\theta + 0) = (1 - \varepsilon)S(n\theta - 0),$$

where $n$ is an integer, $b$ denotes the birth rate, and $\mu$ is the natural death rate. It is relatively easy to find a $\theta$-periodic (disease-free) solution as follows:

$$S(t) = \frac{b}{\mu} + \left(S^* - \frac{b}{\mu}\right) e^{-\mu(t-n\theta)}, \quad t \in [n\theta, (n+1)\theta),$$

where

$$S^* := \frac{(b/\mu)(1 - e^{-\mu\theta})(1 - \varepsilon)}{1 - (1 - \varepsilon)e^{-\mu\theta}}.$$

Let $\gamma$ be the recovery rate of infected individuals. If the infected population in the linear phase is described by

$$\frac{dI(t)}{dt} = \beta S(t)I(t) - (\mu + \gamma)I(t),$$

then the effective reproduction number, denoted by $R_e$, is calculated using (9.99) as

$$R_e = \frac{\beta}{\mu + \gamma}\frac{1}{\theta}\int_0^\theta S(\sigma)d\sigma = R_0\left[1 - \frac{\varepsilon(1 - e^{-\mu\theta})}{\mu\theta(1 - (1 - \varepsilon)e^{-\mu\theta})}\right], \tag{9.100}$$

where

$$R_0 = \frac{b\beta}{\mu(\mu + \gamma)}$$

is the basic reproduction number for the totally susceptible population of size $b/\mu$. The criterion for a disease-free population is then formulated as $R_e < 1$, a result that was derived by Shulgin et al. [62] using Floquet theory.

From (9.100), the critical coverage of immunization for a single pulse, denoted by $\varepsilon^*$, satisfies the relation

$$1 - \frac{1}{R_0} = \frac{\varepsilon^*(1 - e^{-\mu\theta})}{\mu\theta(1 - (1 - \varepsilon^*)e^{-\mu\theta})},$$

where the left-hand side gives the critical coverage of immunization for the continuous vaccination policy applied to newborns. For a sufficiently small $\mu$, the right-hand side is greater than $\varepsilon^*$, so we can expect $\varepsilon^*$ to be less than the critical coverage of immunization given by the traditional control relation. Expression (9.100) is also used to estimate the maximum period of the pulse in which the eradication criterion $R_e < 1$ is satisfied.

If the transmission coefficient $\beta$ is a periodic function of time reflecting the seasonal nature of the epidemic, the effective reproduction number is calculated as

$$R_e = \frac{1}{\mu + \gamma} \frac{1}{\theta} \int_0^\theta \beta(\sigma) S(\sigma) d\sigma \tag{9.101}$$

as long as $\beta$ has a period $\theta$ (that is, the season) and the period of the pulse has the same period $\theta$. This situation occurs if the period of the pulse is an integer-valued number of years. The above argument can be extended to the age-dependent case [41] and could be applied to the hybrid model with birth pulses [21, Exercise 4.14].

## 9.5.2 Control Relations Using T in a Constant Environment

In the context of epidemic models [31], *S-control* means that control (intervention in the infection process) acts primarily to reduce the availability of susceptible individuals of the target type. Therefore, S-control reduces the potential of all types to produce the target type, which is described as a change in the type-specific NGO $K_1$ if state one indicates the target type. If $K_1$ is replaced by $(1 - v)K_1$ under control effort $v \in [0, 1]$ (e.g., $v$ may be the vaccination coverage), the effective TRN is given by

$$T_e = r((1 - v)K_1(I - K_2)^{-1}) = (1 - v)T.$$

The critical effort devoted to the target type alone to prevent an epidemic in the whole population is then calculated as

$$v^* = 1 - \frac{1}{T}.$$

Note that from (9.91), if $R_0 > 1$, we have

$$1 - \frac{1}{R_0} < 1 - \frac{1}{T},$$

which shows that the critical control effort directed at the target host only is always larger than the critical effort directed at the total host.

In contrast to S-control, *I-control* signifies intervention in the survival process of the target type. Therefore, I-control reduces the potential of the target type to produce cases of all types. This changes both $K_1$ and $K_2$, and we cannot obtain a simple control relation such as for S-control under our decomposition $K = K_1 + K_2$. Instead, let us split the NGO $K$ as

$$K = \tilde{K}_1 + \tilde{K}_2,$$

where $\tilde{K}_j = KP_j$. The operator $\tilde{K}_1$ expresses the reproductivity (for all types) of the target type, so applying I-control to the target type changes $\tilde{K}_1$ alone. From Proposition 9.7, if $r(\tilde{K}_2) < 1$, it follows that

$$\text{sign}(r(K) - 1) = \text{sign}(r(\tilde{K}_1(I - \tilde{K}_2)^{-1}) - 1),$$

so we can obtain an alternative definition of the TRN as

$$T = r(\tilde{K}_1(I - \tilde{K}_2)^{-1}).$$

Therefore, if I-control causes $\tilde{K}_1$ to be replaced by $(1 - v)\tilde{K}_1$ under control effort $v$, the effective TRN under I-control is given by

$$T_e = r((1 - v)\tilde{K}_1(I - \tilde{K}_2)^{-1}) = (1 - v)T.$$

Then, the critical effort is again calculated as $v^* = 1 - 1/T$. As discussed in [31], if the control strategy is a mixture of S-control and I-control, we cannot obtain such a simple control relation.

### 9.5.3 *Critical Coverage of Vaccination*

One of most important purposes of the age-structured epidemic model is to calculate the *critical coverage of vaccination*, or the *critical proportion of immunization*.

Let $S(a)$ be the age density of the susceptible host stationary population. Suppose that some fraction $e$ of the host population at age $a_0$ is vaccinated. Then, the stationary age density of susceptible individuals under the continuous mass vaccination policy is given by

$$S^*(a) = S_1(a) + (1 - e)S_2(a),$$

where $S_1(a) := S(a)\chi_1(a)$, $S_2(a) := S(a)\chi_2(a)$, and $\Omega_b = [0, \infty)$, $\Omega_1 = [0, a_0)$ and $\Omega_2 = [a_0, \infty)$.

If we adopt a homogeneous force of infection for the age–duration-structured epidemic model (9.114), the effective NGO under the one-time vaccination policy, denoted by $K_e$, is given by

$$(K_e u)(a) = \frac{S^*(a)}{N} \int_0^\infty \int_0^\infty \beta(a, \tau + \zeta) \frac{\ell(\tau + \zeta)}{\ell(\zeta)} f(\tau)\Gamma(\tau) d\tau u(\zeta) d\zeta, \quad (9.102)$$

where $\beta(a,\sigma)$ is the transmission coefficient between ages $a$ and $\sigma$, $f(\tau)$ is the probability of successful transmission of infective agents from infective individuals with infection-age $\tau$, $N$ is the total size of the host population, which is assumed to be constant, $\Gamma(\tau)$ is the survival probability induced from the recovery rate, and $\ell(a)$ is the survival probability with respect to the natural death rate.

Then, we can decompose $K_e$ as

$$K_e = K_1 + (1-e)K_2, \tag{9.103}$$

where $K_j$ $(j = 1, 2)$ are positive operators given by

$$(K_j u)(a) = \frac{S_j(a)}{N}\int_0^\infty \int_0^\infty \beta(a, \tau+\zeta)\frac{\ell(\tau+\zeta)}{\ell(\zeta)} f(\tau)\Gamma(\tau)d\tau u(\zeta)d\zeta.$$

It is clear from $R_e = r(K_e) \geq r(K_1)$ that we cannot control the disease using a one-time vaccination policy if $r(K_1) \geq 1$. Therefore, we assume that the vaccination age $a_0$ can be chosen such that $r(K_1) < 1$. That is, $\Omega_2$ is the target state. Under this condition, we can define the TRO $M_2$ as

$$M_2 := K_2(I - K_1)^{-1}. \tag{9.104}$$

The TRO $M_2$ produces the next generation of infected individuals in age class $[a_0, \infty)$ intermediated by the infected individuals in age class $[0, a_0)$.

Let $T = r(M_2)$ be the TRN of the post-vaccination age class. Because $R_e = r(K_1 + (1-e)K_2) < 1$ if and only if

$$r((1-e)K_2(I-K_1)^{-1}) = r((1-e)M_2) = (1-e)T < 1,$$

the critical coverage of vaccination at age $a_0$ is given by

$$e^* = 1 - \frac{1}{T}.$$

Then, if $e > e^*$, then $R_e < 1$, and the local disease invasion can be prevented by one-time vaccination. For the special case $a_0 = 0$ in which all newborns are targeted by the immunization policy, we have the well-known control relation $e^* = 1 - 1/R_0$, because $K_1 = 0$ and $T = K_2 = K$.

Using the separable mixing assumption, the type-specific NGO $K_j$ is one dimensional:

$$(K_j u)(a) = \frac{S_j(a)}{N}\beta_1(a)\langle F_0, u\rangle,$$

where $F_0$ is a positive functional defined by

$$\langle F_0, u\rangle := \int_0^\infty \int_0^\infty \beta_2(\tau+\zeta)\frac{\ell(\tau+\zeta)}{\ell(\zeta)} f(\tau)\Gamma(\tau)d\tau u(\zeta)d\zeta.$$

Then, it is easy to see that

$$r(K_1) = \left\langle F_0, \frac{S_1}{N}\beta_1 \right\rangle.$$

If $r(K_1) < 1$, we obtain

$$((I - K_1)^{-1}u)(a) = u(a) + \frac{\langle F_0, u\rangle}{1 - r(K_1)} \frac{S_1(a)}{N}\beta_1(a).$$

The TRO is also a one-dimensional operator given by

$$(M_2 u)(a) = \frac{S_2(a)}{N}\beta_1(a)\left[\langle F_0, u\rangle + \frac{r(K_1)\langle F_0, u\rangle}{1 - r(K_1)}\right]. \tag{9.105}$$

Hence, the TRN is calculated as

$$T = r(M_2) = r(K_2) + \frac{r(K_1)r(K_2)}{1 - r(K_1)} = \frac{r(K_2)}{1 - r(K_1)}, \tag{9.106}$$

where

$$r(K_2) = \left\langle F_0, \frac{S_2}{N}\beta_1 \right\rangle.$$

If we introduce the immunization policy with coverage $e$ at age $a_0$, the effective reproduction number is $R_e = r(K_1) + (1 - e)r(K_2)$ and the critical proportion of immunization is given by

$$e^* = 1 - \frac{1}{T} = \left(1 - \frac{1}{R_0}\right)\frac{1}{\xi}, \tag{9.107}$$

where $R_0 = r(K_1) + r(K_2)$ and

$$\xi := \frac{r(K_2)}{R_0} = \frac{\int_{a_0}^{\infty}\int_0^{\infty}\beta_2(\tau+\zeta)\frac{\ell(\tau+\zeta)}{\ell(\zeta)}f(\tau)\Gamma(\tau)d\tau c_2(\zeta)\beta_1(\zeta)d\zeta}{\int_0^{\infty}\int_0^{\infty}\beta_2(\tau+\zeta)\frac{\ell(\tau+\zeta)}{\ell(\zeta)}f(\tau)\Gamma(\tau)d\tau c_2(\zeta)\beta_1(\zeta)d\zeta}.$$

The proportion $e^*$ gives the critical coverage of immunization, and the proportion $\xi$ is the proportion of secondary cases occurring for susceptible individuals above $a_0$ years of age. If we can assume that $\beta$ is constant and the length of infectiousness is sufficiently short, we have

$$\xi \approx \frac{1}{e_0}\int_{a_0}^{\infty}\ell(\zeta)d\zeta,$$

which is the proportion of individuals above $a_0$ years of age in the host stationary population.

### 9.5.4 Critical Proportion of Case Isolation

To calculate the critical proportion of *case isolation* based on the *onset* observation, let us now consider the (linearized) asymptomatic transmission model with age structure. The ODE version of the asymptomatic transmission model is studied in Sect. 5.2, where we consider the state-reproduction number for the onset cases. Here, we formulate the critical coverage of isolation based on the renewal process of the asymptomatic infected individuals. For simplicity, we neglect the infection-age and the disease-age structure:

$$\begin{aligned} &\frac{\partial E(t,a)}{\partial t} + \frac{\partial E(t,a)}{\partial a} = -(\mu(a) + \eta(a))E(t,a) + S(a)\lambda(t,a), \\ &\frac{\partial I(t,a)}{\partial t} + \frac{\partial I(t,a)}{\partial a} = -(\mu(a) + \gamma(t,a))I(t,a) + \eta(a)E(t,a), \\ &E(t,0) = I(t,0) = 0, \end{aligned} \tag{9.108}$$

where $S(a)$ is the stationary age distribution of susceptible individuals, $E(t,a)$ is the age distribution of asymptomatic infectives at time $t$, $I(t,a)$ is the age distribution of symptomatic infectives, $\eta(a)$ is the age-specific rate of onset, and $\gamma(a)$ is the recovery rate. The force of infection $\lambda(t,a)$ is given by

$$\lambda(t,a) = \int_0^\infty (\beta_E(a,\sigma)E(t,\sigma) + \beta_I(a,\sigma)I(t,\sigma))d\sigma,$$

where $\beta_E$ is the transmission coefficient between susceptibles and asymptomatic infectives, and $\beta_I$ is the transmission coefficient between susceptibles and symptomatic infectives.

Although all newly infected individuals appear as asymptomatic infectives, and so the state-at-infection is the space of the age variable $\Omega_b = \mathbb{R}_+$, we can extend the $h$-state space by dividing the new infectives into two parts:

$$\begin{aligned} b_1(t,a) &= S(a)\int_0^\infty \beta_E(a,\sigma)E(t,\sigma)d\sigma, \\ b_2(t,a) &= S(a)\int_0^\infty \beta_I(a,\sigma)I(t,\sigma)d\sigma, \end{aligned} \tag{9.109}$$

where $b_1(t,a)$ is the incidence of new infectives who are infected by asymptomatic infectives, and $b_2(t,a)$ is the incidence of new infectives infected by symptomatic infectives. Then, $\Omega_b = \mathbb{R}_+ \times \mathbb{R}_+$, $E_+ = L^1_+(\mathbb{R}_+) \times L^1_+(\mathbb{R}_+)$, and we can apply the type-reproduction theory.

If we assume that an epidemic starts at $t = -\infty$, we obtain

$$E(t,a) = \int_0^a \frac{\ell(a)\Xi(a)}{\ell(a-\sigma)\Xi(a-\sigma)}(b_1(t-\sigma,a-\sigma)+b_2(t-\sigma,a-\sigma))d\sigma,$$
$$I(t,a) = \int_0^a \frac{\ell(a)\Gamma(a)}{\ell(a-\sigma)\Gamma(a-\sigma)}\eta(a-\sigma)E(t-\sigma,a-\sigma)d\sigma, \tag{9.110}$$

where $\Xi(a) := \exp(-\int_0^a \eta(\sigma)d\sigma)$. Therefore, we obtain a system of abstract renewal equations for the $E$-valued function $(b_1(t), b_2(t))^{\mathrm{T}}$:

$$\begin{pmatrix} b_1(t,\cdot) \\ b_2(t,\cdot) \end{pmatrix} = \int_0^\infty \begin{pmatrix} \Psi_1(\tau) & \Psi_1(\tau) \\ \Psi_2(\tau) & \Psi_2(\tau) \end{pmatrix} \begin{pmatrix} b_1(t-\tau,\cdot) \\ b_2(t-\tau,\cdot) \end{pmatrix} d\tau, \tag{9.111}$$

where we have omitted the initial data and $\Psi_j(\tau)$ are linear operators from $L^1(\mathbb{R}_+)$ into itself defined by

$$(\Psi_1(\tau)f)(a) := S(a)\int_\tau^\infty \beta_E(a,\sigma)\frac{\ell(\sigma)\Xi(\sigma)}{\ell(\sigma-\tau)\Xi(\sigma-\tau)}f(\sigma-\tau)d\sigma,$$
$$(\Psi_2(\tau)f)(a) := S(a)\int_\tau^\infty \beta_I(a,\sigma) \times \int_{\sigma-\tau}^\sigma \frac{\ell(\sigma)\Gamma(\sigma)}{\ell(x)\Gamma(x)}\frac{\eta(x)\ell(x)\Xi(x)}{\ell(\sigma-\tau)\Xi(\sigma-\tau)}dx f(\sigma-\tau)d\sigma$$

for $f \in L^1(\mathbb{R}_+)$.

For the renewal system (9.111), the NGO $K$ on $E_+$ is given by

$$K = \begin{pmatrix} k_1 & k_1 \\ k_2 & k_2 \end{pmatrix} = K_1 + K_2,$$

where

$$k_j := \int_0^\infty \Psi_j(\tau)d\tau, \quad K_1 := \begin{pmatrix} k_1 & k_1 \\ 0 & 0 \end{pmatrix}, \quad K_2 := \begin{pmatrix} 0 & 0 \\ k_2 & k_2 \end{pmatrix}.$$

Note that $k_j$ denotes a positive integral *operator* on $L^1_+(\mathbb{R}_+)$ and $K_j$ is the type-specific NGO from $E_+$ into $L^1_+(\mathbb{R}_+) \times \{0\}$ or $\{0\} \times L^1_+(\mathbb{R}_+)$. That is, the projection operator $P_j$ is defined as $P_1 : (f_1, f_2) \to (f_1, 0)$ and $P_2 : (f_1, f_2) \to (0, f_2)$, where $(f_1, f_2) \in E$, and the target state-at-infection is given by $\{0\} \times L^1_+(\mathbb{R}_+)$. From the argument in the previous section, if $r(K_1) < 1$, the TRN for the target state is given by $T = r(K_2(I - K_1)^{-1})$.

Moreover, we can calculate $T$ using the operators $k_j$ on $L^1(\mathbb{R}_+)$ as $T = r(k_2(I - k_1)^{-1})$, where $I$ denotes the identity operator on $L^1(\mathbb{R}_+)$. To see this fact, let $\phi = (0, \phi_2)^{\mathrm{T}}$ be the positive eigenvector of $K_2(I - K_1)^{-1}$ associated with $T$, that is, $K_2(I - K_1)^{-1}\phi = T\phi$. Let

$$(I - K_1)^{-1}\begin{pmatrix} 0 \\ \phi_2 \end{pmatrix} = \begin{pmatrix} v_1 \\ v_2 \end{pmatrix}.$$

Because $r(K_1) < 1$, $v := (v_1, v_2)^{\mathrm{T}}$ is a positive vector. Thus, we have

$$v_1 - k_1(v_1 + v_2) = 0, \quad v_2 = \phi_2,$$

from which we obtain $v_1 = (I - k_1)^{-1} k_1 v_2$. In contrast, from

$$K_2 \begin{pmatrix} v_1 \\ v_2 \end{pmatrix} = T \begin{pmatrix} 0 \\ v_2 \end{pmatrix},$$

it follows that $k_2(v_1 + v_2) = T v_2$. Therefore, we have

$$k_2((I - k_1)^{-1} k_1 v_2 + v_2) = k_2(I - k_1)^{-1} v_2 = T v_2,$$

which shows that $v_2$ is the positive eigenvector of $k_2(I - k_1)^{-1}$ and $T = r(k_2(I - k_1)^{-1})$ is its positive eigenvalue.

Let $\varepsilon$ be the case isolation proportion of new onsets of infection. Under a case isolation policy, the number of secondary cases infected by symptomatic infected individuals decreases. In fact, the incidence of new onset $\eta(a)E(t, a)$ is replaced by $(1 - \varepsilon)\eta(a)E(t, a)$, and hence, the effective NGO is given by $K_e = K_1 + (1 - \varepsilon)K_2$, so the effective TRN for state two is calculated as $r((1 - \varepsilon)k_2(I - k_1)^{-1}) = (1 - \varepsilon)T$. Therefore, the critical proportion of case isolation is again calculated as

$$\varepsilon^* = 1 - \frac{1}{T}. \tag{9.112}$$

Let us use the separable mixing assumption that $\beta_E(a, \sigma) = \beta_{E1}(a)\beta_{E2}(\sigma)$ and $\beta_I(a, \sigma) = \beta_{I1}(a)\beta_{I2}(\sigma)$. Then, $k_j$ is a one-dimensional map:

$$(k_1\phi)(a) = \langle F_1, \phi \rangle S(a)\beta_{E1}(a), \quad (k_2\phi)(a) = \langle F_2, \phi \rangle S(a)\beta_{I1}(a),$$

where $F_1$ and $F_2$ are positive functionals defined by

$$\langle F_1, \phi \rangle = \int_0^\infty \int_\tau^\infty \beta_{E2}(\sigma) \frac{\ell(\sigma)\Xi(\sigma)}{\ell(\sigma - \tau)\Xi(\sigma - \tau)} \phi(\sigma - \tau) d\sigma d\tau,$$

$$\langle F_2, \phi \rangle = \int_0^\infty \int_\tau^\infty \beta_{I2}(\sigma) \int_{\sigma-\tau}^{\sigma} \frac{\ell(\sigma)\Gamma(\sigma)}{\ell(x)\Gamma(x)} \frac{\eta(x)\ell(x)\Xi(x)}{\ell(\sigma - \tau)\Xi(\sigma - \tau)} dx \phi(\sigma - \tau) d\sigma d\tau.$$

It is then easy to see that

$$r(k_1) = \langle F_1, S\beta_{E1} \rangle, \quad r(k_2) = \langle F_2, S\beta_{I1} \rangle,$$

where $r(k_1)$ is the average number of secondary infectives produced by a new infective individual in the asymptomatic state, and $r(k_2)$ is the average number of secondary infective individuals produced by a new infective individual in the symptomatic state.

If $r(K_1) = r(k_1) < 1$, we obtain

$$((I - k_1)^{-1}\phi)(a) = \phi(a) + \frac{\langle F_1, \phi\rangle}{1 - r(k_1)} S(a)\beta_{E1}(a).$$

Then, it follows that

$$(k_2(I - k_1)^{-1}\phi)(a) = S(a)\beta_{I1}(a)\left[\langle F_2, \phi\rangle + \frac{\langle F_1, \phi\rangle\langle F_2, S\beta_{E1}\rangle}{1 - r(k_1)}\right].$$

The TRN $T$ for people infected by symptomatic infected individuals is therefore calculated as

$$T = r(k_2(I - k_1)^{-1}) = r(k_2) + \frac{\langle F_1, S\beta_{I1}\rangle\langle F_2, S\beta_{E1}\rangle}{1 - r(k_1)}. \tag{9.113}$$

Finally, note that if the case isolation proportion $\varepsilon$ is age-dependent, we cannot obtain such a simple control relation as (9.112). However, even in such a case, we can compute the TRN for state two as $T = r(k_2(I - k_1)^{-1})$, where

$$\begin{aligned}(k_2 f)(a) = &\int_0^\infty d\tau S(a) \int_\tau^\infty \beta_I(a, \sigma) \\ &\times \int_{\sigma-\tau}^{\sigma} \frac{\ell(\sigma)\Gamma(\sigma)}{\ell(x)\Gamma(x)} \frac{(1 - \varepsilon(x))\eta(x)\ell(x)\Xi(x)}{\ell(\sigma - \tau)\Xi(\sigma - \tau)} dx f(\sigma - \tau) d\sigma,\end{aligned}$$

and $\varepsilon(a)$ denotes the age-specific proportion of case isolation.

## 9.6 Estimation of $R_0$

A most important application of the age-structured epidemic model is the estimation of $R_0$ from the age-specific incidence rate or serological data (fraction of susceptible individuals at each age) at the endemic steady state. When estimating $R_0$, difficulties arise in obtaining reasonable observations of the transmission coefficient $\beta$. Therefore, we present some formulas for estimating $R_0$ without using the transmission coefficient.

To account for the length of infectiousness explicitly, we start our argument from the age–duration-structured SIR model[7]:

[7] In this section, we use the convention that the upper bound of chronological age is infinity. The basic model (9.114) is equivalent to the model introduced by Dietz and Schenzle [22].

$$
\begin{aligned}
&\left(\frac{\partial}{\partial t}+\frac{\partial}{\partial a}\right) S(t,a) = -(\mu(a)+\lambda(t,a))S(t,a), \\
&\left(\frac{\partial}{\partial t}+\frac{\partial}{\partial \tau}\right) J(t,\tau;a) = -(\mu(a+\tau)+\gamma(\tau))J(t,\tau;a), \\
&\left(\frac{\partial}{\partial t}+\frac{\partial}{\partial a}\right) R(t,a) = -\mu(a)R(t,a)+\int_0^a \gamma(\tau)J(t,\tau;a-\tau)d\tau, \\
&S(t,0) = B, \\
&J(t,0;a) = \lambda(t,a)S(t,a), \\
&R(t,0) = 0,
\end{aligned}
\tag{9.114}
$$

where $J(t,\tau;a)$ is the density of infectives with age at infection $a$ at infection-age $\tau$ and time $t$, $B$ is the number of newborns per unit time, $\mu(a)$ is the force of mortality, and $\gamma(\tau)$ is the rate of recovery at infection-age $\tau$. The force of infection $\lambda$ is given by

$$
\lambda(t,a) = \frac{1}{N}\int_0^\infty \beta(a,\sigma)\int_0^\sigma f(\tau)J(t,\tau;\sigma-\tau)d\tau d\sigma,
$$

where $f(\tau)$ is the probability of successful transmission of infective agents from infective individuals with infection-age $\tau$, and $N$ is the total size of the host population, which is assumed to be constant; that is, the total population is a demographic stationary population with age density

$$
B\ell(a) = S(t,a)+\int_0^a J(t,\tau;a-\tau)d\tau + R(t,a),
$$

where $\ell(a) := \exp(-\int_0^a \mu(\sigma)d\sigma)$.

By solving the infection cohort equation for $J(t,\tau;a)$ in (9.114) along the characteristic line, it follows that

$$
J(t,\tau;a) = \begin{cases} \frac{\ell(a+\tau)}{\ell(a)}\Gamma(\tau)J(t-\tau,0;a), & t-\tau>0, \\ \frac{\ell(a+\tau)\Gamma(\tau)}{\ell(a+\tau-t)\Gamma(\tau-t)}J(0,\tau-t;a), & \tau-t>0, \end{cases}
\tag{9.115}
$$

where $\Gamma(\tau)$ is the survival function induced from the recovery rate $\gamma(\tau)$ given by

$$
\Gamma(\tau) := \exp\left(-\int_0^\tau \gamma(\sigma)d\sigma\right).
$$

Here, we introduce the *age-specific incidence rate*, denoted by $v(t,a)$, which is the number of new cases per unit time and per capita at age $a$ and time $t$:

$$
v(t,a) = \frac{\lambda(t,a)S(t,a)}{N} = \frac{J(t,0;a)}{N}.
$$

Inserting (9.115) into the expression for $\lambda$ and changing the order of the integrals, we have

$$\begin{aligned} v(t,a) &= \frac{S(t,a)}{N}\left[g(t,a) + \int_0^t d\tau \int_\tau^\infty \beta(a,\sigma) f(\tau)\Gamma(\tau)\frac{\ell(\sigma)}{\ell(\sigma-\tau)} v(t-\tau,\sigma-\tau)d\sigma\right] \\ &= \frac{S(t,a)}{N}\left[g(t,a) + \int_0^t d\tau \int_0^\infty \beta(a,\tau+\zeta)\frac{\ell(\tau+\zeta)}{\ell(\zeta)} f(\tau)\Gamma(\tau)v(t-\tau,\zeta)d\zeta\right], \end{aligned}$$

where

$$g(t,a) := \int_t^\infty d\tau \int_\tau^\infty \beta(a,\sigma) f(\tau)\frac{\ell(\sigma)\Gamma(\tau)}{\ell(\sigma-t)\Gamma(\tau-t)} i(0,\tau-t;\sigma-\tau)d\sigma.$$

The linearized equation at the disease-free steady state is then given by

$$v(t,a) = c_2(a)\left[g(t,a) + \int_0^t d\tau \int_0^\infty \beta(a,\tau+\zeta)\frac{\ell(\tau+\zeta)}{\ell(\zeta)} f(\tau)\Gamma(\tau)v(t-\tau,\zeta)d\zeta\right]. \tag{9.116}$$

Because we can consider $v(t)(\cdot) := v(t,\cdot)$ as a function from $\mathbb{R}_+$ into $L^1$, we obtain an abstract integral equation in $L^1$:

$$v(t) = c_2 g(t) + \int_0^t \Pi(\tau)v(t-\tau)d\tau, \tag{9.117}$$

where $\Pi(\tau)$ is a nonnegative operator from $L^1_+$ into itself defined by

$$(\Pi(\tau)\phi)(a) := c_2(a)\int_0^\infty \beta(a,\tau+\zeta)\frac{\ell(\tau+\zeta)}{\ell(\zeta)} f(\tau)\Gamma(\tau)\phi(\zeta)d\zeta, \quad \phi \in L^1(\mathbb{R}_+).$$

Therefore, the NGO $K$ is calculated as

$$(K\phi)(a) = \int_0^\infty (\Pi(\tau)\phi)(a)d\tau = \int_0^\infty \Phi(a,\zeta)\phi(\zeta)d\zeta, \quad \phi \in L^1, \tag{9.118}$$

where

$$\Phi(a,\zeta) := c_2(a)\int_0^\infty \beta(a,\tau+\zeta)\frac{\ell(\tau+\zeta)}{\ell(\zeta)} f(\tau)\Gamma(\tau)d\tau.$$

In the following, we develop a method to estimate $R_0 = r(K)$ from the prevalence data at the endemic steady state. Let $S^*$ and $J^*$ be the susceptible and the infective population density functions. From (9.114), we obtain the following expressions:

$$\begin{aligned} S^*(a) &= B\ell(a)\exp\left(-\int_0^a \lambda^*(\sigma)d\sigma\right), \\ J^*(\tau;a) &= \frac{\ell(a+\tau)}{\ell(a)}\Gamma(\tau)\lambda^*(a)S^*(a). \end{aligned} \tag{9.119}$$

Inserting the above expression into the definition of the force of infection and changing the order of the integrals, we have

$$\begin{aligned} \lambda^*(a) &= \frac{1}{N}\int_0^\infty \int_\tau^\infty \beta(a,\sigma)\frac{\ell(\sigma)}{\ell(\sigma-\tau)}f(\tau)\Gamma(\tau)d\tau J^*(0;\sigma-\tau)d\sigma \\ &= \frac{1}{c_2(a)}\int_0^\infty \Phi(a,\zeta)\nu^*(\zeta)d\zeta, \end{aligned} \tag{9.120}$$

where we have used the relation $\nu^*(a) = J^*(0;a)/N$. Because $\nu^* = \lambda^* c_2 s^*$, we have the incidence rate equation at the endemic steady state:

$$\nu^*(a) = s^*(a)\int_0^\infty \Phi(a,\zeta)\nu^*(\zeta)d\zeta, \tag{9.121}$$

where

$$s^*(a) := \exp\left(-\int_0^a \lambda^*(\sigma)d\sigma\right),$$

is the fraction of susceptible individuals at age $a$ in the endemic steady state.

Using the NGO, the incidence rate Eq. (9.121) can be expressed as

$$\nu^* = s^* K \nu^*. \tag{9.122}$$

Although $s^*K$ is a nonlinear operator (as $s^*$ depends on $\nu^*$), we can consider it as a positive linear operator if we think of $s^*$ as a given function. In such a case, we can define its spectral radius (the dominant positive eigenvalue), denoted by $R_e = r(s^*K)$, as the *effective reproduction number*. Some early ideas regarding the effective reproduction number were discussed by Anderson and May [3, p. 17]. We use the term *effective* if the host population is not totally susceptible.

The endemicity condition can then be expressed as $R_e = r(s^*K) = 1$, which has the intuitively clear biological meaning that the infected population should simply reproduce itself in the endemic steady state. In our case, we have $R_e \le R_0$, as $s^* \le 1$. Hence, we can again confirm that there is no endemic steady state if $R_0 < 1$.

From (9.122) and $\nu^* = c_2 s^* \lambda^*$, we obtain

$$(K\nu^*)(a) = \frac{\nu^*(a)}{s^*(a)} = c_2(a)\lambda^*(a). \tag{9.123}$$

Let $K^*$ be the adjoint operator of $K$ and $F_0$ be its eigenfunctional associated with the eigenvalue $R_0$. Observe that

$$\langle K^* F_0, v^* \rangle = R_0 \langle F_0, v^* \rangle = \langle F_0, K v^* \rangle,$$

where $\langle K^*, \psi \rangle$ denotes the value of $K^*$ at $\psi$. Then, we have

$$R_0 = \frac{\langle F_0, K v^* \rangle}{\langle F_0, v^* \rangle} = \frac{\langle F_0, c_2 \lambda^* \rangle}{\langle F_0, c_2 \lambda^* s^* \rangle}. \tag{9.124}$$

This estimation formula was proposed by Dietz and Schenzle [22] and Farrington et al. [24] and was extended to the case in which the host population is in a stable population by Inaba and Nishiura [36].

If the separable mixing assumption $\beta(a, \sigma) = \beta_1(a)\beta_2(\sigma)$ holds, the NGO $K$ becomes a one-dimensional operator as

$$\begin{aligned}(Ku)(a) &= c_2(a)\beta_1(a) \int_0^\infty \int_0^\infty \beta_2(\tau + \zeta) \frac{\ell(\tau + \zeta)}{\ell(\zeta)} f(\tau)\Gamma(\tau) d\tau u(\zeta) d\zeta \\ &= \langle F_0, u \rangle \beta_1(a) c_2(a),\end{aligned} \tag{9.125}$$

where $F_0$ is the eigenfunctional given by

$$\langle F_0, u \rangle := \int_0^\infty \int_0^\infty \beta_2(\tau + \zeta) \frac{\ell(\tau + \zeta)}{\ell(\zeta)} f(\tau)\Gamma(\tau) d\tau u(\zeta) d\zeta, \quad u \in L^1(\mathbb{R}_+).$$

Therefore, we know that $\beta_1 c_2$ is the eigenvector of $K$ associated with $R_0$, and it follows that

$$\begin{aligned}R_0 = \langle F_0, \beta_1 c_2 \rangle &= \int_0^\infty \int_0^\infty \beta_2(\tau + \zeta) \frac{\ell(\tau + \zeta)}{\ell(\zeta)} f(\tau)\Gamma(\tau) d\tau \beta_1(\zeta) c_2(\zeta) d\zeta. \\ &= \frac{1}{e_0} \int_0^\infty \int_0^\infty \beta_2(\tau + \zeta)\ell(\tau + \zeta) f(\tau)\Gamma(\tau) d\tau \beta_1(\zeta) d\zeta,\end{aligned} \tag{9.126}$$

where $e_0 = \int_0^\infty \ell(a) da$ denotes the life expectancy of the host population.

Observe that, for any $u \in L^1$,

$$\langle K^* F_0, u \rangle = \langle F_0, K u \rangle = \langle F_0, \beta_1 c_2 \rangle \langle F_0, u \rangle = R_0 \langle F_0, u \rangle.$$

Thus, we can confirm that $K^* F_0 = R_0 F_0$, that is, $F_0$ is the eigenfunctional of $K^*$ associated with the eigenvalue $R_0$. Moreover, it follows from (9.123) and (9.125) that

$$\lambda^*(a) = \langle F_0, v^* \rangle \beta_1(a). \tag{9.127}$$

From (9.120) and $N = B e_0$, we obtain

$$\lambda^*(a) = \frac{1}{e_0} \int_0^\infty \int_0^\infty d\zeta \beta(a, \tau + \zeta)\ell(\tau + \zeta) f(\tau)\Gamma(\tau) d\tau \lambda^*(\zeta) e^{-\int_0^\zeta \lambda^*(x) dx} d\zeta. \tag{9.128}$$

If we insert relation (9.127) into (9.128), we can see that the coefficient $c := \langle F_0, v^* \rangle$ should be given by a positive root of the characteristic equation:

$$\frac{1}{e_0} \int_0^\infty \int_0^\infty \beta_2(\tau + \zeta)\ell(\tau + \zeta) f(\tau)\Gamma(\tau) d\tau \beta_1(\zeta) e^{-c \int_0^\zeta \beta_1(\sigma) d\sigma} d\zeta = 1. \quad (9.129)$$

The left-hand side of (9.129) gives $R_0$ when $c = 0$. Hence, it is easy to see that under the separable mixing assumption, there exists a unique endemic steady state if and only if $R_0 > 1$.

If we apply (9.124) to the separable mixing case, it follows that

$$\begin{aligned} R_0 &= \frac{\langle F_0, c_2 \lambda^* \rangle}{\langle F_0, c_2 \lambda^* s^* \rangle} = \frac{\langle F_0, c_2 \beta_1 \rangle}{\langle F_0, c_2 \beta_1 s^* \rangle} \\ &= \frac{\int_0^\infty \int_0^\infty \beta_2(\tau + \zeta)\ell(\tau + \zeta) f(\tau)\Gamma(\tau) d\tau \beta_1(\zeta) d\zeta}{\int_0^\infty \int_0^\infty \beta_2(\tau + \zeta)\ell(\tau + \zeta) f(\tau)\Gamma(\tau) d\tau \beta_1(\zeta) s^*(\zeta) d\zeta}. \end{aligned} \quad (9.130)$$

In this separable mixing case, the effective NGO at the endemic steady state is given by

$$(s^* K u)(a) = \langle F_0, u \rangle s^*(a) c_2(a) \beta_1(a),$$

and its spectral radius must be unity:

$$R_e = \langle F_0, \beta_1 c_2 s^* \rangle = 1, \quad (9.131)$$

which is the condition for endemic stationarity. Using (9.131), we find that (9.130) and (9.126) are identical. However, formula (9.130) is advantageous for estimation purposes, because we can expect unknown common factors to be canceled from its numerator and denominator.

If we assume that the length of the infectious period $L$ is sufficiently short compared with the life expectancy of the host population, the following approximation holds:

$$\int_0^\infty \beta_2(\tau + \zeta)\ell(\tau + \zeta) f(\tau)\Gamma(\tau) d\tau \approx \beta_2(\zeta)\ell(\zeta) D,$$

where $D$ is the total infectivity given by

$$D := \int_0^L f(\tau)\Gamma(\tau) d\tau.$$

Then, using the above approximation to (9.131), we arrive at the estimate:

$$R_0 \approx \frac{\int_0^\infty \beta_1(\zeta)\beta_2(\zeta)\ell(\zeta) d\zeta}{\int_0^\infty \beta_1(\zeta)\beta_2(\zeta) s^*(\zeta)\ell(\zeta) d\zeta}. \quad (9.132)$$

Moreover, if we adopt the *proportional mixing assumption*—that is, assume that $\beta_1(\zeta)$ is proportional to $\beta_2(\zeta)$—then $\beta_1$ and $\beta_2$ are both proportional to $\lambda^*$, and we have the estimation formula given by Dietz and Schenzle [22] and Greenhalgh and Dietz [27]:

$$R_0 \approx \frac{\int_0^\infty \lambda^{*2}(\zeta)\ell(\zeta)d\zeta}{\int_0^\infty \lambda^{*2}(\zeta)s^*(\zeta)\ell(\zeta)d\zeta}. \tag{9.133}$$

However, if we assume that $\beta_2$ is constant in (9.132), we can obtain another useful formula:

$$R_0 \approx \frac{\int_0^\infty \lambda^*(\zeta)c_2(\zeta)d\zeta}{\int_0^\infty \nu^*(\zeta)d\zeta}, \tag{9.134}$$

where we have applied (9.127)–(9.132) and used the fact that

$$\nu^*(a) = c_2(a)\lambda^*(a)s^*(a) = \frac{\ell(a)}{e_0}\lambda^*(a)\exp\left(-\int_0^a \lambda^*(\sigma)d\sigma\right).$$

Note that if the incidence rate $\nu^*$ is observed, we can calculate the force of infection as

$$\lambda^*(a) = \frac{\nu^*(a)}{c_2(a)\left(1 - \int_0^a \frac{\nu^*(\zeta)}{c_2(\zeta)}d\zeta\right)}.$$

Then, $R_0$ can be estimated from data of the incidence rate $\nu^*$ using (9.134).

If the force of mortality is assumed to be constant, the survival probability is given by $\ell(a) = e^{-\mu a}$ ($\mu$ is the constant force of mortality) and is referred to as *Type II* in epidemiology, whereas if the survival probability is a step function; that is, there exists a maximal age $\omega$ such that $\ell(a) = 1$ for $a \in [0, \omega)$ and $\ell(a) = 0$ for $a \geq \omega$, then it is referred to as *Type I*. If we use a Type I survival rate, we can establish the *final size equation* as follows:

**Proposition 9.10** *Suppose that $\beta_2$ is constant and the survival probability is of Type I. If the host population is in a demographic stationary state, it follows that*

$$R_0 \approx -\frac{\log(1-p)}{p}, \tag{9.135}$$

*where $p := 1 - s^*(\omega)$ is the proportion of ever removed population in a birth cohort.*

*Proof* If we apply a Type I survival probability such that $\ell(a) = 1$ for $a \in [0, \omega)$ and $\ell(a) = 0$ for $a \geq \omega$ to (9.134), we have

$$R_0 \approx \frac{\int_0^\omega \lambda^*(\zeta)d\zeta}{\int_0^\omega \lambda^*(\zeta)s^*(\zeta)d\zeta} = \frac{1}{1 - s^*(\omega)}\int_0^\omega \lambda^*(\zeta)d\zeta,$$

where $s^*(\omega)$ is the proportion of the population who are susceptible at the end of their life. From $p = 1 - s^*(\omega)$ and $s^*(\omega) = \exp\left(-\int_0^\omega \lambda^*(\zeta)d\zeta\right)$, it follows that

$$\int_0^\omega \lambda^*(\zeta)d\zeta = -\log(1-p).$$

The estimate in (9.135) follows immediately. □

The number $p$ can be interpreted as the *final size* of the epidemic with respect to a birth cohort. At an individual level, $p$ is called a *quantum* of an (infection) event in the demographic terminology, which is the lifelong average number of occurrences of infection. From (9.135), we know that in this case, the well-known *final size equation*

$$1 - p = e^{-pR_0}$$

holds approximately for a birth cohort.

From expression (9.132), we can induce the well-known reciprocal relation between $R_0$ and the proportion of the susceptible population at the endemic steady state.

**Proposition 9.11** *Suppose that the transmission rate $\beta$ is constant and the average duration of infectiousness L is sufficiently short. Then, it follows that*

$$R_0 \approx \frac{1}{\int_0^\infty s^*(\zeta)c_2(\zeta)d\zeta} = \frac{1}{x^*}, \tag{9.136}$$

*where*

$$x^* := \int_0^\infty c_2(\zeta)s^*(\zeta)d\zeta = \frac{1}{N}\int_0^\infty S^*(a)da$$

*is the proportion of susceptible individuals in the host population at the endemic steady state.*

**Corollary 9.1** *Suppose that the survival probability is of either Type I or Type II. The basic reproduction number can then be approximated as*

$$R_0 \approx \frac{e_0}{A}, \tag{9.137}$$

*where A is the average age at the occurrence of new infection in the endemic steady state:*

$$A = \frac{\int_0^\infty a\lambda^*(a)S^*(a)da}{\int_0^\infty \lambda^*(a)S^*(a)da}.$$

*Proof* If we use a Type II survival function $\ell(a) = e^{-\mu a}$, then we can calculate

$$x^* = \frac{\mu}{\lambda^* + \mu}, \quad A = \frac{1}{\lambda^* + \mu}.$$

Therefore, from (9.136) and $e_0 = 1/\mu$, we have

$$R_0 \approx \frac{1}{x^*} = \frac{e_0}{A}.$$

Next, if we use a Type I survival function—that is, there exists a maximal age $\omega$ such that $\ell(a) = 1$ for $a \in [0, \omega)$ and $\ell(a) = 0$ for $a \geq \omega$—then we have

$$x^* = \frac{1 - e^{-\lambda^* \omega}}{\omega \lambda^*}, \quad A = \frac{1}{\lambda^*} - \frac{\omega e^{-\lambda^* \omega}}{1 - e^{-\lambda^* \omega}}.$$

Thus, it follows that

$$R_0 \approx \frac{1}{x^*} = \frac{\omega \lambda^*}{1 - e^{-\lambda^* \omega}} = \frac{e_0(1+\varepsilon)}{A},$$

where $e_0 = \omega$ and

$$\varepsilon := \frac{e^{-\lambda^* \omega}}{1 - e^{-\lambda^* \omega}} - \frac{\lambda^* \omega e^{-\lambda^* \omega}}{(1 - e^{-\lambda^* \omega})^2}.$$

For common childhood diseases, $\varepsilon$ is very small, so we conclude that (9.137) also holds for the Type I survival function. □

In reality, demographic or epidemiological data are given as discrete, age-specific values. Therefore, it is reasonable to estimate $R_0$ based on piecewise constant parameter functions. Let $\omega$ be an upper bound of the chronological age and divide the age interval into $n$ age classes $[0, \omega] = \cup_{k=1}^{n} [a_{k-1}, a_k]$, where $0 = a_0 < a_1 < a_2 < \cdots < a_n = \omega$. Suppose that the transmission coefficient $\beta$ is given by the piecewise constant function

$$\beta(a, \sigma) = \sum_{i,j} \beta_{ij} I_i(a) I_j(\sigma),$$

where $I_i(a) = 1$ for $a \in [a_{i-1}, a_i)$, $I_i(a) = 0$ for $a \notin [a_{i-1}, a_i)$, and $\beta = 0$ for $a, \sigma > \omega$. Then, we have

$$\begin{aligned}
(Ku)(a) &= \frac{S(a)}{N} \int_0^\infty \int_0^\infty \beta(a, \tau + \zeta) \frac{\ell(\tau + \zeta)}{\ell(\zeta)} f(\tau) \Gamma(\tau) d\tau u(\zeta) d\zeta \\
&= \frac{S(a)}{N} \sum_{i,j} \beta_{ij} I_i(a) \int_0^\infty \int_0^\infty I_j(\tau + \zeta) \frac{\ell(\tau + \zeta)}{\ell(\zeta)} f(\tau) \Gamma(\tau) d\tau u(\zeta) d\zeta \\
&= \sum_{i,j} \beta_{ij} \Theta_i(a) \langle F_j, u \rangle \\
&= \sum_i \left( \sum_j \beta_{ij} \langle F_j, u \rangle \right) \Theta_i(a),
\end{aligned}$$

where $\Theta_i(a) := S(a) I_i(a)/N$ and

$$\langle F_j, u \rangle := \int_0^\infty \int_0^\infty I_j(\tau + \zeta) \frac{\ell(\tau + \zeta)}{\ell(\zeta)} f(\tau) \Gamma(\tau) d\tau u(\zeta) d\zeta.$$

The range of the NGO becomes $n$-dimensional, so its eigenfunction associated with $R_0$ is expressed by a linear combination of $\Theta_i$: $f = \sum_i \alpha_i \Theta_i$. Then, we obtain

$$\begin{aligned} Kf &= \sum_i \left( \sum_j \beta_{ij} \langle F_j, \sum_k \alpha_k \Theta_k \rangle \right) \Theta_i(a) \\ &= \sum_i \left( \sum_j \sum_k \beta_{ij} \langle F_j, \Theta_k \rangle \alpha_k \right) \Theta_i(a) = R_0 \sum_i \alpha_i \Theta_i. \end{aligned}$$

Therefore, we arrive at a finite-dimensional eigenvalue problem

$$\tilde{K} \alpha = R_0 \alpha,$$

where $\alpha = (\alpha_1, \ldots, \alpha_n)^{\mathrm{T}}$ and $\tilde{K} = (k_{ij})_{1 \le i, j \le n}$ is a $n \times n$ matrix whose $(i, j)$th entry is given by

$$k_{ij} = \sum_k \beta_{ik} \langle F_k, \Theta_j \rangle.$$

The basic reproduction number $R_0$ is then given by the spectral radius of the NGM $\tilde{K}$, so the main problem is how to estimate the entries of $\tilde{K}$ from the available data. For estimation purposes, we can again make use of the endemic relation in (9.123) (see [27]).

## References


1. Agur, Z., Cojocaru, L., Mazor, G., Anderson, R.M., Danon, Y.L.: Pulse mass measles vaccination across age cohorts. Proc. Natl. Acad. Sci. USA **90**, 11698–11702 (1993)
2. Allen, L.J.S., Bolker, B.M., Lou, Y., Nevai, A.L.: Asymptotic properties of the steady states for an SIS epidemic reaction-diffusion model. Discret. Contin. Dyn. Syst. **21**(1), 1–20 (2008)
3. Anderson, R.M., May, R.M.: Infectious Diseases of Humans: Dynamics and Control. Oxford UP, Oxford (1991)
4. Anita, S.: Analysis and Control of Age-Dependent Population Dynamics. Kluwer, Dordrecht (2000)
5. Bacaër, N., Guernaoui, S.: The epidemic threshold of vector-borne diseases with seasonality. J. Math. Biol. **53**, 421–436 (2006)
6. Bacaër, N., Ouifki, R.: Growth rate and basic reproduction number for population models with a simple periodic factor. Math. Biosci. **210**, 647–658 (2007)
7. Bacaër, N.: Approximation of the basic reproduction number $R_0$ for vector-borne diseases with a periodic vector population. Bull. Math. Biol. **69**, 1067–1091 (2007)
8. Bacaër, N., Abdurahman, X.: Resonance of the epidemic threshold in a periodic environment. J. Math. Biol. **57**, 649–673 (2008)

9. Bacaër, N.: Periodic matrix population models: growth rate, basic reproduction number, and entropy. Bull. Math. Biol. **71**, 1781–1792 (2009)
10. Bacaër, N., Ait Dads, E.H.: Genealogy with seasonality, the basic reproduction number, and the influenza pandemic. J. Math. Biol. **62**, 741–762 (2011)
11. Bacaër, N.: The model of Kermack and McKendrick for the plague epidemic in Bombay and the type reproduction number with seasonality. J. Math. Biol. **64**(3), 403–422 (2012)
12. Bacaër, N., Ait Dads, E.H.: On the biological interpretation of a definition for the parameter $R_0$ in periodic population models. J. Math. Biol. **65**(4), 601–621 (2012)
13. Bacaër, N., Khaladi, M.: On the basic reproduction number in a random environment. J. Math. Biol. **67**, 1729–1739 (2013)
14. Birkhoff, G., Varga, R.S.: Reactor criticality and nonnegative matrices. J. Soc. Indust. Appl. Math. **6**(4), 354–377 (1958)
15. Cantrell, R.S., Cosner, C.: Spatial Ecology via Reaction-Diffusion Equations. Wiley, Chichester (2003)
16. Caswell, H.: Matrix Population Models, 2nd edn. Sinauer, Sunderland (2001)
17. Diekmann, O., Heesterbeak, J.A.P., Metz, J.A.J.: On the definition and the computation of the basic reproduction ratio $R_0$ in models for infectious diseases in heterogeneous populations. J. Math. Biol. **28**, 365–382 (1990)
18. Diekmann, O., Gyllenberg, M., Metz, J.A.J., Thieme, H.R.: On the formulation and analysis of general deterministic structured population models I. Linear Theory. J. Math. Biol. **36**, 349–388 (1998)
19. Diekmann, O., Heesterbeek, J.A.P.: Mathematical Epidemiology of Infectious Diseases: Model Building Analysis and Interpretation. Wiley, Chichester (2000)
20. Diekmann, O., Heesterbeek, J.A.P., Roberts, M.G.: The construction of next-generation matrices for compartmental epidemic models. J. Roy. Soc. Interface 6, **7**(47), 873–885 (2010)
21. Diekmann, O., Heesterbeek, J.A.P., Britton, T.: Mathematical Tools for Understanding Infectious Disease Dynamics. Princeton University Press, Princeton (2013)
22. Dietz, K., Schenzle, D.: Proportionate mixing models for age-dependent infection transmission. J. Math. Biol. **22**, 117–120 (1985)
23. Dublin, L.I., Lotka, A.J.: On the true rate of natural increase. J. Am. Stat. Ass., New Series, No. 150 **20**, 305–339 (1925)
24. Farrington, C.P., Kanaan, M.N., Gay, N.J.: Estimation of the basic reproduction number for infectious diseases from age-stratified serological survey data. Appl. Stat. **50**, Part 3, 251–292 (2001)
25. Gantmacher, F.R.: The Theory of Matrices, vol. 1960. Chelsea Publishing Company, New York (1960)
26. Grassly, N.C., Fraser, C.: Seasonal infectious disease epidemiology. Proc. R. Soc. B **273**, 2541–2550 (2006)
27. Greenhalgh, D., Dietz, K.: Some bounds on estimates for reproductive ratios derived from the age-specific force of infection. Math. Biosci. **124**, 9–57 (1994)
28. Heesterbeek, J.A.P., Roberts, M.G.: Threshold quantities for helminth infections. J. Math. Biol. **33**, 415–434 (1995)
29. Heesterbeek, J.A.P., Roberts, M.G.: Threshold quantities for infectious diseases in periodic environments. J. Biol. Sys. **3**(3), 779–787 (1995)
30. Heesterbeek, J.A.P.: A brief history of $R_0$ and a recipe for its calculation. Acta Biotheor. **50**, 189–204 (2002)
31. Heesterbeek, J.A.P., Roberts, M.G.: The type-reproduction number $T$ in models for infectious disease control. Math. Biosci. **206**, 3–10 (2007)
32. Heffernan, J.M., Smith, R.J., Wahl, L.M.: Perspectives on the basic reproductive ratio. J. Roy. Soc. Interface **2**, 281–293 (2005)
33. Heijmans, H.J.A.M.: The dynamical behaviour of the age-size-distribution of a cell population. In: Metz, J.A.J., Diekmann, O. (eds.), The Dynamics of Physiologically Structured Populations, Lecture Notes Biomathematics, vol. 68, pp. 185–202. Springer, Berlin (1986)

34. Hethcote, H.W.: Asymptotic behavior in a deterministic epidemic model. Bull. Math. Biol. **35**, 607–614 (1973)
35. Inaba, H.: Threshold and stability results for an age-structured epidemic model. J. Math. Biol. **28**, 411–434 (1990)
36. Inaba, H., Nishiura, H.: The basic reproduction number of an infectious disease in a stable population: the impact of population growth rate on the eradication threshold. Math. Model. Nat. Phenom. **3**(7), 194–228 (2008)
37. Inaba, H. and Nishiura, H.: The state-reproduction number for a multistate class age structured epidemic system and its application to the asymptomatic transmission model, Math. Biosci. **216**, 77–89 (2008)
38. Inaba, H.: The net reproduction rate and the type-reproduction number in multiregional demography. Vienna Yearb. Popul. Res. **2009**, 197–215 (2010)
39. Inaba, H.: The Malthusian parameter and $R_0$ for heterogeneous populations in periodic environments. Math. Biosci. Eng. **9**(2), 313–346 (2012)
40. Inaba, H.: On a new perspective of the basic reproduction number in heterogeneous environments. J. Math. Biol. **65**, 309–348 (2012)
41. Inaba, H.: On the definition and the computation of the type-reproduction number $T$ for structured populations in heterogeneous environments. J. Math. Biol. **66**, 1065–1097 (2013)
42. Jagers, P., Nerman, O.: Branching processes in periodically varying environment. Ann. Prob. **13**, 254–268 (1985)
43. Kuniya, T., Inaba, H.: Endemic threshold results for age-structured SIS epidemic model with periodic parameters. J. Math. Anal. Appl. **402**, 477–492 (2013)
44. Kuniya, T., Iannelli, M.: $R_0$ and the global behavior of an age-structured SIS epidemic model with periodicity and vertical transmission. Math. Biosci. Eng. **11**, 929–945 (2014)
45. Kuniya, T.: Existence of a nontrivial periodic solution in an age-structured SIR epidemic model with time periodic coefficients. Appl. Math. Lett. **27**, 15–20 (2014)
46. Li, C.K., Schneider, H.: Applications of Perron-Frobenius theory to population dynamics. J. Math. Biol. **44**, 450–462 (2002)
47. Li, J., Blakeley, D., Smith?, R.J.: The failure of $R_0$. Comput. Math. Methods Med. **2011**, Article ID 527610
48. Lotka, A.J.: Théorie Analytique des Associations Biologiques. Deuxième Partie: Analyse Démographique avec Application Particulière è l'Espèce Humaine, Actualités Scientifiques et Industrielles, No. 780, Hermann et Cie, Paris (1939)
49. Lotka, A.J.: Analytical Theory of Biological Populations, the Plenum Series on Demographic Methods and Population Analysis. Plenum Press, New York (1998)
50. Marcati, P., Serafini, R.: Asymptotic behaviour in age dependent population dynamics with spatial spread. Bollettino U. M. I. **16-B**(5), 734–753 (1979)
51. Marek, I.: Iterations of linear bounded operators in non self-adjoint eigenvalue problems and Kellogg's iteration process. Czech. Math. J. **12**, 536–554 (1962)
52. Marek, I.: Frobenius theory of positive operators: comparison theorems and applications. SIAM J. Appl. Math. **19**, 607–628 (1970)
53. Nishiura, H., Dietz, K., Eichner, M.: The earliest notes on the reproduction number in relation to herd immunity: Theophil Lotz and smallpox vaccination. J. Theor. Biol. **241**, 964–967 (2006)
54. Nishiura, H., Inaba, H.: Discussion: emergence of the concept of the basic reproduction number from mathematical demography. J. Theor. Biol. **244**, 357–364 (2007)
55. Nokes, D.J., Swinton, J.: The control of childhood viral infections by pulse vaccination. IMA J. Math. Appl. Med. Biol. **12**, 29–53 (1995)
56. Nokes, D.J., Swinton, J.: Vaccination in pulses: a strategy for global eradication of measles and polio? Trends Microbiol. **5**(1), 14–19 (1997)
57. Peng, R., Zhao, X.Q.: A reaction-diffusion SIS epidemic model in a time-periodic environment. Nonlinearity **25**, 1451–1471 (2012)
58. Roberts, M.G., Heesterbeek, J.A.P.: A new method for estimating the effort required to control an infectious disease. Proc. R. Soc. Lond. B **270**, 1359–1364 (2003)
59. Roberts, M.G.: The pluses and minuses of $\mathscr{R}_0$. J. R. Soc. Interface **4**, 949–961 (2007)

60. Rogers, A.: Multiregional Demography: Principles, Methods and Extensions. Wiley, New York (1995)
61. Shuai, Z., Heesterbeek, J.A.P., van den Driessche, P.: Extending the type reproduction number to infectious disease control targeting contacts between types. J. Math. Biol. **67**, 1067–1082 (2013)
62. Shulgin, B., Stone, L., Agur, Z.: Pulse vaccination strategy in the SIR epidemic model. Bull. Math. Biol. **60**, 1123–1148 (1998)
63. Thieme, H.R.: Renewal theorems for linear periodic Volterra integral equations. J. Integral Equ. **7**, 253–277 (1984)
64. Thieme, H.R.: Analysis of age-structured population models with additional structure. In: Arino, O., Axelrod, D.E., Kimmel, M. (eds.), Mathematical Population Dynamics, pp. 115-126. Marcel Dekker, New York (1991)
65. Thieme, H.R.: Spectral bound and reproduction number for infinite-dimensional population structure and time heterogeneity. SIAM J. Appl. Math. **70**(1), 188–211 (2009)
66. van den Berg, F., Bacaër, N., Metz, J.A.J., Lannou, C., van den Bosch, F.: Periodic host absence can select for higher or lower prasite transmission rates. Evol. Ecol. **25**, 121–137 (2011)
67. van den Driessche, P., Watmough, J.: Reproduction numbers and sub-threshold endemic equilibria for compartmental models of disease transmission. Math. Biosci. **180**, 29–48 (2002)
68. Varga, R.S.: Matrix Iterative Analysis, 2nd edn. Springer, Berlin (2000)
69. Wallinga, J., Lipsitch, M.: How generation intervals shape the relationship between growth rates and reproductive numbers. Proc. R. Soc. B **274**, 599–604 (2007)
70. Wang, W., Zhao, X.-Q.: Threshold dynamics for compartmental epidemic models in period environments. J. Dyn. Diff. Equat. **20**, 699–717 (2008)
71. Wang, W., Zhao, X.-Q.: Basic reproduction numbers for reaction-diffusion epidemic models. SIAM J. Appl. Dyn. Syst. **11**(4), 1652–1673 (2012)
72. Wang, B.G., Zhao, X.Q.: Basic reproduction ratios for almost periodic compartmental epidemic models. J. Dyn. Diff. Equat. **25**, 535–562 (2013)
73. Webb, G.F.: Diffusive age-dependent population models and an application to genetics. Math. Biosci. **61**, 1–16 (1982)
74. Xinli, H.: Threshold dynamics for SIR epidemic model in periodic environments. In: 2010 International Conference on Computer Application and System Modeling, vol. V7, pp. 41–45 (2010)
75. Zhao, X.-Q.: Basic reproduction ratios for periodic compartmental models with time delay. J. Dyn. Diff. Equa. 1–16. First online: 21 January 2015, doi:10.1007/s10884-015-9425-2

# Chapter 10
# Mathematical Tools

**Abstract** This last chapter is a collection of mathematical tools to treat population equations. First, we explain a semigroup approach to consider the stable population model. Using the semigroup setting, the idea of strong ergodicity is extended as the *asynchronous exponential growth* of the semigroup, which can be applied to certain nonlinear problems. We then briefly discuss functional analytic methods for nonlinear population problems. Next, we introduce some results for the infinite-dimensional Perron–Frobenius theorem, the contraction mapping principle given by the Hilbert pseudometric, Birkhoff's linear multiplicative process theory, and some properties of nonlinear positive operators, which are useful tools for studying population dynamics. Finally, we summarize some basic results about the Laplace transformation and Volterra integral equation that are used in the previous chapters.

## 10.1 Basics for Semigroup Approach

In comparison with the classical integral equation approach, the abstract differential (and integral) equation approach can be a more powerful tool for treating nonlinear structured population models, because this formulation can make use of mathematical methods developed for infinite-dimensional dynamical systems theory. In fact, the theoretical developments in structured population dynamics since the 1980s have been led by the functional analytic approach. For classical results, readers are referred to the work of Metz and Diekmann [75] and Webb [108, 111]. For more elaborated modern methods, readers are referred to [34–36, 94]. Here, we only sketch an introduction for the semigroup approach to age-dependent population models.

### *10.1.1 Linear Problems*

As a most simple linear case, we now consider the semigroup solution of the scalar-type stable population model. Desch and Schappacher [31], Prüss [85], Song et al. [95–98], and Webb [107, 108] all considered semigroup approaches for the stable

 
H. Inaba, *Age-Structured Population Dynamics in Demography and Epidemiology*, DOI 10.1007/978-981-10-0188-8_10

population model in the early 1980s. Readers are referred to [54, 55] for the case of multistate populations.

The stable population model (1.25) is formulated as an abstract Cauchy problem (initial value problem) on a Banach space $L^1(0, \omega)$:

$$\frac{dp(t)}{dt} = Ap(t), \quad p(0) = p_0, \tag{10.1}$$

where $A$ denotes the population operator introduced in Sect. 1.2:

$$(A\phi)(a) := -\frac{d\phi(a)}{da} - \mu(a)\phi(a),$$

the domain of which is given by

$$\mathscr{D}(A) = \left\{ \phi \in L^1(0, \omega) : \phi(0) = \int_0^\omega \beta(a)\phi(a)da, \ A\phi \in L^1(0, \omega) \right\},$$

where we assume that $\beta \in L^1_+(0, \omega) \cap L^\infty_+(0, \omega)$, $\mu \in L^1_{\mathrm{loc},+}(0, \omega)$ and $\int_0^\omega \mu(\sigma)d\sigma = \infty$.

*Remark 10.1* Note that if we choose an age $a_\dagger \in (\beta_2, \omega)$ where $\beta_2$ is the maximum reproductive age, we can assume that $\mu \in L^\infty_+(0, a_\dagger)$ and the term $-\mu\phi$ can be treated as a bounded perturbation in $L^1$. If individuals with age $a > a_\dagger$ do not have any influence on individuals in the age interval $[0, a_\dagger]$, as in the linear case, we do not lose generality, even if we limit our argument to the age interval $[0, a_\dagger]$. Conversely, it is usual to assume $\mu \in L^\infty_+(\mathbb{R}_+)$ if $\omega = \infty$. Note that the most natural state space for age-density functions is $L^1$. However, $L^2$ is also used in population control problems [95].

We now introduce the concept of a strongly continuous semigroup ($C_0$-semigroup). For the definition and the Hille–Yosida theorem of a one-parameter semigroup, readers are referred to [9, 24, 38, 78, 83].

**Definition 10.1** Let $X$ be a Banach space. A one-parameter family $T(t)$, $t \geq 0$, of bounded linear operators from $X$ to $X$ is a *strongly continuous semigroup of bounded linear operators* on $X$ if

1. $T(0) = I$ where $I$ is the identity operator on $X$.
2. $T(t + s) = T(t)T(s)$ for every $t, s \geq 0$.
3. $\lim_{t\downarrow 0} T(t)x = x$ for every $x \in X$.

The linear operator $A$ defined by

$$Ax = \lim_{t\downarrow 0} \frac{T(t)x - x}{t}, \quad x \in \mathscr{D}(A) := \left\{ x \in X : \lim_{t\downarrow 0} \frac{T(t)x - x}{t} \text{ exists} \right\}$$

is the infinitesimal generator of the semigroup $T(t)$ with the domain $\mathscr{D}(A)$.

**Proposition 10.1** (Hille–Yosida theorem) *A closed linear operator $A$ with dense domain $\mathscr{D}(A)$ is the generator of a strongly continuous semigroup $T(t)$, $t \geq 0$, if and only if real numbers $\zeta$ and $M$ exist such that, for all $\lambda > \zeta$, the operator $\lambda - A$ has a bounded inverse and*

$$\|R(\lambda, A)^n\| \leq \frac{M}{(\lambda - \zeta)^n}, \quad n = 1, 2, 3, \ldots, \tag{10.2}$$

*where $R(\lambda, A) = (\lambda - A)^{-1}$. Then, $T(t)$, $t \geq 0$, is uniquely determined by $A$, and it holds that $\|T(t)\| \leq Me^{\zeta t}$. If $\phi \in \mathscr{D}(A)$, then $T(t)\phi \in \mathscr{D}(A)$ for all $t > 0$, $T(t)\phi$ is strongly differentiable for $t \geq 0$ and*

$$\frac{d}{dt}T(t)\phi = AT(t)\phi = T(t)A\phi. \quad \phi \in \mathscr{D}(A). \tag{10.3}$$

We now mention some basic observations regarding the strongly continuous semigroups and spectral properties of generators. For more detail, readers may consult Nagel et al. [38, 78] and Webb [108]. Let $\sigma(A)$ be the spectral set of linear operators $A$, $P_\sigma(A)$ be the set of point spectra of $A$, $\rho(A) := \mathbb{C} \setminus \sigma(A)$ be the resolvent set and $R(\lambda, A) := (\lambda - A)^{-1}, \lambda \in \rho(A)$ be the resolvent. The *spectral bound* $s(A)$ is defined by $s(A) := \sup\{\Re\lambda : \lambda \in \sigma(A)\}$, the *spectral radius* of a $C_0$-semigroups $T(t)$ is defined by $r(T(t)) := \sup\{|\lambda| : \lambda \in \sigma(T(t))\}$ and the *growth bound* is defined by

$$\omega_0 := \inf\{w \in \mathbb{R} : \|T(t)\| \leq M(w)e^{wt}, \forall t \geq 0, \exists M(w) \in \mathbb{R}_+\}.$$

Then, it holds that

$$\omega_0 = \lim_{t \to \infty} \frac{\log \|T(t)\|}{t}, \tag{10.4}$$

and $-\infty \leq \omega_0 < \infty$.

**Proposition 10.2** *Let $\omega_0$ be the growth bound of a strongly continuous semigroup $T(t), t \geq 0$, and let $A$ be its generator. For any $t \geq 0$, it follows that $r(T(t)) = e^{\omega_0 t}$ and $r(T(t)) \geq e^{s(A)t}$.*

If $A$ is a bounded operator, we know that $e^{s(A)t} = r(T(t)) = e^{\omega_0 t}$ and $\omega_0 = s(A)$. Therefore, in such a case, if the spectral bound of the generator is negative, we can conclude that $\lim_{t\to\infty} \|T(t)\| = 0$. This implies the asymptotic stability of the zero solution of the abstract ordinary differential equation $\dot{x} = Ax$. However, if $A$ is unbounded, this is not generally the case.

If there exists a number $t_0 > 0$ such that $T(t_0)$ becomes a compact operator (hence, $T(t)$ is compact for all $t \geq t_0$), $T(t)$ is said to be *eventually compact*. For eventually compact semigroups, the following *spectral mapping theorem* holds [5, p. 87]:

$$\sigma(T(t)) \setminus \{0\} = e^{t\sigma(A)} := \{e^{t\lambda} : \lambda \in \sigma(A)\}, \quad t \geq 0. \tag{10.5}$$

If (10.5) holds, $\omega_0 = s(A)$, and hence, the zero solution of (10.1) is globally asymptotically stable if and only if the spectral bound of the generator $A$ is negative.

**Proposition 10.3** *Let us define*

$$\hat{\Psi}(\lambda) := \int_0^\omega e^{-\lambda a}\beta(a)\ell(a)da$$

*and* $\underline{\mu} := \inf_{a\in[0,\omega]} \mu(a)$. *Then, the population operator A has the following properties:*

*(1)* $\sigma(A) = P_\sigma(A) = \{\lambda \in \mathbb{C} : \hat{\Psi}(\lambda) = 1\}$,
*(2)* $R(\lambda, A)$ *is a compact operator and*

$$\begin{aligned}(R(\lambda, A)\phi)(a) &= \int_0^a e^{-\lambda(a-z)}\frac{\ell(a)}{\ell(z)}\phi(z)dz \\ &+ \frac{e^{-\lambda a}\ell(a)}{1-\hat{\Psi}(\lambda)}\int_0^\omega \int_z^\omega e^{-\lambda(a-z)}\frac{\ell(a)}{\ell(z)}\beta(a)da\phi(z)dz,\end{aligned} \tag{10.6}$$

*(3) If* $\lambda \in \sigma(A)$, *then the algebraic multiplicity of* $\lambda$ *is finite and the corresponding eigenspace is given by* $N(\lambda - A) = \{ce^{-\lambda a}\ell(a) : c \in \mathbb{C}\}$,
*(4)* $\rho(A) \supset \{\lambda \in \mathbb{C} : \Re\lambda > \bar{\beta} - \underline{\mu}\}$ *and* $R(\lambda, A)$ *is a positive operator if* $\lambda \in \mathbb{R}$ *and* $\lambda > \bar{\beta} - \underline{\mu}$,
*(5) If* $\lambda \in \sigma(A)$, *then* $\bar{\lambda} \in \sigma(A)$,
*(6) For any* $\alpha > -\infty$, *there exist at most a finite number of eigenvalues of A in the half plane* $\Re\lambda \geq \alpha$.

*Proof* Solving the resolvent equation $(\lambda - A)f = \phi,\ f \in \mathscr{D}(A),\ \phi \in L^1$, we obtain expression (10.6) for $\lambda \in \mathbb{C}$ such that $\hat{\Psi}(\lambda) \neq 1$. Because $R(\lambda, A)$ is the sum of a one-dimensional operator and a Volterra integral operator, it is a compact operator for $\lambda \in \rho(A)$ and it follows that $\sigma(A) = \{\lambda \in \mathbb{C} : \hat{\Psi}(\lambda) = 1\}$. As $1 - \hat{\Psi}(\lambda)$ is a holomorphic function, it has at most countably many discrete zeros, which are not accumulated except for infinity. Therefore, $\sigma(A)$ is composed of discrete complex numbers, and we know that if $\lambda \in \sigma(A)$, the eigenfunctions are proportional to $e^{-\lambda a}\ell(a)$ and $\sigma(A) = P_\sigma(A)$. This completes the proof for (1)–(3), and the proofs for (4)–(6) have already been given in Proposition 1.7. □

**Proposition 10.4** *The population operator A is a closed operator with a dense domain in* $L^1(0, \omega)$.

*Proof* To show that $A$ is a closed operator, it is sufficient to prove that, if $u_n \in \mathscr{D}(A)$, $u_n \to u \in L^1$ and $Au_n \to v \in L^1$, then $u \in \mathscr{D}(A)$ and $Au = v$. Because

$$(Au_n)(a) = -\frac{du_n(a)}{da} - \mu(a)u_n(a),$$

it follows that

$$u_n(a) = u_n(0)\ell(a) - \int_0^a (Au_n)(\sigma)\frac{\ell(a)}{\ell(\sigma)}d\sigma,$$

where $\ell(a) := \exp\left(-\int_0^a \mu(\sigma)d\sigma\right)$ denotes the survival function. Then, it is easy to see that

$$|u_n(0) - u_m(0)||\ell|_{L^1} \le |u_n - u_m|_{L^1} + \omega|Au_n - Au_m|_{L^1},$$

and hence, $\{u_n(0)\}_{n=1,2,...}$ is a Cauchy sequence. Then, there exists a number $\alpha \in \mathbb{R}$ such that $\lim_{n\to\infty} u_n(0) = \alpha$. If we define

$$w(a) := \alpha\ell(a) - \int_0^a v(\sigma)\frac{\ell(a)}{\ell(\sigma)}d\sigma,$$

then we obtain $\lim_{n\to\infty} |u_n - w|_{L^1} = 0$, so $u = w$ almost everywhere. Conversely, if we let $t \to \infty$ in the boundary condition

$$u_n(0) = \int_0^\omega \beta(a)u_n(a)da,$$

then we have

$$\alpha = w(0) = \int_0^\omega \beta(a)u(a)da = \int_0^\omega \beta(a)w(a)da.$$

Hence, $w \in \mathscr{D}(A)$ and $Aw = v$. Because $u = w$ in the $L^1$-sense, we conclude that $u \in \mathscr{D}(A)$ and $Au = v$. Next, let us show that the domain is dense in $L^1(0, \omega)$. Consider the resolvent equation

$$(\lambda - A)f = g, \quad f \in \mathscr{D}(A), \quad g \in L^1.$$

It follows from (10.6) that, for $\lambda \in \rho(A)$, $R(\lambda, A) = (\lambda - A)^{-1}$ can be decomposed as follows:

$$(\lambda - A)^{-1}g = J(\lambda)g + K(\lambda)g,$$

where $J(\lambda)$ and $K(\lambda)$ are defined as

$$\begin{aligned}
(J(\lambda)g)(a) &:= \int_0^a e^{-\lambda(a-z)}\frac{\ell(a)}{\ell(z)}g(z)dz, \\
(K(\lambda)g)(a) &:= \frac{e^{-\lambda a}\ell(a)}{1-\hat{\psi}(\lambda)}\int_0^\omega \beta(a)\int_0^a e^{-\lambda(a-z)}\frac{\ell(a)}{\ell(z)}g(z)dzda.
\end{aligned} \tag{10.7}$$

Observe that

$$|\lambda(\lambda - A)^{-1}g - g|_{L^1} \le |\lambda J(\lambda)g - g|_{L^1} + |\lambda K(\lambda)g|_{L^1},$$

where we have $\lim_{\lambda\to\infty}|\lambda K(\lambda)g|_{L^1}=0$. In contrast, $J(\lambda)$ is a resolvent of the closed operator $A_0$ defined by

$$(A_0\phi)(a)=-\frac{d\phi(a)}{da}-\mu(a)\phi(a),$$

with the domain

$$\mathscr{D}(A_0)=\{\phi\in L^1(0,\omega):\phi(0)=0,\ A_0\phi\in L^1(0,\omega)\}.$$

It is clear that $A_0$ is an infinitesimal generator of a contraction semigroup as

$$(T_0(t)\phi)(a)=\begin{cases}\phi(a-t)\frac{\ell(a)}{\ell(a-t)}, & a-t>0,\\ 0, & t-a>0.\end{cases}$$

Hence, the following relation holds for $\lambda>0$ (e.g., see [24, Chap. 3]):

$$(\lambda-A_0)^{-1}\phi=\int_0^\infty e^{-\lambda s}T_0(s)\phi ds.$$

Observe that

$$\begin{aligned}|\lambda J(\lambda)g-g|_{L^1}&=\int_0^\omega\left|\int_0^\infty\lambda e^{-\lambda s}(T_0(s)g)(a)ds-g(a)\right|da\\&\le\int_0^\omega da\int_0^\infty\lambda e^{-\lambda s}|(T_0(s)g)(a)-g(a)|ds\\&\le\int_0^\infty\lambda e^{-\lambda s}|T_0(s)g-g|_{L^1}ds\to0,\quad(\lambda\to\infty).\end{aligned}$$

Thus, $\lim_{\lambda\to\infty}|\lambda J(\lambda)g-g|_{L^1}=0$. Therefore, it follows that, for any $g\in L^1$, $\lambda(\lambda-A)^{-1}g\in\mathscr{D}(A)$ and $\lim_{\lambda\to\infty}\lambda(\lambda-A)^{-1}g=g$. We then conclude that $\mathscr{D}(A)$ is dense in $L^1(0,\omega)$. □

*Remark 10.2* The above proof implies that $\mathscr{D}(A)$ is dense if $\mathscr{D}(A_0)$ is dense and

$$\lim_{\lambda\to\infty}|\lambda K(\lambda)g|_{L^1}=0,$$

where $K(\lambda)=(\lambda-A)^{-1}-(\lambda-A_0)^{-1}$. This is a general result given in [10].

**Proposition 10.5** *The population operator $A$ generates a strongly continuous semigroup $T(t)=e^{tA}$, $t\ge0$ where $T(t)$ is a compact operator for $t>\omega$ that satisfies the following properties:*

$$\|T(t)\|\le e^{(\bar{\beta}-\underline{\mu})t},\tag{10.8}$$

$$T(t)(L^1_+)\subset L^1_+,\quad\forall t\ge0,\tag{10.9}$$

*where* $\|\cdot\|$ *denotes the operator norm.*

*Proof* From (10.7), the following holds for $\Re\lambda > \bar{\beta} - \underline{\mu}$:

$$|J(\lambda)g|_{L^1} \leq \int_0^\omega da \int_0^a e^{-(\lambda+\underline{\mu})(a-z)}|g(z)|dz = \frac{|g|_{L^1}}{\Re\lambda + \underline{\mu}} \leq \frac{|g|_{L^1}}{\Re\lambda - (\bar{\beta} - \underline{\mu})},$$

$$|K(\lambda)g| \leq |(1 - \hat{\Psi}(\lambda))^{-1}| \int_0^\omega e^{-(\Re\lambda+\underline{\mu})a} da\bar{\beta} \int_0^\omega \int_0^a e^{-(\lambda+\underline{\mu})(a-z)}|g(z)|dzda$$
$$\leq \frac{\Re\lambda + \underline{\mu}}{\Re\lambda - (\bar{\beta} - \underline{\mu})} \cdot \frac{\bar{\beta}}{\Re\lambda + \underline{\mu}} \cdot \frac{|g|_{L^1}}{\Re\lambda + \underline{\mu}} \leq \frac{|g|_{L^1}}{\Re\lambda - (\bar{\beta} - \underline{\mu})}.$$

Therefore, we have the Yosida estimate as

$$\|R(\lambda, A)\| \leq \frac{1}{\Re - (\bar{\beta} - \underline{\mu})},$$

and it follows from the Hille–Yosida theorem that $A$ generates a strongly continuous semigroup $S(t),\ t \geq 0$, satisfying (10.8). Moreover, from Hille's formula, it holds that

$$S(t) = \lim_{n\to\infty} \left(\frac{n}{t}\right)^n R(\frac{n}{t}, A)^n$$

in the sense of strong convergence. As $R(n/t, A)$ is a nonnegative operator for $n/t > \bar{\beta} - \underline{\mu}$, $S(t)$ is also nonnegative. From Proposition 1.3, the generator of the population semigroup $T(t)$ is the population operator $A$, so from the uniqueness property of the generator, we know that $T(t) = S(t)$. Thus, it follows from Proposition 1.6 that $S(t) = T(t)$ is compact for $t > \omega$. □

From the above proposition, we know that $T(t)p_0,\ p_0 \in \mathscr{D}(A)$, gives the classical solution of (10.1), the population semigroup on a finite age interval is eventually compact and its generator has a *compact resolvent*. Then, it is *eventually norm continuous* and the spectral mapping theorem is valid [78, A-II, Sect. 1; A-III, Sect. 6]. The Fundamental Theorem of Demography (FTD; strong ergodicity theorem) given in Chap. 1 can then be reformulated as follows [54]:

**Proposition 10.6** *Let* $\lambda_0$ *be the dominant real eigenvalue of the population operator. Then, there exists a one-dimensional projection operator* $P_0$ *and positive numbers* $\varepsilon > 0,\ M(\varepsilon) \geq 1$ *such that*

$$\|e^{-\lambda_0 t}T(t) - P_0\| \leq M(\varepsilon)e^{-\varepsilon t}, \quad P_0\psi = \frac{\langle v_0, \psi\rangle}{\langle v_0, u_0\rangle}u_0,$$

*where* $u_0$ *and* $v_0$ *are the eigenfunction and the adjoint eigenfunctional of* $A$ *associated with eigenvalue* $\lambda_0$.

As we can see from the results obtained in Chap. 1, the population semigroup allows an asymptotic expansion by generalized eigenfunctions. This kind of asymptotic eigenfunction expansion is well known in transport equation theory [63]. If the age interval is infinite, the population operator is not eventually compact (because $T_0(t)$ is not compact for $t > 0$) and the resolvent spectrum of the population operator has the *essential spectrum.*[1] Webb [49, 107, 108] proved that $T(t)$ has a stable age distribution with the intrinsic growth rate $\lambda_0$, that is, Proposition 10.6 holds when $\omega \leq \infty$ if $\lambda_0$ is the strictly dominant, simple eigenvalue of the generator $A$ and $\lambda_0 > \omega_1(A)$ where $\omega_1(A)$ denotes the *essential growth bound* of $T(t)$.[2]

Gyllenberg and Webb [47] extended the idea of strong ergodicity to apply it to more general nonlinear semigroups as follows:

**Definition 10.2** Let $T(t),\ t \geq 0$, be a strongly continuous semigroup in a Banach space $X$. If there exists a constant $\lambda \in \mathbb{R}$ and an operator $P$ such that the range of $P$ is included in a one-dimensional space and, for any $x \in X$,

$$\lim_{t\to\infty} e^{-\lambda t}T(t)x = Px, \tag{10.10}$$

then $T(t)$ is called *asynchronous exponential growth* (AEG) and the constant $\lambda$ is called the *intrinsic growth constant.* If the convergence of (10.10) is uniform in a bounded set of $X$, it is called *uniform*. Moreover, if for any $x \in X$, there exist $M > 0$ and $\delta > 0$ such that

$$|e^{-\lambda t}T(t)x - Px|_X \leq Me^{-\delta t}|x|_X, \tag{10.11}$$

then the convergence is said to be *exponential.* Suppose that $X$ is a Banach space with a positive cone $X_+$. If there is a subset $Y \subset X_+$ such that, for all $x \in Y \setminus \{0\}$, $Px \in X_+ \setminus \{0\}$, then the AEG is called *strictly positive* in $Y$. If $Px$ is a quasi-interior point, it is called *ergodic* in $Y$.[3]

Webb [110] derived a necessary and sufficient condition for a linear semigroup $T(t),\ t \geq 0$, to become AEG. For the linear case, $P$ is a spectral projection to a one-dimensional subspace. Gyllenberg and Webb [47] proved a sufficient condition for a nonlinear solution semigroup of a semilinear equation to become AEG.

*Example 10.1* Let us consider the stable population model with immigration introduced in Sect. 2.2:

---

[1] $\lambda \in \sigma(A)$ is an essential spectrum if either $R(\lambda, A)$ is not closed, $\lambda$ is a limit point of $\sigma(A)$, or the generalized eigenspace $N_\lambda(A)$ is infinite-dimensional.

[2] The essential growth bound of $T(t)$ is defined by $\omega_1(A) := \lim_{t\to\infty} t^{-1}\log(\alpha[T(t)])$ where $\alpha[A]$ is the *measure of non-compactness* of a bounded linear operator $A$.

[3] Let $X^*$ be the adjoint space of $X$. $x \in X_+$ is called a *quasi-interior point* if $\langle x^*, x\rangle > 0$ for any $x^* \in X^*_+ \setminus \{0\}$.

$$\begin{aligned}
&\frac{\partial p(t,a)}{\partial t}+\frac{\partial p(t,a)}{\partial a}=-\mu(a)p(t,a)+f(t,a), \quad t>0, \quad 0<a<\omega,\\
&p(t,0)=\int_0^\omega \beta(a)p(t,a)da, \quad t>0,\\
&p(0,a)=p_0(a) \quad 0<a<\omega.
\end{aligned} \tag{10.12}$$

Then, we can consider (10.12) as an inhomogeneous linear Cauchy problem in $L^1$:

$$\frac{dp(t)}{dt}=Ap(t)+f(t), \quad p(0)=p_0. \tag{10.13}$$

If we assume that $f(t):=f(t,\cdot)$ is an $L^1$-valued function such that, for any $t_0>0$, $f(\cdot)\in L^1_+([0,t_0];L^1(0,\omega))$, then the population semigroup $T(t)=e^{tA}$ can be used to give the mild solution of (10.13) by the following variation of constants formula [83]:

$$p(t)=T(t)p_0+\int_0^t T(t-s)f(s)ds, \quad 0\le t\le t_0, \tag{10.14}$$

where $T(t)$ is the population semigroup generated by $A$. If $f(t)$ is continuously differentiable and $p_0\in\mathscr{D}(A)$, the mild solution given by (10.14) becomes a classical solution of the initial value problem (10.13). In Chap. 2, we only considered the special case where $f$ is time-independent. Here, we treat more general alternative conditions for the immigration term $f(t)$ as follows:

(1) $\sup_{t\ge0}|f(t)|_{L^1(0,\omega)}<\infty$,
(2) There exists a function $f_0\in L^1_+$ such that $\lim_{t\to\infty}f(t)=f_0$,
(3) There exists a function $f_0\in L^1_+$ such that $\int_0^\infty |f(s)-f_0|_{L^1}ds<\infty$.

Then, we can prove the following [55]:

**Proposition 10.7** *Suppose that $p(t)$ is a mild solution given by (10.14) and $T(t)$ is strongly ergodic, that is, there exist a number $\lambda_0\in\mathbb{R}$, a one-dimensional projection $P_0$ and positive numbers $\varepsilon>0$, $M(\varepsilon)\ge1$ such that $\|e^{-\lambda_0 t}T(t)-P_0\|\le M(\varepsilon)e^{-\varepsilon t}$. Then, the following hold:*

*(a) If both conditions (1) and (2) hold, or condition (3) holds and $\lambda_0<0$, then*

$$\lim_{t\to\infty}p(t)=-A^{-1}f_0,$$

*(b) If both conditions (1) and (2) hold, or condition (3) holds and $\lambda_0=0$, then*

$$\lim_{t\to\infty}\frac{p(t)}{t}=P_0f_0,$$

*(c) If condition (1) or condition (3) holds and $\lambda_0>0$, then*

$$\lim_{t\to\infty}e^{-\lambda_0 t}p(t)=P_0\left(p_0+\int_0^\infty e^{-\lambda_0 s}f(s)ds\right).$$

If the migration term $f$ is time-independent, conditions (1)–(3) are each satisfied and Proposition 2.4 is obtained as a special case of the above result. From the above result, if $\lambda_0 > 0$, the asymptotic Malthusian parameter is not affected, even if there is some migration. However, if $\lambda_0 < 0$, a constant immigration stream leads to a stationary population, the age distribution of which is given by

$$((-A)^{-1}f_0)(a) = \int_0^a \frac{\ell(a)}{\ell(z)} f_0(z)dz + \frac{\ell(a)}{1-R_0}\int_0^\omega \int_0^x \beta(x)\frac{\ell(x)}{\ell(z)} f_0(z)dzdx,$$

where $R_0 = \int_0^\omega \beta(a)\ell(a)da$ is the basic reproduction number. It is easy to see that the above stationary population produced by immigration has a more aged structure than the stationary age distribution of a closed population. The above model was studied by [6, 55]. The periodic version of (10.12) or (10.13) is investigated in [3, 4].

*Remark 10.3* If we extend the variation of constants formula in some sense, we can construct the population semigroup as a perturbed aging semigroup $T_0(t) = e^{tA_0}$. Define a linear bounded operator $B : X \to X$ as $(Bx)(a) = -\ell(a)\langle \beta, x\rangle$ where $\langle \beta, x\rangle := \int_0^\omega \beta(a)x(a)da$. We can formally rewrite the basic Eq. (10.1) as

$$\frac{dp(t)}{dt} = A_0(I + B)p(t), \tag{10.15}$$

where $I$ denotes the identity operator. Desch et al. [32] proved that the weak solution of the *abstract boundary control problem* (10.15) is given by a unique continuous solution of

$$p(t) = T_0(t)p_0 + A_0\int_0^t T_0(t-s)Bp(s)ds, \tag{10.16}$$

and $A_0(I + B)$ is the infinitesimal generator of a $C_0$-semigroup $T(t), t \geq 0$, on $X$ such that $p(t) = T(t)p_0$ [31, 32]. Note that the range of $B$ is not included in $\mathscr{D}(A_0)$, but in the Favard class of $A_0$. In general, $B$ is a continuous linear operator from $X$ to $Z$, which is a Banach space with norm $|\cdot|_Z$ continuously embedded in $X$ such that for all $x : [0, t] \to Z$, $\int_0^t T_0(t-s)x(s)ds \in \mathscr{D}(A_0)$ and there exists a continuous non-increasing function $\gamma : [0, \infty) \to [0, \infty)$ such that $\gamma(0) = 0$ and $|A_0\int_0^t T_0(t-s)x(s)ds| \leq \gamma(t)\sup_{0\leq s\leq t}|x|_Z$. This method can be applied to the case where $B$ is nonlinear and non-autonomous [12, 33, 88].

Another powerful method is known as the *sun and star arguments* developed by Clément and coworkers [23, 25–28, 36, 51]. Let $T_0^*(t)$ be the dual semigroup of an unperturbed semigroup $T_0(t)$. Then, it is well known that $T_0^*(t), t \geq 0$ is not strongly continuous on the whole space $X^*$, but only on the subspace $X^{\odot} = \overline{\mathscr{D}(A_0^*)} = \{x^* \in X^* : \lim_{t\downarrow 0}|T_0^*(t)x^* - x^*| = 0\}$. If we define $T_0^{\odot}(t)$ as the restriction of $T_0^*(t)$ to $X^{\odot}$, we obtain a $C_0$-semigroup on $X^{\odot}$. Repeating this procedure, we obtain a semigroup $T_0^{\odot *}(t)$ with weak * generator $A_0^{\odot *}$ on $X^{\odot *}$, which is again not strongly continuous on the whole space, but only on the subspace $X^{\odot\odot} = \{x^{\odot *} \in X^{\odot *} : \lim_{t\downarrow 0}|T_0^{\odot *}(t)x^{\odot *} - x^{\odot *}| = 0\}$. As we can embed $X$ into $X^{\odot *}$ by means of the natural mapping and $T_0(t)$

is strongly continuous, we have $X \subset X^{\odot\odot}$. $X$ is called *sun-reflexive* ($\odot$-reflexive) with respect to $A_0$ if and only if $X = X^{\odot\odot}$. For the stable population model (10.1), $X = L^1(0, \omega)$ is $\odot$-reflexive with respect to the aging operator $A_0$ and $X^{\odot *}$ can be identified with $C_0[0, \omega]^*$ where $X^{\odot} = C_0[0, \omega] = \{\phi \in C[0, \omega] : \phi(\omega) = 0\}$. Define the perturbation $B : X \to X^{\odot *}$ by $(Bx)(a) = H(a)\langle \beta, x \rangle$ where $H \in X^{\odot *}$ is the Heaviside function and $B$ is a linear continuous operator. Then, (10.1) can be formulated as the perturbation problem

$$\frac{dp(t)}{dt} = (A_0^{\odot *} + B)p(t). \tag{10.17}$$

The weak solution with $p(0) = x \in X$ is given by $p(t) = T(t)x$ where $T(t)$ satisfies an extended variation of constants formula

$$T(t)x = T_0(t)x + \int_0^t T_0^{\odot *}(t-s)BT(s)xds, \tag{10.18}$$

$T(t)$ defines a $C_0$-semigroup on $X$, and its infinitesimal generator is $A_0^{\odot *} + B$ with domain $\{x \in \mathscr{D}(A_0^{\odot *}) : A_0^{\odot *}x + Bx \in X\}$. If we identify $X^{\odot\odot}$ with $X$, $T(t), t \geq 0$ is none other than the population semigroup [23, 24, 58]. Readers are referred to [34, 35] for recent developments of the sun and star arguments.

Although we omit the details, Greiner [44] also developed another kind of generation theorem for boundary-perturbed operators [7]. These kinds of perturbation arguments led to developments in the functional analytic approach for structured population dynamics during the 1980s.

### *10.1.2 Nonlinear Problems*

As is seen above, the structured population model can be formulated by an abstract initial–boundary value problem:

$$\begin{aligned}
&\frac{\partial p(t,a)}{\partial t} + \frac{\partial p(t,a)}{\partial a} = C(a)p(t,a) + G(a)p(t,\cdot), \quad t, a > 0, \\
&p(t,0) = F(p(t,\cdot)), \quad t > 0, \\
&p(0,a) = p_0(a), \quad a \geq 0,
\end{aligned} \tag{10.19}$$

where $p(t,a)$ takes values in a Banach space $E = L^1(\Omega)$ (the additional trait state space), $C(a)$ is an age-dependent differential operator acting on $E$ to describe the individual's trait changes (such as size, weight, and spatial distribution) as time evolves, $G(a)$ is the nonlinear multiplication operator on $X$, which describes the mortality rates and the interstate migration rates, and $F$ is a nonlinear birth rate operator from $X = L^1(\mathbb{R}_+, E)$ to $E$. For each time $t$, $p(t,\cdot) \in X$ with norm

$|p(t, \cdot)|_X = \int_0^\infty |p(t, a)|_E da$. Note that $C(\cdot)$ and $G(\cdot)$ are operator-valued functions.

The semigroup approach to nonlinear population problems such as (10.19) was developed by Webb [103, 104, 106, 108, 109] and Prüss [84–87] for the case that $C = 0$ and $E = \mathbb{R}^n$ (see also [49]). In a most simple case, system (10.19) can be formulated as the semilinear Cauchy problem on a Banach lattice $X$:

$$\frac{dp(t)}{dt} = Ap(t) + F(p(t)), \quad t \geq 0, \tag{10.20}$$

where $A$ is the infinitesimal generator of a strongly continuous semigroup of positive linear operators in $X$ and $F$ is Lipschitz continuous on each bounded set of $X_+$ and for all $r > 0$ there exists $h(r) > 0$ such that $z + h(r)F(z) \in X_+$ for all $z \in X_+ \cap B_r := \{z \in X : |z|_X \leq r\}$. Then,

$$p(t) = e^{tA}p(0) + \int_0^t e^{(t-s)A}F(p(s))ds, \tag{10.21}$$

has a unique continuous solution in $X_+$ for each $p(0) \in X_+$ on some maximal interval of existence $[0, \tau)$ with either $\tau = \infty$ or $\lim_{t \to \tau - 0} |p(t)|_X = \infty$, which is the *mild solution* of (10.20).

In more general situations, (10.19) is transformed into the following integral equation to construct a nonlinear solution semigroup $S(t)$, $t \geq 0$:

$$p(t, a) = \begin{cases} F(p(t-a, \cdot)) + \displaystyle\int_0^a G(a-s)p(t-s, \cdot)ds, & a \in (0, t), \\ p_0(a-t) + \displaystyle\int_0^t G(a-s)p(t-s, \cdot)ds, & a \in (t, \infty). \end{cases} \tag{10.22}$$

Then, $p(t, a) = (S(t)p_0)(a)$ and the infinitesimal generator of $S(t)$ is given by

$$(A\phi)(a) = -\frac{d\phi(a)}{da} + G(a)\phi(\cdot), \quad \phi \in \mathscr{D}(A), \tag{10.23}$$

where its domain is given by $\mathscr{D}(A) = \{\phi \in W^{1,1}(\mathbb{R}_+; E) : \phi(0) = F(\phi)\}$.

To construct the solution semigroup, and to analyze the stability of steady states and the asymptotic behavior of the nonlinear system (10.19), it is very advantageous if we can deal with the nonlinear generator as a perturbation of a simple linear generator by excluding the nonlinearity from the domain of the generator. We can then construct the solution semigroup using the variation of constants formula. As we remarked above, various methods to deal with the nonlinear boundary condition as a perturbation term have been developed since the late 1980s.

As an example, we sketch the method developed by Thieme [100] using the generation theorem of the non-densely defined Hille–Yosida operator [30]. For simplicity, we neglect the aspects of positivity and invariance of solutions, which can be overcome by technical operations [59, 60, 100]. Let $X := L^1(\mathbb{R}_+; E)$ be the state space of the population vectors, let $Z := E \times X$ be the extended state space and let $Z_0 := \{0\} \times X$ be its closed subspace. Define an operator $\mathscr{A}$ on $Z$ by

$$\mathscr{A}(0, \psi) := (-\psi(0), A\psi), \quad (0, \psi) \in \mathscr{D}(\mathscr{A}) := \{0\} \times \mathscr{D}(A), \tag{10.24}$$

where $A$ is a differential operator that is densely defined on $X$:

$$(A\psi)(a) := -\frac{d\psi(a)}{da}, \quad \mathscr{D}(A) = W^{1,1}(\mathbb{R}_+; E). \tag{10.25}$$

Define a bounded nonlinear perturbation operator $\mathscr{B} : Z_0 \to Z$ by

$$\mathscr{B}(0, \psi) = (F(\psi), G(\cdot)\psi), \quad (0, \psi) \in Z_0. \tag{10.26}$$

Let $u(t) = (0, p(t)) \in Z_0$. Then, (10.19) can be written as a semilinear Cauchy problem in the space $Z$:

$$\frac{du(t)}{dt} = \mathscr{A}u(t) + \mathscr{B}u(t), \quad u(0) = (0, p_0) \in Z_0. \tag{10.27}$$

The domain of the operator $\mathscr{A}$ is dense in $Z_0$, but is not dense in $Z$. The nonlinear perturbation operator $\mathscr{B}$ is defined only on $Z_0 = \overline{\mathscr{D}(\mathscr{A})}$, but it can take values outside of the space $Z_0$. For $\lambda > 0$, the Hille–Yosida estimate holds:

$$\|(\lambda - \mathscr{A})^{-1}\| \le \frac{1}{\lambda}.$$

In this case, although we cannot apply the classical Hille–Yosida theorem and the variation of constants formula, we can instead consider the following extended formula on $Z_0$:

$$u(t) = \mathscr{T}_0(t)u(0) + \lim_{\lambda \to \infty} \int_0^t \mathscr{T}_0(t-s)\lambda(\lambda - \mathscr{A})^{-1}\mathscr{B}u(s)ds, \tag{10.28}$$

where $\mathscr{T}_0(t)$ is a semigroup on $Z_0$ generated by the *part* $\mathscr{A}_0$. The part $\mathscr{A}_0$ is defined as $\mathscr{A}_0 = \mathscr{A}$ on $\mathscr{D}(\mathscr{A}_0) = \{(0, \psi) \in \mathscr{D}(\mathscr{A}) : \mathscr{A}(0, \psi) \in Z_0\}$. Then, it follows that

$$\mathscr{A}_0(0, \psi) := (0, -\psi'), \quad (0, \psi) \in \mathscr{D}(\mathscr{A}_0) := \{0\} \times \mathscr{D}(A_0),$$

where $A_0$ is a differential operator that is densely defined on $X$:

$$(A_0\psi)(a) := -\psi'(a), \quad \mathscr{D}(A_0) = \{\psi \in W^{1,1} : \psi(0) = 0\}.$$

Let $T_0(t) = e^{tA_0}$ be a shift semigroup on $X$. Then, we have

$$\mathscr{T}_0(t)(0, \psi) = (0, T_0(t)\psi).$$

Note that the integral of the right-hand side of (10.28) is meaningful because of the property of the resolvent $(\lambda - \mathscr{A})^{-1}(Z) \subset Z_0$. Although we cannot exchange the integral and the limit, it can be proved that under appropriate conditions the extended variation of constants formula has a unique continuous solution, which is the *integral solution* of (10.27) such that

$$u(t) = u(0) + \mathscr{A} \int_0^t u(s)ds + \int_0^t \mathscr{B}u(s)ds. \tag{10.29}$$

An integral solution becomes a classical solution if and only if it is differentiable [94, Corollary B.20].

Using the extended variation of constants formula, the principle of linearized stability and a center manifold theory for semilinear equations with non-dense domains can be proved [71]. For concrete examples of the method of non-densely defined Hille–Yosida operators, readers are referred to [59] (two-sex age-structured marriage model), [74] (super-infection model), [60] (infection-age-structured HIV epidemic model), and [72] (evolutionary epidemic model).

Finally, let us consider the case in which $C$ is not zero in (10.19) and $U(a, \sigma)$, $a \geq \sigma$, is the evolutionary family generated by $C(\cdot)$ on $E$. Again, one method of treating (10.19) is to transform it into an integral equation:

$$p(t, a) = \begin{cases} U(a, 0)F(p(t-a, \cdot)) + \int_0^a U(a, a-s)G(a-s)p(t-s, \cdot)ds, & a \in (0, t), \\ U(a, a-t)p_0(a-t) + \int_0^t U(a, a-s)G(a-s)p(t-s, \cdot)ds, & a \in (t, \infty). \end{cases} \tag{10.30}$$

Although it is possible to construct the solution semigroup directly from the integral equation (10.30) [105], Thieme [101] demonstrated that it is possible to formulate problem (10.19) as the integrated formula (10.29), which can again be solved by the extended variation of constants formula (10.28). Readers are referred to the work of Thieme [101] for details of the above approach. One point of interest is how to define the operator $A = -d/da + C(\cdot)$ precisely, which can also be examined in the theory of evolution semigroups [22, p. 63].

**Exercise 10.1** Let us calculate $R_0$ for the stable population (10.1) using the well-known recipe developed for ODE systems. Let $X := L^1(\mathbb{R}_+)$ be the state space of the population vectors, let $Z := \mathbb{R} \times X$ be the extended state space and let $Z_0 := \{0\} \times X$ be its closed subspace. Define an operator $\mathscr{A}$ on $Z$ by

$$\mathscr{A}(0, \psi) := \left(-\psi(0), -\frac{d\psi(a)}{da} - \mu(a)\psi(a)\right), \quad (0, \psi) \in \mathscr{D}(\mathscr{A}),$$

and a bounded linear perturbation operator $\mathscr{B} : Z_0 \to Z$ by

$$\mathscr{B}(0,\psi) = \left(\int_0^\infty \beta(a)\psi(a)da, 0\right), \quad (0,\psi) \in Z_0.$$

Then, the stable population model (10.1) can be written as a linear Cauchy problem in the extended state space $Z$ as (10.27). According to the general recipe formally, calculate the next-generation operator $\mathscr{K} := \mathscr{B}(-\mathscr{A})^{-1}$ and its spectral radius to show that $R_0 = r(\mathscr{K}) = \int_0^\infty \beta(a)\ell(a)da$.

## 10.2 Linear Positive Operators

As we have seen in the previous chapters, positive operator theory is a most useful tool for biological applications. For finite-dimensional models, we can make use of the well-known Perron–Frobenius theorem for positive matrices. However, continuous-time models in structured population dynamics have the natural state space $L^1$, so we have to consider infinite-dimensional positive operators acting in a cone without interior points. For Perron–Frobenius-type theories of infinite-dimensional positive operators, a number of results have been given since the pioneering work of Krein and Rutman [68]. Here, we introduce some basic ideas and results, mainly those derived by Sawashima [90], Marek [73], and Heijmans [50], which are useful for our applications. Therefore, we do not pursue complete generality and mainly consider the $L^1$-space as the state space. For more general formulations, readers can consult Krasnoselskii [65, 66], Birkhoff [18], Schaefer [91], and Guo and Lakshmikantham [46].

Let $E$ be a real or complex Banach space and let $E^*$ be its dual space. Then, $E^*$ is a space of all linear functionals on $E$. In the following, we write the value of $f \in E^*$ at $\psi \in E$ as $\langle f, \psi \rangle$. A closed subset $C \subset E$ is called the *cone* (or *positive cone*) if the following conditions hold: (1) $C + C \subset C$, (2) $\lambda \geq 0 \Rightarrow \lambda C \subset C$, (3) $C \cap (-C) = \{0\}$, and (4) $C \neq \{0\}$. With respect to the cone $C$, we write $x \leq y$ if $y - x \in C$ and $x < y$ if $y - x \in C^+ := C \setminus \{0\}$. If the set $\{\psi - \phi : \psi, \phi \in C\}$ is dense in $E$, the cone $C$ is said to be *total*. If $E = C - C$, $C$ is called a *reproducing* cone. If a cone $C$ has a non-empty interior $C^\circ$, it is called a *solid* cone. Any solid cone is reproducing. We write $x \ll y$ if $y - x \in C^\circ$.

Let $B(E)$ be a set of bounded linear operators from $E$ into itself. Let $r(T)$ be the spectral radius of $T \in B(E)$ and let $P_\sigma(T)$ be the point spectrum of $T$. The dual cone $C^*$ is a subset of $E^*$ composed of all positive linear functionals. $f \in C^*$ is called a positive linear functional if $\langle f, \psi \rangle \geq 0$ for all $\psi \in C$. $\psi \in C$ is called a *quasi-interior point* or *nonsupporting point* if $\langle f, \psi \rangle > 0$ for all $f \in C^* \setminus \{0\}$. A positive linear functional $f \in C^*$ is called *strictly positive* if $\langle f, \psi \rangle > 0$ for all $\psi \in C^+$. A nonzero operator $T \in B(E)$ is called *positive* if $T(C) \subset C$ and is *strictly* *positive* if $T(C^+) \subset C^+$.[4] If $(T - S)(C) \subset C$ for $T, S \in B(E)$, we write $S \leq T$. If $C$ is a solid cone and $T(C^+) \subset C^\circ$, $T$ is called *strongly positive*.

**Proposition 10.8** (Krein–Rutman theorem; [68, 92]) *Suppose that $C$ is total and that the positive linear operator $T : C \to C$ is compact and $r(T) > 0$. Then, $r(T)$ is an eigenvalue of $T$ corresponding to a positive eigenvector $\psi \in C^+$.*

**Proposition 10.9** *Suppose that $C$ is a solid cone and $T : C \to C$ is a compact linear strongly positive operator. Then, it follows that*

*(1) $r(T) > 0$, $r(T)$ is a simple eigenvalue associated with an eigenvector in $C^\circ$ and there is no other eigenvalue with a positive eigenvector.*
*(2) $|\lambda| < r(T)$ for all eigenvalues $\lambda \neq r(T)$.*

**Definition 10.3** ([73, 90]) A positive operator $T \in B(E)$ is called *semi-nonsupporting* if, for any $\psi \in C^+$ and $f \in C^* \setminus \{0\}$, there exists an integer $p = p(\psi, f)$ such that $\langle f, T^p\psi\rangle > 0$. A positive operator $T \in B(E)$ is called *non-supporting* if, for any $\psi \in C^+$ and $f \in C^* \setminus \{0\}$, there exists an integer $p = p(\psi, f)$ such that $\langle f, T^n\psi\rangle > 0$ for all $n \geq p$. A positive operator $T \in B(E)$ is called *strictly nonsupporting* if, for any $\psi \in C^+$, there exists a positive integer $p = p(\psi)$ such that $T^n\psi$ is a quasi-interior point of $C$ for all $n \geq p$.

The idea of being semi-nonsupporting is an infinite-dimensional extension of the indecomposability of nonnegative matrices; Krasnoselskii referred to this concept as being *irreducible*. The idea of nonsupporting is an infinite-dimensional extension of the primitivity of nonnegative matrices.

**Proposition 10.10** ([73, 90]) *Suppose that the cone $C$ is total, $T \in B(E)$ is semi-nonsupporting with respect to $C$ and $r(T)$ is a pole of the resolvent $R(\lambda, T) = (\lambda - T)^{-1}$. Then, the following holds:*

*(1) $r(T) \in P_\sigma(T) \setminus \{0\}$ and $r(T)$ is a simple pole of the resolvent $R(\lambda, T)$;*
*(2) The eigenspace corresponding to $r(T)$ is one-dimensional and its eigenvector $\psi \in C$ is a quasi-interior point. Any eigenvector in $C$ is proportional to $\psi$;*
*(3) The adjoint eigenspace corresponding to $r(T)$ is one-dimensional and its eigenfunctional $f \in C^* \setminus \{0\}$ is strictly positive.*

**Proposition 10.11** ([73, 90]) *Suppose that the cone $C$ is total, $T \in B(E)$ is nonsupporting with respect to $C$ and $r(T)$ is a pole of the resolvent $R(\lambda, T) = (\lambda - T)^{-1}$. Then, (1)–(3) of Proposition 10.10 hold and, moreover, it follows that*

*(1) $r(T)$ is a dominant point of the spectrum $\sigma(T)$, that is, $|\mu| < r(T)$ for all $\mu \in \sigma(T) \setminus \{r(T)\}$;*

[4] Note that in older papers such as Birkhoff's, a positive operator is called *nonnegative* and a strictly positive operator is called *positive*.

(2) *$B_1 := \lim_{n\to\infty} r(T)^{-n}T^n$ converges in the operator norm and $B_1$ is a strictly nonsupporting operator given by*

$$B_1 = \frac{1}{2\pi i}\int_{\Gamma_0} R(\lambda, T)d\lambda,$$

*where $\Gamma_0$ is a positively oriented circle with center at $r(T)$ such that no points of the spectrum $\sigma(T)$ except $r(T)$ lie on and inside the circle $\Gamma_0$.*

From the above statement, we know that $\lim_{n\to\infty} r(T)^{-n}T^n$ converges to a projection operator on the one-dimensional eigenspace spanned by the positive eigenvector associated with the dominant positive eigenvalue $r(T)$. Combining the Krein–Rutman theorem with Sawashima's theorem (Proposition 10.11), we can obtain a useful statement[5]:

**Corollary 10.1** *Suppose that the cone $C$ is total, $r(T) > 0$, $T$ is power compact and nonsupporting with respect to $C$. Then, all statements of Propositions 10.10 and 10.11 hold.*

*Proof* Suppose that $T^n$ is compact. Then, the spectrum $\sigma(T^n)$ is a countable set with no accumulation point different from zero. From the spectral mapping theorem, we have $\sigma(T^n) = \{\sigma(T)\}^n$; in particular, $r(T^n) = r(T)^n > 0$ and $r(T) = r(T^n)^{1/n} \in \sigma(T)$. Because $T$ is power compact, its nonzero eigenvalue $r(T)$ is a pole of the resolvent $R(\lambda, T)$ [37, p. 579]. Therefore, we can apply Proposition 10.10 to $T$ and arrive at the conclusion. □

**Proposition 10.12** ([73]) *Let $E$ be a Banach lattice. Suppose that $S, T \in B(E)$ are positive operators. Then, the following holds:*

(1) *If $S \leq T$, then $r(S) \leq r(T)$.*
(2) *If $S$, $T$ are semi-nonsupporting and compact, $S \leq T$, $S \neq T$ and $r(T) \neq 0$, then $r(S) < r(T)$.*

## 10.3 The Principle of Projective Contraction Mapping

The Hilbert projective metric and Hopf's oscillation ratio are key ideas in the linear multiplicative process (positive evolutionary system) theory by Birkhoff [13–19]. We now introduce some preliminary concepts that are needed to describe the linear multiplicative process theory.

A set $E$ is said to be *partially ordered* if, for two elements $x, y \in E$, there exists a binary relation $\leq$ such that (1) $x \leq x$; (2) if $x \leq y$ and $y \leq x$, then $x = y$; (3) if $x \leq y$ and $y \leq z$, then $x \leq z$. The relation $\leq$ is called the *partial order*. If $x \leq y$ and $x \neq y$, we write $x < y$. For a real linear space $E$ with partial order $\leq$, $E$ is

[5] $T \in B(E)$ is *power compact* if there is a positive integer $n$ such that $T^n$ is compact.

called a *lattice* if, for any two elements $x, y \in E$, there is a least upper bound given by the partial order $\leq$, denoted by $\sup(x, y) =: x \vee y$, and a greatest lower bound, denoted by $\inf(x, y) =: x \wedge y$. Moreover, if the partial order $\leq$ satisfies the following conditions: (1) if $x \leq y$, then $x + z \leq y + z$; (2) if $x \leq y$ and $\lambda \geq 0$, then $\lambda x \leq \lambda y$, then $E$ is called a *vector lattice*. The absolute value of an element of a vector lattice $x \in E$ is defined by $|x| = x \vee (-x)$. A real Banach space $E$ is called a *Banach lattice* if it is a vector lattice and its norm $\| \cdot \|$ satisfies $\|x\| \leq \|y\|$, provided that $|x| \leq |y|$. For the purpose of the applications considered here, $L^1$ and $\mathbb{R}^n$ are all Banach lattices. Moreover, a semiordered linear space $\{E, \leq\}$ is called *Archimedean* if $nx \leq y, n = 1, 2, \ldots$ implies $x \leq 0$. Any Banach lattice is Archimedean.

Let $E$ be a real linear space with a positive cone $C$. For $(x, y) \in E \times C^+$, we define

$$\sup(x/y) := \inf\{\lambda : x \leq \lambda y\}, \quad \inf(x/y) := \sup\{\mu : \mu y \leq x\},$$

where we adopt the conventions that $\inf \emptyset = \infty$ and $\sup \emptyset = -\infty$. If $E$ is Archimedean, $\inf(x/y) \neq \infty$, $\sup(x/y) \neq -\infty$. In fact, if $\inf(x/y) = \infty$, then $ny \leq x$, $n = 1, 2, \ldots$, holds and implies $y \leq 0$, which contradicts our assumption that $y \in C^+$. In a similar manner, we obtain $\sup(x/y) \neq -\infty$. Then, for $(x, y) \in E \times C^+$, their *oscillation* is defined by

$$\operatorname{osc}(x/y) := \sup(x/y) - \inf(x/y).$$

From the Archimedean property, the oscillation $\operatorname{osc}(x/y)$ can be defined for any $(x, y) \in E \times C^+$ and takes a value in $[0, \infty]$. The oscillation is zero if and only if $x$ and $y$ are proportional to each other. In $C^+$, the *Hilbert projective pseudometric* is defined as follows:

$$d(x, y) := \log \left[ \frac{\sup(x/y)}{\inf(x/y)} \right].$$

Then, it is easy to see that $d(x, y)$ has the following properties:

**Lemma 10.1** *If $x, y, z \in C^+$, then*

*(1)* $d(x, x) = 0$, $d(x, y) = d(y, x)$ *and* $d(x, z) \leq d(x, y) + d(y, z)$,
*(2)* $d(x, y) = 0$ *if and only if there exists some* $\lambda > 0$ *such that* $x = \lambda y$,
*(3) For any* $\lambda > 0$ *and* $\mu > 0$, $d(\lambda x, \mu y) = d(x, y)$.

The Hilbert metric is called a *pseudometric* because it does not satisfy the axiom of metrics that $d(x, y) = 0$ implies $x = y$. However, we simply call it a projective metric. By the metric $d$, $\{C^+, d\}$ becomes a pseudometric space. The *connected component* in $\{C^+, d\}$ is an equivalent class composed of elements such that $d(x, y) < \infty$. The *ray* is an equivalent class composed of elements such that $d(x, y) = 0$. Two elements $x, y$ in $C^+$ are called *comparable* if there exist $\mu > 0$ and $\alpha \geq 1$ such that $\mu y \leq x \leq \alpha \mu y$.

**Lemma 10.2** *Two elements $x, y$ of $C^+$ belong to the same component if and only if they are comparable.*

*Proof* If $\mu y \leq x \leq \alpha\mu y$, then it is clear that $d(x, y) \leq \alpha$. Conversely, if we assume that $d(x, y) < \infty$, it follows that

$$\inf(x/y)y \leq x \leq \sup(x/y)y \leq e^{d(x,y)} \inf(x/y)y,$$

which shows that $x$ and $y$ are comparable. □

If two elements in $C^+$ are comparable, the images of those elements given by a strictly positive linear operator $A$ are also comparable, and we have

$$d(Ax, Ay) \leq d(x, y), \quad \forall(x, y) \in C^+ \times C^+.$$

Thus, a strictly positive linear operator is a contraction mapping with respect to the projective metric $d$. The *projective diameter* of a positive operator $A$ is defined by

$$\Delta(A) := \sup\{d(Ax, Ay) : (x, y) \in C^+ \times C^+\}.$$

A linear positive operator $A$ is called *uniformly positive* if $\Delta(A) < \infty$. If a power of $A$ becomes uniformly positive, $A$ is called *uniformly primitive*. The uniform primitivity of a positive linear operator can be characterized as follows:

**Lemma 10.3** *A strictly positive linear operator A is uniformly primitive if and only if there exist an integer r, $e \in C^+$, $\alpha \geq 1$ and a strictly positive functional $\lambda(x)$ such that*

$$\lambda(x)e \leq A^r x \leq \alpha\lambda(x)e, \tag{10.31}$$

*where $\lambda(x) > 0$ for any $x \in C^+$.*

*Proof* If inequality (10.31) holds, then $d(A^r x, e) \leq \log \alpha$. Hence, for any $(x, y) \in C^+ \times C^+$, we have

$$d(A^r x, A^r y) \leq d(A^r x, e) + d(e, A^r y) \leq 2\log\alpha, \tag{10.32}$$

which shows that $\Delta(A^r) \leq 2\log\alpha < \infty$, and so $A$ is uniformly primitive. Conversely, if we assume that $A$ is uniformly primitive, there exists an integer $r$ such that, for any fixed $y \in C^+$, $d(A^r x, A^r y) \leq \Delta(A^r) < \infty$ for all $x \in C^+$. Then, we have

$$\inf(A^r x/A^r y)A^r y \leq A^r x \leq \sup(A^r x/A^r y)A^r y \leq e^{\Delta(A^r)} \inf(A^r x/A^r y)A^r y.$$

Hence, if we let $e = A^r y$, $\lambda(x) = \inf(A^r x/e)$ and $\alpha = \exp(\Delta(A^r))$, inequality (10.31) holds. □

**Lemma 10.4** *Let A and B be strictly positive linear operators. It follows that*

$$\Delta(AB) \leq \min\{\Delta(A), \Delta(B)\}.$$

*Proof* From the definition, we have

$$\begin{aligned}\Delta(AB) &= \sup\{d(ABx, ABy) : (x, y) \in C^+ \times C^+\} \\ &\leq \sup\{d(Ax, Ay) : (x, y) \in C^+ \times C^+\} = \Delta(A),\end{aligned}$$

because $B(C^+) \subset C^+$. It follows from $d(ABx, ABy) \leq d(Bx, By)$ that $\Delta(AB) \leq \Delta(B)$. This completes the proof. □

**Corollary 10.2** *If $A$ is a uniformly primitive operator such that inequality* (10.31) *holds for a quasi-interior point $e$, then $A$ is strictly nonsupporting.*

*Proof* Suppose that, for some integer $n$, $A^n$ satisfies (10.31) with a quasi-interior point $e$. For any $v^* \in C^* \setminus \{0\}$ and $x \in C^+$, we have $\langle v^*, A^n x\rangle \geq \lambda(x)\langle v^*, e\rangle > 0$, which shows that $A^n x$ is a quasi-interior point. □

**Corollary 10.3** *If a strictly positive linear operator $A$ is uniformly primitive, there exists an integer $r$ such that the range of $A^n(C^+)$ $(n \geq r)$ is included in a connected component $K$, and $K$ is invariant with respect to $A$, that is, $A(K) \subset K$.*

*Proof* Let $r$ be an integer such that $A^r$ becomes uniformly positive. For a fixed $y \in C^+$, we have inequality (10.32) for any $x \in C^+$. Because any element of $A^r(C^+)$ is comparable with $A^r y$, $A^r(C^+) \subset K$ for a connected component $K$ that contains $A^r y$. Moreover, for any $z \in K$, there exist $\mu_1 > 0$, $\mu_2 > 0$ such that $\mu_1 A^r y \leq z \leq \mu_2 A^r y$. Therefore, we have

$$\begin{aligned}\mu_1 \inf(A^{r+1}y/A^r y)A^r y \leq \mu_1 A^{r+1}y \leq Az \leq \mu_2 A^{r+1}y \\ \leq \mu_2 e^{\Delta(A^r)} \inf(A^{r+1}y/A^r y)A^r y,\end{aligned}$$

which shows that any element of $A(K)$ is comparable with $A^r y$, that is, it is also an element of $K$. Then, $K$ is invariant with respect to $A$. □

For a positive operator $A$ and elements $x, y$ such that $0 < d(x, y) < \infty$, the *projective norm* $\|A\|_p$ or the *Birkhoff contraction ratio* $k(A)$ is defined by

$$\|A\|_p = k(A) := \sup\left\{\frac{d(Ax, Ay)}{d(x, y)} : 0 < d(x, y) < \infty, (x, y) \in C^+ \times C^+\right\}.$$

The *Hopf oscillation ratio* $N(A)$ for a positive operator $A$ is given by

$$N(A) := \sup\left\{\frac{\operatorname{osc}(Ax/Ay)}{\operatorname{osc}(x/y)} : 0 < \operatorname{osc}(x/y) < \infty, (x, y) \in E \times C^+\right\}.$$

Birkhoff's theorem for positive operators based on the projective metric can be stated as follows:

**Proposition 10.13** *Suppose that $A$ is a positive linear operator in the Archimedean semiordered real linear space $\{E, C, \leq\}$. It follows that*

$$k(A) = N(A) \leq \tanh\left[\frac{\Delta(A)}{4}\right].$$

From the above theorem, a uniformly positive linear operator $A$ becomes a strictly contractive mapping with respect to the projective metric. That is, its contraction ratio is less than $\tanh(\Delta(A)/4)$:

$$d(Ax, Ay) \leq \tanh\left[\frac{\Delta(A)}{4}\right] d(x, y).$$

This result was first proved by Birkhoff [13], with entirely arithmetic simple proofs later given by Ostrowski [82], Bauer [8], and Bushell [20, 21].

**Proposition 10.14** (The projective contraction mapping principle) *If $A$ is a uniformly primitive and connected components of a positive cone $C$ are complete with respect to the projective metric $d$, $A$ has a unique fixed point (positive eigenvector) $v \in C^+$ with respect to $d$, and $A^n x$ $(n = 1, 2, \ldots)$ converges to $v$ for any $x \in C^+$.*

*Proof* From the uniform primitivity of $A$, there exists an integer $r$ such that $\Delta(A^r) < \infty$. Then, we have $k(A^r) < 1$. For any $x \in C^+$, if we let $y = Ax$, then

$$d(A^r x, A^{r+1}x) = d(A^r x, A^r y) \leq \Delta(A^r) < \infty.$$

For integers $n > r$, if we define $q := [n/r]$, then we have

$$d(A^n x, A^{n+1}x) \leq k(A^r)^{q-1} d(A^r x, A^{r+1}x).$$

Therefore, $A^n x$ becomes a Cauchy sequence for sufficiently large $n > r$, and it is included in a connected component of $A^r(C^+)$. From the completeness of connected components, $A^n x$ has a limit with respect to $d$. That is, $d(A^n x, v) \to 0$ $(n \to \infty)$. It is evident from the triangle inequality that such a limit $v$ is unique up to a positive constant coefficient. From $d(Av, v) \leq d(Av, A^{n+1}x) + d(A^{n+1}x, v) \leq d(v, A^n x) + d(A^{n+1}x, v) \to 0$ $(n \to \infty)$, we have $d(Av, v) = 0$. Hence, $v$ is a fixed point with respect to $d$. Then, there exists some $\lambda > 0$ such that $Av = \lambda v$, and hence, $v$ is a positive eigenvector associated with the positive eigenvalue $\lambda$. □

To apply the above result, we must show that connected components of a positive cone $C$ are complete with respect to the projective metric $d$. However, the Banach lattice, which is most important for application purposes, satisfies this condition:

**Proposition 10.15** ([16]) *If $\{E, C, \leq\}$ is a Banach lattice, a connected component of a positive cone $C$ is a complete pseudometric space with respect to the projective metric.*

From the above proposition, we know that a connected component on a unit ball $U$ in a Banach lattice becomes a complete metric space with respect to the distance defined by the projective metric $d$. If $A$ is a uniformly primitive linear map, there exists an integer $n$ such that a map from a unit ball into itself defined by $x \to A^n x/|A^n x|_E$ becomes a contraction map with respect to $d$, in which the contraction ratio is less than unity. Therefore, it has a unique fixed point $v \in U$ in a connected component and $\lim_{n\to\infty} d(A^n x, v) = 0$. Moreover, we can state that $|A^n x/|A^n x|_E - v|_E \to 0$, $(n \to \infty)$. In fact, it follows that:

**Proposition 10.16** *On a unit ball in a Banach lattice E, it holds that*

$$|f - g|_E \le e^{d(f,g)} - 1, \quad (f, g \in U \cap C^+). \tag{10.33}$$

*Proof* Suppose that $|f|_E = |g|_E = 1$. If $d(f, g) = \infty$, our conclusion holds, so we consider the case in which $f$ and $g$ are included in the same connected component. Thus, it follows that

$$\inf(f/g)g \le f \le e^{d(f,g)} \inf(f/g)g.$$

From the monotonicity of the norm with respect to the absolute value, we have

$$\inf(f/g) \le 1 \le e^{d(f,g)} \inf(f/g).$$

Therefore, we obtain

$$\begin{aligned} |f - g|_E &= |f \vee g - f \wedge g|_E \\ &\le |e^{d(f,g)} \inf(f/g)g - \inf(f/g)g|_E = (e^{d(f,g)} - 1) \inf(f/g). \end{aligned}$$

From $\inf(f/g) \le 1$, we have inequality (10.33). □

**Proposition 10.17** *Let $\phi$ be a positive eigenvector of a uniformly primitive operator A associated with the positive eigenvalue $\gamma$. There then exist a strictly positive linear functional $v^*$, a positive constant $M(x)$ and $0 < \rho < \gamma$ that is independent of $x \in C^+$ such that*

$$|A^n x - \langle v^*, x\rangle \gamma^n \phi|_E \le M(x)\rho^n \phi, \tag{10.34}$$

*where $v^*$ is a strictly positive eigenfunctional of the dual operator $A^*$ associated with the positive eigenvalue $\gamma$.*

*Proof* Suppose that $A^k$ is uniformly positive. Observe that $\sup(A^n x/A^n\phi)$, $n = 1, 2, \ldots,$ is monotone decreasing and $\inf(A^n x/A^n\phi)$ is monotone increasing and positive for $x \in C^+$, because $A^n(C^+)$ is a connected component for $n \ge k$, and it holds that

$$\begin{aligned} 0 \le \operatorname{osc}(A^n x/A^n\phi) &= \sup(A^n x/A^n\phi) - \inf(A^n x/A^n\phi) \\ &\le (e^{d(A^n x, A^n\phi)} - 1) \sup(A^k x/A^k\phi) \to 0, \quad (n \to \infty). \end{aligned}$$

Then, we can define a strictly positive functional $v^*$ as

$$\langle v^*, x\rangle := \lim_{n\to\infty} \inf(A^n x/A^n\phi) = \lim_{n\to\infty} \sup(A^n x/A^n\phi).$$

From the definition, we have

$$\inf(A^n x/A^n\phi)A^n\phi \le A^n x \le \sup(A^n x/A^n\phi)A^n\phi.$$

$$\inf(A^n x/A^n\phi)A^n\phi \le \langle v^*, x\rangle A^n\phi \le \sup(A^n x/A^n\phi)A^n\phi.$$

Therefore, it follows that

$$|A^n x - \langle v^*, x\rangle \gamma^n \phi|_E \le \mathrm{osc}(A^n x/A^n\phi)\gamma^n\phi.$$

Let $\alpha := [n/k] - 1$ for $n \ge k$. Then, we obtain

$$\mathrm{osc}(A^n x/A^n\phi) \le \|A^k\|_p^\alpha \mathrm{osc}(A^k x/A^k\phi).$$

If we choose $\rho$ and $M(x)$ as

$$\gamma\|A^k\|_p^{\alpha/n} < \rho < \gamma, \quad M(x) := \mathrm{osc}(A^k x/A^k\phi),$$

then we arrive at inequality (10.34). It follows from (10.34) that $v^*$ is a continuous linear functional, and so we obtain

$$\lim_{n\to\infty} \gamma^{-n}\langle A^* v^*, x\rangle = \langle v^*, x\rangle\langle v^*, \phi\rangle.$$

Observe that

$$\begin{aligned}\lim_{n\to\infty} \gamma^{-n}\langle (A^*)^{n+1} v^*, x\rangle &= \langle v^*, Ax\rangle\langle v^*, \phi\rangle \\ &= \gamma \lim_{n\to\infty} \gamma^{-(n+1)}\langle (A^*)^{n+1} v^*, x\rangle = \gamma\langle v^*, x\rangle\langle v^*, \phi\rangle,\end{aligned}$$

which holds for any $x \in C$, implying that $A^* v^* = \gamma v^*$. Then, $v^*$ is a positive eigenfunctional of $A^*$ associated with eigenvalue $\gamma$. □

*Remark 10.4* If $X$ is a finite-dimensional Euclidean space $\mathbb{R}^n$, a nonnegative linear operator is expressed by a nonnegative matrix and the positive cone $C$ of $X$ is a set of nonnegative vector $x = (x_1, \ldots, x_n) \ge 0$, then the interior $C^\circ$ is a subset composed of strictly positive vectors $x = (x_1, \ldots, x_n) > 0$. Moreover, $C^\circ$ is a connected component. If $x = (x_1, \ldots, x_n)$, $y = (y_1, \ldots, y_n) \in C^\circ$, it follows that

$$d(x, y) = \log\left[\frac{\max_i (x_i/y_i)}{\min_j (x_j/y_j)}\right] = \max_{i,j} \log\left(\frac{x_i y_j}{x_j y_i}\right).$$

If a positive matrix $A$ is primitive, there exists an integer $r$ such that $A^r(C^+) \subset C^\circ$. Then, $A$ is uniformly primitive, and it follows from Propositions 10.14 to 10.15

that there exists a positive eigenvector $v \in C^\circ$ such that, for any $x \in C^+$, $\lim_{r\to\infty} d(A^r x, v) = 0$. From this, we obtain the well-known Perron–Frobenius theorem and the strong ergodicity property. Using a similar argument, we can prove the weak ergodicity theorem for matrix population models [43, 69, 93].

## 10.4 Linear Multiplicative Processes

A *time-inhomogeneous multiplicative process* (positive evolutionary system) for positive [negative] time $J = [s_0, \infty)$ $[J = (-\infty, s_0]]$ on a Banach lattice $(E, C)$ with a positive cone $C$ is a two-parameter family of strictly positive[6] linear operators $U(t, s), t \geq s, t, s \in J$, satisfying the multiplicative property

$$U(t, r)U(r, s) = U(t, s), \quad t \geq r \geq s, \quad t, r, s \in J,$$

and $U(s, s) = I$ where $I$ is the identity operator. A time-inhomogeneous multiplicative process for positive [negative] time is *uniformly primitive for positive [negative] time* when, for some $\alpha > 0$, there exists some $t > s > K$ $[s < t < K]$ for any $K > s_0$ $[K < s_0]$ such that $\Delta(U(t, s)) \leq \alpha$. A function $f(t)$ defined for all $t \in J$ and with values $f(t) \in C$ is *consistent* with the multiplicative process $U(t, s), s, t \in J$ when, for all $t, s \in J$, $f(t) = U(t, s)f(s)$.

Let $C^*$ be the set of linear nonnegative functionals on $E$ and let $E^* := \{f^* - g^* : f^*, g^* \in C^*\}$. Then, $C^*$ is a positive cone of $E^*$ and the vector space $(E^*, C^*)$ is the dual space of $(E, C)$. As the dual of any Banach lattice is again a Banach lattice, we can define the *dual multiplicative process* $U^*(s, t) = U(t, s)^*, s \leq t$, by

$$\langle U^*(s, t)v^*, \phi\rangle = \langle v^*, U(t, s)\phi\rangle,$$

where $\langle v^*, \phi\rangle$ denotes the value of $v^* \in E^*$ at $\phi \in E$. A function $v^*(t)$ defined for all $t \in J$ and with values $v^*(t) \in C^*$ is *consistent* with $U^*(s, t)$ when, for all $s \leq t, s, t \in J$, $v^*(s) = U^*(s, t)v^*(t)$. Then, it is easy to see that:

**Proposition 10.18** *If $f(t)$ is consistent with $U(t, s)$ and $v^*(t)$ is consistent with $U^*(s, t)$, it follows that, for any $t, s \in J$,*

$$\langle v^*(t), f(t)\rangle = \langle v^*(s), f(s)\rangle = const. \tag{10.35}$$

---

[6] Although, for simplicity, we restrict our argument to strictly positive processes, we can relax this condition [16, 17].

### 10.4.1 Weak Ergodicity

Let $U(t,s)$, $t \geq s$, be a time-inhomogeneous multiplicative process for positive time on $(E, C)$. $U$ is called *weakly ergodic* if, for any $\psi, \phi \in C^+$,

$$\lim_{t\to\infty} d(U(t,s)\psi, U(t,s)\phi) = 0.$$

In addition, if there exist positive numbers $\varepsilon > 0$, $M(\varepsilon) \geq 1$ such that, for any $s \in J$,

$$\|U(t,s)\|_p \leq M(\varepsilon)e^{-\varepsilon(t-s)},$$

then $U$ is called *exponentially weakly ergodic.*

**Proposition 10.19** *Let $U(t,s)$, $t \geq s$, be a weakly ergodic time-inhomogeneous multiplicative process for positive time on a Banach lattice $(E, C)$, and let $f(t)$ and $g(t)$, $t \in J$, be consistent with $U(t,s)$. If $d(f(s), g(s)) < \infty$, there exists a positive functional $v^*(s) \in E^*$ such that*

$$f(t) = (\langle v^*(s), f(s)\rangle + \varepsilon(t))g(t), \tag{10.36}$$

*where $\lim_{t\to\infty} \varepsilon(t) = 0$ and $v^*(s)$, $s \in J$, is consistent with the dual process $U^*(s,t)$, $s \leq t$, and is uniquely determined. The positive functional $v^*(s)$ defined above is called the importance functional, and it holds that*

$$\langle v^*(t), g(t)\rangle = 1, \quad \langle v^*(t), \varepsilon(t)g(t)\rangle = 0, \quad \forall t \in J.$$

*Proof* Suppose that $f(t)$ and $g(t)$ are consistent with $U(t,s)$. Then, we have

$$\begin{aligned}\mathrm{osc}(f(t)/g(t)) &= \sup(f(t)/g(t)) - \inf(f(t)/g(t))\\ &\leq \{\exp[d(f(t), g(t))] - 1\}\inf(f(t)/g(t))\\ &\leq \{\exp[d(f(t), g(t))] - 1\}\sup(f(s)/g(s)),\end{aligned}$$

and it follows from our assumption that the right-hand side goes to zero when $t \to \infty$. Because $\sup(f(t)/g(t))$ is monotone decreasing and bounded below, $\inf(f(t)/g(t))$ is monotone increasing and bounded above, and both have a common positive limit. For a fixed function $g(t)$, a positive functional $v^*(s)$ is defined by

$$\langle v^*(s), f(s)\rangle := \lim_{t\to\infty} \inf(f(t)/g(t)) = \lim_{t\to\infty} \sup(f(t)/g(t)).$$

It is easy to see that $v^*(s)$ is a linear map. From

$$\inf(f(t)/g(t))g(t) \leq f(t) \leq \sup(f(t)/g(t))g(t),$$

we have

$$|f(t) - \langle v^*(s), f(s)\rangle g(t)|_E \leq |g(t)|_E \mathrm{osc}(f(t)/g(t)).$$

Equation (10.36) follows immediately from the monotonicity of $|f|_E$ as a function of $|f|$. Furthermore, for any $s < r$, we observe that

$$\begin{aligned}\langle v^*(r), f(r)\rangle &= \lim_{t\to\infty} \inf(U(t,r)f(r)/U(t,r)g(r)) \\ &= \lim_{t\to\infty} \inf(U(t,r)U(r,s)f(s)/U(t,r)U(r,s)g(s)) = \langle v^*(s), f(s)\rangle.\end{aligned}$$

Therefore, it follows that

$$\langle v^*(s), f(s)\rangle = \langle v^*(t), f(t)\rangle = \langle v^*(t), U(t,s)f(s)\rangle = \langle U^*(s,t)v^*(t), f(s)\rangle,$$

which shows that $v^*(s) = U^*(s,t)v^*(t)$, $v^*(t)$ is consistent with $U^*(s,t)$. Although $v^*(s)$ depends on the choice of $g(t)$, any importance functional of $U(t,s)$ is proportional to $v^*(s)$. To see this fact, let $h(t)$ be any consistent function with $U(t,s)$ and let $u^*(s)$ be an importance functional induced from $h(t)$. From (10.36), we have

$$h(t) = (\langle v^*(s), h(s)\rangle + \varepsilon(t))g(t).$$

Then, for any consistent function $f$, it follows that

$$\sup(f(t)/h(t)) = \frac{\sup(f(t)/g(t))}{\langle v^*(s), h(s)\rangle + \varepsilon(t)} \to \langle u^*(s), f(s)\rangle.$$

Thus, we obtain

$$\langle v^*(s), f(s)\rangle = \langle v^*(s), h(s)\rangle\langle u^*(s), f(s)\rangle,$$

which shows that $v^*(s)=\langle v^*(s), h(s)\rangle u^*(s)$. From (10.35), we know that $\langle v^*(s), h(s)\rangle$ is constant, and the importance functional is unique up to a positive constant factor. Finally, taking the duality pairing in (10.36), we have

$$\langle v^*(t), f(t)\rangle = \langle v^*(s), f(s)\rangle\langle v^*(t), g(t)\rangle + \langle v^*(t), \varepsilon(t)g(t)\rangle.$$

Because $\langle v^*(t), f(t)\rangle = \langle v^*(s), f(s)\rangle$, it follows that

$$\langle v^*(t), f(t)\rangle(1 - \langle v^*(t), g(t)\rangle) = \langle v^*(t), \varepsilon(t)g(t)\rangle.$$

From Proposition 10.18, both $\langle v^*(t), f(t)\rangle$ and $\langle v^*(t), g(t)\rangle$ are nonzero constants, whereas $\langle v^*(t), \varepsilon(t)g(t)\rangle$ goes to zero as $t \to \infty$, so we have $\langle v^*(t), g(t)\rangle = 1$ and $\langle v^*(t), \varepsilon(t)g(t)\rangle = 0$ for all $t \in J$. □

**Proposition 10.20** *Let $U(t,s)$, $t \geq s$, be a time-inhomogeneous multiplicative process for positive time on $(E, C)$. If $U(t,s)$, $t \geq s$, is uniformly primitive for positive time, it is weakly ergodic.*

*Proof* Because $U(t,s)$ is uniformly primitive for positive time, for any $s \in J$ there exists a number $\alpha > 0$ and an infinite sequence of positive numbers $s = t_0 < t_1 < t_2 < \ldots, t_n \to \infty$ in $J$ such that $\Delta(U(t_{2n+1}, t_{2n})) \leq \alpha$. Then, it is easily seen that, for $t > t_{2n+1}$, $\psi, \phi \in C^+$,

$$d(U(t,s)\phi, U(t,s)\psi) \leq \left(\tanh \frac{\alpha}{4}\right)^n \alpha \to 0, \quad (n \to \infty),$$

which shows that $U(t,s)$ is weakly ergodic. □

**Proposition 10.21** ([17]) *Let $U(t,s)$, $t \geq s$ be a weakly ergodic time-inhomogeneous multiplicative process for positive time on a Banach lattice $(E, C)$. The importance functional is the one and (essentially) only positive linear functional consistent with the dual process $U^*(s,t)$, $t \geq s$.*

*Proof* Let $v^*(s)$ be the importance functional associated with a positive consistent function $g(t)$. For any $\psi \in C^+$, weak ergodicity implies that we can assume $d(U(t,s)\psi, g(t)) < \infty$ without loss of generality. Let $w^*(s)$ be any positive consistent functional. It follows from (10.36) that

$$|\langle w^*(t), U(t,s)\psi\rangle - \langle v^*(s), \psi\rangle\langle w^*(t), g(t)\rangle| \leq \langle w^*(t), g(t)\rangle \mathrm{osc}(U(t,s)\psi/g(t)).$$

From the weak ergodicity, we have $\lim_{t\to\infty} \mathrm{osc}(U(t,s)\psi/g(t)) = 0$. As $\langle w^*(t), g(t)\rangle$ is constant, we conclude that

$$\langle w^*(s), \psi\rangle = \langle v^*(s), \psi\rangle\langle w^*(s), g(s)\rangle,$$

where we have used (10.35). Therefore, we have

$$w^*(s) = v^*(s)\langle w^*(s), g(s)\rangle,$$

which shows that $w^*(s)$ is proportional to the importance functional $v^*$, because $\langle w^*(s), g(s)\rangle$ is a constant. □

In applications, the consistent function $f(t)$ is a state vector of the physical or biological system and $U(t,s)$ is its time evolution operator. Suppose that there are two time series of state vectors $f(t)$ and $g(t)$ that are consistent with a weakly ergodic multiplicative process $U(t,s)$, $t \geq s$. From (10.33), it follows that

$$\lim_{t\to\infty} \left| \frac{f(t)}{|f(t)|_E} - \frac{g(t)}{|g(t)|_E} \right|_E = 0. \tag{10.37}$$

Moreover, if $v^* \in E^*$, we have

$$\lim_{t\to\infty} \left| \langle v^*, \frac{f(t)}{|f(t)|_E}\rangle - \langle v^*, \frac{g(t)}{|g(t)|_E}\rangle \right| = 0.$$

Because any two state vectors that evolve with time according to the same weakly ergodic multiplicative process become asymptotically proportional, the normalized state vectors and their linear functionals converge to each other.

Moreover, if a multiplicative process is uniformly primitive for negative time, there exists a unique function that is consistent with it:

**Proposition 10.22** ([17]) *Let $U(t,s)$, $t \geq s$ be a uniformly primitive multiplicative process for negative time on $(E, C)$. Then, up to a proportional constant, there exists a unique consistent function $f(t) : J \to C^+$.*

In principle, the above proposition guarantees that we can reconstruct the past time series of state vectors if it is generated by a uniformly primitive multiplicative process.

Again, let $U(t,s)$, $t \geq s$, be a time-inhomogeneous multiplicative process for positive time $J = [s_0, \infty)$ on $(E, C)$. Then, the *growth bound for positive time*, denoted by $\omega_p^+(U)$, is defined by

$$\omega_p^+(U) := \lim_{\tau \to \infty} \frac{\log \eta(\tau)}{\tau}, \tag{10.38}$$

where

$$\eta(\tau) := \sup_{s \geq s_0} \|U(s+\tau, s)\|_p.$$

The existence of the limit in (10.38) follows from the subadditivity of $\log \eta(\tau)$ on $[0, \infty)$. In fact, we can observe that

$$\begin{aligned}\eta(x+y) &= \sup_{s \geq s_0} \|U(s+x+y, s)\|_p \leq \sup_{s \geq s_0} \|U(s+x+y, s+x)\|_p \|U(s+x, s)\|_p \\ &\leq \sup_{s \geq s_0} \|U(s+x+y, s+x)\|_p \sup_{s \geq s_0} \|U(s+x, s)\|_p \leq \eta(y)\eta(x).\end{aligned}$$

Then, $\log \eta(\tau)$ is a subadditive function on $[0, \infty)$, and it is known that there exists a limit $\lim_{\tau \to \infty} \log \eta(\tau)/\tau$ [37, p. 618, Lemma 4].

Because $U(t,s)$ is a contraction map with respect to the projective metric, the growth bound is always less than or equal to zero. Moreover, it follows from the definition that, if $\varepsilon > \omega_p^+(U)$, there exists a number $M(\varepsilon) \geq 1$ such that

$$\|U(t,s)\|_p \leq M(\varepsilon) e^{\varepsilon(t-s)}.$$

Therefore, if $\omega_p^+(U) < 0$, the uniformly primitive process $U(t,s)$ is in fact exponentially weakly ergodic.

**Proposition 10.23** ([56]) *Let $U(t,s)$, $t \geq s$, be a time-inhomogeneous multiplicative process for positive time $J = [s_0, \infty)$ on $(E, C)$. If there exist positive numbers $\xi > 0$ and $\alpha > 0$ such that, for all $s \geq s_0$, $\Delta(U(s+\xi, s)) \leq \alpha$, then $U(t,s)$, $t \geq s$, is exponentially weakly ergodic and*

$$\omega_p^+(U) \leq \frac{\log(\tanh(\alpha/4))}{\xi} < 0.$$

## 10.4.2 Strong Ergodicity

The case in which $U(t,s)$, $t \geq s$, depends only on the time interval $t - s$ is of considerable interest. In this special case, we have a one-parameter family of positive operators $T(t)$, $t \geq 0$, if we set $T(t) := U(t, 0)$. Then, it is clear that $T(t)$, $t \geq 0$, has the *semigroup property*:

$$T(t)T(u) = T(t+u), \quad t \geq 0, \quad u \geq 0.$$

A one-parameter semigroup of positive operators on $(E, C)$ is called a *time-homogeneous multiplicative process*. If there exists a number $t_1 > 0$ such that $T(t_1)$ becomes uniformly positive, $T(t), t \geq 0$ is called *uniformly primitive*.

Let $T(t)$, $t \geq 0$, be a positive semigroup on $(E, C)$. If there exists $\phi_0 \in C^+$ such that

$$d(T(t)\phi_0, \phi_0) = 0, \ \forall t \geq 0, \quad \lim_{t\to\infty} d(T(t)\psi, \phi_0) = 0, \ \forall \psi \in C^+,$$

then $T(t)$, $t \geq 0$, is called *strongly ergodic* and $\phi_0$ is called the *stable distribution*. Moreover, if there exist positive numbers $\varepsilon > 0$ and $M(\varepsilon) \geq 1$ such that

$$\|T(t)\|_p \leq M(\varepsilon)e^{-\varepsilon t},$$

then $T(t), t \geq 0$ is called *exponentially strongly ergodic*.

From the above definition, it is clear that the stable distribution $\phi_0$ is an eigenvector of $T(t)$ and there exists a function $r(t) > 0$ such that

$$T(t)\phi_0 = r(t)\phi_0, \quad \forall t \geq 0.$$

The positive function $r(t)$ is a positive eigenvalue of $T(t)$ associated with $\phi_0$, and it follows from the semigroup property that

$$r(t)r(s) = r(t+s), \quad r(0) = 1.$$

If $T(t)$ is strongly continuous, $r(t)$ becomes a continuous function and there exists $\lambda_0 \in \mathbb{R}$ such that $r(t) = e^{\lambda_0 t}$. That is, $\lambda_0$ is the *intrinsic growth rate* (Malthusian parameter) of this multiplicative process.

**Proposition 10.24** *Let $T(t)$, $t \geq 0$, be a strongly continuous, strongly ergodic positive semigroup on $(E, C)$, let $\phi_0$ be the stable distribution of $T(t)$ and let $r(t) = e^{\lambda_0 t}$ be the eigenvalue of $T(t)$ associated with $\phi_0$. Then, there exists a positive functional $v_0^* \in C^*$ such that, for any $\phi$ satisfying $d(\phi, \phi_0) < \infty$,*

$$\lim_{t+\infty} e^{-\lambda_0 t} T(t)\phi = \langle v_0^*, \phi \rangle \phi_0,$$

*where $v_0^*$ is the eigenvector of the dual process $T^*(t) = T(t)^*$ associated with its eigenvalue $e^{\lambda_0 t}$.*

**Proposition 10.25** *Suppose that $T(t)$, $t \geq 0$, is a time-homogeneous multiplicative process on a Banach lattice $(E, C)$. If $T(t)$, $t \geq 0$ is uniformly primitive, it is exponentially strongly ergodic.*

Let $J$ be an unbounded subset of $\mathbb{R}$. A function $f(t) : J \to C^+$ is *consistent* with $T(t)$, $t \geq 0$, if $f(t) = T(t-s)f(s)$ with $t > s$, $t, s \in J$, and a functional $v^*(t) : J \to C_+^*$ is *consistent* with $T^*(t)$, $t \geq 0$, if $v^*(s) = T^*(t-s)v^*(t)$ with $t > s$, $t, s \in J$. In particular, the *exponential solution* $e^{\lambda_0 t}\phi_0$ is consistent with $T(t)$ for $t \geq 0$ and the exponential solution $e^{-\lambda_0 t} v_0^*$ is consistent with $T^*(t)$ for $t \geq 0$. Applying the strong ergodicity to a time series of distributions starting from infinitely remote past or future, we obtain the following result:

**Proposition 10.26** *Let $T(t)$, $t \geq 0$, be a uniformly primitive time-homogeneous multiplicative process on $(E, C)$. For any $t_0 \in \mathbb{R}$, if $f(t)$, $t \in J = (-\infty, t_0]$, is consistent with $T(\tau)$, $\tau \geq 0$, $f(t)$ is proportional to the exponential solution $e^{\lambda_0 t}\phi_0$ for all $t \in J$, and if $v^*(t)$, $t \in J = [t_0, \infty)$ is consistent with $T^*(t)$, $t \geq 0$, $v^*(t)$ is proportional to the exponential solution $e^{-\lambda_0 t} v_0^*$ of $T^*(t)$ for all $t \in J$.*

The above result was partly recognized by Lotka [70] for the stable population model [89]. Even for the time-inhomogeneous multiplicative process $\{U(t, s)\}_{s \leq t}$, we can define its strong ergodicity as follows:

**Definition 10.4** For the multiplicative process $\{U(t, s)\}_{0 \leq s \leq t < \infty}$, if there exist a one-dimensional projection $P_s$ and a number $\lambda \in \mathbb{R}$ such that, for any $\phi \in C$,

$$\lim_{t \to \infty} e^{-\lambda(t-s)} U(t, s)\phi = P_s \phi,$$

then $\{U(t, s)\}_{0 \leq s \leq t < \infty}$ is said to be *strongly ergodic* with the intrinsic growth rate $\lambda$.

The strong ergodicity of non-autonomous systems can be observed in asymptotically autonomous systems, as shown in Chap. 2 (Sect. 2.6.3, [58]). It follows immediately from the definition that:

**Lemma 10.5** *For the one-dimensional operator $P_s$ in Definition 10.4, there exist $\phi_0 \in C^+$ and $f_s \in C^* \setminus \{0\}$ such that*

$$P_s\phi = \langle f_s, \phi\rangle\phi_0, \quad U^*(s,t)f_t = e^{\lambda(t-s)}f_s.$$

*Proof* For $s < t < u$, observe that

$$e^{-\lambda(u-t)}U(u,t)e^{-\lambda(t-s)}U(t,s) = e^{-\lambda(u-s)}U(u,s).$$

From the definition, if $u \to \infty$, then for any $\phi \in C$,

$$P_t e^{-\lambda(t-s)}U(t,s)\phi = P_s\phi. \tag{10.39}$$

Because $P_s$ and $P_t$ are one-dimensional, their range is spanned by $\phi_0 \in C^+$, which is independent of the time origin. As $P_s$ is linear, there exists $f_s \in C^* \setminus \{0\}$ such that $P_s\phi = \langle f_s, \phi\rangle\phi_0$. From (10.39), we have that, for any $\phi \in C$,

$$\langle f_t, e^{-\lambda(t-s)}U(t,s)\phi\rangle = \langle f_s, \phi\rangle,$$

which shows that $f_s = e^{-\lambda(t-s)}U^*(s,t)f_t$. □

In (10.36), if we set $g(t) = U(t,s)\phi_0 = e^{\lambda(t-s)}\phi_0$, then we have

$$U(t,s)\phi = (\langle v^*(s), \phi\rangle + \varepsilon(t))e^{\lambda(t-s)}\phi_0.$$

Thus, we can state the following:

**Proposition 10.27** *Let $\{U(t,s)\}_{0\le s\le t<\infty}$ be a uniformly primitive multiplicative process. Suppose that there exist $\phi_0 \in C^+$ and $\lambda \in \mathbb{R}$ such that $U(t,s)\phi_0 = e^{\lambda(t-s)}\phi_0$. Then, $\{U(t,s)\}_{0\le s\le t<\infty}$ is strongly ergodic with intrinsic growth rate $\lambda$.*

### 10.4.3 Periodic Evolutionary System

Finally, let us consider the case in which a strictly positive evolutionary system $\{U(t,s)\}_{s\le t}$ on a Banach lattice $(E, C)$ has a periodicity, a scenario that was investigated by Thieme [102] and Inaba [61]. Suppose that, for all $s \le t$, there exists $\theta > 0$ such that

$$U(t+\theta, s+\theta) = U(t,s), \quad s \le t.$$

Then, it is clear that the dual process is also a periodic evolutionary system: $U^*(s+\theta, t+\theta) = U^*(s,t)$. For any $s \in J$, $U(s+\theta, s)$ is called the *monodromy operator*.

**Lemma 10.6** *If $U$ is a uniformly primitive $\theta$-periodic evolutionary system, the monodromy operator $U(s+\theta, s)$ is uniformly primitive for any $s \in J$.*

*Proof* From the periodicity, we have $U(s+\theta,s)^n = U(s+n\theta,s)$ for any positive integer $n$. From the uniform primitivity of $U$, for some $\alpha > 0$ and any $K > s_0$, there exist some $t > s > K$ such that $\Delta(U(t,s)) \le \alpha$. Then, we can choose a sufficiently large $n$ such that there exists an interval $(t_1, t_2) \subset (s, n\theta)$ with $\Delta(U(t_2,t_1)) \le \alpha$. From Lemma 10.4, we have $\Delta(U(s+n\theta,s)) \le \Delta(U(t_2,t_1)) \le \alpha$, which shows that $U(s+\theta,s)$ is uniformly primitive. □

If there exists a consistent function $f(t)$ such that $f(t) = e^{\lambda t}\phi(t)$ where $\lambda \in \mathbb{R}$ and $\phi(t)$ is a $\theta$-periodic positive function, we call $f(t)$ the *exponential solution* [61]. Then, we can prove the following:

**Lemma 10.7** *Suppose that $U$ is a strictly positive periodic evolutionary system on $(E, C)$. There exists a positive exponential solution with exponent $\lambda$ if and only if the monodromy operator $U(s+\theta,s)$ has a positive eigenvector associated with a positive eigenvalue $e^{\lambda\theta}$.*

*Proof* If there exists an exponential solution $e^{\lambda t}\phi(t)$, it follows that $e^{\lambda t}\phi(t) = U(t,0)\phi(0)$. Because $\phi(\theta) = \phi(0)$, we have $e^{\lambda\theta}\phi(0) = U(\theta,0)\phi(0)$, which shows that $U(\theta,0)$ has a positive eigenvector $\phi(0)$ associated with an eigenvalue $e^{\theta\lambda}$. Observe that $U(s+\theta,s)U(s,0) = U(s+\theta,\theta)U(\theta,0) = U(s,0)U(\theta,0)$. Therefore, it follows that $U(s+\theta,s)[U(s,0)\phi(0)] = e^{\lambda\theta}[U(s,0)\phi(0)]$, which shows that $U(s+\theta,s)$ has an eigenvector $U(s,0)\phi(0)$ associated with an eigenvalue $e^{\theta\lambda}$. Conversely, suppose that $U(s+\theta,s)$ has a positive eigenvector $\phi(s)$ associated with a positive eigenvalue $\rho > 0$. Let us define a positive functional $\phi(t)$ by $\phi(t) := e^{-\lambda(t-s)}U(t,s)\phi(s)$ with $\lambda = (\log \rho)/\theta$. Then, it is easy to see that $\phi(t)$ has a period $\theta$ and $e^{\lambda t}\phi(t)$ is a consistent function, which is an exponential solution. □

**Proposition 10.28** *If $U$ is a uniformly primitive periodic evolutionary system on $(E, C)$, it has a positive exponential solution.*

*Proof* From Lemma 10.7, it is sufficient to show that $U(s+\theta,s)$ has a positive eigenvector. Because the monodromy operator $U(s+\theta,s)$ is uniformly primitive (Lemma 10.6), it has a positive eigenvector (Proposition 10.14). □

From Corollary 10.1 and Proposition 10.14, we have another condition for the existence of an exponential solution:

**Proposition 10.29** *Suppose that $E$ is a real Banach space with a total cone $C$ and $U$ is a strictly positive periodic multiplicative process on $(E, C)$. If the monodromy operator $U(s+\theta,s)$ is power compact, nonsupporting and $r(U(s+\theta,s)) > 0$, it has a positive exponential solution.*

**Proposition 10.30** *Suppose that $U$ is a uniformly primitive $\theta$-periodic evolutionary system on $(E, C)$. Let $e^{\lambda t}\phi(t)$ be its exponential solution. Then, there exists a positive linear functional (importance functional) $v^*(s)$ such that*

$$\lim_{t\to\infty} |e^{-\lambda t}U(t,s)x - \langle v^*(s), x\rangle\phi(t)|_E = 0, \tag{10.40}$$

*where $v^*$ is an exponential solution of the dual system $U^*$, that is, there exists a periodic functional $w^*(s)$ such that $v^*(s) = e^{-\lambda s} w^*(s)$ and $w^*$ is a positive eigenfunctional of the dual monodromy operator $U^*(s, s+\theta)$ associated with the eigenvalue $e^{\lambda\theta}$.*

*Proof* It follows from Eq. (10.36) that there exists a positive functional $v^*(s)$ such that, for $x \in C^+$ satisfying $d(\phi(s), x) < \infty$,

$$e^{-\lambda t} U(t, s)x = \langle v^*(s), x\rangle \phi(t) + o(|\phi(t)|_E), \tag{10.41}$$

where $\phi(t)$ is a periodic function, and so $|\phi(t)|_E$ is bounded. Any two consistent functions become comparable after finite time, and then (10.40) holds for any $x \in C$. From (10.41), we have

$$\begin{aligned} e^{-\lambda(t+\theta)} U(t+\theta, s+\theta)x &= \langle v^*(s+\theta), x\rangle \phi(t+\theta) + o(|\phi(t+\theta)|_E) \\ &= \langle v^*(s+\theta), x\rangle \phi(t) + o(|\phi(t)|_E) \\ &= e^{-\lambda(t+\theta)} U(t, s)x \\ &= e^{-\lambda\theta}(\langle v^*(s), x\rangle \phi(t) + o(|\phi(t)|_E)), \end{aligned}$$

which shows that $v^*(s+\theta) = e^{-\lambda\theta} v^*(s)$. If we define a functional $w^*(s)$ by $w^*(s) = e^{\lambda s} v^*(s)$, $w^*$ is $\theta$-periodic. In fact, $w^*(s+\theta) = e^{\lambda(s+\theta)} v^*(s+\theta) = e^{\lambda s} v^*(s) = w^*(s)$ and $v^*(s) = e^{-\lambda s} w^*(s)$ is an exponential solution (consistent functional) of the dual system. Moreover, it follows from $v^*(s) = U^*(s, s+\theta) v^*(s+\theta)$ that

$$e^{-\lambda s} w^*(s) = U^*(s, s+\theta) e^{-\lambda(s+\theta)} w^*(s+\theta) = e^{-\lambda\theta} U^*(s, s+\theta) e^{-\lambda s} w^*(s).$$

Then, we have $e^{\lambda\theta} w^*(s) = U^*(s, s+\theta) w^*(s)$, which shows that $w^*$ is a positive eigenfunctional of the dual monodromy operator $U^*(s, s+\theta)$ associated with the eigenvalue $e^{\lambda\theta}$. Because the importance functional is consistent with the dual system, $v^*(s) = e^{-\lambda s} w^*(s)$ is the exponential solution of the dual system. □

From (10.41), for any function $f(t)$ consistent with a uniformly primitive periodic multiplicative process, we obtain

$$\lim_{t\to\infty} \left| \frac{f(t)}{|f(t)|_E} - \frac{\phi(t)}{|\phi(t)|_E} \right|_E = 0,$$

which shows that the normalized distribution $f(t)/|f(t)|_E$ converges to a periodic distribution $\phi(t)/|\phi(t)|_E$.

Although we have only treated a linear positive evolution system and its ergodicity, there are some classes of nonlinear positive operators to which we can apply the idea of a projective metric and ergodicity [41, 67, 79–81, 99].

## 10.5 Nonlinear Positive Operators

When we consider the existence, uniqueness and stability of steady states in models of structured population dynamics, we must usually deal with infinite-dimensional nonlinear positive operators. In the previous chapters, to avoid a direct treatment of infinite-dimensional problems, we have often considered simplified finite-dimensional cases. Here, we present some useful results that are applicable to infinite-dimensional nonlinear population problems.

**Definition 10.5** Let $C$ be a positive cone of a real Banach space $E$ and let $\leq$ be the partial ordering induced by $C$. A positive operator $A : C \to C$ is called a *concave operator* if there exists some $\psi_0 \in C^+$ that satisfies the following conditions:

(1) For any $\psi \in C^+$, there exist $\alpha = \alpha(\psi) > 0$ and $\beta = \beta(\psi) > 0$ such that $\alpha\psi_0 \leq A\psi \leq \beta\psi_0$. That is, $A\psi$, $\psi \in C$ is *comparable* with $\psi_0$ or $A$ is the $\psi_0$-positive operator [65].
(2) For any $\psi \in C$ that is comparable with $\psi_0$ and $0 \leq t \leq 1$, $A(t\psi) \geq tA\psi$.

Although a one-dimensional monotone increasing concave function has at most one fixed point, for the infinite-dimensional case, we need additional conditions to guarantee the uniqueness of the fixed point of the monotone concave operator.

**Proposition 10.31** *Suppose that a positive operator $A : C \to C$ is monotone and concave. For any $\psi \in C$ satisfying $\alpha_1\psi_0 \leq \psi \leq \beta_1\psi_0$ ($\alpha_1 = \alpha_1(\psi) > 0$, $\beta_1 = \beta(\psi) > 0$) and any $0 < t < 1$, if there exists $\eta = \eta(\psi, t) > 0$ such that*

$$A(t\psi) \geq tA\psi + \eta\psi_0,$$

*then $A$ has at most one positive fixed point.*

*Proof* Suppose that $A$ has two positive fixed points $\psi_1, \psi_2 \in C^+$. From the concavity assumption, we can choose positive constants $\alpha_1 = \alpha_1(\psi_1) > 0$, $\beta_2 = \beta_2(\psi_2) > 0$ such that

$$\psi_1 = A\psi_1 \geq \alpha_1\psi_0 \geq \alpha_1\beta_2^{-1}A\psi_2 = \alpha_1\beta_2^{-1}\psi_2.$$

If we define $k := \sup\{\mu : \psi_1 \geq \mu\psi_2\}$, it follows from the above inequality that $k > 0$. If we assume that $0 < k < 1$, there exists $\eta = \eta(\psi_2, k) > 0$ such that

$$\psi_1 = A\psi_1 \geq A(k\psi_2) \geq kA\psi_2 + \eta\psi_0 \geq k\psi_2 + \eta\beta_2^{-1}A\psi_2 = (k + \eta\beta_2^{-1})\psi_2,$$

which contradicts the definition of $k$. Hence, we know that $k \geq 1$ and $\psi_1 \geq \psi_2$. By changing the numbers, we can repeat the same argument and prove $\psi_2 \geq \psi_1$. Thus, we have $\psi_1 = \psi_2$. □

The following theorem by Krasnoselskii is often useful in proving the existence of positive fixed points for nonlinear operators:

**Proposition 10.32** ([65, Theorem 4.11]) *Let $\Psi$ be a positive operator from a cone $C$ in a real Banach lattice into itself. Suppose that $\Psi(0) = 0$, $\Psi$ has a strong Fréchet derivative $T := \Psi'(0)$ and $T$ has a positive eigenvector $v_0 \in C$ associated with the eigenvalue $\lambda_0 > 1$, but has no eigenvector in $C$ associated with unity. Moreover, $\Psi$ has a strong asymptotic derivative $\Psi'(\infty)$ with respect to a cone $C$ and the spectrum of the operator $\Psi'(\infty)$ lies in the circle $|\lambda| \le \rho < 1$. Then, $\Psi$ has at least one nonzero fixed point in $C$ if $\Psi$ is completely continuous.*

Using the argument of Krasnoselskii to prove the above theorem, we can show a slightly modified statement:

**Proposition 10.33** *Suppose that there exist numbers $M > 0$ and $\varepsilon > 0$ such that $\Psi$ is a compact positive operator from $B_\varepsilon := \{x \in C : |x| \le (1+\varepsilon)M\}$ to $B_0$. Moreover, $\Psi(0) = 0$, $\Psi$ has a strong Fréchet derivative $T := \Psi'(0)$ and $T$ has a positive eigenvector $v_0 \in C$ associated with eigenvalue $\lambda_0 > 1$, but has no eigenvector in $C$ associated with unity. Then, $\Psi$ has at least one nonzero fixed point in $B_0$.*

*Proof* Define an operator $\Psi_\eta$ with a parameter $\eta \in [0, \varepsilon]$ by

$$\Psi_\eta(x) = \begin{cases} \Psi(x), & \text{if } |x| \ge \eta, \\ \Psi(x) + (\eta - |x|)v_0, & \text{if } |x| \le \eta, \end{cases}$$

where $v_0$ is the positive eigenvector of $T = \Psi'(0)$ corresponding to the eigenvalue $\lambda_0 > 1$ such that $|v_0| < M$. Then, $\Psi_\eta$ is also positive, completely continuous, defined on $B_\varepsilon$ and $\Psi_\eta(B_\eta) \subset B_\eta$. Since $B_\eta$ is bounded, convex and closed in $E$, it follows from Schauder's principle that $\Psi_\eta$ has a fixed point $x_\eta \in B_\eta$. Then, the norm of $x_\eta$ is greater than $\eta$ if $\eta$ is sufficiently small. In fact, if every operator $\Psi_\eta$ has a fixed point whose norm does not exceed $\eta$, we can construct a sequence $x_n \in C$ such that $0 < |x_n| \le \eta_n$ where $\varepsilon > \eta_1 > \eta_2 > \cdots > \eta_n \to 0$ when $n \to \infty$ and $\Psi(x_n) + (\eta_n - |x_n|)v_0 = x_n$. Since $T$ is compact, we can assume without loss of generality that $T(x_n/|x_n|)$ converges to some element $z \in C$. Observe that

$$\left(\frac{\eta_n}{|x_n|} - 1\right) v_0 = \frac{x_n}{|x_n|} - \frac{\Psi(x_n) - Tx_n}{|x_n|} - T\frac{x_n}{|x_n|},$$

which implies that

$$|v_0| \limsup_{n\to\infty} \left(\frac{\eta_n}{|x_n|} - 1\right) \le 1 + |z|.$$

Therefore, we can assume without loss of generality that $\eta_n/|x_n| - 1$ converges to some nonnegative number $\gamma \ge 0$. Then, the sequence $x_n/|x_n|$ converges to some vector $u_0 \in C$ such that $|u_0| = 1$ and $u_0 = Tu_0 + \gamma v_0$. Then, $\gamma > 0$, because $T$ does not have eigenvectors in $C$ corresponding to eigenvalue unity. Let $k_0 := \sup \Lambda$ where $\Lambda := \{k \in \mathbb{R}_+ : u_0 \ge kv_0\}$. Then, $k_0 > 0$ because $\gamma \in \Lambda$. On the other hand, we can observe that $u_0 \geq T(k_0 v_0) + \gamma v_0 = (k_0\lambda_0 + \gamma)v_0$, which contradicts the definition of $k_0$, because $k_0\lambda_0 + \gamma > k_0$. Therefore there exists $\eta > 0$ such that $|x_\eta| \geq \eta$ and $x_\eta$ is a positive fixed point of $\Psi$ in $B_0$. □

The condition of the above Proposition 10.33 is simply satisfied if $\Psi$ is a positive compact operator from $C$ into $B_0$, which case is observed in proving the existence of an endemic steady state for the age-structured SIR epidemic model (see Chap. 6, [57]).

If $\Psi$ has a unique positive fixed point, we may expect it to be obtained as the limit of the successive approximation

$$x_n = \Psi x_{n-1}, \quad (n = 1, 2, \ldots).$$

In fact, if $\Psi$ is concave, monotone, and compact in the normal cone, the sequence $\{x_n\}_{n=1,2,\ldots}$ converges to the fixed point regardless of the initial data $x_0 \in C^+$ (Theorem 6.6, [65]). Readers may find applications of the above idea to epidemic models in [52, 62].

For a nonlinear positive operator $\Psi$ with $\Psi(0) = 0$, a *majorant* of the operator $\Psi$ is a linear positive operator $T$ such that $\Psi(x) \leq Tx$ for any $x \in C$. Then, we can easily prove that:

**Proposition 10.34** *Let $\Psi$ be a positive operator on a positive cone $C$ of a real Banach space $E$. Suppose $\Psi$ has a compact and nonsupporting majorant $T$ such that $Tf - \Psi(f) \in C^+$ for any $f \in C^+$. Then, if $r(T) \leq 1$, $\Psi$ has no fixed point in $E_+ \setminus \{0\}$.*

*Proof* Suppose that there exists a nonzero fixed point $x \in C^+$. Then, we have $x = \Psi(x) \leq Tx$. Let $F^* \in C^* \setminus \{0\}$ be the adjoint eigenfunctional of $T$ associated with the positive eigenvalue $r(T)$. Because $F^*$ is strictly positive, it follows that $\langle F^*, x\rangle > 0$ and $\langle F^*, Tx - x\rangle = (r(T) - 1)\langle F^*, x\rangle > 0$. Then, we have $r(T) > 1$, which contradicts our assumption. □

For more details on nonlinear positive operator theory, readers may consult the work of Amann [1, 2], Krasnoselskii [65, 66], Krause [67], and Guo and Lakshmikantham [46].

## 10.6 Volterra Integral Equations

Following the style of Iannelli [53], we now collect together some results about Volterra integral equations. For precise formulations, readers should refer to Miller [77] and Gripenberg et al. [45].

First, we consider a finite-dimensional linear system of Volterra convolution integral equations:

$$B(t) = G(t) + \int_0^t \Psi(t - s)B(s)ds, \tag{10.42}$$

where the unknown function $B(t)$ and the initial data $G(t)$ are $n$-dimensional vectors and $\Psi(t)$ is an $n \times n$ matrix. We assume that $\Psi(\cdot)$ is an $n \times n$ matrix-valued locally integrable function and $G(\cdot)$ is an $n$-vector-valued locally integrable function.

The *resolvent* with respect to the integral kernel $\Psi(t)$ is defined as the solution (matrix-valued function) of the following integral equation:

$$\begin{aligned} R(t) &= \Psi(t) + \int_0^t \Psi(t-s)R(s)ds \\ &= \Psi(t) + \int_0^t R(t-s)\Psi(s)ds. \end{aligned} \tag{10.43}$$

It is easy to see that the resolvent is calculated as

$$R(t) = \sum_{j=1}^{\infty} \Psi^{(j)}(t),$$

where $\Psi^{(j)}$, $j = 1, 2, \ldots$ are iteratively calculated as

$$\Psi^{(1)}(t) = \Psi(t), \quad \Psi^{(j+1)}(t) = (\Psi^{(j)} * \Psi)(t) = \int_0^t \Psi^j(t-s)\Psi(s)ds,$$

where $*$ denotes the convolution operation and the above series converges in the sense of norm convergence. Using the resolvent, it is clear that the solution of the integral equation (10.42) is given by

$$B(t) = G(t) + \int_0^t R(t-s)G(s)ds. \tag{10.44}$$

Equation (10.42) has a trivial solution $B(t) \equiv 0$ corresponding to the trivial initial data $G(t) \equiv 0$. Let us introduce the idea of the stability of the trivial solution:

**Definition 10.6** For any $\varepsilon > 0$, if there exists a positive number $\delta > 0$ such that $|B|_{L^\infty} < \varepsilon$ if $|G|_{L^\infty} < \delta$, then the trivial solution of (10.42) is said to be *stable*. If the trivial solution is stable and $\lim_{t\to\infty} B(t) = 0$ provided that $\lim_{t\to+\infty} G(t) = 0$, then it is called *asymptotically stable*.

From (10.44), it is easy to see that the integrability of the resolvent is related to the stability of the trivial solution:

**Proposition 10.35** *Suppose that the resolvent is integrable on the half line $\mathbb{R}_+$; $R(\cdot) \in L^1(\mathbb{R}_+; L(\mathbb{R}^n))$. Then, if $G(\cdot)$ is bounded and continuous on $\mathbb{R}_+$, it holds that $|B(t)| \leq (1 + |R|_{L^1})|G|_{L^\infty}$ for all $t > 0$. Moreover, if $\lim_{t\to\infty} G(t) = 0$, then $\lim_{t\to+\infty} B(t) = 0$.*

**Proposition 10.36** *The trivial solution of* (10.42) *is stable if and only if the resolvent is integrable on the half line.*

**Corollary 10.4** *The trivial solution of* (10.42) *is stable if and only if it is asymptotically stable.*

Under the condition that the resolvent is integrable, the Laplace transform of the kernel $\Psi(\cdot)$ exists for all $\lambda$ such that $\Re\lambda \geq 0$. Then, we obtain the following well-known theorem [77, Theorem 5.2]:

**Proposition 10.37** (Paley–Wiener theorem) *The resolvent* $R(\cdot)$ *is integrable if and only if the following condition is satisfied:*

$$\det(I - \hat{\Psi}(\lambda)) \neq 0, \quad \forall \Re\lambda \geq 0. \tag{10.45}$$

It is clear that condition (10.45) implies that the characteristic equation of the renewal equation (10.45) has no root in the right half plane. In particular, the initial data and the kernel are nonnegative, and we thus obtain the following theorem for the asymptotic behavior:

**Proposition 10.38** *Suppose that* $G(\cdot)$ *and* $\Psi(\cdot)$ *are nonnegative and integrable on* $\mathbb{R}_+$ *and that* $\hat{\Psi}(0) = \int_0^\infty \Psi(a)da$ *is indecomposable. Let* $R_0 = r(\hat{\Psi}(0))$ *be the spectral radius of* $\hat{\Psi}(0)$*. Then, if* $R_0 < 1$*, it follows that*

$$\int_0^\infty B(t)dt = (I - \hat{\Psi}(0))^{-1}\hat{G}(0). \tag{10.46}$$

*If* $R_0 \geq 1$*, it holds that*

$$\lim_{t\to\infty} \frac{1}{t}\int_0^t e^{-\lambda_0 t} B(t)dt = \frac{\langle v_0, \hat{G}(\lambda_0)\rangle}{\langle v_0, K_1 u_0\rangle} u_0, \tag{10.47}$$

*where* $\lambda_0$ *is a unique real root of* $r(\hat{\Psi}(\lambda)) = 1$, $v_0$ *and* $u_0$ *are the left and the right eigenvectors of* $\hat{\Psi}(\lambda_0)$ *associated with its eigenvalue of unity and*

$$K_1 := \int_0^\infty t\Psi(t)dt.$$

*Proof* Multiplying both sides of (10.42) by $e^{-xt}$, $x \geq 0$, and integrating over the interval $(0, T)$, $T > 0$, we have

$$\int_0^T e^{-xt} B(t)dt \leq \hat{G}(x) + \hat{\Psi}(x)\int_0^T e^{-xt} B(t)dt,$$

where $\hat{G}(x)$ and $\hat{\Psi}(x)$ exist. If $R_0 < 1$, $I - \hat{\Psi}(x)$ is nonnegatively invertible for $x \geq 0$,

$$\int_0^T e^{-xt} B(t)dt \leq (I - \hat{\Psi}(x))^{-1}\hat{G}(x).$$

Because the left-hand side is monotone increasing and bounded above when $T \to \infty$, it converges to $\hat{B}(x)$. Then, for $x \geq 0$, $\hat{B}(x)$ exists, and hence,

$$\hat{B}(x) = (I - \hat{\Psi}(x))^{-1}\hat{G}(x), \tag{10.48}$$

where if we let $x \to +0$, we have (10.46). Next, consider the case where $R_0 = 1$. As $r(\hat{\Psi}(0)) = 1$, (10.48) holds for $x > 0$. Then, we have

$$\lim_{x \to +0} x\hat{B}(x) = \frac{\langle v_0, \hat{G}(0)\rangle}{\langle v_0, K_1 u_0\rangle} u_0,$$

which is the same calculation used in the proof of Proposition 2.2. From the Tauberian theorem [11, Theorem 7.8.], it follows that

$$\lim_{t \to \infty} \frac{1}{t} \int_0^t B(t)dt = \lim_{x \to +0} x\hat{B}(x).$$

Thus, we conclude that (10.47) holds. Finally, if $R_0 > 1$, instead of (10.42), we can consider the equivalent system

$$e^{-\lambda_0 t} B(t) = e^{-\lambda_0 t} G(t) + \int_0^t e^{-\lambda_0 a} \Psi(a) e^{-\lambda_0 (t-a)} B(t-a)da,$$

to which we can apply the result of the case for $R_0 = 1$, because $\int_0^\infty e^{-\lambda_0 a}\Psi(a)da$ has the Frobenius root of unity. □

The following stability result was stated by [53, 76]. Consider an $n$-dimensional nonlinear Volterra integral equation system given by

$$B(t) = \int_0^t \Psi(a)B(t-a)ds + \mathscr{P}[B(\cdot), c(\cdot)](t), \tag{10.49}$$

where $B \in C([0,\infty); \mathbb{R}^n)$, $\Psi \in L^1([0,\infty); \mathscr{L}(\mathbb{R}^n))$ and $c \in L^1([a,b]; \mathbb{R}^m)$ are the given input data (initial data). The nonlinear term

$$\mathscr{P} : C_0([0,+\infty); \mathbb{R}^n) \times L^1([a,b]; \mathbb{R}^m) \to C_0([0,+\infty); \mathbb{R}^n)$$

satisfies $\mathscr{P}[0,0] = 0$. There exists some function $L(x)$, $x \in \mathbb{R}_+$, such that $\lim_{x\to 0} L(x) = 0$ and, for $|u|_{L^\infty}, |\bar{u}|_{L^\infty}, |c|_{L^1} \leq \eta$, it holds that

$$|\mathscr{P}[u(\cdot), c(\cdot)] - \mathscr{P}[\bar{u}(\cdot), c(\cdot)]|_{L^\infty} \leq L(\eta)|u - \bar{u}|_{L^\infty}$$

and there exists a number $\mathscr{K} > 0$ such that

$$|\mathscr{P}[0, c(\cdot)]|_{L^\infty} \leq \mathscr{K}|c|_{L^1}, \quad \forall c \in L^1([a,b]; \mathbb{R}^m).$$

Then, we obtain the following:

**Proposition 10.39** ([53]) *Under the above conditions, if the resolvent $R(\cdot)$ belongs to $L^1([0,\infty); \mathscr{L}(\mathbb{R}^n))$, then, for any $\varepsilon > 0$, there exists $\delta > 0$ such that for any input data $c \in L^1([a,b]; \mathbb{R}^m)$ with $|c|_{L^1} \le \delta$, (10.49) has a unique solution $B \in C_0([0,+\infty); \mathbb{R}^n)$ and $|B|_{L^\infty} \le \varepsilon$.*

As we discussed in the previous chapters, the basic models of structured population dynamics are formulated as abstract Volterra integral equations. Therefore, we now discuss an asymptotic result for abstract Volterra integral equations. Earlier studies about the asymptotic behavior of Volterra integral equations have been reported by Kermack and McKendrick [64] and Feller [39]. Some rigorous treatments of vector-valued equations appeared in the 1960s and 1970s [11, 29, 40, 42, 45, 50, 77].

Let $E$ be a Banach space with norm $\|\cdot\|_E$. A vector-valued function $f : I \to E$ on the interval $I = (a,b)$ $(-\infty \le a < b \le \infty)$ is Bochner integrable if $f$ is strongly measurable and $\|f(t)\|_E$ is integrable. Then, the space of integrable functions $L^1(I;E)$ with norm $\|f\| := \int_I \|f(t)\|_E dt$ becomes a Banach space. Suppose that $G(t),\ t \ge 0$, takes a value in $E$ and $\Psi(a),\ a \ge 0$, takes a value in the space of bounded linear operators from $E$ to itself. Then, we can consider an abstract renewal integral equation as follows:

$$B(t) = G(t) + \int_0^t \Psi(a)B(t-a)da. \tag{10.50}$$

For any $T > 0$, if a vector-valued function $B : \mathbb{R}_+ \to E$ is Bochner integrable on the interval $[0,T]$ and satisfies (10.50), it is called the solution of the renewal equation (10.50).

Just as for the finite-dimensional case, (10.50) can be solved using the Laplace transformation as

$$B(t) = G(t) + \frac{1}{2\pi i}\int_{x-i\infty}^{x+i\infty} e^{\lambda t}(I - \hat{\Psi}(\lambda))^{-1}\hat{\Psi}(\lambda)\hat{G}(\lambda)d\lambda, \tag{10.51}$$

where $x$ is chosen so that $x > \lambda_0 := \sup_{\lambda\in\Lambda} \Re\lambda$ provided that the set of singular points $\Lambda$ is given by $\Lambda := \{\lambda \in \mathbb{C} : 1 \in \sigma(\hat{\Psi}(\lambda))\}$. Moreover, we assume that there exists some $\varepsilon$ such that $\lambda_0 > \varepsilon$ and $\hat{G}(\lambda)$ and $\hat{\Psi}(\lambda)$ exist for $\Re\lambda > \varepsilon$.

The asymptotic analysis in Chap. 1 for a scalar-type Volterra integral equation can then be extended to the infinite-dimensional case (10.50) in the same manner if $\hat{\Psi}(\lambda)$ is a nonsupporting compact operator for $\lambda \in \mathbb{R}$. That is, there exists some number $\eta > 0$ such that

$$B(t) = G(t) + e^{\lambda_0 t}\left[\frac{\langle v_0, \hat{G}(\lambda_0)\rangle}{\langle v_0, \Psi_1 u_0\rangle}u_0 + O(e^{-\eta t})\right], \tag{10.52}$$

where $v_0$ and $u_0$ are the positive eigenfunctional and the positive eigenfunction of the positive linear operator $\hat{\Psi}(\lambda_0)$ associated with the eigenvalue unity and

$$\Psi_1 = -\frac{d}{d\lambda}\hat{\Psi}(\lambda)\bigg|_{\lambda=\lambda_0}.$$

The above result hinges on the existence of the dominant characteristic root $\lambda_0$ in $\Lambda$. Using positive operator theory, we can prove that the elements of $\Lambda$ are distributed in the same manner as Lotka's characteristic roots. The following result was proved by Heijmans [50]:

**Proposition 10.40** *Suppose that $\hat{\Psi}(\lambda)$ is a compact operator for any $\lambda \in \mathbb{C}$ and that for $\lambda \in \mathbb{R}$ there exists a quasi-interior point $e$ and a strictly positive functional $F_\lambda$ with respect to the positive cone $L^1_+$ such that*

$$\hat{\Psi}(\lambda)\psi \geq \langle F_\lambda, \psi\rangle e, \quad \lim_{\lambda\to-\infty}\langle F_\lambda, e\rangle = +\infty. \tag{10.53}$$

*Then, the following holds:*

*(1)* $\Lambda = \{\lambda \in \mathbb{C} : 1 \in P_\sigma(\hat{\Psi}(\lambda))\}$,
*(2) For any $\lambda \in \mathbb{R}$, the operator $\hat{\Psi}(\lambda)$ is strictly nonsupporting,*
*(3) For $\lambda \in \mathbb{R}$, the spectral radius $r(\hat{\Psi}(\lambda))$ is monotone decreasing from $+\infty$ to zero,*
*(4) The characteristic equation $r(\hat{\Psi}(\lambda)) = 1$ has a unique real root $\lambda_0$ such that*

$$\mathrm{sign}(\lambda_0) = \mathrm{sign}(r(\hat{\Psi}(0)) - 1),$$

*(5)* $\lambda_0 > \sup\{\mathrm{Re}\lambda : \lambda \in \Lambda \setminus \{\lambda_0\}\}$.

*Proof* If we assume that $\hat{\Psi}(\lambda)$ is a compact operator for any $\lambda$, $(I - \hat{\Psi}(\lambda))^{-1}$ becomes a meromorphic function of $\lambda$ and $\Lambda$ is the set of its isolated poles, $\Lambda = \{\lambda \in \mathbb{C} : 1 \in P_\sigma(\hat{\Psi}(\lambda))\}$.[7] On the real axis, $\hat{\Psi}(\lambda)$ is a positive operator and its spectral radius $r(\hat{\Psi}(\lambda))$ is non-increasing with respect to $\lambda$. If $\hat{\Psi}(\lambda)$ is a compact operator, it follows from the Krein–Rutman theorem (Proposition 10.8) that its nonzero spectral radius gives a positive eigenvalue. Hence, characteristic roots of the equation $r(\hat{\Psi}(\lambda)) = 1$ become elements of $\Lambda$. Next, let us show (2). From the assumption in (10.53), it follows that, for natural numbers $n = 0, 1, 2, \ldots$,

$$\hat{\Psi}(\lambda)^{n+1}\psi \geq \langle F_\lambda, \psi\rangle\langle F_\lambda, e\rangle^n e.$$

Then, we have $\langle F, \hat{\Psi}(\lambda)^n\psi\rangle > 0$, $n \geq 1$, $\lambda \in \mathbb{R}$ for any nonzero positive functional $F \in L^\infty_+ \setminus \{0\}$, which shows that $\hat{\Psi}(\lambda)$ is strictly nonsupporting. From the comparison theorem of positive operators (Proposition 10.14, the spectral radius $r(\hat{\Psi}(\lambda))$ is monotonically decreasing from $+\infty$ to zero with respect to $\lambda \in \mathbb{R}$ and the characteristic equation has a unique real root $\lambda_0 \in \Lambda$. Thus, we have (3) and (4). Finally, we show (5). For any $\lambda \in \Lambda$, there exists a corresponding eigenfunction $\psi_\lambda$ such that $\hat{\Psi}(\lambda)\psi_\lambda = \psi_\lambda$. Then, we have $|\psi_\lambda| = |\hat{\Psi}(\lambda)\psi_\lambda| \leq \hat{\Psi}(\Re\lambda)|\psi_\lambda|$. Let $f_{\Re\lambda}$ be an

[7] $P_\sigma(A)$ denotes the point spectrum of an operator $A$.

adjoint positive eigenfunctional of $\hat{\Psi}(\Re\lambda)$ corresponding to the eigenvalue unity. It follows that

$$\langle f_{\Re\lambda}, \hat{\Psi}(\Re\lambda)|\psi_\lambda|\rangle = r(\hat{\Psi}(\Re\lambda))\langle f_{\Re\lambda}, |\psi_\lambda|\rangle \geq \langle f_{\Re\lambda}, |\psi_\lambda|\rangle.$$

Therefore, we have $r(\hat{\Psi}(\Re\lambda)) \geq 1$ and $\Re\lambda \leq \lambda_0$, because $r(\hat{\Psi}(x))$ is monotone decreasing with respect to $x \in \mathbb{R}$ and $r(\hat{\Psi}(\lambda_0)) = 1$. If $\Re\lambda = \lambda_0$, we obtain $\hat{\Psi}(\lambda_0)|\psi_\lambda| = |\psi_\lambda|$. In fact, if $\hat{\Psi}(\lambda_0)|\psi_{\lambda_0}| > |\psi_{\lambda_0}|$, applying the eigenfunctional $f_{\lambda_0}$ to $r(\hat{\Psi}(\lambda_0)) = 1$, we have $\langle f_{\lambda_0}, \hat{\Psi}(\lambda_0)|\psi_\lambda|\rangle = \langle f_{\lambda_0}, |\psi_\lambda|\rangle > \langle f_{\lambda_0}, |\psi_\lambda|\rangle$, which is a contradiction. Therefore, we can write $|\psi_\lambda| = c\psi_0$ using the eigenfunction $\psi_0$ corresponding to the eigenvalue $r(\hat{\Psi}(\lambda_0)) = 1$. Without loss of generality, we can assume that $c = 1$ and that there exists a real function $\alpha(a)$ such that $\psi_\lambda(a) = \psi_0(a)\exp(i\alpha(a))$. Substituting this relation into

$$\hat{\Psi}(\lambda_0)\psi_0 = \psi_0 = |\psi_\lambda| = |\hat{\Psi}(\lambda)\psi_\lambda|,$$

we obtain the following:

$$\int_0^\infty e^{-\lambda_0 a}\Psi(a)\psi_0 da = \left|\int_0^\infty e^{-\lambda a}\Psi(a)\psi_0 \exp(i\alpha)da\right|.$$

From Lemma 6.12 of Heijmans [50], there exists a constant $\theta$ such that $-\Im\lambda a + \alpha = \theta$. It follows from $\hat{\Psi}(\lambda)\psi_\lambda = \psi_\lambda$ that $e^{i\theta}\hat{\Psi}(\lambda_0)\psi_0 = \psi_0 e^{i\alpha}$. We then have $\theta = \alpha(a)$, which implies that $\Im\lambda = 0$. Thus, there is no element $\lambda \in \Lambda\setminus\{\lambda_0\}$ such that $\Re\lambda = \lambda_0$, which completes the proof of (5). □

## References


1. Amann, H.: Fixed point equations and nonlinear eigenvalue problems in ordered Banach spaces. SIAM Rev. **18**(4), 620–709 (1970)
2. Amann, H.: On the number of solutions of nonlinear equations in ordered Banach spaces. J. Func. Anal. **11**, 346–384 (1972)
3. Anita, S.: Analysis and Control of Age-Dependent Population Dynamics. Kluwer, Dordrecht (2000)
4. Anita, S., Iannelli, M., Kim, M.-Y., Park, E.-J.: Optimal harvesting for periodic age-dependent population dynamics. SIAM. J. Appl. Math. **58**(5), 1648–1666 (1998)
5. Arendt, W., Grabosch, G., Greiner, G., Groh, U., Lotz, H.P., Moustakas, U., Nagel, R., Neubrander, F., Schlotterbeck, U.: One-Parmeter Semigroups of Positive Operators. Lecture Notes in Mathematics, vol. 1184. Springer, Berlin (1986)
6. Bacaër, N.: The asymptotic behavior of the McKendrick equation with immigration. Math. Popul. Stud. **10**, 1–20 (2003)
7. Banasiak, J., Mokhtar-Kharroubi, M.: Evolutionary Equations with Applications in Natural Sciences. Lecture Notes in Mathematics, vol. 2126. Springer, Cham (2015)
8. Bauer, F.L.: An elementary proof of the Hopf inequality for positive operators. t Numer. Math. **7**, 331–337 (1965)

9. Belleni-Morante, A.: Applied Semigroups and Evolution Equations. Clarendon Press, Oxford (1979)
10. Belleni-Morante, A., Busoni, G.: Some remarks on densely defined streaming operators. Math. Comput. Model. **21**(8), 13–15 (1995)
11. Bellman, R., Cooke, K.: Differential-Difference Equations. Acadimic Press, New York (1963)
12. Bertoni, S.: Periodic solutions for non-linear equations of structured populations. J. Math. Anal. Appl. **220**, 250–267 (1998)
13. Birkhoff, G.: Extensions of Jentzsch's theorem. Trans. Am. Math. Soc. **85**, 219–227 (1957)
14. Birkhoff, G.: Lattices in applied mathematics. In: Lattice Theory. Proceedings of Syposia in Pure Mathematics, vol. 2, pp. 155-184. American Mathematical Society, Providence (1961)
15. Birkhoff, G.: Positivity and criticality. In: Birkhoff, G., Wigner, E.P. (eds.) Nuclear Reactor Theory. Proceedings of Symposia in Applied Mathematics, vol. XI, pp. 116–126. American Mathematical Society, Providence (1961)
16. Birkhoff, G.: Uniformly semi-primitive multiplicative process. Trans. Am. Math. Soc. **104**, 37–51 (1962)
17. Birkhoff, G.: Uniformly semi-primitive multiplicative process II. J. Math. Mech. **14**(3), 507–512 (1965)
18. Birkhoff, G.: Lattice Theory, 3rd edn. American Mathematical Society, Providence (1967)
19. Birkhoff, G., Varga, R.S.: Reactor criticality and nonnegative matrices. J. Soc. Indust. Appl. Math. **6**(4), 354–377 (1958)
20. Bushell, P.J.: Hilbert's metric and positive contraction mappings in a Banach space. Arch. Rat. Mech. Anal. **52**, 330–338 (1973)
21. Bushell, P.J.: On the projective contraction ratio for positive linear mappings. J. Lond. Math. Soc. **2**(6), 256–258 (1973)
22. Chicone, C., Latushkin, Y.: Evolution Semigroups in Dynamical Systems and Differential Equations, Mathematical Surveys and Monographs, vol. 70. American Mathematical Society, Providence (1999)
23. Clément, Ph, Diekmann, O., Gyllenberg, M., Heijmans, H.J.A.M., Thieme, H.R.: Perturbation theory for dual semigroups I. The sun-reflexive case. Math. Ann. **277**, 709–725 (1987)
24. Clément, Ph., Heijmans, H.J.A.M., Angenent, S., van Duijn, C.J., de Pagter, B.: One-Parameter Semigroups, CWI Monograph 5. North-Holland, Amsterdam (1987)
25. Clément, Ph, Diekmann, O., Gyllenberg, M., Heijmans, H.J.A.M., Thieme, H.R.: Perturbation theory for dual semigroups II. Time-dependent perturbations in sun-reflexive case. Proc. R. Soc. Edinb. **109A**, 145–172 (1988)
26. Clément, Ph., Diekmann, O., Gyllenberg, M., Heijmans, H.J.A.M., Thieme, H.R.: Perturbation theory for dual semigroups III. Nonlinear Lipschitz continuous perturbations in the sun-reflexive. In: Da Prato, G., Iannelli, M. (eds.) Volterra Integrodifferential Equations in Banach Spaces and Applications. Pitman Research Notes in Mathematics Series, vol. 190, pp. 67–89. Longman, Harlow (1989)
27. Clément, Ph., Diekmann, O., Gyllenberg, M., Heijmans, H.J.A.M., Thieme, H.R.: Perturbation theory for dual semigroups IV: The interwining formula and the canonical pairing. In: Clément, Ph, Invernizzi, S., Mitidieri, E., Vrabie, I.I. (eds.) Semigroup Theory and Applications. Lecture Notes in Pure and Applied Mathematics, vol. 116, pp. 95–116. Marcel Dekker, New york (1989)
28. Clément, Ph, Diekmann, O., Gyllenberg, M., Heijmans, H.J.A.M., Thieme, H.R.: A Hille-Yosida theorem for a class of weakly * continuous semigroups. Semigroup Forum **38**, 157–178 (1989)
29. Crump, K.S.: On systems of renewal equations. J. Math. Anal. Appl. **30**, 425–434 (1970)
30. Da Prato, G., Sinestrari, E.: Differential operators with non dense domain. Annali della Scuola Normale Superiore di Pisa **14**(2), 285–344 (1987)
31. Desch, W., Schappacher, W.: Spectral properties of finite-dimensional perturbed linear semigroups. J. Diff. Equ. **59**, 80–102 (1985)
32. Desch, W., Lasiecka, I., Schappacher, W.: Feedback boundary control problems for linear semigroups. Isr. J. Math. **51**(3), 177–207 (1985)

33. Desch, W., Schappacher, W., Kang Pei Zhang: Semilinear evolution equations, Houst. J. Math. **15**(4), 527–552 (1989)
34. Diekmann, O., Gyllenberg, M.: Abstract delay equations inspired by population dynamics. In: Amann, H., Arendt, W., Hieber, M., Nuebrander, F., Nicaise, J., von Below, J. (eds.) The Gunter Lumer Volume, pp. 187–200. Birkhauser, Basel (2007)
35. Diekmann, O., Gyllenberg, M.: The second half - with a quarter of a century delay. Math. Model. Nat. Phenom. **3**(7), 36–48 (2008)
36. Diekmann, O., van Gils, S.A., Verduyn Lunel, S.M., Walther, H.-O.: Delay Equations: Functional-, Complex-, and Nonlinear Analysis. Applied Mathematical Sciences, vol. 110. Springer, Berlin (1995)
37. Dunford, N., Schwartz, J.T.: Linear operators. General Theory. Wiley, New York (1957). Part I
38. Engel, K.J., Nagel, R.: One-Parameter Semigroups for Linear Evolution Equations. Springer, Berlin (2000)
39. Feller, W.: On the integral equation of renewal theory. Ann. Math. Stat. **12**, 243–267 (1941)
40. Friedman, A., Shinbrot, M.: Volterra integral equations in Banach space. Trans. Am. Math. Soc. **126**(1), 131–179 (1967)
41. Fujimoto, T., Krause, U.: Asymptotic properties for inhomogeneous iterations of nonlinear operators. SIAM J. Math. Anal. **19**(4), 841–853 (1988)
42. Gohberg, I.C., Fel'dman, I.A.: Convolution Equations and Projection Methods for their Solution. Translations of Mathematical Monographs, vol. 41. American Mathematical Society, Providence (1974)
43. Golubitsky, M., Keeler, E.B., Rothschild, M.: Convergence of the age structure: Applications of the projective metric. Theor. Popul. Biol. **7**, 84–93 (1975)
44. Greiner, G.: Semilinear boundary conditions for evolution equations of hyperbolic type. In: Clément, Ph, Invernizzi, S., Mitidieri, E., Vrabie, I.I. (eds.) Semigroup Theory and Applications. Lecture Notes in Pure and Applied Mathematics, vol. 116, pp. 201–214. Marcel Dekker, New york (1989)
45. Gripenberg, G., Londen, S.-O., Staffans, O.: Volterra Integral and Functional Equations. Cambridge U. P, Cambridge (1990)
46. Guo, D., Lakshmikantham, V.: Nonlinear Problems in Abstract Cones. Academic Press, London (1988)
47. Gyllenberg, M., Webb, G.F.: Asynchronous exponential growth of semigroups of nonlinear operators. J. Math. Anal. Appl. **167**, 443–467 (1992)
48. Hadeler, K.P., Waldstätter, R., Wörz-Busekros, A.: Models for pair formation in bisexual populations. J. Math. Biol. **26**, 635–649 (1988)
49. Hale, J.K.: Asymptotic Behavior of Dissipative Systems. Mathematical Survey and Monographs, vol. 25. American Mathematical Society, Providence (1988)
50. Heijmans, H.J.A.M.: The dynamical behaviour of the age-size-distribution of a cell population. In: Metz, J.A.J., Diekmann, O. (eds.) The Dynamics of Physiologically Structured Populations. Lecture Notes in Biomathematics, vol. 68, pp. 185–202. Springer, Berlin (1986)
51. Heijmans, H.J.A.M.: Semigroup theory for control on sun-reflexive Banach spaces. IMA J. Math. Cont. Inf. **4**, 111–129 (1987)
52. Hethcote, H.W., Thieme, H.R.: Stability of the endemic equilibrium in epidemic models with subpopulations. Math. Biosci. **75**, 205–227 (1985)
53. Iannelli, M.: Mathematical Theory of Age-Structured Population Dynamics, Giardini Editori e Stampatori in Pisa (1995)
54. Inaba, H.: A semigroup approach to the strong ergodic theorem of the multistate stable population process. Math. Popul. Stud. **1**(1), 49–77 (1988)
55. Inaba, H.: Asymptotic properties of the inhomogeneous Lotka-Von Foerster system. Math. Popul. Stud. **1**(3), 247–264 (1988)
56. Inaba, H.: Weak ergodicity of population evolution processes. Math. Biosci. **96**, 195–219 (1989)

57. Inaba, H.: Threshold and stability results for an age-structured epidemic model. J. Math. Biol. **28**, 411–434 (1990)
58. Inaba, H.: Strong ergodicity for perturbed dual semigroups and application to age-dependent population dynamics. J. Math. Anal. Appl. **165**(1), 102–132 (1992)
59. Inaba, H.: Persistent age distributions for an age-structured two-sex population model. Math. Popul. Stud. **7**(4), 365–398 (2000)
60. Inaba, H.: Endemic threshold results in an age-duration-structured population model for HIV infection. Math. Biosci. **201**, 15–47 (2006)
61. Inaba, H.: The Malthusian parameter and $R_0$ for heterogeneous populations in periodic environments. Math. Biosci. Eng. **9**(2), 313–346 (2012)
62. Inaba, H.: On a pandemic threshold theorem of the early Kermack-McKendrick model woth individual heterogeneity. Math. Poul. Stud. **21**, 95–111 (2014)
63. Jörgens, K.: An asymptotic expansion in the theory of neutron transport. Commun. Pure Appl. Math. **11**, 219–242 (1958)
64. Kermack, W.O., McKendrick, A.G.: The solution of sets of simultaneous integral equations related to the equation of Volterra. Proc. Lond. Math. Soc., Ser. 2 **41**, 462–482 (1936)
65. Krasnoselskii, M.A.: Positive Solutions of Operator Equations. Noordhoff, Groningen (1964)
66. Krasnosel'skij, M.A., Lifshits, JeA, Sobolev, A.V.: Positive Linear Systems -The Method of Positive Operators. Heldermann Verlag, Berlin (1989)
67. Krause, U.: Positive Dynamical Systems in Discrete Time: Theory, Models, and Applications. de Gruyter, Berlin (2015)
68. Krein, M.G., Rutman, M.A.: Linear operators leaving invariant a cone in a Banach space. Uspehi. Mat. Nauk. **3**, 3–95 (1948) [in Russian]; English translation: Am. Math. Soc. Transl. **10**(1), 199–325 (1950)
69. Lopez, A.: Problems in Stable Population Theory. Office of Population Research. Princeton University, Princeton (1961)
70. Lotka, A.J.: Population analysis: a theorem regarding the stable age distribution. J. Wash. Acad. Sci. **27**(7), 299–303 (1937)
71. Magal, P., Ruan, D.: Center Manifolds for Semilinear Equations with Non-dense Domain and Applications to Hopf Bifurcation in Age Structured Models. Memoirs of the American Mathematical Society, vol. 951. American Mathematical Society (2009)
72. Magal, P., Ruan, S.: Sustained oscillations in an evolutionary epidemiological model of influenza A drift. Proc. R. Soc. A **466**, 965–992 (2010)
73. Marek, I.: Frobenius theory of positive operators: Comparison theorems and applications. SIAM J. Appl. Math. **19**, 607–628 (1970)
74. Martcheva, M., Thieme, H.R.: Progression age enhanced backward bifurcation in an epidemic model with super-infection. J. Math. Biol. **46**, 385–424 (2003)
75. Metz, J.A.J., Diekmann, O.: The Dynamics of Physiologically Structured Populations. Lecture Notes in Biomathematics, vol. 68. Springer, Berlin (1986)
76. Miller, R.K.: On the linearization of Volterra integral equations. J. Math. Anal. Appl. **23**, 198–208 (1968)
77. Miller, R.K.: Nonlinear Volterra Integral Equations. Benjamin, Menlo Park (1971)
78. Nagel, R. (ed.): One-Parameter Semigroups of Positive Operators. Lecture Notes in Mathematics, vol. 1184. Springer, Berlin (1986)
79. Nussbaum, R.D.: Hilbert's Projective Metric and Iterated Nonlinear Maps. Memoirs of the American Mathematical Society, vol. 75, No. 391. American Mathematical Society, Providence (1988)
80. Nussbaum, R.D.: Iterated Nonlinear Maps and Hilbert's Projective Metric, II. Memoirs of the American Mathematical Society, vol. 79, No. 401. American Mathematical Society, Providence (1989)
81. Nussbaum, R.D.: Some nonlinear weak ergodic theorems. SIAM J. Math. Anal. **21**(2), 436–460 (1990)
82. Ostrowski, A.M.: Positive matrices and functional analysis. In: Schrecher, H. (ed.) Recent Advances in Matrix Theory, pp. 81–101. Univ. of Wisconsin Press, Madison (1964)

83. Pazy, A.: Semigroups of Linear Operators and Applications to Partial Differential Equations. Springer, Berlin (1983)
84. Prüss, J.: Periodic solutions of semilinear evolution equations. Nonlinear Analysis, Theory, Methods and Applications **3**(5), 601–612 (1979)
85. Prüss, J.: Equilibrium solutions of age-specific population dynamics of several species. J. Math. Biol. **1**, 65–84 (1981)
86. Prüss, J.: Stability analysis for equilibria in age-specific population dynamics. Nonlinear Analysis, Theory, Methods and Applications **7**(12), 1291–1313 (1983)
87. Prüss, J.: On the qualitative behaviour of populations with age-specific interactions. Comput. Math. Appl. **9**(3), 327–339 (1983)
88. Prüss, J., Schappacher, W.: Semigroup methods for age-structured population dynamics. In: Chatterji, S.D., Fuchssteiner, B., Kulisch, U., Liedl, R. (eds.) Jahrbuch Uberblicke Mathematik 1994, pp. 74–90. Braunschweig, Viewveg (1994)
89. Samuelson, P.A.: Resolving a historical confusion in population analysis. Hum. Biol. **48**, 559–580 (1976)
90. Sawashima, I.: On spectral properties of some positive operators. Nat. Sci. Rep Ochanomizu Univ. **15**, 53–64 (1964)
91. Schaefer, H.H.: Banach Lattices and Positive Operators. Springer, Berlin (1974)
92. Schaefer, H.H., Wolff, M.P.: Topological Vector Spaces, 2nd edn. Springer, New York (1999)
93. Seneta, E.: Non-negative Matrices and Markov Chain, 2nd edn. Springer, Berlin (1981)
94. Smith, H.L., Thieme, H.R.: Dynamical Systems and Population Persistence. Graduate Studies in Mathematics, vol. 118. American Mathematical Society, Providence (2011)
95. Song, J., Yu, J.: Population System Control. Springer, Berlin (1988)
96. Song, J., Yu, J.Y., Wang, X.Z., Hu, S.J., Zhao, Z.X., Liu, J.Q., Feng, D.X., Zhu, G.T.: Spectral properties of population operator and asymptotic behaviour of population semigroup. Acta Mathematica Scientia **2**(2), 139–148 (1982)
97. Song, J., Tuan, C.H., Yu, J.Y.: Population Control in China: Theory and Applications. Praeger, New York (1985)
98. Song, J., Yu, J., Liu, C., Zhang, L., Zhu, G.: Spectral properties of population evolution and controllability of population system. Scientia Sinica (Series A) **XXIX**(8), 800–812 (1986)
99. Thieme, H.R.: Asymptotic proportionality (weak ergodicity) and conditional asymptotic equality of solutions to time-heterogeneous sublinear difference and differential equations. J. Diff. Equ. **73**, 237–268 (1988)
100. Thieme, H.R.: Semiflows generated by Lipschitz perturbations of non-densely defined operators. Diff. Integr. Equ. **3**(6), 1035–1066 (1990)
101. Thieme, H.R.: Analysis of age-structured population models with additional structure. In: Arino, O., Axelrod, D.E., Kimmel, M. (eds.) Mathematical Population Dynamics, pp. 115–126. Marcel Dekker, New York (1991)
102. Thieme, H.R.: Spectral bound and reproduction number for infinite-dimensional population structure and time heterogeneity. SIAM J. Appl. Math. **70**(1), 188–211 (2009)
103. Webb, G.F.: Comapctness of bounded trajectories of dynamical systems in infinite dimensional spaces. Proc. R. Soc. Edinb. **84A**, 19–33 (1979)
104. Webb, G.F.: Nonlinear semigroups and age-dependent population models. Ann. Mat. Pura Appl. **139**, 43–55 (1981)
105. Webb, G.F.: Diffusive age-dependent population models and an application to genetics. Math. Biosci. **61**, 1–16 (1982)
106. Webb, G.F.: Nonlinear age-dependent population dynamics in $L^1$. J. Integr. Equ. **5**, 309–328 (1983)
107. Webb, G.F.: A semigroup proof of the Sharpe-Lotka theorem. In: Kappel, F., Schappacher, W. (eds.) Infinite-Dimensional Systems. Lecture Notes in Mathematics, vol. 1076, pp. 254–268. Springer, Berlin (1984)
108. Webb, G.F.: Theory of Nonlinear Age-Dependent Population Dynamics. Marcel Dekker, New York (1985)

109. Webb, G.F.: Dynamics of population structured by internal variables. Math. Zeit. **189**, 319–335 (1985)
110. Webb, G.F.: An operator-theoretic formulation of asynchronous exponential growth. Trans. Am. Math. Soc. **303**(2), 751–763 (1987)
111. Webb, G.F.: Structured population dynamics. In: Rudnicki, R. (ed.) Mathematical Modelling of Population Dynamics, vol. 63, pp. 123–163. Banach Center Publication, Warszawa (2004)

# Index




H. Inaba, *Age-Structured Population Dynamics in Demography and Epidemiology*, DOI 10.1007/978-981-10-0188-8

Kolibri 360 GmbH
Ferdinand-Jühlke-Straße 7
99095 Erfurt, Deutschland
produktsicherheit@kolibri360.de